Low Temperature Plasma Technology

Low Temperature Plasma Technology

Methods and Applications

Edited by
Paul K. Chu
XinPei Lu

CRC Press
Taylor & Francis Group
Boca Raton London New York

CRC Press is an imprint of the
Taylor & Francis Group, an **informa** business

CRC Press
Taylor & Francis Group
6000 Broken Sound Parkway NW, Suite 300
Boca Raton, FL 33487-2742

First issued in paparback 2020

© 2014 by Taylor & Francis Group, LLC
CRC Press is an imprint of Taylor & Francis Group, an Informa business

No claim to original U.S. Government works

ISBN 13: 978-0-367-57636-3 (pbk)
ISBN 13: 978-1-4665-0990-0 (hbk)

Library of Congress Cataloging-in-Publication Data

Low temperature plasma technology : methods and applications / edited by Paul K. Chu and XinPei Lu.
  pages cm
  Includes bibliographical references and index.
  ISBN 978-1-4665-0990-0 (hardback)
  1. Low temperature plasmas. 2. Low temperature plasmas--Industrial applications. 3. Low temperature plasmas--Scientific applications. I. Chu, Paul K.

  QC718.5.L6L694 2013
  621.5'6--dc23   2013001659

Visit the Taylor & Francis Web site at
http://www.taylorandfrancis.com

and the CRC Press Web site at
http://www.crcpress.com

Contents

SECTION III Applications

Preface

During the last few decades, low-temperature plasmas, especially atmospheric pressure plasmas, which are driven by several urgent applications including plasma medicine and treatment of biocompatible materials, have attracted much attention. In order to meet the ever-increasing requirements, novel methods have been proposed since the mid-1990s, for example, using nanosecond voltage pulses rather than DC or kHz AC voltages to drive the plasmas, generating plasmas by using confined micro-discharge gaps, and generating plasma in open space (plasma jets). At the same time, in order to better understand the various plasma characteristics, there have been major advances in plasma diagnostics such as cavity ringdown spectroscopy and laser-induced fluorescence methods. On the heels of these developments, applications of low-temperature plasmas have been extended to various fields, including nanomaterials, environment, liquid treatment, biocompatible materials treatment, and plasma medicine.

The objective of this book is to summarize recent technological advances and research in the rapidly growing field of low-temperature plasmas and their applications. The book is intended to provide a comprehensive overview of the related phenomena such as plasma bullets, plasma penetration into biofilms, the discharge mode transition of atmospheric pressure plasmas, self-organization of microdischarges, and so on. It also describes relevant technology and diagnostics such as nanosecond pulsed discharge, cavity ringdown spectroscopy, laser-induced fluorescence measurement, and also fast-developing research on atmospheric pressure nonequilibrium plasma jets. Finally, applications of low-temperature plasmas, including synthesis of nanomaterials, environmental applications, treatment of biomaterials, and plasma medicine will be discussed. All in all, this book will provide a balanced and thorough treatment of the core principles, relevant novel technologies and diagnostics, and state-of-the-art applications of low-temperature plasmas.

Although the book focuses mainly on low-temperature plasmas and related topics, the scope of this book is actually quite wide. In order to ensure high quality, we are very happy to have renowned authors from many different countries including Italy, the Netherlands, Belgium, the United States, France, Slovakia, Singapore, and the People's Republic of China. These authors are pioneers in the respective fields. For example, Prof. Schoenbach, the contributor of Chapter 5 on high-pressure micro-cavity discharges, was the first to report stable high-pressure operation of microdischarges, even in air, in a cylindrical hollow cathode geometry. His group innovated the term "microhollow cathode discharges (MHCDs)" for these discharges, and it is well adopted by others to describe the three-layer configuration. Other contributors such as Profs. Y.N. Wang, D.Z. Wang, C.O. Laux, S. De Benedictis, C. Wang, C. Leys, S.Y. Xu, B.R. Locke, K.P. Yan, and X.Y. Liu have been known for decades for their excellent contributions to their fields.

This book is intended for graduate students and scientists working in low-pressure plasmas, atmospheric pressure plasmas, plasma diagnostics, plasma nanotechnology, and plasma medicine.

It is written at a level appropriate for graduate education in low-temperature plasma physics and materials science and has relevance in biology, chemistry, and engineering. The book also constitutes an excellent advanced reference for senior college students who want to pursue research in these topics on the graduate level.

XinPei Lu and Paul K. Chu
Editors

Editors Note

Paul K. Chu received his BS in mathematics from the Ohio State University in 1977 and his MS and PhD in chemistry from Cornell University in 1979 and 1982, respectively. He is Chair Professor of Materials Engineering in the Department of Physics and Materials Science in City University of Hong Kong. Paul's research activities are quite diverse, encompassing plasma surface engineering and various types of materials and nanotechnology. Professor Chu has coedited 8 books on plasma science, biomedical engineering, and nanotechnology, special issues in *IEEE Transactions of Plasma Science* and *Surface and Coatings Technology*, as well as MRS proceedings. He has coauthored more than 30 book chapters, 1100 journal papers, and  800 conference papers. He has also been granted numerous patents in the United States, Europe, and the People's Republic of China. He is chairman of the Plasma-Based Ion Implantation and Deposition (PBII&D) International Committee and member of the Ion Implantation Technology (IIT) International Committee and IEEE Nuclear and Plasma Science Society Fellow Evaluation Committee. He is fellow of the IEEE, APS, AVS, MRS, and HKIE (Hong Kong Institution of Engineers), senior editor of *IEEE Transactions on Plasma Science*, and associate editor of *Materials Science & Engineering Reports*. Professor Chu is also an editorial board member of international journals that include *Biomaterials, Plasma Sources Science and Technology*, and *Surface and Coatings Technology*, and he has won a number of awards, including the 2007 IEEE NPSS Merit Award.

XinPei Lu received his PhD in electrical engineering from the Huazhong University of Science and Technology, Hubei, People's Republic of China. Upon graduation, he worked at Old Dominion University as a research associate for 4 years. In 2007, he joined Huazhong University of Science and Technology, where he is now a professor (Chang Jiang Scholar) with the College of Electrical and Electronic Engineering. He is a senior member of IEEE. He has also served as a guest editor for *IEEE Transactions on Plasma Science* and as a session chair at the International Conference on Plasma Science since 2007. He has given many invited talks at international conferences, including the IEEE International Conference on Plasma Science.  His research interests include low-temperature plasma sources and their biomedical applications, modeling of low-temperature plasmas, and plasma diagnostics. He is the author or coauthor of about 50 peer-reviewed journal articles and holds six patents in these areas.

Contributors

Peter J. Bruggeman
Department of Applied Physics
Eindhoven University of Technology
Eindhoven, the Netherlands

Huiliang Cao
Shanghai Institute of Ceramics
Chinese Academy of Sciences
Shanghai, People's Republic of China

Santolo De Benedictis
CNR Istituto di Metodologie Inorganiche e dei
    Plasmi
Unità Organizzativa di Supporto di Bari
Bari, Italy

Nathalie De Geyter
Department of Applied Physics
Ghent University
Ghent, Belgium

Giorgio Dilecce
CNR Istituto di Metodologie Inorganiche e dei
    Plasmi
Unità Organizzativa di Supporto di Bari
Bari, Italy

Peter Dubruel
Department of Organic Chemistry
Ghent University
Ghent, Belgium

Mario Janda
Department of Astronomy, Earth Physics
    and Meteorology
Comenius University
Bratislava, Slovakia

Chunqi Jiang
Department of Electrical Engineering
University of Southern California
Los Angeles, California

Wei Jiang
School of Physics
Huazhong University of Science and Technology
Hubei, People's Republic of China

Christophe O. Laux
Department of Astronomy, Earth Physics
    and Meteorology
Comenius University
Bratislava, Slovakia

Christophe Leys
Department of Applied Physics
Ghent University
Ghent, Belgium

DaWe Liu
State Key Laboratory of Advanced
    Electromagnetic Engineering and Technology
Huazhong University of Science and Technology
Hubei, People's Republic of China

Xuanyong Liu
Shanghai Institute of Ceramics
Chinese Academy of Sciences
Shanghai, People's Republic of China

Bruce R. Locke
Department of Chemical and Biomedical
    Engineering
Florida State University
Tallahassee, Florida

XinPei Lu
State Key Laboratory of Advanced
    Electromagnetic Engineering and Technology
Huazhong University of Science and Technology
Hubei, People's Republic of China

Zdenko Machala
Department of Astronomy, Earth Physics
    and Meteorology
Comenius University
Bratislava, Slovakia

Rino Morent
Department of Applied Physics
and
Department of Organic Chemistry
Ghent University
Ghent, Belgium

David Z. Pai
Department of Astronomy, Earth Physics
    and Meteorology
Comenius University
Bratislava, Slovakia

Karl H. Schoenbach
Frank Reidy Research Center for Bioelectrics
Old Dominion University
Norfolk, Virginia

Jizhong Sun
Department of Physics
Dalian University of Technology
Dalian, People's Republic of China

Chuji Wang
Department of Physics and Astronomy
Mississippi State University
Mississippi State, Mississippi

Dezhen Wang
Department of Physics
Dalian University of Technology
Dalian, People's Republic of China

Yanhui Wang
Department of Physics
Dalian University of Technology
Dalian, People's Republic of China

You-nian Wang
School of Physics and Optoelectronic
    Technology
Dalian University of Technology
Dalian, People's Republic of China

Shaoqing Xiao
NIE and Institute of Advanced Studies
Nanyang Technological University
Singapore

Shuyan Xu
NIE and Institute of Advanced Studies
Nanyang Technological University
Singapore

Keping Yan
Department of Chemical and Biological
    Engineering
Zhejiang University
Hangzhou, People's Republic of China

Yu-ru Zhang
School of Physics and Optoelectronic Technology
Dalian University of Technology
Dalian, People's Republic of China

Haiping Zhou
NIE and Institute of Advanced Studies
Nanyang Technological University
Singapore

Fundamentals

I

1

Introduction

1.1 A Historical Perspective of Plasma

Plasmas make up more than 99% of visible matter in the universe. They consist of positive ions, electrons or negative ions, and neutral particles. Plasma is regarded as the fourth state of matter. When a solid (the first state of matter) is heated, the particles in it get sufficient energy to loosen their structure and thus melt to form a liquid (the second state of matter). After obtaining sufficient energy, the particles in a liquid escape from it and vaporize to gas (the third state of matter). Subsequently, when a significant amount of energy is applied to the gas through mechanisms such as an electric discharge, the electrons that escape from atoms or molecules not only allow ions to move more freely but also produce more electrons and ions via collisions after accelerating rapidly in an electric field. Eventually, the higher number of electrons and ions change the electrical property of the gas, which thus becomes ionized gas or plasma (Figure 1.1).

The study of plasmas can be traced to the seventeenth century. Natural plasmas such as lightning and polar lights are often observed and have intrigued people for many centuries. The desire to understand the mechanism governing plasma generation led to the invention of the discharge device by early researchers.

Although the fact that frictional charge induces discharges was first observed by Greek philosophers, the mechanism of plasma production was not understood until the seventeenth and eighteenth centuries. Figure 1.2a shows an early discharge device made by Anders; the apparatus consists of a glass sphere that can be rapidly spun on an axle by a great wheel.[1] The axle is hollow and connected to a globe through a valve on the other end to a vacuum pump. A glow appeared when the sphere was spun in the dark. Later on, the phenomena of static electricity and discharge became a subject for study. In 1734 Priestley described a "pencil of electric light," today known as corona discharge.[2] Around the same time, significant progress in the development of electric charge storage devices such as Leyden batteries made it possible to study spark discharges. Figure 1.2b shows a spark discharge device[3] that consists of two metal electrodes and a charge storage device known as the Leyden jar. The short-circuit connection between the two metal electrodes caused the quick discharge of the stored charge, resulting in sparks. Later in 1800, an electrochemical battery was invented by Volta.[4] Petrov discovered continuous arc discharge in 1803 using a sufficiently powerful electric battery.[5]

During the nineteenth century, significant progress was made with regard to electric energy storage and vacuum systems. Faraday developed the direct current (DC) glow discharge, the forerunner to today's vacuum plasma, by applying voltage up to 1000 V to an evacuated tube (~1 Torr) in 1831–1835.[6]

3

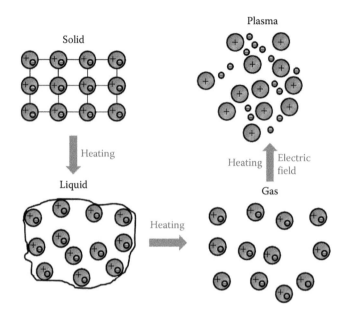

FIGURE 1.1 The transition of states of matter on application of heat.

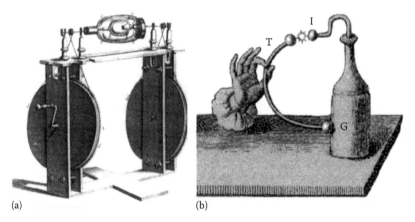

FIGURE 1.2 (a) A glow discharge device developed in 1705. (From Anders A., *IEEE Trans. Plasma Sc.*, 31, 1052, 2003.) (b) An arc discharge device developed in 1775. (From Dibner, B., *Galvani-Volta: A Controversy That Led to the Discovery of Useful Electricity*, Burndy Library, 1952.)

Plasma was first identified as radiant matter by Sir William Crookes in 1879.[7] In the second half of the nineteenth century and the early twentieth century, significant understanding about gas discharges was achieved. J. Townsend studied gas discharge in a uniform electric field and came up with the Townsend discharge theory.[8] He laid the foundation of modern plasma research. His contribution also included the discoveries of cross sections of various electron–atom collisions, drift velocities of electrons and ions, and their recombination coefficients. Later in 1920s, Langmuir not only came up with the term "plasma,"[9] but also invented the Langmuir probe to determine the electron temperature, electron density, and electric potential of a plasma.[10]

The twentieth century witnessed rapid progress in the development, diagnostics, and applications of plasma. Low-pressure radio-frequency (RF) plasma produced in a vacuum chamber is being used intensively in basic processing such as for deposition and etching in the semiconductor industry since the 1970s. Since the 1990s the application of atmospheric pressure plasma eliminated the need for expensive vacuum chamber and pumping systems; as such, atmospheric pressure plasmas are being widely used for environmental applications, surface modification of materials, biomedical applications, and so on.

1.2 Types of Plasma

Plasma is composed of electrons, positive ions, and neutral particles, and can be described based on the ionization degree, density, thermodynamic equilibrium, and so on; therefore, plasma is classified in many different ways.

1.2.1 Plasma Ionization

The ionization degree is defined as $\alpha_i = N_i/(N_i + N_n)$, where N_i is the number density of ions and N_n is the number density of neutrals. The response of any plasma to a magnetic field and the electric conductivity of plasma are determined by α_i. Plasma with $10^{-6} < \alpha_i < 10^{-1}$ is weakly ionized. Because the degree of ionization is determined by the electron temperature in the plasma, weakly ionized plasma is also referred to as low-temperature plasma. In most plasma-processing chambers, the degree of ionization is less than 10^{-4}. The degree of ionization of inductively coupled plasma (ICP) and electron cyclotron resonance is a lot higher, about 10^{-2}. Plasma with $\alpha_i \approx 1$ is fully ionized, and is referred to as "hot" plasma. Examples include fusion plasmas, solar wind (interplanetary medium), and stellar interiors (the Sun's core).

1.2.2 Plasma Densities

1.2.2.1 High-Density Plasma (High-Pressure Plasma)

High-density plasma refers to plasma with particle density $N > 10^{15-18}$ cm^{-3}. The high number of ions and free radicals of high-density plasma not only enhance excitation/ionization collisions but also increase the ion bombardment rate. Therefore, high-density plasmas generated by ICP and capacitively coupled plasma (CCP) are often used for etching in microelectronics, producing nanomaterials, decontaminating plasmas, and so on.

1.2.2.2 Low-Density Plasma (Low-Pressure Plasma)

Low-density plasma refers to plasma with particle density $N < 10^{12-14}$ cm^{-3}. Unlike in the case of high-density plasma, the collision rate between particles of low-density plasma is negligible. Low-density plasma is used in the laser wakefield accelerator,[11] where plasma density must be low enough to allow propagation of lasers of optical or infrared frequencies.

1.2.3 Plasma Thermal Equilibrium

On the basis of relative temperature between electrons, ions, and neutrals, plasmas are classified as *thermal equilibrium*, *local thermal equilibrium*, or *nonthermal equilibrium*.

1.2.3.1 Thermal Equilibrium Plasma

The electron temperature (T_e), ion temperature (T_i), and neutral temperature (T_n) are identical in thermal equilibrium plasma. This is attributed to the frequent collisions between electrons and ions/neutrals inside high-temperature and high-density plasma. Examples include the natural fusion reactor (Sun), a magnetic field (of tokamak design), or inertial (laser) confinement of a plasma.

1.2.3.2 Nonthermal Equilibrium Plasma

In nonthermal equilibrium plasma, the momentum transfer between light electrons and heavy particles (ions and neutrals) is not efficient and the power applied to plasma favors electrons; therefore, the electron

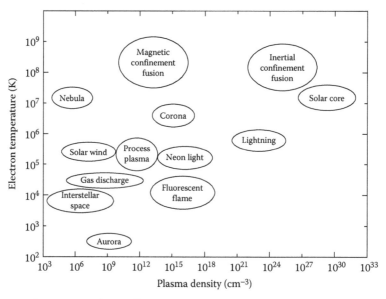

FIGURE 1.3　Typical parameters of naturally occurring and laboratory plasmas.

temperature (T_e) is considerably higher than in ions (T_i) and neutrals (T_n), that is, $T_e \gg T_i, T_n$. Nonthermal equilibrium plasmas are generated by corona discharge, glow discharge, arc discharge, capacitively coupled discharge, inductively coupled discharge, wave heated plasma, and so on. Applications of nonthermal plasma have expanded to cover a large number of fields including environmental engineering, aeronautics and aerospace engineering, biomedicine, textile technology, and analytical chemistry.

1.2.3.3 Local Thermal Equilibrium Plasma

Unlike thermal equilibrium plasma and nonthermal equilibrium plasma, local thermal equilibrium plasma is in quasi-equilibrium: the electron, positive ion, and neutral temperatures are in the same range. The ion temperature of local thermal equilibrium plasma is 3,000–10,000 K (0.4–1 eV), which is much higher than that of nonthermal plasma, but its electron temperature is much lower (0.4–1 eV compared with 2–10 eV of nonthermal plasma). Local thermal equilibrium plasma can be generated by DC and RF arcs, or by an inductively coupled torch. They are used for plasma spraying (coating) and thermal plasma chemical and physical vapor deposition.

Figure 1.3 shows the typical parameters for naturally occurring and laboratory plasmas. Gas discharge and process plasmas have been discussed in this book. Their densities are in the range of 10^7–10^{14} cm^{-3}, and their electron energy is in the range of 1–100 eV.

1.3　Plasma Diagnostics

The oldest and frequently used diagnostic tool for low-temperature plasmas is the Langmuir probe. Since the 1920s, when the probe was invented by Irving Langmuir and his coworkers to measure DC plasma properties,[10] it has been developed for measurement under more general conditions such as for pulsed DC,[12] RF,[13] and microwave plasmas.[14] The Langmuir measurement is based on I–V characteristics of the Debye sheath, that is, the current density flowing to the surface of a plasma as a function of the voltage drop across the sheath. Detailed analysis of the I–V characteristics can yield primary plasma parameters: plasma density, plasma potential, floating potential, electron temperature, and electron energy distribution function (EEDF). More details on special arrangements and practical considerations of the Langmuir probe can be found in Chung et al.,[15] Huddlestone and Leonard,[16] Hutchinson,[17] and Lieberman and Lichtenberg.[18]

Active spectroscopy such as cavity ring-down spectroscopy (CRDS) and laser-induced fluorescence (LIF) provide highly sensitive ways to measure absolute density of plasma species. CRDS involves a laser that is used to illuminate a high-finesse optical cavity, which consists of a simple setup of two highly reflective mirrors. After the intensity of the laser builds up in the cavity due to constructive interference, the laser is turned off, so light is reflected between the two mirrors thousands of times and decays exponentially. After the plasma is introduced inside the cavity, the plasma species absorbs light, so the intensity of light decreases faster. The "ring-down time" is obtained by measuring the time needed by the light to decrease to 1/e of its initial intensity, and is used to calculate the density of plasma species.[19] After decades of development, plasma-CRDS (P-CRDS) has evolved into a powerful plasma diagnostic tool. Different laser sources, namely, continuous and pulsed wave lasers, are used to analyze various plasmas such as an ICP and a microwave-induced plasma; the absolute number density of many plasma species in time and in space are determined thus.[20-23] More details on configurations of an experimental P-CRDS and coupling of P-CRDS with OES for plasma diagnostics can be found in Chapter 8.

The LIF technique provides an efficient way to measure the ground state and long-lived, nonradiative, excited atoms, molecules, or radicals.[24] This technique has been used for many years for diagnostics in plasmas. It includes two major steps. The first one is to excite the atoms or molecules in the ground state (E_1) to a higher energy level (E_3) through resonant absorption of laser photons, with $E_3 - E_1 = h\upsilon_{laser}$. Afterward, the excited state comes down to a lower energy state E_2, by emitting fluorescence, with $E_2 - E_1 = h\upsilon_{LIF}$. The LIF system consists of a laser source, an arrangement of lenses, a fluorescent medium (plasma), collection optics, and a detector.[25] The popular laser sources for LIF are the Nd:YAG laser, dye lasers, excimer lasers, and ion lasers. The laser light passes through a set of lenses and mirrors to illuminate the plasma. Subsequently, the signal is captured by the charged coupled device (CCD) camera. A timing trigger device is usually used to synchronize the laser source and detectors. Two-photon absorption LIF (TALIF), as the name suggests, is the absorption of two photons followed by fluorescence of the third photon.[26,27] Due to the second-order effect, two-photon absorption is much weaker than single-photon absorption, but forbidden transition probabilities and vacuum ultraviolet (VUV) laser requirements make TALIF an alternative to single-photon absorption to investigate the ground-state species.[28-31] A quantitative description of the role of the collision process in plasma, calibration of LIF, and LIF measurement of active species in thermal plasma can be found in Chapter 9.

1.4 Applications of Plasma

The most common man-made plasmas in our daily life are plasma lamps. The fluorescent lamp and high-intensity arc lamp are primarily two types of plasma light sources.[32] The fluorescent lamp is a gas discharge lamp that uses electricity to excite mercury vapor. The UV light produced by mercury atoms causes a phosphor to fluoresce and produces visible light. The energy conversion efficiency of fluorescent lamps is a lot higher than that of incandescent lamps, so they are used as an energy-saving alternative in homes.[33] The high-intensity arc lamp produces light by means of an electric arc between tungsten electrodes housed inside a transparent quartz tube. Its color characteristics depend on the gaseous elements in the tube. They are widely used in commercial settings, especially for advertisement lighting in public areas. The plasma display panel is a technique based on fluorescent lamps. A panel typically has millions of tiny cells in the compartmentalized space between two panels of glass. The UV photons produced by the plasma strike on the phosphor that is painted on the inside of the cell. Different colors can be obtained depending on the type of phosphor used. Three cells comprising the primary colors of visible light constitute each pixel in the display panel. Varying the signal voltage to the cells thus results in different color outputs.[34] Compared with light-emitting diode (LED) display and liquid crystal display (LCD), plasma display has better color fidelity and wider viewing angles. A quantitative description on plasma light sources can be found in Chapter 5.

Plasma application is also an effective, cheap, and environmentally friendly process for the disinfection and degradation of organic pollutants in water. Compared with the traditional chlorination process, the ozonation process has stronger oxidization efficiency and no side effects. Siemens' ozonizer based on air corona discharge has been successfully used for ozone synthesis in many industrial fields for more than 150 years without any major modification.[35,36] Instead of *ex situ* discharge for ozone synthesis, the *in situ* discharge inside water or in close proximity to a water surface can produce more chemically active species such as H_2O_2, $O^\bullet$, $OH^\bullet$, $HO_2^\bullet$, $O_3^\bullet$, $N_2^\bullet$, e^-, O_2^-, O^-, and O_2^+.[36] Most of these species have even stronger oxidation potential than that of ozone. Therefore, dielectric barrier discharge (DBD), contact glow discharge electrolysis, and silent discharges are used for water treatment through direct electrical discharge. Furthermore, the strong electric field and UV radiation caused by these discharges are also lethal to several kinds of microorganisms present in water.[37] More details on plasma water purification and discharges in liquids can be found in Chapters 11 and 12.

The removal of NO_x from a mobile device is of growing concern in nonthermal plasma processing for air pollution control. Unlike removal of NO_x from a stationary source, this involves chemical reduction of NO_x to N_2 molecule rather than oxidation to HNO_2 or HNO_3. The plasma-driven catalysis reactor with a TiO_2 catalyst greatly enhances NO_x removal efficiency to 95% under proper plasma power.[35] Moreover, large-scale facilities for nonthermal plasma processing have been set up in Poland, China, Korea, and Japan. For the waste gas with flow rate of 35,000 $Nm^3 h^{-1}$, 70% removal rate of NO_x and 99% removal rate of SO_2 were achieved through plasma driven by 160 kW pulse power. DC + AC driving plasmas are also being considered for large-scale applications.[35] A quantitative description of environmental applications can be found in Chapter 11.

Due to the high bactericidal effectiveness and ease of access into narrow and confined spaces, plasma has been used for decades for packaging in the food industry, sterilization of surgery equipment, and blood coagulation. Recently, the development of low-temperature (<40°C) atmospheric pressure plasma sources extended plasma treatment applications.[38] Low-temperature plasma can combat fungal diseases efficiently and propagate through socks; therefore, 25–40% of the population with tinea pedis infection can be treated effectively with appropriate plasma devices.[39] Contact-free plasma is also an ideal candidate for normal dental care, because it can clean out bacteria in the teeth cavity without drilling, which reduces the patient's suffering a lot.[40] Moreover, for chronic wounds caused by venous diseases, arterial diseases, diabetes mellitus, and carcinoma, although plasma cannot cure the underlying disease, by eliminating bacterial and fungal infection, it can support the treatment and speed up recovery. Finally, the development of hand plasma sterilization devices provides a fast and efficient way to sterilize public buildings including hospitals, children nurseries, nursing homes, and so on.[38] More details on plasma medicine can be found in Chapter 14.

1.5　Organization of This Book

There are three major sections of this book. After the introductory chapter, Chapter 2 proceeds to provide the reader with necessary theoretical foundations in nonthermal plasmas at atmospheric pressure. The next section, "Processing and Characterization" consisting of seven chapters (Chapters 3–9), educates readers about an advanced plasma laboratory. Phenomena such as plasma bullets, plasma penetration into biofilms, discharge mode transition of atmospheric pressure plasmas, self-organization of microdischarges are introduced. The numerical diagnostics by particle-in-cell/Monte Carlo (PIC/MC) and fluid/hybrid models and experimental diagnostics through CRDS and the LIF measurement are presented. The final section, "Applications of Plasmas," consisting of six chapters (Chapters 10–15), not only discusses important practical applications such as environmental protection, treatment of biomaterials, and plasma medicine, but also summarizes research challenges and future trends associated with these applications. The book is a suitable degree-level text for students of engineering and science, and a research monograph for practicing engineers and scientists. In the remainder of this section, each chapter is examined one by one.

Chapter 3 presents general properties of atmospheric pressure thermal plasmas, as well as methods for their generation are discussed. Then, a complete description of atmospheric pressure nonthermal plasmas, including their general properties, instabilities of diffuse discharges and stabilization mechanisms, and different types of discharges, is presented.

Chapters 3 and 4 illustrate how the theoretical formulations on plasma physics and plasma chemistry presented in Chapter 2 are incorporated into the computational model. For the modeling of low-pressure plasmas, the principles, capabilities, and limitations of fluid/hybrid and PIC/MC models are reviewed and compared. For the modeling of atmospheric pressure plasmas, besides a detailed discussion on theoretical models and computation details, the studies on discharge mode transition and nonlinear behaviors through the modeling are also presented.

Chapter 5 begins with a review of electrode geometries and materials and fabrication techniques. The modes of operation, electrical characteristics, and microplasma parameters including gas temperature, electron density, and electron energy are highlighted. The chapter then turns to parallel operation, series operation, and OES diagnostics of microplasmas. The large surface-to-volume ratio of microplasmas not only results in higher energy coupling efficiency with the surroundings than the other discharge methods but also promotes a series of applications, including light sources, plasma reactors, plasma cathodes, thrusters, detectors, and biomedicals, which are described in the final section.

Chapter 6 reviews a pulsed power strategy for plasma generation. The mechanism of pulsed discharge plasmas and their advantages over conventional DC and AC discharge plasmas are highlighted. Different kinds of pulsed discharge sources such as pulsed corona, DBD, spark, and discharges in water are discussed. A detailed description of nanosecond repetitively pulsed (NRP) discharges, including NRP discharge regimes and their transition from corona and glow to spark, gas heating mechanism of NRP, and its applications on plasma-assisted combustion, is presented. A limitation of NRP is the necessity of expensive high-voltage pulse generators with high repetitive rates. The chapter then turns to more economic self-pulsed DC-driven discharges with discussions on streamer mechanism, transient spark–repetitive streamer-to-spark transition, self-pulsed discharge with water, and its applications on flue gas cleaning and decontamination.

Chapter 7 begins with a review of the application driving the development of plasma jet during the last two decades. Then, the key parameters such as driven source, electrode structures, and gas temperatures governing the reactivity of a plasma plume are highlighted. A dominant theme in this chapter is the change in the working gas from noble gas to air, in order to promote applications of plasma jet. The modifications in plasma jet structure based on working gas change are also presented. Then, the measurements of reactive species, such as O, OH, NO, and O_3, are discussed, followed by a section detailing plasma plume propagation characteristics.

Chapter 8 presents the experimental configuration details of P-CRDS and prospective applications of P-CRDS. First, the concept CRDS is covered, which is followed by a discussion on the principles of P-CRDS. Configurations of an experimental P-CRDS system, including laser sources, coupling of a plasma with CRDS, and electronics and data acquisition, are discussed. Descriptions of P-CRDS measurements including absolute number density measurements of plasma species and coupling of P-CRDS with OES for plasma diagnostics are presented. The final section in this chapter summarizes P-CRDS applications such as in plasma medicine, plasma-assisted combustion, and materials processing.

Chapter 9 starts with an overview of fluorescence excitation spectroscopy and LIF diagnostics in plasmas at various pressures. Background mathematical expressions of one-photon LIF, two-photon LIF, and optical–optical double-resonance LIF (OODR-LIF) are all presented. The measurements of internal state distributions and absolute density, including data set of quenching, VET and RET design LIF scheme, calibration of LIF for radical detection, calibration by sources of known radical density, calibration by noble gases, and Rayleigh scattering, are discussed. In the end, the applications of LIF measurement are presented, with emphasis on detection of $N_2(A)$ in pulsed DBD and nanosecond discharges by OODR-LIF and one-photon LIF, detection of N and O in DBD and atmospheric pressure plasma jet by TALIF, and drawbacks of LIF measurements.

Chapter 10 reviews the development of plasma nanotechnology in the past decades, highlights the basic principles of plasma nanotechnology, and discusses the condition and the methods required to obtain high-performance nanomaterials by plasma.

Chapter 11 examines gaseous discharges for pollution emission control, including electrostatic precipitation, ozone generation, and streamer corona discharges. Due to the advantage of air plasma generation, corona discharge modes and chemical reactivity are discussed taking into consideration initial radical generation, fine particle charging, and global chemical kinetics and multiple processing. A thorough discussion on power source and streamer corona plasma reactor, including energization methods and circuit technology, matching between power source and reactors, and system development and technical elements, is presented. The state of the art in flue gas cleaning and volatile organic compounds (VOCs) and odor emission abatement are then explained, with emphasis on NO_x, SO_2, and Hg oxidation and removal, plasma-initiated catalysis intensification, and integrated system and industrial demonstration of flue gas cleaning and plasma catalyst hybrid system. The chapter concludes with three sections detailing indoor air cleaning, fuel gas cleaning, and water cleaning, respectively.

Chapter 12 illustrates different types of discharges, including direct discharges in water, discharges in bubbles, discharges with water electrodes, and discharges with added water vapor or aerosol sprays. The physical properties and elementary water-based chemistry of the discharges in liquids are also discussed. Moreover, their applications are reviewed with emphasis on water and gas treatment, high-voltage switching, biomedical applications, and material synthesis and nanoparticle production.

Chapter 13 starts with a general introduction of surface engineering, including the reason for surface modification, surface modification strategies, and different methods to change polymers. A dominant theme discussed in this chapter is the ability of nonthermal plasmas to modify the polymeric surface. A complete description on the principles of plasma surface interactions and recent achievements on plasma surface modifications are presented, which is followed by techniques employed in polymeric surface modification including plasma postirradiation grafting, plasma syn-irradiation grafting, and plasma polymerization. The chapter concludes with a section on trends and future prospects of these techniques.

Chapter 14 presents the historical progression and development of applications of atmospheric pressure plasma on medicine. The plasma-mediated mechanisms for each application, including instrument sterilization, food decontamination, dental disinfection treatment, wound treatment, and dermatological therapy, are discussed. The research challenges and opportunities associated with these applications are also summarized in the end.

Chapter 15 discusses the surface modifications in implantable biomaterials via plasma-based processes. While several materials (316L stainless steel, cobalt–chromium alloys, and titanium-based alloys) are currently in use, titanium alloys are fast emerging as the first choice for the majority of applications. Hence, this chapter focuses on plasma processing of titanium-based materials, which are an ideal choice for orthopedic applications. The chapter is divided into five sections, starting with the requirements to be met in orthopedic implants, surface modification of titanium-based materials by liquid discharging processes, surface modification of titanium-based materials via plasma spraying processes, surface modification of titanium-based materials using plasma-ion implantation processes, and future prospects.

References

1. A. Anders, *IEEE Transactions on Plasma Science* 31, 1052 (2003).
2. J. Priestley, *The History and Present State of Electricity: With Original Experiments* (Dodsley, Johnson and Davenport, Cadell, 1767).
3. B. Dibner, *Galvani-Volta: A Controversy That Led to the Discovery of Useful Electricity* (Burndy Library, 1952).
4. A. Volta, *Philosophical Transactions of the Royal Society of London* 90, 403 (1800).
5. A. Anders, *IEEE Transactions on Plasma Science* 31, 1060 (2003).
6. M. Faraday, Abstracts of the papers printed in the *Philosophical Transactions of the Royal Society of London* 4, 54 (1837).
7. A.V. Engel, *Ionized Gases* (Oxford University Press, London, 1965).
8. Y.P. Raizer, *Gas Discharge Physics*, corrected edition (Springer, Germany, 1991).

9. I. Langmuir, *PNAS* 14, 627 (1928).

10. H.M. Mott-Smith and I. Langmuir, *Physical Review* 28, 727 (1926).

11. E. Liang, *Physics of Plasmas* 13, 064506 (2006).

12. C. Corbella, M. Rubio-Roy, E. Bertran, and J.L. Andújar, *Journal of Applied Physics* 106, 033302 (2009).

13. J. Hopwood, C.R. Guarnieri, S.J. Whitehair, and J.J. Cuomo, *Journal of Vacuum Science Technology A: Vacuum, Surfaces, and Films* 11, 152 (1993).

14. F. Werner, D. Korzec, and J. Engemann, *Plasma Sources Science and Technology* 3, 473 (1994).

15. P.M.-H. Chung, L. Talbot, and K.J. Touryan, *Electric Probes in Stationary and Flowing Plasmas: Theory and Application* (Springer-Verlag, New York, 1975).

16. R.H. Huddlestone and S.L. Leonard, *Plasma Diagnostic Techniques* (Academic Press, New York, 1965).

17. I.H. Hutchinson, *Principles of Plasma Diagnostics* (Cambridge University Press, New York, 2005).

18. M.A. Lieberman and A.J. Lichtenberg, *Principles of Plasma Discharges and Materials Processing* (John Wiley and Sons, Hoboken, 2005).

19. M.J. Thorpe, K.D. Moll, R.J. Jones, B. Safdi, and J. Ye, *Science* 311, 1595 (2006).

20. W.M.M. Kessels, J.P.M. Hoefnagels, M.G.H. Boogaarts, D.C. Schram, and M.C.M. van de Sanden, *Journal of Applied Physics* 89, 2065 (2001).

21. G. Berden, R. Peeters, and G. Meijer, *International Reviews in Physical Chemistry* 19, 565 (2000).

22. R. Engeln, K.G. Letourneur, M.G. Boogaarts, M.C. van de Sanden, and D. Schram, *Chemical Physics Letters* 310, 405 (1999).

23. Y. Duan, C. Wang, and C.B. Winstead, *Analytical Chemistry* 75, 2105 (2003).

24. J. Amorim, G. Baravian, and J. Jolly, *Journal of Physics D: Applied Physics* 33, R51 (2000).

25. G.P. Davis and R.A. Gottscho, *Journal of Applied Physics* 54, 3080 (1983).

26. E. Caussé, P. Malatray, R. Calaf, P. Charpiot, M. Candito, C. Bayle, P. Valdiguié, R. Salvayre, and F. Couderc, *Electrophoresis* 21, 2074 (2000).

27. J.P. Booth, G. Hancock, N.D. Perry, and M.J. Toogood, *Journal of Applied Physics* 66, 5251 (1989).

28. R.E. Walkup, K.L. Saenger, and G.S. Selwyn, *The Journal of Chemical Physics* 84, 2668 (1986).

29. K. Niemi, V.S. der Gathen, and H.F. Döbele, *Plasma Sources Science and Technology* 14, 375 (2005).

30. D.J. Bamford, L.E. Jusinski, and W.K. Bischel, *Physical Review A* 34, 185 (1986).

31. G.D. Stancu, F. Kaddouri, D.A. Lacoste, and C.O. Laux, *Journal of Physics D: Applied Physics* 43, 124002 (2010).

32. D.R. Demers and C.D. Allemand, *Analytical Chemistry* 53, 1915 (1981).

33. J.F. Waymouth and F. Bitter, *Journal of Applied Physics* 27, 122 (1956).

34. T. Jüstel, J.-C. Krupa, and D.U. Wiechert, *Journal of Luminescence* 93, 179 (2001).

35. H. Kim, *Plasma Processes and Polymers* 1, 91 (2004).

36. M.A. Malik, A. Ghaffar, and S.A. Malik, *Plasma Sources Science and Technology* 10, 82 (2001).

37. A. Bogaerts, E. Neyts, R. Gijbels, and J. van der Mullen, *Spectrochimica Acta Part B: Atomic Spectroscopy* 57, 609 (2002).

38. M.G. Kong, G. Kroesen, G. Morfill, T. Nosenko, T. Shimizu, J. van Dijk, and J.L. Zimmermann, *New Journal of Physics* 11, 115012 (2009).

39. G.E. Morfill, T. Shimizu, B. Steffes, and H.-U. Schmidt, *New Journal of Physics* 11, 115019 (2009).

40. X. Lu, Y. Cao, P. Yang, Q. Xiong, Z. Xiong, Y. Xian, and Y. Pan, *IEEE Transactions on Plasma Science* 37, 668 (2009).

2

Atmospheric Pressure Plasmas

Peter J. Bruggeman

2.1 Introduction

Atmospheric pressure plasmas (APPs) are very diverse and have widespread applications such as in waste reduction and gas purification, chemical conversion and syntheses such as ozone production, nanoparticle synthesis, surface functionalization of materials, and disinfection.

It is relatively easy to produce nonequilibrium* gas discharges at low pressure with gas temperatures close to room temperature. With increasing pressure, however, gas discharges have the tendency to become unstable and constricted: typically a glow-to-spark/arc transition occurs. At atmospheric pressure it is thus necessary to use special geometries, electrodes, or excitation methods to obtain nonequilibrium plasmas for which the gas temperature remains significantly smaller than the electron temperature.

In this chapter a description of the basic properties of different APPs is given. We start with a discussion on energy transfer and heating in APPs, which have direct implications for the wide range of gas temperatures occurring in APPs. Next the typical timescales of important chemical and physical processes are presented. High-pressure (atmospheric pressure) discharges also have some specific chemical and physical properties that differ from those of low-pressure discharges, which are outlined

---

* In the case of a nonequilibrium discharge the electron temperature (T_e) is significant larger than the gas temperature (T_g).

in a separate section. This is followed by a discussion on the electrical breakdown at atmospheric pressure and by transitions and instabilities of high-pressure plasmas. The different APPs are classified according to the excitation method and/or geometry. First DC and low-frequency AC capacitively coupled discharges that often have a high gas temperature and can even be in local thermal equilibrium (LTE), for which $T_e = T_g$, are discussed. The same can be said for inductively coupled plasmas (ICPs) and microwave (MW)-excited plasmas. All other discharge types are used when one of the goals is to reduce gas heating as much as possible. These discharges include corona discharges, radiofrequency (RF) and pulsed glow discharges, dielectric barrier discharges (DBDs), and so-called microplasmas.

It is not the goal to address all the specific topics in detail but rather to discuss and illustrate the general working mechanisms of APPs. More details on specific topics can be found in the dedicated chapters in this book or in references provided in the text.

2.2 Energy Transfer and Heating: Fundamental Relations

As most APPs are produced by applying a voltage across two electrodes, the electrical energy that is contained in the electric field in the electrode gap is transferred to the plasma by accelerating charged species. In general at high pressure there is limited direct energy transfer of the electric field to the heavy particles (ions) and the electrons are mainly heated by the electric field. The energy transfer of the electrons to the heavy particles (gas molecules) occurs through ionization, excitation, dissociation of molecules, and elastic collisions. In many steady-state high-pressure discharges the electron temperature is between 1 and 3 eV, and in this case, in atomic gases, the energy transfer from the electrons to the heavy particles is mainly dominated by elastic collisions. In the case of molecular gases vibrational excitation is normally dominant in this range of electron temperatures. An example of the relative energy transfer from electrons to the different processes is shown in the case of Ar and N_2 in Figure 2.1. Note that the energy transfer from the electric field to the electrons can be much more efficient than the subsequent collisional energy transfer between electrons and heavy particles mainly because only a fraction of about 10^{-4} of the energy difference between the electron and the heavy

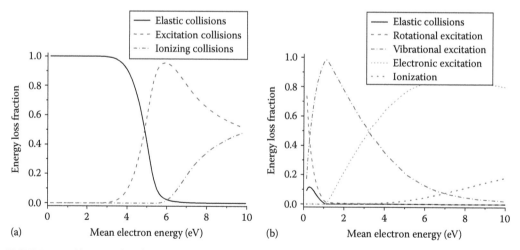

FIGURE 2.1 (a) Energy loss fractions of electron collisions in Ar consisting of elastic, excitation, and ionization collisions. (b) Energy loss fractions of electron collisions in N_2 consisting of elastic collisions, rotational, vibrational, electronic excitation, and ionization collisions. The data are obtained by the program Bolsig +. (Data from Hagelaar, G.J.M. and Pitchford, L.C., *Plasma Sources Sci. Technol.*, 14, 722–733, 2005.)

particle is transferred in each collision due to the large mass difference. Thus, the gas temperature can be significantly lower than the electron temperature.

Let us now consider a plasma in an atomic gas with an electron temperature below 3 eV. The energy transfer of the electrons to the gas by elastic collisions is determined by the elastic collision rate coefficient. Typical elastic collision rates are of the order of 10^{-13} m^3 s^{-1}. With increasing pressure, the collision frequency increases, which causes a more efficient energy transfer and thus gas heating. The difference between the electron temperature and gas temperature can be estimated by equating the average energy gained by an electron in the electric field between two collisions and the average energy transferred in a collision between an electron and a neutral. This leads to the following equation (Finkelnburg and Maecker 1956):

$$\frac{T_e - T_g}{T_e} = \frac{m_g}{4m_e} \frac{(\lambda_e e E)^2}{\left(\frac{3}{2} k_B T_e\right)^2}, \tag{2.1}$$

where
  m_g is the heavy neutral particle mass
  m_e is the electron mass
  k_B is the Boltzmann constant
  E is the electric field
  e is the elementary charge
  λ_e is the electron mean free path

Equation 2.1 shows that the difference between T_e and T_g increases with increasing pressure (decreasing main free path) and increasing electric field. Figure 2.2 illustrates a typical pressure dependence of T_e and T_g for a steady-state arc plasma. Of course the dependence in this figure is only valid when the gas temperature is mainly determined by the balance between the electron energy transfer to the heavy particles and not by strong heat removal which, for example, occurs in regions of large plasma gradients such as sheath regions or when a strong gas flow is applied.

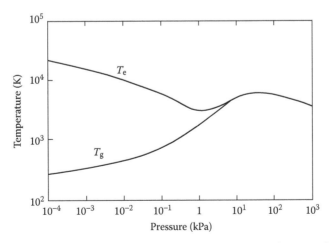

FIGURE 2.2 Electron temperature and gas temperature in a mercury arc as a function of pressure. (Reprinted from *Gaseous Electronics: Electrical Discharges*, 1, Hirsh, M.N. and Oskam, H.J., Electric arcs and arc gas heaters, 291–398, Copyright 1978, with permission from Elsevier.)

A lot of additional insight can be obtained from a power balance between the energy input in a plasma and the heat removal. Assuming that the energy used per ionization in the bulk is known and equal to $E(T_e)$,* the power dissipation in the plasma can be written as

$$\frac{P_{\text{heat}}}{V} = n_e n_g k_e(T_e) \varepsilon(T_e),$$

where

n_e is the electron density
n_g is the neutral gas density
$k_e(T_e)$ is the (effective) ionization rate
P_{heat} is the power
V is the plasma volume

One can estimate the heat loss (P_{loss}) in the plasma as follows by assuming that the heat loss is mainly due to thermal conduction in the gas phase:

$$\frac{P_{\text{loss}}}{V} = -\nabla \lambda \nabla (T_g - T_{\text{wall}}) \propto \frac{n_g T_g^{3/2}}{L^2},$$

where λ is the thermal conductivity of the gas. On the right-hand side of the equation the gradient is estimated by a plasma dimension L, the wall temperature (T_{wall}) is assumed to be much smaller than T_g, and $\lambda \propto n_g \sqrt{T_g}$ as is known from kinetic gas theory. By assuming that the energy dissipation in the plasma equals the heat loss in the steady-state condition, the following relation can be deduced:

$$n_e L^2 \propto \frac{T_g^{3/2}}{\varepsilon(T_e) k_e(T_e)}. \tag{2.2}$$

This relation clearly shows that the gas temperature increases with increasing electron density and increasing plasma size. Low-density small-sized plasmas will thus always have a smaller T_g than their large-density large-sized counterparts.

Equations 2.1 and 2.2 show that to obtain nonthermal plasmas for which $T_e \gg T_g$, the following approaches can be used:

- Preventing thermalization/equilibrium by pulsing the plasma
- Increasing the electric field by using sharp electrodes as in corona discharges
- Reducing n_e or the current by introducing dielectric or resistive barriers
- Improving the heat transfer by
  - Forced convection
  - Using gases with a large thermal conductivity such as helium
  - Reducing the plasma size

It is important to note that plasmas produced by continuous DC excitation that have a relatively large volume at atmospheric pressure often have an elevated T_g that can approach T_e. The most natural high-pressure plasma is thus a hot plasma, and to obtain plasmas that operate close to room temperature one or more of the approaches summarized above can be employed.

2.3 Typical Timescales

The different physical and chemical processes that are of importance or directly influence the plasma properties at atmospheric pressure span a time range of about 12 orders of magnitude. It is thus very important to have a good understanding of which processes are relevant especially for time-modulated plasmas.

---

* In this energy the average energy losses by all elastic and inelastic collisions are included. For more details the reader is referred to Lieberman (2005).

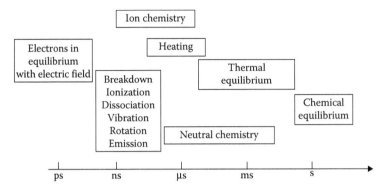

FIGURE 2.3 Schematic overview of the typical important timescales in APPs.

An overview of the most important physical and chemical processes is shown in Figure 2.3. The fastest timescale is of the order of picoseconds and determines the equilibrium of the electrons in the applied electric field. The typical neutral collision frequency at atmospheric pressure is of the order of 10^9 s^{-1} at room temperature. Ionization typically occurs at nanosecond timescales. Heavy particle heating takes a significantly longer time, that is, about 100 ns–1 μs. For chemical reactions (e.g., radical formation) three different timescales need to be considered: (1) electron-induced dissociation reactions for larger T_e values typically occurs on timescales of a few nanoseconds; (2) radical chemistry due to interaction with neutrals is mainly on a microsecond timescale; and (3) ionic reactions that are important in higher density plasmas occur at an intermediate timescale, which is typically in the range 10 ns–1 μs.

It normally takes several milliseconds for thermalization to occur between the electron and gas temperature. As thermalization is a necessary condition for chemical equilibrium it only occurs at a timescale of seconds. The above-mentioned timescales are the reason why plasma chemistry is in most cases strongly nonequilibrium chemistry. Note that reaction rates can strongly depend on T_e and n_e, so different plasmas can have different timescales for ionic and electronic processes.

2.4 High-Pressure Plasma Chemistry

As we have seen above, at low pressure the (elastic) collision frequency is much smaller compared to that at atmospheric pressure. Apart from an increase in collision frequency a shift in dominant chemical reactions can occur.

A very straightforward example is three-body reactions that are of increasing importance at high pressure. Two examples that are the basis of two major applications of atmospheric pressure are (Kogelschatz 2003):

ozone formation

$$O + O_2 + O_2 \rightarrow O_3 + O_2$$

and

excimer formation (e.g., in xenon)

$$Xe_m + 2Xe \rightarrow Xe_2^* + Xe,$$

in which the Xe_2^* decay causes the excimer band emission in the UV at around 172 nm. Both ozone generators and excimer lamps (and lasers) are based on high-pressure discharges because the above processes are very slow at reduced pressures as the rate constant has a quadratic dependence with the gas density. Note that in general the heterogeneous (surface) reactions that often play an important role at low pressure become significantly less important at high pressure, and apart from a few exceptions bulk chemistry becomes dominant.

A second difference at atmospheric pressure is the ion chemistry. Even in the case of atomic noble gases, at high pressure and not very elevated temperatures atomic ions are quickly converted into molecular ions:

$$Ar^+ + 2Ar \rightarrow Ar_2^+ + Ar.$$

Typical rates of dissociative electron–ion recombination such as for the following reaction

$$e + Ar_2^+ \rightarrow Ar_m + Ar,$$

are of the order of 10^{13}–10^{14} m^3 s^{-1}, which is several orders of magnitude faster compared to three-body electron–ion recombination or radiative electron–ion recombination for which a second electron or photon is necessary to simultaneously conserve momentum and energy (Fridman 2009):

$$Ar^+ + 2e \rightarrow 2Ar + e \quad \left[k \approx 10^{-39} \text{ m}^6\text{s}^{-1} \ (T_e = 1 \text{ eV}) \right],$$

$$Ar^+ + e \rightarrow Ar + h\nu \quad \left[k \approx 3 \times 10^{-19} \text{ m}^3\text{s}^{-1} \ (T_e = 1 \text{ eV}) \right].$$

Because of the low recombination rates of atomic ions in low-pressure plasmas, diffusion is often the dominant particle-loss mechanism, while at atmospheric pressure (for which the electron density is often in the range of 10^{20}–10^{21} m^{-3}) dissociative electron–ion recombination and thus bulk recombination often becomes dominant compared to wall losses. Additionally, at high pressure dissociative recombination and in general ion chemistry can play an important role in radical production when the ion density is significantly large (10^{20}–10^{21} m^{-3} or larger). A typical example is OH production on the edge of a plasma filament, which can be explained by charge exchange and dissociative electron–ion recombination reactions when the core of the filament is assumed to consist mainly of atomic ions and the water is fully ionized (Verreycken et al. 2012):

$$H^+ + H_2O \rightarrow H_2O^+ + H \quad (k \approx 7 \times 10^{-15} \text{ m}^3\text{s}^{-1}),$$

$$H_2O^+ + H_2O \rightarrow H_3O^+ + OH \quad (k \approx 10^{-15} \text{ m}^3\text{s}^{-1}),$$

$$H_3O^+ + e \rightarrow OH + 2H \quad \left[k \approx 10^{-13} \text{ m}^3\text{s}^{-1} \ (T_e = 1 \text{ eV}) \right].$$

In electronegative gases, negative ion chemistry can be significantly different at high-pressure compared to low-pressure plasmas. At low pressure attachment mainly proceeds through dissociation, which requires bridging the threshold energy for the dissociation. A typical example is

$$O_2 + e \rightarrow O + O^-.$$

However, at atmospheric pressure three-body attachment plays a key role for balancing the charged particles

$$O_2 + e + M \rightarrow O_2^- + M,$$

where M is a third-body heavy particle. This is particularly important for pulsed discharges in air for which the above three-body attachment reaction is often a dominant charge-loss mechanism in the recombination phase, because of which when the next pulse is applied negative ions can increase the production of electrons in the newly developing ionization front. For a more detailed discussion on ion chemistry in APPs, the reader is referred to the book *Plasma Chemistry* (Fridman 2009).

An additional difference at high pressure compared to at low pressure is the strong electronic quenching of excited states, which in some cases significantly reduces the emission. Many examples exist such as for the excited OH(A), which causes the typical OH(A–X) emission band at around 306 nm. The excited state can be strongly quenched (and correspondingly the emission drastically reduced) due to the competitive reaction

$$OH(A) + M \rightarrow OH(X) + M^*,$$

where M is a colliding heavy particle. This phenomenon not only plays an important role in applications such as in the production of efficient gas discharge lamps but is also of importance for fundamental investigations and diagnostics when probing the OH density by laser-induced fluorescence.

2.5 Breakdown, Transitions, Instabilities, Constrictions, and Streamer Formation

A typical figure that is discussed in almost every textbook is the current–voltage (*I–V*) characteristic of a DC discharge in a range from nanoamperes up to several kiloamperes, as shown in Figure 2.4. The graph is of course not stably realizable in one single setup, but it illustrates the different stages of a discharge. Basically when a voltage is applied to the electrodes, the charges produced by external influences (cosmic radiation and so on) are collected at the electrodes. At some value of the voltage the electric field is large enough to accelerate electrons, which can cause ionizing collisions. This process continues with the help of secondary electron emission at the cathode and develops into a Townsend discharge for which the space charge is still negligible. With increasing current, space charge becomes important and sheaths are formed at the electrode so a transition to a glow discharge occurs. This transition is often called a breakdown as a glow discharge is a self-sustained discharge. With a further increase in current, the current density starts to increase when the transition from a normal glow to an abnormal glow occurs and plasma heating becomes important. At a critical point this triggers a transition to an arc discharge.

At low pressure, a common breakdown mechanism is the Townsend breakdown. Basically avalanches run between two metal electrodes in a homogeneous applied field, and space charge effects are negligible. During this process the electron density builds up by bulk ionization, and secondary electron emission occurs by ionic impact at the cathode (Figure 2.5a). However, with increasing pressure ionization can occur on much smaller length scales, and space charge effects are sometimes no longer negligible.

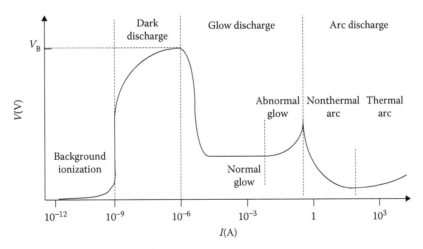

FIGURE 2.4 *I–V* characteristic of a parallel-plate discharge and the transitions from Townsend to glow and arc discharge. (Modified from Roth, J.R., *Industrial Plasma Engineering*, Institute of Physics Publishing, Bristol and Philadelphia, 1995.)

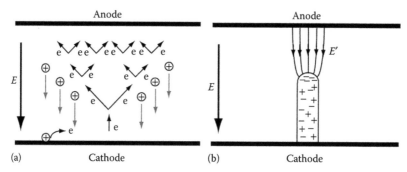

FIGURE 2.5 Schematic representation of (a) Townsend breakdown and (b) streamer breakdown.

The amount of charge in the gap can, at sufficient charge density, produce its own electric field that equals the applied homogeneous electric field, and it can effectively sustain its own ionization front and a transition to a filamentary self-propagating ionization front which is called a streamer occurs (Figure 2.5b). The criterion for the conditions under which the transition to a streamer occurs is called the Meek criterion. In the case of air at atmospheric pressure the criterion can be estimated as follows (Raizer 1987):

$$\int_0^d \left[\alpha\left(\frac{E}{N}\right) - \beta\left(\frac{E}{N}\right) \right] dx \approx 18 - 20, \tag{2.3}$$

where
  α is the Townsend ionization coefficient
  β is the attachment coefficient
  E/N is the reduced electric field
  d is the gap length

Note that from this criterion a critical electron density can also be estimated, which is of the order of 10^{17} m^{-3} in the case of air at atmospheric pressure. When the Meek criterion is satisfied, the Townsend discharge changes into a streamer, which can subsequently change into a spark when reaching the second electrode. In the spark significant gas heating occurs, and depending on the power input of the power supply it turns into an arc- or a glow-like discharge and even quench.

If a diffuse discharge is required, one needs to ensure that the Meek criterion is not reached. Apart from obvious techniques such as varying the geometry and excitation voltages, streamer formation can be prevented by preionization and slowing down the ionization in the avalanche by using well-chosen gas mixtures that strongly influence the ionization process.

The approach of using preionization is perhaps not self explaining but is relative easy to understand (Levatter and Liu 1980). Two avalanches that are close to each other are impacted by each other's electric field. The electric field of these avalanches is the same but has a different sign in the overlapping region of the field lines. This causes an effective reduction of the electric field in the avalanche. Increasing the preionization will increase the number of avalanches in the discharge gap, and thus the average distance between the avalanches is reduced so the electric field buildup by one avalanche is reduced by the surrounding avalanches. Reducing the electric field and thus the ionization frequency directly reduces the integral in Equation 2.3 and ensures that the Meek criterion is not fulfilled and the avalanche to streamer transition is postponed.

Another more chemical approach is introducing a two-step ionization such as Penning ionization:

$$Ar_m + M \rightarrow Ar + M^+.$$

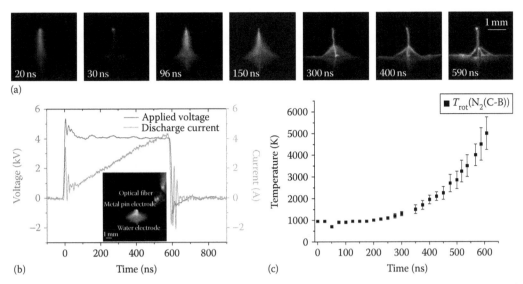

FIGURE 2.6 (a) 10 ns resolved images of a 600-ns pulsed discharge in air between a metal pin electrode and a water surface. (b) The corresponding *I–V* waveform. A time-averaged image (integration time of 1/30 s) of the plasma is inserted in the graph. (c) The evolution of the gas temperature during the plasma pulse. The constriction of the discharge around 300 ns can be clearly seen with a corresponding temperature increase. (Reprinted with permission from Bruggeman, P., et al., *Plasma Sources Sci. Technol.*, 18, 045023, 2009, © IOP.)

Basically one adds a trace gas to the bulk gas (e.g., Ar). The production of Ar metastables (11 eV) can be done at smaller electric fields compared to the ionization of Ar (15.6 eV). The metastables then easily ionize molecules or atoms (M) with an ionization energy ≤11 eV, and if this Penning ionization is the dominant mechanism, then the two-step process is typically slower than a direct ionization and can proceed at smaller electric fields. An example is the generation of Ar discharges with a trace gas such as ammonia, which is diffuse for conditions for which a pure Ar discharge is filamentary (Massines et al. 2009).

In order to understand the typical timescales of glow-to-spark/arc transition, it is illustrative to discuss the example of a DC pulsed discharge, as shown in Figure 2.6. The discharge is produced by applying a 600 ns voltage pulse (with amplitude of 4 kV) to a needle electrode positioned 1.5 mm above a water surface that serves as a grounded electrode. The plasma is operated in air (containing water vapor evaporated from the liquid water electrode) at a frequency of 6 kHz. From Figure 2.6 it is clear that a transition occurs at 300 ns when the plasma constricts. This constriction coincides with the start of a close to exponential increase in the gas temperature. The exact timescale of the increase depends on the specific conditions (applied voltage, energy injected in the discharge, electrode material, etc.) but a typical timescale for excessive heating is of the order of 100 ns. It shows that pulsing the plasma to prevent instabilities (contraction) from occurring has to be done at typical timescales of 10–100 ns.

The above example also clearly illustrates the tendency of APPs to constrict and not be diffuse but rather filamentary. This is generally valid as can also be seen in naturally occurring plasmas such as the aurora high in the atmosphere (at low pressure), which is diffuse, and lightning close to the earth's surface at atmospheric pressure, which is strongly filamentary.

A constriction of an initially diffuse (glow) discharge is a common instability. This instability is induced and coincides with an increase in electron density and gas temperature. When a discharge is stable, electron production is balanced by electron losses. The electron density demonstrates runaway behavior if the plasma operates under conditions such that an increase in the electron density cannot be compensated by electron loss. Mathematically, the balance of the electron density can be written as

$$\frac{dn_e}{dt} = k_{ion}(T_e)n_e n_g - \frac{D_a}{\Lambda^2}n_e - k_{dr}n_e^2, \tag{2.4}$$

where

k_{dr} is the dissociative recombination losses
D_a is the ambipolar diffusion coefficient
Λ is the effective diffusion length

If a change in electron density does not influence the electron temperature, the ionization rate scales linear with n_e. Thus, diffusion and dissociative recombination contribute to stabilizing the discharge when a fluctuation of the ionization rate causes a small linear increase in n_e.

However, any process that increases the ionization rate faster than the linear dependence of n_e on the diffusion losses can cause potentially runaway behavior. This occurs with an increase in the gas temperature. Indeed, an increase in the gas temperature leads to an increase in the reduced electric field (E/N) (assuming that the electric field remains constant as in the positive column of a glow discharge). This leads to a subsequent increase of the electron temperature and ionization rate, which increases the electron density. A larger electron density in turn leads to more Ohmic losses and thus an increase in gas temperature. It is clear that this process causes an increase in the ionization rate in Equation 2.4 and causes a stronger effective dependence on the electron density than the linear dependence when T_e remains constant. The heating and increase in T_e preferentially happen in the core of the discharge, which also promotes the radial shrinking of the discharge. The above instability is referred to as thermal instability and is common in high-pressure gas discharges.

Apart from thermal instability, stepwise ionization due to a stronger accumulation of metastable species can significantly increase the ionization rate and also cause instabilities. This is why in some cases strong gas flow is used to stabilize the discharge. An additional instability can also be triggered in electronegative discharges when detachment (which is independent of T_e) balances a significant part of the attachment. In this case an increase in n_e will lead to a reduction in T_e and consequently a reduction in the attachment rate. In this case the detachment is faster compared to the effect of gas composition on the attachment, and the detachment causes a further increase in n_e.

An example of constriction of a high-pressure atomic plasma column is studied by Castaños-Martínez et al. (2009) for an atmospheric pressure MW discharge in Ar–Ne mixtures. It has been shown that both pure Ne and Ar discharges are radially constricted, while a Ne discharge with an addition of 0.5–1% Ar fills the entire tube and is more diffuse (Figure 2.7). Note that the ionization energy of Ne and Ar is 21.6 eV and 15.6 eV, respectively. The charge conservation in the case of an Ar discharge with Ar_2^+ as the dominant ion is governed by the following reactions:

$$e + Ar_2^+ \rightarrow Ar_m + Ar,$$

$$e + Ar_m \rightarrow Ar^+ + 2e,$$

$$Ar^+ + 2Ar \rightarrow Ar_2^+ + Ar.$$

Note that in this case the ionization is very fast and depends nonlinearly on the electron density as the electron loss process produces metastables that can easily be ionized (energy threshold of about 4 eV) in a plasma with a T_e of 1–2 eV (which is typical for these MW discharges). The ionic and electronic losses are also bulk losses and not determined by diffusion.

In the case of the diffuse plasma, however, the dominant ion is Ar^+ and the following reactions dominate the charge balance:

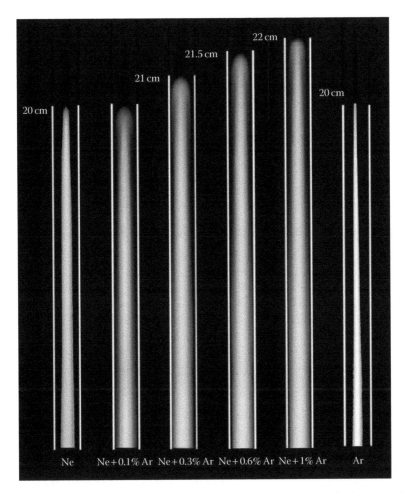

FIGURE 2.7 Images of a surface wave tubular discharge in Ne–Ar mixtures at atmospheric pressure. The diameter of the dielectric tube (indicated in the image by the white lines) is 12 mm. Both the pure Ne and Ar discharges are contracted but a mixture of Ne with about 1% of Ar fills the entire tube with a diffuse discharge. (Reprinted with permission from Castaños-Martínez, E., et al., *J. Phys. D Appl. Phys.*, 42, 012003, 2009, © IOP.)

$$e + Ar \rightarrow Ar^+ + 2e,$$

$$e + Ne \rightarrow Ne_m + e,$$

$$Ne_m + Ar \rightarrow Ar^+ + Ne,$$

$$e + Ar^+ (+Ar) \rightarrow Ar (+Ar).$$

The ionization rate is clearly much smaller as it is dominated by Penning ionization and direct electron ionization from the ground state of Ar. This is consistent with the above discussion on ionization instabilities. Additionally, the recombination rate of atomic ions is about 5 orders of magnitude smaller than the dissociative recombination rate, which makes ambipolar diffusion the dominant loss mechanism for charged species in this diffuse discharge.

Note that the above mechanism seemingly has similarities with the Penning mixtures used to create diffuse discharges and prevent streamer formation. However, it is important to realize that in high-density diffuse discharges, processes at timescales that are significantly slower compared to the few nanoseconds in which an avalanche crosses the electrode gap can be important for ionization. The charge density is also significantly larger; thus recombination products from the produced ions can indeed facilitate the charge creation, while this is not the case in an ionization front.

The above case is a typical example of APP instabilities. In several discharges that are contracted dissociative recombination is the dominant loss mechanism. This is mainly due to the fact that atomic ion recombination is normally significantly slower and in this case ambipolar diffusion is often important, which also causes the discharge to become more diffuse. The above illustrates that a requirement for contraction is that the electrons or ions have to be produced and lost locally. This can be important even in atomic noble gases for which the dominant ion can be a dimer (e.g., Ar_2^+).

2.6 DC and Low-Frequency AC Capacitively Coupled Discharges

2.6.1 DC and Low-Frequency AC-Excited Glow Discharges

The classical low-pressure glow discharge has been studied extensively for several decades. The light emission pattern of a low-pressure glow discharge is described in most standard gas discharge textbooks and includes a cathode glow, a cathode dark space, a negative glow, Faraday dark space, a positive column, an anode dark space, and an anode glow (e.g., Lieberman and Lichtenberg 2005). The main characteristic of the DC glow discharge is that the discharge is maintained by electron production at the cathode by secondary electron emission due to ion impact. At high pressure ionization in the sheath also occurs. The electrons originating from the cathode are accelerated in the sheath and cause significant ionization, which maintains the discharge. The positive column just connects both sheath regions. The constant electric field in the positive column ensures that the electron losses are compensated by an equal amount of electron production.

In spite of the fact that it is easy to produce DC glow discharges at low pressure, with increasing pressure the glow discharge has the tendency to become unstable and constricted: a glow-to-arc/spark transition then occurs. At atmospheric pressure it is thus necessary to use special geometries, electrodes, or excitation methods to obtain diffuse glow discharges. As the timescale on which a glow-to-arc/spark transition occurs is typically of the order of (a few) 100 ns (see previous section), switching off the discharge before the transition to a spark/arc occurs is an often used approached to produce diffuse atmospheric pressure glow discharges. A lot of studies on the glow-to-spark transition are reported. Unfortunately, the instability (constriction) can occur in the cathode, the anode, and the positive column region. This illustrates that the exact discharge geometry, electrode properties, and even gas composition have a strong influence on the instabilities of the glow discharge. Nonetheless, there needs to be a minimum power input for the glow-to-spark transition to occur. This is related to a large electron density and an increase in the gas temperature, conditions under which thermal instability can occur.

In some literature the term glow discharges is used for a diffuse-looking discharge. However, even diffuse-looking atmospheric pressure discharges can consist of a large number of filaments. It is therefore more correct to reserve the term glow discharge for a discharge with properties similar to those of a glow discharge at low pressure.

The typical properties of a normal glow discharge that are also valid at atmospheric pressure include the following:

- The reduced current density (J/N) is independent of the density and applied voltage.
- The characteristic light emission pattern, especially the cathode dark space, is visible, although the size is significantly smaller compared to the low-pressure cases as it scales with density.
- The discharge voltage is independent of the current when corrected for temperature rise, constriction of the positive column, and current dependence of the cathode–anode voltage drop.
- The burning voltage and cathode voltage drop is significantly larger than for an arc discharge.

Nonetheless, apart from the similarities with low-pressure glow discharges, there are several differences. The electron temperature is still significantly larger compared to the gas temperature, but due to the higher pressure the collisionallity is much larger and the gas temperature can reach 1000 K or more in a glow discharge. Therefore, the typical scaling laws valid at low pressure have to be used as a function of density and not pressure.

Additionally, the sheath is highly collisional at atmospheric pressure, which means that the ion energies impacting on the electrode or substrate are significantly lower (typically not more than a few electron volts under standard conditions) compared to lower-pressure discharges, which can have ion energies of several tens of electron volts and are often used for material sputtering applications.

Diffuse atmospheric pressure glow discharges have an electron density in the range 10^{16}–10^{19} m^{-3} (e.g., Barinov et al. 1998). The main difference between low-pressure and atmospheric pressure glow discharges is the loss mechanisms for electrons in the positive column. At low pressure the electron losses are due to radial ambipolar diffusion, while at higher pressure electron–ion dissociative recombination becomes more important. Because of this difference in electron loss mechanisms, at larger electron densities/powers atmospheric pressure glow discharges are often contracted and resemble arc discharges. In electronegative gases such as air, attachment losses can be important.

A recent study of the glow-to-spark transition in a metal pin–water electrode geometry showed that broadened sparks occur close to the water electrode (Bruggeman et al. 2008b). This illustrates the stabilizing effect of resistive (water) electrodes on the constriction of diffuse glow discharges even after a contraction has occurred in the bulk of the discharge. Stable DC and low-frequency AC-excited glow discharges in air at atmospheric pressure have been obtained by Laroussi et al. (2003) by using liquid electrodes. An example of a DC-excited glow discharge at atmospheric pressure with a liquid electrode that clearly illustrates the different characteristic zones of a glow discharge is shown in Figure 2.8.

A second approach to stabilize DC or low-frequency AC-excited glow discharges is by applying a fast gas flow. Significant gas flow not only can hasten heat removal in the discharge but also can contribute in enhancing charge and radical transport, which additionally alter the discharge properties. Stable DC glow discharges have also been produced at atmospheric pressure between two metal electrodes in microgaps (Staack et al. 2007).

The term abnormal glow discharge is normally used for glow discharges that show an increase in voltage with increasing current. At low pressure this occurs when the current density increases in the case that the electrode area of the cathode is fully covered by the discharge. At high pressure the radial

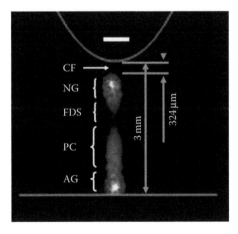

FIGURE 2.8 Glow discharge at atmospheric pressure between a metal pin cathode and water anode. The glow discharge is sustained at 1.25 mA and 1 kV. The structure of a glow discharge is clearly visible. CF, cathode fall; NG, negative glow; FDS, Faraday dark space; PC, positive column; AG, anode glow. (Reprinted with permission from Bruggeman, P., et al., *Plasma Sources Sci. Technol.*, 17, 045014b, 2008a, © IOP.)

diameter of the glow discharge is less and this specific phenomenon is mainly an issue when dealing with small electrodes.

2.6.2 Arc Discharges

The main difference between an arc and a glow discharge at atmospheric pressure is that the current of an arc is significantly larger (1 up to several kA) and the cathode sheath region is significantly different. Unlike in glow discharges in which the secondary electrons from the cathode are produced by ion impact, the secondary electron emission in arc discharges can be triggered by both thermal emission and field emission. As such, the cathode voltage drop is smaller (typically 15 V) because there is no need to accelerate the ions toward the cathode and the voltage drop only serves to accelerate the electrons up to the ionization threshold of the gas. Thermal emission of electrons can be calculated from the well-known Richardson equation for current density J (e.g., Boulos et al. 1994):

$$J = AT^2 \exp\left(-\frac{e\varphi}{k_B T}\right) (\text{A m}^{-2}),$$

where the theoretical value of the constant A is equal to 1.2×10^6 A m^{-2}K^{-2}. However, the practical value of A depends on the electrode metal. As the work function φ is typically between 2 and 5 eV, thermal electron emission becomes important only at elevated temperatures typically present in an arc discharge.

The bulk region of an arc consists of a positive column that is normally more contracted than its counterpart in the glow discharge and also has a significantly higher electron density of 10^{22}–10^{25} m^{-3}. Arc discharges are typically categorized as thermal and nonthermal arcs. Low-pressure arc discharges are often nonthermal, as can be clearly seen in the steady-state T_e–T_g variation with pressure in Figure 2.2. In the case of thermal arcs it is often a significant simplification to assume that local thermodynamic equilibrium (LTE) exists. LTE conditions require that the following three conditions are fulfilled locally:

1. The velocity distributions of particles follow a Maxwell–Boltzmann distribution.
2. The population density of excited states follows a Boltzmann distribution.
3. The particle densities and ionization degrees follow the Saha equation.

The above three requirements are met when collision processes (and not radiative processes) govern transitions and reactions in plasma. Additionally, as mentioned, the gradients in plasma properties should be small, or, more accurately, the diffusion time should be larger than the time necessary to obtain equilibrium.

It is instructive to treat the third condition in more detail. The Saha equation (e.g., Boulos et al. 1994) is as follows:

$$\frac{n_i n_e}{n_a} = \frac{g_i g_e}{g_a}\left(\frac{2\pi m_e k_B T}{h^2}\right)^{3/2} \exp\left(-\frac{E_{ion}}{k_B T}\right),$$

where
E_{ion} is the ionization energy
h is the Planck constant
g and n are the degeneracy and densities of the ion, electron, and atom, respectively

This equation is obtained from the following reaction by assuming equilibrium

$$e + A \leftrightarrow A^+ + 2e,$$

and is a specific form of detailed balancing or a mass action law that can also be applicable for the dissociation degree in a molecular plasma. However, in the case of ionization, electron impact ionization

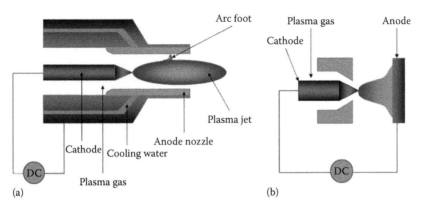

FIGURE 2.9 Schematic representation of (a) transferred arc and (b) nontransferred arc. (Reprinted from *Spectrochim. Acta Part B At. Spectrosc.*, 61, Tendero, C., et al., Atmospheric pressure plasmas: A review, 2–30, Copyright 2006, with permission from Elsevier.)

needs to be balanced by two electron recombinations. Thus, this equation is not applicable when molecular ions are present in the plasma or at low electron densities when other recombination losses such as dissociative molecular ion dissociation or diffusive losses are dominant.

In the context of applications one often refers to transferred and nontransferred arcs. In the case of transferred arcs, the arc is drawn between an electrode and a conducting workpiece. The energy transfer efficiency to the workpiece is very high, and the discharge has a large gas temperature. Transferred arcs are typically used in welding and cutting applications. Nontransferred arcs are basically remote plasma torches and are typically used for spraying and waste destruction. Many designs exist and the reader can refer to Boulos et al. (1994) and Roth (1995) for a more extensive overview. Two typical examples of a transferred and a nontransferred arc are shown in Figure 2.9.

It is important to remember that arc discharges have to be ignited with an additional voltage apart from the standard power supply. From Figure 2.4 it is clear that the discharge has to pass all the discharge stages that require a significantly larger voltage compared to that for an arc discharge. Often an arc is drawn by making contact with the two electrodes, or an inductor is used to produce a voltage spike as in a "lamp starter."

Arc discharges can also become unstable due to many prevalent phenomena. At currents of several kiloamperes and higher it is possible that the kinetic pressure of an arc column may not be able to withstand the inward magnetic pressure induced by the electric current through the column. In this case pinching occurs. This phenomenon can produce a very strong plasma that yields multiple ionized atomic ions and is used in applications for extreme UV generation (Kieft et al. 2003).

2.6.3 Gliding Arc Discharges

Gliding arcs have properties of both thermal and nonthermal plasmas. They are highly reactive and often have a high selectivity for chemical processes. The main reason why gliding arcs are often used for chemical-oriented studies is because they have properties of both thermal (large electron densities, currents, and power) and nonthermal plasmas (low gas temperature).

A gliding arc is normally generated between two diverging electrodes in a gas flow. An example of a configuration is shown in Figure 2.10.

The discharge ignites where the electric field is maximum, that is, at the smallest distance between the curved electrodes. This distance is typically of the order of a (few) millimeters. At this position a hot quasi-thermal plasma that moves upward is formed (due to the gas flow or the buoyancy force). As the electrodes are curved the length of the plasma column increases, increasing the heat losses in the column,

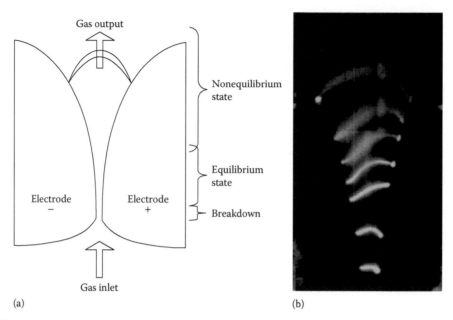

(a)   (b)

FIGURE 2.10 (a) Gliding arc electrode configuration with indication of the LTE and non-LTE region. (b) Superposition of snapshot images of a gliding arc discharge that illustrates the dynamic movement of the arc. (Reprinted with permission from Fridman, A., *Plasma Chemistry*, Cambridge University Press, Cambridge, 2008, © Cambridge University.)

which exceed the input energy of the power supply. Consequently, ionization reduces and quasi-thermal plasma is converted into nonthermal plasma. Eventually, the plasma extinguishes as the power supply cannot maintain such a long plasma column. Thereafter, recombination of plasma starts, and a reignition of the discharge occurs at the minimum distance between the electrodes. This self-pulsing nature is characteristic of a gliding arc discharge and typically occurs at a timescale of the order of 10 ms.

The exact plasma properties depend on the input power, which can range from about 100 W up to the order of 40 kW, corresponding to a gas temperature range from 2,500 K up to 10,000 K in the initial "quasi-thermal" state of the plasma. For the high power end, the gas temperatures in the nonthermal zone reduce to 1000–2000 K or even lower. The initial electron density is as high as in a typical arc, while the downstream or average electron density is of the order of 10^{17}–10^{19} m^{-3} (Kalra 2005).

The chemical efficiency of these discharges is due to the two distinctive phases in the plasma. In the thermal phase the molecules introduced in the discharge are strongly dissociated. The fast transition from a thermal to a nonthermal discharge (and a corresponding drop in the gas temperature) allows a fast recombination of the dissociation products to molecules and freezing of the reaction products, which are of key importance for applications requiring selective chemistry. As the flow is rather large it allows large throughputs with residence times of the reagents of the order of a millisecond. The above clearly illustrates why these discharges are typically used in applications related to chemical plasma conversion.

2.7 Inductively Coupled Plasmas

ICPs are used extensively at low pressure but can also operate at atmospheric pressure. They are often used as so-called analytical plasmas that ionize gases subjected to mass spectrometry (ICP-MS) or for spectroscopic analyses of trace components in gases and liquids. Also, ICPs are used for toxic gas/waste water treatment and coating applications. In an ICP the electrical power has a completely different coupling as compared to in a capacitively coupled plasma (see Figure 2.11). It basically consists of a helical

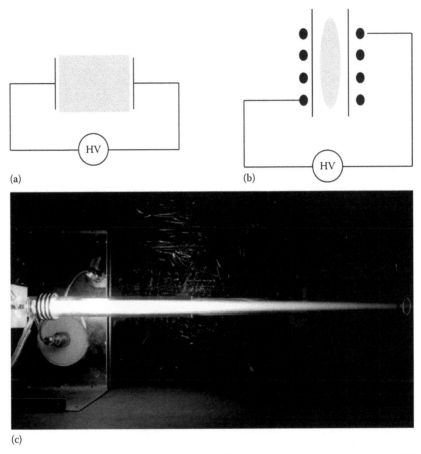

(a)　　　　　　(b)

(c)

FIGURE 2.11 Comparison between a capacitively coupled discharge (a) and an inductively coupled discharge (b). (c) ICP discharge at atmospheric pressure operating at 1600 W in a mixture of N_2O, O_2, and air. The coil outside the tube produces the electric field inside the dielectric tube, which has an internal diameter of 16 mm. (Reprinted with permission from Tamura, T., et al., Direct decomposition of anesthetic gas exhaust using atmospheric pressure multigas inductively coupled plasma. *IEEE Trans. Plasma Sci.*, 39, 1684–1688, © 2011 IEEE.)

coil wrapped around a dielectric tube in which the plasma is generated. By driving the coil with an RF field (typical in the megahertz range) the current in the coil induces a time-varying magnetic field in the tube, which causes a resulting electric field that accelerates the electrons and sustains the discharge. As ICPs are plasmas that are not directly in contact with electrodes or walls and no ionic sputtering of the wall occurs, they can be used for a large range of processing gases. Mostly they operate at high power (several kilowatts up to megawatts) and have a large electron density (10^{21} m^{-3} or larger) and a gas temperature close to the electron temperature (van de Sande et al. 2003). In the case of higher power the discharge is close to LTE, although for high flow rates, with the introduction of a liquid spray or low-power input, a stronger deviation from LTE can be found.

2.8 Wave-Heated Plasmas

Electromagnetic (EM) waves generated near a plasma surface can propagate into the plasma or along the surface and be absorbed by the plasma. EM waves can also lead to the ignition of a discharge when a local electric field that is high enough to cause breakdown is produced. The heating of the electrons in the plasma is caused by an EM wave. A typical example of wave-heated plasmas is the so-called

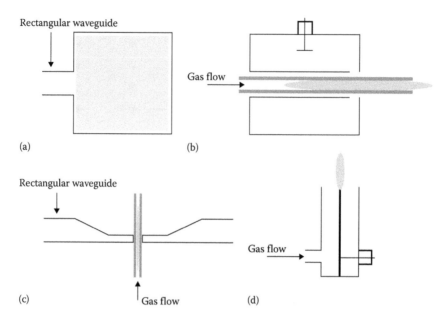

FIGURE 2.12 Four configurations that are commonly used to produce MW-excited discharges: (a) MW cavity, (b) surfatron, (c) surfaguide, and (d) antenna/electrode MW discharge. Many variations in geometries exist.

MW-excited plasma. Basically in the case of wave-heated plasmas one of the critical dimensions of the MW plasma should be comparable with the wavelength of the EM wave used for excitation, which implies that they should be generated at gigahertz frequencies. The coupling of MW power to the plasma can be done with different geometries based on different principles. An overview of four different geometries is shown in Figure 2.12. The conceptually easiest configuration is the MW cavity that is brought into resonance and the standing wave efficiently couples its energy into the plasma. As the geometry of the cavity and the plasma density determine the resonance frequency, the plasma is very sensitive to small changes in frequency (Figure 2.12a). The dependence of the resonance frequency on the plasma density confines application to a limited range of plasma conditions.

A second approach is using surface waves that travel along the plasma–dielectric boundary interface (Moisan and Zakrzewski 1991). Surface wave launching devices exist in different forms and can operate from frequencies below 1 MHz up to 40 GHz and more. It is based on the property that EM surface waves propagating along a cylindrical plasma column can be efficiently absorbed by the plasma, thus sustaining the plasma. A typical commonly used surface wave plasma device is the surfatron (see Figure 2.12b), which consists of a launcher cavity that has a small opening that allows the formation of a local high field in the dielectric tube. At this location ionization occurs and plasma is created. This plasma is then able to support a surface wave and it grows along the length of the column. This in turn creates specific conditions for long plasma columns. The surface wave attenuates due to the losses as it propagates along the plasma column, which causes a decrease in electron density with increasing distance from the launcher.

It is also possible to run a surface wave plasma in the gigahertz region by passing the dielectric discharge tube through the broad face of a waveguide (as shown in Figure 2.12c). Tapering the waveguide causes the electric field to increase and assists in breakdown. The launcher in this case is basically the gap between the waveguide walls. Also, for MW discharges it is possible to use an electrode/antenna to produce the plasma. Some of the MW jets are based on this principle as shown in Figure 2.12d.

Typical wave-heated plasmas at atmospheric pressure operate at an elevated gas temperature (above 500 K) and have electron temperatures around 1–2 eV and electron densities in the range of 10^{20}–10^{21} m^{-3} (van der Mullen et al. 2007), although the electron density can be larger as it depends strongly on the power input.

2.9 Corona Discharges

Corona discharges are localized discharges in the neighborhood of a pin or a thin wire, where the field is significantly enhanced. Ionization and emission thus occurs locally around the pin or the wire. Corona discharges normally do not extend up to the counter electrode. For this reason a corona discharge is also known as a partial discharge. Two typical configurations of a corona discharge are shown in Figure 2.13a and b. A corona driven by a positive (negative) pin/wire electrode is called a positive (negative) corona. In the case of DC coronas the electrode gap can be divided into two zones: the high electric field region in which high-energetic electrons are produced and the ion drift region in which the electric field is low and the continuity of current toward the second electrode is guaranteed by a flux of positive ions (in the case of positive corona) or negative ions produced by electron attachment to, for example, oxygen (in the case of a negative corona). The morphology of the discharge is highly dependent on polarity and the applied voltage. This can also be seen in the current waveforms, which show that several of these corona discharges are highly dynamic in nature.

Negative DC corona can exhibit pulses. These pulses are called Trichel pulses and are caused by the space charge buildup produced by the corona. As the removal of the space charge (negative ions) is necessary to allow the next pulse, the frequency of the current pulses is typically of the order of 1–100 kHz. At a certain time-averaged current amplitude (or voltage) the negative corona becomes a continuous glow corona and the pulsation of the discharge stops. The current can be increased during this regime until a glow discharge is formed. This glow discharge is normally unstable and constricts to a spark. The stability region of the negative corona is thus determined by the threshold of the glow-to-spark transition.

Positive coronas can operate in a glow-like regime but are often intermittent and pulsed. For the larger current range the positive corona is streamer-like and the stability region is limited by the streamer-to-spark transition, which occurs when a streamer reaches the second (grounded) electrode.

In spite of the fact that the morphology of the negative and positive corona can be rather different, the time-averaged *I–V* characteristic is very much the same and is determined by the drift zone.

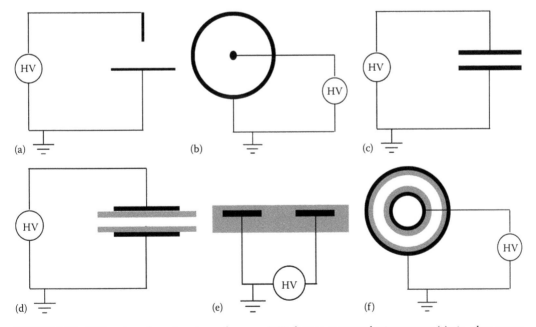

FIGURE 2.13 Different configurations to produce non-LTE plasmas at atmospheric pressure: (a) pin–plate corona geometry, (b) coaxial wire–cylinder corona geometry, (c) parallel-plate capacitively coupled RF discharge geometry, (d) parallel-plate DBD geometry, (e) surface DBD geometry, and (f) coaxial DBD geometry as used in ozone production applications.

The *I–V* characteristic can be analytically determined by solving the Poisson equation and the continuity of the current in the drift region in a wire–cylinder geometry assuming that space charge effects are small. Even in a general geometry the following *I–V* relation holds (Raizer 1987):

$$I = KV(V - V_c),$$

where

K is a constant depending on the geometry and the mobility of the charge carriers in the drift zone
V_c is the threshold voltage for corona onset

The ignition voltage of the DC corona discharge depends on the polarity. This is easy to understand as in the case of the negative polarity the criterion for Townsend breakdown can be applied, while this is not the case for a positive corona as secondary electron emission does not occur at the pin/wire electrode. In this case the Meek criterion is considered for corona onset.

DC coronas are extensively used in electrostatic precipitators. Basically the ions in the drift region charge the dust particles, which are attracted by the grounded electrode and as such transported to the grounded electrode and removed from the exhaust gases.

Corona discharges are also used for air treatment such as removal of NO_x or volatile organic compounds (VOCs) (Chang et al. 1991). The energetic electrons in the corona produce radicals that can convert unwanted products in the gas stream into less harmful products or products that are easier to remove from the gas phase. For air treatment nanosecond pulsed discharges are mostly used. Nanosecond puls-ing has several advantages that are correlated to the fact that the corona can be operated at much higher voltages compared to the DC case without spark formation, given that the voltage is switched off before the streamer-to-spark transition can occur. Higher voltages also leads to longer streamers that have a better filling ratio of the electrode gap and enables a more homogeneous treatment. Larger voltages lead to higher electric fields and electron temperatures, which increases ionization and dissociation rates. Additionally, due to the short pulse only electrons are significantly accelerated in a few nanoseconds, and gas heating can be kept to an absolute minimum.

Pulsed coronas or streamers have been investigated in detail and the basic mechanisms are now under-stood. In the case of negative streamers electron avalanches can cause the propagation of the streamer as the streamer develops in the same direction as the avalanches. Also, electron avalanches cause the propagation of positive streamers but it is generally accepted that photoionization in gas mixtures such as air plays an important role in generating electrons in front of the streamer head. Most streamers are highly branched. Inhomogeneities can explain branching (Babaeva and Kushner 2009), while a Laplacian instability of

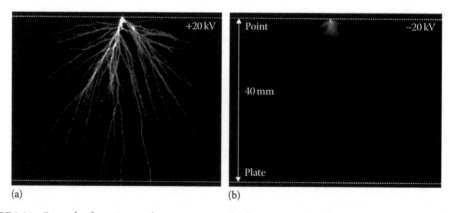

FIGURE 2.14 Example of a positive and negative corona discharge in a point–plate geometry with a gap of 40 mm. The needle is at the top of the image and a clear difference is seen for positive and negative streamers. (Reprinted with permission from Briels, T.M.P., et al., *J. Phys. D Appl. Phys.*, 41, 23400, 2008, © IOP.)

the streamer head can theoretically explain branching (Arrayas et al. 2002) without the necessity of a inhomogeneity, which almost always occurs. As shown in Figure 2.14 a positive and a negative corona have a significantly different morphology even when excited by the same voltage pulse. The electron temperature is very large during the streamer propagation time up to the ionization energy of the gas and more and the typical electron density in streamers is often 10^{20}–10^{21} m^{-3} (Spyrou et al. 1992; Liu et al. 2007). Note that the chemistry in transient streamer discharges is a two-step process. During streamer propagation, excitation, ionization and dissociation occur, and excited species, metastables, radicals, electrons, and ions are produced. These highly reactive or energetic species subsequently recombine and react with molecules on a much longer timescale as the lifetime of a streamer. At this timescale also the main chemistry that is important for the applications happens.

2.10 RF Glow Discharges and Pulsed Diffuse Discharges

As discussed above, increasing the preionization in the gas leads to a larger avalanche density and interaction avalanches, because of which the conditions of the Meek criterion are not met. In relative high-frequency discharges or nanosecond pulsed discharges with a repetition frequency of more than 1 kHz this phenomenon can be important as the time between two subsequent discharges does not allow a full relaxation or recombination of the charge.

Diffuse glow discharge in atmospheric pressure helium with small admixtures (typically <2%) of molecular gases can be obtained in a capacitively coupled parallel-plate geometry even when bare metal electrodes are used (Figure 2.13c). Operating the discharge in the same reactor in other gases normally leads to a contraction of the discharge to a single filament with a significant increase in gas temperature. The electron density of the discharge is of the order of 10^{16}–10^{17} m^{-3} (Waskoenig et al. 2010), for which diffusive losses are still important. The discharge operates in the so-called alpha mode, which means that the main electron production/heating is in the gas phase (sheath region) rather than due to secondary electron production (gamma mode). The electron temperature in these discharges is typically in the range of 2–3 eV. This discharge configuration is very much appreciated by modelers as it can be well described by a one-dimensional code. Indeed, it is diffuse and behaves very much like a low-pressure glow discharge.

Nanosecond pulsed glow discharge are often produced at kilohertz frequencies in pin–pin geometries. Intrinsically, they can be produced at lower voltages compared to "single shot" glow discharges due to the significant memory effect, but they can still operate at voltages that would lead long pulses to a glow-to-spark transition, as shown in Figure 2.4. This is because these glow discharges have a higher power density compared to their DC counterparts and also often a larger electron density (typically 10^{19} m^{-3} or larger) (Pai et al. 2009). They are often investigated in the context of plasma-assisted combustion. However, there is a trend in the plasma-enhanced combustion field to move toward shorter and shorter voltage pulses, which does not allow time for the formation of a sheath, and only an ionization front travels across the electrode gap.

2.11 Dielectric Barrier Discharges

As already mentioned, the introduction of dielectric barriers between the metal electrode and the plasma can be used to reduce the current. The displacement current through the dielectric causes the continuity of the current in the circuit so in this case the discharge has to be driven by AC or pulsed voltage excitation. In the case of a DBD in general at least one of the metal electrodes is covered by a dielectric barrier and often both electrodes are. DBDs have been generated in parallel-plate or in coaxial cylindrical reactor geometries (Figure 2.13 d and f). Surface DBDs propagating over a dielectric surface are produced when, for example, both metal electrodes are encapsulated in the same dielectric material as shown in Figure 2.13e.

DBDs can operate in a diffuse mode or a filamentary mode (Figure 2.15). The most common morphology is a filamentary mode as in air or O_2. In this case a small filament, very much like a streamer, is locally produced and bridges the gap. Typically, because of the dielectric nature of the electrode,

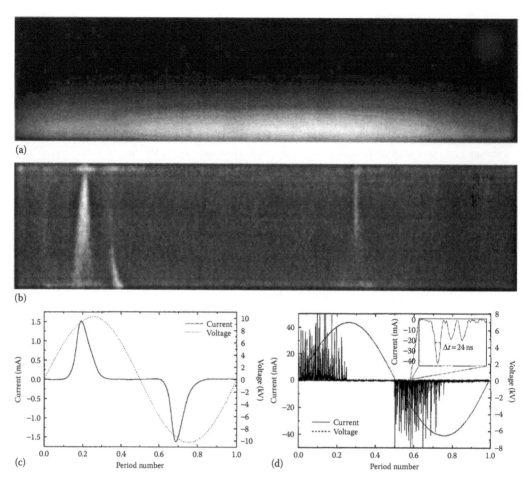

FIGURE 2.15 (a) and (b) Diffuse and filamentary DBD in N_2 with a gap of 4 mm and a voltage of 11 and 14 kV, respectively. (c) and (d) Typical current and voltage waveform of a diffuse and filamentary DBD. In the case of the filamentary discharges the current peaks produced by the microdischarges are clearly visible. (Reprinted with permission from Gherardi, N., et al., *Plasma Sources Sci. Technol.*, 9, 340–346, 2000, © IOP.)

the filament propagates partly over the electrode surface and often has a large contact area with the dielectric. In this process the filament transfers charge from one dielectric to the other and reduces the field locally. This leads to a memory effect, which causes the next filament to be produced at a different location. This phenomenon causes a spreading of the filaments, making the DBDs look diffuse with the naked eye even when they consist of many filaments.

These filaments behave as microchemical reactors. One of the most successful and oldest applications of DBDs is in ozone formation. Werner von Siemens reported the first experimental investigations more than 150 years ago (Siemens 1857). Ozone formation requires that three criteria be met: energetic electrons that are able to efficiently dissociate O_2 (4–5 eV), high pressure because of the three-body reaction, which is responsible for the ozone production, and low gas temperatures because of the reduced lifetime of O_3 at elevated gas temperatures. DBDs can deliver all these requirements as the gas temperature of the bulk gas can be maintained close to room temperature (for larger powers cooling of the electrodes is required), while the electron temperature is typically 2–5 eV. The electron density is reported to range from 10^{17} m^{-3} in the diffuse mode up to 10^{21} m^{-3} in the filamentary mode (Dong et al. 2008; Zhu et al. 2009).

The stabilization by the dielectric barriers makes the discharge very robust to instabilities, which lead to constriction, heating, and large current sparks/arcs even in electronegative gases such as oxygen.

An additional advantage is that the stabilizing component (the dielectric) is a capacitor that in the ideal case does not consume energy. This is a major advantage compared to discharges that are stabilized by resistors. Moreover, DBDs can easily be scaled up from small laboratory reactors to large industrial installations.

As stated above, most DBDs consist of filaments. However, there exist diffuse DBDs for specific conditions (see Figure 2.15b). The current waveform gives a good indication if one deals with a filamentary or diffuse discharge (Figure 2.15c and d). Diffuse discharges are easily obtained at reduced pressures or in He-rich mixtures. Plasma chemistry has a major effect on filamentation. As discussed above, a reduction in the ionization coefficient leads to diffuse discharges. Note that the high energy of He metastables causes most impurities to be ionized by Penning ionization. Many of the plasma jets used in biomedical applications are DBDs that are converted into a jet by applying a gas flow.

2.12 Microplasmas

Microplasmas are plasmas with dimensions of a few micrometers up to a few millimeters. As the typical cathode sheath thickness of a glow discharge at 1 mbar is of the order of 1 cm and at atmospheric pressure of the order of 100 µm, it is clear that microplasmas mostly operate under atmospheric pressure conditions. Unlike large-scale discharges at atmospheric pressure, microplasmas are strongly influenced by boundary-dominated phenomena. This behavior is very similar to that of low-pressure discharges given that microplasmas operate at similar pressure–distance (pd) values.

The effect that "microcavities" have on a plasma can be illustrated by the well-known microhollow cathode discharge. The microhollow cathode discharge is basically a DC-excited discharge in which the cathode is a metal tube (inner diameter D) and the anode a normal plate. When it operates at small currents, the discharge has the cathode sheath outside the tube, such as when using two parallel-plate electrodes. However, for increasing currents in the pressure range $pD \sim$ 0.1–10 Torr cm, the cathode sheath enters the inner tube and a cathode sheath is formed along the inner tube wall. This causes the electrons to move back and forth between the sheaths at the inner walls (pendulum effect). The electrons are trapped significantly longer and cause more ionization before exiting the tube and moving toward the anode. As a result the plasma density increases for this type of cathode.

The main reason behind using microscale geometries at atmospheric pressure is that high-pressure plasmas, which typically have a tendency to constrict or undergo filamentation and excessive heating, are stabilized. The surface-to-volume ratio is increased in microgeometries and thermal gradients also increase, which leads to an increased heat transfer to the walls and electrodes. This consequently reduces the risk of thermal instabilities significantly. Additionally, as the size of the discharge is of the same order of magnitude as the cathode sheath (~100 µm), glow discharges can be operated without a (fully developed) positive column. As a positive column has the tendency to contract at large currents, the discharges can often be operated at higher densities without the risk of instabilities.

Amazingly, microplasma arrays are used to create large-volume diffuse APPs (Kunhardt 2000). With a microplasma in a three-electrode geometry, it is possible to generate a glow discharge with dimensions of a few cubic centimeters even in atmospheric pressure air (Mohamed et al. 2002). Figure 2.16 shows the conceptual geometry. A microplasma is used as a virtual cathode and the electrons are extracted from the microplasma by the anode, creating a diffuse low-density glow discharge.

Many different configurations of microplasmas exist but they are a scaled-down version of one of the types of discharges treated in the previous sections. Plasma densities can span a large parameter range as both plasma jets used for biomedical applications, which are often DBD-like plasma, and continuous micro DC-excited glow-like discharges are considered. The latter discharges operate with electron densities of the order of 10^{21} m^{-3} and with temperatures reaching 1000 K (Belostotskiy et al. 2010). One of the most popular applications of microplasmas are plasma display panels and UV (excimer) sources (Kogelschatz 2003).

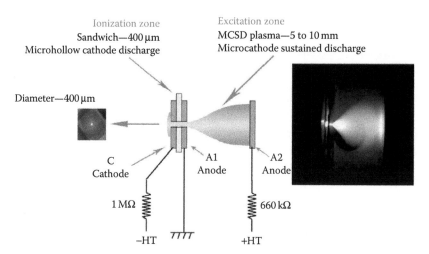

FIGURE 2.16 Microdischarge in three-electrode arrangement to produce large-volume diffuse atmospheric pressure discharges. (With kind permission from Springer Science+Business Media: *Eur. Phys. J. Appl. Phys.*, Microplasmas: Physics and application to the production of singlet oxygen $O_2(a^1\Delta_g)$, 42, 2008, 17–23, Puech, V.)

2.13 Conclusion

A wide variety of different APPs exist, which range from Townsend discharges up to thermal arcs, which span a range of ionization degrees of almost 10 orders of magnitude. The mean electron energies can be equal to the ionization energy of the gas in streamers up to close to room temperature in recombining room temperature plasmas. The gas temperatures typically range from ambient room temperature up to more than 1 eV. Because of the wide range of plasma properties, APPs have widespread applications ranging from cutting steel plates by a thermal arc to using cold APPs for wound treatment. At atmospheric pressure significant heating is most common although LTE conditions are an exception rather than the rule. To prevent heating, time-modulated plasmas have to be used. The large variety of plasma properties, chemistry, and applications make APPs a very fertile topic for both fundamental and application-oriented studies. Additional complexity is even obtained when a liquid phase is added (Bruggeman and Leys 2009b). The complex plasma chemistry that normally exists at application conditions in combination with the self-organizing nature and filamentation of APPs provides us with many interesting and challenging unresolved scientific questions.

References

Arrayas, M., Ebert, U., and Hundsdorfer, W. 2002. Spontaneous branching of anode-directed streamers between planar electrodes. *Phys. Rev. Lett.* 88 (17): 174502.

Babaeva, N.Y. and Kushner, M.J. 2009. Effect of inhomogeneities on streamer propagation: I. Intersection with isolated bubbles and particles. *Plasma Sources Sci. Technol.* 18 (3): 035009.

Barinov, Y.A., Kaplan, V.B., Rozhdestvenskii, V.V., and Shkol'nik, S.M. 1998. Determination of the electron density in a discharge with nonmetallic liquid electrodes in atmospheric pressure air from the absorption of microwave probe radiation. *Tech. Phys. Lett.* 24 (12): 929–931.

Belostotskiy, S.G., Ouk, T., Donnelly, V.M., et al. 2010. Gas temperature and electron density profiles in an argon DC microdischarge measured by optical emission spectroscopy. *J. Appl. Phys.* 107: 053305.

Boulos, M.I., Fauchais, P., and Pfender, E. 1994. *Thermal Plasmas: Fundamentals and Applications.* New York: Plenum.

Briels, T.M.P., Kos, J., Winands, G.J.J., et al. 2008. Positive and negative streamers in ambient air: Measuring diameter, velocity and dissipated energy. *J. Phys. D Appl. Phys.* 41: 234004.

Bruggeman, P., Guns, P., Degroote, J., et al. 2008a. Influence of the water surface on the glow-to-spark transition in a metal-pin-to-water electrode system. *Plasma Sources Sci. Technol.* 17: 045014.

Bruggeman, P. and Leys, C. 2009. Non-thermal plasmas in and in contact with liquids. *J. Phys. D Appl. Phys.* 42: 053001.

Bruggeman, P., Liu, J.J., Degroote, J., et al. 2008b. DC excited glow discharges in atmospheric pressure air in pin-to-water electrode systems. *J. Phys. D Appl. Phys.* 41: 215201.

Bruggeman, P., Walsh, J.L., Schram, D.C., et al. 2009. Time dependent optical emission spectroscopy of sub-microsecond pulsed plasmas in air with water cathode. *Plasma Sources Sci. Technol.* 18: 045023.

Castaños-Martínez, E., Moisan, M., and Kabouzi, Y. 2009. Achieving non-contracted and non-filamentary rare-gas tubular discharges at atmospheric pressure. *J. Phys. D Appl. Phys.* 42: 012003.

Chang, J.S., Lawless, P.A., and Yamamoto, T. 1991. Corona discharge processes. *IEEE Trans. Plasma Sci.* 19 (6): 1152–1166.

Dong, L., Qi, Y., Zhao, Z., and Li, Y. 2008. Electron density of an individual microdischarge channel in patterns in a dielectric barrier discharge at atmospheric pressure. *Plasma Sources Sci. Technol.* 17: 015015.

Finkelnburg, W. and Maecker, H. 1956. Electric arcs and thermal plasmas. In *Encyclopedia of Physics XXII*, ed. S. Flügge, Berlin: Springer, 307.

Fridman, A. 2008. *Plasma Chemistry*. Cambridge: Cambridge University Press.

Gherardi, N., Gouda, G., Gat, E., et al. 2000. Transition from glow silent discharge to micro-discharges in nitrogen gas. *Plasma Sources Sci. Technol.* 9: 340–346.

Hagelaar, G.J.M. and Pitchford, L.C. 2005. Solving the Boltzmann equation to obtain electron transport coefficients and rate coefficients for fluid models. *Plasma Sources Sci. Technol.* 14: 722–733.

Kalra, C., Gutsol, A., and Fridman, A. 2005. Gliding arc discharges as a source of intermediate plasma for methane partial oxidation. *IEEE Trans. Plasma Sci.* 33 (1, Part 1): 32–41.

Kieft, E.R., van der Mullen, J.J.A.M., Kroesen, G.M.W., and Banine, V. 2003. Time-resolved pinhole camera imaging and extreme ultraviolet spectrometry on a hollow cathode discharge in xenon. *Phys. Rev. E.* 68: 056403.

Kogelschatz, U. 2003. Dielectric-barrier discharges: Their history, discharge physics and industrial applications. *Plasma Chem. Plasma Process.* 23 (1): 1–46.

Kunhardt, E.E. 2000. Generation of large-volume, atmospheric-pressure, nonequilibrium plasmas. *IEEE Trans. Plasma Sci.* 28 (1): 189–200.

Laroussi, M., Lu, X., and Malott, C.M. 2003. A non-equilibrium diffuse discharge in atmospheric pressure air. *Plasma Sources Sci. Technol.* 12: 53–56.

Levatter, J.J. and Liu, S.-C. 1980. Necessary conditions for the homogeneous formation of pulsed avalanche discharges at high gas pressures. *J. Appl. Phys.* 51 (1): 210–222.

Lieberman, M.A. and Lichtenberg, A.J. 2005. *Principles of Plasma Discharges and Materials Processing*. New York: John Wiley and Sons, Inc.

Liu, N., Celestin, S., Bourdon, A., and Pasko, V.P. 2007. Photo ionization and optical emission effects of positive streamers in air at ground pressure. *Appl. Phys. Lett.* 91: 211501.

Massines, F., Gherardi, N., Naudé, N., and Ségur, P. 2009. Recent advances in the understanding of homogeneous dielectric barrier discharges. *Eur. Phys. J. Appl. Phys.* 47: 22805.

Mohamed, A.-A.H., Block, R., and Schoenbach, K.H. 2002. Direct current glow discharges in atmospheric air. *IEEE Trans. Plasma Sci.* 30 (1): 182–183.

Moisan, M. and Zakrzewski, Z. 1991. Plasma sources based on the propagation of electromagnetic surface-waves. *J. Phys. D Appl. Phys.* 24 (7): 1025–1048.

Pai, D.Z., Stancu, G.D., Lacoste, D.A., and Laux, C.O. 2009. Nanosecond repetitively pulsed discharges in air at atmospheric pressure—The glow regime. *Plasma Sources Sci. Technol.* 18 (4): 045030.

Pfender, E. 1978. Electric arcs and arc gas heaters in *Gaseous Electronics: Electrical Discharges vol. 1.* eds. M.N. Hirsh and H.J. Oskam, New York: Academic, 291–398.

Puech, V. 2008. Microplasmas: Physics and application to the production of singlet oxygen $O_2(a^1\Delta_g)$. *Eur. Phys. J. Appl. Phys.* 42: 17–23.

Raizer, Y.P. 1987. *Gas Discharge Physics.* Berlin: Springer-Verlag.

Roth, J.R. 1995. *Industrial Plasma Engineering.* Bristol and Philadelphia: Institute of Physics Publishing.

Siemens, W. 1857. Ueber die elektrostatische Induktion un die Verzögerung des Stroms in Flaschendrähten. Poggendorffs. *Ann. Phys. Chem.* 102: 66–122.

Spyrou, N., Held, B., Peyrous, R., et al. 1992. Gas temperature in a secondary streamer discharge—An approach to the electric wind. *J. Phys. D Appl. Phys.* 25: 211.

Staack, D., Farouk, B., Gutsol, A.F., and Fridman, A. 2007. Spatially resolved temperature measurements of atmospheric pressure normal glow microplasmas in air. *IEEE Trans. Plasma Sci.* 35 (5): 1448–1455.

Tamura, T., Kaburaki, Y., Saski, R., et al. 2011. Direct decomposition of anesthetic gas exhaust using atmospheric pressure multigas inductively coupled plasma. *IEEE Trans. Plasma Sci.* 39 (8): 1684–1688.

Tendero, C., Tixier, C., Tristant, P., et al. 2006. Atmospheric pressure plasmas: A review. *Spectrochim. Acta Part B At. Spectrosc.* 61: 2–30.

van de Sande, M.J., van Eck, P., Sola, A., et al. 2003. Electron production and loss processes in a spectrochemical inductively coupled argon plasma. *Spectrochim. Acta Part B At. Spectrosc.* 58 (5): 783–795.

van der Mullen, J.J.A.M., van de Sande, M.J., de Vries, N., et al. 2007. Single-shot Thomson scattering on argon plasmas created by the microwave plasma torch: Evidence for a new plasma class. *Spectrochim. Acta B At. Spectrosc.* 62: 1135–1146.

Verreycken, T., van der Horst, R.M., Baede, A.H.F.M., van Veldhuizen, E.M., and Bruggeman, P.J. 2012. Time and spatially resolved LIF of OH in a plasma filament in atmospheric pressure He–H$_2$O. *J. Phys. D Appl. Phys.* 45: 045205.

Waskoenig, J., Niemi, K., Knake, N., et al. 2010. Atomic oxygen formation in a radio-frequency driven micro-atmospheric pressure plasma jet. *Plasma Sources Sci. Technol.* 19: 045018.

Zhu, X.M., Pu, Y.K., Balcon, N., and Boswell, R. 2009. Measurement of the electron density in atmospheric pressure low temperature argon discharges by line-ratio method of optical emission spectroscopy. *J. Phys. D Appl. Phys.* 42: 142003.

II

Processing and Characterization

3

Modeling of Low-Pressure Plasmas

Wei Jiang,
Yu-ru Zhang,
and You-nian Wang

3.1 Introduction

3.1.1 Basic Structure of Low-Pressure Plasma Sources

Low-temperature plasma has been widely used for surface treatment of solid materials and can be classified under two categories—cold plasma and thermal plasma—based on thermal equilibrium between electrons and ions [1,2]. All the plasmas used in the microelectronics industry are cold plasmas. In most low-pressure plasmas, the pressure is lower than 1 mTorr–1 Torr. Correspondingly, the electron density n_e is of the order of 10^8–10^{13} cm^{-3} and electron temperature $T_e = 1$–10 eV, and the ions remain at room temperature of $T_i \approx 300$ K. Because $T_e \gg T_i$, electrons and ions are not in thermal equilibrium, and the ionization rate is as low as 10^{-5}–10^{-1} cm^{-3}. For cold plasma discharge, the chemical bonds of the molecular gas can mostly be formed at 1–10 eV. This discharge produces electrons and destroys chemical bonds, thus generating a variety of free radicals and positive ions, in order to remove part of the material (etching) or form a surface film (deposition). Low-temperature plasmas mentioned in this chapter refer to cold plasmas.

As is well known, all laboratory plasmas are not self-sustaining. Therefore, to produce a stable plasma, it must be heated. Plasma heating is the most important topic covered in all fields of plasma physics and can be achieved by charged or neutral beam injection and by the interaction between electromagnetic waves and plasmas, which is a common and practical method for plasma heating. In low-temperature and low-pressure plasmas, the discharge is sustained by electron impact ionization, and the energy of electromagnetic waves impacts electrons.

Direct current (DC), radio frequency (RF), or microwave sources can all be used to generate low-temperature plasmas. Because the coupling efficiency of electromagnetic energy into the plasma is low, high frequency (HF) is not suitable for low-pressure plasmas. In addition, the electromagnetic wave frequency should not be too high or too low when compared with the plasma frequency to ensure that the skin depth of electromagnetic waves δ_{em} is comparable to the plasma character length L, which thus ensures effective resonance absorption.

At the same time, ions must be kept cold in low-temperature plasmas so that only electron heating rather than ion heating takes place. This requires that the electron density not be too high because if the density is too high, even if the electromagnetic wave does not heat ions, electrons can transfer energy to them through Coulomb collisions, which will ultimately heat them. This means that the electron–ion Coulomb collision rate must be much lower than that between electrons/ions and neutral gas. So the electron density in plasmas is generally lower than 10^{13} cm^{-3} with cutoff frequency in the microwave band. Thus, the driving frequency is correspondingly lower than in the microwave band.

In addition, the most commonly used sources in practical reactors are 1 kHz–1 MHz low-frequency sources, 1–100 MHz RF sources, HF sources, as well as 2.45 GHz microwave sources. Electromagnetic waves can be applied to the plasmas in three ways, that is, through the electrode structure, the coil structure, and the antenna structure, as shown in Figure 3.1. The coil structure here is also a kind of antenna structure.

If the wavelength of the electromagnetic wave is much larger than the size of the device, all three structures can be used; however, if the wavelength of the electromagnetic wave is less than the size of the device, only the antenna structure can be adopted. Because the actual size of the device is generally about 0.1–1 m, only electrode structure can be used for DC sources. However, for sources with frequency lower than 100 MHz, all three structures, that is, the corresponding electrode structure of the capacitively coupled plasma (CCP), the coil structure of the inductively coupled plasma (ICP), and the antenna structure of the helicon, can be used. Because the wavelength of the microwave is generally ~1 cm, only the antenna structure can be used, which corresponds to electron cyclotron resonance (ECR). Regardless of the different configurations and sizes, discharge devices usually consist of a cylindrical vacuum chamber and a DC, RF, or microwave source. The vacuum chamber usually consists of prepumped high vacuum with pressure of less than 10^{-3} Pa and is filled with working gas, such as Ar and CF$_4$, at 1 mTorr–1 Torr pressure. The gas flows continually into the chamber and the reacted gas is pumped out, but the gas pressure is kept constant.

In most low-pressure plasmas, the discharge current is usually very small, so the magnetic field is much weaker than the electric field and can even be completely ignored in some cases, such as in DC, CCP, or ICP discharges. A weak external magnetic field, which is generated by permanent magnets or coils, can magnetize electrons and thus drastically increase the electron heating efficiency and change the nature of the discharge. However, the ions are still unmagnetized. For the electrode structure and the coil structure, the external magnetic field is optional, but for the antenna structure, an external magnetic field is mandatory.

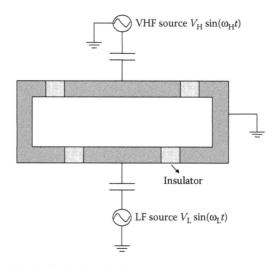

FIGURE 3.1 Schematics of electrode structure device.

3.1.1.1 Direct Current and Capacitively Coupled Plasma

DC and CCP discharges have similar electrode structure, in which the external power is directly applied to the two parallel disk-like metal electrodes, as shown in Figure 3.1. The difference between the two sources is only the frequency, that is, DC discharge uses a DC power supply, while CCP employs RF sources (typically 13.56 MHz). Due to more efficient heating caused by an oscillating field, the plasma density in the CCP discharge is usually an order higher than in the DC discharge.

In DC and CCP, the longitude fields are directly applied to the plasma. As a result, a thin (a few Debye length λ_d) positively charged layer builds up near the electrode, which is often referred to as a sheath. In DC discharges, the sheath is stationary, but it oscillates in RF discharges. Because ions are much heavier than electrons, they can only respond to the time-averaged electric fields and are accelerated to the substrates in the sheath.

DC discharges, which often work under very low pressure (~1 Pa), have been widely used in sputtering devices for depositing metal films. The secondary electrons produced by ion bombardment to the substrates are essential for sustaining these discharges. Besides, weak external magnetic fields are generally used to increase the density and sputtering rate.

CCP has been commonly used for etching and deposition of thin films. For anisotropic etching, the plasmas are always operated at low gas pressure, for example, 10–100 mTorr, and the frequency is usually set at 13.56 MHz. Collisionless processes play an important role in the energy transfer and introduce the nonlocal effects. Nevertheless, in plasma-enhanced chemical vapor deposition (PECVD) processes, the pressure is usually in the range of 0.1–10 Torr.

Very high frequency (VHF) sources have recently attracted a lot of attention and are widely used in silicon wafer and flat panel display processing due to the deposition of good-quality films at high rates. In the discharge sustained at HF (i.e., tens of megahertz to hundreds of megahertz) in large-area reactors, the electromagnetic effects, such as standing-wave effect and skin effect, have a significant influence on the plasma characteristics [3].

Moreover, dual-frequency (DF) CCP has been proposed to control the ion flux and ion bombardment energy to the wafer independently [4] under the condition when $\omega_{hf} \gg \omega_{lf}$ and $V_{hf} \ll V_{lf}$. Frequency combinations of 2/27, 2/60, and 13.56/160 MHz are used commercially.

3.1.1.2 Inductively Coupled Plasma

ICP is a discharge with a coil structure, and the applied RF current flows through single-turn or multiturn metal coils rather than through the plasma. For a cylindrical plasma, the coils can be placed on the top of the cylinder or around the cylindrical sidewall. The former structure is known as planar ICP or transformer coupled plasma (TCP), and the latter structure is referred to as coaxial ICP (see Figure 3.2).

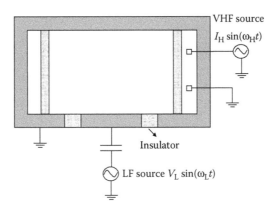

FIGURE 3.2 Schematics of coil structure device.

Usually, coils and plasmas must be separated by quartz glass and other media to prevent the plasma current from flowing directly into the coil. Sometimes an encaged metal disk is placed between the coil and quartz window to shield the longitude electric field, which is called the Faraday shield.

ICPs have three main advantages: (1) ICP discharge can achieve higher plasma density with lower RF power than a CCP discharge and therefore decrease substrate-level contamination; (2) because no DC magnetic field is required, the equipment structure is simpler; and (3) the energy of ions incident on the substrate and the ion flux can be independently controlled by applying a bias power on the substrate.

ICP discharge is usually sustained at a lower pressure of 0.5–50 mTorr, with high density of 10^{11}–10^{12} cm^{-3}. Because the ions undergo collision less frequently when they are accelerated across thin sheaths, ICP has been commonly used in anisotropic etching of metals.

It is well known that there are two distinct discharge modes in inductive discharges, that is, the E mode and the H mode. In the E mode, the discharge is sustained by a longitude electric field, the power deposition is low, as is the plasma density. As the current in the coil increases, the discharge is sustained by a transverse electric field, and higher density plasmas with lower energy electrons are obtained.

3.1.1.3 Helicons and ECR

The structures of ECR and helicon discharge devices are similar, as shown in Figure 3.3. Electromagnetic waves from the waveguide at the top of the chamber are fed into the vacuum chamber along the axial direction. The plasma is generated in the region of the vacuum chamber, which is surrounded by coils below the antenna, and the coils are used to produce an axial magnetic field in the discharge. In most of the ECR and helicon discharge devices, plasmas are generated at the top of the reactor, and diffuse continuously into a larger chamber at the bottom. The excitation wavelength of the electromagnetic field is comparable to the device size in both of them. The differences are the electromagnetic wave frequency and the intensity of the external magnetic field, that is, 2.45 GHz microwave and 845 G magnetic field are used in ECRs, but helicon discharges are often driven by a 10–100 MHz RF source with the external magnetic field set to be 100–1000 G.

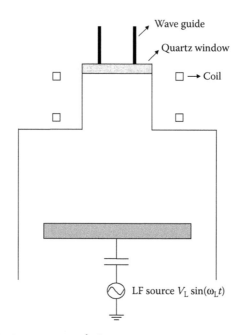

FIGURE 3.3 Schematics of antenna structure device.

The structure of helicon is very similar to that of an ICP, in which the ring-shaped RF coils (antenna) are located at the top or around the side wall of the discharge chamber. The main differences between them are the coil length, aspect ratio, and magnetic field strength. The ICP discharge is sustained in the reactor with smaller aspect ratio ($R \sim L$), the magnetic field is as weak as 0–100 G, and the electromagnetic wave does not propagate along the axial direction. For helicon discharges, magnetic fields are as strong as 100–1000 G, and some of the electromagnetic waves excited in plasmas propagate along the axial direction in larger aspect ratio geometries ($R < L$). By changing the magnetic field strength and other parameters, a continuous variation in the E mode, the H mode, and the wave mode (helicon mode) can be observed.

Another common feature of ECRs and helicons is that the electromagnetic wave vector **k** is parallel to the magnetic field B and perpendicular to the plasma surface. But wave vector of the electromagnetic wave can also be parallel to the plasma surface sometimes, which is usually called the surface wave plasma (SWP). Because the energy deposition length is long in the SWP, it is difficult to be incorporated in cylindrical chambers for material processing.

In ICP, ECR, and helicon discharges, the plasma density is high, so they are collectively referred to as high-density plasma sources. In these sources, an additional DC or RF power is often applied to the electrode where the materials are processed. Note that the additional DC/RF bias source has almost no effect on the plasma properties, but can only be used to control the ion flux and energy to the substrate surface.

3.1.2 Modeling of Low-Pressure Plasmas

Modeling of low-pressure plasmas is a typical multiscale problem, which includes the transport of neutral molecules and charged particles, electromagnetic fields, and surface processes. Variations are there with respect to physical and chemical processes with different spatial scales, such as the device size (~1 m), the sheath size (~10^{-2} m), the surface scale (10^{-6} m), and the atomic scale (10^{-9} m); also, different temporal scales coexist: the surface evolutionary time (~10 s), the neutral particle relaxation time (10^{-3} s), the electromagnetic wave cycle (10^{-6} s), the electron oscillation period (10^{-9} s), and the atomic collision time (10^{-15} s). The ultimate goal of the theoretical and experimental study is to quantitatively predict the etching or deposition process under a given discharge condition, which provides a basis for optimizing the adjustable parameters.

There are three kinds of controlling parameters in a low-pressure plasma discharge:

1. RF source parameters, such as drive frequencies and amplitudes of the applied power (or voltages and currents). The power sources can be either continuous or pulse-modulated. In some low-pressure discharges, the plasmas are unmagnetized; therefore, an external static magnetic field (in the order of 10–100 G) can significantly impact the transportation of electrons and enhance the plasma density. The external magnetic field can be produced by either a permanent magnet or a coil with DC current.
2. Gas parameters, such as gas type, pressure, and temperature. The feedstock gases for etching devices are usually C_xF_y, Ar, O_2, and so on, with the pressure typically being several tens of mTorr. The feedstock gases for deposition devices are usually SiH_y, NH_3, H_2, and so on, with the pressure being about several hundreds of mTorr or higher.
3. Reactor dimensions, such as gap length and reactor radius. Based on whether neutrality is maintained, low-pressure plasmas can be divided into the bulk and sheath region. In the bulk region, we have $n_e \approx n_i$, where n_e and n_i are the electron and ion density, respectively; the sheath region is located between the bulk plasma and the chamber wall, where $n_e < n_i$.

Generally, physical processes can be divided into plasma processes and surface processes, and we consider only plasma processes here. They can be studied either analytically or numerically, of which the numerical model can be classified into the fluid model and the particle model, as we have done later in this chapter.

3.2 Fluid Model

Plasma parameters such as density, mean velocity, and mean energy can be obtained by solving a set of moments of the Boltzmann equation, which are formed by multiplying the Boltzmann equation by v^0, v^1, v^2, and integrating over velocity. Because the fluid models used in the simulation are similar, we only take the CCP discharge as an example here.

3.2.1 Model Description

3.2.1.1 Electron Equations

For electrons, the continuity equation can be expressed as

$$\frac{\partial n_e}{\partial t} + \nabla \cdot \Gamma_e = S_e, \tag{3.1}$$

where
n_e is the electron density
Γ_e is the electron flux

The source term S_e is given by $S_e = \sum_a k_a n_1 n_2 - \sum_b k_b n_3 n_e$, where k is the reaction coefficient. The electrons are created during the reaction type a between the species with density n_1 and n_2 and are lost in the collision b between them and other species with density n_3.

Because the electron mass is very small, we can ignore the inertia term, so the electron flux can be presented in the drift–diffusion form

$$\Gamma_e = -\frac{1}{m_e \nu_e} \nabla \left(n_e k_B T_e \right) - \frac{e n_e}{m_e \nu_e} E, \tag{3.2}$$

where
m_e and T_e are the mass and temperature of electrons, respectively
ν_e is the electron–neutral collision frequency
E is the static electric field

The energy balance equation is used to describe the electron temperature

$$\frac{\partial \left(\frac{3}{2} n_e k_B T_e \right)}{\partial t} = -\nabla \cdot q_e - eE \cdot \Gamma_e - W_e, \tag{3.3}$$

where k_B is the Boltzmann constant. Here, the flux of energy q_e is given by

$$q_e = \frac{5}{2} k_B T_e \Gamma_e - \frac{5}{2} \frac{n_e k_B T_e}{m_e \nu_e} \nabla \left(k_B T_e \right), \tag{3.4}$$

where the first term on the right-hand side represents energy transport due to electron convention, and the second term represents energy transport due to the electron thermal conduction. The energy exchange during all the collisional processes is $W_e = \sum_j \varepsilon_j k_j n_1 n_2$. Positive values of ε_j imply energy loss by electrons (e.g., ionization), and negative values imply energy gain (e.g., superelastic collisions).

For ICP discharges, an additional term P_{tot} should be added to the right-hand side of the electron energy equation, which is used to describe the energy deposition caused by the inductive and capacitive electromagnetic fields:

$$P_{tot} = \frac{1}{2} Re \left(\sigma_p |E_r|^2 \right) + \frac{1}{2} Re \left(\sigma_p |E_z|^2 \right) + \frac{1}{2} Re \left(\sigma_p |E_\theta|^2 \right), \tag{3.5}$$

TABLE 3.1 Reactions in Argon Discharges

Reaction	Rate Coefficient ($cm^{-3}s^{-1}$)	Reference
$e + Ar \rightarrow e + Ar$	$10^{-9}(0.514 + 5.51T_e + 22.9T_e^2 - 6.42T_e^3 + 6T_e^4)$, $T_e \geq 4eV$	[5]
	$10^{-9}(39.9\,T_e - 1.86\,T_e^2 - 0.4672\,T_e^3 + 0.0006429\,T_e^4)$, $T_e > 4$ eV	
$e + Ar \rightarrow 2e + Ar^+$	$8.7 \times 10^{-9}(1.5T_e - 5.3)\exp(-4.9/(1.5T_e - 5.3)^{0.5})$, $1.5T_e > 5.3$ eV	[6]
$e + Ar \rightarrow e + Ar(4s)$	$5.0 \times 10^{-9}\,T_e^{0.74}\exp\left(\dfrac{-11.56}{T_e}\right)$	[1]
$e + Ar(4s) \rightarrow e + Ar$	$4.3 \times 10^{-10}\,T_e^{0.74}$	[1]
$e + Ar(4s) \rightarrow 2e + Ar^+$	$6.8 \times 10^{-9}\,T_e^{0.67}\exp\left(\dfrac{-4.2}{T_e}\right)$	[1]
$e + Ar_m \rightarrow e + Ar_r$	2.0×10^{-7}	[1]

Note: T_e represents the electron temperature in eV.

where

$\sigma_p = \varepsilon_0\omega_{pe}^2/(\nu_e - i\omega)$ is the plasma conductivity

ω_{pe} is the electron plasma frequency

The electromagnetic fields can be obtained by solving the Maxwell equations.

For an argon discharge, the reactions and the rate coefficients are given in Table 3.1.

3.2.1.2 Ion Equations

The continuity equation, momentum balance equation, and Equation 3.5 for ions are as follows:

$$\frac{\partial n_i}{\partial t} + \nabla \cdot n_i u_i = S_i, \tag{3.6}$$

$$\frac{\partial n_i m_i u_i}{\partial t} + \nabla\left(n_i m_i u_i u_i\right) = -\nabla p_i + Z_i e n_i E + M_i, \tag{3.7}$$

$$\frac{\partial\left(n_i C_{v,i} T_i\right)}{\partial t} + \nabla(n_i C_{v,i} T_i u_i) = -\nabla \cdot q_i - p_i(\nabla \cdot u_i) - W_i, \tag{3.8}$$

where

n_i, u_i, m_i, Z_i, p_i, and T_i are the density, velocity, mass, charge, pressure, and temperature for ions, respectively

$C_{v,i}$ is the ion heat capacity

W_i and S_i are the transfer of energy caused by collisions and the source term, respectively, which are similar as mentioned above

The collisional transfer of momentum to ions is handled explicitly through the term

$$M_i = \sum_n -\frac{m_i m_n}{m_i + m_n} n_i u_i \nu_i, \tag{3.9}$$

where

m_n is the mass of the neutral species

ν_i is the ion–neutral collision frequency

The ion heat flux can be expressed as

$$q_i = -\frac{5}{2}\frac{n_i k_B T_i}{m_i \nu_i}\nabla\left(k_B T_i\right).$$

(3.10)

3.2.1.3 Neutral Equations

For neutral species, the continuity equation and the momentum equation are included

$$\frac{\partial n_n}{\partial t} + \nabla \cdot n_n u_n = S_n,$$

(3.11)

$$\frac{\partial n_n m_n u_n}{\partial t} + \nabla\left(n_n m_n u_n u_n\right) = -\nabla p_n - \nabla \cdot \overrightarrow{\pi_n} + M_n,$$

(3.12)

$$\frac{\partial\left(n_n C_{v,n} T_n\right)}{\partial t} + \nabla\left(n_n C_{v,n} T_n u_n\right) = -\nabla \cdot q_n - p(\nabla \cdot u_n) - \sum W_n,$$

(3.13)

where

n_n, u_n, p_n, T_n, q_n, and $C_{v,n}$ are the density, velocity, pressure, temperature, energy flux, and heat capacity of neutral species, respectively

S_n and W_n are the source term and energy exchange term, respectively

$\overrightarrow{\pi_n}$ is the neutral viscous stress tensor

The transfer of momentum to neutral species is

$$M_n = \sum_\alpha -\frac{m_n m_\alpha}{m_n + m_\alpha}n_n u_\alpha \nu_n.$$

(3.14)

3.2.1.4 Electromagnetic Equations

In the range of low frequency (LF) (i.e., 13.56 MHz), the electromagnetic effects have no important influence on the plasma characteristics. Therefore, the electrostatic field due to the space charge within the plasma can be determined by solving Poisson's equation

$$\nabla^2\Phi = \frac{e}{\varepsilon_0}\left(n_e - \sum_+ Z_+ n_+ + \sum_- Z_- n_-\right),$$

(3.15)

where

Φ is the electric potential

ε_0 is the vacuum permittivity

Recently, special attention has been paid to VHF sources due to their higher ion flux and lower ion bombarding energy. However, when the excitation wavelength becomes comparable to the electrode dimension at high frequencies, the electromagnetic effects start to have a profound influence on the plasma properties. Therefore, the electromagnetic effects should be taken into account in the simulations for obtaining realistic results.

The electromagnetic fields in the plasmas are governed by the full set of Maxwell equations,

$$\nabla \times E = -\frac{\partial B}{\partial t},$$

$$\nabla \times B = \mu_0 J + \mu_0 \varepsilon_0 \frac{\partial E}{\partial t},$$

$$\nabla \cdot E = \frac{e}{\varepsilon_0}\left(n_e - \sum_+ Z_+ n_+ + \sum_- Z_- n_-\right),$$

(3.16)

$$\nabla \cdot B = 0,$$

where μ_0 is the vacuum permeability, and the current density is expressed as

$$J = e\left(\sum_+ Z_+ n_+ u_+ - \sum_- Z_- n_- u_- - \Gamma_e\right).$$

(3.17)

Introducing the electric potential Φ and the magnetic vector potential $\mathbf{A}$, the electric field can be given as $E = -\nabla\Phi + E_T$, where $E_T = -\partial \mathbf{A}/\partial t$ is the vortex electric field. Furthermore, we can arrange such that $\nabla \cdot E_T = 0$ by making use of the Coulomb gauge $\nabla \cdot \mathbf{A} = 0$. This reduces the full set of Maxwell equations to

$$\nabla^2 \Phi = \frac{e}{\varepsilon_0}\left(n_e - \sum_+ Z_+ n_+ + \sum_- Z_- n_-\right),$$

(3.18)

$$\nabla^2 E_T - \mu_0 \varepsilon_0 \frac{\partial^2 E_T}{\partial t^2} = \mu_0 \frac{\partial J}{\partial t} - \mu_0 \varepsilon_0 \frac{\partial^2 \nabla \Phi}{\partial t^2}.$$

(3.19)

3.2.2 Boundary Condition

In the simulation, boundary conditions must be specified in order to complete the problem. The electron flux at electrodes and side walls is given by using the kinetically limited Maxwellian flux condition

$$\Gamma_e = \pm \frac{1}{4} n_e u_{e,th},$$

(3.20)

where
  The $\pm$ sign corresponds to the direction of the electron flux
  $u_{e,th}$ is the electron thermal velocity, expressed as $u_{e,th} = (8k_B T_e/\pi m_e)^{1/2}$

By considering the electron reflection at the boundaries, the reflection coefficient Θ is fixed at 0.25. Ignoring the electron thermal conduction, the energy flux to the electrodes and walls is set to

$$q_e = \frac{5}{2} k_B T_e \Gamma_e.$$

(3.21)

Meanwhile, we assume that the ion density and ion velocity do not change substantially near the wall in the steady state, so the gradients are set to zero at the walls, that is, $\nabla n_\pm = 0$ and $\nabla \cdot u_\pm = 0$.

The boundary condition for neutral species density is similar to that of ions, that is, $\nabla n_n = 0$.

The potential boundary condition is obvious: $\Phi = 0$ when the electrodes or the walls are grounded, and $\Phi = \sum_i V_i \sin(2\pi f_i t)$ on the powered electrode, where V_i and f_i are the peak voltage and frequency of sources, respectively.

In addition, all external materials are assumed to be perfect electric conductors; so the boundary conditions of the vortex electric field are $\partial E_{T_z}/\partial z = 0$ and $\partial(rE_{T_r})/\partial r = 0$.

3.2.3 Numerical Method

The simulation of plasma discharges poses significant challenges; for instance, it has a wide range of timescales (e.g., from picoseconds for atom collisions to milliseconds for fluid transport) and steep unsteady gradients in charged particle densities and includes electric field, magnetic field, and chemical reactions Therefore, a robust numerical algorithm is needed to solve the problem accurately and efficiently.

Because a staggered grid provides conservative properties for nonlinear fluid equations, vector quantities, such as Γ_e, q_e, $u_\pm$, and E, are located halfway between the location of scalar parameters, such as n_e, n_i, T_e, and Φ.

In 2D simulation, the finite volume method is employed to discretize the electron continuity equation and electron energy equation. For electron flux, a first-order upwind discretization is used as a more robust method than the standard central difference scheme. The electrons can be advanced in time with the Crank–Nicolson method.

The flux-corrected transport (FCT) algorithm is a high-order, monotone, conservative, positivity-preserving algorithm and performs accurately and efficiently in solving nonlinear equations, such as the ion continuity equation and the ion momentum equation. Because the ion density can never be negative, there must be numerical diffusion to a certain extent. However, positivity and accuracy seem to be mutually exclusive. Therefore, an additional antidiffusion stage is taken into consideration in the FCT algorithm to tackle the strong diffusion and increase the accuracy of the calculation. Meanwhile, the antidiffusive fluxes should be modified to avoid generating new maxima or minima in the solution to ensure monotonicity. More details about FCT can be found in the works of Boris et al. [7].

3.2.4 Results

In our simulation, the ion pressure gradient $-\nabla p_i$ can be ignored based on the cold fluid approximation. Because ions are much heavier than electrons, they can hardly gain any energy from the field. Besides, ions are assumed to exchange energy rapidly in elastic collisions, so they are at the same temperature as the neutral species (i.e., at room temperature). Therefore, no energy balance equation is needed for ions. For neutral particles, only the continuity equation is taken into account.

The plasma density has a different distribution when the sources are applied at different electrodes. This study is carried out at 500 mTorr in a (DF) Ar CCP discharge; LF and HF are set at 2 and 60 MHz, respectively, with a constant power of 100 W. In the reactor, the upper and lower electrodes have a radius of 7 and 6 cm, respectively, and both of them are surrounded by a 1-cm-thick dielectric. It is clear that the peak of the plasma density appears near the radial edge of the bottom electrode when it is driven by the HF source (see Figure 3.4b and d). However, when the HF source is applied at the top electrode, the peak takes place near the dielectric that covers the bottom electrode (see Figure 3.4a and c). In addition, the plasma density becomes higher when the two sources are applied at different electrodes.

The second example is a comparison of the results obtained from the electrostatic model (with only the static electric field taken into account) and the electromagnetic model (which includes the electromagnetic effects) in an Ar CCP. The discharge is sustained at 100 mTorr in a reactor with a radius of 20 cm, and the gap between the two electrodes is 3 cm. Figure 3.5 shows that at LF (i.e., 13.56 MHz), no obvious difference is detected between the plasma density profiles obtained from the two models. However, as the frequency increases, the difference between the two density profiles becomes evident, for example, at 60 and 100 MHz. Therefore, it is essential to simulate the VHF discharge with electromagnetic effects taken into account for more realistic results.

In addition, the mode transition in ICPs has also attracted a considerable amount of interest during recent years. It is well known that an inductive discharge involves two sustaining mechanisms, an

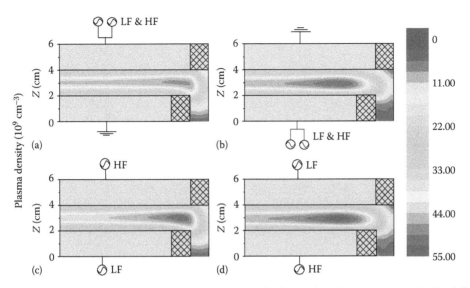

FIGURE 3.4 The distributions of plasma density in an argon discharge when the sources are applied at different electrodes ($P_{HF} = P_{LF} = 100$ W, HF = 60 MHz, LF = 2 MHz, 500 mTorr).

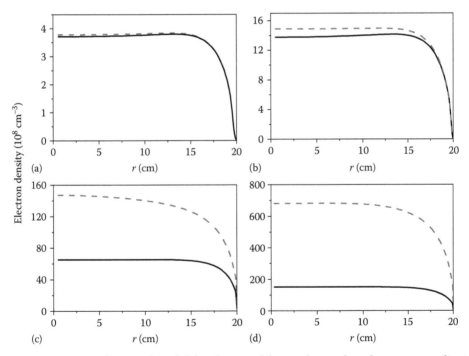

FIGURE 3.5 Comparison between the radial distributions of electron density along the reactor centerline in the electrostatic model (solid line) and the electromagnetic model (dashed line) at different frequencies: (a) 13.56 MHz, (b) 27 MHz, (c) 60 MHz, and (d) 100 MHz, for an argon discharge sustained at 100 mTorr and 30 V.

electrostatic (E) mode with low plasma density and an electromagnetic (H) mode with high plasma density, as we have mentioned above. Evolutions of plasma density with coil current under different pressures when the discharge transforms from the E mode to the H mode are shown in Figure 3.6. An argon discharge is sustained by driving a four-turn planar coil at 13.56 MHz. The results are obtained from the pure fluid model (PFM) and hybrid model (HM), respectively. Because nonlocal property of

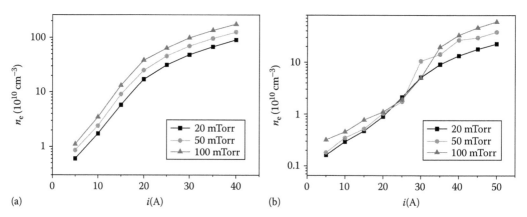

FIGURE 3.6 The evolution of electron density peaks with RF coil currents under different pressures: (a) simulations of PFM and (b) simulations of HM.

electrons is introduced in the HM, the density evolution when the mode transition occurs exhibits different characteristics at various pressures. When the pressure is low, the mode transition obtained from HM is more continuous, which is similar to that calculated by the PFM. However, as the pressure increases, a discontinuous transition takes place, and meanwhile the current range of mode transition becomes wider at 100 mTorr (see Figure 3.6b).

3.3 Particle Model: PIC and MC Method

In most laboratory plasmas, the particle spacing is about 10^{-4}–10^{-7} m, the radius of the particle generally in the range of 10^{-9}–10^{-10} m, and the speed 10^3–10^7 m s^{-1}. Based on the different characteristic scales, the interactions between particles can be divided into two parts: short-range interactions and long-range interactions. Thus the particle motion comprises two phases: the movement and the collision. In the movement phase, the long-range interactions are accomplished on the cells, which correspond to the particle-in-cell (PIC) model. In the collision phase, the interaction time is relatively short, and only the particle velocity rather than the coordinate changes, which corresponds to the Monte Carlo (MC) model. It should be noted that the collision process is completely random.

The researchers who used the particle model were from two different backgrounds: The first group comprised mechanics researchers, who consider the behavior of rarefied neutral gas without taking into account the charged particles. In this case, the Boltzmann equation is solved, and the method is often referred to as direct simulation Monte Carlo method (DSMC) [8]. The second group comprised plasma physicists, who consider the charged particles in the Vlasov–Maxwell system, and this method is known as the PIC method [9,10]. The typical steps are as follows:

1. In accordance with the coordinate and velocity of the particles, charge and current distributions on the grids are calculated.
2. According to the charge and current distributions on the grids, combined with the corresponding field boundary conditions, the electromagnetic fields on the grids are calculated.
3. Taking into account the electromagnetic fields on the grids, the coordinates and the speed of the particles are updated. The particle boundary conditions are also processed (particles leave the simulation area or new particles are generated).
4. Some or all of the particle velocity is changed by means of the appropriate MC process, while keeping the coordinate unchanged.

The processes mentioned above are with charged and neutral particles. For neutral particles, only the last two steps need to be carried out. For fully ionized plasma, the MC process is not necessary.

However, the DSMC/MC and PIC models should be adopted at the same time for typical low-pressure plasmas, with an ionization degree of 10^{-5} to 10^{-1} cm^{-3}.

3.3.1 PIC Method

The PIC method is used for long-range interactions in the particle model. In this method, discrete particles in a Lagrangian frame are tracked in continuous phase space, and moments of the distribution, such as densities and currents, are computed simultaneously on Eulerian mesh points.

Compared to the fluid model, the advantages of the PIC/MC method are (1) no arbitrary particle distribution function is assumed; all nonlocal and nonequilibrium processes can be included self-consistently; (2) the PIC/MC method uses differential cross sections as the input parameters, which can be directly measured experimentally and therefore the results are very reliable; (3) the PIC/MC method can calculate all the discharge parameters in one run, and the results are generally more accurate; and (4) unlike the fluid model, the PIC/MC method is free from numerical dissipation, and the results converge and are reliable only if the stability criterion is satisfied. However, the shortcomings of the PIC/MC method are also very clear: (1) the space step of the explicit scheme is required to resolve the Debye length, and the time step must resolve the electron oscillation frequency; these lead to very small space and time steps, so the computation cost is very high, typically 10–100 times or more that of the fluid model and (2) the PIC/MC method must use a regular grid; however, there are still many problems that need to be resolved for the simulations in the arbitrary region. In contrast, fluid model mesh generation is more mature, and complex shapes can be simulated for actual industrial reactors.

As we have discussed before, plasma includes both fields and particles. For electrode structures, because the electromagnetic wavelength is much larger than the size of the discharge device, while the plasma current density is very small, the self-generated magnetic field can be ignored, so a static model is sufficient to calculate the electric field by solving Poisson's equation. The electrostatic model has been well studied and successfully used in many researches on DC and CCP discharges. On the other hand, for coil and antenna structures, Darwin or full electromagnetic model is necessary. However, due to the significant computation cost, PIC simulations for these discharges are still an unsolved problem. Therefore, we use the electrostatic model as an example here.

3.3.1.1 Weighting

Weighting is the operation that plasma macro-quantities (number density, current density, etc.) are assigned to or from simulation particles. The macroparticle coordinate x_g is mapped to the density function $\rho(x_p)$, where x_g is the grid point coordinate. By introducing the shape function $S(x)$, the interpolation process can be written as

$$\rho(x_g) = \sum_p q_p S(x_g - x),$$
(3.22)

where $\sum_p$ is the sum of all particles; obviously, $S(x)$ must satisfy the normalization condition: $\int_{-\infty}^{\infty} S(x - x_p) dx = 1$. In addition, because the particle size is limited, the range of $S(x) \neq 0$ should be small enough. To avoid anisotropic problems, the condition $S(-x) = S(x)$ must be met. The most simple function for meeting these requirements is the use of the polynomial form of spline interpolation. The first third-order sample interpolations are often referred to as the nearest-grid-point (NGP), PIC, and quadratic spline weighting (QS) methods, respectively.

For the NGP method, macroparticles give rise to a lot of random noises when crossing the grid boundaries. Higher order interpolation methods can partially eliminate this problem, but the aliasing effect still needs to be handled with care. A more advanced approach implies the need for more complex interpolation, but it is rarely used in practice, and most of the studies adopt the one-order PIC method.

3.3.1.2 Poisson Equations

The electric field can be obtained by solving Poisson's equation $\nabla^2\varphi = -\rho/\varepsilon_0$. Although Poisson equation is mostly studied thoroughly in the numerical methods of partial differential equations, applications of these methods are still not very satisfactory. This is mainly because of the global nature of the Poisson equation: Any solution needs to use all solved parameters within the source function and to rely on the field of the boundary points, regardless of how far the boundary points are. In general, the computational cost is much more for solving Poisson's equation than for solving the Maxwell equations. In most higher dimensional cases, the cost for solving Poisson's equation is even greater than that for advancing the particles. Besides, the global nature of Poisson's equation makes it very difficult to parallelize.

For 1D cases, by Gauss' law on the grid, Poisson's equation can be discretized into

$$-\rho_i = \frac{\phi_{i-1} - 2\phi_i + \phi_{i+1}}{\Delta z^2}. \tag{3.23}$$

For 2D axis-symmetric structure, the discrete Poisson equation is in the form of

$$-\rho_{j,k} = \frac{2r_{j+1/2}}{\left(\Delta r^2\right)_j \Delta r_{j+1/2}}\left(\phi_{j+1,k} - \phi_{j,k}\right) - \frac{2r_{j-1/2}}{\left(\Delta r^2\right)_j \Delta r_{j-1/2}}\left(\phi_{j-1,k} - \phi_{j,k}\right)$$

$$+ \frac{1}{\Delta z^2}\left(\phi_{j,k+1} - 2\phi_{j,k} + \phi_{j,k-1}\right), \tag{3.24}$$

$$-\rho_{0,k} = \frac{2r_{1/2}}{\Delta r_0^2}\frac{\phi_{1,k} - \phi_{0,k}}{\delta r_{1/2}} + \frac{\phi_{0,k+1} - 2\phi_{0,k} + \phi_{0,k-1}}{\Delta z^2}.$$

The above equations are solved by coupling the appropriate boundary conditions, usually Dirichlet boundary conditions. Then the electric field is obtained from $E = -\partial\varphi/\partial x$. To avoid nonphysical effects, the electric potential, electric field, and charge density are usually at the same grid; therefore, the electric field is calculated by using the central difference. For the 1D case, we have

$$E_z = \frac{\phi_{i-1} - \phi_{i+1}}{2\Delta z}. \tag{3.25}$$

In some RF discharges, the RF source is loaded through a capacitor to the electrode, so the DC current cannot flow through the electrode, and this results in a self-bias voltage, which should be considered to be self-consistent in the simulation. There are two ways to include such an effect: (1) The most general approach is to include the charges of the RF electrode in Poisson's equation. However, iterations are often needed, and it is much easier to divide the potential into two linear superpositions. (2) Assuming the voltage waveform is $V = V_0 \sin \omega_{rf} t + V_{dc}$, adjust V_{dc} to make sure that the positive and negative charge flows to the electrode is equal.

3.3.1.3 Particle Pushing

Under the influence of the electromagnetic field, the macroparticle moves in space. In the stationary background coordinate system, the motion equations of the nonrelativistic case can be written as

$$m\frac{dv}{dt} = q(E + v \times B),$$

$$\frac{dx}{dt} = v. \tag{3.26}$$

Without taking the external magnetic field into consideration, the motion equations can be expressed by the explicit leapfrog scheme,

$$m \frac{v_z^{t+\Delta t/2} - v_z^{t-\Delta t/2}}{\Delta t} = qE^t,$$

$$\frac{z^{t+\Delta t} - z^t}{\Delta t} = v_z^{t+\Delta t/2}, \tag{3.27}$$

where E is given by

$$E(z_p) = \left[\frac{z_{i+1} - z_p}{z_{i+1} - z_i} \right] E_i + \left[\frac{z_p - z_i}{z_{i+1} - z_i} \right] E_{i+1}. \tag{3.28}$$

3.3.1.4 New PIC Model

In conventional electrostatic PIC simulations, the spatial and temporal steps must be chosen to resolve the fastest temporal and finest spatial behavior of the electrons, namely, $\Delta x \leq \lambda_D$, $\omega_p \Delta t \leq 2$, and $\Delta x/\Delta t \leq v_t$, where v_t is the electron thermal velocity. This requires hundreds to thousands of cells per centimeter, and the time step is 10^{-12}–10^{-11} s in discharge modeling. On the other hand, more than 100 particles per cell are needed to get rid of the stochastic errors in MC simulation. Therefore, the computational costs are very high.

By damping some HF modes, implicit PIC/MC method can eliminate the major constraints of grid spacing and time steps in explicit PIC codes, with most of the kinetic effects maintained, and thus it becomes a better approach to solve these problems. In the implicit particle-in-cell (IPIC) scheme, the field in the next step E^{n+1}, which depends on the future charge density ρ^{n+1}, must be known at the n step. There are two kinds of algorithms, namely, direct implicit simulation (DIPIC) and implicit movement method simulation (IMMPIC). In DIPIC, the field equations are derived from direct summation and extrapolation of the particles' moving equations. In the IMMPIC method, the field equations are derived from the Vlasov movement equations.

The direct implicit scheme has been successfully used in several simulations, but some issues need to be addressed. For instance, a practical electromagnetic and implicit model for low-pressure plasma simulations needs to be developed.

3.3.2 MC Method

For short-range interactions in the particle model, the MC method must be employed. Because electrons, positive ions, and neutral particles (radicals or molecules) coexist in low-pressure plasmas, there are three types of collision processes:

1. Collisions between charged particles
   This type of collision includes the small-angle Coulomb collisions between electron–electron (e–e), electron–ion (e–i), and ion–ion (i–i). Besides, recombination processes between electrons/ negative ions and positive ions are also included.
   All particles within the Debye length participate in Coulomb collisions. During the past few decades, significant work has been undertaken to study the effect of e–e collisions on the electron energy distribution function. However, there are only two models based on the two-body method: the first is the TA77 model developed by T. Takizuka, and the second is the NY97 model developed by K. Nanbu. In these two models, all charged particles are divided into pairs, and collisions occur between them. In the TA77 model, the time step is limited by the collision frequency. However, in the NY97 model, many small-angle scattering processes are cumulated

to a large angle, and thus a much larger time step can be used, which makes this method more suitable at high density.

For recombination processes, although the cross section is about 100–10,000 times the molecular radius, the exact values are still unknown. Therefore, it is more convenient to adopt the recombination rate instead of the cross section. The most general collision method based on the recombination rate was first proposed by M. J. Kushner and generalized by K. Nanbu in which the temperature relied on in recombination processes can be incorporated in the MC model.

2. Collisions between neutral molecules, namely, m–m collision
In the molecular–molecular (m–m) collision, the interaction distance is small, and the collision takes place near the molecular radius. This method, often referred to as the DSMC model, has been well developed by G. A. Bird and K. Nanbu. The hard sphere (HS) model, the variable hard sphere (VHS) model, and the variable soft sphere (VSS) model are the three commonly used models in the DSMC method that can be used to describe chemical reactions occurring in the gas phase.
These two types of collisions are less important or can even be ignored completely in most cases. Due to the limited space here, more details are not mentioned in this chapter. Interested readers can refer to the original papers.

3. Collisions between charged particles and neutral gas
This type of collision, which includes reactions between electron–molecular (e–m) and ion–molecular (i–m), is the most important process in the discharge. Several approximations need to be made before we can have further discussions: (i) The neutral gas is in thermal equilibrium, and the density or temperature of neutral particles does not change with time. This assumption is based on the fact that the plasma ionization rate is very low. So the momentum exchange between charge particles and neutrals is small, and thus the neutral gas is in its own thermal equilibrium. (ii) Only the nonpolar molecules are considered. (iii) Only the single-charge states generated by the electron impact ionization and dissociation process are considered, without taking into consideration the ion impact ionization process. This is valid because the electron energy is lower than 10 eV in most cases. Besides, Auger electrons are not taken into account.
A typical MC process involves the following steps: (1) randomly select collision particles (the first random number); (2) calculate the energy and the corresponding cross section, depending on the speed of selected particles, where the cross-sectional data are usually experimentally determined; (3) calculate the collision type according to the cross section (the second random number); (4) calculate the scattering (the third random number) and azimuthal angle (the fourth random number); and (5) calculate the postcollision speed.

3.3.2.1 Sampling and Calculation of the Postcollision Velocity

When the pressure is low, and the time step small, the collision probability is very low. Under this condition, the null-collision and constant time step method (also known as Nanbu's method) are the two ways used to accelerate the sampling process of particles significantly. Because only the e–m or i–m process is considered in the simulation, the null-collision method is more efficient.

Before starting the simulation, the energy range of interest is divided equally into N_ε parts by the energy interval $\Delta\varepsilon$. The collision frequencies for process j at energy ε_i between electrons and particles with density n_j can be expressed as

$$\nu_{i,j} = \left(\frac{2e\varepsilon_i}{m_e}\right)^{1/2} \sigma_{i,j} n_j, \tag{3.29}$$

where $\sigma_{i,j}$ is the electron impact cross section. The total collision frequencies at energy ε_i are $\nu_i = \sum_{j=1}^{N_{type}} \nu_{i,j}$, where N_{type} is the number of collision types. Then the probability arrays are given by $p_{i,j} = \sum_{l=1}^{j} \nu_{i,l} / \nu_{max}$, and $v_{max} = max\{v_i\}$ is the maximum of total collision frequencies. For each simulated electron at energy ε_i, the collision type can be determined as follows:

$$r \leq p_{i,1} \quad \text{(Collision type 1)},$$

$$p_{i,1} \leq r \leq p_{i,2} \quad \text{(Collision type 2)},$$

$$\cdots$$

$$r \geq p_{i,type} \quad \text{(Null collision)},$$

(3.30)

where r is a random number $r \in [0, 1]$.

Once an elastic collision occurs, the velocities of particles with masses m and M after collision v' and V' are determined by

$$v' = v - \frac{M}{m + M}\left[g(1 - \cos\chi) + h\sin\chi\right],$$

$$V' = V + \frac{m}{m + M}\left[g(1 - \cos\chi) + h\sin\chi\right],$$

(3.31)

where

$g = v - V$ is the relative velocity
v and V are the precollision velocities

The deflection angle of the relative velocity is given by $\cos\chi = [\varepsilon + 2 - 2(1 + \varepsilon)^R]/\varepsilon$, where R is a random number between 0 and 1, and $\varepsilon = mv^2/2$ is the energy of the incident electron before collision. The Cartesian components of h are as follows:

$$h_x = g_\perp \cos\varphi,$$

$$h_y = -\frac{g_x g_y \cos\varphi + g g_z \sin\varphi}{g_\perp},$$

$$h_z = -\frac{g_x g_z \cos\varphi - g g_y \sin\varphi}{g_\perp},$$

(3.32)

where

$$g = \sqrt{g_x^2 + g_y^2 + g_z^2}$$

$$g_\perp = \sqrt{g_y^2 + g_z^2}$$

The azimuthal angle φ is uniformly distributed in the interval $[0, 2\pi]$

For excitation collisions, the precollision velocity is modified as $\tilde{v} = v\left(1 - E_{th}/\varepsilon\right)^{1/2}$, where E_{th} is the threshold energy of the inelastic collision. For ionization collisions, $\tilde{v} = v\left[1 - \left(E_{th} + E_{ej}/\varepsilon\right)\right]^{1/2}$, where E_{ej} is the energy of ejected electrons. More details can be found in the works of Nanbu [11] and Georgieva [12].

3.3.2.2 Electron–Molecule Collision Process

All e–m collision processes are given in Table 3.2. Collisions between electrons and monatomic molecules include elastic, excitation, and ionization reactions. In addition, collisions between electrons and polyatomic molecules include dissociation and adsorption reactions. Differences between the monatomic and polyatomic molecule collisions are based on the fact that only the outermost electrons are involved in monatomic molecule collisions, but for polyatomic molecules, excited, rotational, and vibrational states are also important.

TABLE 3.2 Electron–Molecule Collision Process

Collision Process	Process	Example
Elastic	$e + AB \rightarrow e + AB$	$e + Ar \rightarrow e + Ar$
Excitation	$e + AB \rightarrow e + AB^*$	$e + Ar^* \rightarrow e + Ar^*$
Ionization	$e + AB \rightarrow 2e + AB^+$	$e + Ar \rightarrow 2e + Ar^+$
Attachment	$e + AB \rightarrow AB^-$	$e + SF_6 \rightarrow SF_6^-$
Dissociation	$e + AB \rightarrow e + A + B$	$e + CF_4 \rightarrow e + F + CF_3$
Polar dissociation	$e + AB \rightarrow e + A^+ + B^-$	$e + CF_4 \rightarrow e + F^+ + CF_3^-$
Dissociative ionization	$e + AB \rightarrow 2e + A + B^+$	$e + CF_4 \rightarrow 2e + F + CF_3^+$
Dissociative attachment	$e + AB \rightarrow A + B^-$	$e + CF_4 \rightarrow F + CF_3^-$

Electrons lose energy during collisions except the elastic reaction, in which the total momentum and energy of electrons are conserved. Excitation collisions include two types: the first type occurs when the outermost electron of the monatomic molecule is excited to a high-energy state, which has a larger cross section; the second type takes place when polyatomic molecules are excited to their rotational and vibrational excited states. For ionization reactions, electrons collide with molecules during the processes and produce new electrons and positive ions. In addition, the chemical bond of polyatomic molecules is broken and a number of free radicals are produced in dissociation processes. Then the radicals or molecules capture electrons and form negative ions in the adsorption collision, which only takes place in electronegative gas discharges.

Ionization, dissociation, and adsorption processes can take place simultaneously during one reaction, which are known as dissociation–ionization collision, desorption–ionization collision, and polarization collision.

The electron scattering is anisotropic, and the scattering angle is usually fitted from an integral cross section. The fitting formula is

$$\cos \chi = \frac{2 + \varepsilon - 2(1 + \varepsilon)^R}{\varepsilon}, \tag{3.33}$$

where ε is the electron energy or in a more general form

$$\cos \chi = 1 - \frac{2R(1 - \xi)}{1 + \xi(1 - 2R)}, \tag{3.34}$$

where

ξ is the fitting function and determined by the electron energy
R is a random number between 0 and 1

3.3.2.3 Ion–Molecule Collision Process

All e–m collision processes are given in Table 3.3. Ion–monatomic molecule collisions include two types: elastic and charge exchange. For ion–polyatomic molecule collisions, two more types are added: dissociation and desorption (the reverse process of adsorption).

The HS model and the induced polarization model can be used for i–m collisions. Assume that the incident ion is A and the neutral particle is B, then the postcollision velocity of elastic collision can be simplified as

$$v_{a'} = \frac{1}{2} \left(m_a v_a + m_b v_b + m_b \left| v_a - v_b \right| R \right),$$

$$v_{b'} = \frac{m_a}{m_a + m_b} \left(m_a v_a + m_b v_b - m_a \left| v_a - v_b \right| R \right). \tag{3.35}$$

TABLE 3.3 Ion–Molecule Collision Process

Collision Process	Process	Example
Elastic	$A^{\mp} + B \rightarrow A^{\mp} + B$	$Ar^{+} + Ar \rightarrow Ar^{+} + Ar$
Charge exchange	$A^{\mp} + B \rightarrow A + B^{\mp}$	$Ar^{+} + Ar \rightarrow Ar + Ar^{+}$
Detachment	$A^{-} + B \rightarrow e + A + B$	$O^{-} + O_{2} \rightarrow e + O + O_{2}$
Ion dissociation	$A^{\mp} + AB \rightarrow A^{\mp} + A + B$	$CF_{3}^{+} + CF_{4} \rightarrow CF_{3}^{+} + F + CF_{3}$

For reactive gases like SF_6, ion collisions are very complex. Due to a lack of adequate cross-sectional data, a theoretical cross section and treatment proposed by K. Nanbu can be used.

References

1. Lieberman M A and Lichtenberg A J 2005 *Principles of Plasma Discharges and Materials Processing,* 2nd edn (New York: Wiley).
2. Makabe T and Petrovic Z L 2006 *Plasma Electronics: Applications in Microelectronic Device Fabrication* (New York: Taylor and Francis Group).
3. Lieberman M A, Booth J P, Chabert P, Rax J M, and Turner M M 2002 *Plasma Sources Sci. Technol.* **11** 283.
4. Goto H H, Lowe H D, and Ohmi T 1992 *J. Vac. Sci. Technol. A* **10** 3048.
5. Bukowski J D and Graves D B 1996 *J. Appl. Phys.* **80** 2614.
6. Gogolides E and Sawin H H 1992 *J. Appl. Phys.* **72** 3971.
7. Boris J P, Landsberg A M, Oran E S, and Gardner J H 1993 *LCPFCT—A Flux-Corrected Transport Algorithm for Solving Generalized Continuity Equations* (No. 6410-93-7192, Washington DC: NRL).
8. Bird G A 1994 *Molecular Gas Dynamics and the Direct Simulation of Gas Flows* (Oxford: Clarendon Press).
9. Hockney R W and Eastwood J W 1988 *Computer Simulation Using Particles* (New York: Adam-Hilger).
10. Birdsall C K and Langdon A B 1985 *Plasma Physics via Computer Simulation* (New York: McGraw-Hill).
11. Nanbu K 2000 *IEEE Trans. Plasma Sci.* **28** 971.
12. Georgieva V 2006 Computer modeling of low-pressure fluorocarbon-based discharges for etching purposes (Ph.D. diss., University of Antwerp).

4

Modeling of Atmospheric Pressure Plasmas

Yanhui Wang,
Jizhong Sun, and
Dezhen Wang

4.1 Introduction

Atmospheric pressure discharges (APDs) are useful because of their specific advantages over low-pressure discharges. They do not need expensive vacuum equipment, and generate nonthermal plasmas, which are more suitable for assembly line processes. Hence, this category of discharges has significant industrial applications (Fridman et al. 2005).

APDs exhibit rich physical phenomena and, as such, significant experimental and theoretical work has been done in this area. However, nonthermal atmospheric pressure plasmas suffer from severe limitations in the field of experimental diagnostics. Commonly used diagnostic techniques suitable for low-pressure discharges cannot be used in APDs because of more frequent collisions in APDs. Also, there is no reliable technique for measuring basic plasma parameters such as energy distributions and densities of both electrons and ions (let alone their spatiotemporal variations). For example, the Langmuir probe cannot be used for this category of discharges. To understand the evolution of important physical quantities, we resort to numerical models.

In the following sections, we mainly concentrate on dielectric-barrier discharges (DBDs) but limit our numerical models to the cases where the applied electric field is below the Meek's criteria to avoid streamer development. The use of a dielectric barrier in the discharge gap helps prevent spark formation. Although DBDs have been used for over 100 years, there has been renewed interest with the discovery of atmospheric glow discharge by Kanazawa et al. (1988) and later its confirmation by Massines et al. (1998). DBDs exhibit two major discharge modes: filamentary and glow (homogeneous). The glow discharge mode has obvious advantages over the filamentary one for applications such as treatment of surfaces and deposition of thin films. Glow mode discharges with average power densities comparable to those of filamentary discharges are of enormous interest for applications in which reliable control is required. A theoretical study of glow discharge mode would be of great help to scientists. Further, by assuming that noble gases represent a model system, nonlinear phenomena, such as chaos, can be studied. This would also be of interest to people working in other related fields.

In this chapter, we first introduce numerical models, and then discuss several applications in detail.

4.2 Theoretical Models and Computational Details

4.2.1 Fluid Model

4.2.1.1 Basic Equations

The fluid model is the most commonly used approach to study atmospheric pressure plasmas in which plasma particles are represented by macroscopic quantities such as density, flux, and mean energy. In APDs, the evolution of particle positions and velocities can generally be described by Boltzmann's equation:

$$\frac{\partial f}{\partial t} + \boldsymbol{v} \cdot \nabla f + \frac{q}{m}\boldsymbol{E} \cdot \nabla_v f = \left(\frac{\delta f}{\delta t}\right)_{col}, \tag{4.1}$$

where

f is the distribution function of the considered particles in six-dimensional phase space
$\boldsymbol{v}$ is the velocity coordinate
q is the charge
m is the mass of charged particles
$\boldsymbol{E}$ is the electric field
∇_v is the velocity gradient operator
$\left(\delta f / \delta t\right)_{col}$ represents the change rate of the distribution function caused by collisions

Successive integrations of Boltzmann's equation over velocity space can lead to the velocity moments, also known as the fluid equations (transport of mass, transport of momentum, and transport of energy). The first moment of Boltzmann's equation (i.e., the continuity equation for each particle) is obtained by integrating Equation 4.1 over the velocity space:

$$\frac{\partial n}{\partial t} + \nabla \cdot \boldsymbol{J} = S, \tag{4.2}$$

where

n is the particle density
$\boldsymbol{J}$ is the particle flux
S is the source term

By multiplying Equation 4.1 with $m\boldsymbol{v}$ and integrating it over velocity space, the momentum conservation equation is obtained. Under the drift–diffusion approximation, it is written as

$$\boldsymbol{J} = \left(\pm\mu\boldsymbol{E}n\right) - D\nabla n. \tag{4.3}$$

In the first term on the right-hand side of Equation 4.3, the symbols "+" and "−" are for positively and negatively charged particles, respectively, where μ is the mobility of the particles and D the diffusion coefficient.

Because of the short energy relaxation time of electrons in APDs, local field approximation is usually used to close the system of hydrodynamic equations. In this approximation, the energy transfer for electrons and ions is ignored, and the transfer and reaction rate coefficients of charged particles are expressed as a function of the local reduced electric field. Thus the simplified model has only the continuity and momentum transfer equations. When the discharge current density is moderate, the fluid model based on local field approximation can accurately predict many behaviors of APDs such as discharge current and voltage. Therefore, it is a practical tool for studying atmospheric plasmas.

For heavy particles with low mobility the local field approximation is always valid. However, electrons may not always be in equilibrium with the local electric field. In regions with a strong electric field, such as the sheath region, if the mean free path of electrons is not small with respect to the characteristic distance of field variations, electrons can gain energy at one position and dissipate it at another position. When this occurs, it is more appropriate to calculate another moment of Boltzmann's equation describing the transport of the mean energy.

Multiplying Equation 4.1 by $1/2\left(mv^2\right)$ and integrating over velocity, the energy conservation equation for an electron is obtained:

$$\frac{\partial n_e \bar{\varepsilon}}{\partial t} = -\nabla \cdot \Gamma_\varepsilon - e\boldsymbol{E} \cdot \boldsymbol{J}_e - P(\bar{\varepsilon}), \tag{4.4}$$

where

$\bar{\varepsilon} = 3/2\left(k_\mathrm{B} T_e\right)$ is the mean electron energy (k_B and T_e represent the Boltzmann constant and the electron temperature, respectively)

$\boldsymbol{J}_e$ is the electron flux

$e\boldsymbol{E} \cdot \boldsymbol{J}_e$ represents the energy gained by electrons in the electric field

$P(\bar{\varepsilon})$ is the energy lost in various collisions

Here, Γ_ε is the electron mean energy flux given by

$$\Gamma_\varepsilon = \left(\frac{5}{3}\boldsymbol{J}_e\bar{\varepsilon}\right) - \left(\frac{5}{3}n_e D_e \nabla\bar{\varepsilon}\right). \tag{4.5}$$

The fluid model with the electron energy conservation equation, i.e., Equations 4.2 through 4.4 treats the electron transport coefficients as functions of the local mean electron energy. It can describe the nonlocal behavior of electrons more accurately than the fluid model without the electron energy equation.

In general, the electric field is calculated from Poisson's equation:

$$\nabla \cdot \boldsymbol{E} = \frac{\rho}{\varepsilon_0}, \tag{4.6}$$

where

$\rho = e\left(n_+ - n_- - n_e\right)$ is the total space charge density (n_+, n_-, and n_e are number densities of positive ions, negative ions, and electrons, respectively)

ε_0 is vacuum permittivity

The numerical solution for Poisson's equation is complicated and very time-consuming, especially in two- or three-dimensional space. Thus it is sometimes proposed to replace Poisson's equation by the current balance equation (Kulikovsky 1994a):

$$\frac{\partial \boldsymbol{E}}{\partial t} + \frac{e}{\varepsilon_0}\boldsymbol{J}_c = \frac{e}{\varepsilon_0}\boldsymbol{J}_\mathrm{T}, \tag{4.7}$$

where $\boldsymbol{J}_\mathrm{T}$ is the vector for total current density, which consists of conductivity current density $\boldsymbol{J}_c$ and displacement current density $\left(\varepsilon_0/e\right)\partial\boldsymbol{E}/\partial t$. The only condition that $\boldsymbol{J}_\mathrm{T}$ should obey is $\nabla \cdot \boldsymbol{J}_\mathrm{T} = 0$. Obviously, Equation 4.7 is much simpler to solve.

4.2.1.2 Example of One-Dimensional Fluid Simulation of Atmospheric Pressure DBDs in Helium

A simplified DBD system is shown schematically in Figure 4.1. It comprises two parallel-plate metal electrodes. Each electrode is coated with a $d_\mathrm{B} = 0.2$-cm-thick dielectric layer with relative permittivity $\varepsilon_\mathrm{B} = 7.5$. The left electrode is grounded, while the right electrode is driven by an ideal sinusoidal voltage. The discharge gap is $d_\mathrm{C} = 1.0$ cm and is filled with atmospheric helium. Assuming that the atmospheric

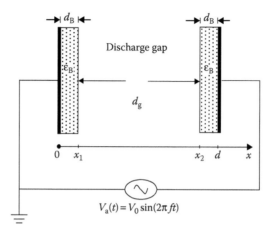

FIGURE 4.1 The electrode configuration used in the model.

DBD under study is uniform in the direction parallel to the electrodes, the plasma dynamics can be described by a one-dimensional fluid model.

In the simplest helium discharge model in which He$^+$ ions are produced by direct impact ionization, under local field approximation plasma evolution can be described by the following equations:

$$\frac{\partial n_e(x,t)}{\partial t} + \frac{\partial J_e(x,t)}{\partial x} = S(x,t),\qquad(4.8)$$

$$\frac{\partial n_p(x,t)}{\partial t} + \frac{\partial J_p(x,t)}{\partial x} = S(x,t),\qquad(4.9)$$

$$J_e(x,t) = -\mu_e E(x,t)n_e(x,t) - D_e\frac{\partial n_e(x,t)}{\partial x},\qquad(4.10)$$

$$J_p(x,t) = \mu_p E(x,t)n_p(x,t) - D_p\frac{\partial n_p(x,t)}{\partial x},\qquad(4.11)$$

$$\varepsilon_0\varepsilon(x)\frac{\partial E(x,t)}{\partial t} + i_c(x,t) = i_T(t).\qquad(4.12)$$

Here x and t are the interelectrode axial distance and time, respectively, subscripts "e" and "p" represent electrons and positive ions, respectively, and $s(x,t)$ is the source term. Only direct ionization by electron impact and electron–ion recombination are considered, and the source term is given by

$$s(x,t) = \alpha\mu_e|E(x,t)|n_e(x,t) - \eta n_e(x,t)n_p(x,t),\qquad(4.13)$$

where

  α is the Townsend ionization coefficient obtained using Ward's formula
  η is the electron recombination coefficient, which is taken from the literature (Ward 1962, Deloche et al. 1976). They are all functions of the local reduced electric field.

In Equation 4.12, $\varepsilon(x)$ is the relative permittivity of either the dielectric layers or the discharge gas, depending on x, and $i_c(x,t) = e\left[J_p(x,t) - J_e(x,t)\right]$ is the conduction current density. Secondary electron emission from the instantaneous cathode is considered here for ion bombardment alone, and therefore the electron flux leaving the cathode is taken as $\gamma J_p(x,t)$. Here γ is the secondary emission coefficient. The discharge current density is i_T. Because the electric field satisfies the condition $\int_0^d E(x,t)dx = V_a(t)$, the expression of i_T can be obtained by integrating Equation 4.12 from $x = 0$ to $x = d$:

$$i_T(t) = \left(\frac{2d_B}{\varepsilon_0\varepsilon_B} + \frac{d_g}{\varepsilon_0}\right)^{-1}\left[\int_{x_1}^{x_2}\frac{i_c(x,t)}{\varepsilon_0}dx + \frac{\partial V_a(t)}{\partial t}\right].\qquad(4.14)$$

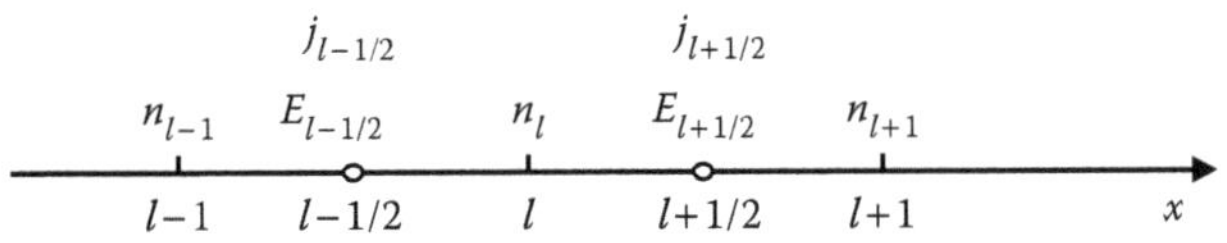

FIGURE 4.2 The computational cell.

This method, in which the local electric field is determined from the current conservation equation instead of Poisson's equation related with space charge density, can yield the electric field from i_T and i_c directly, thus avoiding the complex boundary conditions associated with dielectric–gas interfaces.

The above set of equations is often numerically solved with the semi-implicit Scharfetter–Gummel scheme (Scharfetter et al. 1969, Boeuf 1987, 1988, Kulikovsky 1995), which has already been extensively used in the modeling of semiconductor devices and gas discharges. This scheme is based on the following idea.

Introduce a uniform (or nonuniform) grid with N nodes $\{x_l, l = 1, 2,..., N\}$ in the computational domain. The number densities are given at the nodes, while fields and currents are given at the half-integer points, as shown in Figure 4.2. Between two nodes l and $l + 1$, E, D, and j are assumed to be constant and equal to $E_{l+1/2}$, $D_{l+1/2}$, and $j_{l+1/2}$, respectively. Thus, Equations 4.10 and 4.11 can be considered as differential equations with respect to n over the interval (x_l, x_{l+1}). Solving these two equations can yield the expressions for $j_{el+1/2}$ and $j_{pl+1/2}$.

Therefore, the numerical solutions of the system equations are obtained from the following finite-difference equations:

$$\frac{n_l^{t+1} - n_l^t}{\tau} + \frac{j_{l+1/2}^{t+1} - j_{l-1/2}^{t+1}}{h} = s_l^t, \tag{4.15}$$

$$\varepsilon_0 \frac{E_{l+1/2}^{t+1} - E_{l+1/2}^t}{\tau} + i_{cl+1/2}^{t+1} = i_T^{t+1}, \tag{4.16}$$

$$i_T^{t+1} = \left(\frac{2d_B}{\varepsilon_B} + d_g\right)^{-1} \left[h \sum_l i_{cl+1/2}^{t+1} - \varepsilon_0 \frac{v_a^{t+1} - v_a^t}{\tau}\right]. \tag{4.17}$$

In the above equations, τ is the time step, h the space step, and n and j stand for number density and flux of electrons or ions, respectively. The expression of j at half-integer points is given by

$$j_{l+1/2}^{t+1} = \frac{D}{h R_{l+1/2}^t} \left(n_l^{t+1} - e^{\alpha_{l+1/2}^t} n_{l+1}^{t+1}\right), \tag{4.18}$$

where

$$R_{l+1/2}^t = \frac{e^{\alpha_{l+1/2}^t} - 1}{\alpha_{l+1/2}^t},$$

$$\alpha_{l+1/2}^t = \pm \frac{\mu h}{D} E_{l+1/2}^t.$$

Here the symbol "+" is for electrons and "−" is for ions.

This scheme is accurate only on the condition that the potential drop between two adjacent nodes is much less than the electron temperature. To maintain the calculation accuracy, fine grids are often required. To simplify the complexity in computation, Kulikovsky proposed an improved scheme: In this scheme a pair of additional nodes is inserted between two adjacent nodes to make the potential drop between these *virtual* nodes less than the electron temperature, and particle densities at virtual nodes are obtained by interpolation. The current density is then calculated from Equation 4.18, where values at virtual nodes are used as parameters. This procedure can dramatically improve the accuracy of the Scharfetter–Gummel scheme; Figures 4.3 and 4.4 present some numerical results of Atmospheric pressure DBDs in Helium.

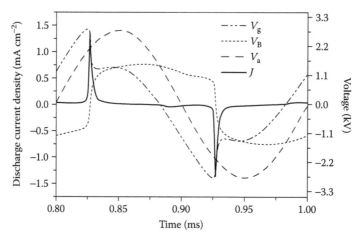

FIGURE 4.3 Temporal evolutions of the current density (solid curve), applied voltage (dashed curve), gas voltage (dot–dashed curve), and memory voltage (dotted curve) over one cycle.

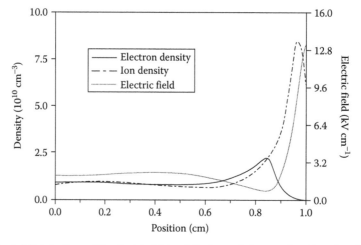

FIGURE 4.4 Spatial distributions of the electric field and the ion and electron densities when the discharge current reaches its maximum.

4.2.2 Particle-in-Cell Monte Carlo Collision Model

Atmospheric discharges exhibit strong nonlinear effects, which the fluid models cannot easily represent; besides, fluid models are not capable of modeling nonlocal particle kinetics because they assume that the specific velocity distribution is based on the local value of the electric field (Kim et al. 2005). In reality, a combination of factors plays an important role in many plasmas of interest. Therefore, a kinetic model is needed. The particle-in-cell (PIC) method allows the statistical representation of general distribution functions in phase space and handles charge effects well and is therefore a good alternative. Because the PIC method employs fundamental equations, most physical parameters can be determined.

In general, the PIC method is simple and is therefore used for analyzing the physical parameters of an electrostatic plasma system. Here we assume the system contains only electrons and one type of ions with one elementary charge for simplicity. PIC simulation tracks superparticles for both electrons and ions in phase space by integrating a set of Newton's equations of motion (Equation 4.19) (assuming N superparticles for both ions and electrons) for their positions and velocities at the next time step, and by solving Poisson's equation for the new electric field resulting from the redistribution of charged particles; the new electric field is then substituted into the set of Newton's equations of motion for the positions and velocities of these charged superparticles at the later time step. This process is repeated until the predefined conditions are satisfied:

$$\frac{\mathrm{d}\boldsymbol{v}_{ek}}{\mathrm{d}t} = -\frac{e}{m_e}\big(\boldsymbol{E} + \boldsymbol{v}_{ek} \times \boldsymbol{B}\big), \quad \frac{\mathrm{d}\boldsymbol{v}_{ik}}{\mathrm{d}t} = \frac{e}{m_i}\big(\boldsymbol{E} + \boldsymbol{v}_{ik} \times \boldsymbol{B}\big), \tag{4.19}$$

$$\nabla^2 \phi(\boldsymbol{r}, t) = -\frac{e}{\varepsilon_0}\big(n_i - n_e\big). \tag{4.20}$$

This series of Newton's equations of motion and Poisson's equation constitutes a set of self-consistent equations. With certain boundary conditions, the set of equations can be numerically solved. Here $\boldsymbol{v}$ is the velocity of superparticle k, $\boldsymbol{E}$ the electric field, and $\boldsymbol{B}$ the magnetic field. A superparticle represents a group of particles of one type, either electrons or protons. The number of particles contained in a superparticle is called the *weighting factor*. Superparticles are introduced because the product of the number of superparticles of a specific type and its weighting factor represents the real number of this type of species contained in the system of interest. It is worth noting that superelectrons have a specific feature; for instance, their ratio of charge to mass is equal to e/m_e, as in the case of electrons. In other words, the trajectory of a superparticle is identical to the species it stands for. In the course of movement in the background gas, all superparticles are tracked because they mimic the collision events. The collision processes are treated by the Monte Carlo (MC) technique. The typical flow diagram for a PIC simulation is shown in Figure 4.5. If certain criteria are satisfied, a production run takes place for final statistical treatment. A detailed review and recent advances in PIC simulations can be found in the works of Verboncoeur (2005).

Despite having many advantages, the PIC model has a number of weaknesses. Because the PIC model is a statistical sampling from the real system of interest, its numerical fluctuations of the PIC model converge as $1/N^{1/2}$ for N superparticles. Consequently, the number of particles tracked must fulfill certain requirements; for example, the number of each type of tracked particle is usually around or over 100 to avoid large statistical fluctuation and self-heating. Other conditions should also be fulfilled; for instance, the time step should generally be smaller than the shortest characteristic timescale of the plasma, such as the inverse of plasma frequency, radio frequency (RF), driving frequency, and maximum collision frequency. Also, the cell size should generally be smaller than the shortest characteristic length scale of the plasma, that is, the

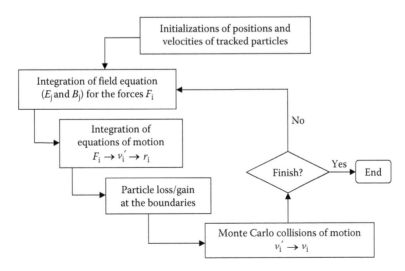

FIGURE 4.5 Flow chart of a basic PIC program.

Debye length, sheath length, and Larmor radius. For more information on its limitations, please refer to Verboncoeur (2005). Due to very small cell size, a characteristic of the discharge system of interest is that it unusually contains many cells; thus a large number of superparticles needs to be followed. Further, the time step is very short. Hence, the most challenging issue with regard to the PIC method is computational efficiency. In general, the PIC method suffers huge limitations from both timescales and space scales. In most applications, one-dimensional position and three-dimensional velocity (1D3V) PIC models are used. For a 2D3V PIC model, it is difficult to carry out these simulations and one has to resort to advanced numerical algorithms and parallel computing to enhance the computational efficiency. When this method is applied to atmospheric discharges, these limitations are much more severe.

4.2.3 Hybrid Model (PIC-MCC+Fluid)

Hybrid models that have more precise kinetics than fluid models and can run faster than a pure PIC model can be built by combining fluid simulations with PIC models. Also, depending on the physics to be modeled, various hybrid models can be derived. Ions can be modeled as a fluid, while electrons are treated in a PIC scheme (Sommerer and Kushner 1992). Alternatively, only high-energy electrons can be tracked in a PIC scheme, while ions and low-energy electrons are treated as fluids (Kushner 2004).

It is critical to guarantee free movement between the fluid and particle parts in hybrid simulations. Depending on the issues that need to be addressed, various strategies can be undertaken (Kim et al. 2005). So far, not much work has been done on atmospheric discharges (Iza et al. 2009).

4.3 Discharge Modes

4.3.1 Glow and Townsend Discharge Modes

Under different conditions the uniform atmospheric DBDs can operate in two modes: glow and Townsend. These two discharge modes have completely different electrical characteristics and spatial structures. The transition from one to the other can be controlled by external parameters. In this section, taking the atmospheric DBD in helium as an example, we discuss the properties of the two discharge modes.

The discharge considered here is generated between two parallel-plate electrodes. Assuming that the discharge is homogeneous along the electrode surfaces of electrodes, we made a one-dimensional fluid model.

The ionization and excitation species included in the model are electron, atomic ion He^+, molecular ion He_2^+ and the main atomic metastable states He (2^3S) and He (2^1S). Because the metastable states He (2^3S) and He (2^1S) exhibit very similar behavior, we group them together into one effective species: He^*. The elementary processes included in the source term are listed in Table 4.1. The mobility and diffusion coefficients have been taken from the works of Xu and Chu (1996) and Deloche et al. (1976). The secondary emission coefficient is assumed to be constant and equal to 0.01 for the two kinds of ions.

Figures 4.6 and 4.7 illustrate the electrical characteristics and spatial structures of two discharge modes obtained computationally under the following conditions: the discharge gap width is equal to 0.7 cm, the amplitude and frequency of applied sinusoidal voltage are 2.4 kV and 5 kHz, respectively, and the permittivity of the dielectric barrier is 7.5. The thickness of the dielectric barrier is 0.1 cm in Figure 4.6 and 0.2 cm in Figure 4.7. In the glow discharge mode (Figure 4.6), the discharge current increases abruptly to a higher amplitude, the densities of charged particles are high, and peak value densities occur near the cathode, which means high-level excitation and ionization occur near the cathode. The space charge effect plays a dominant role in the glow mode. The electric field is strongly distorted by spatial charges. Three specific regions that a classical normal DC glow discharge possesses are formed: the cathode fall, the Faraday dark space, and the positive column. In the Townsend mode, the ionization level is relatively low, and thus the discharge current and the charged particle densities are much smaller than in the glow discharge mode, as shown in Figure 4.7. The maximum electron density occurs at the anode (Figure 4.7). This indicates that the maximum production rate of excited or ionized species

TABLE 4.1 The Elementary Processes in the Model of Helium Plasma

Reaction	Threshold Energy (eV)	Rate Coefficient	Reference
$e + He \rightarrow He^+ + 2e$	24.6	α cm^3 s^{-1}	Shon and Kushner (1994)
$e + He \rightarrow He^* + 2e$	20.2	$4.2 \times 10^{-9} T_e^{0.31} e^{-19.8/T_e}$ cm^3 s^{-1}	Stevefelt et al. (1982)
$e + He^* \rightarrow He^+ + 2e$	4.4	$1.28 \times 10^{-7} T_e^{0.6} e^{-4.78/T_e}$ cm^3 s^{-1}	Stevefelt et al. (1982)
$He^+ + 2He \rightarrow He_2^+ + He$	0.0	2.0×10^{-31} cm^6 s^{-1}	Stevefelt et al. (1982)
$He^* + He^* \rightarrow e + He^+ + He$	−15.0	8.7×10^{-10} cm^3 s^{-1}	Pouvesle et al. (1982)
$He^* + He^* \rightarrow e + He_2^+$	0.0	2.03×10^{-9} cm^3 s^{-1}	Pouvesle et al. (1982)
$He_2^+ + e \rightarrow 2He^* + He$	0.0	5.9×10^{-9} cm^3 s^{-1}	Ward (1962)
$He^+ + e \rightarrow He$	0.0	$8.1 \times 10^{20} \dfrac{n_e}{1 + 0.079p}(T_e/T_g)^{-4.4}$ cm^3 s^{-1}	Deloche et al. (1976)
$He^* + e \rightarrow He + e$	−20.2	2.9×10^{-9} cm^3 s^{-1}	Pouvesle et al. (1982)

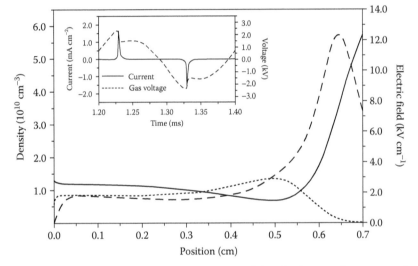

FIGURE 4.6 Electrical characteristics and spatial structures of glow discharge at the moment of current peak occurrence. The solid curve is an electric field, the dotted curve is electron density, and the dashed curve is ion density. Cathode is at $x = 0.7$ cm. Inset shows current density and gas voltage versus time. (From Wang Y. H. and Wang D. Z., *Chin. Phys. Lett.*, 21, 2235, 2004. Chinese Physical Society and IOP Publishing Ltd.)

by electron impact is close to the anode. The low ionization level makes the space charge effect too small to affect the applied electric field. The electric field decreases linearly from the cathode to the anode, and the localized high-field cathode fall region and the positive column regions are not formed. In fact, the discharge in Figure 4.6 changes from glow to Townsend mode because the increase in dielectric layer thickness results in a decrease of the electric field strength in the gas gap, thus leading to a drop in the ionization level. It is clear that, in Figure 4.7, while the external voltage is identical, the electric field strength is only half of that in Figure 4.6. Generally, in atmospheric DBD, Townsend discharge occurs under conditions of relatively low driving frequency, small peak voltage, wide discharge gap, or thick dielectric layer.

In the discharges discussed above, there is only one discharge current peak per half cycle of the external voltage. But in some cases, such as small gap width, multiple current peaks can be formed during a half cycle of the applied voltage. This multiple peak discharge can also operate in two different modes. For the discharge conditions described in Figure 4.6, when the gap width is decreased to 0.5 cm and the peak voltage is increased to 3.0 kV, two current peaks per half period can be formed, as shown in Figure 4.8. It can be seen from the spatial distributions of electron density, ion density, and electric field that both breakdowns have the typical features of glow discharge. If the gap width is decreased to 0.3 cm, more current peaks are

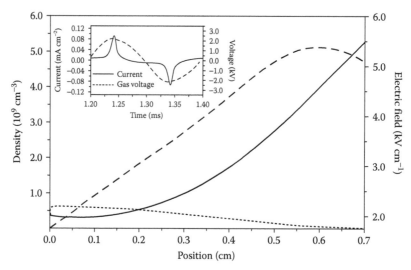

FIGURE 4.7 Electrical characteristics and spatial structures of glow discharge at the moment of current peak occurrence. The solid curve is an electric field, the dotted curve is electron density, and the dashed curve is ion density. Cathode is at $x = 0.7$ cm. Inset shows current density and gas voltage versus time. (From Wang Y. H. and Wang D. Z., *Chin. Phys. Lett.*, 21, 2236, 2004. Chinese Physical Society and IOP Publishing Ltd.)

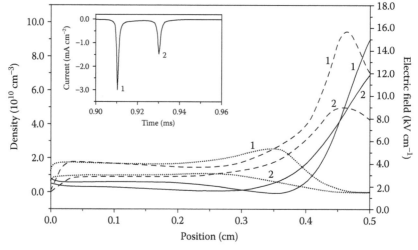

FIGURE 4.8 Glow discharge with two current peaks. The solid curve is the electric field, the dotted curve is the electron density, and the dashed curve is ion density. The cathode is at $x = 0.5$ cm. (From Wang Y. H. and Wang D. Z., *Chin. Phys. Lett.*, 21, 2236, 2004. Chinese Physical Society and IOP Publishing Ltd.)

formed in the growth phase of the external voltage, but the discharge structures show a qualitative change, as shown in Figure 4.9. For each breakdown, the maximum ion density is near the cathode, whereas the electron density is much lower than the ion density and increases from cathode to anode following the Townsend law. Though the electric field is slightly distorted, the localized high-field cathode fall region and positive column are not formed. These suggest that the discharge has changed to Townsend type.

Not only different multiple peak discharges have different operation modes, but even in the same multiple peak discharge, different breakdowns taking place in the same half period of the applied voltage can operate in two discharge modes. This can be seen from Figure 4.10, where the discharge is driven at 10 kHz frequency, other conditions remaining the same as in Figure 4.9. It is clear that the behavior of

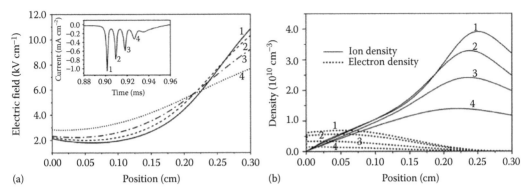

(a) (b)

FIGURE 4.9 Townsend discharge with multiple current peaks (a) Spatial variations of electric field corresponding to four current peaks; (b) spatial distributions of ion and electron densities corresponding to four current peaks. The cathode is at $x = 0.3$ cm. The numbers 1, 2, 3, and 4 correspond to four current peaks. (From Wang Y. H. and Wang D. Z., *Chin. Phys. Lett.*, 21, 2236, 2004. Chinese Physical Society and IOP Publishing Ltd.)

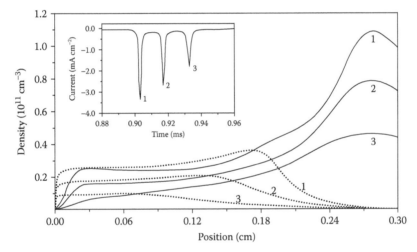

FIGURE 4.10 Multiple peak discharge with two modes. The solid curve is ion density and the dotted curve is electron density. The cathode is at $x = 0.3$ cm. (From Wang Y. H. and Wang D. Z., *Chin. Phys. Lett.*, 21, 2237, 2004. Chinese Physical Society and IOP Publishing Ltd.)

the first discharge is similar to the evolution of a glow DBD. But with the decrease of the current peak value, the densities of the charged particles drop, and the electron density peak is gradually displaced to the anode. In the third discharge, the maximum electron density is close to the anode, and the positive space charge density near the cathode becomes too low to induce a quasi-neutral plasma domain. Thus, the third discharge operates in the Townsend mode. On further increasing the driving frequency, the total series completely transforms to the glow mode.

Wang et al. (2011) studied how atmospheric pressure barrier discharge in helium was influenced by external frequency using a modified two-dimensional fluid model, which is discussed below. The discharge configuration was as in Figure 4.1, with a discharge gap of 0.3 cm, an electrode width of 1.2 cm, and a dielectric layer thickness of 0.1 cm. The driving sinusoidal voltage is 1.5 kV at the frequency ranging from 1 to 100 kHz. They found that helium discharge exhibited three operation modes: Townsend, homogeneous glow, and local glow discharges (from 1 to 100 kHz); the discharge operated in the Townsend mode when the driving frequency ranged from 1 to about 7 kHz; it exhibited homogeneous glow characteristics in the range from 7 to 65 kHz, and turned into a local glow discharge when the external frequency exceeded 65 kHz. For detailed information, readers can refer to the works of Wang et al. (2011).

4.4 Characteristics of Atmospheric Pressure Glow Discharges

4.4.1 Analysis of Townsend and Glow Discharge Modes with a Modified Model

For simplicity, theoretical studies often explicitly or implicitly make an approximation that the average electron energy is constant throughout the discharge space at any time; also, the value of electron energy is chosen empirically regardless of subtle discharge details. The first approximation is correct because the Townsend ionization coefficient reflects basic features of atmospheric discharges, those energy-dependent reactions do not play key role in the discharges. However, to obtain quantitative information, it is necessary to take into account the fact that the external voltage imposed on the discharge gas varies periodically. Consequently, the average electron energy varies with time. Many reaction rate coefficients are energy-dependent. By incorporating the energy conservation equation into the original set of equations (Shi and Kong 2000), these models can be made self-consistent. Even so, self-consistent models cannot give the electron energy distribution function (EEDF). Without EEDF information, it is impossible to quantify the energy dependence of reaction rate coefficients, and thus it is necessary to develop a simple model.

4.4.1.1 Modified Model

At atmospheric pressure, the frequency of an electron colliding with background species is of the order of 10^{12} Hz, much higher than that of externally applied voltage (in which case the frequency is of the order of 10^4 Hz). Therefore, when an external voltage is applied, electrons can relax fully. MC simulations also show that the time it takes the electron to reach equilibrium at 1 bar is of the order of 10^{-11} s or less for O_2/N_2 mixtures (Eliasson and Egli 1987), while the time constant of the inherent changes in the electric field due to space charge buildup is of the order of 10^{-9} s. Hence, a local equilibrium can be reached among electrons under a constant field (Eliasson 1991). Thus, we can solve the stationary Boltzmann's equation under a fixed electric field to obtain the EEDF $f(\varepsilon)$:

$$-\frac{2eE^2}{3m_e}\frac{\mathrm{d}}{\mathrm{d}\varepsilon}\left(\frac{\varepsilon^{3/2}}{\upsilon_e(\varepsilon)}\frac{\mathrm{d}f(\varepsilon)}{\mathrm{d}\varepsilon}\right)=I_{ea}^e+I_{ea}^i, \qquad (4.21)$$

where
  ε is the kinetic energy of electrons
  E is the electric field strength
  I_{ea}^e and I_{ea}^i are the terms for elastic and inelastic collisions between electrons and background species
  υ_e is the elastic collision frequency of an electron with background neutrals

A total of 18 processes involving electrons colliding with helium atoms were included in the solution to Boltzmann's equation. These consist of momentum transfer, 16 excitations [from the ground state up to excited states (5^1S, 5^3S, 5^1P, and 5^3P)], and ionization. Equation 4.21 does not include the term for electron–electron collisions because the ionization rate in barrier discharges is less than 10^{-6}.

The reaction rate coefficients for excited neutrals and other new species generated by the energetic electron impact can be evaluated by the following integration:

$$k_i=\int_0^\infty\left(\sum_i Q_i(\varepsilon)\right)\sqrt{\frac{2e\varepsilon}{m}}f(\varepsilon)\sqrt{\varepsilon}\mathrm{d}\varepsilon, \qquad (4.22)$$

where
  Q_i is the cross section for one of the inelastic processes of an atom due to collisions with electrons
  $\sqrt{2e\varepsilon/m}$ is the electron velocity

Thus, a specific $f(\varepsilon)$ is obtained with the electric field at a fixed value. Consequently, the average electron energy obtained according to $f(\varepsilon)$ is a function of the electric field strength, and reaction rate coefficients for species generation and consumption are functions of the electron energy. We can therefore express these coefficients as functions of the local electric field strength, which can be obtained by solving the fluid model. Thus, the set of Equations 4.8 through 4.13 and 4.21 is self-consistent.

By combining the Boltzmann equation with the fluid model, we can represent the energy evolution of electrons. In solving the Boltzmann equation for EEDF, only the processes directly involved in direct electron impact collisions are considered. Now the question is whether the model describes the electron energy evolution reasonably accurately. Taking helium discharge as an example, we include a total of 18 collision processes in the Boltzmann equation (Alkhazov 1970). All these collision processes are fast interactions (Emmert et al. 1988), and the effects of an external electric field on them are therefore negligible. In reality, electrons can also gain or lose energy by other processes. For He discharge, electrons can gain energy from He* + e $\rightarrow$ He + e; Penning ionization also produces electrons with certain energy (Table 4.1). To take into consideration the contribution from these processes, we include an extra equation for the time evolution of the resulting electron energy to the set of Equations 4.1 through 4.5 and solve the new set of equations. In the fluid model, we consider the process summarized in Table 4.1, where all electronic excited states are not discerned and labeled as He*. For computational details, please refer to the works of Wang et al. (2009).

The evolution of electron energy at location 3 (see Figure 4.11) is shown in Figure 4.12a without taking into consideration the above-mentioned processes. The contribution from these extra processes to the

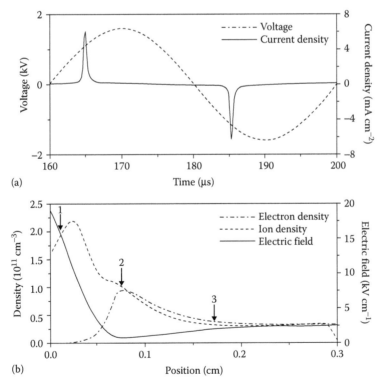

(a)

(b)

FIGURE 4.11 Typical characteristics of glow discharge in helium: (a) waveform of the externally applied voltage and discharge current during one cycle; (b) spatial distributions of electric field and ion and electron densities from the cathode to the anode at the moment when the discharge current intensity reaches its maximum. (Reprinted with permission from Wang Q., et al., *Phys. Plasmas*, 16, 3, 2009. Copyright 2009, American Institute of Physics.)

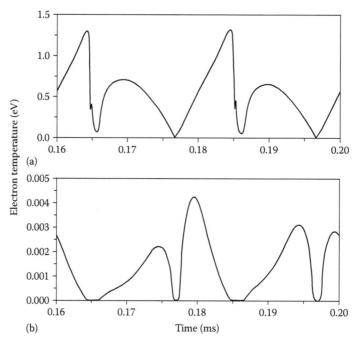

FIGURE 4.12   Electron temperature results from the local electric field (a) and from three processes that are not taken into account in solving Boltzmann's equation (b). (From Sun J. Z., et al., *Chin. Phys. Lett.*, 25, 4056, 2008. Chinese Physical Society and IOP Publishing Ltd.)

electron energy is shown in Figure 4.12b. These extra processes actually affect the electron energy very little. This conclusion is useful when handling complex discharge gases such as O_2, N_2, and mixtures of gases.

With the modified model, further details of the different modes of homogeneous discharge can be obtained. For convenience of presentation, we plot a typical Townsend discharge (Figure 4.13) as in Figure 4.11 (glow discharge).

4.4.1.2 Electron Energy Distribution Function

To measure the difference in electron energy, we have a look at the EEDFs at different locations in Figures 4.11b and 4.13b. Figure 4.14 presents the EEDFs at different locations in Figure 4.11b. The EEDFs look considerably different in shape: the EEDF at location 1 spreads over the widest range of energy, and the EEDF at location 2 contains the largest percentage of electrons with low energy. We can see from Figure 4.11b that the major part of the external voltage falls on the cathode fall (at location 1), and the electrons gain more energy mainly in this region; then the energy of the electrons are consumed in the processes of exciting and ionizing the particles in the negative glow region. Consequently, the EEDF at location 2 shows that the majority of electrons are in the low-energy part. In the positive column (location 3), there are more high-energy electrons than in the negative glow layer owing to a slightly higher electric field formed in the former region. This can explain why three noticeable regions exist in the discharge. With the EEDF, we can calculate the electron-related reaction rate coefficients self-consistently.

The EEDFs at the three locations marked in Figure 4.13b are shown in Figure 4.15. The three curves are similar. The EEDF at location 4 has a higher percentage of high-energy electrons than that at any of the other locations because the electric field at location 4 is more intense.

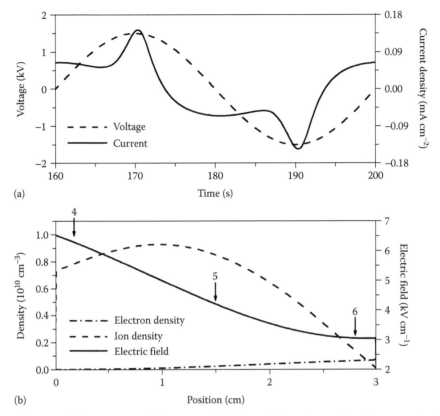

(a)

(b)

FIGURE 4.13 Typical characteristics of the Townsend discharge: (a) waveform of the voltage upon the discharge gas and discharge current density during one cycle; (b) distributions of the particle densities and the electric field when the discharge current is at its maximum. (Reprinted with permission from Wang Q., et al., *Phys. Plasmas*, 16, 3, 2009. Copyright 2009, American Institute of Physics.)

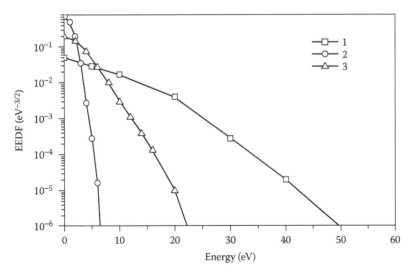

FIGURE 4.14 EEDFs for the three locations in Figure 4.11b. (Reprinted with permission from Wang Q., et al., *Phys. Plasmas*, 16, 4, 2009. Copyright 2009, American Institute of Physics.)

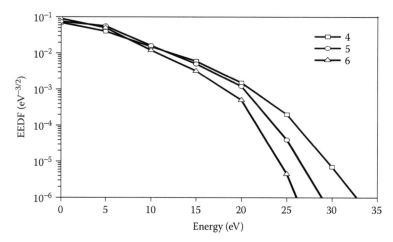

FIGURE 4.15 EEDFs for the three locations in Figure 4.13b. (Reprinted with permission from Wang Q., et al., *Phys. Plasmas*, 16, 4, 2009. Copyright 2009, American Institute of Physics.)

4.4.1.3 Electron Temperature

It is important to have the evolution information of electron temperature both temporally and spatially because the rate coefficients of some reactions are temperature-dependent. Figure 4.16 shows the spatiotemporal distribution of electron temperature in the glow discharge. When the breakdown takes place, the electron temperature reaches 6 eV in the domain of the cathode fall, which is higher than in the other three regions, and after the breakdown, the electron temperature in all the regions decreases almost to zero. To compare with Figure 4.15, we plot the electron temperature distribution versus space and time in Figure 4.17 for the Townsend discharge. The electron temperature near the cathode is 1.5 times higher than near the anode when the discharge current reaches its maximum, but there is no dominant peak appearing near the cathode. As in the case of temporal evolution of the glow discharge,

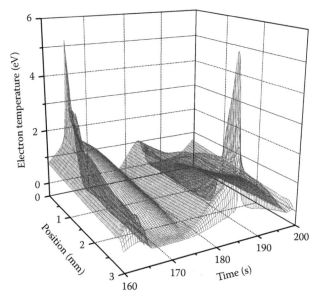

FIGURE 4.16 Spatiotemporal distributions of electron temperature in glow discharge. (Reprinted with permission from Wang Q., et al., *Phys. Plasmas*, 16, 4, 2009. Copyright 2009, American Institute of Physics.)

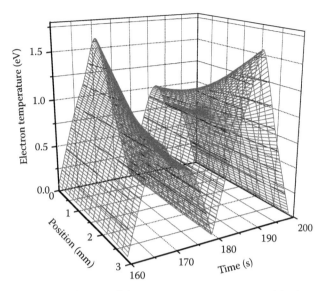

FIGURE 4.17 Spatiotemporal distributions of electron temperature in Townsend discharge. (Reprinted with permission from Wang Q., et al., *Phys. Plasmas*, 16, 4, 2009. Copyright 2009, American Institute of Physics.)

the electron temperature drops nearly to zero after the discharge. Moreover, the maximum electron temperature (1.6 eV) is much lower than that of the glow discharge (6.0 eV) and drops slowly over time. As can be seen, the electron temperature is in general a function of space and time. It would be an error to assume that the electron temperature is constant no matter what the mode. However, one can design an analytical function to approximate the electron temperature in the Townsend discharge.

4.4.1.4 Ignition Voltage for Mode Transition

With the modified model, we can predict the ignition voltage at which the transition of the discharge modes takes place. Figure 4.18 shows the maximum of the discharge current density and the corresponding voltage drop on the discharge gas (we name it the gas voltage in the following text) versus applied voltage. At lower voltage (below 1.5 kV), the discharge current density is very low, and it increases slowly with the applied voltage, but the gas voltage is higher. This is the typical Townsend discharge. When the applied voltage exceeds a critical value, the discharge current density jumps abruptly to a much higher value, but the gas voltage drops suddenly to a much lower value. We define the critical value as the ignition voltage. At the applied voltage of 1.6 kV, the peak value of the discharge current density ascends to 6 mA cm^{-2}. This then increases linearly with the applied voltage, but the gas voltage is nearly constant. This is in agreement with the characteristics of a normal glow DC discharge at low pressure.

To see how the ignition voltage for the transition from the Townsend discharge to the glow discharge at atmospheric pressure varies with the spacing of the gas gap, we evaluated the values of ignition voltage at different spaces by drawing the curves of the current density against the external voltage, as in the curve in Figure 4.18. The ignition voltages sought are the voltage values corresponding to the sudden jumps of the current density. As can be seen from Figure 4.19, the minimum of the ignition potential occurs when the discharge gap spacing is 2.5 mm. Narrowing the gap spacing below 2 mm results in the ignition voltage rising drastically, and increasing the gap spacing leads to the ignition voltage going up slowly but steadily. Shortening the gap space means that electrons have a lesser chance of colliding with helium atoms at the same pressure. To maintain the discharge, a stronger electric field is needed to accelerate electrons in a shorter distance to have enough energy to ionize the neutrals.

Although certain improvements have been made in the model to represent the evolution of electron temperature, it is worth noting that the application of the Boltzmann equation to the electrons in the

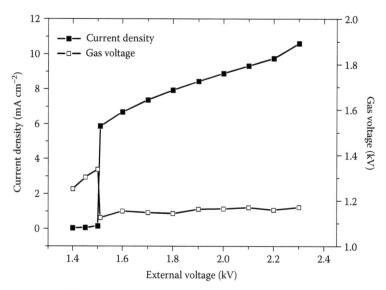

FIGURE 4.18   Maximum of discharge current density and its corresponding voltage imposed upon the discharge gas (i.e., gas voltage in the graph) versus applied voltage. (From Wang Q., et al., *Phys. Plasmas*, 16, 5, 2009. Copyright 2009, American Institute of Physics.)

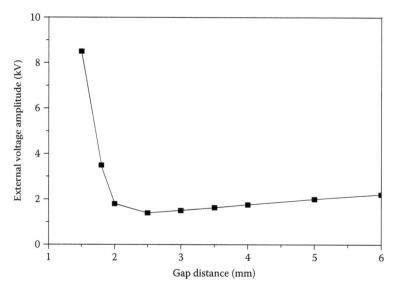

FIGURE 4.19   Ignition voltage for glow discharge versus the discharge gap distance. (From Wang Q., et al., *Phys. Plasmas*, 16, 5, 2009. Copyright 2009, American Institute of Physics.)

sheath is questionable because the local field approximation may not hold there because of a strong electric field gradient and electron density. For more details, readers can refer to the works of Sun et al. (2008).

4.4.2   Radial Evolution of Atmospheric DBD

Given the apparent homogeneous appearance of atmospheric glow DBD, a numerical simulation is usually based on a one-dimensional model that implicitly assumes radial uniformity. Although the one-dimensional model can provide significant insights on atmospheric glow DBDs, some experimental and numerical results have indicated that they possess certain radial structures (Mangolini 2002,

Anderson et al. 2004, Zhang et al. 2005a, 2005b, Zhang and Kortahagen 2006). The motion of the charged particles is influenced by the radial space charge field, resulting in the radial propagation of the discharge. It is therefore important to investigate the discharge behaviors in the radial directions.

Here we consider an atmospheric DBD in pure helium gap between two dielectrically coated electrodes. The particles taken into account include electrons, ions (He$^+$, He$_2^+$), and neutral particles (He, He*). The simulation is based on a two-dimensional fluid model. Within the scope of the drift–diffusion approximation, the charged particles can be described by the continuity equations in cylindrical geometry:

$$\frac{\partial n_k}{\partial t} + \frac{1}{r}\frac{\partial \left(r j_{kr}\right)}{\partial r} + \frac{\partial j_{kz}}{\partial z} = S_k, \tag{4.23}$$

$$j_{kr} = \pm \mu_k E_r n_k - \left(D_k \frac{\partial n_k}{\partial r}\right), \tag{4.24}$$

$$j_{kz} = \pm \mu_k E_z n_k - \left(D_k \frac{\partial n_k}{\partial z}\right), \tag{4.25}$$

where
  The indices r and z correspond to the radial and axial components, respectively
  n_k, j_k, and S_k represent the density, the flux, and the source term, respectively, of the corresponding particles
  μ_k and D_k are the mobility and diffusion coefficients, respectively

The minus signs in Equations 4.24 and 4.25 represent electrons, and the plus signs positive particles. As for the neutral particle, only the diffusion term is considered.

As in the one-dimensional model, in two-dimensional simulation the electric field is also derived from the current conservation equation. In the cylindrical coordinate system, the radial and axial components of the electric field r, respectively, are formulated as

$$\frac{\partial E_r}{\partial t} + \frac{e}{\varepsilon_0} j_r = \frac{e}{\varepsilon_0} j_{0r}, \tag{4.26}$$

$$\varepsilon(z)\frac{\partial E_z}{\partial t} + \frac{e}{\varepsilon_0} j_z = \frac{e}{\varepsilon_0} j_{0z}, \tag{4.27}$$

where
  j_{0r} and j_{0z} are the radial and axial components of the total current density, respectively
  $\varepsilon(z)$ is the permittivity of either the dielectric or the gas gap, depending on the position z

The axial total current, including the displacement current and the conduction current, can be represented as

$$j_{0z} = -\frac{\varepsilon_0}{e\left(d + \dfrac{2d_b}{\varepsilon_b}\right)}\frac{\partial V_0(t)}{\partial t} + \frac{1}{d + \dfrac{2d_b}{\varepsilon_b}}\int_0^d j_z dz, \tag{4.28}$$

where
  d_b is the thickness of the dielectric barrier
  ε_b is its permittivity
  d is the distance between the two dielectric barriers
  $V_0(t)$ is the applied voltage

The radial component of the total current density is given by Kulikovsky (1994a, 1994b).

$$j_{0r} = 0. \tag{4.29}$$

The total current densities of j_{0r} and j_{0z} satisfies

$$\nabla \cdot \boldsymbol{j}_0 = 0. \tag{4.30}$$

Through the entire space between the two electrodes (Kulikovsky 1994a, 1994b), their solutions are given analytically in Equations 4.28 and 4.29, and do not require boundary conditions. The elementary processes included in the source are given in Table 4.1. The emission coefficient of secondary electrons is 0.01.

The above set of equations is solved numerically using the Scharfetter–Gummel scheme on a set of uniform rectangular grids. The calculation begins with a spatially uniform density for ions and electrons, both set at 10^7 cm^{-3}.

4.4.2.1 Radial Evolution of Atmospheric DBD with One Current Peak in Half Cycle of Applied Voltage

For all numerical examples presented here, the gap between the two electrodes is 3 mm and the thickness of each dielectric layer is 1 mm with their permittivity at 7.6. The electrode radius is set to 5 mm and the driving frequency of the applied sinusoidal voltage is 20 kHz. The working gas is helium at 760 Torr. The amplitude of the applied voltage is 1.6 kV. Figure 4.20 shows the profile of the space-averaged current density and gap voltage over the whole electrode surface, together with the applied voltage. This discharge is typical atmospheric DBD with a single current peak during each half cycle. To study the radial behavior of this discharge, the radial distribution of the current density at different radial positions is presented in Figure 4.21. Two features of the radial evolution of atmospheric pressure glow discharge (APGD) are shown in Figure 4.21. First, the discharge current densities at different r do not reach their

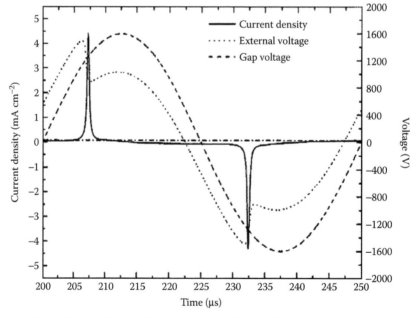

FIGURE 4.20 The profiles of the overall average current density, the gas voltage, and the applied voltage. (Reprinted with permission from Zhang Y. T., et al., *J. Appl. Phys.*, 98, 3, 2005a. Copyright 2005, American Institute of Physics.)

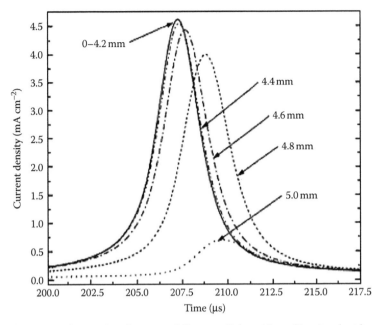

FIGURE 4.21 The current density distribution at different radial positions. (Reprinted with permission from Zhang Y. T., et al., *J. Appl. Phys.*, 98, 3, 2005a. Copyright 2005, American Institute of Physics.)

peaks simultaneously. In the central region of $r = 0$–4.2 mm, the discharge current density reaches its peak earlier than in the boundary region of $r = 4.6$–5 mm, which indicates that gas breakdown takes place first in the central region. Second, the amplitude of the current density decays radially, meaning the discharge is stronger in the central region than in the boundary. These radial characteristics obtained computationally are consistent with the experimental results reported by Mangolini et al. (2002).

In order to understand the radial evolution of APGD better, we plot the spatial distributions of ion and electron densities at two key moments in Figure 4.22. When the gas breakdown occurs in the central region at $t = 207.25$ μs, corresponding electron and ion densities are much larger in the central region, which are shown in Figure 4.22a and c. As can be seen, gas breakdown is confined to the central region and is not uniform radially. After the discharge is formed from this centrally confined gas breakdown, it begins to expand toward the electrode boundary. Consequently, a new gas breakdown occurs outside the central region, and this propagates progressively, radially from 4 to 5 mm. This radial propagation is associated with a progressive reduction of the electron and ion densities in the central region and a simultaneous increase in particle densities in the boundary region. In Figure 4.22b and d, the electron and ion densities are largest near $r = 4.7$ mm in the boundary region at $t = 207.4$ μs.

The mechanism responsible for the radial evolution of APGD in Figures 4.21 and 4.22 can be understood from the radial electric field distribution in Figure 4.23. At $t = 207.25$ μs the electric field in Figure 4.23a is clearly larger in the central region and this is directly related to the large current density for $r = 0$–4.2 mm in Figure 4.21. The small electric field in the boundary region of Figure 4.24a is also consistent with the very low current density for $r = 4.6$–5 mm at the same time. Similarly, at $t = 207.4$ μs, the electric field in Figure 4.24b shows a clear maximum of almost 18 kV cm^{-1} at $r = 4.6$ mm and this large electric field triggers gas breakdown in the boundary region as identified by the current density peak at $t = 207.4$ μs for $r = 4.6$ mm in Figure 4.21. It is evident that the electric field in the boundary region has a radial profile that decays rapidly away from its peak. As a result gas breakdown is likely to be triggered within a radially confined region, and so the space average of the induced discharge current for the boundary region is much smaller than the current density at the location of the maximum electric field.

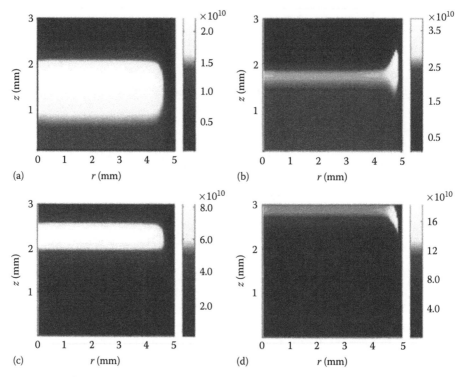

FIGURE 4.22 Ion density profiles at (a) 207.25 μs and (b) 207.4 μs. Electron density profiles at (c) 207.25 μs and (d) 207.4 μs. The momentary cathode is located at the top in each picture. (Reprinted with permission from Zhang Y. T., et al., *J. Appl. Phys.*, 98, 4, 2005a. Copyright 2005, American Institute of Physics.)

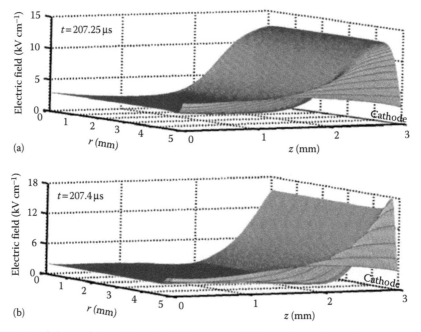

FIGURE 4.23 Total electric field profiles at (a) 207.25 μs and (b) 207.4 μs. (From Zhang Y. T., et al., *J. Appl. Phys.*, 98, 4, 2005. Copyright 2005, American Institute of Physics.)

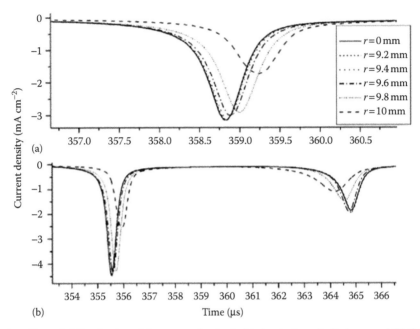

FIGURE 4.24 The radial distribution of the current density in (a) the single-peak discharge and (b) the two-peak discharge. (From Zhang Y. T., et al., *Chin. Phys. Lett.*, 22, 172, 2005b. Chinese Physical Society and IOP Publishing Ltd.)

4.4.2.2 Radial Evolution of Atmospheric DBD with Multicurrent Peaks per Half Cycle of Applied Voltage

As the discharge with one current peak per half voltage cycle, the multiple peak discharges also possess some radial structures. Take the discharge with two current peaks per half voltage cycle as an example. Based on the above two-dimensional fluid model, when the electrode radius is 1 cm, the driving frequency is 10 kHz, the amplitude of the applied voltage is 2.3 kV, and other parameters are the same as those in Section 4.4.1, a discharge with two current peaks is obtained. Figure 4.24b illustrates the radial distribution of the current density in two-peak discharge; we also give the radial distribution of the current density in single-peak discharge at the applied voltage 2.0 kV in Figure 4.24a as a comparison. Figure 4.25 illustrates spatial distributions of ion and electron densities at two key moments of the second discharge in two-peak discharge. It is evident from Figure 4.24b that the radial evolution behaviors are different between the first peak and the second. The radial evolution of the first peak is similar to that of the single peak in Figure 4.24a. The discharge takes place first in the central region between the electrodes, and then expands to the boundary, accompanied by a decrease in the current amplitude. However, in the second discharge, the current density in the boundary of the electrode reaches a peak earlier than that in the central region of $r < 9.2$ mm, and the maximal densities of charged particles occur first in the boundary region (Figure 4.25a and c). This also indicates that gas breakdown takes place first at the edge of the electrode, and then spreads rapidly toward the center area (Figure 4.25b and d).

The formation mechanism of the first current peak is also similar to that of the single peak. The calculations indicate that radial electric fields and radial sheath are important factors in the evolution of the second peak. Figure 4.26 gives the spatiotemporal distributions of total electric fields at two key moments corresponding to Figure 4.25a and b. Because of the ambipolar diffusion and velocity difference between electrons and ions, radial electric fields are formed, and electrons play a dominant role in this process. The detailed descriptions of radial fields and radial sheath were reported in the works of Boeuf (1988),

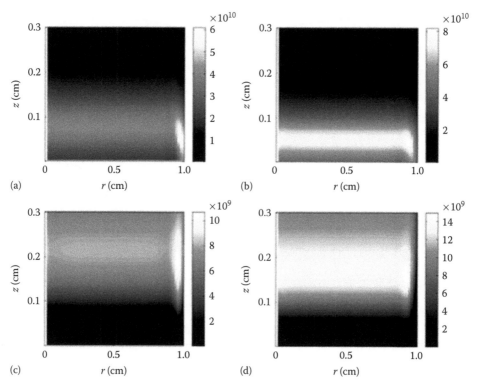

FIGURE 4.25   Ion density profiles at (a) 364.5 μs and (b) 364.8 μs. Electron density profiles at (c) 364.5 μs and (d) 364.8 μs. The momentary cathode is located at the bottom in every graph. (From Zhang Y. T., et al., *Chin. Phys. Lett.*, 22, 173, 2005b. Chinese Physical Society and IOP Publishing Ltd.)

Tsai and Wu (1990), and Young and Wu (1993). After the first breakdown is complete, with the increase in the applied voltage, electrons and ions are forced toward the anode and the cathode by the axial fields, respectively, resulting in the deposition of a great number of particles on the dielectric surface. At the same time, due to the effects of radial fields, the higher ion densities are formed near the cathode at the edge of the discharge gap, and higher electron densities also occur near the anode at the periphery. This distribution of ion densities strengthens the electric fields near the cathode sheath, as shown in Figure 4.26a. Thus, higher densities and stronger fields make the breakdown easier at the edge of the discharge gap. While the discharge is developing, the fields are enhanced in the wide central region (Figure 4.26b), and the breakdown ignites in the center, as shown in Figures 4.25b and 4.26d.

Further simulations show that when several current peaks are formed in each half cycle of the applied voltage, except the first current peak, other discharges all start at the electrode periphery.

It is worth mentioning that, under simulation conditions, both the single-peak and the multiple peak atmospheric DBD are radially uniform in the wide central region.

4.4.3 Characteristics of APDs Driven by Combined RF/Pulse Sources

We will now present some characteristics of APDs driven by combined RF/pulse sources. The work was originally motivated by the idea that a low RF power sustains the discharge and keeps the working gas at a low temperature, and one can apply a low duty ratio of nanosecond pulses to the discharge to enhance the plasma density. Thus, it would be interesting to explore the possibility of applying nanosecond pulses to the RF discharge to achieve high plasma density but consume relatively low power. For further details, readers can refer to the works of Wang et al. (2010).

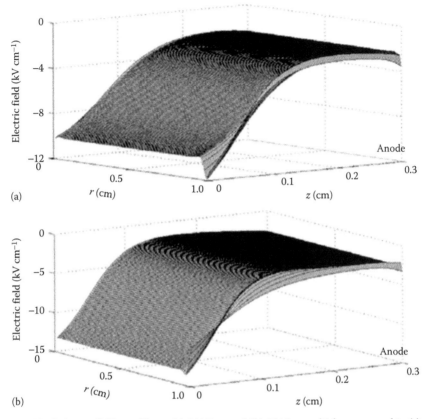

(a)

(b)

FIGURE 4.26 Total electric fields profiles at (a) 364.5 μs and (b) 364.8 μs, which correspond to (a) and (b) in Figure 4.25, respectively. (From Zhang Y. T., et al., *Chin. Phys. Lett.*, 22, 174, 2005. Chinese Physical Society and IOP Publishing Ltd.)

An atmospheric discharge is generated between two parallel metal electrodes, the powered one at $y = 0$ mm and the grounded one at $y = 2$ mm. Both RF voltage and pulse power sources are applied to the electrode. The pulse voltage $V_i(t)$ is chosen to have a smoothed trapezoidal wave form to eliminate the discontinuity of $dV(t)/dt$ (Stewart and Lieberman 1991). The mathematical form is expressed as follows:

$$\frac{V_i(t)}{V_{i0}} = \begin{cases} \dfrac{t/t_r}{\left[1+(t/t_r)^8\right]^{1/8}}, & 0 < t \leq t_r + \left(\dfrac{t_p}{2}\right) \\ \dfrac{(t_t - t)/t_f}{\left\{1+\left[(t_t - t)/t_f\right]^8\right\}^{1/8}}, & t_r + \left(\dfrac{t_p}{2}\right) < t < t_t. \end{cases}$$

Here the pulse is characterized by three different times: rise time t_r, plateau time t_p, and fall time t_f, with the pulse-on period t_t. For simplicity, we set $t_r = t_p = t_f$. The modified fluid model introduced in Section 4.2.2 is used here.

The detailed simulation parameters are as follows: the pressure of helium is 760 Torr, and its temperature is set to 300 K as usual. The electrodes absorb all the incoming charged particles and their secondary electron emission coefficient caused by incident ions is assumed to be 0.05. The amplitude and the frequency of the RF source are 400 V and 15 MHz, respectively. The peak value of the pulse voltage V_{i0} is 800 V unless stated otherwise, and the frequency of the pulse is 500 kHz, that is, 1 pulse every 30 RF cycles. The pulse-on period t_t is chosen to be 25 ns, and the pulse and RF sources are applied to the electrode with the initial phase angle. For presentation convenience, we define the moment when the data starts to be extracted as $t = 0$, and, accordingly, the first RF cycle starts at $t = 0$; capital T represents one RF period.

4.4.3.1 Discharges Driven by Single RF Power Source

Before analyzing the discharges driven by combined pulse/RF power sources, we have a look at the discharges driven by a single RF power source, which will then be the comparative reference for a subsequent session. The simulation is run until the RF discharge reaches a steady phase and the discharge information is later extracted. The typical waveforms of external voltage and RF discharge current density as a function of time are shown in Figure 4.27. From the figure we can see that both external voltage and the RF current density have a smooth, sinusoidal wave form, implying a linear response of the discharge to external power and a continuous discharge mode. In addition, the capacitive nature of the discharge is clearly shown as the current waveform leads the voltage waveform by about 63°. The maximal amplitude of the current density is about 45 mA cm^{-2}.

The typical spatiotemporal profiles of the electron and the ion densities in the RF discharge in one cycle are shown in Figure 4.28. In Figure 4.28a, we can see that most of the electrons are confined in the central region of the discharge gap, and a few electrons in the vicinity of the electrodes are governed by the electric field in the sheath. In contrast, as shown in Figure 4.28b, almost all of the ions are restricted in the central region of the gap, and the ions next to the electrodes are not visibly affected by the external voltage. This clearly indicates that only electrons can respond to the rapidly varying RF voltage. The density of neutral plasma in the central part of discharge gap is 4.4×10^{11} cm^{-3}.

4.4.3.2 Discharges Driven by Combined Pulse/RF Power Sources

When an extra pulse with a duty ratio of 1–30 is applied to the RF discharge as stated above, what changes will be introduced to the discharge? Figure 4.29a shows the time evolutions of the external voltage and the discharge current density in the 30 RF cycles. To see the picture more clearly, we zoom in on the waveforms of the external voltage and discharge current density in the time between $t/T = 0$ and $t/T = 2$ and plot them in Figure 4.29b. The discharge current density can reach a value as high as 200 mA cm^{-2},

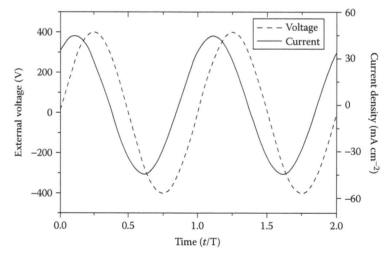

FIGURE 4.27 Waveforms of external voltage and the RF current density after the discharge has already stabilized fully. The symbol T in this figure (as well as in the following figures) is the time of one RF period. The time is set to zero when the data starts to be extracted (this definition is applicable for Figures 4.27 through 4.32). (From Wang Q., et al., *Phys. Plasmas*, 17, 3, 2010. Copyright 2010, American Institute of Physics.)

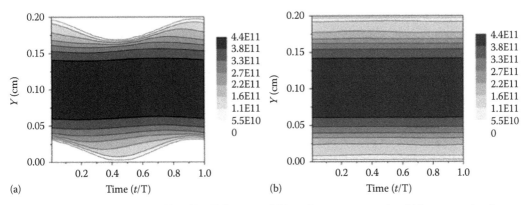

FIGURE 4.28 Spatiotemporal profiles of the (a) electron and (b) ion densities in one cycle, which correspond to the case shown in Figure 4.27. (From Wang Q., et al., *Phys. Plasmas*, 17, 3, 2010. Copyright 2010, American Institute of Physics.)

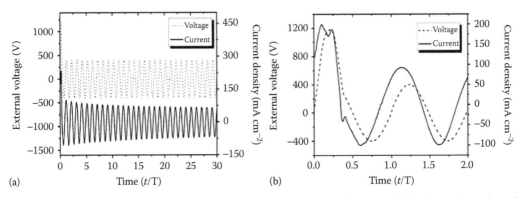

FIGURE 4.29 Waveforms of the external voltage and discharge current density in (a) the first 30 RF cycles and (b) the waveforms in the first two cycles, in which the time axis is magnified. (From Wang Q., et al., *Phys. Plasmas*, 17, 4, 2010. Copyright 2010, American Institute of Physics.)

against 45 mA cm^{-2} in the discharge driven by a single RF power source; in the second RF cycle, the peak value of the current density is about 100 mA cm^{-2}, still twice as high as that of current density in the RF discharge without applied pulses (see Figure 4.27). The two peaks of current density occurring during $t/T = 0$ and $t/T = 5$ are induced, respectively, by the pulse rise time and the pulse fall time. The drop in the current density between the two peaks suggests that the plateau part of the pulse contributes relatively less to neutral ionization. As the pulse is off, the current density gradually relaxes toward the current density in the pure RF discharge until the next pulse is on.

To see the evolution of the ion number density clearly, we plot the spatiotemporal evolutions of ions number density in Figure 4.30, in which Figure 4.30a shows the ion density curve in the first two RF cycles and Figure 4.30b displays the ion density evolution in the 30th RF cycle. In the first two RF cycles, the ion density in the cathode region is as high as 3.5×10^{12} cm^{-3}; while the ion density in the central part of the discharge space is 1.0×10^{12} cm^{-3}. In the 30th RF cycle just before the next pulse is switched on, the peak ion density is around 9.0×10^{11} cm^{-3}, still twice that in the pure RF discharge. Clearly, the short pulse can enhance the plasma density quite remarkably, although the amplitude of the pulse is 800 V, only twice the RF voltage, and the duty ratio is only 1–30.

In the following we will find that the plasma density and current density do not decay at a similar rate. The peak values of current density and the spatial average values of the positive ion densities versus time are shown in Figure 4.31. Both of these values can be fitted perfectly by the second-order

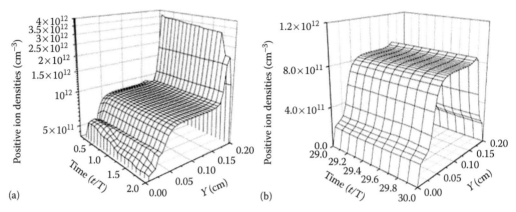

FIGURE 4.30 Spatial distributions of ion density in the (a) first two RF cycles and (b) the 30th RF cycle. (From Wang Q., et al., *Phys. Plasmas*, 17, 4, 2010. Copyright 2010, American Institute of Physics.)

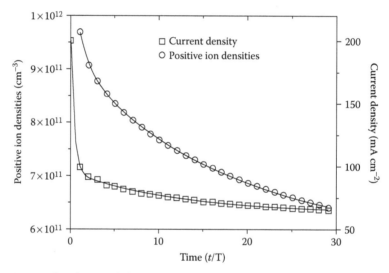

FIGURE 4.31 Temporal evolutions of the current density peaks (cycles) and spatial average value of the positive ion density (squares); solid lines represent the fitting curves. (From Wang Q., et al., *Phys. Plasmas*, 17, 4, 2010. Copyright 2010, American Institute of Physics.)

exponential functions $y = y_0 + A_1 e^{-x/t_1} + A_2 e^{-x/t_2}$, where $y_0 = 63.04$ mA cm^{-2}, $A_1 = 136.63$ mA cm^{-2}, $A_2 = 32.45$ mA cm^{-2} $t_1 = 0.38$, and $t_2 = 11.90$ for the current density; and $y_0 = 5.78 \times 10^{11}$ cm^{-3}, $A_1 = 1.86 \times 10^{11}$ cm^{-3}, $A_2 = 3.38 \times 10^{11}$ cm^{-3}, $t_1 = 1.19$, and $t_2 = 17.61$ for the positive ion density. Figure 4.31 indicates that the current density drops very rapidly just after the pulse is switched off, having a characteristic time length of 0.38 T, and then decreases gradually, having a longer characteristic time. In contrast, the ion density falls more mildly, which can be approximately represented by one exponential decay function $3.38 \times 10^{11} \times e^{-x/17.61}$ cm^{-3} with a half-life time of 12.2 T.

This result is very important, which suggests that applying short pulses can enhance the plasma density, and does not consume much more power. The physical reason behind this phenomenon can be explained as follows: the RF voltage cannot sustain the high current density generated by the short pulse, and the diffusion and other mechanisms responsible for reducing the plasma density have long characteristic times, hence plasma density decreases slowly. More detailed explanation can be obtained from an analysis of the evolutions of the electric field and EEDF (Wang et al. 2010). It was found that the

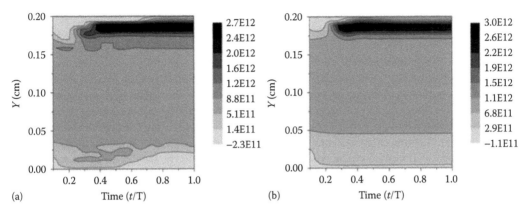

FIGURE 4.32 The contour maps of (a) electron and (b) ion densities (in unit per cubic centimeter). (From Wang Q., et al., *Phys. Plasmas*, 17, 5, 2010. Copyright 2010, American Institute of Physics.)

short voltage pulse generates a very strong electric field next to the powered electrode, because of which a large number of electrons transition to high-energy electrons, which function as the seed electrons and produce active particles and ion–electron pairs, and newly generated electrons and the seed electrons subsequently generate more ion–electron pairs.

Another interesting issue is whether the pulsed RF discharge alters the discharge mode. We can reasonably assume that the RF discharge in helium gas operates in a glow discharge mode because it is well-known that a He APGD is readily realized with RF power. Now we check whether the applied pulse will change the discharge mode. Figure 4.32 displays the contour maps of electron and ion densities on the time–space plane. In the first two-fifths of the RF period, the discharge in the regions next to the electrodes is noticeable. Next to the powered electrode, a wider sheath exists, which is similar to that in pure RF discharge (see Figure 4.32); next to the other electrode, a much narrower sheath forms after the pulse-off period begins. Higher charge density exists next to the narrow sheath. We also note that the electron density in this sheath varies. In contrast with the ion density, the electron density varies much more drastically. Clearly, the plasma density in the central part of the discharge gap still remains homogeneous, which is consistent with the slow decrease in ion density shown in Figure 4.30. This implies that the discharge mode does not change.

Sang et al. (2010) employed the PIC method to study the effects of additional nanosecond power on a similar discharge, except that the working gas was argon and the grounded metal electrode was covered with a thin dielectric layer. They found that the plasma density could be enhanced by three orders of magnitude. The dielectric layer played an important role in plasma density enhancement.

4.5 Nonlinear Behaviors

As a spatially extended dissipated system, gas discharge plasmas could in principle possess complex spatiotemporal nonlinear behaviors. Many studies on gas discharge systems have shown the transitions to chaos through period multiplication, quasi-periodicity, and intermittency (Cheung et al. 1988, Ding et al. 1993, Jonas et al. 2000, Strumpel et al. 2000, Bruhn 2004, Šijačiæ et al. 2004, Wang et al. 2007, Wilson IV and Podder 2007). Among these systems, glow discharges, having the advantage of relatively simple configuration and convenience with regard to experimental handling, have become a medium for exploring the universal characteristics of chaos (Bruhn 2004, Wilson IV and Podder 2007). Also, glow DBDs at atmospheric pressure have been shown to possess very rich nonlinear behaviors (Wang et al. 2007, Qi et al. 2008, Shi et al. 2008, Zhang et al. 2010). The studies on these complex nonlinear behaviors are important, which not only offer valuable guidance to effective control of atmospheric pressure plasma stability, but also provide a platform to develop and utilize nonlinear science.

To demonstrate the nonlinear behaviors of homogeneous atmospheric DBDs, a one-dimensional fluid simulation is performed in atmospheric argon between two parallel-plate electrodes and over a very broad frequency range from kilohertz to RF. Discharge characteristics are computationally studied in the direction perpendicular to the electrode plane by solving the continuity equation, the momentum transfer equations, and the current balance equation. Only two plasma species, namely, electrons and argon ion, are considered in the model. For the cases presented here, the amplitude of the voltage is fixed at 1.8 kV and the discharge gap is 1 mm. The thickness of the dielectric layers covering the electrodes is 1 mm with a relative permittivity of 7.5. The secondary electron emission coefficient is set to 0.01. By altering the excitation frequency the complex dynamic behaviors of DBD in atmospheric argon can be observed. When the frequency of the applied voltage is below 25.0 kHz, the discharge is a well-known glow DBD with one current pulse every half voltage cycle. For simplicity we use 1P to signify this state. As the frequency increases, the discharge undergoes period multiplication and a clear period-doubling sequence to chaos is observed. Figure 4.33 illustrates a series of period-doubling bifurcations from the

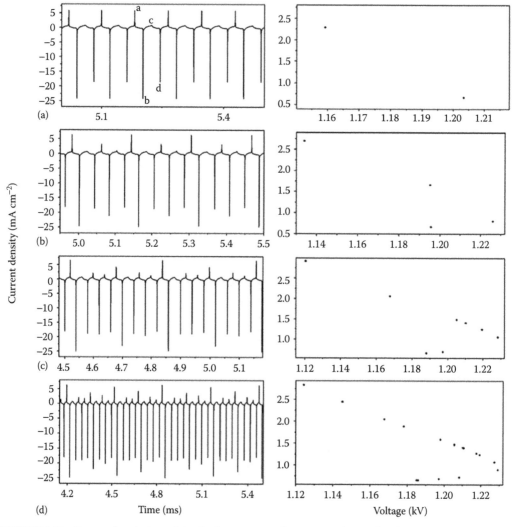

FIGURE 4.33 Temporal evolutions of the discharge current density in 2P (at 26.0 kHz) (a), 4P (at 26.2 kHz) (b), 8P (at 26.37 kHz) (c), and 16P (at 26.378 kHz) (d), together with their corresponding Poincare sections. (From Shi H., et al., *Phys. Plasmas*, 15, 2, 2008. Copyright 2008, American Institute of Physics.)

period 2 to the period 16 state, which can be expressed as $2nP$ ($n = 1, 2,...$). The left wave forms in Figure 4.33a–d are the temporal oscillations of discharge current densities, and the right ones are corresponding Poincare sections that are formed through the intersections of phase–space trajectories with a plane at a fixed phase. When the frequency is equal to 26.0 kHz, the discharge first bifurcates into a period 2 state from the common 1P mode, denoted by 2P, as shown in Figure 4.33a. In the 2P state, the temporal profile of the current pulses repeats every two cycles of the applied voltage, and the amplitude of the current pulse is obviously different in each oscillation period, which is composed of four current pulses. As the driving frequency is increased to 26.2 kHz, the discharge bifurcates again and the period 4 state with a period of eight different pulses appears (see Figure 33b), denoted by 4P. If the frequency keeps on increasing, the bifurcation speed of the discharge becomes faster. At 26.36 kHz, period 8 discharge occurs, and then it further bifurcates into the 16P state under a frequency of 26.378 kHz. The Poincare sections confirm further the periodicity of the above discharges. In Figure 4.33a, the Poincare map contains two fixed points, suggesting that the discharge operates in the 2P state. By increasing the driving frequency, the number of points doubles gradually and at 26.378 kHz, there are 16 scattered points appearing in the Poincare section, which indicates the discharge is in the 16P state.

When the frequency is further increased over 26.385 kHz, the discharge undergoes a transition to the chaotic state, in which the discharge becomes out of order and the discharge current amplitude fluctuates stochastically, the resultant Poincare map consisting of randomly scattered points with some points almost connected to a continuous line segment, as shown in Figure 4.34.

In the chaotic region, with an increase in the driving frequency, more nonlinear characteristics are found in argon atmospheric DBD. At around 26.48 kHz a narrow frequency window of period 5 first emerges, which is then replaced by a period 10 state. The period 10 state is regarded as a secondary bifurcation from the 5P state via period doubling. The 10P state is maintained over a small frequency range, and then is quickly replaced by the chaotic state. If we continue to increase the driving frequency, an obviously inverse bifurcation sequence 8P–4P–2P–1P can be observed in the frequency range between 26.795 and 27.1 kHz, which means that the DBD in atmospheric argon can be converted from the chaotic state to common single-period motion through inverse period-doubling bifurcation. After the driving frequency is above 27.1 kHz, the discharge is kept at the single-period state until the period 2 bifurcation occurs under a frequency of about 70 kHz. This period 2 state lasts with frequency of about 5 kHz, and then is directly replaced by period 3 discharge, or 3P (see Figure 4.35). According to Li–York theory, the 3P state indicates the existence of chaotic motion (Huang and Huang 2005). Therefore, the appearance of the 3P state further confirms that the discharge system under study can be driven to the chaotic state. If we continue to increase the frequency, it is found that the 3P state occupies a rather wider range of frequency, from 75 to 133 kHz.

Furthermore, period 2 discharges, in which the temporal profiles of current pulses repeat every two cycles of the applied voltage, are shown to occur at different excitation frequencies and exhibit

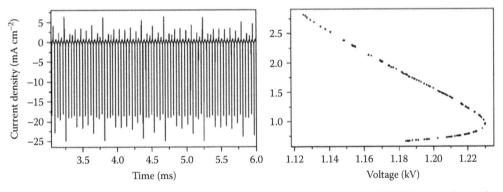

FIGURE 4.34 Temporal evolutions of the discharge current density for chaos at 26.385 kHz, together with its corresponding Poincare section. (From Shi H., et al., *Phys. Plasmas*, 15, 3, 2008. Copyright 2008, American Institute of Physics.)

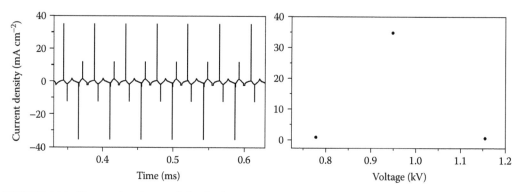

FIGURE 4.35 Temporal evolution of the discharge current in 3P state and corresponding Poincare sections. (From Shi H., et al., *Phys. Plasmas*, 15, 5, 2008. Copyright 2008, American Institute of Physics.)

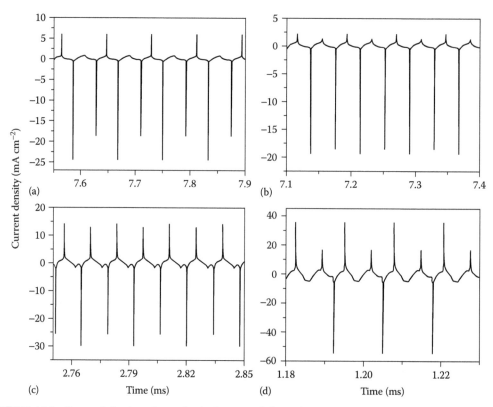

FIGURE 4.36 Current behaviors of period 2 discharges at different frequencies of 26.0 (a), 26.88 (b), 72.0 (c), and 160.1 kHz (d). (From Wang Y. H., et al., *Phys. Plasmas*, 16, 2, 2009. Copyright 2009, American Institute of Physics.)

different current and voltage behaviors. Figure 4.36a–d shows the temporal behaviors of the discharge current in period 2 discharges at 26.0, 26.88, 72.0, and 160.1 kHz, respectively. It is clear that the peak amplitudes of current pulses are very different in each period 2 discharge. The stability of the period 2 discharge is associated with the symmetry of the discharge current between positive and negative half cycles. When the discharge current becomes highly symmetrical, the period 2 discharge can reach a steady state and can sustain over a broad frequency range. The stable period 2 discharge is observed from 200 kHz to about 11.5 MHz, and, finally, it is replaced by a stable single period. Figure 4.37a–d shows the temporal evolutions of the discharge current density and the gas voltage in stable period 2 discharge under four different frequencies of 250 kHz, 500 kHz, 1 MHz, and 2 MHz.

When the excitation frequency is increased to 13.56 MHz, namely, the RF, the nonlinear behaviors of atmospheric DBD are also observed by changing the amplitude of the applied voltage. Figure 4.38 gives the bifurcation diagram of the discharge, in which the existing regions of different discharge behaviors can be easily identified. In the low applied voltage region, only one current peak value is produced at

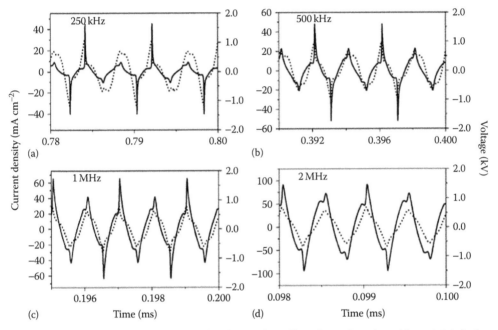

FIGURE 4.37 Discharge current (solid curve) and gas voltage (dotted curve) in the stable period 2 discharge: (a) at 250 kHz, (b) at 500 kHz, (c) at 1 MHz, and (d) at 2 MHz. (From Wang Y. H., et al., *Phys. Plasmas*, 16, 3, 2009. Copyright 2009, American Institute of Physics.)

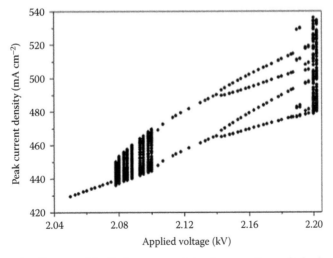

FIGURE 4.38 Bifurcation diagram of the discharge current density versus the applied voltage amplitude. (From Zhang J., et al., *Phys. Plasmas*, 17, 3, 2010. Copyright 2010, American Institute of Physics.)

TABLE 4.2 The Largest Lyapunov Exponent Calculated at Different Voltages

Applied Voltage (kV)	Discharge Behavior	Lyapunov Exponent
1.700	1P	−0.0078
2.120	2P	−0.0077
2.150	4P	−0.0091
2.195	8P	−0.0061
2.200	Chaos	1.782
2.240	Chaos	2.062

one fixed voltage amplitude, meaning the discharge is in the common 1P state. As the applied voltage amplitude is increased above 2.08 kV, the discharge becomes irregular and bifurcates into an unstable period 2 state in which the discharge fluctuates intermittently from one kind of period 2 state to another with the discharge time being prolonged. This unstable period 2 discharge can be seen as the transition state between stable 1P and 2P discharge. At 2.1 kV, the discharge settles into a strict period 2 state with two current peak values at one voltage amplitude. The region where two pitchfork doubling sequences are observed corresponds to the period 4 state. Between 2.19 and 2.2 kV, the discharge operates in 8P manner. Notice that in the 8P region, the current peak values do not fluctuate following the same tendency, but have a sudden change at 2.193 kV, which indicates that there exist two different 8P discharges in this voltage region. After the 8P state the discharge enters the chaotic domain with a large number of current peak values appearing at one voltage amplitude.

To detect the presence of chaos in a dynamical system quantitatively, Lyapunov exponents, which quantify the exponential divergence of initially closed state space trajectories and estimate the amount of chaos in a system, are usually calculated. For time series produced by dynamical systems, the presence of a positive characteristic exponent indicates chaos. At present, there are many methods for calculating the Lyapunov exponent. Among these, the method for calculating the largest Lyapunov exponent from small data sets presented by Rosenstein et al. (1993) is relatively easy to implement and is accurate because it takes advantage of all the available data. Table 4.2 gives the largest Lyapunov exponents calculated at different voltages. It is clear that in periodic states, the largest Lyapunov exponent is approximately zero. While the applied voltage reaches 2.2 kV, the largest Lyapunov exponent becomes positive, which confirms the presence of a chaotic state.

In addition, the Feigenbaum universal constant δ is also calculated to confirm the period-doubling behavior. The obtained δ value is about 4.7, which agrees well with the result given by Feigenbaum.

Apart from the rich temporal nonlinear behaviors, the atmospheric DBDs also possess complex spatial nonlinear behaviors, such as a filamentary, self-organized pattern. Though it is difficult to simulate these behaviors, a great variety of patterns have been revealed by a series of experimental studies (Guikema et al. 2000, Kogelschatz 2002, Schoenbach et al. 2004, Dong et al. 2005, Ra'hel and Sherman 2005, Stollenwerk et al. 2006).

It should be pointed that so far the studies on nonlinear behaviors of atmospheric DBDs are limited to the findings of various nonlinear phenomena. The physical mechanism causing these nonlinear behaviors is far from being understood. Further studies are thus needed.

References

Alkhazov G. D. 1970. *Sov. Phys. Tech. Phys.* 15: 66.
Anderson C., Hur M., Zhang P., Mangolini L., Kortahagen U. 2004. *J. Appl. Phys.* 96: 1835.
Boeuf J. P. 1987. *Phys. Rev. A.* 36: 2782–2792.
Boeuf J. P. 1988. *J. Appl. Phys.* 63: 1342–1347.
Bruhn B. 2004. *Phys. Plasmas* 11: 4446.

Cheung P. Y., Donovan S., Wong A. Y. 1988. *Phys. Rev. Lett.* 61: 1360.
Deloche R., Monchicourt P., Charlet M., Lampert F. 1976. *Phys. Rev. A* 13: 1140–1176.
Ding W. X., Huang W., Wang X. D., Yu C. X. 1993. *Phys. Rev. Lett.* 70: 170.
Dong L. F., Liu F. C., Liu S. H., He Y. F., Fan W. L. 2005. *Phys. Rev. E* 72: 046215.
Eliasson B. 1991. *IEEE Trans. Plasma Sci.* 19: 309.
Eliasson B., Egli W. 1987. *Helv. Phys. Acta* 60: 241.
Emmert J., Angermann H. H., Dux R., Langhoff H. 1988. *J. Phys. D Appl. Phys.* 21: 667.
Fridman A., Chirokov A., Gutsol A. 2005. *J. Phys. D Appl. Phys.* 38: R1–R24.
Guikema J., Miller N., Niehof J., Klein M., Walhout M. 2000. *Phys. Rev. Lett.* 85: 3817.
Huang R. S. H., Huang H. 2005. *Chaos and Its Application*. Wuhan: Wuhan University Press.
Iza F., James L., Kong M. G. 2009. *IEEE Trans. Plasma Sci.* 37: 1289.
Jonas P., Bruhn B., Koch B. P., Dinklage A. 2000. *Phys. Plasmas* 7: 729.
Kanazawa S., Kogoma M., Moriwaki T., Okazaki S. 1988. *J. Phys. D Appl. Phys.* 21: 838.
Kim H. C., Iza F., Yang S. S., Radmilovic-Radjenovic M., Lee J. K. 2005. *J. Phys. D Appl. Phys.* 38: R283–R301.
Kogelschatz U. 2002. *IEEE Trans. Plasma Sci.* 30: 1400.
Kulikovsky A. A. 1994a. *J. Phys. D Appl. Phys.* 27: 2556.
Kulikovsky A. A. 1994b. *J. Phys. D Appl. Phys.* 27: 2564.
Kulikovsky A. A. 1995. *J. Comp. Phys.* 119: 149–155.
Kushner M. J. 2004. *J. Appl. Phys.* 95: 846.
Mangolini L., Orlov K., Kortahagen U., Heberlein J., Kogelschatz U. 2002. *Appl. Phys. Lett.* 80: 1722.
Massines F., Rabehi A., Decomps P., et al. 1998. *J. Appl. Phys.* 83: 2950.
Pouvesle J. M., Bouchoule A., Stevefelt J. 1982. *J. Chem. Phys.* 77: 817.
Qi B., Hang J. J., Zhang Z. H., Wang D. Z. 2008. *Chin. Phys. Lett.* 25: 3323.
Ra'hel J., Sherman D. M. 2005. *J. Phys. D Appl. Phys.* 38: 547.
Rosenstein M. T., Collins J. J., Luca C. J. D. 1993. *Physica D* 65: 117.
Sang C. F., Sun J. Z., Wang D. Z. 2010. *J. Phys. D Appl. Phys.* 43: 045202 (6pp).
Scharfetter D. L., Gummel H. K. 1969. *IEEE Trans. Elec. Dev.* ED-16(1): 64–77.
Schoenbach K. H., Moselhy M., Shi W. 2004. *Plasma Sources Sci. Tech.* 13: 177.
Shi H., Wang Y. H., Wang D. Z. 2008. *Phys. Plasmas* 15: 122306.
Shi J. J., Kong M. G. 2000. *Trans. Plasma. Sci.* 33: 624.
Shon J. W., Kushner M. J. 1994. *J. Appl. Phys.* 75: 1883.
Šijaèiæ D. D., Ebert U., Rafatov I. 2004. *Phys. Rev. E* 70: 056220.
Sommerer T. J., Kushner M. J. 1992. *J. Appl. Phys.* 71: 1654.
Stevefelt J., Pouvesle J. M., Bouchoule A. 1982. *J. Chem. Phys.* 76: 4006.
Stewart R. A., Lieberman M. A. 1991. *J. Appl. Phys.* 70: 3481.
Stollenwerk L., Amiranashvili S., Boeuf J. P., Purwins H. G. 2006. *Phys. Rev. Lett.* 96: 255001.
Strumpel C., Astrov Y. A., Purwins H. G. 2000. *Phys. Rev. E* 62: 4889.
Sun J. Z., Wang Q., Zhang J. H., Wang Y. H., Wang D. Z. 2008. *Chin. Phys. Lett.* 25: 4054–4057.
Tsai J. H., Wu C. 1990. *Phys. Rev. A* 41: 5626.
Verboncoeur J. P. 2005. *Plasma Phys. Control Fusion* 47: A231.
Wang Q., Sun J. Z., Wang D. Z. 2009. *Phys. Plasmas* 16: 043503.
Wang Q., Sun J. Z., Wang D. Z. 2011. *Phys. Plasmas* 18: 103504.
Wang Q., Sun J. Z., Zhang J. H., Ding Z. F., Wang D. Z. 2010. *Phys. Plasmas* 17: 053506.
Wang Y. H., Shi H., Sun J. Z., Wang D. Z. 2009. *Phys. Plasmas* 16: 063507.
Wang Y. H., Wang D. Z. 2004. *Chin. Phys. Lett.* 21: 2235–2237.
Wang Y. H., Zhang Y. T., Wang D. Z., Kong M. G. 2007. *Appl. Phys. Lett.* 90: 071501.
Ward A. L. 1962. *J. Appl. Phys.* 33: 2789–2794.
Wilson IV R. B., Podder N. K. 2007. *Phys. Rev. E* 76: 046405.
Xu X. J., Chu D. C. 1996. *Discharge Physics of Gas*. Shanghai: Fudan University Press, p. 277 (in Chinese).

Young F. F., Wu C. 1993. *Appl. Phys. Lett.* 62: 473.
Zhang J., Wang Y. H., Wang D. Z. 2010. *Phys. Plasmas* 17: 043507.
Zhang P., Kortahagen U. 2006. *J. Phys. D Appl. Phys.* 39: 153.
Zhang Y. T., Wang D. Z., Kong M. G. 2005a. *J. Appl. Phys.* 98: 113308.
Zhang Y. T.,Wang D. Z., Wang Y. H., Liu C. S. 2005b. *Chin. Phys. Lett.* 22: 171–174.

5

High-Pressure
Microcavity Discharges

Karl H. Schoenbach

5.1 Toward Micrometer-Sized Glow Discharges

Glow discharges are likely the most studied and widely used gas discharges. They are found in numerous applications ranging from light sources to plasma reactors for materials processing (Liebermann and Lichtenberg 1994, Roth 2001). The simplest version is the direct current (DC), self-sustaining glow discharge. The discharge is established when a sufficiently high voltage is applied between the two electrodes, that is, the cathode and the anode. After the electrical breakdown, space charge-dominated boundary layers develop at the cathode and anode, that is, the cathode and anode fall regions. They are connected by a uniform plasma, the positive column. How small can we make such a discharge before conditions in voltage and current density are reached that make it difficult to sustain the discharge? An estimate can be obtained from data on glow discharges between two plane-parallel electrodes published in a book by one of the pioneers in discharge physics, von Engel (1965). Because we are interested in high-pressure glow discharges, we will extrapolate his results to discharges at 1 atmosphere—a high-pressure microglow discharge.

In his book, von Engel discusses an experiment in a hydrogen glow discharge between iron electrodes, which was based on measurements by Gůntherschulze (1930). When the electrodes are brought close together, the positive glow is first eliminated, then the anode fall disappears when the anode reaches the cathode fall layer or the cathode dark space with the dimension d. Further reduction of the electrode distance, D, can only be obtained by increasing the voltage, in this case, the cathode fall voltage, drastically.

The minimum axial dimension of a glow discharge, defined as the length that corresponds to the minimum in cathode fall voltage, is consequently of the order of the cathode fall length, d. The cathode fall length, d_n, for normal glow varies with the inverse of pressure, p:

$$pd_n = \frac{\ln(1 + 1/\gamma)}{\alpha/p} \tag{5.1}$$

with p in Torr and d_n in centimeters. The first Townsend coefficient, α, is defined as the number of ionizing collisions per path length, and γ is the number of electrons emitted from the cathode for each incoming ion. They are constants that depend on the gas and the electrode material, respectively. For hydrogen discharges between iron electrodes, the left side of the equation was determined to be 0.9 Torr cm. Accordingly, the cathode fall thickness, and, consequently, the minimum axial dimension, of a glow discharge at atmospheric pressure (760 Torr) in hydrogen between iron electrodes is 10 μm. A further reduction in distance would lead to an "obstructed discharge" with a strongly increasing voltage. So if we assume that a steep rise in voltage limits the axial distance, we can determine the current density for such a microdischarge (von Engel 1965). The reduced normal current density, J_n/p^2, is 72×10^{-6} A cm^{-2} Torr^2. For atmospheric pressure this corresponds to a current density of approximately 40 A cm^{-2}. The average power density, $P = J_n V_c/d_n$, in the cathode fall, assuming that the cathode fall voltage, V_c, is 150 V, is of the order of 10^6 W cm^{-3} for atmospheric pressure normal glow discharges. Typical cross-sectional areas for stable discharges are of the order of square millimeters (Kunhardt 2000).

Whereas this estimate for the minimum axial dimension is based on the assumption that even for small gaps the generation of electrons is determined by the impact of positive ions, more recent studies have shown that for very small gaps this assumption does not hold. Studies by Ono et al. (2000) on metal-to-silicon and silicon-to-silicon microgaps have shown that for metal–silicon gaps of less than 7 μm, the breakdown voltage actually decreased, instead of increasing as expected from the Paschen law. Measurements by Slade and Taylor have confirmed this observation (2002). This deviation from Paschen's curve for very small gaps (>2 μm) is likely caused by pure field emission (Fowler–Nordheim field emission) and in the transition region (up to about 10 μm) through ion-enhanced field emission caused by the presence of positive ions near the cathode surface. The breakdown in small gaps has been simulated by means of particle-in-cell and Monte Carlo techniques (Zhang et al. 2004). Recently, David B. Go's group (Go and Pohlman 2010, Tirumala and Go 2010) has developed an analytical model that provides an excellent description of microscale breakdown and the modified Paschen's curve (Figure 5.1). It must be noted that the breakdown for very small gaps is no longer a function of pressure, p, times distance, d, but depends on p and d separately.

Independent of the breakdown and sustaining mechanism, the power density in these small gap discharges can be very high. Discharges at high power densities are prone to instabilities: in particular, thermal instabilities [which for atmospheric pressure discharges are discussed in Raizer (1997), Kunhardt (2000), and Garscadden (2008)] lead to a radial constriction of the discharge column and to glow-to-arc transition. Stable high-pressure operation in self-sustained, atmospheric pressure glow discharges between parallel-plate electrodes can be reached by limiting the current density to values below the instability threshold current density or by pulsing the discharge with pulse durations that are short compared to the characteristic times for the development of the dominant instability (Kunhardt 2000).

Another way to stabilize high-pressure glow discharges is to confine the plasma in small dielectric cavities. Diffusion to the walls effectively damps small fluctuations in the plasma that would otherwise lead to filamentation. For high-pressure discharges, these cavity dimensions are in the 100 μm range. In addition, for these microdischarges conduction cooling of the electrodes helps in keeping the gas temperature in the glow discharge relatively low. With this approach, using cavities with dimensions

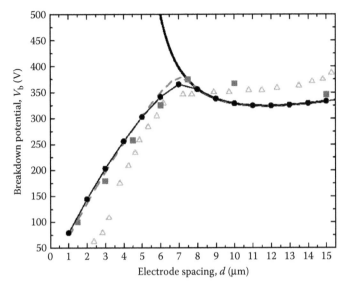

FIGURE 5.1 Breakdown potential as a function of the electrode spacing. The bold solid curve shows the Paschen's curve for atmospheric pressure air between electrodes with a secondary emission coefficient of $\gamma = 0.0075$. Experimental studies by Slade and Taylor (triangles) (From Slade, G.P. and E.D. Taylor, *IEEE Trans. Compon. Packaging Technol.*, 25, 390–396, 2002.) show clearly the strong deviation of the breakdown potential from Paschen's law at very small gaps. The results of particle-in-cell and Monte Carlo simulations (squares) (From Zhang, W., et al., *J. Appl. Phys.*, 96, 6066–6072, 2004.) and that of recently developed analytical models (dashed line) (From Tirumala, R. and D.B. Go, *Appl. Phys. Lett.*, 97, 151502, 2010.), (solid line with filled circles) (From Go, D.B. and D.A. Pohlman, *J. Appl. Phys.*, 107, 103303, 2010.), which include ion-enhanced field emission as a major charge generation mechanism in small gaps, agree well with the experimental results. (Figure courtesy of David Go, University of Notre Dame.)

ranging from about twice the cathode fall distance to about ten times this distance, it was possible to extend the pressure and current density range and to generate stable glow discharges at pressures up to and above 1 atmosphere, even in molecular gases.

5.2 Microcavity Discharges

The stability of glow discharges confined in a small cavity was first reported in an article by White (1959) who studied "hollow cathode" discharges with the cathode containing a nearly spherical cavity of 0.75 mm diameter. Stable discharges in neon were observed at pressures of 100 Torr. In the mid to late 1990s there were efforts to apply this concept to high-pressure discharges. Schoenbach et al. (1996) at Old Dominion University were the first to report stable high-pressure (350 Torr) operation of microdischarges, even in air, in a cylindrical hollow cathode geometry such as the one shown in Figure 5.1a, with a cathode diameter of 75 μm. Electrode materials for these discharge devices (and generally in the studies of the group at Old Dominion University) were molybdenum as electrode material with mica (in the earlier studies) or alumina as the dielectric. It was also shown that a positive slope in the current–voltage (*I–V*) characteristics of these discharges allowed the formation of microdischarge arrays. These authors coined the term "microhollow cathode discharges (MHCDs)" for these discharges, a connotation that was subsequently used by other groups to describe the three-layer configuration. The paper by Schoenbach et al. in 1996 was followed shortly thereafter by a paper from Gary Eden's group (Frame et al. 1997), who used a similar electrode geometry, but rather than using metal electrodes, they used a cathode made of silicon. The use of silicon opened the possibility of semiconductor microfabrication as a way to generate large arrays of microdischarges, a method that was successfully employed by the Eden group at the University of Illinois

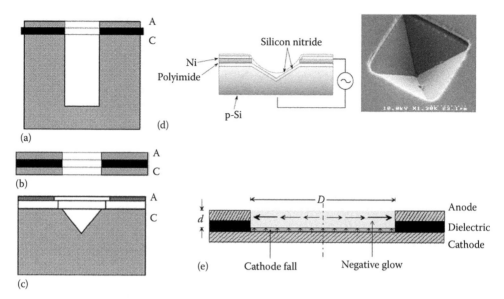

FIGURE 5.2 Examples of microdischarge electrode configurations (From Schoenbach, K.H. and W.D. Zhu, *IEEE Trans. Quantum Electron.*, 48, 768–782, 2012.): (a) and (b) are microhollow cathode structures (From Schoenbach, K.H., et al., *Appl. Phys. Lett.*, 68, 13–15, 1996.); structure (b) exists also with the anode being closed; (c) inverted pyramid device (From Chen, J., et al., *J. Microelectromech. Syst.*, 11, 536–543, 2002.); (d) *left*: inverted Si pyramid device with a dielectric structure that allows AC and bipolar operation (figure provided by Gary Eden for the review on microdischarges by Becker, K.H., et al. 2006. *J. Phys., D Appl. Phys.*, 39: R55–R70.); *right*: SEM image of a single device with an emitting aperture of 50 μm by 50 μm (From Becker, K.H., et al., *J. Phys. D Appl. Phys.*, 39, R55–R70, 2006.); (e) shows the electrode structure of a CBL discharge geometry (arrows indicate the electric field distribution). (From Schoenbach, K.H., et al., *Plasma Sources Sci. Technol.*, 13, 177–185, 2004.)

(Park et al. 2002, Eden and Park 2005). Similarly, at the Universities of Frankfurt and Braunschweig in Germany, micromachining techniques were employed to generate so-called micro-structured electrode (MSE) arrays using glass, polyimide, or ceramics as the dielectric (Penache et al. 2002a). Whereas the planar MSE arrays generate discharges in planar microgaps, the three-dimensional MSE arrays use a microcavity structure, as shown in Figure 5.2b, as the base structure. Other microcavity geometries, not shown in Figure 5.2, include microtube cathodes made of stainless steel, with the anode at the orifice (Sankaran and Giapis 2002, Yokoyama et al. 2005) and microslots (Yu et al. 2003, Rahul et al. 2005). An overview of the variety of microcavity and other microdischarge geometries in general can be found in the works of Iza et al. (2008).

In the following I will focus mainly on microcavity discharges and discharges that develop in a cathode hollow and only include as special cases the microhollow cathode sustained (MCS) glow discharge and the cathode boundary layer (CBL) discharge, which is defined by a planar cathode with the plasma confined by a "hollow" anode (Figure 5.2e). The phrase "hollow cathode" historically refers to a specific mode of discharge operation, where the cathode fall regions in the interior of the cylindrical cathode are so close that high-energy electrons emitted from one side of the cathode can enter the opposite cathode fall, where they are accelerated back toward the axis. The "pendulum" motion of such electrons leads to increased ionization on the hollow cathode axis, which appears in the *I–V* curve as negative differential resistance (referred to as hollow cathode mode or negative glow mode). Nowadays, MHCDs are often operated as normal or abnormal glow discharges, rather than in the hollow cathode mode. Therefore, the inclusion of the phrase "hollow cathode" in the name MHCD might be misleading for the larger holes. As a consequence, some groups have referred to these microcavity discharges simply as microdischarges.

High-pressure microcavity plasmas, generated by using electrode geometries as shown in Figure 5.2, have unique properties. Most importantly, they are nonequilibrium discharges. Because the cathode fall largely determines the discharge characteristics, the electron energy distribution contains electrons at energies comparable to the cathode fall (which is of the order of 150 V). Also, because they can be stably operated at high pressure, three-body reactions, such as those important for the generation of excimers, become dominant. Microcavity discharges and their applications are the topic of a number of review papers published in the past decade (Schoenbach et al. 2003, Kogelschatz et al. 2005, Becker et al. 2006, Kogelschatz 2007, Becker and Schoenbach 2008, Iza et al. 2008, Mariotti and Sankaran 2010).

5.3 Microcavity (Microhollow Cathode) Discharge Modes of Operation

DC, pulsed DC, alternating current (AC), radio frequency (RF), and microwave (μw) sources can be and have been used to excite microplasmas in various configurations. However, many, if not most, of the studies, both modeling and experimental, have been performed on discharges in cavity-type geometries, as shown in Figure 5.2a–c.

5.3.1 DC Operation

In the DC operation, sustaining voltages are in the range from 150 to 500 V, depending on the discharge current, the type of gas, and the electrode material. Lowest voltages are obtained with noble gases, highest voltages are measured for attaching gases or mixtures that contain attaching gases, such as air. The DC I–V characteristics of MHCDs show distinct regions. An example of such a characteristic is shown at the bottom of Figure 5.3 for a discharge in xenon at 750 Torr (Schoenbach et al. 2000). Vacuum ultraviolet (VUV) images obtained via a band-pass spectral filter at characteristic current values are also shown in this figure. In the predischarge mode (lowest current, positive slope in the I–V characteristics), the plasma is confined to the hole. When the plasma expands beyond the microhole as shown in Figure 5.3b, the current voltage characteristic shows a negative slope and the plasma begins to cover the cathode area outside the hole. When the cathode surface is limited, for example, by a cylindrical dielectric layer placed on the outer, planar cathode surface, the discharge enters an abnormal glow mode (characterized by a positive slope in the I–V curve) after reaching the boundaries of the cathode surface (Moselhy et al. 2002).

The initial assumption that the negative differential conductivity range represents a hollow cathode discharge mode (Schoenbach et al. 1996, 1997) was found to hold only for openings with diameters of the order of twice the cathode fall length. With a hole diameter of 250 μm, at a xenon pressure of 760 Torr, this condition is not satisfied. Modeling results by Boeuf et al. (2005) have shown that for cavity diameters exceeding the required distance for "hollow cathode operation," such a negative differential conductivity can be explained by an increase in the discharge area when the plasma expands from the inside of the cavity to the flat outside area of the cathode. The measured voltage maximum, as shown in Figure 5.3, can be attributed to the transition from the low-current, abnormal discharge inside the hollow cathode to a higher-current, normal glow discharge at the outer surface of the cathode.

A two-dimensional computational study of neon microdischarges in pyramidal structures, such as shown in Figure 5.2c, was published in 2004 (Kushner 2004), followed by a computational study of argon microdischarges (Kushner 2005) at varying pressure and current in an electrode configuration similar to that used by Boeuf et al. (2005), such as shown in Figure 5.2b. For both types of microdischarges it was shown that they do not operate in the typical hollow cathode mode, but ionization by beam electrons, accelerated in the cathode fall, as in hollow cathode discharges, is, nevertheless, critical to the operation of such a microdischarge. This points to the importance of the electrode material, or more exactly, the sensitivity of the electron density and gas temperature to changes in the secondary electron emission coefficient, γ. The electron and ion kinetics in a microcavity discharge were studied by Lee's group at Pohang University. Computational results confirm the nonequilibrium character of such discharges: the electron energy distributions are non-Maxwellian with a high-energy tail (Kim et al. 2006).

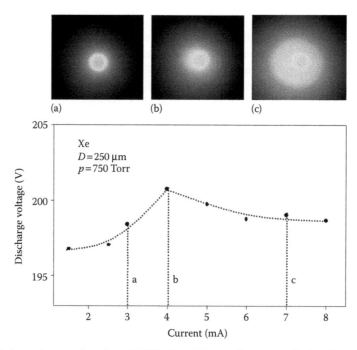

FIGURE 5.3 End-on photographs of an MHCD in a 250 μm diameter, cylindrical cavity (as shown in Figure 5.2b) at a pressure of 750 Torr in xenon at various currents. The current values are indicated in the *I–V* characteristic of the MHCD, shown below the photographs. The photographs were taken via a CCD camera through a band-pass spectral filter that centered around 171 nm (FWHM: 26 nm). This filter allowed only the second continuum emission of xenon excimer radiation to pass. The photographs show clearly that with increasing current the plasma extends from the cathode hole along the plane surface of the cathode.

5.3.2 Pulsed DC Operation

Operation of DC MHCDs with currents exceeding approximately 8 mA leads to thermal damage to the electrodes. However, by operating the discharge in a pulsed DC mode, the current can be increased considerably, as demonstrated with pulses of 0.7 ms duration and a repetition rate of 10 Hz (Moselhy et al. 2002). The pulsed operation allowed us to increase the current from 8 to 80 mA, limited only by the onset of instabilities. This increase in current is related to a strong increase in the radial dimension of the excimer source, which means the current density stays approximately constant. This is expected because the microdischarge develops from a discharge inside the cathode hole to a normal glow discharge that expands onto the outer planar cathode surface (Figure 5.3).

Whereas the use of pulsed DC with pulse durations in the millisecond range allows us to increase the current, the use of ultrashort pulses allows us to alter the electron energy distribution. This was demonstrated by recording the electron density and the excimer emission in nanosecond pulsed discharges (Moselhy et al. 2001a, Schoenbach and Moselhy 2003). By superimposing 20-ns-long pulses on a DC microdischarge, the electron density could be increased by more than an order of magnitude. This strong increase in electron density is assumed to be due to pulsed electron heating (Stark and Schoenbach 2001).

5.3.3 AC Operation

Although most of the basic studies and modeling efforts have been carried out with DC microcavity discharges, it must be noted that efforts at many laboratories have been devoted to developing AC microcavity discharges. Studies of RF-driven microdischarges in hollow slot electrode geometries were reported by

Collins' team at Colorado State University (Yalin et al. 2003, Rahul et al. 2005). RF microdischarge arrays were studied at the National Cheng Kung University in Taiwan (Guo and Hong 2003). The use of AC voltages allows coating of the electrodes with insulating materials and consequently leads to an increased lifetime of the discharge systems. With Si as material for microcavity cathodes, and coated with a thin film of silicon nitride, as shown in Figure 5.1d, large arrays of microdischarges could be fabricated using micromachining technologies (Eden and Park 2005). More recently, the use of polymer-based replica molding processes by the same group has even allowed fabrication of flexible arrays of microcavity plasma devices (Readle et al. 2009). Tachibana et al. have used two aluminum meshes coated with plasma-sprayed Al_2O_3 with millimeter and submillimeter rectangular openings as electrodes to operate arrays of discharges (2005). Although not belonging to the same class of microcavity discharges as were discussed so far in this chapter, it must be mentioned that planar MSE arrays operating at RF have been successfully used to generate atmospheric pressure microplasmas (Penache et al. 2002a, Schrader et al. 2006).

5.4 Microhollow Cathode Sustained (MCS) Glow Discharges

An extension of the microplasma two-electrode structure has been studied by the group at Old Dominion University. By adding a third electrode on the anode side of a microcavity structure as shown in Figure 5.3b, it was possible to extend atmospheric pressure plasmas beyond the microhole (Stark and Schoenbach 1999). The microcavity discharge acts as a plasma cathode. Small variations in the microcavity discharge voltage cause large changes in the microcavity discharge current and consequently in the current to the third external electrode. Figure 5.4 shows a stable DC microhollow cathode sustained discharge in atmospheric pressure argon.

This concept has also been used to generate glow discharges in atmospheric pressure air with dimensions up to cubic centimeters. As in the case of argon, electrons were extracted through the anode opening at moderate electric fields when the microdischarge was operated in the hollow cathode mode. These electrons support a stable plasma between the anode and the third electrode with sustaining voltages (in air) in the range of 200–400 V, depending on the current, gas pressure, and gap distance. The microcavity plasma current was limited to values of less than 30 mA DC to prevent overheating of the sample. Electron densities and temperatures have been measured by means of heterodyne laser interferometry and were found to be of the order of 10^{13} cm^{-3} and 2000 K, respectively (Leipold et al. 2000). The air plasma can be extended in size by placing microdischarges in parallel (Mohamed et al. 2002) or by using split third electrodes (Shin and Rahman 2011).

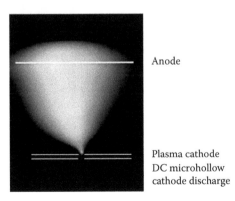

Anode

Plasma cathode
DC microhollow
cathode discharge

FIGURE 5.4 MCS glow discharge in argon at atmospheric pressure with an anode potential of 213 V and a current of 0.5 mA. The MHCD voltage is 169 V and the current is 0.47 mA. The gap between the plasma cathode and the third electrode (anode) is 2 mm. (From Stark, R.H. and K.H. Schoenbach, *J. Appl. Phys.*, 85, 2075–2079, 1999.)

5.5 Cathode Boundary Layer (CBL) Discharges

The CBL discharge is a high-pressure glow discharge that belongs (from the discharge physics point of view) to the family of microcavity discharges discussed in Section 5.3. However, the discharge does not develop in a cathode microcavity but between a planar cathode and a ring-shaped anode, which are separated by a thin, ring-shaped dielectric layer (typically of 100–300 μm thickness). The circular openings on the dielectric layer and the anode are generally of the same diameter (from less than a millimeter to several millimeters) and aligned concentrically (Figure 5.1e) (Schoenbach et al. 2004). This electrode configuration had been studied with respect to the discharge stability of plane cathode microdischarges (Biborosch et al. 1999). It was also considered by the group at the University of Illinois to be an electrode configuration that could be fabricated using micromachining techniques but was found to be less attractive than electrode geometries similar to those shown in Figure 5.1a and c (Chen et al. 2002) and was, to my knowledge, not explored further.

However, this discharge (as shown in the following) has some exciting features, not just the physics, but also its applications as simple, large-area excimer emitters (Zhu et al. 2007). Due to the small interelectrode distance, the discharge is reduced to cathode fall and negative glow. In such a discharge geometry, the current flow changes from axially directed at the cathode surface to radially directed via the negative glow to the ring-shaped anode (Figure 5.2e). The thickness measurement of the plasma layers has been elaborated in detail by Moselhy et al. (2002) and Takano and Schoenbach (2006), and a comparison with published data on the thickness of the cathode region (Cobine 1958) supported the assumption that the plasma layer in the CBL discharges consists of cathode fall and negative glow only. Side-on photographs of a CBL in a discharge geometry in which the surrounding dielectric (quartz) served as an optical window are shown in Figure 5.5. For high-pressure operation in xenon and argon, the sustaining voltage in the normal glow mode was approximately 200 V (Moselhy and Schoenbach 2004), which is of the order of reported cathode fall voltages in noble gases (Cobine 1958).

Homogeneous CBL discharges were observed in xenon and argon at elevated pressure (>75 Torr) and/or at current densities larger than 50 mA cm^{-2} (Schoenbach et al. 2004). However, at pressures below 200 Torr, particularly in high-purity xenon, when lowering the discharge current, the discharge

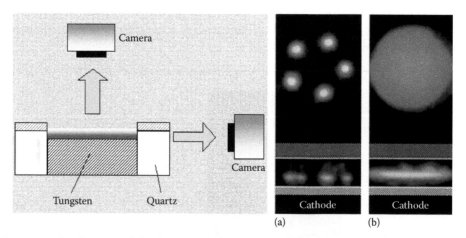

FIGURE 5.5 *Left*: Schematics of the electrode configuration of a CBL discharge. The cathode is formed of a tungsten wire of 750 μm diameter, and the ring-shaped anode is approximately 400 μm away from the planar cathode. The cylindrical, 250-μm-thick quartz tube surrounding the cathode allows us to observe the discharge side on. *Right*: Photographs of the discharge end on and side on are shown for a discharge in xenon at 75 Torr at a current of 92 μA and a voltage of 279 V (a) and a current of 224 μA and 346 V (b). The cathode fall length, determined by the dark space between the cathode and the negative glow, was 50 μm (a) and 70 μm (b). (From Schoenbach, K.H. and W. Zhu, *IEEE Trans. Quantum Electron.*, 48, 768–782, 2012.)

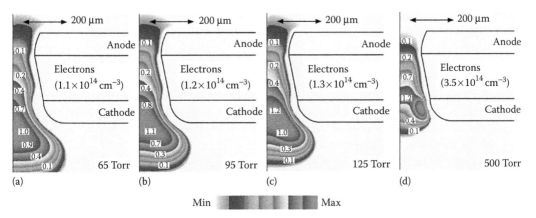

FIGURE 5.6 Electron density for microdischarges sustained in Ar (2 mA) for (a) 65 Torr, (b) 95 Torr, (c) 125 Torr, and (d) 500 Torr. All images are two-decade log-scale plots with maximum values indicated. The electron density peaks on axis at pressures <125 Torr. The progressively off-axis peak in density at higher pressure is due to the more confined beam ionization and a larger proportion of Ar_2^+ producing volumetric recombination. (From Kushner, M.J., *J. Phys. D Appl. Phys.*, 38, 1633–1643, 2005.)

became filamentary with the filaments forming distinctive patterns on the planar cathode surface. An example of such a pattern is shown in Figure 5.5a. The geometric arrangement and the number of filaments depend on the gas pressure, the discharge current, and the diameter of the opening. It is possible to form a single plasma filament (~50 μm in diameter) in a CBL device with a 300 μm opening at a xenon pressure of 50 Torr, when the current is reduced to values below 50 μA (Takano and Schoenbach 2006). Coulomb interaction between space charges and surface charges accumulated on the insulator walls are considered to be the cause of self-organized pattern formations in xenon. Another explanation for the self-organization was presented by Benilov's group at the University of Madeira (Almeida et al. 2011). It was based on a self-consistent model of the near-cathode region in a DC glow discharge. Multiple steady-state solutions were shown to describe different modes of current transfer to the cathodes.

The *I–V* curve of CBL discharges is characterized by a positive slope (except for the low current range), making it possible to expand the area of these discharges or to operate them in parallel without individual ballasting. This allows the construction of large-area, thin (of the order of 100 μm) plasma sources. Almost all of the measurements on plasma parameters, such as electron temperature, electron density, and gas temperature, have been performed on cavity-type discharges. However, it can be assumed that the plasma parameters of CBL discharges are of the same order of magnitude as those of DC MHCD.

5.6 Plasma Parameters

5.6.1 Gas Temperature

MHCDs are nonthermal discharges, which means that the gas temperature is low compared to the electron temperature even at high pressures. Measurements of the gas temperature have been performed mainly in DC microcavity plasmas, but over a wide range of currents, pressures, and for different gases. One of the most common methods is based on emission spectroscopy: rotational temperature measurements in discharges with gases that contain trace mixtures of nitrogen (Kurunczi et al. 2004) or in air plasmas (Block et al. 1999a). In this diagnostic method, the measured spectrum of the (1, 2) band of the N_2 second positive band is compared with model spectra that vary with temperature. For an MHCD in atmospheric pressure air, the temperature range obtained with this diagnostic method rises from 1700 to 2000 K when the current is increased from 4 to 12 mA. The temperature in microdischarges operating in noble gases and/or lower pressure is considerably smaller. For a 400 Torr neon MHCD at 1 mA the temperature was determined to be approximately 400 K, and increasing with discharge current (Kurunczi et al. 2004).

These low-temperature data for noble gas operation are consistent with results obtained with absorption spectroscopy (Penache et al. 2002b). The gas temperature in MSE arrays (where the single discharge geometry resembles microhollow cathode geometries shown in Figure 5.2c) in argon at pressures ranging from 50 to 400 mbar was derived from the Doppler broadening of argon lines in the near-infrared region. The gas temperature was found to increase with pressure from 380 to 1100 K in the studied pressure range at a constant discharge current of 0.5 mA.

The results indicate that the gas temperature depends strongly on the type of gas and pressure: highest values of approximately 2000 K were obtained in atmospheric pressure air discharges (Block et al. 1999a, 1999b) and the lowest with noble gases at pressures below 100 mbar. The temperature increases with discharge current. So, for noble gas discharges at relatively low pressure and low current, the microdischarge plasma is "cold," approaching values close to room temperature.

5.6.2 Electron Density

Electron densities in MHCDs in argon have been measured using either Stark broadening of the hydrogen Balmer-β line at 486.1 nm (Moselhy et al. 2003) or Stark broadening and shift of argon lines in the near-infrared region (Penache et al. 2002b). In both cases the measured electron densities for DC microdischarges were of the order of 10^{15} cm^{-3}, showing a slight increase with current. When operated in the pulsed mode, applying 10-ns electrical pulses of 600 V, the electron densities increased to 5×10^{16} cm^{-3} (Moselhy et al. 2003). Electron densities in MHCDs in atmospheric pressure air have also been measured using heterodyne infrared interferometry (Leipold et al. 2000). In an MHCD with a hole diameter of 200 μm, with a current of 12 mA at a voltage of 380 V, the electron density was found to be 10^{16} cm^{-3} (Block et al. 1999b).

The spatial distribution of the electron density in a microdischarge in argon with an electrode configuration similar to that shown in Figure 5.2b was modeled by Kushner (2005). Figure 5.6 shows the results dependent on pressure. The electron density increases with pressure. It peaks for low pressure (<200 Torr) on the axis and for pressures exceeding 200 Torr in an annulus at the cathode.

5.6.3 Electron Temperature

The electron temperature in DC MHCDs in argon was found to be of the order of 1 eV (Frank et al. 2001). The value was obtained from line intensity measurements, assuming a Maxwellian electron energy distribution. However, because the electron energy distribution is highly nonthermal (the discharge is to a large extent determined by the processes in the cathode fall and negative glow), a large concentration of electrons with energies up to cathode fall energy (beam electrons) is expected. This presence of beam electrons has been proven through modeling (Kushner 2005, Kim et al. 2006, Choi et al. 2007), but, to our knowledge, it has been verified experimentally only for lower pressure discharges (Gill and Webb 1977). According to modeling studies (Kushner 2005), beam electrons accelerated in the cathode fall clearly play a major role in sustaining DC microdischarges.

An even more pronounced increase in the concentration of high-energy electrons can be expected when the discharge is pulsed. A pulsed electric field causes the electron energy distribution function to shift temporarily toward higher energies. For a 10-ns pulsed discharge the electron temperature was estimated to reach values of 2.25 eV, assuming that the electron energy distribution is a Boltzmann distribution (Moselhy et al. 2003). This is a rather crude approximation. More accurate values of the electron temperature, or, rather, the electron energy probability functions (EEPFs), are obtained by means of particle-in-cell and Monte Carlo collision models (Kim et al. 2006). The electron and ion kinetics in a 300-Torr argon discharge in a microhollow cathode geometry (radius of the cathode hollow: 50 μm) was simulated for the ignition phase, which is of the order of nanoseconds, and therefore describes the changes in the electron energy

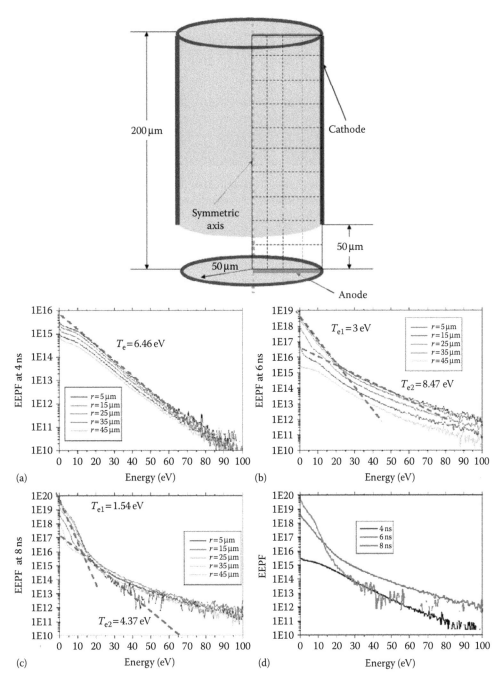

FIGURE 5.7 *Top*: Geometry and simulation domain of the MHCD; *bottom*: evolution of the EEPF for an argon discharge at 300 Torr, 0.1 mA. (a) 4 ns; (b) 6 ns; (c) 8 ns; (d) $r = 5$ μm. (From Kim, G.J., et al., *J. Phys. D Appl. Phys.*, 39, 4386–4392, 2006.)

distribution of a 10-ns pulsed discharge well. Figure 5.7 shows the EEPF at 4, 6, and 8 ns after ignition. At 4 ns the function is Druyvesteyn. As electrons drift toward the anode, the low-energy part of the EEPF is filled up with electrons generated by ionizing electron collision, generating a bi-Maxwellian distribution. Later in time, at 8 ns, the cathode fall region is established. As a result, the electron temperature of the low-energy electrons decreases to values of approximately 1.5 eV, but the high-energy tail persists.

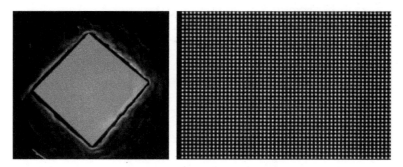

FIGURE 5.8 Photographs of the 500 × 500 array. *Left*: the entire array operating in 700 Torr of Ne; *right*: a 54 × 40 segment of the array depicting the pixel-to-pixel emission uniformity. The Ne pressure is 700 Torr. (From Eden, J.G. and S.J. Park, *Plasma Phys. Control. Fusion*, 47, B83–B92, 2005.)

5.7 Scaling

Scaling of single microdischarges has been the topic of experimental studies (see Becker et al. 2006, Becker and Schoenbach 2008) and modeling studies (Kushner 2005, Greenan et al. 2011). Generally, the gas temperature increases with pressure, as does the electron density. The electron density is less affected by current because with increasing current the plasma volume increases. As clearly seen in Figure 5.3, with increasing current, the plasma extends from the cathode hole to cover the planar cathode surface. This allows us to generate relatively large-area plasmas with a single microhollow discharge.

5.7.1 Parallel Operation of Microcavity Discharges—Microdischarge Arrays

However, rather than increasing the plasma volume by increasing current in a single discharge, large-area plasmas of various shapes and amplitudes have been generated by operating microdischarges in series or in parallel, or, in some cases, both. Parallel operation of microdischarge arrays in the negative glow mode or the hollow cathode mode (corresponding to a negative slope in *I–V* characteristics) and the normal glow mode (corresponding to a zero slope in *I–V* characteristics) are made possible by using distributed resistive ballast. Shi et al. (1999) used a semi-insulating silicon wafer (with a resistivity of 1200 Ω cm) as the anode material and successfully sustained a 4 × 4 array of 100 μm microdischarges in argon at pressures up to 300 Torr. von Allmen et al. (2003a) used thick film resistors to individually ballast an array of microdischarges.

Parallel operation of MHCDs is possible without individual ballasting for any modes where the discharge current increases with voltage. By limiting the cathode surface by means of a layer of dielectric material, the discharge enters into an abnormal glow mode when reaching the boundaries of the cathode surface. Because in this mode the *I–V* characteristic has a positive slope, operation of multiple discharges in parallel without ballasting each discharge becomes possible (Stark et al. 2000). This concept also seems to be the basis for the demonstrated parallel operation of CBL discharges, where the cathode surface is confined by the surrounding dielectric (Zhu et al. 2007).

Parallel operation of multiple microdischarges in the predischarge mode, which also has a positive *I–V* characteristic (see Figure 5.3), was successfully demonstrated for DC and AC operation by several groups (Schoenbach et al. 1996, Frame et al. 1998, Penache et al. 2000, Park et al. 2001, Guo and Hong 2003). In particular, Gary Eden's group at the University of Illinois excelled in the formation of large arrays by focusing on Si-based devices. The most extensively characterized Si-based device structure is that presented in Figure 5.1d, an inverted Si pyramid device with a dielectric structure that allows AC and bipolar operation.

The operation of a 500 × 500 array (or one-fourth of a million devices) of microplasmas has been successfully demonstrated by the group at the University of Illinois. Photographs of the entire device and a

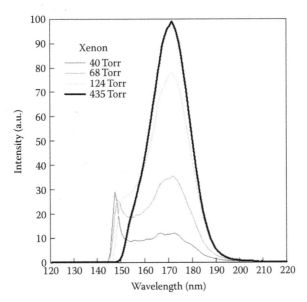

FIGURE 5.9 Pressure dependence of the VUV emission spectrum of a high-pressure microdischarge in xenon. The discharge voltage and current were kept in the range of 215–235 V and 3–3.3 mA, respectively. (From El-Habachi, A. and K.H. Schoenbach, *Appl. Phys. Lett.*, 73, 885–887, 1998b.)

segment of the device operating in neon at a pressure of 700 Torr are shown in Figure 5.8 (Eden and Park 2005). The devices have square cross sections with an area of 50 μm² and a device pitch of 100 μm. A 15-kHz, sinusoidal AC voltage waveform was used to drive the array. More recently, the same research group extended the formation of arrays fabricated in plastic by replica molding (Readle et al. 2009). Noteworthy for these structures is the excellent uniformity of emission from device to device over the entire array.

5.7.2 Series Operation of Microdischarges

The series operation of microdischarges was mainly motivated by the increase in excimer emission based on the consideration that excimer molecules do not reabsorb the emitted excimer radiation. Consequently, except for a few absorbers (such as neutral excimer and dimer ions), any background gas is essentially transparent to excimer emission. Theoretically, if *n* circular microdischarge excimer devices are lined up concentrically along their central axis, the excimer irradiance measured when all devices are turned on should be approximately equal to the sum of the irradiances from *n* devices. The first experiments of microdischarges in series were demonstrated by El-Habachi et al. (2000) using a tandem MHCD structure generating XeCl excimers. The excimer irradiance was doubled in this arrangement compared to in a single discharge. A ceramic structure with three microdischarge cells in series was successfully operated in neon by Vojak et al. (2001).

5.8 Applications of High-Pressure Microcavity Discharges

The growth of applications of microplasmas over the past decade, both in terms of the diversity of applications and in the breadth and depth in a given area of application, makes it impossible to provide a detailed account of all microplasma applications in a single chapter. I have, therefore, selected a few applications that will be discussed at some level of detail below. In all other cases, extensive references, and if available, references on review papers on applications of microcavity discharges will be provided.

5.8.1 Light Sources

From the discussion in Section 5.6.1 it is obvious that one of the major applications of microcavity arrays is their use as microdisplays. The luminance and luminous efficacy of silicon microplasmas in neon and neon–xenon mixtures has been measured (Eden and Park 2005). It was found that the efficacy of a microdischarge array phosphor assembly exceeded that of commercially available plasma display panels considerably. Another application of microdischarges as light sources is their use as UV lamps. A review on high-pressure microdischarges as sources of UV radiation has been published by Schoenbach and Zhu (2012).

5.8.1.1 Excimer Light Sources

Efficient excimer formation requires the following two conditions: (1) a sufficiently large number of electrons with energies high enough for metastable (or ion) formation and (2) a high pressure to ensure a sufficiently high rate of three-body collisions (Becker et al. 2002b). Microcavity discharges as well as CBL discharges satisfy these conditions. Because they are restricted to the cathode fall and negative glow, the electron energy distribution is non-Maxwellian and contains a large number of high-energy electrons (see Section 5.5.3). The high-energy electrons in such discharges in rare gases and rare gas–halogen mixtures provide a high concentration of excited or ionized rare gas atoms. The high-pressure operation of these microcavity discharges favors excimer formation, making them efficient excimer radiation sources.

The initial studies on excimer emission from microdischarges with 100 μm cathode hole diameters were performed with argon and xenon (Schoenbach et al. 1997, El-Habachi and Schoenbach 1998a, 1998b). The efficiency (VUV power/electrical power) of the xenon excimer radiation source was 7.5% ± 1.5% at pressures of approximately 400 Torr; it was less for argon discharges (6%) (Moselhy et al. 2003). Thus, a single microdischarge in xenon, with a forward voltage slightly more than 200 V and a current slightly higher than 3 mA, generated about 50 mW of VUV radiation peaking at 172 nm. Higher efficiencies can be obtained with nanosecond pulsed discharges. Figure 5.9 shows that the emission spectrum is dependent on pressure. At low pressures (<50 Torr), the spectrum is dominated by the Xe 147 nm resonance line. With an increase in pressure, there are indications of the first

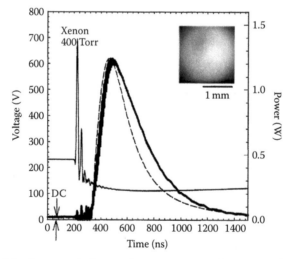

FIGURE 5.10 Temporal development of voltage and excimer emission for a 20-ns pulsed excitation. The voltage before the pulse application is the sustaining voltage of the DC discharge (230 V). After pulse application, which raises the voltage to 700 V, for the 20 ns pulse, the discharge voltage drops to 110 V, and approaches the DC value after several microseconds. The solid bold line represents the temporal development of the excimer emission. The dashed line represents modeling results. An end-on image of the excimer source with a 100 μm diameter cathode hole in the center at an applied pulsed voltage of 700 V is shown in the inset picture. (From Moselhy, M., et al., *Appl. Phys. Lett.*, 79, 1240–1242, 2001a.)

continuum of Xe extending from the resonance level toward longer wavelengths. At even higher pressures (>100 Torr) the second continuum, peaking at 172 nm, dominates the spectrum.

The efficiency of xenon excimer radiation was found to increase to approximately 20% when, instead of applying a DC voltage, a 20-ns pulsed voltage was raised to approximately 800 V. Further increasing the voltage actually caused a decrease in efficiency, although the optical power was still rising (Moselhy and Schoenbach 2003). The excimer response to a 20 ns voltage pulse in a xenon microdischarge is shown in Figure 5.10. The optical response to the voltage pulse, which peaks a few hundred nanoseconds after the pulse application, was modeled using a rate-equation model (Adler and Mueller 2000). The insert in Figure 5.9 shows that this relatively high voltage, even if applied only for 20 ns, causes the area of the source to expand from the cathode hole over the planar surface of the cathode (Moselhy et al. 1999).

In 1999, a group at Stevens Institute of Technology reported excimer emission from neon and helium DC microdischarges (Kurunczi et al. 1999, Becker et al. 2002a, 2002b). Helium has the highest excitation energy (about 20 eV for the first excited state manifold) and ionization energy (>24 eV) among all atoms. Nonetheless, helium excimer emission could readily be obtained in the range from 60 to 100 nm with MHCD plasmas at 400–600 Torr (Kurunczi et al. 1999).

The emission from DC microdischarges in Ar/F$_2$ (Schoenbach et al. 2000) and Xe/Cl$_2$ mixtures (El-Habachi et al. 2000) at 193 and 308 nm, respectively, was studied by the group at Old Dominion University. At atmospheric pressure for the ArF discharge and 1000 Torr for the XeCl discharge, measured excimer efficiencies were 1% and 3%, respectively. Continuous wave (CW) emission in the mid- to deep-UV region from diatomic excimers was reported by the group at the University of Illinois (Frame et al. 1998). Emission from microdischarges with electrode geometries similar to that shown in Figure 5.1a, with Si as the cathode material, was studied for several gas mixtures (Kr/I$_2$, Xe/I$_2$, Xe/O$_2$) at gas pressures of 300 Torr. Particular emphasis was placed on XeI (B→X) emission, which appears to make such microdischarges potential replacements for mercury resonance line radiation at 253.7 nm.

As important as high excimer power is for many applications, such as photolithography, the radiant emittance, that is, the optical (excimer) power per surface element, is even more important. As shown qualitatively in the photographs of Figure 5.3, the discharge plasma, and consequently, the area of the excimer source, expands with increasing current along the plane surface of the cathode, limiting the increase in radiant emittance. If the cathode surface is confined, and the plasma has reached these limits,

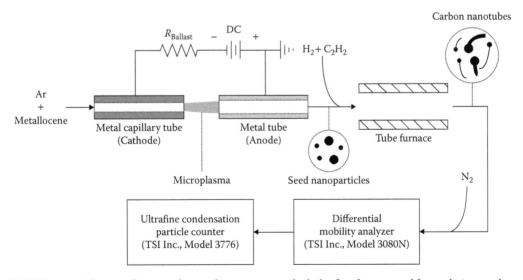

FIGURE 5.11 Schematic diagram of microplasma reactor and tubular flow furnace used for catalytic, gas-phase growth of carbon nanotubes. As-grown nanostructures are continuously monitored by a cylindrical differential mobility analyzer (DMA) and an ultrafine condensation particle counter (CPC). (From Chiang, W.-H. and R.M. Sankaran, *Appl. Phys. Lett.*, 91, 121503, 2007.)

the normal discharge transfers into an abnormal discharge where the excimer intensity and (due to the confined cathode area) the radiant emittance increase linearly with current, but it is limited by instabilities at higher currents (Schoenbach et al. 2003).

5.8.1.2 Line Emission from Microdischarges

Impurities, even in small concentrations, are detrimental to excimer formation. It is therefore important to evacuate the discharge chamber thoroughly before operation and use only highly purified gases and materials with low gas emission inside the discharge chamber. On the other hand, impurities such as O_2 have a place in VUV line emission sources. The addition of small amounts of O_2 to Ar at high pressure was shown to increase the intensity of the 130.5 nm O line at the expense of excimer emission to values of 13 mW at discharge currents of 10 mA (Moselhy et al. 2001b). This strong line emission is assumed to be due to resonant energy transfer from Ar excimers or their precursors to oxygen. Similarly, resonant energy transfer from Ne excimers to H_2 resulting in intense H Lyman-α line emission at 121.6 nm was observed (Kurunczi et al. 1999).

5.8.1.3 Toward Microexcimer Lasers

A simple estimate of the power density in a microdischarge indicated that these discharges may serve as an excimer laser medium (El-Habachi et al. 2000). That optical gain can actually be achieved with microplasmas was demonstrated by the group at the University of Illinois (von Allmen et al. 2003b). In a discharge arrangement comprising seven sets of microdischarges that were arranged optically in series with an overall active length of ~1 cm, indications of gain on the 460.30 nm transition of Xe+ were observed. This makes this the first example of a microdischarge-pumped optical amplifier. If a cavity and reflectors are properly designed, it seems possible to use these discharges as the medium for a miniaturized excimer laser.

5.8.2 Microdischarges as Plasmareactors

The energetic electrons in high-pressure microdischarges assist in the production of radicals and ions, species that are needed for materials processing and surface modification. Hsu and Graves (2003) explored the use of an MHCD as a flow reactor. Flowing of molecular gases through the microplasma was found to induce chemical modifications by processes such as molecular decomposition. Becker and coworkers (Qiu et al. 2004) used a single flow-through DC-excited microplasma reactor to generate H_2 from an atmospheric pressure mixture of NH_3 and Ar for use in small, portable fuel cells. They observed NH_3 conversion rates of up to 20% for residence times of 5 µs for the gas mixture in the plasma.

MCS atmospheric pressure microdischarges (see Section 5.4) in air with admixtures of benzene and methane were used by Jiang et al. (2005) to study the remediation of these volatile organic compounds. The measured removal fraction of 300 ppm methane was 80% and that of benzene 90%, with a residence time in the microdischarge plasma of less than 0.5 ms. In a similar MCS discharge sustained in pure oxygen and in mixtures of oxygen and rare gases, the production of singlet oxygen, with its potential application of pumping the oxygen iodine laser, was studied by teams at the CNRS–University Paris–Sud and CNRS University Paul Sabatier in Toulouse, France (Bauville et al. 2007). Singlet oxygen relative yields of 7.6% were measured in an afterglow with a flowing argon–oxygen mixture (1.15% O_2) downstream at a distance of 23 cm from the discharge.

Amorphous carbon films were fabricated by Guo and Hong (2003) by adding 1% hexamethyldisiloxane (HMDSO) to atmospheric pressure helium in a microcavity plasma array with a third biased electrode. Maskless etching of silicon and diamond deposition on a heated Mo substrate have been demonstrated by Sankaran and Giapis (2001, 2002). Nanoparticle growth in microdischarges was studied by the Sankaran group at Case Western University. Metal nanoparticles were grown in a DC, atmospheric pressure microplasma reactor, as shown in Figure 5.11. The microplasma was generated between two stainless-steel capillary tubes with a 180 µm inside diameter as the cathode and a stainless-steel mesh as the anode. The use of tubes rather than cylindrical openings as microhollow cathodes, as shown in Figure 5.2b, was introduced by Sankaran and Giapis (2001, 2002). With the system shown in Figure 5.11, the controlled synthesis of Fe and Ni catalyst particles by such a microplasma was demonstrated for gas-phase growth of carbon nanotubes (Chiang and Sankaran 2007). An overview of such material applications of high-pressure microplasmas, focusing on nanoparticle synthesis with microplasmas, has been published by Mariotti and Sankaran (2010). A review of material applications with a broader scope is presented in a chapter by Sankaran and Giapis (2008).

5.8.3 Microdischarges as Sensors

Microdischarges have also been used as detectors. MHCDs were studied by Miclea and collaborators for their application as analytical plasmas (Miclea et al. 2002, 2005, Miclea and Franzke 2007). Their small size allows the miniaturization of sensors used in analytical spectrometry. The discharge is able to excite and efficiently ionize organic molecules that flow through the hollow cathode structure. Based on this concept, a high-pressure microplasma has been used as a detector of halogenated hydrocarbons (Miclea et al. 2002). Detection limits of 20 ppb for chlorine as well as fluorine were measured. Miclea and collaborators (Miclea et al. 2005, Miclea and Franzke 2007) also used a microcavity plasma at atmospheric pressure as a miniaturized ionization source for mass spectrometry. The microplasma jet source consists of a sandwich metal/insulator/metal structure with a central hole 100–300 µm in diameter. In the jet mode, atmospheric pressure is maintained on the anode side with a much lower pressure (0.2 mbar) on the cathode side. For mass spectrometric applications, the samples are introduced into the He gas flow. Detection limits for halogenated organic compounds of the order of picograms per second were observed. In the case of ferrocene, a detection limit of about 500 ppb was reported for Fe (Miclea et al. 2005). MHCDs are not the only microdischarges that were studied with respect to their applications as detectors. An overview of analytical detectors based on microplasmas is given in a 2007 review article (Miclea and Franzke 2007) and very recently in a review article by Gianchandini's group at the University of Michigan (Eun and Gianchandini 2012).

5.8.4 Microplasma Thrusters

The effect of gas heating through deposition of electrical energy in microdischarges, particularly in the MHCDs (Figure 5.2b), and the resulting flow dynamics in such microcavities has been studied by Kushner (2005). The heated gas can be expanded in a nozzle to generate thrust: the concept of a microthruster with applications in altitude and orbit control of small satellites. Modeling studies of DC microdischarges for applications in small satellite propulsion devices by a team at the University of Texas, Austin, showed that in the case of helium discharges approximately 40% of the electrical power input is converted into thermal energy and that the gas temperature increases linearly to values over 1000 K (Kothnur and Raja 2007). Another computational study published in 2008 by the same group showed that the thrust produced by ions exiting the discharge is only a small fraction of the total thrust (Deconinck et al. 2007). In a computational study, Kushner's group (Arakoni et al. 2008) found that if such discharges are sustained in argon with tens of Torr backpressure and with the discharge confined to the nozzle, thrusts from 50 to 200 µN for power depositions of 0.25–2 W can be expected.

5.8.5 Biological Applications of Microplasmas

The interaction of plasmas with biological systems generally falls under two categories. Most of the work has focused on applications such as inactivation, biodecontamination, and sterilization, where lethal plasma intensities are applied to initiate cell death. However, there is now more emphasis on nonlethal effects that allow us to control the response of biological cells. The topic of biological applications of cold plasmas, including microplasmas, has generated significant interest in the scientific community and has led to the establishment of a new scientific discipline: "plasma medicine." All of the biological applications of microplasmas depend on the use of plasma jets, plasma ejected from microdischarges, which, at the target site, has actually turned into a gas (afterglow plasma) with high concentrations of excited species and radicals. Plasma jets and their medical and biological applications are topics covered in Chapters 7 and 14. Therefore, I will only refer to two review articles that deal with biomedical applications of microplasmas (Iza et al. 2008, Kong et al. 2009).

5.9 Summary

Microcavity discharges have allowed the generation of stable plasmas at pressures up to and above 1 atmosphere. They have been operated in a DC mode, at RF and μw frequencies, and pulsed with pulse durations as short as ten nanoseconds. They have been operated in rare gases and in molecular gases and mixtures such as atmospheric pressure air. Despite running at high pressures, microdischarges are nonthermal discharges. Their electron energy distribution contains a tail of high-energy electrons, which causes high excitation and ionization rates. Electron densities exceeding 10^6 cm^{-3} have been measured in pulsed MHCDs. Gas temperatures, on the other hand, are low compared to temperatures in arcs and sparks. They can be close to room temperature at low currents in rare gases and generally do reach not more than 2000 K in atmospheric pressure air discharges. With discharge voltages of a few hundred volts, and current densities of the order of 10 A cm^{-2} (at currents of 1 mA in an MHCD with a 100 μm cathode opening), the power densities in such a discharge with typical dimensions of 100 μm can reach values of the order of 10^5 Wcm^{-3}.

The nonthermal characteristics of microdischarges together with the high power density opens up the possibility of a large number of applications, ranging from plasma microreactors to sources for microplasma jets for medical applications or as microthrusters. The possibility to arrange them in parallel allows us to generate large displays, or, if placed in series, possibly even microlasers. The nonthermal electron energy distribution in such high-pressure microdischarges favors the generation of excimers through three-body collisions. Microdischarges in rare gases and rare gas–halogen mixtures have been shown to generate intense, deep UV radiation in a CW mode.

The physics of microdischarges need to be better understood for widening the applications of plasmas. The measurements of plasma parameters, although not yet spatially resolved, in combination with computational studies, not only will allow us to expand and optimize the presently pursued applications, but likely open up pathways to new applications. Besides applications, the study of microplasmas will also provide a window into fundamental plasma physics issues at high power density in plasmas with extremely small dimensions. As noted in Becker et al. (2006), "the pursuit of microplasmas affords the possibility to explore new realms of plasma science."

References

Adler, F. and S. Mueller. 2000. Formation and decay mechanisms of excimer molecules in dielectric barrier discharges. *J. Phys. D Appl. Phys.* 33: 1705–1715.

Almeida, P.G.C., M.S. Benilov, and M.J. Faria. 2011. Three-dimensional modeling of self-organization in DC glow discharges. *IEEE Trans. Plasma Sci.* 39: 2190–2191.

Arakoni, R.A., J.J. Ewing, and M.J. Kushner. 2008. Microdischarges for use as microthrusters: Modeling and scaling. *J. Phys. D Appl. Phys.* 41: 105208 (12 pp).

Bauville, G., B. Lacour, L. Magne, V. Puech, J.P. Boeuf, E. Munoz-Serrano, and L.C. Pitchford. 2007. Singlet oxygen production in a microcathode sustained discharge. *Appl. Phys. Lett.* 90: 031501-1–031501-3.

Becker, K.H., P.F. Kurunczi, M. Moselhy, and K.H. Schoenbach. 2002a. Vacuum ultraviolet spectroscopy of microhollow cathode discharge plasmas. In *Spectroscopy of Nonequilibrium Plasmas at Elevated Pressures*, Vol. 4460, pp. 239–250, ed. V.N. Ochkin, *Proceedings of SPIE*, SPIE Press, Bellingham, WA.

Becker, K.H., P.F. Kurunczi, and K.H. Schoenbach. 2002b. Collisional and radiative processes in high-pressure discharge plasmas. *Phys. Plasmas* 9: 2399–2404.

Becker, K.H. and K.H. Schoenbach. 2008. High-pressure microdischarges. In *Low Temperature Plasmas*, Chapter 17, pp. 463–493, eds. R. Hippler, H. Kersten, M. Schmidt, and K.H. Schoenbach. Wiley-VCH, Weinheim.

Becker, K.H., K.H. Schoenbach, and J.G. Eden. 2006. Microplasmas and applications. *J. Phys. D Appl. Phys.* 39: R55–R70.

Biborosch, L.D., O. Bilwatsch, S. Ish-Shalom, E. Dewald, U. Ernst, and K. Frank. 1999. Microdischarges with plane cathodes. *Appl. Phys. Lett.* 75: 3926–3928.

Block, R., M. Laroussi, F. Leipold, and K.H. Schoenbach. 1999b. Optical diagnostics of non-thermal high pressure discharges. *Proceedings of the 14th International Symposium on Plasma Chemistry*, Vol. II, pp. 945–950. IEEE, Prague, Czech Republic.

Block, R., O. Toedter, and K.H. Schoenbach. 1999a. Gas temperature measurements in high pressure glow discharges in air. *Proceedings of the 30th AIAA Plasma Dynamics and Lasers Conference*, Norfolk, VA, Paper no. AIAA 99-3434.

Boeuf, J.P., L.C. Pitchford, and K.H. Schoenbach. 2005. Predicted properties of microhollow cathode discharges in xenon. *Appl. Phys. Lett.* 86: 070151.

Chen, J., S.J. Park, Z. Fan, J.G. Eden, and C. Liu. 2002. Development and characterization of micromachined cathode plasma display devices. *J. Microelectromech. Syst.* 11: 536–543.

Chiang, W.-H. and R.M. Sankaran. 2007. Microplasma synthesis of metal nanoparticles for gas-phase studies of catalyzed carbon nanotube growth. *Appl. Phys. Lett.* 91: 121503.

Choi, J., F. Iza, J.K. Lee, and C.-M. Ryu. 2007. Electron and ion kinetics in a DC microplasma at atmospheric pressure. *IEEE Trans. Plasma Sci.* 35: 1274–1278.

Cobine, J.D. 1958. *Gaseous Conductors: Theory and Engineering Applications*, Dover Publications, Inc., New York, pp. 218–225.

Deconinck, T., S. Mahadevan, and L.L. Raja. 2008. Simulation of a direct-current microdischarge for the micro plasma thruster. *IEEE Trans. Plasma Sci.* 36: 1200–1201.

Eden, J.G. and S.J. Park. 2005. Microcavity plasma devices and arrays: A new realm of plasma physics and photonic applications. *Plasma Phys. Control. Fusion* 47, B83–B92.

El-Habachi, A. and K.H. Schoenbach. 1998a. Emission of excimer radiation from direct-current, high-pressure hollow cathode discharges. *Appl. Phys. Lett.* 72: 22–24.

El-Habachi, A. and K.H. Schoenbach. 1998b. Generation of intense excimer radiation from high-pressure hollow cathode discharges. *Appl. Phys. Lett.* 73: 885–887.

El-Habachi, A., W. Shi, M. Moselhy, R.H. Stark, and K.H. Schoenbach. 2000. Series operation of direct current xenon chloride excimer sources. *J. Appl. Phys.* 88: 3220–3224.

Eun, C.K. and Y. Gianchandani. 2012. Microdischarge-based sensors and actuators for portable microsystems: Selective examples. *IEEE Trans. Quantum Electron.* 48: 814–826.

Frame, J.W., P.C. John, T.A. DeTemple, and J.G. Eden. 1998. Continuous wave emission in the ultraviolet from diatomic excimers in a microdischarge. *Appl. Phys. Lett.* 72: 2634–2636.

Frame, J.W., D.J. Wheeler, T.A. DeTemple, and J.G. Eden. 1997. Microdischarge devices fabricated in silicon. *Appl. Phys. Lett.* 71: 1165–1167.

Frank, K., U. Ernst, I. Petzenhauser, and W. Hartmann. 2001. Spectroscopic investigations of high-pressure microhollow cathode discharges. *Conference Record IEEE International Conference on Plasma Science*, p. 381. IEEE, Las Vegas, NV.

Garscadden, A. 2008. Atmospheric pressure glow discharges. In *Low Temperature Plasmas*, 2nd ed., Vol. 2, Chapter 15, pp. 411–437, eds. R. Hippler, H. Kersten, M. Schmidt, and K.H. Schoenbach. Wiley-VCH, Weinheim.

Gill, P. and C.E. Webb. 1977. Electron energy distribution in the negative glow and their relevance to hollow cathode lasers. *J. Phys. D Appl. Phys.* 10: 299–311.

Go, D.B. and D.A. Pohlman. 2010. A mathematical model of the modified Paschen's curve for breakdown in microscale gaps. *J. Appl. Phys.* 107: 103303.

Greenan, J., C.M.O. Mahony, D. Mariotti, and P.D. Maguire. 2011. Characterization of hollow cathode and parallel plate microplasmas: Scaling and breakdown. *Plasma Sources Sci. Technol.* 20: 025011.

Güntherschulze, A. 1930. Die behinderte Glimmentladung. *Z. Phys.* 61: 1–14.

Guo, Y.-B. and F.C.-N. Hong. 2003. Radio-frequency microdischarge arrays for large-area cold atmospheric plasma generation. *Appl. Phys. Lett.* 82: 337–339.

Hsu, D.D. and D.B. Graves. 2003. Microhollow cathode discharge stability with flow and reaction. *J. Phys. D Appl. Phys.* 36: 2898–2907.

Iza, F., G.J. Kim, S.M. Lee, J.K. Lee, J.L. Walsh, Y.T. Zhang, and M.G. Kong. 2008. Microplasmas: Sources, particle kinetics, and biomedical applications. *Plasma Process Polym.* 5: 322–344.

Jiang, C., A.-A.H. Mohamed, R.H. Stark, J.H. Yuan, and K.H. Schoenbach. 2005. Removal of volatile organic compounds in atmospheric pressure air by means of direct current glow discharges. *IEEE Trans. Plasma Sci.* 33: 1416–1425.

Kim, G.J., F. Iza, and J.K. Lee. 2006. Electron and ion kinetics in a microhollow cathode discharge. *J. Phys. D Appl. Phys.* 39: 4386–4392.

Kogelschatz, U. 2007. Applications of microplasmas and microreactor technology. *Contrib. Plasma Phys.* 47: 80–88.

Kogelschatz, U., Y.S. Akishev, K.H. Becker, E.E. Kunhardt, M. Kogoma, S. Kuo, M. Laroussi, A.P. Napartovich, S. Okazaki, and K.H. Schoenbach. 2005. DC and low frequency air plasma sources. In *Nonequilibrium Air Plasma at Atmospheric Pressure*, Chapter 6, pp. 276–361, eds. K.H. Becker, U. Kogelschatz, K.H. Schoenbach, and R.J. Barker, Institute of Physics Publishing, Bristol.

Kong, M.G., G. Groesen, G. Morfill, T. Nosenko, T. Shimizu, J. van Dijk, and J.L. Zimmermann. 2009. Plasma medicine: An introductory review. *New J. Phys.* 11: 115012 (35 pp).

Kothnur, P.S. and L.L. Raja. 2007. Simulation of direct-current microdischarges for application in electro-thermal class of small satellite propulsion devices. *Contrib. Plasma Phys.* 47: 9–18.

Kunhardt, E.E. 2000. Generation of large-volume, atmospheric pressure, nonequilibrium plasmas. *IEEE Trans. Plasma Sci.* 28: 189–200.

Kurunczi, P., N. Abramzon, M. Figus, and K. Becker. 2004. Measurement of rotational temperatures in high-pressure microhollow cathode (MHC) and Capillary Plasma Electrode (CPE) discharges. *Acta Phys. Slovaca* 54: 115–124.

Kurunczi, P., H. Shah, and K. Becker. 1999. Hydrogen Lyman-α and Lyman-β emissions from high-pressure microhollow cathode discharges in Ne-H2 mixtures. *J. Phys. B At. Mol. Opt. Phys.* 32: L651–L658.

Kushner, M.J. 2004. Modeling of microdischarge devices: Pyramidal structures. *J. Appl. Phys.* 95: 846–849.

Kushner, M.J. 2005. Modelling of microdischarge devices: Plasma and gas dynamics. *J. Phys. D Appl. Phys.* 38: 1633–1643.

Leipold, F., R.H. Stark, A. El-Habachi, and K.H. Schoenbach. 2000. Electron density measurements in an atmospheric pressure air plasma by means of infrared heterodyne interferometry. *J. Phys. D Appl. Phys.* 33: 2268–2273.

Liebermann, M.A. and A.J. Lichtenberg. 1994. *Principles of Plasma Discharges and Materials Processing*. John Wiley and Sons, Inc., New York.

Mariotti, D. and R.M. Sankaran. 2010. Microplasmas for nanoparticle synthesis. *J. Phys. D Appl. Phys.* 43: 323001 (21 pp).

Miclea, M. and J. Franzke. 2007. Analytical detectors based on microplasma spectrometry. *Plasma Chem. Plasma Process* 27: 205–224.

Miclea, M., K. Kunze, J. Franzke, and K. Niemax. 2002. Plasmas for lab-on-the-chip applications. *Spectrochim. Acta* B57: 1585–1592.

Miclea, M., K. Kunze, U. Heitmann, S. Florek, J. Franzke, and K. Niemax. 2005. Diagnostics and application of the microhollow cathode discharge as an analytical plasma. *J. Phys. D Appl. Phys.* 38: 1709–1715.

Mohamed, A.-A.H., R. Block, and K.H. Schoenbach. 2002. Direct current glow discharges in atmospheric air. *IEEE Trans. Plasma Sci.* 30: 182–163.

Moselhy, M., A. El-Habachi, and K.H. Schoenbach. 1999. Pulsed operation of microhollow cathode excimer sources. *Bull. Am. Phys. Soc.* 44: 29.

Moselhy, M., I. Petzenhauser, K. Frank, and K.H. Schoenbach. 2003. Excimer emission from microhollow cathode argon discharges. *J. Phys. D Appl. Phys.* 36: 2922–2927.

Moselhy, M. and K.H. Schoenbach. 2003. Nanosecond pulse generators for microdischarge excimer lamps. Digest of technical papers. *14th IEEE International Pulsed Power Conference*, p. 1317. IEEE, Dallas, TX.

Moselhy, M. and K.H. Schoenbach. 2004. Excimer emission from cathode boundary layer discharges. *J. Appl. Phys.* 95: 1642–1649.

Moselhy, M., W. Shi, R.H. Stark, and K.H. Schoenbach. 2001a. Xenon excimer emission from pulsed microhollow cathode discharges. *Appl. Phys. Lett.* 79: 1240–1242.

Moselhy, M., W. Shi, R.H. Stark, and K.H. Schoenbach. 2002. A flat glow discharge excimer radiation source. *IEEE Trans. Plasma Sci.* 30: 198–199.

Moselhy, M., R.H. Stark, K.H. Schoenbach, and U. Kogelschatz. 2001b. Resonant energy transfer from argon dimers to atomic oxygen in microhollow cathode discharges. *Appl. Phys. Lett.* 78: 880–882.

Ono, T., D.Y. Sim, and M. Esachi. 2000. Micro-discharge and electric breakdown in a micro-gap. *J. Micromech. Microeng.* 10: 445–451.

Park, S.-J., J. Chen, C. Liu, and J.G. Eden. 2001. Silicon microdischarge devices having inverted pyramidal cathodes: Fabrication and performance of arrays. *Appl. Phys. Lett.* 78: 419–421.

Park, S.-J., J. Chen, C.J. Wagner, N.P. Ostrom, C. Liu, and J.G. Eden. 2002. Microdischarge arrays: A new family of photonic devices. *IEEE J. Sel. Top. Quantum Electron.* 8: 139–147.

Penache, C., A. Braeuning-Demian, L. Spielberger, and H. Schmidt-Boecking. 2000. Experimental study of high pressure glow discharges based on MSE arrays. *Proceedings of Hakone VII*, Vol. 2, pp. 501–505. Johann Wolfgang Goethe Universität, Greifswald, Germany.

Penache, C., A. Braeuning-Demian, O. Hohn, S. Schoessler, T. Jahnke, K. Niemax, and H. Schmidt-Boecking. 2002b. Characterization of high-pressure microdischarge using laser-diode atomic absorption spectroscopy. *Plasma Sources Sci. Technol.* 11: 476–483.

Penache, C., C. Gessner, A. Bräuning-Demian, P. Scheffler, L. Spielberger, O. Hohn, S. Schössler, T. Jahnke, K.-H. Gericke, and H. Schmidt-Böcking. 2002a. Micro-structured electrode arrays: A source of high-pressure non-thermal plasma. In *Spectroscopy of Nonequilibrium Plasma at Elevated Pressures*, Vol. 4460, pp. 17–25, ed. V.N. Ochkin. *Proceedings of SPIE*, SPIE Press, Bellingham, WA.

Qiu, H., K. Martus, W.Y. Lee, and K. Becker. 2004. Hydrogen generation in a microhollow cathode discharge in high-pressure ammonia-argon gas mixtures. *Int. J. Mass Spectrom.* 233: 19–24.

Rahul, R., O. Stan, A. Rahman, E. Littlefield, K. Hoshimiya, A.P. Yalin, A. Sharma, et al. 2005. Optical and RF electrical characteristics of atmospheric pressure open-air hollow slot microplasmas and application to bacterial inactivation. *J. Phys. D Appl. Phys.* 38: 1750–1759.

Raizer, Y.P. 1997. *Gas Discharge Physics.* Springer-Verlag, Berlin, 2nd printing.

Readle, J.D., K.E. Tobin, K.S. Kim, J.K. Yoon, J. Zheng, S.K. Lee, S.-J. Park, and J.G. Eden. 2009. Flexible, light weight arrays of microcavity plasma devices: Control of cavity in thin substrates. *IEEE Trans. Plasma Sci.* 37: 1045–1053.

Roth, J.R. 2001. *Industrial Plasma Engineering*, Vol. 2. IOP Publishing, Bristol.

Sankaran, R.M. and K.P. Giapis. 2001. Maskless etching of silicon using patterned microdischarges. *Appl. Phys. Lett.* 79: 593–595.

Sankaran, R.M. and K.P. Giapis. 2002. Hollow cathode sustained plasma microjets: Characterization and application to diamond deposition. *J. Appl. Phys.* 92: 2406–2411.

Sankaran, R.M. and K.P. Giapis. 2008. Materials applications of high-pressure microplasmas. In *Low Temperature Plasmas—Fundamentals, Technologies and Techniques*, Vol. 2, Chapter 18, pp. 495–524, eds. R. Hippler, H. Kersten, M. Schmidt, and K.H. Schoenbach. Wiley-VCH, Verlag.

Schoenbach, K.H., A. El-Habachi, M.M. Moselhy, W. Shi, and R.H. Stark. 2000. Microhollow cathode discharge excimer lamps. *Phys. Plasmas* 7: 2186–2191.

Schoenbach, K.H., A. El-Habachi, W. Shi, and M. Ciocca. 1997. High-pressure hollow cathode discharges. *Plasma Sources Sci. Technol.* 6: 468–477.

Schoenbach, K.H., M. Moselhy, and W. Shi. 2004. Self-organization in cathode boundary layer microdischarges. *Plasma Sources Sci. Technol.* 13: 177–185.

Schoenbach, K.H., M. Moselhy, W. Shi, and R. Bentley. 2003. Microhollow cathode discharges. *J. Vac. Sci. Technol. A* 21: 1260–1265.

Schoenbach, K.H., R. Verhappen, T. Tessnow, F.E. Peterkin, and W.W. Byszewsi. 1996. Microhollow cathode discharges. *Appl. Phys. Lett.* 68: 13–15.

Schoenbach, K.H. and W.D. Zhu. 2012. High-pressure microdischarges—Sources of ultraviolet radiation. *IEEE Trans. Quantum Electron* 48: 768–782.

Schrader, C., L. Bars-Hippe, K.-H. Gericke, E.M. van Veldhuizen, N. Lucas, P. Sichler, and S. Büttgenbach. 2006. Micro-Structured electrode array: Glow discharges in Ar and N₂ at atmospheric pressure using a variable radio-frequency generator. *Vacuum* 80: 1144–1148.

Shi, W., R.H. Stark, and K.H. Schoenbach. 1999. Parallel operation of microhollow cathode discharges. *IEEE Trans. Plasma Sci.* 27: 16–17.

Shin, J. and M.T. Rahman. 2011. Formation of microhollow cathode sustained discharge with split third electrodes. *IEEE Trans. Plasma Sci.* 39: 2676–2677.

Slade, G.P. and E.D. Taylor. 2002. Electrical breakdown in atmospheric pressure air between closely spaced (0.2 μm–40 μm) electrical contacts. *IEEE Trans. Compon. Packaging Technol.* 25: 390–396.

Stark, R.H., A. El-Habachi, and K.H. Schoenbach. 2000. Parallel operation of microhollow cathode discharges. *Conference Record IEEE International Conference on Plasma Science*, p. 111. IEEE, New Orleans, LA, Paper no. 1 P24.

Stark, R.H. and K.H. Schoenbach. 1999. Direct current high-pressure glow discharge. *J. Appl. Phys.* 85: 2075–2079.

Stark, R.H. and K.H. Schoenbach. 2001. Electron heating in atmospheric pressure glow discharges. *J. Appl. Phys.* 89: 3568–3572.

Tachibana, K., Y. Kishimoto, S. Kawai, T. Sakaguchi, and O. Sakai. 2005. Diagnostics of microdischarge-integrated plasma sources for displays and materials processing. *Plasma Phys. Control. Fusion* 47: A167–A177.

Takano, N. and K.H. Schoenbach. 2006. Self-organization in cathode boundary layer discharges in xenon. *Plasma Sources Sci. Technol.* 15: S109–S117.

Tirumala, R. and D.B. Go. 2010. An analytical formulation for the modified Paschen's curve. *Appl. Phys. Lett.* 97: 151502.

Vojak, B.A., S.-J. Park, C.J. Wagner, J.G. Eden, R. Koripella, J. Burdon, F. Zenhausern, and D.L. Wilcox. 2001. Multistage, monolithic ceramic microdischarge device having an active length of ~0.27 mm. *Appl. Phys. Lett.* 78: 1340–1342.

von Allmen, P., S.T. McCain, N.P. Ostrom, B.A. Vojak, J.G. Eden, F. Zenhausern, C. Jensen, and M. Oliver. 2003a. Ceramic microdischarge arrays with individually ballasted pixels. *Appl. Phys. Lett.* 82: 2562–2564.

von Allmen, P., D.J. Sadler, C. Jensen, N.P. Ostrom, S.T. McCain, B.A. Vojak, and J.G. Eden. 2003b. Linear, segmented microdischarge array with an active length of ~1 cm: Cw and pulsed operation in the rare gases and evidence of gain on the 460.30 nm transition of Xe⁺. *Appl. Phys. Lett.* 82: 4447–4449.

von Engel, A. 1965. *Ionized Gases*, American Vacuum Society Classics, American Institute of Physics, New York, reprint by arrangement with Oxford University Press.

White, A.D. 1959. New hollow cathode glow discharge. *J. Appl. Phys.* 30: 711–719.

Yalin, A.P., Z.Q. Yu, O. Stan, K. Hoshimiya, A. Rahman, V.K. Surla, and G.J. Collins. 2003. Electrical and optical emission characteristics of radio-frequency-driven hollow slot microplasmas operating in open air. *Appl. Phys. Lett.* 83: 2766–2768.

Yokoyama, T., S. Hamada, S. Ibuka. K. Yasuoka, and S. Ishii. 2005. Atmospheric DC discharges with miniature gas flow as microplasma generation method. *J. Phys. D Appl. Phys.* 38: 1684–1689.

Yu, Z., K. Hoshimiya, J.D. Williams, S.F. Polvinen, and G.J. Collins. 2003. Radio-frequency-driven near atmospheric pressure microplasma in a hollow slot electrode configuration. *Appl. Phys. Lett.* 83: 854–856.

Zhang, W., T.S. Fisher, and S.V. Garimella. 2004. Simulation of ion generation and breakdown in atmospheric air. *J. Appl. Phys.* 96: 6066–6072.

Zhu, W., N. Takano, K.H. Schoenbach, D. Guru, J. McLaren, J. Heberlein, R. May, and J.R. Cooper. 2007. Direct current planar excimer source. *J. Phys. D Appl. Phys.* 40: 3896–3906.

6

Atmospheric Pressure Nanosecond Pulsed Discharge Plasmas

Zdenko Machala,
David Z. Pai,
Mario Janda, and
Christophe O. Laux

6.1 Introduction

Nonthermal atmospheric pressure plasma can induce chemical reactions in various gas mixtures [1–3] without significantly heating them, thus keeping energy consumption relatively low. The desired chemical effect is achieved by efficient production of reactive species resulting from the collision of high-energy electrons with neutral species. To break chemical bonds, the mean energy of electrons should be at least around 1 eV and their density high enough to ensure efficient conversion of reactants to products. These conditions can be met without heating the neutral species by using an external electric field.

Electrons easily gain sufficient energy from the applied electric field, because they lose only a small fraction of it in elastic collision with neutral species. Once they have enough energy, their density starts growing due to electron impact ionization processes that resemble the development of an avalanche. At atmospheric pressure, an electron avalanche transforms into a streamer once the space charge exceeds a critical value of 10^8 electrons, which is the so-called Meek's criterion [4].

A streamer is an ionization wave that propagates the region with a strong field from a high voltage (HV) electrode toward the grounded electrode. Because of space charge, the electric field in the streamer's head can reach more than 200 kV cm^{-1} [5,6], so that the chemical and ionization processes are very efficient there. Streamers are thus considered to be crucial for the efficiency of plasma-induced chemistry at atmospheric pressure.

However, a streamer may also lead to an electrical breakdown if it reaches the opposite electrode and creates a conductive plasma bridge between the electrodes. The electrical breakdown is a crucial issue for the generation of electrical discharges and the design of HV devices. The first studies of this phenomenon appeared decades ago [7–9] and the streamer breakdown theory was introduced by Meek, Raether, and Loeb [4,10,11] in the middle of the twentieth century. However, because of the complexity of this problem, the study of the breakdown mechanism, streamers, and their propagation still continues [5,6,12–18].

Electrical breakdown typically leads to the formation of spark or arc discharge, depending on the power supply and the external electric circuit parameters. Arc and spark usually generate thermal plasma, where the temperature of all species is equilibrated and the gas is heated to a very high temperature. For many applications, very high gas temperature and waste of energy for heating gas are not desirable. Several strategies were therefore developed to avoid the breakdown and streamer-to-spark transition. Two of the most common ways are based on covering at least one electrode by an insulator (the concept of dielectric barrier discharges [2,19,20] or by using very short HV pulses [3,21–26]), which will be discussed in detail in Section 6.4 that is dedicated to nanosecond repetitively pulsed (NRP) discharges. It will be shown that NRP discharges have many specific features different from just nanosecond discharges repeated more often, especially the cumulative effects of the energy and chemical species, and change of the streamer breakdown mechanism.

Restriction of thermal plasma generation can be also achieved with a simple direct current (DC) power supply by adding a ballast resistor ($R > 1$ MΩ) limiting the discharge current. Several types of electrical discharges producing nonthermal plasma can be generated in this way, depending on the electric circuit parameters, gas composition, flow rate, and geometry of electrodes [27–30]. Despite the DC power supply used, some of these discharges have a pulsed character, such as prevented spark developed by Marode et al. [27,31–34], or its more general case transient spark (TS) described in Section 6.3. In these discharges, the streamer-to-spark transition is not stopped like in the pulsed discharges generated by the pulsed HV power supplies, but the spark current peak is limited by an additional resistor between the discharging external capacity and the gap. In this way it is possible to prevent the thermalization of the plasma during the spark phase, and even to enhance the chemical effect because of the so-called secondary streamer [33].

6.2 Mechanisms of Filamentary Plasma Growth in Air at Atmospheric Pressure

At atmospheric pressure in air, the ionization growth of plasma discharges develops generally under two basic forms. One is more diffuse and continuous in time, and is often defined as a "glow" [not to be confused with what is called "glow discharge" (GD), characterizing the structure of usually low-pressure discharges typical with a cathode fall]. In a nonuniform discharge gap typically for corona discharge, a "glow" often surrounds the highly stressed electrode with a low radius of curvature (e.g., point). This discharge form has been understood as a self-sustained discharge in the "Townsend" regime, where each electron avalanche, developing under the applied electrical field with little influence from the space charge, creates another avalanche initiated in the surrounding space by one secondary electron.

The other form is much more concentrated and develops filamentary plasmas which may extend far away from the electrodes. It appears when the avalanche reaches a critical threshold of the number of electrons where the head of the avalanche suddenly extends in filamentary forms in both directions of the field (this threshold is often considered as being of around some 10^8 electrons). In such a case, steep moving ionizing fronts are set up, which are referred to as cathode- or anode-directed streamers. They also need secondary electrons to be self-sustained, but the continuous ionization growth inside the front avalanches are under the space-charge field left by successive avalanches within the streamer front, which predominates over the geometrical field set by the electrodes.

These two basic discharge forms may exist simultaneously or sequentially, which leads to the development of new structures and a large number of visual discharge aspects [33].

6.2.1 The Positive Point-to-Plane Discharge

In positive point-to-plane 1 cm gap, the filamentary structure appears as shown in the still photograph given in Figure 6.1. This discharge is obtained with a coaxial HV pulser delivering a pulse with a short rise time, and a long plateau decreasing in the millisecond range. After applying this HV pulse it can be observed that the current is composed of many pulses appearing in sequence. The streak photograph in Figure 6.1 of this discharge shows that the mean streamer crossing time is on the order of 26 ns. The streak consists of very quick displacement of the still image of the discharge on the film from the left to the right, the temporal axis also being the radial axis of the discharge, so that the time sequence must be corrected by this superimposed space and time axis [33].

If a continuous positive DC voltage is applied instead of a pulsed voltage, the first current pulse is followed by many others with a variable period between them from 20 to about 100 μs. All current pulses after the first one are much smaller in size. Indeed, the visual shape of the discharges associated with these smaller pulses present in only one streamer filament following the discharge gap axis, so that the current is much smaller. More precisely, the next discharge after the first one may contain two filaments, but after this second pulse only one. This means that the preceding discharges leave some memory inside the gas, which gives them the axial path as a preferred one. The physical reason is not completely elucidated, but it is probable that it is due to the temperature imprint inside the gas [33]. More details may be found in the works of Marode [12] and Bessiers et al. [35].

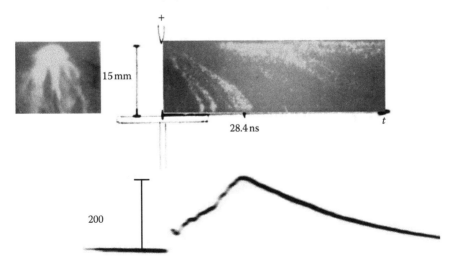

FIGURE 6.1 Still photograph and streak recordings of the positive point-to-plane discharge. (From Marode, E., *Plasma for Environmental Issues.* eds. E. Tatarova, V. Guerra, and E. Benova, ArtGraf, Sofia, 2012.)

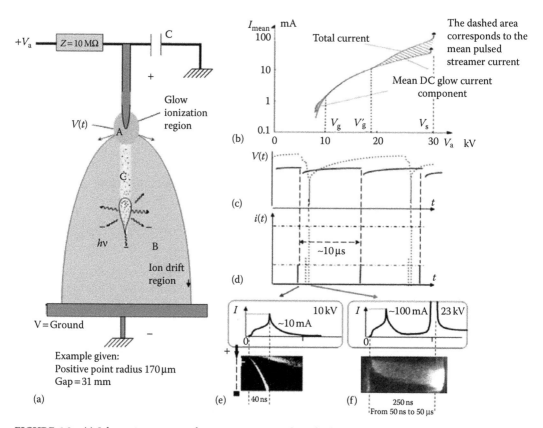

FIGURE 6.2 (a) Schematic structure of a positive point-to-plane discharge in air at atmospheric pressure during the streamer propagation toward the negative plane; (b) mean current as a function of applied voltage I_{mean} (V_a) plot; (c) time evolution of the voltage $V(t)$; (d) time evolution of the current $i(t)$; (e) enlarged current pulse with the associated streak photograph of the streamer and (f) streamer-to-spark transition. (From Marode, E., *Plasma for Environmental Issues*. eds. E. Tatarova, V. Guerra, and E. Benova, ArtGraf, Sofia, 2012.)

In spite of the influence of the preceding discharges, the single streamer filaments obtained with the DC potential may be taken as representative of the filamentary discharge structure. The fact that the DC discharge follows the discharge axis is obviously a considerable advantage for studying the properties of this discharge under DC potential, which are summarized in Figure 6.2. The schematic structure of a positive point-to-plane discharge at a given time during the streamer propagation toward the negative plane is shown in Figure 6.2a. The superposition of the two discharge forms already introduced, the glow in ionizing region around point (A), and the filamentary plasma discharge (C) produced as the trail of the ionizing streamer fronts at the end of the filament are sketched in Figure 6.2a. The secondary electrons produced by the photo-ionizing radiations from avalanches inside the streamer front sustain the propagation of the streamer.

The whole structure of the streamer filament is similar to a low-pressure GD (again not to be confused with the "glow" around the point which is also a discharge but not space charge dominated). The streamer is similar to the cathode fall, using the gas as a source of electrons instead of the cathode, while the filament corresponds to the positive column. The positive ions, produced by the continuous glow and the repetitive streamer, drift slowly toward the cathode (B). Figure 6.2b–d comprises plots and sketches, as an example of measured voltages and currents reported in Goldman and Goldman [36]. In the $I_{mean}(V_a)$ plot (b), the lower curve is the continuous current uniquely due to the glow, and it increases with the applied potential. The upper curve is the total measured mean current I_{mean}. The difference between these two curves corresponds to the mean time-integrated streamer filament current pulses. As seen, the glow current is not negligible but the visual appearance may be misleading, because the ordinate is logarithmic. The streamer current may largely dominate at large applied potentials between 20 and 30 kV. Only a continuous current without any streamers exists in between two specific potentials V_g and V_g'. These $I(V)$ curves are similar for a large range of point radii and gaps. Here, above 30 kV, the spark breakdown occurs.

Sketches of the time evolution of the voltage $V(t)$ in Figure 6.2c and the currents $i(t)$ in Figure 6.2d for two different applied voltages V_a are shown. Let us take an applied potential V_a below the sparking potential V_s. A regular sequence of streamer filament development appears, inducing small regular repetitive decrease $\Delta V(t)$ on the gap potential, for instance each 30 μs, as seen on the blue plain curve in (c), with associated current pulses seen as plain current spikes in (d). The $\Delta V(t)$ on the sketch are increased a lot, while they are negligible compared to the applied voltage. For $V_a = 28$ kV, a capacity $C = 20$ pF and a streamer pulse of 10 mA, $\Delta V(t) \sim 25$ V. Figure 6.2e shows the enlarged current pulse with the associated streak photograph of the discharge. If now $V_a > V_s$, then the gap voltage shows sharp decreases of $\Delta V(t)$ as seen on (c) on the dotted potential curve, reaching practically zero potential, while the associated discharge current (d) shows two spikes: the streamer filament spike followed by the strong TS spike (dotted lines). It is clearly seen that the streamer filament system triggers the spark formation.

It is now easy to understand the two streamer groups. Indeed, if the applied HV pulse is between V_g' and V_s then the decrease of the HV plateau will sweep along the voltage axis shown in Figure 6.2b, and no streamer will be launched during the crossing of the $(V_g' - V_g)$ region.

Enlarging Figure 6.2f, we see that some already known properties of the streamer-to-spark transition can be summarized [12,37,38], and the different plasma regions of this transition which are used for various applications can be identified. The current pulse shape in the figure is time synchronized with the various sequences of the discharge development. The slow current increase at the beginning corresponds to the streamer propagation. Within the streamer front, while the electrons probably do not have a Maxwellian distribution, their energy may be taken as equivalent to the electron temperature T_e between 10 and 30 eV (with 1 eV = 7735 K in the $\varepsilon = 3/2 \, kT_e$ definition), while the neutral gas temperature T_g is slightly increased to some 380 K. Streamer arrival at the cathode plane is indicated by a sharp current increase. The following current decrease is due to electron attachment along the filamentary plasma left behind by the streamer ionizing front. The streak photograph also shows that the light near the positive point, called the secondary streamer, decreases in intensity (this light is actually improperly designated as "secondary streamer," because it is not an ionizing region, but a plateau of high-field region [38]. The electron temperature is more pertinent here, because the electrons should be thermalized and have a Maxwellian distribution function, and $T_e \sim 1.4$ eV with $T_g = 400$ K. If the applied voltage is under V_a, this decrease falls down and the discharge stops until complete clearing of the ions, either due to ion recombination or to their collection by the electrodes. Above V_a, the current decrease slows down, reaches a minimum and restarts again, along with the light from the secondary streamer. The interpretation of this restart is that there is a decrease of the neutral density N due to hydrodynamic expansion, and a corresponding increase of the reduced field E/N, which restarts the electron ionization, and leads to spark formation. Playing with the applied potential shapes and amplitudes (DC, pulsed, or alternative), with the electrode nature (metallic or dielectric), and with the geometrical shapes of the electrodes, a specific part of the plasma state during this transition may be intensified dominantly for specific applications. The prevented spark regime is already indicated in the figure with dotted lines [33].

6.2.2 Theory of Streamer-to-Spark Transition (in Air)

The ionization processes are very efficient in the streamer's head. However, the electric field behind this head is weak and the produced plasma is in decay phase. The electron density is initially decreased by a factor of 10 through dissociative recombination of electrons with O_4^+ ions [16]. Afterward, about 100 ns later, the three-body attachment on O_2 becomes the most important electron sink mechanism. The O_4^+ ions are built up according to the following scheme:

$$N_2 \xrightarrow{e} N_2^+ \xrightarrow{N_2+M} N_4^+ \xrightarrow{O_2} O_2^+ \xrightarrow{O_2+M} O_4^+. \tag{6.1}$$

This conversion is fast (within a few nanoseconds) and it is necessary for decrease of electron density, because the rate coefficient of electron recombination with O_4^+ is about 2 orders of magnitude higher than rate coefficients of reactions with O_2^+ and N_2^+ ions [39,40]. There must be another process that balances the electron loss processes and finally enables the breakdown by accelerating the ionization processes. According to Marode [33], there are three possible mechanisms.

6.2.2.1 Gas Density Decrease

This mechanism is based on the heating of gas in the plasma channel [41]. The increase of the gas temperature T_g in the channel leads to steep increase of the pressure. This is followed by a hydrodynamic channel expansion that empties the core of the channel, and so, decreases the gas density N. For this reason, a reduced electric field strength E/N in the plasma channel increases and accelerates the electron impact ionization reactions and spark formation follows [33].

6.2.2.2 Chemical and Stepwise Ionization

The accumulation of excited metastable species can lower the threshold electron-impact ionization energy (stepwise ionization) [15]. Lowke [13] stressed the role of molecular oxygen metastable species $O_2(a^1\Delta_g)$. Their collisions with O_2^- ions can induce electron detachment even at room temperature [33]:

$$O_2\left({}^1\Sigma_g^+\right) + O_2^- \rightarrow 2O_2 + e \quad k = 2 \times 10^{-10}\,\text{cm}^3\text{s}^{-1}. \tag{6.2}$$

6.2.2.3 Attachment Control Processes

The distribution of the attachment rate along the plasma filament produced by the streamer produces an increase of the reduced electric field E/N near the anode [38,42,43]. At least 80 Td prevails here and the plasma-induced chemistry may be efficiently activated [44]. This region was also called a secondary streamer by Loeb, who supposed that it was a new ionization wave [10]. A decrease of gas density leads to the extension of this region that finally reaches the cathode, as was shown in a model by Bastien and Marode [43]. Then the spark may start because E/N becomes high enough along the whole plasma filament [33].

The increase of T_g may also have a direct effect on the gas-phase chemistry [38]. The rate coefficients usually depend strongly on T_g. Elementary processes causing the loss of electrons include electron–ion recombination reactions and electron attachment reactions producing negative molecular ions with much lower mobility. The regeneration of electrons from these ions by thermal detachment is actually considered by many authors as the most important mechanism responsible for maintaining the conductivity of the streamer for a period of 1–10 μs [45–48]. To overcome the electron affinity, the gas in the channel must be heated above a critical value of 1000–2000 K so that the rate of these thermal detachment reactions is large enough. Moreover, the rate coefficients of electron–ion recombination reactions also decrease with the increasing gas temperature [39,40].

6.3 Transient Spark

In this section we will focus on a filamentary self-pulsed discharge named the TS. TS is initiated by a streamer, followed by a short (~10–100 ns) high current (~1–10 A) pulse. A transition to a typical arc or a spark discharge is inhibited by the components of the electric circuit: a large external resistor $R \sim 5$–10 MΩ and a small internal capacity $C \sim 10$–40 pF being discharged. Unlike in the case of the prevented spark proposed by Marode [32,33], the current peak is not necessarily limited in its amplitude by an additional resistor but is always limited in its duration.

The process of periodical charging and discharging of C repeats with a characteristic frequency f of a few kilohertz and it can be controlled by the applied DC voltage. HV pulsed devices working at repetition frequencies above 1 kHz appeared only recently [24–26,49], and the importance

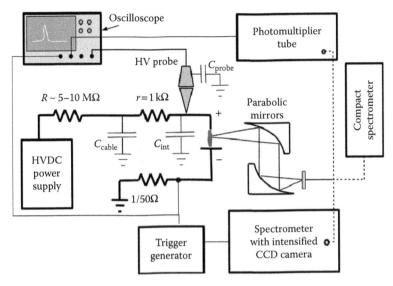

FIGURE 6.3 Schematic of the experimental setup: HV, high voltage; R and r, resistors; C_{cable}, C_{probe}, and C_{int}, capacities. (From Janda, M. et al., *Plasma Sources Sci. Technol.*, 21, 045006, 2012, with permission of IOP Publishing Ltd.)

of the high repetition frequency on the power efficiency of plasma generation was emphasized [25,26,49]. High repetition frequency of TS can certainly play an important role in its chemical efficiency. However, we observed changes of some electrical and optical characteristics of TS with increasing f [50,51]: smaller current pulses and a disappearance of atomic lines.

To understand changes of TS characteristics with f, we performed a time-resolved optical emission spectroscopic (OES) study, using a fast photomultiplier tube, as well as a fast iCCD camera coupled with a spectrometer. The schematic of the experimental setup is shown in Figure 6.3. The typical visual appearance of TS is shown in Figure 6.4. The OES technique can provide valuable information on excited atomic and molecular states. It enables determining the rotational, vibrational, and electronic excitation temperatures of the plasma, and thus the level of nonequilibrium and gas temperature [52,53].

Our results presented in this section are also interesting for deeper understanding of the streamer-to-spark transition process, which is not fully explained yet. TS enables studying two interesting features—the influence of a large external resistor and the influence of the increasing repetition frequency on the spark breakdown. The influence of a large external resistor R on the breakdown was already experimentally studied by Larsson [14].

The major influence of R was found to be the extension of streamer-to-spark transition delay time. This was later theoretically approved by Naidis [15]. However, Larsson [14] used a pulsed voltage generator with low repetition frequency, so that breakdowns did not influence each other, that is, each breakdown occurred in "virgin" gas. Moreover, his applied voltage typically highly exceeded the threshold breakdown voltage. In TS, the breakdown occurs when the applied voltage reaches the threshold breakdown value.

6.3.1 Evolution of the TS

When an HV V_a applied to the needle electrode from the generator is progressively increased, a glow corona develops into a streamer. As the voltage V achieves a breakdown value V_{TS}, a transition to TS may occur. This happens when a streamer crosses the whole gap and creates a relatively conductive plasma bridge, through which the capacity C of the circuit can be discharged. This capacity is composed of several components—internal capacity of the discharge chamber C_{int}, capacity of the HV probe C_{probe}

FIGURE 6.4 Photograph of TS in positive needle–plane gap of 4 mm, $f = 2$ kHz, $R = 6.6$ MΩ, and exposure 0.05 s. (From Janda, M. et al., *Plasma Sources Sci. Technol.*, 20, 035015, 2011, with permission of IOP Publishing Ltd.)

(when this probe is used for HV measurement) and the capacity of the HV cable C_{cable} connecting the needle electrode with the ballast resistor R (Figure 6.3).

When C is discharged, the current is given approximately by

$$I(t) \approx -C\frac{\mathrm{d}V}{\mathrm{d}t} \tag{6.3}$$

reaches a high value (~1 A) and the voltage V drops to zero (Figure 6.5). Then, during the quenched phase, V grows in time t and C gets recharged, according to the following equation:

$$V(t) = V_a\left[1 - \exp\left(\frac{-t}{RC}\right)\right], \tag{6.4}$$

where

V_a is the generator voltage

As soon as V reaches the characteristic TS breakdown voltage V_{TS}, a new TS pulse appears. It occurs in time $t = T$, from which we get the characteristic repetition frequency f of this process:

$$f = \frac{1}{T} = \frac{1}{RC\ln\left[\dfrac{V_a}{V_a - V_{TS}}\right]}. \tag{6.5}$$

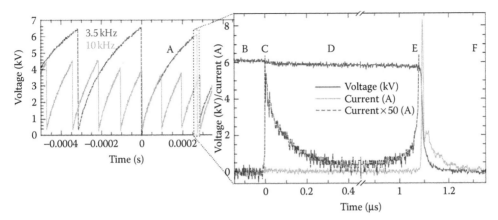

FIGURE 6.5 Typical waveforms of TS on (a) ms timescale, (b) ns timescale, $R = 6.5$ MΩ, $f \approx 2$ kHz, $r = 0$ Ω, $C = 43 \pm 4$ pF, $d = 5$ mm, and TS pulse parameters: FWHM ≈ 12 ns, rise and decay time ~5 ns. A, recharging phase; B, prebreakdown phase (corona and streamers); C, streamer; D, streamer-to-spark transition phase; E, spark pulse; F, discharged phase. (From Janda, M. et al., *Plasma Sources Sci. Technol.*, 21, 045006, 2012, with permission of IOP Publishing Ltd.)

For typical R and C, the repetition frequency f is on the order of several kilohertz, and it can be controlled by the generator voltage V_a. The control of TS by other external circuit parameters was described in detail recently [51]. The typical voltage and current waveforms and TS phases are shown in Figure 6.5.

6.3.2 TS Control by Electric Circuit Parameters

The RC term determines which frequencies can be achieved and how fast f grows with V_a, but R and C also influence other TS properties. The major role of R is to avoid the transition to arc or the pulseless GD, because increase in f is accompanied by increase of mean current I_{mean}.

As I_{mean} exceeds approximately 1.5 mA, TS tends to transform into a pulseless GD regime with a constant current up to 2 mA [phase (F), Figure 6.5b]. However, due to the high value of R and the electronegativity of air, this regime is not stable and the discharge randomly switches between the glow regime and the high frequency TS regime. Typically, R should be above 5 MΩ to avoid transition to a stable GD. GD is a continuous discharge, previously also named "high-pressure GD" (HPGD) and the continuous current may be in the range of 2–10 mA or even more. This discharge regime has all common parts of the low-pressure GD [28,54,55] and its plasma remains out of thermal equilibrium at low currents and approaches equilibrium at high currents (>100 mA). On the other hand, if $R > 10$ MΩ, the energy losses on the external resistor become too high for operating at high f and very high I_{mean}.

The influence of C on TS is even more significant. Because TS is based on charging and discharging of C, total charge and the energy delivered to the discharge gap per pulse are functions of C. We can therefore affect the shape of TS current pulses by changing C. Larger C typically means larger current pulses.

The question is how to control C? It consists of several components, from which we can easily change only the value of C_{cable} by changing the cable length. Longer cable means larger C_{cable}. However, it is not very practical to change C using different HV cables with various lengths. So we prefer using long cables ($C_{cable} > 20$ pF) and place a small separating resistor r between the HV electrode and the HV cable. This r separates C_{cable} from $C_{int} + C_{probe}$. This allows us to control the shape of the current pulse by changing r, without changing C_{cable}.

The circuit shown in Figure 6.3 with a large resistor R and a small resistor r between the HV cable and the gap, as well as a capacity C_{cable} between them and C_{int} of the discharge chamber practically represents the configuration of the prevented spark, proposed by Marode [37]. The idea of the prevented spark is that the stored energy inside C can only be delivered through the resistance r, so that the maximum current will be $I_{max} = V/r$; for example, for 10 kV and 5 kΩ, the maximum current is limited to $I_{max} = 200$ mA. Then, when a potential V_a is applied it charges not only C_{int} but also C_{ext} (in our circuit of Figure 6.3 equal to C_{cable}). Thus, when a discharge occurs at the applied potential above V_{TS}, the current evolution will depend on the values of the couple (C_{int}, C_{ext}). The sustained "prevented spark plasma" is created, whose properties may be changed by defining the cutoff current maximum with the value of r, and the duration with the value of rC_{ext}. The values of r and C_{ext} can be set so as to limit the amplitude and the length of the current pulse and thereby control the plasma properties. TS presented here is a more general case, where r can be $r = 0$; thus, $C = C_{int} + C_{ext}$ and the current pulse amplitude is only limited by the energy stored in this total C. It also comprises the case when $r \neq 0$ and the current amplitude can be controlled by C_{ext} (prevented spark case).

6.3.3 The Influence of Separating Resistance r

We expected that the increasing value of r would lead to a gradual decrease in I_{max}. This tendency was experimentally confirmed (Figure 6.6), but the dependence of the pulse shape and width on r was not the expected monotonic increase. We found that full width at half maximum (FWHM) is not a proper parameter to describe TS pulses, because they are obviously formed by a superposition of two independent current pulses (Figure 6.7). The first one (I_{C1}) is due to discharging of the capacitance $C_1 = C_{int} + C_{probe}$ in the gap. The second one (I_0) is due to discharging of the HV cable through the resistor r.

These two currents, I_{C1} and I_0, can explain the dependences of I_{max} and pulse shape on r. While I_{C1} does not change significantly with r (Figure 6.6, first peak), I_0 becomes smaller (Figure 6.6, second peak) and broader. As a result, I_{max} also decreases with r, but when the maximum of I_0 becomes small compared with the maximum of I_{C1}, I_{max} remains almost constant for a further increase in r (Figure 6.6, $r > 10$ kΩ). Here we obtained I_0 from I after subtraction of I_{C1}. C_1 necessary to calculate I_{C1} was obtained by fitting measured current waveforms by Equation 6.6 for large r.

We expected that I_0 can also be derived from the measured potential drop on r (ΔV_r):

$$I(t) \approx -C_1 \frac{dV}{dt} + \frac{\Delta V_r}{r}. \tag{6.6}$$

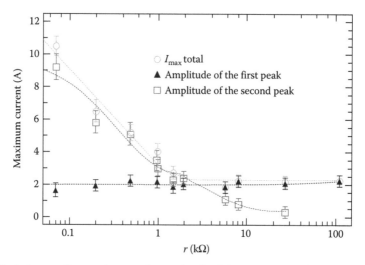

FIGURE 6.6 The influence of separating r on the amplitude of the TS current pulse, and its two components, $R = 9.84$ MΩ, $f \approx 1.1$ kHz, and $C = 43 \pm 4$ pF. (From Janda, M. et al., *Plasma Sources Sci. Technol.*, 20, 035015, 2011, with permission of IOP Publishing Ltd.)

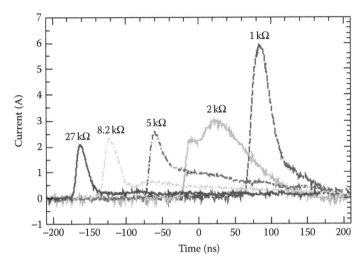

FIGURE 6.7 The influence of separating r on the shape of TS current pulses, $R = 9.84$ MΩ, $f \approx 1.1$ kHz, and $C = 43 \pm 4$ pF. (From Janda, M. et al., *Plasma Sources Sci. Technol.*, 20, 035015, 2011, with permission of IOP Publishing Ltd.)

We experimentally tested the validity of Equation 6.6 by measuring current I_0 flowing through r. However, the agreement between the measured current and the current calculated from ΔVr was good enough only for $r > 1.5$ kΩ. For smaller r, experimental results were in agreement only with I_0 derived from I after subtraction of IC_1. To find a better formula for $I(t)$, which would enable us to calculate the measured voltage waveforms for all values of r from it, we performed a detailed analysis of the electric circuit representing our experimental setup, which can be found in the works of Janda et al. [51]. A more accurate version of Equation 6.3, which takes into account the influence of r, was also provided therein.

6.3.4 Changes of TS Electrical Characteristics with f

The increase of the generator voltage V_a leads to a monotonous increase of f, as can be also derived from Equation 6.4. However, the repetition frequency is not absolutely regular. It strongly depends on V_{TS}, and each TS pulse may appear at a slightly different value of V_{TS}. We can therefore only define an average frequency, determined by an average value of V_{TS}.

Even if we know the precise value of V_{TS}, obtained at low TS repetition frequency, further increase of f cannot be easily calculated from the external electrical parameters (V_a, C, R) according to Equation 6.5. We found that V_{TS} itself is a function of f: it tends to decrease as f increases (Figure 6.8).

The increase of f is associated with changes of other TS properties as well. Current pulses get smaller and broader with increasing f (Figure 6.8). The broadening is mostly due to the slower current fall after the peak value (Figure 6.9). This could be explained by the decrease of the I_{max} with f. Smaller I_{max} also means lower peak electron density and higher impedance of the generated plasma. Because TS is based on discharging of C, higher impedance means longer discharging time.

6.3.5 Streamer-to-Spark Transition Time

An average streamer-to-spark transition time τ shortens significantly with increasing f. This is analogous with the shortening of τ with increasing mean discharge current in the prevented spark [12]. We also observed the prolongation of τ with increasing R as reported by Larsson and Naidis [14,15]. With further increase of f, the influence of R becomes negligible and the average streamer-to-spark transition time shortens down to ~100 ns.

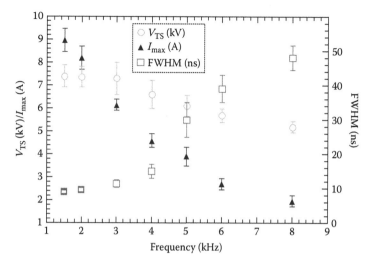

FIGURE 6.8 The dependence of peak current (I_{max}), breakdown voltage V_{TS} and FWHM of the current pulses on frequency, $C = 32 \pm 4$ pF, $r = 1$ kΩ, $d = 5.5$ mm, and $R = 6.6$ MΩ. (From Janda, M. et al., *Plasma Sources Sci. Technol.*, 21, 045006, 2012, with permission of IOP Publishing Ltd.)

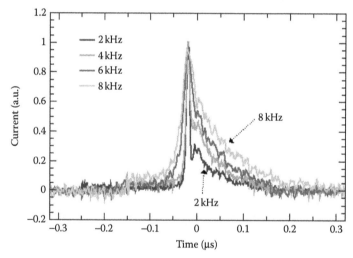

FIGURE 6.9 Changes of the normalized current waveforms measured on 1 Ω with increasing frequency, $C = 32 \pm 4$ pF, $r = 1$ kΩ, $d = 5.5$ mm, and $R = 6.6$ MΩ. (From Janda, M. et al., *Plasma Sources Sci. Technol.*, 21, 045006, 2012, with permission of IOP Publishing Ltd.)

The shortening of τ and changes of the streamer versus spark peak current ratio are also partly responsible for the modification of the shape of the spark current pulse rising edge (Figure 6.9). At lower TS frequencies (<3 kHz), τ is very random and it can vary from a few hundred nanoseconds up to several microseconds. The streamer and spark current pulses are well separated, and the streamer peak current (~100 mA) is a lot smaller compared to the spark pulse (~10 A). Moreover, because τ is long enough, the current after the streamer pulse falls down to a few milliampere before the transition to the spark. At higher frequencies, the maximum spark current is lower (<1 A), whereas the maximum streamer current does not change. The current decrease during the streamer-to-spark transition phase is also less significant. Sometimes, we even observed almost instantaneous formation of spark after the streamer with no current decrease in between. Finally, the resulting current pulse looks like two merging peaks.

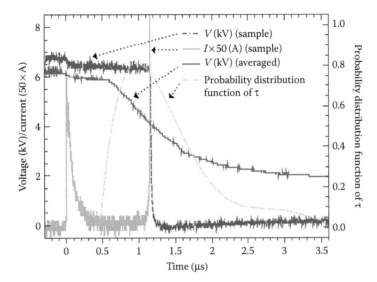

FIGURE 6.10 The estimate of probability distribution function of streamer-to-spark transition delay τ, derived from the averaged voltage waveform (128 samples), $f \sim 2.5\,kHz$, $C = 32 \pm 4\,pF$, $r = 1\,k\Omega$, $d = 5.5\,mm$, and $R = 8.2\,M\Omega$. (From Janda, M. et al., *Plasma Sources Sci. Technol.*, 21, 045006, 2012, with permission of IOP Publishing Ltd.)

Because of the random character of the streamer-to-spark transition, τ must be characterized by an average value, or even better by a probability distribution function $P(\tau)$. To obtain $P(\tau)$ we synchronized the acquisition of voltage and current waveforms by the rising edge of the streamer current measured on 50 Ω resistor shunt. This moment was defined as the beginning of the streamer-to-spark transition event. The end of the transition was defined as the moment of maximum spark current.

If we acquire averaged current waveform from a few hundred TS pulses synchronized by the rising edge of the streamer, we get a well-smoothed streamer pulse followed by a distribution function of spark current after the streamer. Based on our definition of τ, this spark current distribution function corresponds in the first approximation to $P(\tau)$.

We had to cut off the current during the spark phase to protect the oscilloscope, when we measured the current on the 50 Ω shunt. Thus, we had to use voltage waveforms to calculate the spark current according to Equation 6.2. To obtain more accurate derivative of V, we first fitted the region with the voltage drop by a polynomial function.

For example, Figure 6.10 shows $P(\tau)$ at 2.5 kHz ($d = 5.5$ mm, $R = 8.2$ MΩ). A difference between single pulse and averaged waveforms is also demonstrated here. It might be more appropriate to define τ as a moment when the spark current starts to grow, not when it reaches the peak value. However, the resulting difference is negligible at low TS frequencies, because of the steep spark current rise time (a few nanosecond). This difference grows with increasing f, but we estimate that even at ~10 kHz, it is only about 10% of τ as it is defined. More detailed discussion of the definition of the spark pulse probability distribution function $P(\tau)$ can be found in the works of Janda et al. [56].

6.3.6 Optical Emission Diagnostics of TS

Significant differences between lower and higher frequency regimes of TS were observed in time-integrated emission spectra in the VIS region (Figure 6.11). The emission of O, N, and N$^+$ atomic lines dominated in the spectra at lower frequencies (<3 kHz). At higher frequencies these atomic lines almost disappeared, and N$_2$ first positive system ($B^3\Pi_g - A^3\Sigma_u^+$) was much stronger. In the UV region, N$_2$ second positive system ($C^3\Pi_u - B^3\Pi_g$) dominated at all frequencies, but its relative intensity compared to atomic lines in VIS region also increased significantly with f.

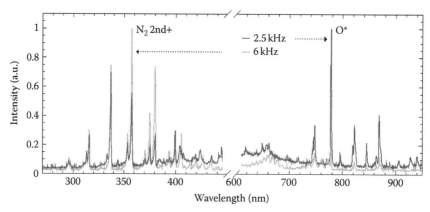

FIGURE 6.11   Time-integrated spectra of TS discharge. (From Janda, M. et al., *Plasma Sources Sci. Technol.*, 21, 045006, 2012, with permission of IOP Publishing Ltd.)

6.3.7 PMT Measurements of Time Evolution of Emission Intensity

To explain the disappearance of atomic lines with increasing f, we used PMT with appropriate narrow-band interference filters to measure the time evolution of the emission from $O(^5P)$ species at 777 nm (the strongest atomic line), and from $N_2(C)$ species (0–0 band of N_2 second positive system at 337 nm). At lower frequencies (Figure 6.12a), we can clearly see two peaks of total emission. The first one is produced by the streamer, whereas the second one corresponds to the short spark. As f increases, these two emission peaks merge and cannot be easily distinguished at higher f (Figure 6.12b). This can be explained by the shortening of the streamer-to-spark transition time τ with growing f.

Time-integrated PMT signal from $O(^5P)$ species decreases with f quite significantly, whereas the signal from $N_2(C)$ does not change much. This confirms the previous observation from time-integrated spectra (Figure 6.11). New information is that the $N_2(C)$ species are produced mainly during the streamer phase and $O(^5P)$ species during the spark phase (Figure 6.12). The electrical properties of streamers, the rise time, and the maximum current do not change significantly with f (Figure 6.6). In contrast, the spark current pulses are smaller and broader with f. For this reason we suppose that changes of the spark phase of TS at higher f are responsible for the disappearance of atomic lines in time-integrated emission spectra.

Based on the work [15], we suppose that the highest E/N is reached in the moment of the maximum spark current. After the maximum, E/N as well as the electron mean energy decline. The electron

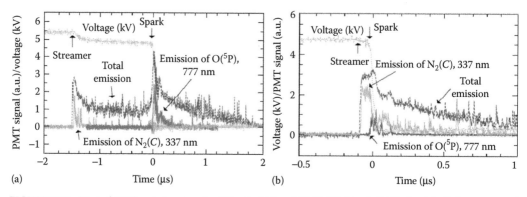

FIGURE 6.12   Typical PMT emission signals of TS at 2.5 (a) and 6 kHz (b), $R = 6.6\ M\Omega$, $C \approx 26$ pF, $d = 5$ mm, and $r = 0\ k\Omega$. (From Janda, M. et al., *Plasma Sources Sci. Technol.*, 21, 045006, 2012, with permission of IOP Publishing Ltd.)

impact excitation processes after the current peak are therefore certainly less efficient than before. The broadening of TS spark pulses with f is mostly due to the longer current decay time. This means that the overall efficiency (or capability) of TS pulses to excite ground state atomic oxygen species decreases with increasing f. From this we can deduce an overall weaker chemical effect of TS with smaller and broader pulses at higher f, which is in agreement with the observed decreasing bio-decontamination efficiency [58].

Another possible explanation for the disappearance of oxygen atomic lines with increasing f might be a more complicated mechanism of O(^{5}P) generation. We suggest the following three-step mechanism, based on experimental study of O(^{3}P) generation by NRP discharge [59]:

$$e + N_2(X) \rightarrow e + N_2(B,C) \quad \text{streamer phase,} \tag{6.7}$$

$$N_2(B,C) + O_2 \rightarrow N_2(X) + 2O(^3P) \quad \text{streamer-to-spark transition phase,} \tag{6.8}$$

$$e + O(^3P) \rightarrow e + O(^5P) \quad \text{spark phase.} \tag{6.9}$$

As the streamer-to-spark transition phase shortens with the growing f, less O(^{3}P) atoms accumulate for the production of O(^{5}P) during the high current phase.

6.3.8 Changes of the Gas Temperature with f

We used the emission spectra of 0–0 band of N_2 second positive system at 337 nm, obtained by a spectrometer coupled with the iCCD camera, for the calculation of time-resolved rotational temperature T_r of $N_2(C)$ species. We obtained T_r by fitting the experimental spectra with the simulated ones using Specair program [52]. We further assumed that in atmospheric pressure air plasma $T_r \approx T_g$, where T_g is the gas temperature.

We can synchronize the iCCD camera either with the beginning of the streamer, or beginning of the spark phase. We were thus able to measure the initial temperature (5–15 ns time window) at which streamers start (T_{str}), as well as the temperature in the initial phase of spark (T_{spark}). T_{spark} approximately represents the breakdown temperature to the spark, while T_{str} approximately represents a "steady-state" temperature to which gas is heated by the previous TS pulse. Figure 6.13 shows dependences of these temperatures on f.

The initial streamer temperature T_{str} obviously increased with f, although the overall heating of the gas was not very significant. Even at 10 kHz, the "steady-state" temperature at which every new streamer initiating the TS starts is only around 550 K.

However, this slight increase of temperature can explain the decrease of V_{TS} with growing f. At higher temperature, lower V_{TS} is required to keep E/N constant, due to the decrease of N resulting from increase of the gas temperature.

Although it seems that T_{spark} also slightly increases with f, the uncertainty of these data is quite high. Within the experimental error, we can conclude that the breakdown temperature at the beginning of the spark is constant: $T_{spark} = 1000 \pm 125$ K. We thus suppose that the increase of T_g to ~1000 K is crucial for the streamer-to-spark transition.

There are two possible reasons for this: an increase of E/N in the plasma column generated by the streamer due to the decrease of N, or the changes of the gas-phase chemistry.

Under an assumption of the constant atmospheric pressure, we estimated that an averaged E/N in the plasma channel would increase from about 60 Td to about 170 Td, while going from streamer to spark due to the increase of T_g from around 350 to 1000 K.

In actuality, we should also take into account changes of pressure inside the plasma channel. A pulsed discharge similar to TS was recently simulated by Naidis [15]. During the spark phase, in the moment of the highest E/N, the pressure on the axis of the plasma channel was almost 2 bar. If we use this value,

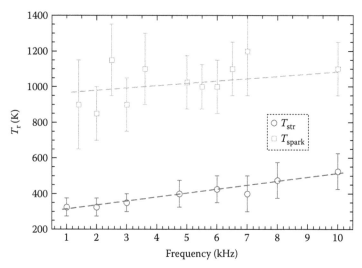

FIGURE 6.13 Changes of the rotational temperature of $N_2(C)$ species with f, T_{str}—measured during the initial 5–20 ns of the streamer, T_{spark}—measured during the initial 20 ns of the spark, $R = 8.2$ MΩ, $C = 32 \pm 4$ pF, $d = 5$ mm, and $r = 1$ kΩ. (From Janda, M. et al., *Plasma Sources Sci. Technol.*, 21, 045006, 2012, with permission of IOP Publishing Ltd.)

we get an increase of average E/N to only about 100 Td instead of 170 Td. The spark breakdown could be therefore also related to the changes of chemistry at higher T_g. The observed breakdown temperature corresponds very well with the critical temperature necessary to overcome the electron affinity by the thermal detachment reactions. Probably both the increase of E/N and the acceleration of thermal electron detachment play a certain role. Further research of TS, including kinetic modeling is necessary to resolve this question.

6.3.9 Changes of the Gas Temperature after the Streamer

It is obvious from T_{spark} and T_{str} that the temperature increases after the streamer. We therefore measured the time evolution of T_r during the transition time for various discharge frequencies (Figure 6.14). A decreasing intensity of $N_2(C)$ emission enabled us to follow T_r only for about 300 ns from the beginning of the streamer. At higher frequencies, where the transition time $\tau < 200$ ns, we also measured time evolution of T_r with the camera triggered by the beginning of the spark pulse, although we still observed light from $N_2(C)$ species generated by the streamer. Below ~4 kHz, where $\tau > 300$ ns, it was not possible to follow the evolution of T_r during the whole streamer-to-spark transition phase. Here, we were therefore able to measure T_{spark} only due to the small amount of $N_2(C)$ species produced during the initial phase of spark.

At both $f = 2.5$ and 6 kHz (Figure 6.14), we observed approximately linear increase of T_r with time before the breakdown. The heating is faster at 6 kHz, although in both cases the streamer-to-spark transition occurs when $T_r \sim 1000$ K. The acceleration of the gas heating with growing f is probably the major reason for the shortening of τ with f, though it can be partially attributed to the increase of T_{str}.

We suppose that the heating of the gas is caused by the electric current flowing through the generated plasma channel (Joule heating). The acceleration of the gas heating with the growing f can be therefore explained by the increase of an average current during the streamer-to-spark transition phase. At 3 kHz, the current after the streamer peak value decreases quickly to around 10 mA and it stays at this value during the remaining streamer-to-spark transition phase. At 5 kHz, the current decrease after the streamer was a lot slower. The current only decreased to ~130 mA before it started to slowly climb up again. At higher frequencies, the current sometimes does not decrease at all and starts to grow slowly right after the streamer.

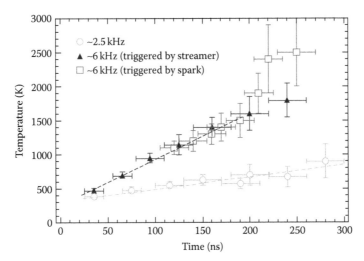

FIGURE 6.14 Changes of rotational temperature of $N_2(C)$ species with time, $R = 8.2\ M\Omega$, $C = 32 \pm 4\ pF$, $d = 5\ mm$, and $r = 1\ k\Omega$. (From Janda, M. et al., *Plasma Sources Sci. Technol.*, 21, 045006, 2012, with permission of IOP Publishing Ltd.)

The reaction rate is given by the product of their rate coefficient with the concentrations of the reactants. The slower current decay after streamers at higher TS frequencies could be therefore explained by the decrease of the gas density N due to the increase of T_{str}, even if the rate coefficients were not influenced. Because of the temperature dependence of their rate coefficients, the recombination reactions could be even more suppressed by the increase of the gas temperature to ~500 K. However, this cannot explain why we sometimes observed no current decay at all above 5 kHz. It means that processes generating electrons must be accelerated with increasing f. Once the conductive plasma channel bridging the whole gap is formed, the axial reduced electric field quickly becomes nearly uniform [15]. We estimated the average E/N to be around 60–70 Td. This is certainly not sufficient for direct electron-impact ionization processes to play an important role. We can also exclude the effect of increased E/N near the needle electrode by the secondary streamer, because its development takes several tens of nanoseconds [12], and we observed almost instantaneous increase of the current. The most probable explanation is therefore a chemical or stepwise ionization owing to any species accumulated from previous TS pulses, that is, a memory effect.

6.3.10 Streamer and TS Diameter Measurement

A set of experiments of single TS pulse imaging by iCCD camera (Figure 6.15a), with varying TS repetition frequency was performed to obtain an estimate of the plasma channel diameter D. We assumed that the plasma diameter can be well defined by the volume from which the emission can be observed. The iCCD camera was triggered directly by the current signal and we were able to synchronize it either with the beginning of the streamer or the spark event. We found that D of a plasma channel generated by a streamer is approximately 300 μm. More accurately, the FWHM of the radial channel profile after the Abel inversion is ~155 μm (Figure 6.15b). This agrees with the results of Gibert and Bastin [60] or van Veldhuizen et al. [61] who also applied optical emission measurement. However, it is much larger than the streamer diameter 40 μm measured by Bastin and Marode [62] who measured it from the broadening of H lines that is related to current density, so it may be closer to real plasma "electrical" diameter.

The subsequent TS channel was found narrower (at least for $f < 4$ kHz): less than 100 μm (FWHM after the Abel inversion being ~55 μm), but it slightly expanded close to the planar electrode. The contraction of the plasma channel is in agreement with the calculations of Naidis [57]. This measured D together with the complex analysis of the electric circuit representing TS, then enabled calculating the plasma conductivity leading to electron density.

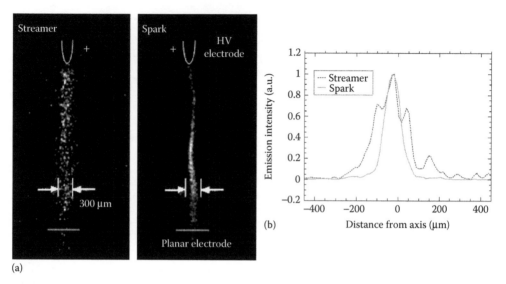

(a)

FIGURE 6.15 (a) Images of the streamer and spark of a single TS pulse taken by iCCD camera, exposure 25 ns, acquisition started ~25 ns after the beginning of the streamer and spark, respectively, $r = 0.9$ kΩ, $f \approx 2$ kHz, $R = 6.6$ MΩ, $C = 32 \pm 4$ pF, and $d = 4$ mm. (From Janda, M. et al., *Plasma Sources Sci. Technol.*, 20, 035015, 2011, with permission of IOP Publishing Ltd.) (b) Radial intensity profiles of single streamer and spark pulses.

6.3.11 Estimation of the Electron Density

We used the dependence of the electron number density on the plasma conductivity σ_p to calculate n_e:

$$n_e = \frac{\sigma_p m_e \upsilon_c}{e^2}.$$ (6.10)

Here e and m_e are the electron charge and mass, respectively, and υ_c is the electron–heavy particle collision frequency. The plasma conductivity σ_p is related to the plasma resistance R_p by

$$R_p = \frac{\rho_p d}{A},$$ (6.11)

where d and A are the gap length and the cross-sectional area of the plasma channel, respectively,

$$A = \frac{1}{4}\pi D^2.$$ (6.12)

The value of υ_c depends on electron temperature T_e and heavy particle density N, which both can change significantly in TS due to changes of E/N and gas temperature T_g inside the plasma channel. We calculated υ_c in air for T_g from 300 to 1500 K and for E/N from 10 to 200 Td using Monte Carlo simulation of electron dynamics [63]. We found that within this region of T_g and E/N, υ_c can vary approximately from 4×10^{11} to 4×10^{12} s^{-1}. Let us take the value 10^{12} s^{-1} to estimate the electron number density in TS. This approximation introduces the uncertainty of less than a factor of 4. This enabled us to estimate the temporal evolution of the electron density during the TS pulse, as shown in Figure 6.16. Further details can be found in the works of Janda et al. [51].

The calculated n_e peaks at ~10^{14} cm^{-3} in the streamer and ~10^{17} cm^{-3} in the spark phase, which was experimentally confirmed by the measurements of H$_\alpha$ Stark broadening in humid air in the spark current peak and 150 ns after this peak. Calculation of n_e from FWHM of H$_\alpha$ line is difficult due to sensitivity

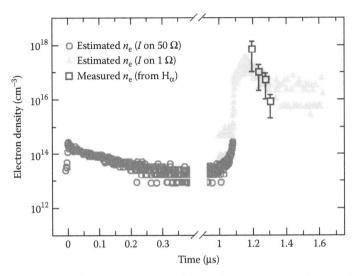

FIGURE 6.16 Electron density during a streamer and TS pulse in microsecond timescale, calculated and measured by H_α broadening. $f \approx 2$ kHz and $C = 32 \pm 4$ pF.

on electron temperature and ion dynamics. However, Gigosos et al. [64] showed that we can avoid these difficulties by using FW at the half area (FWHA) instead of FWHM and to calculate n_e by formula:

$$\Delta\lambda^{H\alpha}_{\text{FWHA}} = 0.549\,\text{nm} \times \left(\frac{n_e}{10^{23}\,\text{m}^{-3}}\right)^{0.67965}. \tag{6.13}$$

Here, $\Delta\lambda^{H\alpha}_{\text{FWHA}}$ is the FWHA of the H_α line in nanometer. To obtain reasonable values of n_e, the FWHA must be calculated from the line profile corrected with respect to Doppler, pressure, and instrumental broadening. However, the minimum value of FWHM we experimentally observed was 0.14 ± 0.03 nm (time 150 ns after the spark current pulse peak). This value is so large that Doppler and pressure broadening can be neglected. Even after the deconvolution of the measured line profile with the slit function describing the instrumental broadening, the value of this FWHM decreased to 0.12 nm, so the influence of instrumental broadening is within the experimental uncertainty of the measured FWHM. In time windows closer to the current peak, with even stronger Stark broadening, the instrumental broadening can be neglected completely.

6.3.12 Influence of Separating Resistor r on Electron Density n_e

As discussed above, by changing the value of the separating resistor r we can shape the amplitude and duration of the TS current pulse. The current pulse shape is associated with n_e evolution. Increasing the value of r leads to a decrease in I_{max} and therefore also to lower maximum n_e (Figure 6.17). On the other hand, increasing r causes the C_{cable} to take longer to get discharged: up to 20 μs for $r = 111$ kΩ (Figure 6.17). This prolongation of the current tail leads to relatively high n_e for longer time. Typical n_e at the end of discharging of C_{cable} was 10^{13} cm^{-3}.

We also estimated that it takes up to 100 μs ($r = 111$ kΩ) for n_e to drop below 10^{12} cm^{-3}. Thus, higher r enables us to merge the characteristics of TS and GD together, that is, relatively high current pulse (2 A) with high efficiency of production of radicals are followed by a long period with the current above 1 mA, during which the plasma reaches characteristics typical for GD. On the other hand, the voltage does not drop to low values for larger r, for example $r = 27$ kΩ, minimum voltage is 1.5 kV, and so C_{cable} cannot be discharged completely. So it is not suitable to use larger values of r. The separating resistor r

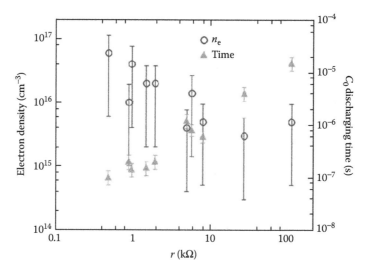

FIGURE 6.17 Dependence of maximum electron density and C_{cable} discharging time on r, $R = 9.84$ MΩ, $f \approx 2$ kHz, $C_0 = 37 \pm 3$ pF, and $C_1 = 7 \pm 1$ pF. (From Janda, M. et al., *Plasma Sources Sci. Technol.*, 20, 035015, 2011, with permission of IOP Publishing Ltd.)

must remain much smaller than the ballast resistor R, otherwise the discharging of C_{cable} has a negligible effect on the discharge current and $r + R$ can be treated as one resistor.

6.3.13 TS Summary

A new concept of a DC-driven self-pulsed discharge was described. The TS is a repetitive streamer-to-spark transition discharge of very short pulse duration (~10–100 ns) and with very limited energy so that the generated plasma remains highly in nonequilibrium. This discharge can be maintained at low energy conditions (up to 1 mJ per pulse) by an appropriate choice of resistances and capacitances in the electrical circuit, and its frequency can be controlled by the applied voltage.

The activity of TS is comparable to the nanosecond repetitive pulsed discharges but its advantage is the ease of the DC operation and does not need special and expensive HV pulsers with high repetition frequency and nanosecond rise times. Temporal evolution of electron density in TS was calculated based on the detailed analysis of the electrical circuit and single pulse imaging with fast iCCD. It showed that $n_e \approx 10^{17}$ cm^{-3} at maximum and 10^{11} cm^{-3} are reached on average. The maximum n_e during the spark pulse was confirmed by measurements of H_α Stark broadening.

The influence of a resistor r separating the HV cable from the HV electrode on the TS properties was presented. The increasing r causes longer discharging of the charge stored in the capacity represented by the HV cable and the tail of the current pulse can thus be longer than several tens of μs. This additional resistor r enables controlling the plasma properties and tailoring them to the specific application. For example, its appropriate choice enables combining relatively high current pulses (~1 A) with high efficiency of radical production in the following long periods with the current above 1 mA during which the plasma reaches characteristics typical for GD.

We also investigated time-resolved optical emission of TS. TS can be maintained at low energy conditions (0.1–1 mJ per pulse) and the generated plasma cannot therefore reach thermal equilibrium conditions, although the current pulse can lead to a very short temporary increase of gas temperature up to ~2500 K. The steady-state temperature, however, remains relatively low, because even at high repetition frequencies (~10 kHz) each streamer-to-spark process starts as low as ~550 K. Subsequent increase of the gas temperature to ~1000 K governs the streamer-to-spark transition. This can be explained either by the increase of E/N due to decreasing N, or by thermal electron-detachment reactions.

The shortening of the average streamer-to-spark transition time with increasing TS frequency can be partly explained by the slight increase of the streamer temperature T_{str}, and mostly by an acceleration of the temperature growth. The acceleration of the heating is caused by the changes of current waveforms during the streamer-to-spark transition phase. At 3 kHz, the current after the streamer peak decays quite quickly to around 10 mA before reforming the spark pulse, whereas at 6 kHz it sometimes starts growing right after the streamer. Just a few tens of nanoseconds later, this increase accelerates and the breakdown occurs, almost like a continuous transition from streamer to spark. We suppose that this behavior can be partly explained by the decrease of the gas density due to the increase of gas temperature. It is also necessary to consider the changes of the gas-phase chemistry due to the increase of gas temperature, and a memory effect—an accumulation of some species [NO, $O_2(a)$, etc.] that favors the ionization processes and thus shortens the transition phase. Further research is required, including new analytic techniques and a kinetic modeling, to explain the role of the increasing initial gas temperature and the increasing TS frequency on the breakdown mechanism.

6.4 NRP Discharges

6.4.1 Scope of This Section

Nanosecond pulsed breakdown in gases has been studied for decades, including at atmospheric pressure, with the last dedicated overview of this field by Korolev et al. [65]. Since then, the most distinctive development has been the emergence of nanosecond discharges at high pulse repetition frequency (PRF), referred to here as NRP discharges. To be clear, NRP discharges are generated by the application of HV pulses across the discharge gap. This is in contrast to the self-pulsed nanosecond discharges discussed in Section 6.3. Becker et al. [66] (see Section 7.4 of [66]) was the first review of NRP discharges, and this chapter can be considered an update.

This chapter discusses work at high PRF (i.e., ~10 kHz and above) and at atmospheric pressure. The interest in using high PRF extends well beyond simply packing in more discharges per unit time; qualitative differences in the physics of discharge transitions (Sections 6.4.5 through 6.4.9), in plasma chemistry (Sections 6.4.14 through 6.4.16), and in energy efficiency (Section 6.4.13) lead us to distinguish NRP discharges from "nanosecond discharges," which in this chapter refers to those at low PRF (i.e., less than 1 kHz). Nanosecond discharges will be discussed only insofar as they contribute to the understanding of NRP discharges. Likewise, work at low pressure is only mentioned when necessary for aiding the discussion of atmospheric pressure discharges.

6.4.2 Research on Nanosecond Discharges at Atmospheric Pressure Leading up to NRP Discharges

Here, we highlight the key steps in nanosecond discharge research that led to the development of NRP discharges. The first study dedicated to nanosecond gas discharges at atmospheric pressure came over half a century ago [67] as part of the effort to explain the formative time lag of spark discharges. Much of the early work on nanosecond discharges focused on determining fundamental breakdown mechanisms [68–71], running parallel to similarly nascent research in streamers and runaway electrons, two phenomena that arise when HV is applied, often very rapidly, across a discharge gap.

The first major application for nanosecond discharges was for pulsed gas lasers starting from the 1960s [72]. The advent of transversely excited atmospheric pressure (TEA) lasers led to much interest in GDs at high pressure because uniform large-volume plasmas were required for the lasing medium [73–75]. Short HV pulses were used to prevent the glow-to-spark (G-S) transition. Much work was done on developing preionization methods to further suppress the localization of the discharge by streamers [76], and we will soon see that NRP discharges represent a means of preionization that was not available at that time. In contrast to TEA lasers, many current applications require volume but not uniformity, and therefore streamer formation is oftentimes permissible. Furthermore, the gas in TEA lasers is circulated to provide fresh gas in the discharge gap for each pulse. With NRP discharges, we seek instead to accumulate active species generated by previous pulses.

Following the development of gas lasers, research in nanosecond discharges at atmospheric pressure was driven by applications in the decomposition of pollutant gases such as SO_x and NO_x. This field gained momentum starting from the 1980s, when Mizuno et al. [21] found that depollution using corona discharges generated using submicrosecond HV pulses was more efficient than with DC or AC coronas as well as more practical than with electron beams. For the first time, nanosecond discharges were primarily used for their ability to direct energy toward the production of useful chemical species instead of heating the gas [77].

In the meantime, there was continued progress in understanding breakdown at high overvoltage starting from the 1990s [78–80]. In particular, at voltages several times that of static breakdown, some electrons reach energies at which more energy is gained from the field than is lost through collisions, thus continuously gaining energy and becoming runaway electrons. Breakdown at high overvoltage can occur via fast ionization waves (FIW), which are distinguished from streamers by spatial uniformity and much higher speeds of propagation (~10^9 cm s^{-1} compared to ~10^8 cm s^{-1} for streamers in air at 1 atm). This form of nanosecond breakdown is distinct from the streamer- or Townsend avalanche-based breakdown mechanisms, which occur at low to moderate overvoltage.

The above research trends had already been fairly developed by the time that nonthermal atmospheric pressure plasmas were being widely studied for the applications discussed in this book. Starting from the 1990s, there was much interest in developing large-volume diffuse plasmas in atmospheric pressure air that maintain high average electron density but also at low volumetric power. GDs would be ideal except that the glow-to-arc transition at atmospheric pressure typically restricts operation to low reduced electric fields, where the vast majority of the discharge energy is channeled into gas heating, leading to high power budgets on the order of kW cm^{-3} to produce 10^{13} cm^{-3} of electrons. As was the case decades before with TEA lasers, nanosecond discharges of ~10 ns duration were suggested as a means of preventing the glow-to-arc transition such that GDs could operate at high reduced electric field and thereby improve the energy efficiency of electron generation [81].

Very high efficiency and large volume could also theoretically be achieved using electron beams [82], but there are still practical difficulties with generating electron beams at atmospheric pressure, mostly because it is difficult to build transparent windows for the electron beam to get from the vacuum region, where it is generated into ambient air. Runaway electrons generated by subnanosecond HV pulses were suggested as the next best alternative. However, runaway electrons were considered to be incapable of penetrating deeply into air at atmospheric pressure, leaving NRP discharges as the most practical alternative.

Unlike previous research in nanosecond discharges, the importance of using high PRF to maintain high average electron density was by this point emphasized. Fortunately, research on NRP discharges was facilitated by the timely advance of HV pulse generators with high PRF capability in the 10-kHz range and above. The key advance enabling this technology was a diode capable of switching simultaneously at high power and high speed, developed at the Ioffe Physico-Chemical Institute [83]. The Ioffe team eventually commercialized their high-PRF HV nanosecond pulse generators (www.fidtechnology.com), which have been used since the first study of NRP discharges in 2000 [84].

The Stanford group [84–86] proposed and demonstrated NRP plasmas as a way to significantly reduce the power budget required for sustaining high electron number densities in atmospheric pressure air. PRFs of 10–100 kHz were chosen to match the recombination times of active species such as electrons. Power budget reductions of 2 to 3 orders of magnitude with respect to DC-GDs were possible. This work was conducted from 1997 to 2002 with support from the "Air Plasma Ramparts Program" of the Air Force Office of Scientific Research (AFOSR) and the Department of Defense (DoD). Using a 10-ns, 10-kV, 100-kHz repetitive nanosecond pulse generator, they produced stable GDs in atmospheric pressure air at 2000 K with average electron densities of 1×10^{12}–2×10^{12} cm^{-3} and a measured power budget of 12 W cm^{-3}, which is close to the theoretical value of 9 W cm^{-3}. The electrode gap was 1 cm and the pulsed electric field was ~5 kV cm^{-1}. Macheret et al. [82] confirmed the performed calculations to quantify the power requirements of NRP glow discharges and the values determined from these experiments.

6.4.3 Nanosecond versus Submicrosecond Discharges

In this section we define what is considered a "nanosecond discharge" for this chapter. As mentioned in the previous section, a HV pulse width of around 10 ns is necessary to prevent the G-S transition at atmospheric pressure. However, minimizing pulse width is also relevant for spark discharges. Energy budgets would be minimized if the pulse were only long enough to induce the desired effect, which for some industrial applications such as plasma-assisted combustion can be as short as about 10 ns [87]. Furthermore, the efficiency of fundamental processes in nanosecond discharges is closely related to the mean electron energy, for which an effective electron temperature can be assigned (to be discussed further in Section 6.4.4), which has been simulated to reach a maximum within a few nanoseconds of applying the electric field and then decreasing dramatically within several tens of nanoseconds [88], as shown in Figure 6.18. This is due to the redistribution of the electric potential upon plasma formation that shields electrons from the strong applied field necessary for high electron temperature. Therefore, the time-averaged electron temperature increases as the pulse duration is reduced, because the phase of comparatively low electron temperature is progressively decreased. The efficiency improvement with decreasing pulse width was remarked upon early on in pulsed corona research [89] and applies to a number of plasma chemical species, as demonstrated in Figure 6.18.

In this chapter, we consider "nanosecond" discharges to be those whose pulse widths are on the 10-ns scale. This is shorter than the characteristic time for gas expansion at atmospheric pressure, and therefore constant gas density is typically assumed. Furthermore, ions react relatively slowly to very fast changes in the electric field and thus may not flow into the cathode by the time the applied field is on the falling edge, creating a temporary collection of positive charge near the cathode and reversal of the electric field [82]. As mentioned previously, very high electron temperatures can be maintained over the course of ~10 ns but not much more, reflecting the time required for the discharge to form a fully developed plasma column. There is a large body of work on "submicrosecond" plasmas that are typically much longer than 100 ns for which the physics is different due to various processes that occur at later times.

Nonetheless, we briefly mention here recent research on submicrosecond discharges that have been used in generating plasma jets/plumes, which are covered in detail in Chapter 7, according to the outline from XinPei Lu

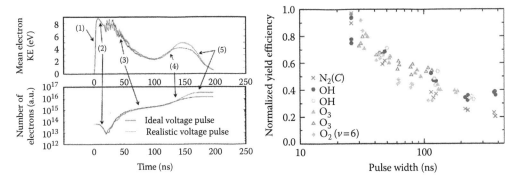

FIGURE 6.18 Generic evolution of the space-averaged mean electron kinetic energy (KE) and total number of electrons during a 150-ns pulsed discharge (left) (From Iza, F. et al., *IEEE Trans. Plasma Sci.*, 37, 1289–1296, 2009.), where the numbers (1)–(5) describe the following stages of discharge development: (1) avalanche of seed electrons that remain from the previous pulse, (2) loss of energetic electrons to the anode, (3) quasi-neutral plasma formation, confinement of low-energy electrons, (4) new avalanches initiated by secondary electrons, and (5) pulse end. Normalized yield efficiency of various species for different gas mixtures as a function of discharge pulse width (right) (From Ono, R. et al., *J. Phys. D Appl. Phys.*, 44, 485201, 2011.), where the gas mixture compositions can be found in the original figure from the reference.

and discussed here only with respect to pulse width. For plasma plumes generated in atmospheric pressure helium and launched into surrounding ambient air, Xiong et al. [91] found that pulse widths much longer than ~1 µs did not lengthen the plume, and that applying pulses of length much longer than ~100 µs actually shortened the plume. For pulse widths between 200 and 800 ns, it was found that the charge carried by the plume continues to increase in this range but levels off at larger values, with the longest plumes generated using 800-ns pulses. It has also been noted that pulsed jets require about 100 ns before significant current occurs [92].

6.4.4 NRP Discharges

In this section we present the reasoning for the pulsing scheme used in generating NRP discharges. An illustrative example of a typical NRP discharge is shown in Figure 6.19, showing pulsing parameters and time-resolved images of the development of the NRP glow regime. For atmospheric pressure, pulses of ~10 ns duration are typically used. The rise and fall times are kept as short as possible, with a few nanoseconds being the realistic limit for most generators capable of reaching ~10 kV in pulse amplitude. The theoretical lower limit for pulse duration is dictated by the electron-neutral collision frequency, which at atmospheric pressure is on the order of 10^{10} s^{-1}, because shorter pulses do not allow enough collisions to occur. Breakdown, as measured by the current and optical emission, typically begins a few

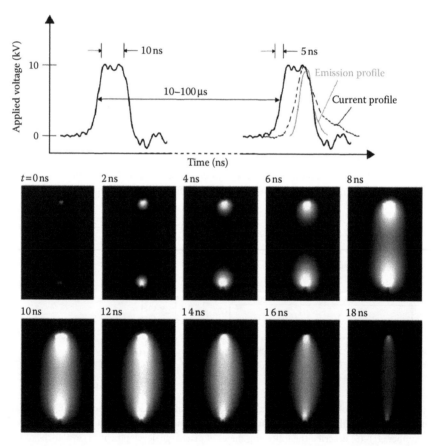

FIGURE 6.19 *Top*: schematic of nanosecond repetitive pulsing with typical parameters of applied voltage amplitude, pulse width, rise time, and period for 1–10 mm electrode gaps in air at atmospheric pressure. *Bottom*: time-resolved images of the NRP glow plasma in atmospheric pressure air, preheated to 2000 K, in terms of the time after the beginning of HV pulse. (From Packan, D. M., *Repetitive Nanosecond Glow Discharge in Atmospheric Pressure Air*, Stanford University, 2003.)

nanoseconds after the rising edge of the pulse and may last several nanoseconds after the falling edge, with a strong dependence on experimental conditions.

The NRP pulsing scheme consists of two parts:

1. The use of HV nanosecond-duration pulses
2. The application of these pulses at high PRF. State-of-the-art generators are currently able to achieve PRFs up to several hundred kilohertz.

The first part of the NRP pulsing scheme is the choice of using HV nanosecond pulses to ionize efficiently without inducing the glow-arc transition. Applying a high electric field causes the electron energy distribution function (EEDF) to develop a large high-energy tail, such that significant numbers of electrons are accelerated to energies at which the electron-impact ionization cross section (σ_i) is much greater than the cross section for vibrational excitation (σ_v). The top graph of Figure 6.20 (left) shows electron-impact cross sections for various processes in N_2.

The bottom graph of Figure 6.20 (left) shows the EEDF at various times following the application of a 100-Td-reduced electric field, calculated using the Boltzmann solver ELENDIF. Comparison of the graphs shows that after the field has been applied for just 100 ps, a much greater proportion of the electrons have energies greater than 10 eV, for which $\sigma_i > \sigma_v$. Using the electron-impact cross sections for different modes of excitation and the EEDF, we can calculate the fractions of the total power dissipated in each mode as a function of the reduced electric field (E/N), as shown in Figure 6.20. Relatively efficient ionization occurs for $E/N > 100$ Td.

Figure 6.20 (left) also shows that the EEDF changes only slightly between 100 ps and 1 ns, indicating that it requires only about 100 ps for the EEDF to reach equilibrium with the electric field. This is much shorter than the characteristic rise time of the pulse shown in Figure 6.19 (top), which means that the evolution of the EEDF toward the desired state with a prominent high-energy tail is limited by the rise time. By minimizing the rise time, we therefore also minimize the time that the EEDF spends in suboptimal states. Note that the EEDF maintains a quasi-Maxwellian profile for which we can assign an effective electron temperature corresponding to the mean electron energy, with the understanding that non-Maxwellian behavior can occur.

The NRP strategy involves applying a strong field to ionize efficiently and then turn it off after about 10 ns. If the field is not switched off, as is the case with DC discharges, the glow–arc transition quickly

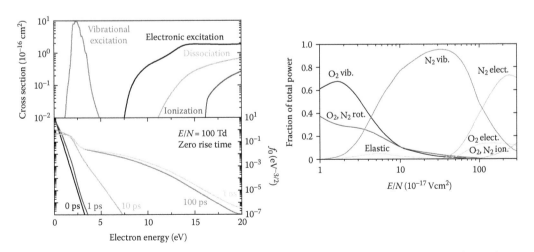

FIGURE 6.20 *Left*: cross sections for electron-impact processes in N_2 (From Raizer, Y. P., *Gas Discharge Physics*, Springer, Berlin, 1991.) and EEDF at times after the application of a 100-Td reduced electric field. *Right*: Fractions of power dissipated in different modes of excitation for air as a function of reduced electric field. (From Aleksandrov, N. L. et al., *High Temp.*, 19, 17–21, 1981; Nighan, W. L., *Phys. Rev. A*, 2, 1989, 1970.)

occurs. To avoid this in DC discharges, the electric field strength must be reduced, to the detriment of the ionization efficiency, as vibrational excitation of N_2 then consumes more than 99% of the input power at reduced fields around 30 Td (Figure 6.20, right).

Thus, to generate an average electron number density of 10^{12} cm^{-3} in 1-atm air at 2000 K, the DC field is 1.6 kV cm^{-1}, whereas the NRP field at PRF = 30 kHz is 5.3 kV cm^{-1} [86]. Consequently, the volumetric power consumption for the NRP plasma to produce 10^{12} cm^{-3} of electrons is 10 W cm^{-3}, much less than the 3 kW cm^{-3} dissipated by the DC plasma for the same electron number density.

The second part of the NRP pulsing scheme is to choose the appropriate PRF such that the time interval between pulses is shorter than the characteristic recombination time of active species. This feature distinguishes the NRP discharge from nanosecond discharges operating in single-shot or low PRF (1–10 Hz) mode. By using high PRF of 10–100 kHz, the accumulative and synergetic effect of repeated pulsing achieves the desired steady-state behavior. This depends strongly on the electron recombination rate following a pulse, which determines the number of seed electrons available to the upcoming pulse for ionization. The electron recombination rate is the net rate of all the individual electron processes. For example, recombination in atmospheric pressure air at 2000 K is effectively due to a single reaction, the dissociative recombination of electrons with NO^+ [86]. It can be shown that for target maximum and minimum electron number densities of 10^{12} and 10^{11} cm^{-3}, respectively, the electron recombination time is 90 μs in atmospheric pressure air at 2000 K. Hence, PRFs of 30 and 100 kHz were chosen in this case.

6.4.5 NRP Discharge Regimes

We will now discuss the three types of NRP discharge regimes that can be observed between pin electrodes. These regimes are termed the NRP spark, NRP glow, and NRP corona. The following discussion is based on a series of studies performed in similar experimental conditions that facilitate comparison [49,95,96]. In all cases, a bare metal electrode configuration was used. While these studies are not comprehensive insofar that they do not consider other reactor configurations such as DBD, surface, or microplasma discharges, they nonetheless provide a useful guide to the range of behavior of NRP discharges. NRP DBDs will be discussed in Section 6.4.10.

Due to their transient nature, NRP discharges challenge classification into the same categories as DC discharges [46]. Visually, however, they resemble their DC counterparts, arising in the same corona-glow-arc regime sequence order with increasing applied voltage, as shown in Figure 6.21a for NRP discharges in atmospheric pressure air at 1000 K. The corona, glow, and spark regimes of NRP discharges have been experimentally found to exist in air at atmospheric pressure, in the gas temperature range of 300–2000 K. A regime of a type of discharge is defined here as the applied voltage range over which it occurs, with all other experimental conditions fixed. The corona discharge emits light only near the anode, whereas the glow regime fills the entire gap with diffuse emission. Finally, the spark regime is characterized by its intense emission and high conduction current. The glow and spark regimes are also distinguished by the dynamics of discharge formation, as shown in Figure 6.21b and c, which will be discussed in detail shortly.

For each regime, Table 6.1 summarizes the characteristic conduction current, energy per pulse, and degree of gas heating. The corona and spark discharge regimes are easily obtained under almost all experimental conditions, whereas the glow regime exists only over a limited range of conditions. The most important observation concerning the existence of the NRP glow regime is the relationship between interelectrode gap distance and ambient gas temperature. When the gap distance is fixed, the NRP GD can only be generated above a minimum gas temperature. Conversely, when the gas temperature is fixed, GD can only be generated in a range of gap distances.

Here we would also like to mention that oftentimes in the literature the notions of "diffuse" and "filamentary" discharges are used, which strictly speaking refer to the spatial character of the discharge. Glow and spark discharges are often associated with being diffuse and filamentary, respectively. However, these discharge regimes are not *defined* by their spatial descriptions but rather by several basic properties such as gas heating and conduction current, as shown in Table 6.1. There are exceptions to

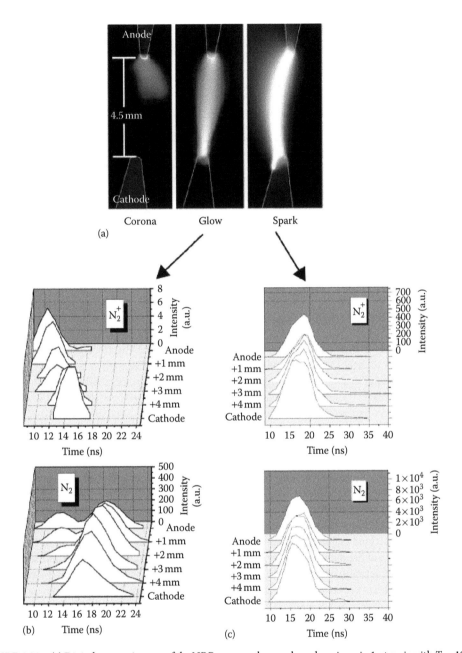

FIGURE 6.21 (a) Digital camera images of the NRP corona, glow, and spark regimes in 1-atm air with $T_g = 1000$ K, PRF $= 10$ kHz, $d = 4.5$ mm. From *left* to *right*, the images are in order of increasing applied voltage (5, 5.5, and 6 kV, respectively). (b) Measured $N_2^+(B\text{-}X)$ (0,0) emission intensity (top) and $N_2(C\text{-}B)$ (0,0) emission intensity (bottom) along the interelectrode axis for a NRP glow discharge and (c) for the NRP spark discharge. (From Pai, D. Z. et al., *Plasma Sources Sci. Technol.*, 18, 045030, 2009; Pai, D. Z. et al., *Plasma Sources Sci. Technol.*, 19, 065015, 2010b.)

the glow-diffuse and spark-filamentary associations. Tardiveau et al. and Shao et al. [97,98] observed a diffuse discharge in atmospheric pressure air whose current density of hundreds of ampere per square centimeter (A cm^{-2}) is more characteristic of a spark or even arc discharge than a glow. Furthermore, the diffuse discharge observed by Tardiveau et al. [97] occurs at a higher voltage than the filamentary regime, which is the opposite tendency of the G-S transition.

TABLE 6.1 Classification and Measured Thermal and Electrical Characteristics of Observed NRP Discharge Regimes for the Conditions $T_g = 1000$ K, PRF $= 30$ kHz, $d = 5$ mm, and $\upsilon = 2$ m/s

NRP Regime Name	Appearance	Discharge Formation Mechanism	Maximum Electron Density	Energy Deposited Per Pulse	Gas Heating $(+\Delta T_g)$	Conduction Current (A)
Corona	Corona			$<1\ \mu J$	<200 K	<2
Glow	Diffuse	Streamer	10^{13} cm^{-3}	$1-10\ \mu J$	<200 K	<2
Spark	Filamentary	Streamerless	10^{15} cm^{-3}	$200\ \mu J-1$ mJ	$2000-4000$ K	$20-40$

Source: Pai, D. Z. et al., *Plasma Sources Sci. Technol.*, 18, 045030, 2009.

Notes: The formation mechanisms shown may not apply in all cases, as streamer formation depends on preionization level, which can be determined by factors other than the discharge regime (see Sections 6.4.6.3, 6.4.7, and 6.4.11).

6.4.6 Dynamics of NRP Glow Discharges

Here, we discuss how the NRP glow discharge develops in detail. Figure 6.21b shows spatially and temporally resolved emission intensity from excited N_2 and N_2^+ that illustrates how breakdown occurs in this regime. The discharge is initiated by a cathode-directed wave that propagates at a speed of about 10^8 cm s^{-1}, followed immediately by a return wave of potential redistribution that travels much faster, at about 10^9 cm s^{-1}. These events are typical of streamer-initiated discharges in atmospheric pressure air [12].

First, the HV pulse initiates an anode-directed Townsend avalanche. Eventually, a space-charge layer builds up sufficiently to meet Meek's criterion for the avalanche-to-streamer transition (AST), whereby the space-charge field is approximately equal in strength to the Laplacian field. Once the streamer head reaches the cathode, it can no longer extract electrons from the gas, use them for ionization, and maintain a high space-charge field. To continue existing, the streamer head must instead extract electrons from the cathode via secondary emission. Given enough time, this streamer–cathode interaction would eventually lead to the formation of a cathode fall typical of DC discharges, which is characterized by secondary electron emission via ion–electron emission. However, Figure 6.21b shows that the return wave occurs immediately following the arrival of the streamer at the cathode. Therefore, the streamer–cathode interaction time is at most 2 ns, the camera gate width. As will be discussed in Section 6.4.19, ion–electron emission is limited in this situation. Instead, faster processes such as field and photo-emission are the predominant mechanisms of secondary emission.

The secondary electrons partially neutralize the positive space charge in the streamer head, which causes a rapid potential drop at the streamer head. This potential drop is then communicated down the discharge gap, generating a return wave of potential redistribution that typically travels at about 10^9 cm s^{-1} in atmospheric pressure air [42]. This agrees well with Figure 6.21b, where it can be discerned that the return wave speed must be significantly greater than the streamer speed of 10^8 cm s^{-1}. Following the return wave, the discharge enters a "conduction" phase in which current builds up for the remainder of the pulse duration. To remain in the glow regime within a given pulse, the voltage must be switched off before strong current growth, that is, the streamer-to-spark transition [12]. There is also the issue of how to maintain the glow regime over the course of many pulses, but this is an issue that will be discussed in Section 6.4.8.

We now introduce a simple model of the NRP glow regime using the above description of discharge dynamics.

6.4.6.1 Streamer Phase

The NRP glow discharge is initiated by the AST. To analyze the AST, we assume that the electric field at the moment of initiation is well-approximated by the Laplacian electric field E_L. For a pin–pin electrode configuration, we approximate the electrodes as two hyperboloid surfaces with foci separated by $a = [(d/2)(d/2 + R)]^{1/2}$, where d is the interelectrode gap distance and R is the radius of curvature of the electrode tip. The field along the interelectrode (z) axis relating the two pins can be approximated as:

$$E_L(z) \approx -\frac{V_p}{\ln\left[(1+\eta_0)/(1-\eta_0)\right]}\frac{a}{a^2-z^2}, \tag{6.14}$$

where V_p is the potential of the anode pin, the cathode pin is at zero potential, and $\eta_0 = \left[(d/2)/(d/2+R)\right]^{1/2}$. The AST occurs when the space-charge field (E_{SC}) generated by an electron avalanche becomes comparable with E_L. One approximation for E_{SC} can be found in Bazelyan and Raizer [99]:

$$E_{SC} = \frac{2eQ\alpha^2}{9\pi\varepsilon_0} \equiv K\alpha^2, \tag{6.15}$$

where

  e is the charge of an electron
  Q is the number of charges required for the AST (also known as the ionization integral)
  ε_0 is the permittivity of free space
  α is the first Townsend (or ionization) coefficient

Combining Equations 6.14 and 6.15 by setting the electric field at the anode $E_L(z = d/2) = E_{SC}$ and using $\alpha = \alpha_{AST}$, where α_{AST} is the ionization coefficient necessary for the AST, we obtain the applied voltage necessary for the AST, V_{AST}:

$$V_{AST} = K\alpha_{AST}^2\left[\frac{a^2-(d/2)^2}{a}\right]\ln\left(\frac{1+\eta_0}{1-\eta_0}\right) \approx K\alpha_{AST}^2 R\ln\left(\frac{1+\eta_0}{1-\eta_0}\right), \tag{6.16}$$

where the last term on the right-hand side is the approximate expression for V_{AST} in the case of sharp tip electrodes, $R \ll d$. It is well known that sharp tips generally lower the voltage required for inducing many discharge phenomena, and Equation 6.16 quantifies this effect approximately for the AST in terms of the geometric parameters R and η_0. In Section 6.4.8, Equation 6.16 will be used for describing NRP discharge transitions.

6.4.6.2 Conduction Phase

Following the streamer and return wave, the NRP glow discharge enters the conduction phase, in which the remaining energy of the pulse is dissipated resistively in the conducting channel created by the streamer. During the return wave phase, the space charge in the channel adjusts to redistribute the potential, and thus the field in the bulk of the plasma becomes approximately uniform. This smoothing of the field represents the formation of the positive column and has been simulated for the conditions of the NRP discharges discussed here [100]. The field in the positive column E_{pc} is

$$E_{pc} = \frac{V_p - V_{cf}}{d}, \tag{6.17}$$

where

  V_p is the applied voltage
  V_{cf} is the cathode fall voltage

For NRP discharges in atmospheric pressure air at 750–2000 K, V_{cf} has been measured and modeled to be 1–2 kV [82,86,95], which indicates that the NRP glow is in the abnormal regime [46]. This conclusion is also supported by the observation that the current density increases with applied voltage.

It has been found that $E_{pc} \approx E_{br}$, the static breakdown field, that is, the electric field necessary to balance ionization (ν_i) and attachment (ν_a) rates in air. In contrast, $E_{pc} \ll E_{br}$ for normal DC-GD in air at atmospheric pressure [101]. The greater relative value for E_{pc} for the NRP glow is also characteristic of the abnormal regime. However, this is also consistent with the fact that DC discharges require a lower electric field to sustain the positive column than pulsed discharges. This is because in DC-GD the species enabling electron detachment (ν_d) and stepwise (ν_{sw}) ionization reach high concentrations over time and lower the electric field needed to overcome electron losses due to attachment, whereas pulsed discharges lack the time for such a build up [46]. For a poststreamer channel in atmospheric pressure air, simulations show that in an electric field of 19 kV cm^{-1}, well over 100 ns is required before the net ionization rate $\left(\nu_i + \nu_{sw} + \nu_d - \nu_a \right)$ is greater than zero [15]. With an electric field of 24 kV cm^{-1}, the same result is reached in about 10 ns. Given that $E_{br} \approx 30$ kV cm^{-1} in atmospheric pressure air, it is plausible that $E_{pc} \approx E_{br}$ for the NRP glow for this reason. This would also imply, however, that few species that participate in detachment and stepwise ionization, that is, electrons, survive between pulses.

6.4.6.3 Streamerless Glow

The foregoing discussion of the streamer-initiated glow is based on experiments in air at temperatures between 300 and 1000 K. In previous experiments at Stanford University at $T_g = 2000$ K, a significant level of preionization existed in the glow regime, because the PRF had been chosen high enough to maintain an electron number density greater than 10^{11} cm^{-3} between pulses. In these experiments, it is therefore likely that the NRP glow occurred through a streamerless process, with uniform ionization in volume. This is confirmed by the NRP glow images shown in Figure 6.19, where no streamers are observed.

6.4.7 Dynamics of NRP Spark Discharges

In this section, we discuss how the NRP spark discharge develops, and we will see that the basic description is simpler than that for the NRP glow. Due to its usefulness in plasma-assisted combustion, the NRP spark has been studied extensively for its thermal and plasma-chemical properties. These areas will be discussed in Sections 6.4.15 and 6.4.17.

Figure 6.21c shows spatially and temporally resolved emission intensity from excited N_2 and N_2^+ that illustrates how breakdown occurs in this regime. Unlike the NRP glow, spark breakdown occurs simultaneously throughout the gap without any indication of streamers or other wave propagation. In contrast, typical spark formation is customarily preceded by a streamer [12]. The streamerless breakdown of NRP spark discharges can be explained by the preionization level. The NRP spark shown in Figure 6.21c maintains $T_g > 2000$ K at all times, and as mentioned in Section 6.4.4, electron recombination in this case is primarily due to the dissociative recombination of NO$^+$. Given this reaction and a maximum electron density of 10^{15} cm^{-3} (see Table 6.1), it is straightforward to show that $n_e(t = 33\ \mu\text{s}) = 3 \times 10^{11}$ cm^{-3} at the end of the recombination period, which is also the preionization electron number density n_{e0} for the following pulse. Depending on whether n_{e0} is below or above a critical level n_{crit}, high-pressure discharges form via the streamer mechanism or via Townsend avalanches in volume, respectively [76]. For the experimental conditions here, it can be shown that n_{crit} has an upper limit of 10^8 cm^{-3} [96]. The condition for streamerless discharge formation is therefore satisfied with $n_{e0} > n_{crit}$. Instead of streamers, avalanche ionization occurs in volume, resulting in the uniform emission observed experimentally.

Streamerless breakdown may not be universally true for NRP sparks; it is possible that at low PRF, the preionization level could decrease such that $n_{e0} < n_{crit}$ and allow streamer initiation. Conversely, the NRP glow discharge may form without streamers at sufficiently high PRF such that a high preionization level is maintained between pulses. In any case, streamerless breakdown provides a more realistic test for models that assume a poststreamer channel such as in the works of Naidis [102], where for NRP sparks in steady state the thermal discharge radius (r_T) varies as follows:

$$r_T \sim \left(\frac{Q}{p_0 d} \right)^{1/2}, \tag{6.18}$$

where

Q is the input energy
p_0 is the ambient pressure
d is the gap distance

If the applied voltage amplitude U is high such that the plasma resistance becomes negligible compared to the external resistance connected in series with the discharge gap, R_{ext}, then the current approaches the value U/R_{ext}. In this case, the gas temperature on the axis T_{ax} varies according to

$$T_{ax} \sim \left(\frac{U p_0 f \tau_{pulse}}{R_{ext}} \right)^{0.2}, \tag{6.19}$$

where

f is the PRF
τ_{pulse} is the pulse duration in this case

Again, if the current approaches U/R_{ext}, then Equation 6.19 for T_{ax} is independent of d until $d \gg r_T$.

6.4.8 NRP G-S Transition

In this section we summarize the principal results of a theory for describing the NRP G-S regime transition based primarily on the AST (see Section 6.4.6.1) and the thermal ionization instability, which is essentially a positive feedback loop between heating of the gas and ionization [46]. The standard treatment of the G-S transition applies to the case of a DC electric field or a pulsed field at low repetition frequency in flow. For the latter, the heated gas is removed from the discharge gap by the flow, and therefore the G-S transition must be induced by a single pulse [12,103]. For NRP discharges, however, high PRF results in incremental heating over a sequence of pulses for as long as the gas remains in the discharge gap. Although ionization and heating both occur during the pulse, substantial heat is released afterward via processes such as vibrational–translational relaxation, in the case of N_2. The NRP G-S transition is thus qualitatively different from those previously studied.

The NRP G-S transition occurs after a number of pulses, indicating that it is a relatively slow process at the millisecond timescale. We can simplify the analysis by considering the NRP heating process to be thermodynamically equivalent to a continuous heating process with the same time-averaged volumetric power. The thermal instability therefore occurs when the time-averaged volumetric power input exceeds the heat loss rate due to conduction, assuming that convection is negligible. It is possible to derive an expression of the gas temperature increase for the NRP glow discharge just below the G-S transition [95]:

$$T - T_0 = \frac{-\lambda + \sqrt{\lambda_0 + 2A\Theta}}{A}, \tag{6.20}$$

where

T is the gas temperature in the plasma
T_0 is the ambient gas temperature

λ_0 is the thermal conductivity of the ambient gas

$A = 6.2 \times 10^{-5}$ W m^{-1} K^{-2} is the linear slope of the variation of the thermal conductivity of air from 300 to 2000 K

$\pi\Theta$ is the time-averaged input power per unit length

A key experimental finding is that the maximum value of Θ in the NRP glow regime is a constant under for a given PRF and gap distance d. Given that the G-S transition occurs at a constant value of $\Theta \sim 10$ W m^{-1} (for the conditions of Pai et al. [95]), it is possible to show that the reduced field for the transition $(E/N)_{GS}$ also remains constant with the ambient gas temperature T_0. The field in the positive column is then $E_{pc} = (E/N)_{GS} \cdot N$, and it is now possible to derive an expression for the G-S transition voltage:

$$V_{GS} = \left[\left(\frac{E}{N}\right)_{GS} N\right] d + V_{cf},\tag{6.21}$$

where N is the gas density in the plasma calculated using Equation 6.20. For the experiments discussed here, the value for $(E/N)_{GS}$ is 144 Td (1 Td = 10^{-17} V cm^2). Note that the above treatment is performed in a time-averaged fashion to obtain semianalytical results from which the physics is relatively transparent.

In addition to the time-averaged treatment over many pulses shown above, there is also a complementary model of the NRP G-S transition resolved in time for a single pulse, with an assumed preionization level to approximate the effect of previous repetitive pulsing [104]. Within the framework of a single-pulse model, the G-S transition is considered primarily from the perspective of streamer propagation. As shown in Figure 6.22, when the streamer reaches the cathode to form a conducting plasma channel, the discharge energy rapidly increases and causes the G-S transition. To obtain the NRP glow regime, the applied voltage pulse needs to last only as long as the streamer propagation time across the gap to prevent energy deposition that would lead to the thermal ionization instability.

Figure 6.22 also demonstrates that, for the same pulse characteristics, the G-S transition occurs sooner at 300 K than at 1000 K, which is consistent with the experimental observation that the NRP glow is much easier to attain at higher temperatures. It follows that experimental parameters must be adjusted to regulate the energy deposition such that the NRP glow can be generated at lower temperatures, which will now be discussed.

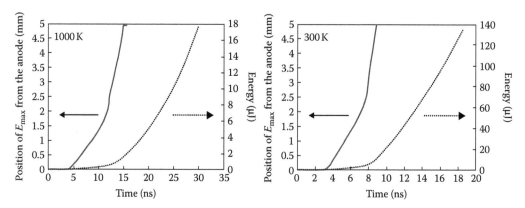

FIGURE 6.22 Position of the streamer head as defined by the position of the maximum electric field E_{max} and the discharge energy as a function of time, for streamer propagation leading to the NRP G-S transition at 1000 K (left) and 300 K (right). (From Tholin, F., and A. Bourdon, *J. Phys. D Appl. Phys.*, 44, 385203, 2011.)

6.4.9 Domain of Existence of the NRP Glow Regime

The breakdown condition and thermal ionization instability discussed in Sections 6.4.6.2 and 6.4.7, respectively, suffice to explain the NRP glow and spark regimes at a fixed interelectrode gap distance. However, when the gap distance is varied, these two conditions cannot explain the range of gap distance over which the NRP glow regime was experimentally found to exist, as shown in Figure 6.23. The maximum gap distance (d_{max}) can be explained as the maximum distance that a streamer can travel for a given pulse duration. The minimum gap distance (d_{min}) represents the condition in which the applied voltage produces a Laplacian field at the anode that is strong enough to initiate the streamer yet does not result in enough heat during the conduction phase to lead to the thermal ionization instability. For $d < d_{min}$, the applied voltage is so high that as soon as the streamer crosses the gap, Joule heating is sufficient to trigger the G-S transition after several pulses. The NRP glow regime can only exist when $d_{min} < d_{max}$. The quantity d_{max} can be easily obtained as $d_{max} = \upsilon_{streamer}\tau$, where τ is the pulse duration and $\upsilon_{streamer} \approx 10^8$ cm s^{-1}.

The quantity d_{min} can be determined as follows. By setting $V_{AST} \leq V_{GS}$ and using Equations 6.16 and 6.21, we derive a criterion for the existence of the NRP glow regime in terms of experimental parameters:

$$\alpha_{AST}^2 KR\ln\left[\frac{1+\eta_0(d)}{1-\eta_0(d)}\right] \leq \left(\frac{E}{N}\right)_{GS} Nd + V_{cf}. \tag{6.22}$$

The PRF and pulse duration are not explicit in Equation 6.22 but influence Θ in Equation 6.20 and thereby modify N. It is possible to calculate theoretical values for d_{min} as a function of the gas temperature using Equation 6.22 on the condition of equality, as shown in Figure 6.23 (left).

It can be shown (Pai et al. [95]) from Equation 6.22 that d_{min} can be reduced by decreasing the radius of curvature of the electrodes R. From Figure 6.23 (left), it can be seen that the stable glow regime narrows with decreasing gas temperature, and choosing sharper electrodes to decrease d_{min} or lengthening the pulse to increase d_{max} are two straightforward ways to enable the existence of the glow regime below 750 K. Figure 6.23 (right) demonstrates that streamer propagation speeds vary by only a factor of 1.2 over a range of electrode radii of curvature from 50 to 300 μm. Using sharper electrodes leaves the curve for d_{max} in Figure 6.23 (left) relatively unchanged while moving the curve for d_{min} downward leads to increasing the domain of the glow regime in a straightforward fashion. On the other hand, changing the

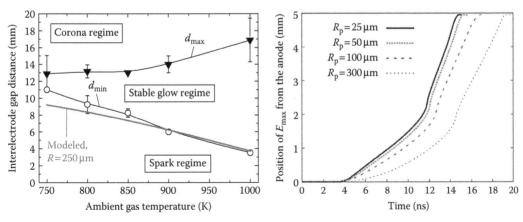

FIGURE 6.23 *Left*: measured gap distance ranges for existence of NRP glow regime and modeled d_{min} as a function of the ambient gas temperature. (From Pai, D. Z. et al., *J. Appl. Phys.*, 107, 093303, 2010a.) *Right*: position of the streamer head as defined by the position of the maximum electric field E_{max} as a function of time, for different values of the radius of curvature of the electrodes for T_g = 1000 K, an applied voltage of 5 kV, and a 5-mm gap. (From Tholin, F., and A. Bourdon, *J. Phys. D Appl. Phys.*, 44, 385203, 2011.)

pulse duration would alter both d_{max} and d_{min} by way of changing the time required for the streamer to cross the gap and changing the discharge energy, respectively, leaving the overall effect on the domain of the glow regime unclear. Thus, sharpening the electrodes is the most transparent way to obtain the NRP glow in atmospheric pressure air at lower temperatures than that shown in Figure 6.23 (left), down to room temperature. However, by doing so, streamer branching may occur if the Laplacian field in the gap falls below $E^* = 8–9$ kV cm^{-1}, although this problem can be avoided through judicious engineering of the Laplacian field [104]. Note that E^* is higher than the external field of about 5 kV cm^{-1} required for the stable propagation of a positive streamer [46].

Other length scales common in gas discharge physics cannot explain d_{min}. A hypothetical pd scaling law based on Townsend breakdown would predict increasing d_{min} as a function of T_0 instead of the decreasing trend shown in Figure 6.23. In any case, the timescale of NRP discharges is far too short for establishing the ion-driven secondary electron emission that is assumed in Townsend breakdown. Also, a hypothetical scaling law based on $\alpha\xi \approx 20$ (where ξ denotes the distance necessary for building up the space-charge layer), the Meek–Raether criterion for streamer inception [46], could be imagined if it could be shown that $\alpha = \alpha_{AST}$ increases with T_0. Not only does α_{AST} decrease with T_0, but also $d_{min} \gg 100$ μm, the characteristic length of the space-charge layer in streamer heads in air at atmospheric pressure that should be close in value to ξ.

For gliding arc discharges, a critical length (l_{crit}) exists for the equilibrium (i.e., arc) to nonequilibrium (i.e., glow) transition [105]. The energy balance between electrical energy deposited into the discharge and heat loss from the plasma column plays a role in determining both l_{crit} and d_{min} discussed here. However, l_{crit} differs from d_{min} because the NRP G-S transition studied here is *from* the glow *to* the spark regime, whereas the gliding arc G-S transition is in the opposite direction. As the gliding arc is initially thermal, streamers are not involved in the generation of the glow phase. On the other hand, streamers initiate the NRP glow, leading to the AST being essential to defining d_{min}. Therefore, d_{min} for NRP glows is of a different nature than l_{crit} for gliding arcs.

6.4.10 NRP Dielectric Barrier Discharges

Besides NRP discharges generated in electrode configurations involving bare conducting electrodes, as discussed in Section 6.4.5, NRP DBD discharges are also increasingly studied. The charging of the dielectric capacitance has two principal effects that are illustrated in Figure 6.24. First, as is the case with AC-driven DBDs, the voltage across the discharge gap is different from that applied across the electrodes. This dielectric screening of the electric field can be put to use when submicrosecond pulses are applied across the electrodes, because the charging of the dielectric results in pulses of ~10 ns duration across the gap [106,107]. The second effect is that during the falling edge of the applied voltage, the energy and surface charge stored in the dielectric is released to generate a secondary discharge.

Unlike AC-DBDs, NRP DBDs are capable of generating diffuse discharges in a wider range of conditions that take on the character of a single discharge with a current density of ~10 A cm^{-2}. This is in contrast to a series of microdischarges with current densities of ~10–100 mA cm^{-2} in the case of AC-DBDs. Furthermore, NRP DBDs have higher energy deposition, electron density, and electron temperature than their AC-DBD counterparts. This enhancement of plasma activity is due to the fact that breakdown in NRP DBDs occurs at higher reduced electric field than in AC-DBDs. As discussed in Section 6.4.4, such a situation improves the efficiency of many plasma processes.

NRP DBDs can also be generated using applied voltage pulses that are themselves ~10 ns long [108]. In this case, there is practically no delay between the primary and secondary discharges because the rising and falling edges of the pulse are only nanoseconds apart. As with submicrosecond pulsed DBDs, applied voltage pulses of 10-ns duration can also generate diffuse discharges, although only a filamentary discharge regime is possible beyond a certain gap distance, as shown in Figure 6.24 (top).

Residual surface charge on the dielectric and preionization is thought to alter the character of NRP DBDs from filamentary to diffuse after a sufficient number of pulses. The assumption of uniform breakdown has

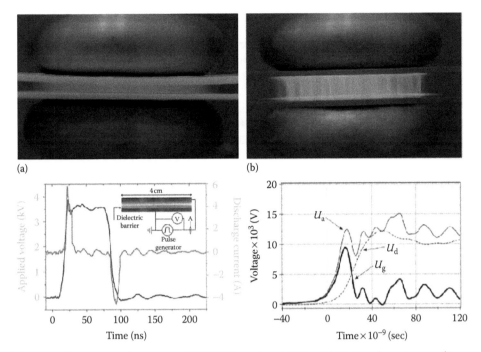

(a)   (b)

FIGURE 6.24 *Top*: images of two typical NRP DBD modes at PRF = 1 kHz for (a) an air gap of 2 mm and (b) 8 mm. (From Shao, T. et al., *J. Phys. D Appl. Phys.*, 41, 215203, 2008.) *Bottom left*: traces of the applied voltage and discharge current for the NRP DBD shown in the inset. (From Walsh, J. L., and M. G. Kong, *Appl. Phys. Lett.*, 91, 251504, 2007.) *Bottom right*: measured and calculated voltage for a surface NRP DBD. Trace labeled U_a is the measured voltage, U_d is the voltage across the dielectric, and U_g is the voltage across the discharge. (From Williamson, J. M. et al., *J. Phys. D Appl. Phys.*, 39, 4400–4406, 2006.)

enabled the formulation of a one-dimensional model that illustrates the role of the dielectric layer in controlling the energy coupling to the plasma [109]. Essentially, about as much energy is dissipated by the plasma as is stored in the capacitance that constitutes the DBD, which is analogous to the fact that in an RC circuit, about as much energy is stored in the capacitor as is dissipated in the resistance when voltage is rapidly applied. An analytic expression is derived in Adamovich et al. [109] for the efficiency (η) energy coupling to the plasma:

$$\eta \approx \frac{\left[\left(V_0/V_{\text{peak}} \right)^2 + \left(\sqrt{2\pi}/\nu_{\text{RC}}\tau \right) \right]}{1 + \left[\left(V_0/V_{\text{peak}} \right)^2 + \left(\sqrt{2\pi}/\nu_{\text{RC}}\tau \right) \right]}, \tag{6.23}$$

where
  V_0 is the breakdown voltage
  V_{peak} is the amplitude of the applied voltage pulse
  ν_{RC} is the time constant of discharge circuit following breakdown
  τ is the pulse duration

From Equation 6.23, it can be seen that η decreases if high overvoltage ($V_0 \ll V_{\text{peak}}$) is combined with a long pulse width $\left(\nu_{\text{RC}}\tau \gg 1 \right)$, with most of the power reflected back to the generator. On the other hand, it can be shown that combining high V_{peak} with a short pulse width $\left(\nu_{\text{RC}}\tau \ll 1 \right)$ increases V_0, leading to increased efficiency.

6.4.11 Preionization

As mentioned throughout this chapter, a number of effects in NRP discharges are suspected to result from high levels of residual charges left from previous pulses. It is also possible that the preconditioned gas contains species that reduce electron attachment as well as increase detachment or stepwise ionization. Experimental evidence of the effect of preionization is usually indirect, such as the streamerless formation of NRP sparks discussed in Section 6.4.7. Direct measurement of plasma species at the end of the recombination period is often difficult due to low concentrations. Instead, both analytical and numerical models have been important in illustrating the effects of preionization.

As discussed in Section 6.4.7, Levatter and Lin [76] developed a criterion for streamerless discharge formation. Briefly, this theory considers the Meek–Raether criterion for the AST, derived from the ionization integral for determining the number of electrons N_e generated in an electron avalanche:

$$\int_0^{\xi_c} \alpha(\xi')d\xi' = N_e, \qquad (6.24)$$

where ξ_c is the critical avalanche track length for the AST, that is, the distance required for ionization originating with a single seed electron to generate an electron cloud of radius r_c with enough charge to initiate a streamer, as shown in Figure 6.25 (left). It is generally accepted that N_e must be on the order of 10^8 to generate a strong enough space-charge field for the AST to occur, leading to the often-used form of the Meek–Raether criterion, $\alpha\xi_c \approx 20$ [46]. For an applied electric field E, it is possible to calculate r_c:

$$r_c = \sqrt{\left(\frac{2\xi_c}{E}\right)\left(\frac{D_T}{\mu_e}\right)}. \qquad (6.25)$$

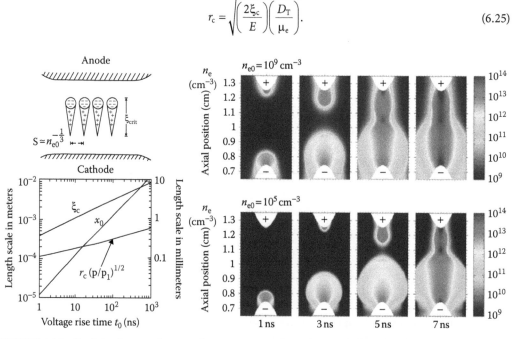

FIGURE 6.25 *Top left*: schematic diagram showing one row of simultaneously formed primary avalanches propagating before their respective space-charge fields reach the AST. *Bottom left*: calculated values of critical avalanche track length (ξ_c), critical avalanche head radius (r_c), and electron-deficient layer thickness (x_0) as a function of the voltage rise time and of the initial gas pressure/standard atmospheric pressure ratio p/p_1 for a He:Xe:F$_2$ = 200:8:1 mixture. (From Levatter, J. I., and S. C. Lin., *J. Appl. Phys.*, 51, 210–222, 1980.) *Right*: cross-sectional view of the electron density at different times for streamers with preionization levels of 10^5 cm^{-3} and 10^9 cm^{-3}. (From Bourdon, A. et al., *Plasma Sources Sci. Technol.*, 19, 034012, 2010.)

If the initial electron density n_{e0} exceeds a critical level $n_{crit} = 1/r_c^3$, then the seed electrons are spaced closely enough that their avalanches will overlap in space, thus canceling the space-charge field necessary for streamer formation. The above analysis applies to the part of the discharge gap far from the cathode region. Near the cathode, seed electrons are first swept away via drift before the local field reaches the breakdown threshold, and to prevent streamer formation here we must have $x_0 < r_c$, where x_0 is the thickness of the electron deficit region near the cathode at the moment t_0 at which the applied field reaches the breakdown threshold:

$$x_0 = \frac{1}{2} N\mu_e \left(\frac{E}{N}\right)_0 t_0, \tag{6.26}$$

where $(E/N)_0$ is the reduced field for breakdown. The applied field is assumed to ramp up from zero.

Numerical modeling of NRP discharges supports and adds to the above analytical theory. For nanosecond pulses, model streamerless breakdown originate from seed electrons [88], but secondary electrons do not play a role. Increasing n_{e0} is found to decrease the applied voltage necessary for such breakdown, which is predicted by the above theory. Furthermore, the delay time of breakdown for sub-microsecond pulses is found to be much shorter in experiment than in the model, and this is attributed to preionization. In this case, breakdown occurs in typical Townsend fashion and requires secondary electrons to build up the discharge.

Lo et al. [110] performed simulations of NRP discharges assuming a high degree of preionization from previous pulses (~10^{10} cm^{-3} of seed electrons) that demonstrate the sweeping out of seed electrons from the cathode region. As expected, space charge forms near the cathode, causing an ionization front to propagate from the edge of the space-charge region to the cathode. Preionization appears to prevent the build-up of space charge in the rest of the discharge gap. Although this model appears to follow the above theory qualitatively, one important difference is that the ionization front results from secondary electrons instead of seed electrons.

Even when streamers are present during breakdown, preionization exerts an influence on their characteristics. For NRP discharges in atmospheric pressure air, streamers have been simulated with preionization levels of $n_{e0} = 10^4$–10^9 cm^{-3} [18,100]. Preionization strongly influences the electron density, electric field, and propagation delay of positive but not negative streamers, as shown in Figure 6.25 (right). For low n_{e0}, electrons generated by photoionization ahead of the negative streamer aid the propagation of the positive streamer. However, at high preionization levels ($n_{e0} > 10^7$ cm^{-3}), photoionization is no longer necessary for positive streamer propagation. The model of Likhanskii et al. [111] exhibits the same behavior for NRP discharges in a surface DBD configuration, although in this case photoionization is not explicitly calculated.

6.4.12 Electron Number Density and Temperature

As mentioned in Section 6.4.4, two outstanding properties of NRP discharges are efficient generation of electrons and high average electron temperature. Recall that the electron temperature is approximative for the quasi-Maxwellian EEDF of NRP discharges. Measurements of the electron number density (n_e) and recombination time (τ_{recomb}) are summarized in Table 6.2. Current–voltage waveforms can provide spatially averaged values of n_e during the discharge but not during recombination, except when a DC probe current is imposed [85]. In this case $n_e = J/(e\mu_e E)$, where J is the current density, e is the electron charge, μ_e is the electron mobility, and E is the average electric field. Optical and electromagnetic diagnostics measure n_e during recombination: atomic line ratios, Stark broadening, laser Thomson scattering, CARS-based electric field measurement combined with current measurements, and millimeter-wave interferometry.

Broadly speaking, measurements of n_e using different techniques agree within an order of magnitude for the same discharge regime. NRP glow and DBDs attain maximum values of 10^{12}–10^{13} cm^{-3}, whereas n_e for NRP sparks reaches up to 10^{15} cm^{-3}. NRP microplasmas are capable of $n_e \gg 10^{16}$ cm^{-3}, all the more

TABLE 6.2 Measured Values of Electron Number Density (n_e) and Recombination Time (τ_{recomb}) for Various Nanosecond and NRP Discharges at or near Atmospheric Pressure

Gas (atm)	T_g (K)	Pulse Width (ns)	PRF (kHz)	Discharge	n_e (cm^{-3})	τ_{recomb}	Method	Reference
Air 1	2000	10	100	Glow	1.7×10^{12} max. 7×10^{11} min.	12 μs	Current–voltage	[85]
H$_2$ 0.2–0.4	300	10	12	Glow	2.5×10^{12} max. 4.5×10^{10} min.		CARS, current	[112]
Air 1	300	10	5	DBD	1.4×10^{13} avg.		Current–voltage	[107]
Air 1	300	12–13	1	DBD	2.3×10^{12} avg.		Current–voltage	[108]
He 1	300	90	1	DBD	8×10^{12} max.	~500 μs	mm-wave interferometry	[113]
Air 1	1000	10	30	Spark	3×10^{15} max.		Current–voltage	[96]
Air 1	300	75	0.1	Spark	10^{15} max.	~40 ns	Laser Thomson scattering	[114]
Ar 1	300	3	10	Microplasma	$\gg 10^{16}$ max.	300 ns	Stark/atomic line ratios	[115]

Notes: Here, T_g refers to the ambient gas temperature. For DBDs, the pulse width shown is the voltage across the discharge gap rather than that applied across the electrodes.

interesting that the gas temperature during the discharge has been measured to be only 480 K [115]. By comparison, arc discharges that are also generated in atmospheric pressure argon are only able to achieve such high values of n_e in the thermal regime where the gas temperature in the plasma is ~10,000 K [46].

As mentioned in Section 6.4.3, NRP discharges should generate very high initial electron temperatures that decrease once the plasma forms and screens the externally applied electric field. Time-resolved measurements of the electron temperature (T_e) have verified this effect for relatively long pulses. Using millimeter-wave interferometry, Lu and Laroussi [113] measured T_e for a submicrosecond DBD in atmospheric pressure helium, although the voltage pulse appearing in the gap is only 90 ns long, an effect of pulsed DBDs discussed in Section 6.4.10. The measured electron temperature reaches a peak value of 8.9 eV very soon after the beginning of the pulse and decreases to 2 eV after about 150 ns, which agrees well with T_e simulated in similar conditions [88]. For a 70-ns pulsed discharge in atmospheric pressure air, Grisch et al. [114] utilized laser Thomson scattering to find that between 30 and 70 ns from the beginning of the pulse, the electron temperature decreases from 2.6 to 1 eV. These measurements of T_e are consistent with the observation that, for the same reduced electric field, T_e is much lower in air than in helium [116]. Also, $T_e = 2.6$ eV is measured at the peak of the current pulse, but Iza et al. [88] showed that the maximum value of T_e should occur before the maximum of the current pulse, that is, before the formation of the plasma that screens the applied electric field.

6.4.13 Energy Cost of Ionization

In Section 6.4.4, we discussed that one of the core principles of the NRP pulsing scheme is to enable efficient ionization. It would therefore be useful to calculate the average energy cost per electron generated (ε_{elec}), which can be viewed as a measure of the energy efficiency of ionization. A reasonable approximation for ε_{elec} is eU_{plasma}/Q_{plasma}, where e is the electron charge, U_{plasma} is the discharge energy calculated from current–voltage measurements, and Q_{plasma} is the total charge calculated by integrating the current in time. Under steady-state conditions, Q_{plasma} would represent the integration of both electron and ion currents. The situation is very different for NRP and nanosecond discharges. Even under extremely high-field conditions of hundreds of Td, ion drift velocities are still only ~10^5 cm s^{-1}. In 100 ns, ions could drift only about 100 μm, which is only a small fraction of the gap distances used for the studies of concern here. Therefore, Q_{plasma} would only contain at most a small fraction of the total ions in the discharge gap, allowing us to assume that Q_{plasma} represents only the integration of the electron current.

Figure 6.26 shows values of ε_{elec} as a function of PRF for some of the studies shown in Table 6.2, as well as from several additional experiments performed at atmospheric pressure. The key observation is that ε_{elec} decreases by at least one order of magnitude as the PRF is increased from 10 Hz to 100 kHz. The horizontal fit from single-shot triggering up to PRF = 10 Hz symbolizes that there is little change in ε_{elec} in this PRF range. The log–log linear fit for PRF = $10-10^5$ Hz is only for the data points concerning air. For a given PRF, Figure 6.26 shows several data points from Pai et al. [95] for the NRP spark regime acquired at different applied voltages, wherein ε_{elec} always decreases with increasing applied voltage.

Note that the data shown in Figure 6.26 are for experiments conducted at different gas temperatures and different applied fields. The E/N thus varies between experiments, and it is known that the energy cost depends on E/N (see Figure 6.5 in Macheret et al. [82]). The decreasing trend of ε_{elec} may therefore not be fully attributable to increasing PRF alone. For example, it is possible that high PRF serves simply to keep the average gas temperature in the discharge gap appreciably above the ambient temperature, thereby reducing N. There does not, however, appear to be an obvious relationship between the discharge regime and ε_{elec}. There is also no clear link between ε_{elec} and pulse duration for studies at PRF $\leq$ 100 Hz, at which the effects of high PRF should not be important. Air discharges appear to have higher ε_{elec} than those in He, Ar, and N_2, but this is to be expected as the breakdown field and electron recombination time are much higher in air.

Thus, there is evidence that the ionization efficiency improves with increasing PRF. This is another fundamental difference between NRP discharges and other nanosecond discharges, aside from the ability to accumulate active species and generate streamerless spark discharges. It would appear that PRF > 10 kHz is necessary to obtain the roughly tenfold improvement in efficiency over nanosecond discharges. Even then, the lowest energy cost that has been reported for air thus far is ε_{elec} = 2 keV, which is about 30 times

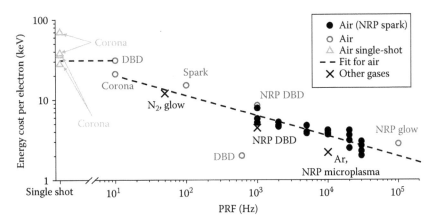

FIGURE 6.26 Average energy cost per electron generated (ε_{elec}) for nanosecond and NRP discharges of various regimes at atmospheric pressure, as a function of PRF. (From Packan, D. M., *Repetitive Nanosecond Glow Discharge in Atmospheric Pressure Air*, Stanford University, 2003; Tardiveau et al., *J. Phys. D Appl. Phys.*, 42, 175202, 2009; Williamson, J. M. et al., *J. Phys. D Appl. Phys.*, 39, 4400–4406, 2006; Udagawa, K. T. et al., *Plasma Sources Sci. Technol.*, 20, 055009, 2011; Pancheshnyi, S. et al., *Phys. Rev. E*, 71, 016407, 2005; Wang, D. Y. et al., *IEEE Trans. Plasma Sci.*, 35, 1098–1103, 2007; Takaki, K. et al., *Plasma Process. Polym.*, 3, 734–742, 2006; as well as the works cited in Table 6.2.) Single-shot measurements are shown to the left of the break in the PRF axis. Discharge regimes are indicated, with corona referring to a corona electrode configuration, that is, pin–plate geometry, in which various discharge regimes can occur: a corona localized near the pin, or diffuse or filamentary discharges bridging the gap.

above the value of 66 eV at Stoletov's point that is theoretically possible if electron beams could be used [82]. However, ε_{elec} shown in Figure 6.26 is a spatial and temporal average. Near the moment of maximum T_e or in the cathode fall region where the electric field is strong, we can expect lower ε_{elec}.

6.4.14 Accumulation of Active Species

In Section 6.4.4, we discussed that one of the core principles of the NRP pulsing scheme is to accumulate active species through the use of high PRF. At low PRF, certain species have long been known to accumulate, as illustrated in Figure 6.27 for the case of ozone, which is possible to accumulate with PRF as low as 10 Hz. Due to its long lifetime in air, a background density of ozone extending well beyond the discharge zone is established through diffusion.

Rather than present an extended discussion of plasma chemistry, in Table 6.3 we summarize measurements of species densities and lifetimes for air-like noncombustible mixtures at atmospheric pressure because there is a large body of work that can be assembled into a coherent overview of plasma chemistry. Fuel–air mixtures for plasma-assisted combustion are not considered here because the chemistry of combustion would obscure that strictly of the plasma. Several studies shown in Table 6.3 concern "pulsed coronas" generated in bare metal pin–plate configurations whose current densities are between those of NRP glow and spark discharges, hence their separate classification. Unlike the NRP corona shown in Figure 6.21, pulsed coronas in this context occupy the entire discharge gap, and the applied voltage extends for several hundred nanoseconds at less than 50% of the peak pulse voltage.

Broadly speaking, from Table 6.3 we see that the NRP glow/DBD produces up to ~10^{17} cm^{-3} of O, ~10^{15} cm^{-3} of O$_3$, and ~10^{14} cm^{-3} of OH in air at atmospheric pressure. NRP sparks/pulsed coronas produce up to ~10^{18} cm^{-3} of O, ~10^{17} cm^{-3} of O$_3$, ~10^{16} cm^{-3} of N and NO, and ~10^{15} cm^{-3} of OH. However, these studies vary greatly in PRF as well as in the time and location of measurement (i.e., in the plasma, afterglow, or downstream from the discharge). The lifetimes of various species range from the nanosecond to millisecond timescales. Most electronically excited species have lifetimes under 100 ns due to rapid quenching at atmospheric pressure of nearly kinetic rates. Other species such as OH and O have lifetimes in air-like mixtures that are long enough to enable accumulation

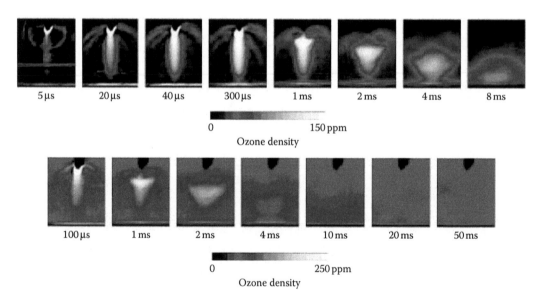

FIGURE 6.27 Time evolution of ozone density distribution after a nanosecond pulsed corona discharge in air at atmospheric pressure for PRF = 0.15 Hz (top) and 10 Hz (bottom). (From Ono, R., and T. Oda, *J. Phys. D Appl. Phys.*, 40, 176–182, 2007.)

TABLE 6.3 Maximum Densities and Lifetimes of Plasma Species in NRP and Nanosecond Discharges at Atmospheric Pressure

Gas	T_g (K)	Pulse Width (ns)	PRF (kHz)	Discharge	Species	Max Density (cm^{-3})	$1/e$ Lifetime	References
N$_2$	300	<1 μs	0.001	Pulsed corona	N	10^{16}	>300 μs	[121]
Air	300	<1 μs	Single-shot	Pulsed corona	O$_3$	10^{17}	>0.1 s	[122]
Air–H$_2$O	300	<1 μs	0.01	Pulsed corona	OH	10^{15}	~20 μs	[123]
Ar–H$_2$O	300	50	0.001–0.01	DBD	OH	10^{15}	900 μs	[124]
Air–H$_2$O						10^{14}	200 μs	
Air	300	14	0.05–0.6	DBD	O$_3$	8.5×10^{15}		[106]
Air	300	<1 μs	1.5×10^{-4}	DBD	O$_3$	$2–5 \times 10^{15}$	>0.02 s	[44]
He–N$_2$	300	90	1–10	DBD	NO	$1.6–2.2 \times 10^{14}$		[125]
					NO$_2$	$5.4–8.1 \times 10^{13}$		
Air	300	200–300	0.01	DBD	O		40 μs	[126]
Air	300	300	0.005	Spark	NO	3×10^{16} (right after discharge)		[126]
N$_2$/O$_2$/Ar					NO$_2$	2×10^{14} (downstream from discharge)		
Air	1000	10	10	Spark	O	1×10^{18}	25 μs	[59, 128]
Air	1000	10	10	Spark	N$_2$(A)	5×10^{14} (avg.)	<100 ns	[129]
N$_2$	1000	10	10	Spark	N$_2$(A)	2.5×10^{15}	600 ns	[129]
Air	1000	10	10	Spark	N$_2$(B)	3×10^{16}	3.8 ns	[130]
Air	1000	10	10	Spark	N$_2$(C)	3×10^{15}	2.0 ns	[130]
Air	1000	10	30	Glow	NO(A)	3×10^{10}	42 ns	[49]
Air	1000	10	30	Glow	N$_2$(B)	3×10^{13}	6 ns	[49]
Air	1000	10	30	Glow	N$_2$(C)	3×10^{12}	<4 ns	[49]
Air	1000	10	10	Glow	N$_2^+$(B)	9×10^{9}	<2 ns	[49]
Air	1000	10	30	Glow	O(3p ^{5}P)	9×10^{9}	6 ns	[49]
Air	1000	10	10	Glow	O	8×10^{16}		[59]

using PRF in the 10-kHz range, which is well within the capability of NRP discharges. As such, OH and O will now be discussed in detail as specific examples of the accumulation of active species made possible by NRP discharges.

6.4.15 Atomic Oxygen Concentration

The recent interest in atomic oxygen stems primarily from its role in plasma-assisted combustion, one of the most successful applications of NRP discharges in atmospheric pressure air thus far [86], both as a radical in the combustion process and as a key piece of evidence in support of the two-step mechanism for ultrafast heating to be discussed in Section 6.4.17.

Two-photon absorption laser-induced fluorescence (TALIF) has been used to measure the population and/or lifetime of ground-state atomic oxygen in nanosecond and NRP discharges in a wide range of experimental conditions [59,126,131]. Several comparisons can be made on the basis of these studies, which were all performed in air. First, for DBDs at low PRF generated at room temperature, the decay times of O atoms after the discharge were measured to be 2 ms and 10–40 μs at 60 Torr and 1 atm, respectively. The range of values in the latter case is due to humidity levels ranging from 0% to 2.4%, with higher humidity resulting in faster decay. The decay of atomic oxygen is often associated with three-body ozone creation:

$$O + O_2 + M \rightarrow O_3 + M. \tag{6.27}$$

The rate of this reaction scales as the square of the pressure, which is consistent with the aforementioned measured lifetimes if the air is assumed to be rather humid.

NRP discharges pulsed at PRF = 10 kHz have much higher densities of O atoms, which are produced about 20 ns after the discharge and increase rapidly in concentration for the following 100 ns. NRP glow discharges in atmospheric pressure at 1000 K can generate up to ~10^{17} cm^{-3} of O atoms compared to 10^{14} cm^{-3} in the case of a "single-pulse" DBD at 60 Torr. Note that there is only a difference in gas density by a factor of 4 between these two cases, and that the volumetric energy of NRP glows and DBDs are similar. Although the NRP glow in this case was operated at 1000 K instead of at room temperature, reaction (Equation 6.27) should still be the dominant loss mechanism at 1000 K. NRP spark discharges are capable of dissociating an even greater fraction (up to 50%) of O$_2$ in air, resulting in an O atom density of ~10^{18} cm^{-3} with a long lifetime of 25 μs. Thus, the accumulation of O atoms due to high PRF results in the exceptionally high O atom density of NRP discharges. Atomic oxygen in NRP spark discharges also plays an important role in the ultrafast gas heating effect, which will be discussed in Section 6.4.17.

6.4.16 OH Concentration

The accumulative effect has also been observed for OH in an air–H$_2$ mixture at low pressure [132]. The application of a single pulse leads to the OH density reaching ~10^{12} cm^{-3} after ~100 μs and then decaying with a lifetime of ~1 ms. On the other hand, for repetitive pulsing at PRF = 40 kHz only about ten pulses are required to reach OH densities of ~10^{13}–10^{14} cm^{-3}, which is a 10- to 100-fold increase over the single-pulse case. For low PRF (1–10 Hz) nanosecond discharges in saturated air–H$_2$O and Ar–H$_2$O mixtures at atmospheric pressure, Hibert et al. [124] measured OH(X) densities of ~10^{14} and ~10^{15} cm^{-3}. Given OH lifetimes of ~200 μs and ~1 ms for the cases of air–H$_2$O and Ar–H$_2$O mixtures, respectively, it is reasonable to expect that OH densities could be increased by over ten times by choosing 1/PRF that is shorter than these lifetimes.

6.4.17 Ultrafast Heating

In strong electric fields, discharges in molecular gases such as N$_2$–O$_2$ mixtures have been found to experience remarkably fast heating [133]. This effect occurs early in discharge development and is much faster than other modes of heating, such as vibrational–translational energy transfer in N$_2$. Instead, "ultrafast heating" in N$_2$–O$_2$ mixtures is attributed to a number of processes that depend on

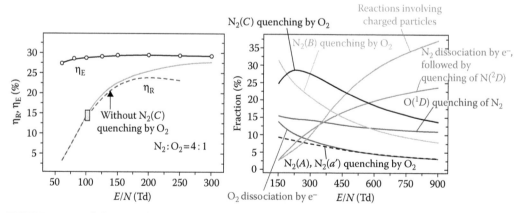

FIGURE 6.28 *Left*: fraction of energy of discharge in air converted to gas heating (η_R) and the fraction of energy spent on ionization, dissociation, and electronic excitation that is converted to gas heating (η_E) as a function of E/N. *Right*: contribution of various processes to the overall energy balance of ultrafast gas heating in air at 1 atm as a function of E/N. (From Popov, N. A., *J. Phys. D Appl. Phys.*, 44, 285201, 2011.)

E/N [134–136]. As shown in Figure 6.28 (left), the fraction of the discharge energy in air channeled into ultrafast heating (η_R) increases significantly with E/N and is not expected to exceed 40% for $E/N < 1000$ Td, although this upper limit requires further experimental verification. On the other hand, of the energy expended on ionization, dissociation, and electronic excitation, the fraction that is converted to ultrafast heating (η_E) remains about constant at 30% for a wide range of E/N and gas pressure. Figure 6.29 (right) shows how the importance of various processes to ultrafast heating changes significantly with E/N. At $E/N \leq 200$ Td, the dissociation of O_2 is the main route for heating, and this occurs through electron impact and the quenching of electronically excited nitrogen molecules $N_2\left(B^3\Pi_g, C^3\Pi_u, a'^1\Sigma_u^-\right)$ by O_2. The quenching of N_2 excited states by excited $O(^1D)$ atoms is also important. At $E/N > 400$ Td, ultrafast heating is primarily due to the electron-impact dissociation of N_2 and processes involving charged particles (e.g., ion–molecule reactions, electron–ion and ion–ion recombination), although the contribution of the latter remains an estimate at this point.

Ultrafast heating has been observed experimentally in NRP spark discharges in air at atmospheric pressure [95]. Gas temperature increases of 900 K in the first 20 ns after the rising edge of the pulse have been measured [59,130]. This is a heating rate of 4.5×10^{10} K s^{-1}, which is the highest ever reported for

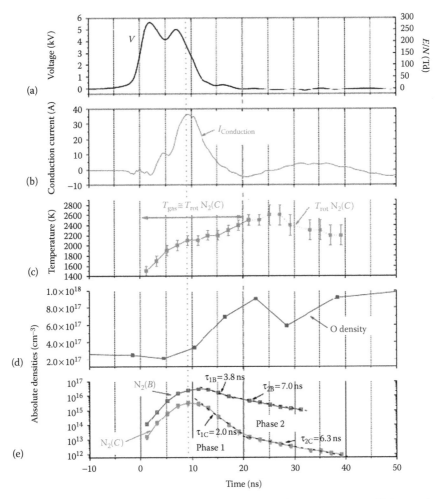

FIGURE 6.29 Measured (a) applied voltage, (b) conduction current, (c) gas temperature, (d) O atom density, and (e) $N_2(B)$ and $N_2(C)$ densities as a function of time for an NRP spark discharge in air at atmospheric pressure. (From Rusterholtz, D. L. et al., *50th AIAA Aerospace Sciences Meeting*, Nashville, TN, 2012.)

gas discharges and is only achievable otherwise through the use of lasers. In the case of NRP discharges, ultrafast heating has been proposed to proceed via a two-step mechanism:

$$N_2(X) + e^- \rightarrow N_2^* + e^-,$$

$$N_2^* + O_2 \rightarrow N_2(X) + 2O + \text{heat}. \tag{6.28}$$

First, the pulse rapidly populates electronically excited states N_2^* via electron impact. Second, the newly formed N_2^* are immediately quenched via O_2 dissociation at near-kinetic rates. The quenching is also exothermic, resulting in ultrafast heating wherein the contributions from the various states of N_2^* are expected to be in the proportions shown in Figure 6.28 (right). Figure 6.29 shows that the increase in O atom density coincides with decreases in the densities of $N_2(B)$ and $N_2(C)$, which is consistent with the two-step mechanism. Furthermore, the increase in gas temperature closely tracks that of atomic oxygen.

6.4.18 Electric Field

For both nanosecond and NRP discharges, measuring the electric field is challenging due to its highly nonuniform and dynamic behavior. This complicates modeling efforts because numerous electron-driven processes are highly sensitive to the reduced electric field. Often, approximations are made by dividing the applied voltage by the gap distance, with perhaps a correction for the cathode fall voltage. In Reference 85, the cathode fall voltage of an NRP glow discharge of 6 kV was determined by measuring the voltage across the gap as a function of gap distance and determining the intercept at null gap. The inferred cathode fall voltage was 1.5 kV. When nitrogen is present, it is possible to deduce the electric field using the intensity ratio of the second positive system of N_2 and the first negative system of N_2^+ [137]. Unfortunately, inaccuracies arise if the electric field is not spatially uniform and not in steady state. If the space- and time-integrated emission is considered, that is, if all of the emitted light is collected, then it is possible to obtain an accurate measurement of the peak electric field value [138].

Recently, a technique based on coherent anti-Raman scattering (E-CRS) has been applied to perform space- and time-resolved measurements of the electric field in NRP discharges in atmospheric pressure N_2, H_2, and air [112,139,140], as shown in Figure 6.30. E-CRS requires only a Raman-active component gas of sufficient density and virtually no assumptions about the state of the gas. The electric field is simply the intensity ratio of two scattered light signals multiplied by a constant found by calibration against a known electric field.

E-CRS therefore provides direct experimental evidence that the local electric field in the discharge is significantly less than the Laplacian field, showing the extent of screening by space charge. In the case of bare metal electrodes, the screening is attributed to the formation of a cathode sheath. As discussed

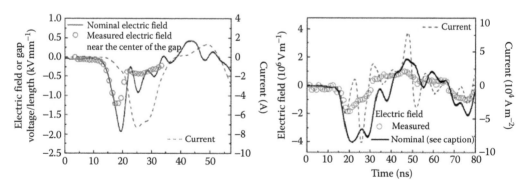

FIGURE 6.30 Temporal evolution of the nominal electric field (the electric field expected without any space charge: solid curve), electric field at the center of the gap measured by E-CRS (empty circles), and the current (dashed curve) for NRP discharges in (left) a bare metal electrode configuration (From Ito, T. et al., *J. Phys. D Appl. Phys.*, 43, 062001, 2010.), and (right) for an NRP DBD. (From Ito, T. et al., *Phys. Rev. Lett.*, 107, 065002, 2011.)

in Section 6.4.3, Macheret et al. [82] modeled the development of the cathode sheath of NRP discharges and found that nearly half of the applied voltage fell over the cathode fall region, which is consistent with Figure 6.30. For NRP DBDs, this space-charge effect is observed before breakdown, indicating the presence of significant residual charge stored on the surface of the dielectric or from preionization.

Another phenomenon associated with space charge is electric field reversal, which has thus far been modeled in NRP discharges [82,111,141]. Space-charge regions can develop either upon the formation of the cathode fall or from charge deposited on a dielectric layer. During the falling edge of a positive pulse, the voltage at the anode may decrease below the positive potential of the space-charge region. In this case, the space-charge region becomes a virtual anode, thus reversing the direction of the electric field in the gap. The potential difference may become large enough to cause reverse breakdown. For a negatively pulsed DBD, the situation is similar except that the virtual electrode is the cathode.

6.4.19 Cathode Fall and Secondary Electron Emission Processes

Due to fast breakdown and preionization, certain aspects of cathode fall dynamics and secondary emission are expected to be particular to NRP discharges. Few experimental data exist due to the difficulty of directly probing a region that is typically at the micron scale at atmospheric pressure and next to a surface. Numerical and analytical models of the development of the cathode fall region in response to nanosecond pulses have been performed for a wide range of pressures, secondary emission processes, preionization levels, pulse widths, and gas compositions [110,142–144], but the results share several common features.

Cathode fall voltages in the kilovolt range are consistently found. As shown in Figure 6.31a–c, in all cases both the cathode fall voltage and length increase up to the moment of breakdown (2–100 ns,

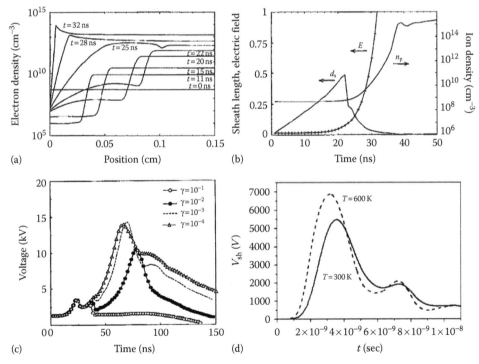

FIGURE 6.31 (a) Electron density in the cathode region as a function of position for different times showing expansion, then collapse of the sheath upon breakdown, (b) sheath length, electric field, and ion density as a function of time, and (c) cathode sheath voltage for different secondary emission coefficients as a function of time. (From Belasri, A. et al., *J. Appl. Phys.*, 74, 1553–1567, 1993.) (d) Cathode sheath voltage as a function of time for a much shorter pulse (~5 ns) than in (a)–(c), where the pulse duration was ~100 ns. (From Macheret, S. O. et al., *Phys. Plasmas*, 13, 023502, 2006.)

depending on the conditions), when the space-charge field from the ion layer at the plasma-sheath boundary becomes strong enough to screen the electric field in the plasma bulk yet strongly enhance the field in the cathode region. At this moment, strong ionization due to the space-charge field in the cathode region causes the plasma-sheath boundary to move rapidly toward the cathode. This phenomenon is otherwise termed an ionization front and is similar to a cathode-directed streamer in terms of its initiation mechanism (i.e., the AST, see Section 6.4.6.1) and propagation speed. The length of the cathode sheath decreases while its electric field increases. As shown in Figure 6.31c, the cathode fall voltage generally remains in the kilovolt range but is very dynamic, with its evolution in time influenced by factors such as secondary emission. If the falling edge of the applied voltage pulse arrives soon after breakdown, then the cathode fall voltage is forced downward, as shown in Figure 6.31d. This is the same situation in which field reversal can occur, as discussed in Section 6.4.18.

Because of the short pulse duration of NRP discharges, the vast majority of ions do not have time to cross discharge gaps on the millimeter scale and greater before the pulse is finished. Therefore, unlike in typical Townsend breakdown, secondary emission is not dominated by ion–electron emission, although ions near the cathode can still contribute to this process [110]. When the breakdown occurs via a cathode-directed streamer, photoemission and ion–electron emission do not become significant until 10 and 20 ns, respectively, after contact of the streamer with the cathode [145]. Thus, the typical NRP pulse width of ~10 ns is too short for these processes, especially when the time required for the AST and streamer propagation across the gap is taken into account. Field emission governs secondary emission for pulses with extremely short rise times (0.5–2.5 ns), whereas explosive emission also plays a role in somewhat longer pulses [146].

6.4.20 Runaway Electrons and X-Rays

As mentioned in Section 6.4.2, there has been a lot of work on the role of runaway electrons in the breakdown on nanosecond discharges, particularly at atmospheric pressure. For high overvoltages, runaway electrons and the x-rays that they produce via bremsstrahlung have long been known to preionize the gas and enable the generation of diffuse discharges [147]. The electric field threshold for electron runaway is much higher than that for the Townsend and streamer breakdown mechanisms. The considerable literature on this subject has already been well summarized [148], and therefore the focus here is on recent work at high PRF.

The possibility of using nanosecond discharges in air at atmospheric pressure as practical sources of electron beams and x-rays has motivated work at high PRF of up to 3 kHz [149], as shown in Figure 6.32. Depending on the geometry of the reactor, the detected x-ray intensity can increase or decrease as the PRF is increased from 1.5 Hz to 1 kHz [98], with the former tendency attributed to the fact that open

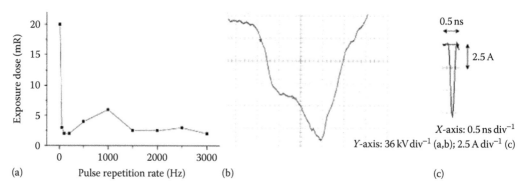

FIGURE 6.32 (a, b) Measured x-ray dose as a function of PRF. (From Tarasenko, V. F., *Appl. Phys. Lett.*, 88, 081501, 2006.) (c) Example of applied voltage pulse and current from runaway electrons. For the axis scale information for (b) and (c), "div" represents divisions on the oscilloscope display. (From Shao, T. et al., *Phys. Plasmas*, 18, 053502, 2011.)

electrode geometry permits heated gas to escape the discharge gap, thereby preventing the initially diffuse discharge from constriction. On the other hand, when the heated gas does not have as much room for escape, the breakdown voltage decreases, and therefore any generation of runaway electrons must occur at lower field strength.

The role of runaway electron in breakdown at yet higher PRF, that is, greater than 10 kHz, has not yet been studied experimentally. Up to now, the streamer and Townsend mechanisms of breakdown have sufficed to explain the dynamics of NRP discharges. Preionization from previous pulses may cause immediate and severe screening of the electric field in the plasma bulk, limiting runaway electrons to the high-field cathode region (see Section 6.4.11), and rapid screening preceding breakdown has been observed experimentally (see Figure 6.30). However, it is not at all out of the question that runaway electrons and x-rays may be present in NRP discharges, especially given the fact that streamer heads in such discharges have been modeled to have maximum electric field values of up to about 250 kV cm^{-1} in atmospheric pressure air at room temperature despite a high preionization electron density of 10^9 cm^{-3} [104]. The reduced electric field in this case is 329 V cm^{-1} Torr^{-1}, which is many times the static breakdown voltage and approaches the runaway threshold of 590 V cm^{-1} Torr^{-1} for nitrogen [65].

6.4.21 NRP Discharges—Conclusions and Outlook

In Section 6.4, we have provided an overview of a number of major areas of recent research in NRP discharges at atmospheric pressure. It has included mention of work at lower PRF and/or pressure whenever necessary for illustrating the same behavior that should be expected in NRP discharges. The overall picture that emerges is that NRP discharges do, or are expected to, exhibit a wide range of phenomena, some of which are common to nanosecond discharges in general:

- Elevated electron temperature and density
- Efficient generation of plasma chemical species
- Ultrafast heating
- Electric field reversal
- Rapid expansion and collapse of the cathode sheath
- Presence of runaway electrons and x-rays

On the other hand, it is important to understand that NRP discharges are not simply nanosecond discharges repeated more often, any more than RF or microwave discharges are the same as sinusoidally driven discharges but at higher frequency. A number of distinguishing characteristics are due to preionization, and an NRP discharge might be considered its own source of initial electrons. Phenomena unique to NRP discharges include

- G-S transition that depends on the cumulative effect of energy deposition.
- Accumulation of plasma chemical species.
- Changing breakdown from the streamer to Townsend mechanisms.
- Increased energy efficiency of ionization.
- Changes in streamer initiation and propagation, including the elimination of the necessity of photoionization for positive streamer propagation.
- Much faster screening of the electric field in the plasma bulk.

Thus, the field of NRP discharge physics has advanced considerably since the publication of Becker et al. [66]. However, there are still a number of fundamental problems remaining:

- Does the ionization efficiency continue to decrease for PRF > 100 kHz?
- Clarify the importance of ion–electron emission at the nanosecond scale
- Determine the presence and importance of runaway electrons and x-rays

- Plasma–fluid interaction in NRP discharges, particularly with regard to how the postdischarge channel affects the upcoming breakdown event
- Determine how the plasma column cools following ultrafast heating
- Combining NRP discharges with microplasmas to maximize electron temperature
- Direct comparison of the characteristics of NRP discharges with those of self-pulsed high-PRF nanosecond discharges (i.e., TSs), discussed in Section 6.3
- Demonstrate that the NRP glow regime can be generated stably in air at atmospheric pressure at 300 K

The NRP glow in ambient air at 300 K and 1 atm was very recently obtained [150] using electrodes of radius of curvature in the range 20–50 microns, with pulse and electrode geometry characteristics verifying the conditions of Equation 6.22. A planar metal electrode located closely behind the cathode tip was used to prevent streamer branching at the sharp cathode tip. Without this plane, a multi-channel glow discharge consisting of several single channel glow discharges was obtained.

Acknowledgments

This work was supported by Slovak Research and Development Agency SK-FR-0038-09 and Slovak grant agency VEGA Grant Nos. 1/0668/11 and 1/0998/12, and also by the ANR project PLASMAFLAME, by a Grant-in-Aid for Young Scientists B, Grant No. 23760688, a Grant-in-Aid for Scientific Research on Innovative Areas (Frontier Science of Interactions between Plasmas and Nano-Interfaces, Grant No. 21110002) from the Ministry of Education, Culture, Sports, Science, and Technology (MEXT) of Japan, a Grant-in-Aid for JSPS Fellows from the Japan Society for the Promotion of Science (JSPS), and JSPS Postdoctoral Fellowship for Foreign Researchers. Zdenko Machala would like to greatly appreciate Dr. Emmanuel Marode of LPGP Paris-Sud for extensive discussions and providing his manuscript of the chapter referenced here as [33] that was a valuable review of streamer-to-spark transition mechanisms. David Z. Pai thanks Professor Kazuo Terashima of the University of Tokyo for allowing the time necessary for the writing of his portion of this chapter.

References

1. Penetrante, B. M., and S. E. Schultheis, ed. 1993. *Non-Thermal Plasma Techniques for Pollution Control*. New York: Springer.
2. Kogelschatz, U. 2003. Dielectric-barrier discharges: Their history, discharge physics, and industrial applications. *Plasma Chem. Plasma Process.* 23:1–46.
3. Lowke, J. J., and R. Morrow. 1995. Theoretical analysis of removal of oxides of sulphur and nitrogen in pulsed operation of electrostatic precipitators. *IEEE Trans. Plasma Sci.* 23:661–671.
4. Meek, J. M. A. 1940. Theory of spark discharge. *Phys. Rev.* 57:722–728.
5. Morrow, R., and J. J. Lowke. 1997. Streamer propagation in air. *J. Phys. D Appl. Phys.* 30:614–627.
6. Kulikovsky, A. A. 1998. Analytical model of positive streamer in weak field in air: Application to plasma chemical calculations. *IEEE Trans. Plasma Sci.* 26:1339–1346.
7. Peek, F. W. 1929. *Phenomena in High Voltage Engineering*. New York: McGraw-Hill.
8. Llewellyn-Jones, F. 1966. *Ionization and Breakdown in Gases*. London: Metheun.
9. Meek, J. M., and J. D. Craggs. 1978. *Electrical Breakdown of Gases*. New York: Wiley.
10. Loeb, L. B. 1965. *Electrical Coronas: Their Basic Mechanisms*. Berkeley, CA: University of California Press.
11. Raether, H. 1964. *Electron Avalanches and Breakdown in Gases*. Washington, DC: Butterworths.
12. Marode, E. 1975. Mechanism of spark breakdown in air at atmospheric pressure between a positive point and a plane. 1. Experimental—Nature of streamer track. *J. Appl. Phys.* 46:2005–2015.
13. Lowke, J. J. 1992. Theory of electrical breakdown in air—The role of metastable oxygen molecules. *J. Phys. D Appl. Phys.* 25:202–210.

14. Larsson, A. 1998. The effect of a large series resistance on the streamer-to-spark transition in dry air. *J. Phys. D Appl. Phys.* 31:1100–1108.

15. Naidis, G. V. 1999. Simulation of streamer-to-spark transition in short non-uniform air gaps. *J. Phys. D Appl. Phys.* 32:2649–2654.

16. Aleksandrov, N. L., and E. M. Bazelyan. 1999. Ionization processes in spark discharge plasmas. *Plasma Sources Sci. Technol.* 8:285–294.

17. Kulikovsky, A. A. 2001. The efficiency of radicals production by positive streamer in air: The role of Laplacian field. *IEEE Trans. Plasma Sci.* 29:313–317.

18. Bourdon, A., Z. Bonaventura, and S. Celestin. 2010. Influence of the pre-ionization background and simulation of the optical emission of a streamer discharge in preheated air at atmospheric pressure between two point electrodes. *Plasma Sources Sci. Technol.* 19:034012.

19. Efremov, N. M., B. Y. Adamiak, V. I. Blouchin, S. J. Dadashev, K. J. Dmitriev, O. P. Gryaznova, and V. F. Jusbashev. 2000. Action of a self-sustained glow discharge in atmospheric pressure air on biological objects. *IEEE Trans. Plasma Sci.* 28:238–241.

20. Wagner, H. E., R. Brandenburg, K. V. Kozlov, A. Sonnenfeld, P. Michel, and J. F. Behnke. 2003. The barrier discharge: Basic properties and applications to surface treatment. *Vacuum* 71:417–436.

21. Mizuno, A., J. S. Clements, and R. H. Davis. 1986. A method for the removal of sulfur dioxide from exhaust gas utilizing pulsed streamer corona for electron energization. *IEEE Trans. Ind. Appl.* 22:516–522.

22. Penetrante, B. M., J. N. Bardsley, and M. C. Hsiao. 1997. Kinetic Analysis of non-thermal plasmas used for pollution control. *Jpn. J. Appl. Phys. Part 1—Regul. Pap. Brief Commun. Rev. Pap.* 36:5007–5017.

23. Masuda, S., and H. Nakao. 1990. Control of NO_x by positive and negative pulsed corona discharges. *IEEE Trans. Ind. Appl.* 26:374–383.

24. Walsh, J. L., J. J. Shi, and M. G. Kong. 2006. Submicrosecond pulsed atmospheric glow discharges sustained without dielectric barriers at kilohertz frequencies. *Appl. Phys. Lett.* 89:161505.

25. Pancheshnyi, S. V., D. A. Lacoste, A. Bourdon, and C. O. Laux. 2006. Ignition of propane-air mixtures by a repetitively pulsed nanosecond discharge. *IEEE Trans. Plasma Sci.* 34:2478–2487.

26. Pai, D., D. A. Lacoste, and C. O. Laux. 2008. Images of nanosecond repetitively pulsed plasmas in preheated air at atmospheric pressure. *IEEE Trans. Plasma Sci.* 36:974–975.

27. Bastien, F., and E. Marode. 1979. The determination of basic quantities during glow-to-arc transition in a positive point-to-plane discharge. *J. Phys. D Appl. Phys.* 12:249–264.

28. Machala, Z., M. Morvova, E. Marode, and I. Morva. 2000. Removal of cyclohexanone in transition electric discharges at atmospheric pressure. *J. Phys. D Appl. Phys.* 33:3198–3213.

29. Machala, Z., E. Marode, C. O. Laux, and C. H. Kruger. 2004. DC glow discharges in atmospheric pressure air. *J. Adv. Oxid. Technol.* 7:133–137.

30. Yu, L., C. O. Laux, D. M. Packan, and C. H. Kruger. 2002. Direct-current glow discharges in atmospheric pressure air plasmas. *J. Appl. Phys.* 91:2678–2686.

31. Akishev, Y., M. Grushkin, V. Karalnik, A. Petryakov, and N. Trushkin. 2010. Non-equilibrium constricted dc glow discharge in N2 flow at atmospheric pressure: Stable and unstable regimes. *J. Phys. D Appl. Phys.* 43:075202.

32. Marode, E., A. Goldman, and M. Goldman. 1993. High pressure discharge as a trigger for pollution control. In *Non-Thermal Plasma Techniques for Pollution Control* (NATO ASI Series G34, vol. 1). eds. B. M. Penetrante, and S. E. Schultheis. Berlin/Heidelberg: Springer-Verlag. pp. 167–190.

33. Marode, E. 2012. The prevented spark plasma in atmospheric air: From field to thermal equilibrium. In *Plasma for Environmental Issues*. eds. E. Tatarova, V. Guerra, and E. Benova. Sofia: ArtGraf.

34. Hafez, R., S. Samson, and E. Marode. 1995. A prevented spark reactor for pollutant control: Investigation of NO_x removal. In *12th International Symposium on Plasma Chemistry*, 21–25 August 1995. University of Minnesota, Minneapolis, MN. pp. 855–861.

35. Bessieres, D., J. Paillol, A. Gibert, and L. Pecastaing. 2005. Positive corona ignition and development in air at atmospheric pressure under a heaviside voltage pulse. *Plasma Process. Polym.* 2:183–187.

36. Goldman, M., and A. Goldman, eds. 1978. *Corona Discharges*. New York: Academic Press.

37. Marode, E. 1983. The glow-to-arc transition. In *Electrical Breakdown and Discharges in Gases.* eds. E. E. Kunhardt, and L. Luessen. New York: Plenum, p. 119.

38. Marode, E., D. Djermoune, P. Dessante, C. Deniset, P. Segur, F. Bastien, A. Bourdon, and C. Laux. 2009. Physics and applications of atmospheric non-thermal air plasma with reference to environment. *Plasma Phys. Control. Fusion* 51:124002.

39. Mitchell, J. B. A. 1990. The dissociative recombination of molecular ions. *Physics Reports* 186:215–248.

40. Johnsen, R. 1993. Electron temperature dependence of the recombination of $H_3O+(H_2O)_n$ ions with electrons. *J. Chem. Phys.* 98:5390.

41. Marode, E., F. Bastien, and M. Bakker. 1979. A model of the streamer-induced spark formation based on neutral dynamics. *J. Appl. Phys.* 50:140–146.

42. Sigmond, R. S. 1984. The residual streamer channel—Return strokes and secondary streamers. *J. Appl. Phys.* 56:1355–1370.

43. Bastien, F., and E. Marode. 1985. Breakdown simulation of electronegative gases in non-uniform field. *J. Phys. D Appl. Phys.* 18:377–394.

44. Ono, R., and T. Oda. 2007. Ozone production process pulsed positive dielectric barrier discharge. *J. Phys. D Appl. Phys.* 40:176–182.

45. Gallimberti, I. 1979. The mechanism of the long spark formation. *J. Phys. Colloq.* 40:C7-193–C197-250.

46. Raizer, Y. P. 1991. *Gas Discharge Physics.* Berlin: Springer.

47. Bondiou, A., and I. Gallimberti. 1994. Theoretical modelling of the development of the positive spark in long gaps. *J. Phys. D Appl. Phys.* 27:1252–1266.

48. Goelian, N., P. Lalande, A. Bondiou-Clergerie, G. L. Bacchiega, A. Gazzani, and I. Gallimberti. 1997. A simplified model for the simulation of positive-spark development in long air gaps. *J. Phys. D Appl. Phys.* 30:2441–2452.

49. Pai, D. Z., G. D. Stancu, D. A. Lacoste, and C. O. Laux. 2009. Nanosecond repetitively pulsed discharges in air at atmospheric pressure—The glow regime. *Plasma Sources Sci. Technol.* 18:045030.

50. Janda, M., and Z. Machala. 2008. Transient-spark discharge in $N^{-2}/CO^2/H_2O$ mixtures at atmospheric pressure. *IEEE Trans. Plasma Sci.* 36:916–917.

51. Janda, M., V. Martisovits, and Z. Machala. 2011. Transient spark—A DC driven repetitively pulsed discharge and its control by electric circuit parameters. *Plasma Sources Sci. Technol.* 20:035015.

52. Laux, C. O., T. G. Spence, C. H. Kruger, and R. N. Zare. 2003. Optical diagnostics of atmospheric pressure air plasmas. *Plasma Sources Sci. Technol.* 12:125–138.

53. Fantz, U. 2006. Basics of plasma spectroscopy. *Plasma Sources Sci. Technol.* 15:S137–S147.

54. Machala, Z., E. Marode, M. Morvova, and P. Lukac. 2005. DC glow discharges in atmospheric air as a source for VOC abatement. *Plasma Process. Polym.* 2:152–161.

55. Machala, Z., M. Janda, K. Hensel, I. Jedlovsky, L. Lestinska, V. Foltin, V. Martisovits, and M. Morvova. 2007. Emission spectroscopy of atmospheric pressure plasmas for bio-medical and environmental applications. *J. Mol. Spectrosc.* 243:194–201.

56. Janda, M., Z. Machala, A. Niklova, and V. Martisovits. 2011. Streamer-to-spark transition in transient spark—A DC driven nanosecond pulsed discharge in atmospheric air. *Plasma Sources Sci. Technol.* 20:035015.

57. Naidis, G. V. 2009. Simulation of streamer-induced pulsed discharges in atmospheric-pressure air. *Eur. Phys. J. Appl. Phys.* 47:22803–22808.

58. Machala, Z., I. Jedlovsky, L. Chladekova, B. Pongrac, D. Giertl, M. Janda, L. Sikurova, and P. Polcic. 2009. DC discharges in atmospheric air for bio-decontamination—Spectroscopic methods for mechanism identification. *Eur. Phys. J. D* 54:195–204.

59. Stancu, G. D., F. Kaddouri, D. A. Lacoste, and C. O. Laux. 2010a. Atmospheric pressure plasma diagnostic by OES, CRDS and TALIF. *J. Phys. D Appl. Phys.* 43:124002.

60. Gibert, A., and F. Bastien. 1989. Fine structure of streamers. *J. Phys. D Appl. Phys.* 22:1078–1082.

61. van Veldhuizen, E. M., P. C. M. Kemps, and W. R. Rutgers. 2002. Streamer branching in a short gap: The influence of the power supply. *Plasma Sources Sci. Technol.* 30:162–163.

62. Bastien, F., and E. Marode. 1977. Stark broadening of Hα and Hβ in ionized gases with space-charge field. *J. Quant. Spectrosc. RA* 17:453–469.

63. Janda, M., V. Martisovits, M. Morvova, Z. Machala, and K. Hensel. 2007. Monte Carlo simulations of electron dynamics in N_2/CO_2 mixtures. *Eur. Phys. J. D* 45:309–315.

64. Gigosos, M. A., M. A. González, and V. Cardeñoso. 2003. Computer simulated Balmer-alpha, -beta and -gamma Stark line profiles for non-equilibrium plasmas diagnostics. *Spectrochim Acta B* 58:1489–1504.

65. Korolev, Y. D., and G. A. Mesyats. 1998. *Physics of Pulsed Breakdown in Gases.* Yekaterinburg: UD RAS.

66. Becker, K. H., U. Kogelschatz, K. H. Schoenbach, and R. J. Barker, eds. 2005. *Non-equilibrium Air Plasmas at Atmospheric Pressure.* Bristol: IOP Publishing.

67. Fletcher, R. C. 1949. Impulse breakdown in the 10(-9)-sec range of air at atmospheric pressure. *Phys. Rev.* 76:1501–1511.

68. Felsenthal, P., and J. M. Proud. 1965. Nanosecond-pulse breakdown in gases. *Phys. Rev.* 139:1796.

69. Mesyats, G. A., Y. I. Bychkov, and V. V. Kremnev. 1972. Impulse nanosecond electric-discharge in a gas. *Uspekhi Fiz. Nauk* 107:201.

70. Kunhardt, E. E., and W. W. Byszewski. 1980. Development of overvoltage breakdown at high gas pressure. *Phys. Rev. A* 21:2069–2077.

71. Dickey, F. R. 1952. Contribution to the theory of impulse breakdown. *J. Appl. Phys.* 23:1336–1339.

72. Heard, H. G. 1963. Ultra-violet gas laser at room temperature. *Nature* 200:667.

73. Beaulieu, A. J. 1970. Transversely excited atmospheric pressure CO_2 lasers. *Appl. Phys. Lett.* 16:504.

74. Reilly, J. P. 1972. Pulser sustainer electric-discharge laser. *J. Appl. Phys.* 43:3411.

75. Hill, A. E. 1973. Continuous uniform excitation of medium-pressure CO_2-laser plasmas by means of controlled avalanche ionization. *Appl. Phys. Lett.* 22:670–673.

76. Levatter, J. I., and S. C. Lin. 1980. Necessary conditions for the homogeneous formation of pulsed avalanche discharges at high gas pressures. *J. Appl. Phys.* 51:210–222.

77. Hackam, R., and H. Akiyama. 2000. Air pollution control by electrical discharges. *IEEE Trans. Dielectr. Electr. Insul.* 7:654–683.

78. Vasilyak, L. M., S. V. Kostiouchenko, N. N. Koudriavtsev, and I. V. Filiouguine. 1994. High-speed ionization waves at an electric break down. *Uspekhi Fiz. Nauk* 164:263–286.

79. Starikovskaia, S. M., N. B. Anikin, S. V. Pancheshnyi, D. V. Zatsepin, and A. Y. Starikovskii. 2001. Pulsed breakdown at high overvoltage: Development, propagation and energy branching. *Plasma Sources Sci. Technol.* 10:344–355.

80. Babich, L. P., T. V. Loiko, and V. A. Tsukerman. 1990. High-voltage nanosecond discharge in dense gases developing with runaway electrons regime under high overvoltages. *Uspekhi Fiz. Nauk* 160:49–82.

81. Stark, R. H., and K. H. Schoenbach. 2001. Electron heating in atmospheric pressure glow discharges. *J. Appl. Phys.* 89:3568–3572.

82. Macheret, S. O., M. N. Shneider, and R. B. Miles. 2002. Modeling of air plasma generation by repetitive high-voltage nanosecond pulses. *IEEE Trans. Plasma Sci.* 30:1301–1314.

83. Grekhov, I. V., V. M. Efanov, A. F. Kardosysoev, and S. V. Shenderey. 1985. Power drift step recovery diodes (DSRD). *Solid-State Electron.* 28:597–599.

84. Nagulapally, M., G. V. Candler, C. O. Laux, L. Yu, D. M. Packan, C. H. Kruger, R. Stark, and K. H. Schoenbach. 2000. Experiments and simulations of DC and pulsed discharges in air plasmas. In *31st AIAA Plasmadynamics and Lasers Conference*, 19–22 June 2000, Denver, CO, AIAA-2000-2417.

85. Kruger, C. H., C. O. Laux, L. Yu, D. M. Packan, and L. Pierrot. 2002. Nonequilibrium discharges in air and nitrogen plasmas at atmospheric pressure. *Pure Appl. Chem.* 74:337–347.

86. Packan, D. M. 2003. *Repetitive Nanosecond Glow Discharge in Atmospheric Pressure Air.* Stanford, CA: Stanford University.

87. Pilla, G., D. Galley, D. A. Lacoste, F. Lacas, D. Veynante, and C. O. Laux. 2006. Stabilization of a turbulent premixed flame using a nanosecond repetitively pulsed plasma. *IEEE Trans. Plasma Sci.* 34:2471–2477.

88. Iza, F., J. L. Walsh, and M. G. Kong. 2009. From submicrosecond- to nanosecond-pulsed atmospheric-pressure plasmas. *IEEE Trans. Plasma Sci.* 37:1289–1296.

89. van Veldhuizen, E. M., W. R. Rutgers, and V. A. Bityurin. 1996. Energy efficiency of NO removal by pulsed corona discharges. *Plasma Chem. Plasma Process.* 16:227–247.

90. Ono, R., Y. Nakagawa, and T. Oda. 2011. Effect of pulse width on the production of radicals and excited species in a pulsed positive corona discharge. *J. Phys. D Appl. Phys.* 44:485201.

91. Xiong, Q., X. Lu, K. Ostrikov, Z. Xiong, Y. Xian, C. Zhou, J. Hu, W. Gong, and Z. Jiang. 2009. Length control of He atmospheric plasma jet plumes: Effects of discharge parameters and ambient air. *Phys. Plasmas* 16:043505.

92. Walsh, J. L., and M. G. Kong. 2011. Portable nanosecond pulsed air plasma jet. *Appl. Phys. Lett.* 99:081501.

93. Aleksandrov, N. L., F. I. Vysikailo, R. S. Islamov, I. V. Kochetov, A. P. Napartovich, and V. G. Pevgov. 1981. Electron-distribution function in 4-1 N-2-O-2 mixture. *High Temp.* 19:17–21.

94. Nighan, W. L. 1970. Electron energy distributions and collision rates in electrically excited N_2, CO, and CO_2. *Phys. Rev. A* 2:1989.

95. Pai, D. Z., D. A. Lacoste, and C. O. Laux. 2010a. Transitions between corona, glow, and spark regimes of nanosecond repetitively pulsed discharges in air at atmospheric pressure. *J. Appl. Phys.* 107:093303.

96. Pai, D. Z., D. A. Lacoste, and C. O. Laux. 2010b. Nanosecond repetitively pulsed discharges in air at atmospheric pressure—The spark regime. *Plasma Sources Sci. Technol.* 19:065015.

97. Tardiveau, P., N. Moreau, S. Bentaleb, C. Postel, and S. Pasquiers. 2009. Diffuse mode and diffuse-to-filamentary transition in a high pressure nanosecond scale corona discharge under high voltage. *J. Phys. D Appl. Phys.* 42:175202.

98. Shao, T., V. F. Tarasenko, C. Zhang, Y. V. Shut'ko, and P. Yan. 2011. X-ray and runaway electron generation in repetitive pulsed discharges in atmospheric pressure air with a point-to-plane gap. *Phys. Plasmas* 18:053502.

99. Bazelyan, E. M., and Y. P. Raizer. 1998. *Spark Discharge*. Boca Raton, FL: CRC Press.

100. Celestin, S., Z. Bonaventura, B. Zeghondy, A. Bourdon, and P. Segur. 2009. The use of the ghost fluid method for Poisson's equation to simulate streamer propagation in point-to-plane and point-to-point geometries. *J. Phys. D Appl. Phys.* 42:065203.

101. Staack, D., B. Farouk, A. Gutsol, and A. Fridman. 2008. DC normal glow discharges in atmospheric pressure atomic and molecular gases. *Plasma Sources Sci. Technol.* 17:025013.

102. Naidis, G. V. 2008. Simulation of spark discharges in high-pressure air sustained by repetitive high-voltage nanosecond pulses. *J. Phys. D Appl. Phys.* 41:234017.

103. Akishev, Y., O. Goossens, T. Callebaut, C. Leys, A. Napartovich, and N. Trushkin. 2001. The influence of electrode geometry and gas flow on corona-to-glow and glow-to-spark threshold currents in air. *J. Phys. D Appl. Phys.* 34:2875–2882.

104. Tholin, F., and A. Bourdon. 2011. Influence of temperature on the glow regime of a discharge in air at atmospheric pressure between two point electrodes. *J. Phys. D Appl. Phys.* 44:385203.

105. Kuznetsova, I. V., N. Y. Kalashnikov, A. F. Gutsol, A. A. Fridman, and L. A. Kennedy. 2002. Effect of "overshooting" in the transitional regimes of the low-current gliding arc discharge. *J. Appl. Phys.* 92:4231–4237.

106. Williamson, J. M., D. Trump, P. Bletzinger, and B. N. Ganguly. 2006. Comparison of high-voltage ac and pulsed operation of a surface dielectric barrier discharge. *J. Phys. D Appl. Phys.* 39:4400–4406.

107. Walsh, J. L., and M. G. Kong. 2007. 10 ns pulsed atmospheric air plasma for uniform treatment of polymeric surfaces. *Appl. Phys. Lett.* 91:251504.

108. Shao, T., K. H. Long, C. Zhang, P. Yan, S. C. Zhang, and R. Z. Pan. 2008. Experimental study on repetitive unipolar nanosecond-pulse dielectric barrier discharge in air at atmospheric pressure. *J. Phys. D Appl. Phys.* 41:215203.

109. Adamovich, I. V., M. Nishihara, I. Choi, M. Uddi, and W. R. Lempert. 2009. Energy coupling to the plasma in repetitive nanosecond pulse discharges. *Phys. Plasmas* 16:113505.

110. Lo, C. W., and S. Hamaguchi. 2011. Numerical analyses of hydrogen plasma generation by nanosecond pulsed high voltages at near-atmospheric pressure. *J. Phys. D Appl. Phys.* 44:375201.

111. Likhanskii, A. V., M. N. Shneider, S. O. Macheret, and R. B. Miles. 2007. Modeling of dielectric barrier discharge plasma actuators driven by repetitive nanosecond pulses. *Phys. Plasmas* 14:073501.

112. Muller, S., D. Luggenholscher, and U. Czarnetzki. 2011. Ignition of a nanosecond-pulsed near atmospheric pressure discharge in a narrow gap. *J. Phys. D Appl. Phys.* 44:165202.

113. Lu, X. P., and M. Laroussi. 2008. Electron density and temperature measurement of an atmospheric pressure plasma by millimeter wave interferometer. *Appl. Phys. Lett.* 92:051501.

114. Grisch, F., G. A. Grandin, D. Messina, and B. Attal-Tretout. 2009. Laser-based measurements of gasphase chemistry in non-equilibrium pulsed nanosecond discharges. *C. R. Mec.* 337:504–516.

115. Walsh, J. L., F. Iza, and M. G. Kong. 2010. Characterisation of a 3 nanosecond pulsed atmospheric pressure argon microplasma. *Eur. Phys. J. D* 60:523–530.

116. Siglo. 1998. *The Siglo Data Base, CPAT and Kinema Software*, http://www.siglo-kinema.com.

117. Udagawa, K. T., Y. Zuzeek, W. R. Lempert, and I. V. Adamovich. 2011. Characterization of a surface dielectric barrier discharge plasma sustained by repetitive nanosecond pulses. *Plasma Sources Sci. Technol.* 20:055009.

118. Pancheshnyi, S., M. Nudnova, and A. Starikovskii. 2005. Development of a cathode-directed streamer discharge in air at different pressures: Experiment and comparison with direct numerical simulation. *Phys. Rev. E* 71:016407.

119. Wang, D. Y., M. Jikuya, S. Yoshida, T. Namihira, S. Katsuki, and H. Akiyama. 2007. Positive- and negative-pulsed streamer discharges generated by a 100-ns pulsed-power in atmospheric air. *IEEE Trans. Plasma Sci.* 35:1098–1103.

120. Takaki, K., H. Kirihara, C. Noda, S. Mukaigawa, and T. Fujiwara. 2006. Production of an atmospheric-pressure glow discharge using an inductive energy storage pulsed power generator. *Plasma Process. Polym.* 3:734–742.

121. Ono, R., Y. Teramoto, and T. Oda. 2009. Measurement of atomic nitrogen in N_2 pulsed positive corona discharge using two-photon absorption laser-induced fluorescence. *Jpn. J. Appl. Phys.* 48:122302–122304.

122. Ono, R., and T. Oda. 2004. Spatial distribution of ozone density in pulsed corona discharges observed by two-dimensional laser absorption method. *J. Phys. D Appl. Phys.* 37:730–735.

123. Nakagawa, Y., R. Ono, and T. Oda. 2011. Density and temperature measurement of OH radicals in atmospheric-pressure pulsed corona discharge in humid air. *J. Appl. Phys.* 110:073304.

124. Hibert, C., I. Gaurand, O. Motret, and J. M. Pouvesle. 1999. OH(X) measurements by resonant absorption spectroscopy in a pulsed dielectric barrier discharge. *J. Appl. Phys.* 85:7070–7075.

125. Lu, X. P., and M. Laroussi. 2005. Optimization of ultraviolet emission and chemical species generation from a pulsed dielectric barrier discharge at atmospheric pressure. *J. Appl. Phys.* 98:023301.

126. Ono, R., Y. Yamashita, K. Takezawa, and T. Oda. 2005. Behaviour of atomic oxygen in a pulsed dielectric barrier discharge measured by laser-induced fluorescence. *J. Phys. D Appl. Phys.* 38:2812–2816.

127. Ono, R., and T. Oda. 2002. NO formation in a pulsed spark discharge in $N_2/O_2/Ar$ mixture at atmospheric pressure. *J. Phys. D Appl. Phys.* 35:543–548.

128. Stancu, G. D., F. Kaddouri, D. A. Lacoste, and C. O. Laux. 2009. Investigations of the rapid plasma chemistry induced by nanosecond discharges in atmospheric pressure air using advanced optical diagnostics. In *40th AIAA Plasmadynamics and Lasers Conference*, 22–25 June 2009, San Antonio, TX, AIAA-2009-3593.

129. Stancu, G. D., M. Janda, F. Kaddouri, D. A. Lacoste, and C. O. Laux. 2010b. Time-resolved CRDS measurements of the N2(A3Σ+u) density produced by nanosecond discharges in atmospheric pressure nitrogen and air. *J. Phys. Chem. A* 114:201–208.

130. Rusterholtz, D. L., D. Z. Pai, G. D. Stancu, D. A. Lacoste, and C. O. Laux. 2012. Ultrafast heating in nanosecond discharges in atmospheric pressure air. In *50th AIAA Aerospace Sciences Meeting*, 9–12 January 2012, Nashville, TN, AIAA-2012-5059.

131. Uddi, M., N. B. Jiang, E. Mintusov, I. K. Adamovich, and W. R. Lempert. 2009. Atomic oxygen measurements in air and air/fuel nano second-pulse. discharges by two photon laser induced fluorescence. *Proc. Combust. Inst.* 32:929–936.

132. Choi, I., Z. Y. Yin, I. V. Adamovich, and W. R. Lempert. 2011. Hydroxyl radical kinetics in repetitively pulsed hydrogen-air nanosecond plasmas. *IEEE Trans. Plasma Sci.* 39:3288–3299.

133. Baranov, V. I., F. I. Vysikailo, A. P. Napartovich, V. G. Nizev, S. V. Pigulskii, and A. N. Starostin. 1978. Contraction of the decaying plasma in a nitrogen discharge. *Sov. J. Plasma Phys.* 4:201–205.

134. Popov, N. A. 2001. Investigation of the mechanism for rapid heating of nitrogen and air in gas discharges. *Plasma Phys. Rep.* 27:886–896.

135. Popov, N. A. 2011. Fast gas heating in a nitrogen-oxygen discharge plasma: I. Kinetic mechanism. *J. Phys. D Appl. Phys.* 44:285201.

136. Aleksandrov, N. L., S. V. Kindysheva, M. M. Nudnova, and A. Y. Starikovskiy. 2010. Mechanism of ultra-fast heating in a non-equilibrium weakly ionized air discharge plasma in high electric fields. *J. Phys. D Appl. Phys.* 43:255201.

137. Paris, P., M. Aints, F. Valk, T. Plank, A. Haljaste, K. V. Kozlov, and H. E. Wagner. 2005. Intensity ratio of spectral bands of nitrogen as a measure of electric field strength in plasmas. *J. Phys. D Appl. Phys.* 38:3894–3899.

138. Bonaventura, Z., A. Bourdon, S. Celestin, and V. P. Pasko. 2011. Electric field determination in streamer discharges in air at atmospheric pressure. *Plasma Sources Sci. Technol.* 20:035012.

139. Ito, T., T. Kanazawa, and S. Hamaguchi. 2011. Rapid breakdown mechanisms of open air nanosecond dielectric barrier discharges. *Phys. Rev. Lett.* 107:065002.

140. Ito, T., K. Kobayashi, U. Czarnetzki, and S. Hamaguchi. 2010. Rapid formation of electric field profiles in repetitively pulsed high-voltage high-pressure nanosecond discharges. *J. Phys. D Appl. Phys.* 43:062001.

141. Unfer, T., and J. P. Boeuf. 2009. Modelling of a nanosecond surface discharge actuator. *J. Phys. D Appl. Phys.* 42:194017.

142. Macheret, S. O., M. N. Shneider, and R. C. Murray. 2006. Ionization in strong electric fields and dynamics of nanosecond-pulse plasmas. *Phys. Plasmas* 13:023502.

143. Nikandrov, D. S., L. D. Tsendin, V. I. Kolobov, and R. R. Arslanbekov. 2008. Theory of pulsed breakdown of dense gases and optimization of the voltage waveform. *IEEE Trans. Plasma Sci.* 36:131–139.

144. Belasri, A., J. P. Boeuf, and L. C. Pitchford. 1993. Cathode sheath formation in a discharge-sustained XeCl laser. *J. Appl. Phys.* 74:1553–1567.

145. Odrobina, I., and M. Cernak. 1995. Numerical simulation of streamer-cathode interaction. *J. Appl. Phys.* 78:3635–3642.

146. Levko, D., S. Yatom, V. Vekselman, and Y. E. Krasik. 2012. Electron emission mechanism during the nanosecond high-voltage pulsed discharge in pressurized air. *Appl. Phys. Lett.* 100:084105.

147. Stankevi.Yl, and V. G. Kalinin. 1967. Fast electrons and x-ray radiation in initial stage of impulse spark discharge development in air. *Doklady Akademii Nauk SSSR* 177:72.

148. Korolev, Y. D., and G. A. Mesyats. 1991. *Physics of Pulse Breakdown in Gases*. Moscow: Nauka.

149. Tarasenko, V. F. 2006. Nanosecond discharge in air at atmospheric pressure as an x-ray source with high pulse repetition rates. *Appl. Phys. Lett.* 88:081501.

150. Rusterholtz, D. 2012. Nanosecond repetitively pulsed discharges in atmospheric pressure air. Ph.D. thesis, Ecole Centrale Paris.

On Atmospheric Pressure Nonequilibrium Plasma Jets

XinPei Lu

7.1 Introduction

The electron temperature is much higher than the temperature of the heavy particles in atmospheric pressure nonequilibrium plasmas. Due to the high collision frequency between electrons and heavy particles, the electrons lose their energy in a short period. If molecular gas is presented, the electrons quickly transfer their energy to molecular rotational and vibrational states because the energy levels of the latter can be much lower than that for electrons' excitation and ionization [1–3]. This makes it difficult to obtain atmospheric pressure nonequilibrium plasmas with high electron energy. Thus the ionization efficiency in such case is low. Furthermore, when electronegative gases, such as O$_2$ and SF$_6$, are present, electrons could be absorbed by the gas in a timescale of tens of nanoseconds, or even shorter, which makes it even harder to obtain atmospheric pressure nonequilibrium plasma with electronegative gases [4].

Nevertheless, for traditional discharges, a plasma is generated as long as the applied electric field across the discharge gap is high enough to initiate a breakdown. However, at a pressure of 1 atmosphere, the electric field required to initiate the discharge is quite high. For example, when air is used, the required electric field is about 30 kV cm^{-1}. That is why the discharge gaps for most atmospheric pressure discharges are from millimeters to several centimeters [5–13]. On the other hand, from an applications point of view, the short discharge gaps significantly limit the size of the objects to be treated if direct treatment (when object is placed between the gaps) is desired. If indirect treatment (object is placed next to the gaps and the active radicals of the plasma reach the object by flowing the gas) is applied, the active radicals with short lifetimes and the charged particles may disappear before reaching the sample to be treated. To overcome the shortcomings of the traditional atmospheric pressure nonequilibrium plasmas, plasmas generated in an open space rather than in a confined discharge gap are needed. However, when a plasma is to be launched in an open space where the applied electric field is normally quite low, it is extremely difficult to sustain the plasma.

Briefly, as pointed above, there are two factors that make it a big challenge to generate atmospheric pressure nonequilibrium plasma jets (APNP-Js), one is the high electron–heavy particles' collision frequency and the other is the low applied electric field. Fortunately, various methods were developed to overcome these challenges and several sources based on different designs were subsequently reported [14–78]. APNP-Js are generated in an open space rather than in confined gaps. Thus, they can be used for direct treatment and there is no limitation on the size of the object to be treated. This is extremely important for applications such as plasma medicine [79–108].

In this chapter, the development of APNP-Js will be reviewed. The chapter is organized as follows: First, because the working gas is one of the main factors that affects the characteristics of APNP-Js, the latter are grouped based on the type of gas used to ignite them. Second, one of the most interesting phenomena of noble gas and nitrogen APNP-Js, the "plasma bullet," is discussed and its behavior described. Third, parameter measurements on various plasma jet devices are reviewed. Finally, perspectives on research on APNP-Js are presented.

7.2 Plasma Jets

7.2.1 Noble Gas Plasma Jet

Various types of APNP-Js with different configurations in which most of the jets are with noble gas mixed with reactive gases, such as O_2, have been reported. Plasma jets operating with noble gases can be classified into four categories, that is, dielectric-free electrode (DFE) jets, dielectric barrier discharge (DBD) jets, DBD-like jets, and single electrode (SE) jets, as shown in Figures 7.1–7.4.

7.2.1.1 Dielectric-Free Electrode Jets

One of the early APNP-Js, developed by Hicks' group, is a DFE jet, as shown in Figure 7.1 [14,15]. The jet is driven by a radiofrequency (RF) power source at 13.56 MHz. It consists of an inner electrode, which is coupled to the power source, and a grounded outer electrode. A mixture of Helium (He) with reactive gases is fed into the annular space between the two electrodes. Cooling water is needed to keep the jet from overheating and the gas temperature of the plasma jet varies from 50°C to 300°C, depending on the RF power. The window of stable operation without apparent arcing of the device is as given: He flow rates greater than 25 l min^{-1}, O_2 concentrations of up to 3.0% by volume, CF_4 concentrations of up to 4.0% by volume, and RF power between 50 and 500 W.

Several notable characteristics of the DFE jets are, first, arcing is unavoidable when stable operation conditions are not met. Second, compared to DBD and DBD-like jets that will be discussed below, the power delivered to the plasma for the DFE jet is much higher. Third, due to the high power delivered, the gas temperature of the plasma is quite high and out of the acceptable range for biomedical applications. Fourth, for this DFE jet, which is driven by an RF power supply, the peak voltage is only a few hundred volts, so the electric field within the discharge gap is relatively low and its direction is radial (perpendicular to the gas flow direction). The electric field in the plasma plume region is even lower, especially along the plasma plume propagation direction (gas flow direction). Finally, because the electric field along the plasma plume propagation direction is very low, the generation of this plasma plume is probably gas flow driven rather than electrically driven.

On the other hand, because a relatively high power can be delivered to the plasma and the gas temperature is relatively high, the plasma is very reactive. This kind of plasma jet is suitable for applications such as material treatment as long as the material to be treated is not very sensitive to high temperatures.

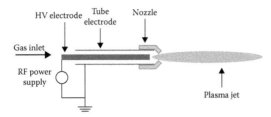

FIGURE 7.1 Schematic of a dielectric-free electrode (DFE) jet. (From Babayan, S.E., et al., *Plasma Sources Sci. Technol.*, 7, 286, 1998.)

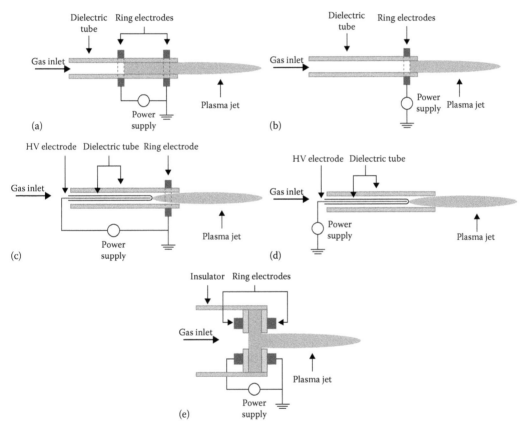

FIGURE 7.2 Schematic of a DBD plasma jet. (a) two ring electrodes DBD jet; (b) single electrode DBD jet; (c) electric field-enhanced two electrode DBD jet; (d) electric field-enhanced single electrode DBD jet; (e) plasma pencil.

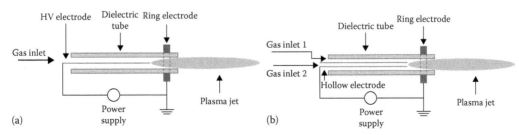

FIGURE 7.3 Schematic of a DBD-like plasma jet. (a) solid needle electrode with dielectric tube DBD-like plasma jet; (b) capillary metal tube inside dielectric tube DBD-like plasma jet.

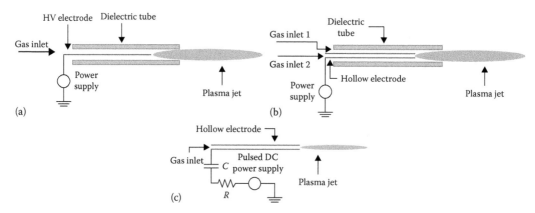

FIGURE 7.4 Schematic of an SE plasma jet. (a) solid needle electrode with dielectric tube single electrode plasma jet; (b) capillary metal tube inside dielectric tube single electrode plasma jet; (c) capillary metal tube single electrode plasma jet.

7.2.1.2 DBD Jets

For DBD jets, as shown in Figure 7.2a–e, there are many different configurations. As shown in Figure 7.2a, which was first reported by Teschke et al. [25], the jet consists of a dielectric tube with two metal ring electrodes on the outer side of the tube. When a working gas (He, Ar) is flowing through the dielectric tube and a kilohertz-high-voltage (HV) power supply is turned on, a cold plasma jet is generated in the surrounding air. The plasma jet consumes a power of several watts. The gas temperature of the plasma is close to room temperature. The gas flow velocity is less than 20 m s^{-1}. The plasma jet, which looks homogeneous to the naked eye, is actually a "bullet"-like plasma volume with a propagation speed of more than 10 km s^{-1}. The applied electric field plays an important role in the propagation of the plasma bullet. More discussion on this phenomenon is given in Section 7.3.

Figures 7.2b eliminates one ring electrode [26]. So the discharge inside the dielectric tube is weakened. Figure 7.2c replaces the HV ring electrode with a centered pin electrode, which is covered by a dielectric tube with one end closed [27]. With this configuration, the electric field along the plasma plume is enhanced. Walsh and Kong's [28] studies show that a high electric field along the plasma plume is favorable for generating long plasma plumes and more active plasma chemistry. Figure 7.2d further removes the ground ring electrode shown in Figure 7.2b [29]. So the discharge inside the tube is also weakened. On the other hand, a stronger discharge inside the discharge tube (Figure 7.2a and c) helps the generation of more reactive species. Along with the gas flow, the reactive species with relatively long lifetimes may also play an important role in various applications. The configuration of Figure 7.2e, developed by Laroussi and Lu [30], is different from the previous four DBD jet devices. The two ring electrodes are attached to the surface of two centrally perforated dielectric disks. The holes in the center of the disks are about 3 mm in diameter. The distance between the two dielectric disks is about 5 mm. With this device, a plasma plume up to several centimeters in length can be obtained.

All the DBD jet devices discussed above can be operated either by kilohertz alternating current (AC) power or by pulsed DC power. The length of the plasma jet can easily reach several centimeters or even longer than 10 cm as reported by Lu et al. [27]. This capability makes the operation of these plasma jets easy and practical. There are several other advantages of DBD jets. First, due to the low power density delivered to the plasma, the gas temperature of the plasma remains close to room temperature. Second, because of the use of the dielectric, there is no risk of arcing whether the object to be treated is placed far away or close to the nozzle. These two characteristics are very important for applications such as plasma medicine, where safety is a strict requirement.

7.2.1.3 DBD-Like Jets

All the plasma jet devices shown in Figure 7.3 are DBD-like jets. This is based on the following facts. When the plasma plume is not in contact with any object, the discharge is more or less like in a DBD. However, when the plasma plume is in contact with an electrically conducting (a nondielectric material) object, especially a ground conductor, the discharge is between the HV electrode and the object to be treated (ground conductor). In such instances, it does not function as a DBD. The devices shown in Figure 7.3 can be driven by a kilohertz AC power, by an RF power, or by a pulsed DC power.

Figure 7.3b replaces the solid HV electrode in Figure 7.3a with a hollow electrode [22,31]. The benefit of this kind of configuration is that two different gases can be mixed in the device. Normally, gas inlet 2 is used for a reactive gas such as O_2 and gas inlet 1 for a noble gas. It was found that the plasma plume is much longer with this kind of gas control than with a premix gas mixture with the same percentage [22]. The role (and advantage) of the ring electrode in Figure 7.3a and b is as in the case of DBD jets.

When the DBD-like plasma jets are used for plasma medicine applications, the object to be treated could be cells or a whole tissue. In such cases, these types of jet devices should be used carefully because of the risk of arcing. On the other hand, if it is used for treatment of conductive material, because there is no dielectric; more power can be easily delivered to the plasma. Thus, as long as arcing is carefully avoided, DBD-like jets have their own advantages.

7.2.1.4 SE Jet

The schematics of SE jets are shown in Figure 7.4a–c. The jets shown in Figure 7.4a and b are like DBD-like jets except there is no ring electrode on the outside of the dielectric tube. The dielectric tube guides the gas flow. These two jets can be driven by direct current (DC), kilohertz AC, RF, or pulsed DC power.

Because of the risk of arcing, the plasma plumes generated in Figure 7.4a and b are not the best for biomedical applications because of safety concerns [32]. To overcome this problem, Lu et al. [33] developed an SE jet like the one in Figure 7.4c. The capacitance C and resistance R are about 50 pF and 60 kΩ, respectively. The resistor and capacitor are used for controlling the discharge current and voltage on the hollow electrode (needle). This jet is driven by pulsed DC power supply with pulse width of 500 ns, repetition frequency of 10 kHz, and amplitude of 8 kV. The advantage of this jet is that the plasma plume or even the hollow electrode can be touched without any danger of harm, making it suitable for plasma medicine applications.

One of the potential applications is in dentistry such as root canal treatment. Due to the narrow channel geometry of a root canal, which typically has a length of a few centimeters and a diameter of 1 mm or less, the plasma generated by a plasma jet is not efficient for delivering reactive agents into the root canal for disinfection. Therefore, to have a better killing efficacy, a plasma needs to be generated inside the root canal. When plasma is generated inside the root canal, reactive agents, including the short-lifetime species, such as charge particles, could play a role in killing bacteria. By using the device shown in Figure 7.4c, a cold plasma could be generated inside a root canal, as shown in Figure 7.5.

7.2.2 N$_2$ Plasma Jet

As pointed out in the introduction section, it is difficult to generate an atmospheric pressure nonequilibrium nitrogen (N_2) plasma jet. Up to now, only few N_2 plasma jets have been reported [34–36]. Figure 7.6 shows the schematic of two N_2 plasma jets. The N_2 plasma jet as shown in Figure 7.6a was reported by Hong and Uhm [34]. A 20 kHz AC power supply is connected to two electrodes with a thickness of 3 mm and a center hole 500 μm in diameter. The two electrodes are separated by a dielectric disk with a center hole of the same diameter. With this configuration, an N_2 plasma up to 6.5 cm long can be generated. When the N_2 gas flow rate is 6.3 SLM (standard liters per minute), the gas is ejected out from the hole at a speed of about 535 m s^{-1}. The gas temperature of the plasma plume at 2 cm from the nozzle is below 300 K. Figure 7.6b shows a slightly different N_2 plasma jet device, which replaces the inner perforated HV electrode of Figure 7.6a with a pin electrode [35]. The inner electrode can also be replaced by a tube as was done by Hong et al. [36].

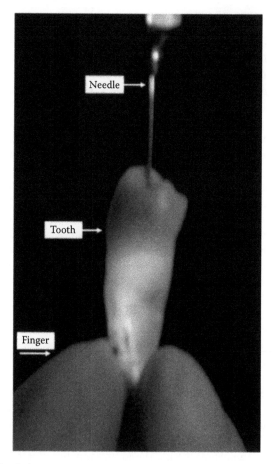

FIGURE 7.5 Photograph of plasma generated in root canal of tooth by using the jet shown in Figure 7.4c. (From Lu, X., et al., *IEEE Trans. Plasma Sci.*, 37, 668, 2009. © 2009 IEEE.)

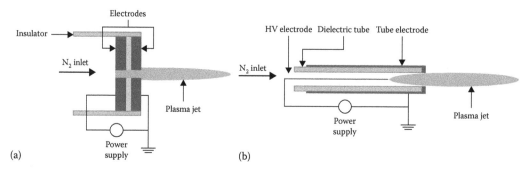

FIGURE 7.6 Schematic of an N_2 plasma jet. (a) two disk electrodes N_2 plasma jet; (b) needle–ring electrode N_2 plasma jet.

7.2.3 Air Plasma Jet

Due to the presence of electronegative oxygen, O_2, in air, it is difficult to sustain an atmospheric pressure nonequilibrium air plasma jet. Nevertheless, several different air plasma jets were reported [37–41]. Mohamed et al. [38] reported a microplasma jet device that can operate in various gases including air. The schematic is shown in Figure 7.7a. A discharge channel through an insulator with a

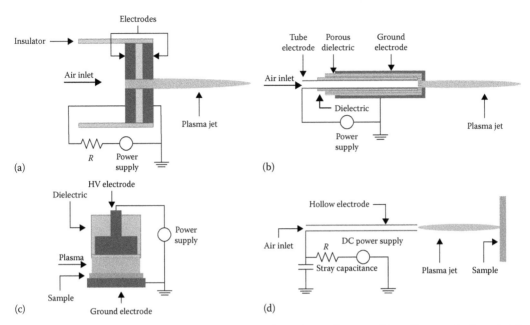

FIGURE 7.7 Schematic of an air plasma jet. (a) two disk electrodes air plasma jet; (b) capillary metal tube with disk ground electrode air plasma jet; (c) floating electrode air plasma jet; (d) capillary metal single electrode air plasma jet.

thickness of about 0.2–0.5 mm and a diameter of 0.2–0.8 mm separates the anode from the cathode, which have a center hole with the same diameter. The ballast resistor is 51 kΩ. When air is flowing through the hole and a DC voltage of a few hundred volts (up to kilovolts) is applied between the anode and the cathode (depending on the thickness of the insulator separating the electrodes), a relatively low-temperature air plasma is generated in the surrounding air with a length of up to 1 cm, depending on the gas flow rate and discharge current. However, the gas temperature of the plasma can still be quite high. The gas temperature within the microgap is about 1000 K. However, it drops quickly as it propagates in the surrounding air. It is about 50°C at 5 mm from the nozzle for an air flow rate of 200 ml min^{-1} and discharge current of 19 mA.

Hong et al. [39] reported another type of air plasma jet device as shown in Figure 7.7b. One of the notable characteristics of this device is that a porous alumina dielectric is used to separate the HV stainless-steel (typical injection needle) electrode and the outer ground electrode. The alumina used in this device has approximately 30 vol% porosity and an average pore diameter of 100 μm. The ground electrode is fabricated from stainless steel and has a centrally perforated hole 1 mm in diameter through which the plasma jet is ejected to the surrounding ambient air. When a 60 Hz HV power supply is applied and the flow rate of air is at several SLMs, an APNP-J up to about 2 cm is generated in the surrounding air. During one voltage cycle, there are multiple discharges. The increase in the input power results in more current pulses. The shortcoming of this device is that the gas temperature of the plasma is quite high. It is about 60°C 10 mm away from the nozzle for an air flow rate of 5 SLM. For slower flow rate, the gas temperature is even higher.

Figure 7.7c and d is the schematics of two "floating" electrode air plasma jets [40,41]. Strictly, they are not plasma jets because the plasmas are generated within a gap. However, because the secondary electrode (ground electrode) can be a human body, they are categorized as plasma jets in this chapter. Both jets could generate room-temperature air plasmas. They are completely safe electrically and do not cause damage to animals or human beings.

In Figure 7.7c, a kilohertz AC or pulsed DC voltage with amplitude of 10–30 kV is used to drive the device. The discharge ignites when the powered electrode approaches the surface to be treated at a distance (discharge gap) of less than 3 mm, depending on the form, duration, and polarity of the driving voltage. This jet is suitable for treating large smooth surfaces.

On the contrary, the jet shown in Figure 7.7d is more suitable for localized three-dimensional treatments. This jet is driven by a homemade DC power supply. The output voltage of the power supply can be adjusted up to 20 kV. The output of the power supply is connected to a stainless-steel needle (typical injection needle) electrode through a resistor R of 120 MΩ, which is several orders of magnitude higher than those reported [42]. When a counterelectrode, such as a finger, is placed close to the needle, a plasma is generated, as shown in Figure 7.8. The plasma is similar to the positive corona discharge. However, this jet can be touched by the human body directly, which is not the case with traditional corona discharge. There is no risk of glow-to-arc transition. The maximum length of the plasma is about 2 cm. The plasma is kept at room temperature. It should be noted that the discharge is actually pulsed. It appears periodically with a pulse frequency of tens of kilohertz, depending on the applied voltage and the distance between the tip of the needle and the object to be treated, as shown in Figure 7.9.

These air plasma jets can also be operated with N_2 gas. On the other hand, the device shown in Figure 7.6a can be operated with air too. But the maximum length of the air plasma plume operated with the jet shown in Figure 7.6a is about 2 cm. In addition, the jet shown in Figure 7.4c can also be operated with air as working gas [43]. But the length of the plasma is only several millimeters.

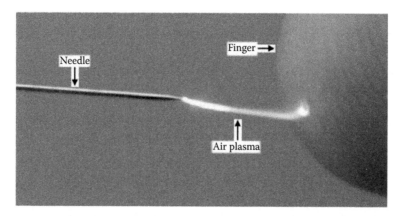

FIGURE 7.8   Plasma generated by DC power supply touched by a finger. (From Wu, S., et al., *IEEE Trans. Plasma Sci.*, 38, 3404, 2010. © 2010 IEEE.)

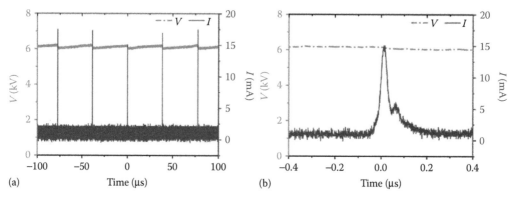

FIGURE 7.9   (a) Typical *I–V* characteristics of the plasma and (b) a close look at the *I–V* characteristics of a typical single pulse. (From Wu, S., et al., *IEEE Trans. Plasma Sci.*, 38, 3404, 2010. © 2010 IEEE.)

7.2.4 Brief Summary

Although noble gas plasma jets are relatively easy to be generated, they are not as reactive as air plasma jets. Therefore a small quantity of reactive gases is added to a noble gas when plasma jets are used for various applications. The noble gas serves as the carrier gas to generate the plasma. For biomedical applications, O_2 or H_2O_2 is usually added. For etching, CF_4 or O_2 is used. It should be noted that N_2 plasma jets are not as reactive as air plasma jets. Adding a small quantity of reactive gases to the carrier gas is also recommended.

7.3 Plasma Bullet of Noble Gas Plasma Jets

The discrete nature associated with the structure of noble plasma jets was first observed by Teschke et al. [25] using an RF-driven plasma jet and by Lu and Laroussi [109] using a pulsed DC plasma jet (the plasma pencil). Using fast imaging, these investigators found that the plasma plume, which appeared continuous to the naked eye, was in fact made up of fast-moving plasma structures. Teschke et al. [25] found that the small volume of plasma, or plasma bullet, travels at a velocity of about $1.5 \times 10^4 \, \mathrm{m \, s^{-1}}$, while Lu and Laroussi [109] measured velocities as high as $1.5 \times 10^5 \, \mathrm{m \, s^{-1}}$. Comparatively, the estimated upper limit of the drift velocity of electrons under the external applied electric field is only $1.1 \times 10^4 \, \mathrm{m \, s^{-1}}$, and the estimated upper limit of the N_2^+ drift velocity is $2.2 \times 10^2 \, \mathrm{m \, s^{-1}}$. Because these speeds are far slower than the measured bullet-like plume velocity, Lu and Laroussi invoked a streamer propagation model based on photoionization in the manner proposed by Dawson and Winn [110] for streamers to explain the properties of these so-called plasma bullets.

However, there are some notable differences between the streamer-like APNP-J and positive corona discharges that are typically used to study cathode-directed streamers. For example, streamers developed in pulsed positive corona discharges are typically not very repeatable due to the stochastic nature of their initiation. In contrast, plasma bullet behavior is mostly very repeatable. In addition, experiments revealed ring-shaped profiles of plasma bullet radiation of nitrogen ions and metastable He atoms, with the maxima shifted from the jet axis. Such a pattern is different from that of a typical streamer in uniform media, which does not have a donut shape. Besides, after the plasma plume propagates, a dark channel is created between the plume head and the electrode. It is not clear yet whether the conductivity of the dark channel is high enough to affect the propagation of the plasma plume as in a streamer discharge. Furthermore, photoionization plays an important role in positive streamers. However, it is not clear whether photoionization plays a similar role in the propagation of the plasma bullet either. Finally, in a streamer discharge, the discharge behaves differently when polarity of the voltage is changed from positive to negative. How plasma jets behave for different polarities needs to be investigated. Because of significant differences between streamers and bullet-like plasma plumes, many studies have been carried out in the past several years to investigate the behavior of a so-called plasma bullet [111–147]. Some of these questions are much better understood now. In Section 7.3.1, recent studies are discussed.

7.3.1 Stochastic or Repeatable Characteristics of Plasma Bullets

It was believed that plasma plumes propagate in a repeatable fashion: they consistently propagate the same distance after the same propagation time. Xian et al. found that under certain conditions this may not be the case. The jet used by Xian et al. is similar to the one shown in Figure 7.4c [109]. When a pulsed DC voltage is applied, as shown in Figure 7.10, the discharge exhibits a chaotic mode when the applied voltage is 8 kV. On the other hand, when the applied voltage is increased to 9 kV, the discharge becomes repeatable.

The setup used by Walsh et al. [112] is similar to that in Figure 7.2b except that there is a ground electrode in front of the plasma jet. Through detailed electrical and optical characterization, Walsh et al. also found that immediately following breakdown the plasma jet operates in a chaotic mode. By increasing the applied voltage, the discharge becomes periodic and the jet plasma is found to produce at least one strong plasma bullet every cycle of the applied voltage. These results show that under certain conditions plasma plumes can also exhibit a stochastic behavior.

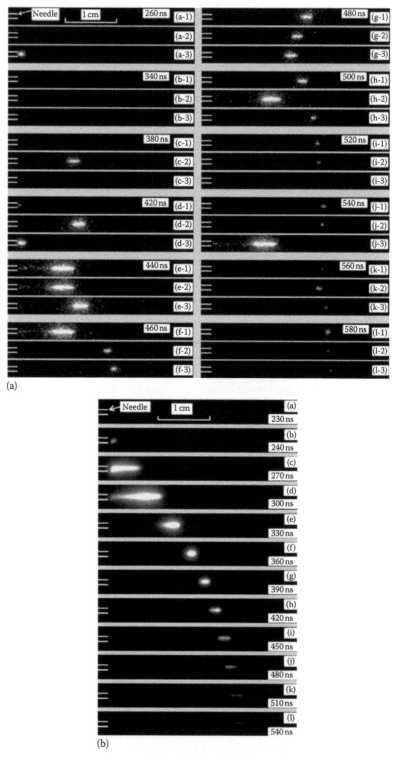

FIGURE 7.10 High-speed photographs of the plasma plume for (a) 8 kV and (b) 9 kV. For (a), due to the randomicity of the discharges, three photographs are taken for every delay time. Pulse frequency, 10 kHz; pulse width, 500 ns; working gas, He/O$_2$ (20%); total flow rate, 0.4 l min^{-1}. (From Xian, Y., et al., *IEEE Trans. Plasma Sci.*, 37, 2068, 2009. © 2009 IEEE.)

7.3.2 "Donut" Shape of the Plasma Bullet

Teschke et al. [25] and Mericam-Bourdet et al. [113] observed that the plasma bullet is, in fact, hollow and has a "donut," or ring-shaped, structure as shown in Figure 7.11.

These observations were subsequently supported by Sakiyama et al. [114] and Naidis [115] through computer simulation. Naidis [115] showed that for He jet propagating in air, the propagation is mainly due to a confinement of the discharge front in the channel where the air molar fraction x is less than 0.01 as the direct ionization rate in He–air mixtures decreases sharply for higher values of x. Naidis [115] showed that the discharge has a ring shape as it propagates at the interface where $x = 0.01$.

Figure 7.12 shows the radial profiles of the number densities of electrons n_e and $N_2(C^3\Pi)$ molecules $n_{N2(C)}$ at axial positions $z = 1$, 2, and 4 cm just after the arrival of the streamer front. It clearly shows that the maxima of n_e and $N_2(C^3\Pi)$ are shifted from the axis to some distance, which decreases with an increase in z. The distance nearly corresponds to the position where the air molar fraction is equal to 10^{-2} [115]. The figure also shows that the Penning process does have a quantitative effect on streamer parameters, though the patterns of streamer structure in both cases, with and without the Penning process, are nearly the same. This is consistent with the experimental observation by Zhu et al. [116].

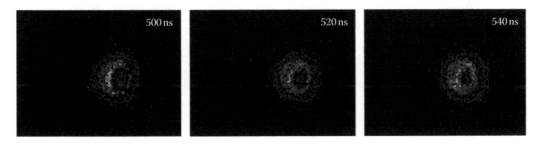

FIGURE 7.11 Photographs of the bullets illustrating their donut shape. (From Mericam-Bourdet, N., et al., *Plasma Sources Sci. Technol.*, 21, 034005, 2012.)

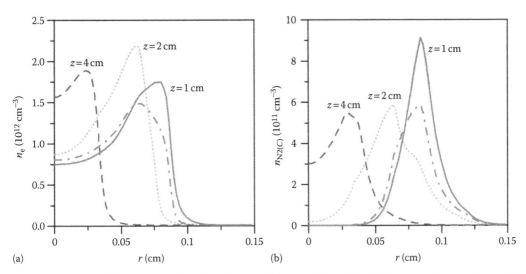

FIGURE 7.12 Radial distributions of number densities of electrons (a) and N_2 ($C^3\Pi$) molecules (b) at various axial positions, for $z = 1$, 2, and 4 cm. Dotted–dashed lines show the results for $z = 1$ cm obtained without taking into account the Penning reactions. (From Naidis, G.V., *J. Phys. D*, 44, 215203, 2011.)

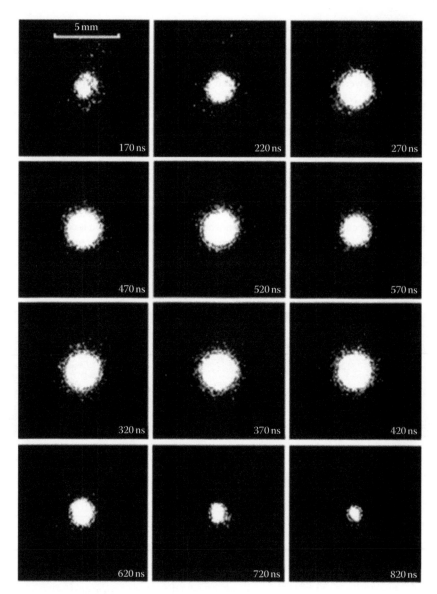

FIGURE 7.13 High-speed photographs of the plasma bullet in the surrounding air taken head on with respect to the plasma plume. The exposure time is 5 ns. The internal diameter of the nozzle is 2.5 mm. The working gas is He/N_2 (He: 1 l min^{-1}; N_2: 0.015 l min^{-1}) mixture. (From Wu, S., et al., *IEEE Trans. Plasma Sci.*, 2123912, 2011. © 2011 IEEE.)

Because the "donut" shape is related to the gas composition caused by diffusion of air, Wu et al. add 1.5% N_2 into the He gas flow and found that the "donut" shape is replaced by a "solid disk" shape (see Figure 7.13). This further confirms that the "donut" shape is due to air diffusion.

7.3.3 Dark Channel Left by the Plasma Bullet

It was noticed that with the propagation of the bullet-like plasma plume a dark channel between the head of the plasma plume and the electrode was created. Lu et al. and Karakas and Laroussi used a plasma jet like the one shown in Figure 7.2e to study the effect of the dark channel [118,119]. Pulsed DC voltage was used to drive the plasma. Lu et al. found that as long as the pulse width is more than

500 ns, the adjustment of the pulse width does not affect the shape and peak value of the discharge current pulses. In other words, when the pulse width is increased to more than 500 ns, it no longer affects the discharge current except by changing the zero current duration. Therefore, the increase of the plasma plume length with the increase of the pulse width from 500 ns to 1 μs can be attributed to the electric field. Thus, although the channel left by the ionization front (plasma bullet) appears dark, based on the high-speed photographs of the plasma, the conductivity of this dark channel is not negligible; it affects the propagation of the plasma bullet. Karakas and Laroussi measured the temporal emission behavior of excited N_2, N_2^+, and He from the plasma plume 2 cm away from the nozzle. It was found that the magnitude of the emission intensity first increases and eventually reaches its highest value. This region corresponds to the point where the ionization front propagates forward. Then, the emission decreases exponentially until it reaches low emission levels. The ionization front leaves behind a channel with low concentrations of short-lived reactive species. This channel initially appears as an extension of the plasma bullet, but if the applied voltage is sufficiently high, the plasma bullet eventually breaks off this extension. Even when the plasma bullet breaks off, the secondary discharge ignition is still able to inhibit the propagation of the plasma bullet. This further confirms that the conductivity of the dark channel left by the ionization front is not negligible. Sands et al. [120] used a unipolar pulsed voltage to drive a jet device like the one shown in Figure 7.2b. They used He mixed with 5% Ar as the working gas. It was found that the $Ar(1s_5)$ column density remained greater than 10^{11} cm^{-2} for up to 10 μs after the discharge was initiated, as shown in Figure 7.14. This confirms that there are some long lifetime species that are present in the channel that could affect plasma plume propagation.

Naidis [51] simulated the distribution of the absolute value of the electric field along the plasma plume. He found that the maximum electric field values for the positive streamer are much higher than those for the negative one. However, the electric field in the channel of the negative streamer is much higher than that in the positive one. Naidis [51] also pointed out that, while at positive polarity, radiation is emitted mainly from a small region adjacent to the streamer head, at negative polarity the whole channel radiates. This feature is related to the difference between electric field values in the channels of positive and negative streamers. At positive polarity the electric field in the channel is low and production

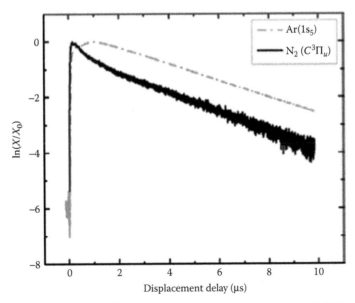

FIGURE 7.14 Decay of the Ar $(1s_5)$ metastable line-integrated column density and N_2 $(C^3\Pi_u)$ emission intensity at 337 nm in the residual streamer channel 1 mm from the capillary tip. (From Sands, B., et al., *J. Phys. D*, 43, 282001, 2010.)

of $N_2(C^3\Pi)$ molecules is insignificant. As the lifetime of $N_2(C^3\Pi)$ molecules is much smaller than the time of streamer propagation, $N_2(C\text{-}B)$ radiation is localized in a small region near the head. At negative polarity the values of the electric field are much higher, so that $N_2(C^3\Pi)$ molecules are generated effectively in the whole channel [51]. In this respect plasma jets driven by negative voltage pulses are similar to glow plasma.

7.3.4 Role of Photoionization

It is difficult to study the role of photoionization in the propagation of the plasma plume. In order to study the effect of photoionization, Naidis [115,121] used two two-dimensional axially symmetric positive streamer models to simulate the plasma plume. In one of the models, the photoionization effect was included [121] and in the other model photoionization was neglected [115]. It is assumed instead that the streamers propagate in a uniform weakly preionized gas. The initial background electron density is at the level of 10^{10} cm^{-3}. In both the models, the external field (the field in the absence of volume charges inside streamers) is assumed to be constant in time and is evaluated as that produced by a positively charged sphere. The simulation results show that the streamer front moves along the jet with a mean velocity of about 10^7 cm s^{-1}. This is close to the experimental observations. Because photoionization effects could be replaced by a preionization channel, so it is still not clear whether photoionization plays an essential role in the propagation of the plasma plume. Figure 7.15 shows profiles of the electric field and number density of electrons along the symmetry axis at various times based on the photoionization model. The line at $t = 0$ shows the external electric field and the initial electron density in Figure 7.15a and b, respectively. As can be seen, the electron density in the discharge channel reaches values higher than 10^{12} cm^{-3}.

7.3.5 Effect of Polarity of the Applied Voltage

Jiang et al. [122] and Xiong et al. [123] did comparative studies on the effects of the polarities of the applied voltages on the propagation of the plasma plumes. They found that the plasma plume is much longer when positive pulsed voltages are used. In addition, the maximum velocities of the plasma plumes driven by positive voltages are higher than that driven by negative voltages.

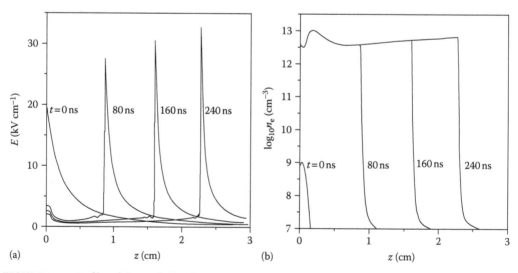

FIGURE 7.15 Profiles of electric field and number density of electrons along the symmetry axis at various times. The jet radius is 0.25 cm. (a) electric field distribution for different time; (b) electron density distribution for different time. (From Naidis, G.V., *J. Phys. D*, 43, 402001, 2010.)

Naidis [51] calculated the dependence of streamer propagation velocity on time for both polarities. The simulation results show that the calculated velocity over time is in good agreement with experiments reported by Jiang et al. and Xiong et al. The velocities of positive and negative streamers, being nearly equal initially, vary with time quite differently. While at positive polarity the velocity changes with time rather slowly, at negative polarity a steep decrease takes place over time. As a result, the propagation length for the positive streamer is much larger than that for the negative one, which is also in agreement with experimental data reported.

7.3.6 Multiple Plasma Bullets Behavior

Xian et al. [148] studied the plasma plume propagation behavior in a controlled gas environment. They found that multiple plasma bullets appeared per voltage pulse for both polarities of the applied voltage. Figure 7.16 is the schematic of the experimental setup. The plasma plume is generated by a SE plasma jet device. The HV wire electrode, which is made of a copper wire, is inserted into a 4-cm-long quartz tube with the right end closed. Details about the plasma jet device can be found in Lu et al. [29]. However, in this study, the plasma plume is generated in a glass container rather than in the surrounding air. The glass container is cylindrical-shaped. The inner diameter of the container is 16 cm, and the depth of the container is about 25 cm. There is an exhaust tube 2 mm in diameter at the lower part of the stopper. So the pressure in the container is at atmospheric pressure. Because the glass container is not vacuum sealed, there is always a trace amount of air inside the glass bottle. Throughout these studies, He/N_2 (1%) mixture was used as the working gas with total flow rate of 2 l min^{-1}. HV pulsed DC power supply (amplitude up to ±10 kV, pulse width from 200 ns to DC adjustable, pulse repetition rate up to 10 kHz) is used to drive the plasma jet device. The He/N_2 (1%) flow is started 10 min before turning on the power supply and the gas flow maintained during the experiment. When HV pulsed DC voltage is turned on, a plasma plume is generated in the glass container, as shown in Figure 7.16.

Figure 7.17a and b shows the current–voltage characteristics of the discharge with different pulse widths for the positive and negative pulse, respectively. It should be pointed out that the current I is the total current. It is the sum of the displacement current and the actual discharge current, where the actual discharge current is much smaller than the displacement current.

The pulse width was increased from 200 ns to 125 µs. When the pulse width is in the range of 200 ns to 1.5 µs, only one plasma bullet is observed, which is as reported in Lu and Laroussi [109]. However, by further increasing the pulse width, two plasma bullets start to appear per voltage pulse. When the pulse width is 2.8 µs, the interval between the two plasma bullets is the longest. Figure 7.18 is the high-speed photographs of the plasma taken at different delay times for a pulse width of 2.8 µs. The time labeled on each photograph corresponds to the time in Figure 7.17a.

As shown in Figure 7.18c, the second plasma bullet starts to appear in the nozzle, while the first plasma bullet propagates in the surrounding gas. The second plasma bullet exits the nozzle at about 1.2 µs, as

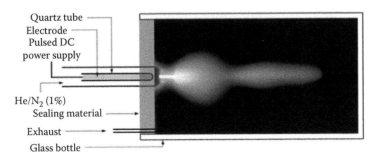

FIGURE 7.16 Schematic of the experimental setup. (From Xian, Y., et al., *Plasma Sources Sci. Technol.*, 21, 034013, 2012.)

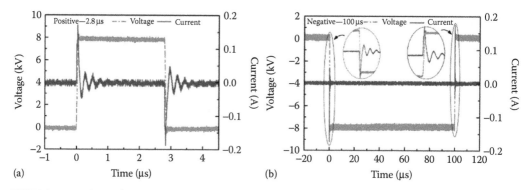

FIGURE 7.17 The total current versus voltage characteristics of the discharge for (a) positive and (b) negative pulses. (From Xian, Y., et al., *Plasma Sources Sci. Technol.*, 21, 034013, 2012.)

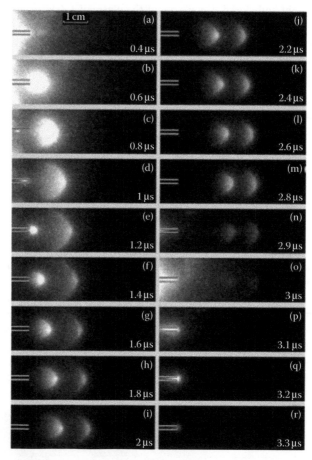

FIGURE 7.18 (a)–(r) High-speed photographs of the plasma plume for the positive pulse. The time labeled on each photograph corresponds to the time in Figure 7.17a. (From Xian, Y., et al., *Plasma Sources Sci. Technol.*, 21, 034013, 2012.)

shown in Figure 7.18e. Then both bullets propagate in the surrounding gas with approximately the same velocity, as shown in Figure 7.18e–n. It is worth mentioning that when the first bullet exits the nozzle, both the dimension and the brightness of the bullet increase dramatically. On the other hand, when the second plasma bullet exits the nozzle, the dimension and the brightness of the bullet increases only slightly. But they keep increasing. As the brightness of the first bullet decreases, both reach similar dimensions. Besides,

Figure 7.18o–q shows that there is another discharge. But the plasma does not propagate further in the surrounding gas. This discharge appears at the falling edge of the applied voltage and it is as reported in Lu and Laroussi [109].

To further understand if the multiple bullet behavior is polarity related, a negative HV pulsed DC voltage is used to drive the plasma device. The amplitude and the repetition frequency of the pulsed DC voltage are the same. The pulse width is increased from 200 ns to 125 μs. It is found that only one plasma bullet is captured for each voltage pulse when the pulse width is shorter than 80 μs or longer than 105 μs. On the other hand, when the pulse width is increased from 80 to 105 μs, two plasma bullets are observed per voltage pulse. When the pulse width is in the range from about 98 to 102 μs, even three plasma bullets are captured per voltage pulse. Figure 7.19 shows the plasma bullet behavior for the pulse width of 100 μs. The first, second, and third bullet exit the nozzle at 101.73, 108.9, and 115.35 μs, respectively, as shown in Figure 7.19a, f, and j, which corresponds to the voltage rising from −8 kV to 0 V, as shown in Figure 7.17b. As per Figure 7.19, the distances between the bullets are almost the same. It is also interesting to note that the brightness of the three bullets decreases on exiting the nozzle. Similarly, as reported in Lu and Laroussi [109], plasma is also generated at the falling edge of the voltage pulse (from 0 V to −8 kV), as shown in Figure 7.19q–u.

As mentioned earlier, for the positive pulse, the two plasma bullets are generated at and after the rising edge of the voltage pulse (from 0 V to 8 kV). So HV serves as the anode and the surrounding gas as the cathode when the two plasma bullets are observed. On the other hand, for the negative pulse, the three plasma bullets are generated at and after the rising edge of the voltage pulse (from −8 kV to 0 V) rather than at or after the falling edge of the voltage pulse (from 0 V to −8 kV). Therefore, when the three plasma bullets are generated, the HV electrode serves as the anode because the dielectric tube covering the HV electrode is positively charged during the primary discharge at the falling edge of the voltage pulse (from 0 V to −8 kV). Thus it is the electrode polarity rather than the applied voltage polarity that plays an important role in the generation and propagation of multiple plasma bullets.

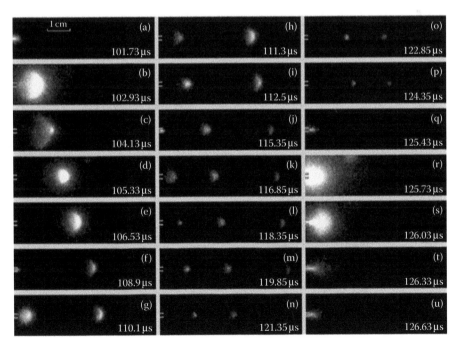

FIGURE 7.19 (a)–(u) High-speed photograph of the plasma plume for the negative pulse. The time labeled on each photograph corresponds to the time in Figure 7.17b. (From Xian, Y., et al., *Plasma Sources Sci. Technol.*, 21, 034013, 2012.)

7.4 Dynamics of N₂ Plasma Jet

As discussed in Section 7.3, previous reports point out that noble gas plasma plumes generated by APNP-Js are electrically driven and propagate at a speed of 10^4–10^5 m s⁻¹. However, there was no report on the dynamics of plasma jets with molecular gas such as N₂ or air as the working gas. Only recently, Xian et al. [149] investigated the dynamics of N₂ plasma plumes. Figure 7.20 shows the schematic of the device together with a photograph showing the N₂ plasma plume. The plasma jet is made of a quartz tube with an inner diameter of 1 mm and a stainless-steel needle with a diameter of 0.2 mm is placed at the center of the quartz tube to serve as the HV electrode. A copper ring with a diameter of 1 mm is placed next to the front end of the quartz tube to serve as the ground electrode. The distance between the HV electrode and ground electrode is about 2.5 mm and the HV electrode is connected to a DC power supply via a 20 MΩ resistor. When N₂ flows through the quartz tube and the HV power supply is on, a cold plasma plume is generated in the surrounding air with a length of up to about 2 cm as shown in Figure 7.20. The gas flow rate is estimated to be about 126 m s⁻¹.

When a grounded steel cone is placed several millimeters away from the plasma jet nozzle, the N₂ plasma plume does not stop propagating and flows around the steel cone, as shown in Figure 7.21. The steel cone does not affect the discharge current or voltage, nor does it have any effects on the intensity of the plasma plume. The observation is different from that reported before and shown in Figure 7.22 [29]. The device in Figure 7.22 consists of an HV wire electrode inserted into a quartz tube with one closed end. The quartz tube along with the HV electrode is inserted into the hollow barrel of a syringe. When He is injected into the hollow barrel at a flow rate of 2 l min⁻¹ with a pulsed DC HV voltage of 6 kV, repetition rate of 10 kHz, and pulse width of 500 ns, a 3-cm-long cold He plasma plume is generated, as shown in Figure 7.22a. When the grounded steel cone is placed close to the plasma plume (but not in direct contact), the plasma plume bends toward the cone-shaped steel and stops propagating, as shown in Figure 7.22b. The plasma plume in Figure 7.22 is electrically driven. The grounded steel cone close to the plasma plume affects the electric field distribution and consequently the propagation of the plasma plume. The plasma plume also stops at the steel cone even if it is grounded via a resistor of several megaohms. On the other hand, for the N₂ plasma plume generated by the device as shown

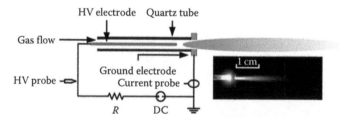

FIGURE 7.20 Schematic of the plasma jet together with a photograph on the right showing the nonequilibrium N₂ plasma plume ($R = 20$ MΩ, applied voltage $V = 10$ kV, and N₂ flow rate = 5 SLM). (From Xian, Y., et al., *Appl. Phys. Lett.*, 100, 123702, 2012.)

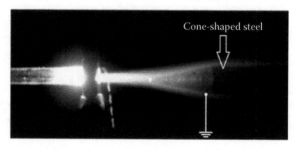

FIGURE 7.21 Photograph of the plasma plume with a grounded steel cone placed in front of the plasma plume. The plasma plume flows around the cone. (From Xian, Y., et al., *Appl. Phys. Lett.*, 100, 123702, 2012.)

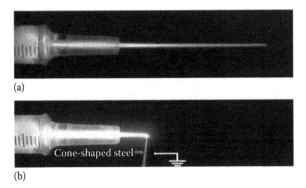

(a)

(b)

FIGURE 7.22 Photographs of a He plasma plume: (a) The plasma plume propagates freely in ambient air. (b) A steel cone close to the plasma plume stops the propagation. (From Xian, Y., et al., *Appl. Phys. Lett.*, 100, 123702, 2012.)

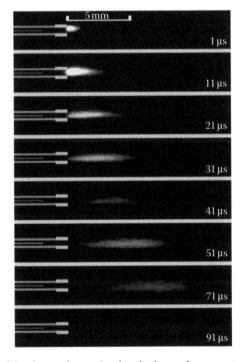

FIGURE 7.23 Photographs of the plasma plume taken by a high-speed camera at an exposure time of 5 μs. (From Xian, Y., et al., *Appl. Phys. Lett.*, 100, 123702, 2012.)

in Figure 7.20, when a grounded steel cone is placed in contact with the plasma plume similar to that shown in Figure 7.22b, propagation of the N_2 plasma plume is not affected at all. Therefore, the propagation mechanism must be different and is perhaps not electrically driven.

To investigate whether plasma plume dynamics are dictated by another factor such as the gas flow velocity, a fast intensified charge-coupled device (ICCD) camera (Princeton Instruments, Model: PIMAX2) is employed. The exposure time is set to 5 μs for all the photographs taken at different times in Figure 7.23. Each picture is an integrated picture of over 50 shots with the same delay time. As seen in Figure 7.23, the plasma plume resembles a bullet too.

The plume propagation velocity is plotted against time in Figure 7.24. Because the actual plume location fluctuates slightly albeit for the same delay time, five shots are taken for each delay time. The initial

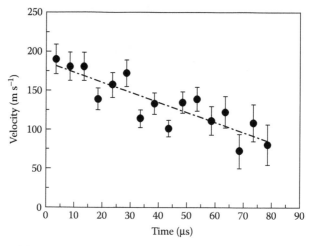

FIGURE 7.24   Velocity of the plasma plume versus time.

propagation velocity is about 180 m s^{-1} and decreases with time, dropping to about 80 m s^{-1} after 80 μs. The velocities are several orders of magnitude lower than those for noble gas-operated jets reported and are in fact close to the estimated gas flow velocity. In the beginning, the propagation velocities are slightly higher than the gas flow velocity, probably due to air diffusion into the N$_2$ channel. Because the ionization energy of O$_2$ is smaller than that of N$_2$, some excited state N$_2$ such as N$_2$(C) states have energy closer to the ionization energy of O$_2$, and so they may affect the plasma propagation. Only a small O$_2$ concentration has this effect because too much of O$_2$ results in attachment.

7.5 Parameter Diagnostics of Plasma Plumes

As discussed before, APNP-Js generate plasma in open spaces rather than in confined gaps. When a bright plume is generated, we call it APNP-J. However, we know little about the bright plume in the beginning, such as what the electron density is, what the various reactive species concentrations are. Fortunately, effort has been made to understand more about several different types of APNP-Js in recent years. In this section, the electron density measurement on a microplasma jet, O atom density distribution in the effluent of a microscale atmospheric pressure plasma jet (μ-APPJ), OH radical concentration in an atmospheric pressure He microwave plasma jet, O$_2$($^1\Delta_g$) of two different APPJs, He and Ar metastable states of a plasma jet are reviewed.

7.5.1 Electron Density

Electron density is one of the most important parameters of plasmas. However, only few measurements have been carried out on the electron density of APNP-Js. This is due to the lack of a diagnostic method for atmospheric pressure nonequilibrium plasmas. In addition, APNP-Js normally have a relatively small dimension (diameter in the millimeter range) and their electron densities are relatively low, which makes measurement even more difficult.

Choi et al. [150] measured the electron density of a microplasma jet by using CO$_2$-laser heterodyne interferometry. Figure 7.25 shows the plasma jet device. He gas was fed into the inlet of the micro-HC and effused out of the anode hole at a flow rate of 6 l min^{-1}. A homemade pulsed DC power supply was used to generate the discharge. The power supply provided repetitive positive pulses with amplitude up to 1.2 kV and pulse width of 400–500 μs. Figure 7.26 shows the obtained electron density. The peak value of the electron density of the plasma jet reaches more than 2 × 10^{14} cm^{-3}.

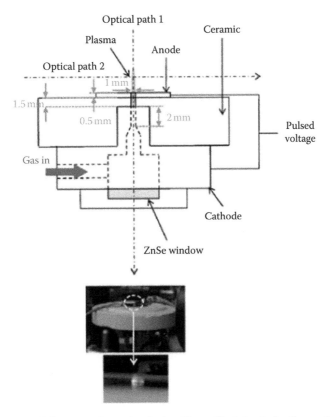

FIGURE 7.25 Schematic of the microplasma jet device. (From Choi, J., et al., *Plasma Sources Sci. Technol.*, 18, 035013, 2009.)

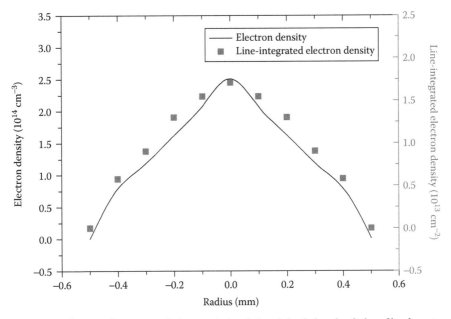

FIGURE 7.26 Distribution of n_e measured along optical path 2 and the deduced radial profile of n_e using inverse Abel transformation. (From Choi, J., et al., *Plasma Sources Sci. Technol.*, 18, 035013, 2009.)

The other method to obtain the electron density of APNP-Js is through simulation. Naidis reported the simulation results of the electron density as shown above in Figures 7.12 and 7.15. It is of the order of 10^{12-13} cm^{-3}.

7.5.2 O Atom Density

O atoms play the most important role in many applications of APNP-Js, such as decontamination, surface treatment. To optimize the discharge parameters and configuration of APNP-Js, the absolute concentration of O atoms has to be measured.

Knake et al. [151] measured the absolute atomic oxygen density distribution in the effluent of a μ-APPJ by two-photon absorption laser-induced fluorescence (TALIF) spectroscopy. The coplanar μ-APPJ is a capacitively coupled RF discharge (13.56 MHz, 15 W RF power) designed for optimized optical diagnostic access as shown in Figure 7.27. It is operated in a homogeneous glow mode with a noble gas flow (1.4 SLM He) containing a small admixture of molecular oxygen (0.5%). Ground-state atomic oxygen densities in the effluent up to 2×10^{14} cm^{-3} are measured as shown in Figure 7.28, which provides space-resolved density maps. Figure 7.29 shows that a maximum of the atomic oxygen density is observed for 0.6% molecular oxygen admixture. In addition, the absolute O atom density profiles in the discharge core of the plasma jet were also measured. Figure 7.30 shows the ground-state atomic oxygen densities up to 3×10^{16} cm^{-3} [66].

Reuter et al. [52] measured the O atom concentration in the effluent of an APPJ by TALIF spectroscopy. The O atoms were generated by a planar plasma jet device, which is driven by a 13.56 MHz RF power supply. The O atom density at the nozzle reaches a value of ~10^{16} cm^{-3}. Even at several centimeters distance

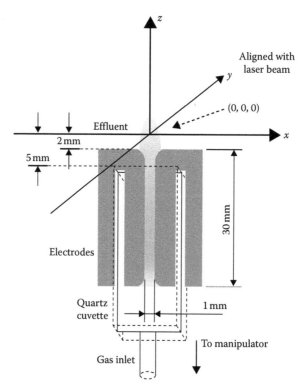

FIGURE 7.27 Schematic sketch of the μ-APPJ and defined coordinate system. (From Knake, N., et al., *J. Phys. D*, 41, 194006, 2008.)

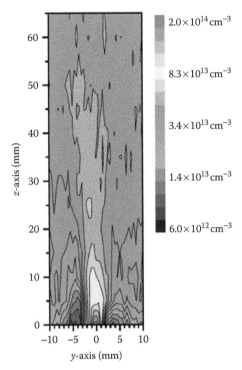

FIGURE 7.28 Spatial y–z density distribution of oxygen atoms in the effluent of the μ-APPJ measured at a base gas flow of 1.4 SLM and a 1% oxygen admixture (logarithmic scale). (From Knake, N., et al., *J. Phys. D*, 41, 194006, 2008.)

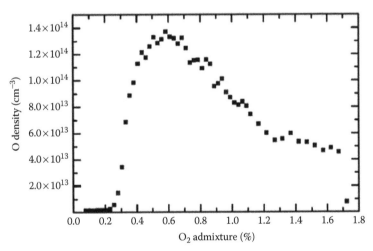

FIGURE 7.29 Effect of O_2 admixture variation on the atomic oxygen density measured at position (0, 0, 0) 1.4 SLM helium flow, and 13 W RF power. (From Knake, N., et al., *J. Phys. D*, 41, 194006, 2008.)

1% of the initial O atom density can be detected. These results are actually consistent with the results of Knake et al. In addition, Reuter et al. found that there are short-lived excited oxygen atoms up to a distance of 10 cm from the jet's nozzle based on optical emission measurements. Figure 7.31 shows that energetic vacuum ultraviolet radiation, measured by OES down to 110 nm, reaches far into the effluent where it is responsible for the generation of atomic oxygen.

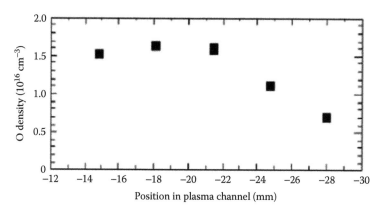

FIGURE 7.30　The absolute atomic oxygen density profiles in the discharge core of the plasma jet. (From Knake, N., et al., *Appl. Phys. Lett.*, 93, 131503, 2008.)

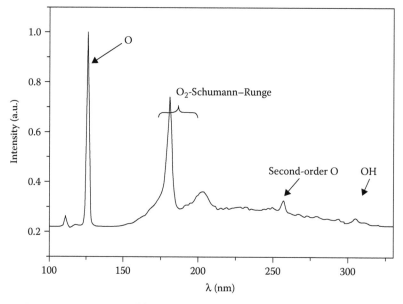

FIGURE 7.31　VUV emission spectrum of the APPJ operated at 100 W RF power with a He gas flux of 1 m³ h⁻¹ and 0.3 vol% O_2 admixture. The spectrum is recorded at a distance of 5 mm from the jet's nozzle. (From Reuter, S., et al., *Plasma Sources Sci. Technol.*, 18, 015006, 2009.)

Jeong et al. [152] obtained the O, $O_2(^1\Delta_g)$, $O_2(^1\Sigma_g^+)$, and O_3 concentrations in the afterglow of a non-equilibrium, capacitive discharge, operated at 600 Torr total pressure with $(0.5\text{–}5.0) \times 10^{17}$ cm⁻³ of oxygen in He by absorption spectroscopy, and numerical modeling. Figure 7.32 shows the evolution of oxygen atoms, metastable oxygen molecules, and ozone over time after turning off the power to the electrode. Excellent agreement is achieved between theory and the experimental results. At time $t = 10$ μs after the RF power is shut off, the concentrations of the five oxygen species are 1.3×10^{17} cm⁻³ ground-state O_2, 6.0×10^{15} cm⁻³ O atoms, 5.0×10^{15} cm⁻³ O $(^1\Delta_g)$, 1.0×10^{15} cm⁻³ $O_2(^1\Sigma_g^+)$, and 2.5×10^{15} cm⁻³ O_3. Assuming that the concentrations of the neutral species are continuous from the on to the off state, these

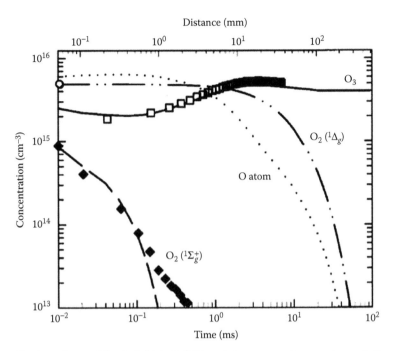

FIGURE 7.32 The dependence of the O, $O_2(^1\Delta_g)$, $O_2(^1\Sigma_g^+)$, and O_3 concentrations on time and distance for initial plasma operation at 24.4 W cm^{-3}, 6.0 Torr of O_2, and 120°C (symbols are the experimental data, whereas the lines are the model prediction, 400 W applied power). (From Jeong, J., et al., *J. Phys. Chem.*, A104, 8027–8032, 2000.)

results indicate that the plasma converts 2% and 3.5% of the oxygen fed into O atoms and singlet-delta metastable O_2 molecules, respectively.

7.5.3 OH Concentration

OH is very reactive. It also plays an important role in various applications of APNP-Js. Srivastava et al. measured the OH radical concentration at atmospheric pressure He microwave plasma jet by cavity ring-down spectroscopy and optical emission spectroscopy [153]. The absolute number densities of OH radicals were measured along the plasma jet column from the jet orifice to far downstream, and the OH concentrations were found to vary from 9.6×10^{15} near the jet orifice to 7.3×10^{12} molecules cm^{-3} 16 mm away from the jet orifice, as shown in Figure 7.33.

7.5.4 $O_2(^1\Delta_g)$ Concentration

$O_2(^1\Delta_g)$ is one of the most important reactive oxygen species (ROS), which is well known as an important agent in numerous biophysical and biochemical processes. $O_2(^1\Delta_g)$ not only generates oxidative damage in a variety of biological targets, but is also a primary active species for killing tumor cells in the emerging cancer therapy known as photodynamic therapy. The most remarkable feature of $O_2(^1\Delta_g)$ is its extremely long and unique radiative lifetime of more than 75 min in the gas phase. Having an excitation energy of 0.98 eV, $O_2(^1\Delta_g)$ is a highly reactive chemical molecule. As such, $O_2(^1\Delta_g)$ plays an important role in gas and liquid chemistry and has therefore attracted the attention of many scientists working in virtually every field of the natural sciences, from physics to medicine, through chemistry and biology.

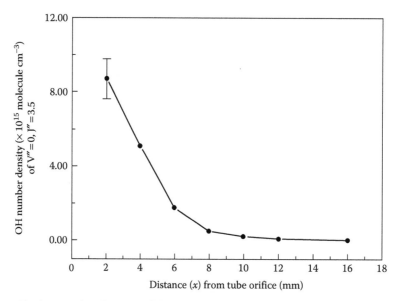

FIGURE 7.33 Absolute number densities of the OH radicals at different locations along the plasma jet axis. Plasma power = 38 W, He gas flow rate = 0.58 SLM, and plasma column length = 2.5 mm. (From Srivastava, N. and Wang, C., *IEEE Trans. Plasma Sci.*, 39, 918, 2011. © 2011 IEEE.)

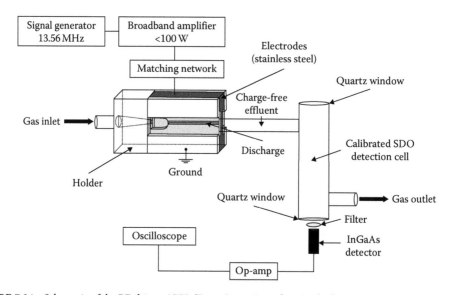

FIGURE 7.34 Schematic of the RF-driven APPJ. (From Sousa, J., et al., *J. Appl. Phys.* 109, 123302, 2011.)

Sousa et al. [50] measured the absolute densities of singlet delta oxygen (SDO) molecules by using infrared optical emission spectroscopy in the flowing effluents of two different APPJs, that is, a capacitively coupled RF-driven jet (RF-APPJ) and a lower frequency kilohertz-driven DBD jet, as shown in Figures 7.34 and 7.35. The plasma jets were operated in He, with small admixtures of molecular oxygen ($O_2 < 2\%$). Figure 7.36a shows the evolution of the SDO density as a function of the applied power, measured in the effluent of the RF-APPJ, for two different O_2 admixtures. An overlinear increase of the $O_2(^1\Delta_g)$ density of 1 order of magnitude was observed. Only a small fraction of the

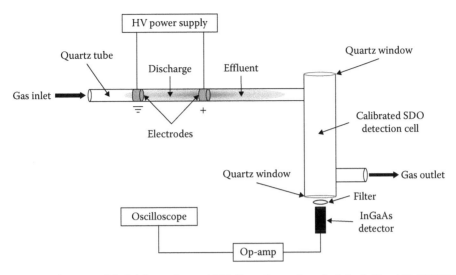

FIGURE 7.35   Schematic of the kilohertz-driven APPJ. (From Sousa, J., et al., *J. Appl. Phys.* 109, 123302, 2011.)

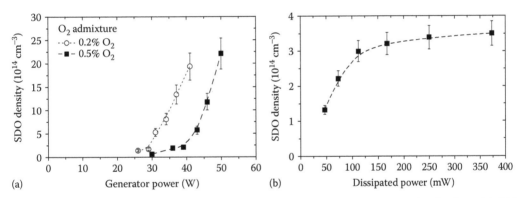

FIGURE 7.36   (a) $O_2(^1\Delta_g)$ density as a function of the applied 13.56 MHz RF power with a He flow of 1 SLM and different O_2 admixtures. (b) $O_2(^1\Delta_g)$ density as a function of the power dissipated into the kilohertz plasma jet when operated at different applied voltages (5–8 kV) and at a 20 kHz pulse repetition rate, with a He flow of 2 SLM and an O_2 admixture of 0.5%. The interelectrode distance is 25 mm. (From Sousa, J., et al., *J. Appl. Phys.* 109, 123302, 2011.)

generator power is actually coupled into the plasma, as there are losses in the connecting cables and matching network, heating of the electrodes, as well as RF radiation of the jet device acting as an antenna. Figure 7.36b shows the evolution of the SDO density measured in the effluent of the kilohertz APPJ as a function of the dissipated power. The SDO density increases with an increase of the dissipated power in the beginning. But it tends to saturate when the dissipated power is higher than 250 mW.

Figure 7.37a shows the evolution of the $O_2(^1\Delta_g)$ density as a function of the O_2 admixture, measured in the effluent of the RF-APPJ for different applied powers. It shows that $O_2(^1\Delta_g)$ density decreases with increasing O_2 fraction. Figure 7.37b shows the evolution of $O_2(^1\Delta_g)$ density as a function of the O_2 fraction, measured in the effluent of the kilohertz plasma for different interelectrode distances. Trends similar to those shown in Figure 7.37a are observed.

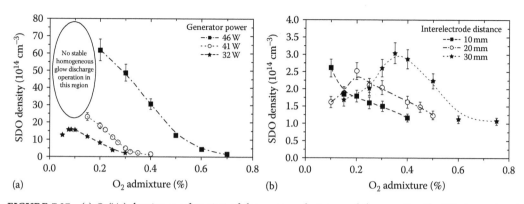

FIGURE 7.37 (a) $O_2(^1\Delta_g)$ density as a function of the oxygen admixture while operating the RF plasma jet at 13.56 MHz, with a He flow of 1 SLM. (b) $O_2(^1\Delta_g)$ density as a function of the oxygen fraction while operating the kilohertz plasma jet at 6 kV and at 20 kHz, with a He flow of 2 SLM. (From Sousa, J., et al., *J. Appl. Phys.* 109, 123302, 2011.)

7.5.5 He and Ar Metastable States' Concentration

He and Ar metastable states play an important role in the plasma generation and plasma chemistry processes. To understand more about their role in APNP-Js, Urabe et al. [138] measured spatio-temporal structures of excited He species by laser spectroscopic methods in a plasma jet, which was driven by a bipolar impulse voltage pulse train of the order of kilohertz repetition rate applied across a pair of electrodes wrapped around a glass tube with a He gas flow, as shown in Figure 7.38.

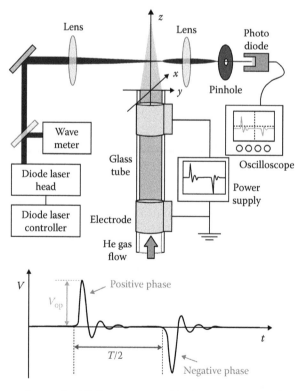

FIGURE 7.38 Experimental setup for APPJ generation and LAS measurement with applied voltage waveform. V_{op}, peak value; T, repetition cycle. (From Urabe, K., et al., *Appl. Phys. Express*, 1, 066004, 2008.)

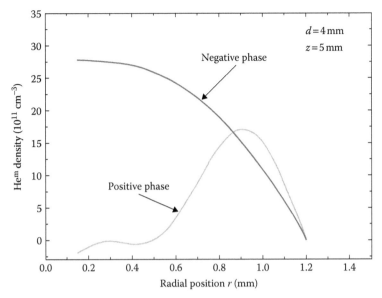

FIGURE 7.39 Radial distributions of absolute He^m density for both voltage phases in the DBD mode measured with a 4 mm bore tube at $V_{op} = 6.0$ kV and $z = 5$ mm at the respective timings of 1.87 μs (positive phase) and 2.08 μs (negative phase) after the increase in the applied voltage. (From Urabe, K., et al., *Appl. Phys. Express*, 1, 066004, 2008.)

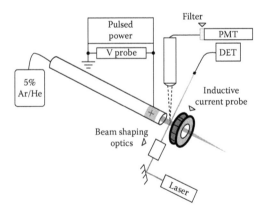

FIGURE 7.40 Schematic of experimental apparatus and diagnostics used in this experiment. (From Sands, B., et al., *J. Phys. D*, 43, 282001, 2010.)

Figure 7.39 shows the measured He^m concentration. It is interesting to note that the radial distribution of the excited species had a hollow shape at the center in the positive voltage phase, while it had a more uniform shape in the negative phase. The peak density of the He metastable atom is about 10^{12} cm^{-3}.

Sands et al. [120] measured the Ar(^{1}s$_5$) metastable state concentration in a He/Ar (5%) APPJ by using diode laser absorption spectroscopy. The schematic of the experimental setup is shown in Figure 7.40. The plasma was driven by a positive unipolar voltage pulse with 20 ns rise time. Figure 7.41 shows the contributions from each production process as a function of position. The total Ar metastable column density reaches a value as high as 10^{12} cm^{-2} near the jet nozzle.

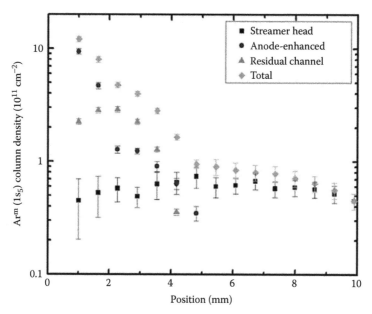

FIGURE 7.41 Contributions from each of the time-resolved production regimes to the total Ar ($^1s_5$) metastable line-integrated column density as a function of position, shown on a semilog plot. The data are shown at 10 kV. (From Sands, B., et al., *J. Phys. D*, 43, 282001, 2010.)

7.6 Summary

After about a decade of development, great progress was made in understanding the operation of APNP-Js. Not only have various types of noble gas APNP-Js been reported, but APNP-Js using N_2 and air have also been introduced. In addition, the behavior of the "plasma bullet" is better understood. Some characteristics of traditional streamers and "plasma bullet" are compared, such as the stochastic nature of their initiation, the ring-shaped profiles of plasma bullet radiation, and the role of the dark channel left by the propagation of the plasma plume. However, there are still some features that are not well understood. For example, it is not clear whether the dynamic behavior of the air/N_2 plasma plume is similar.

Further studies are needed to study the chemical activity of APNP-Js, so that the treatment time can be further shortened. In addition, APNP-Js with air as the working gas need to be developed because they will make many potential applications possible. Finally, for some applications, such as surface treatment, large area treatment is needed. How to make APNP-Js suitable for such applications is a topic of ongoing and future research efforts.

References

1. Y. P. Raizer, *Gas Discharge Physics*, Springer-Verlag, Berlin, 1991.
2. M. A. Lieberman, *Principles of Plasma Discharges and Materials Processing*, John Wiley & Sons, Inc. New York, 1994.
3. A. Fridman and L. A. Kennedy, *Plasma Physics and Engineering*, Taylor & Francis Group, New York, 2004.
4. K. H. Becker, U. Kogelschatz, K. H. Schoenbach, and R. J. Barker, *Non-Equilibrium Air Plasmas at Atmospheric Pressure*, Institute of Physics Publishing Ltd., Bristol, 2005.
5. U. Kogelschatz, *Pure Appl. Chem.* 62, 1667 (1990).
6. A. Chirokov, A. Gutsol, A. Fridman, K. D. Dieber, J. M. Grace, and K. S. Robinson, *Plasma Sources Sci. Technol.* 13, 623 (2004).

7. S. Okazaki, M. Kogoma, M. Uehara, and Y. Kimura, *J. Phys. D Appl. Phys.* 26, 889 (1993).
8. J. Shin and L. Raja, *Appl. Phys. Lett.* 88, 021502 (2006).
9. G. Nersisyan and W. G. Graham, *Plasma Sources Sci. Technol.* 13, 582 (2004).
10. M. Laroussi, X. Lu, V. Kolobov, and R. Arslanbekov, *J. Appl. Phys.* 96, 3028 (2004).
11. P. Xu and M. J. Kushner, *J. Appl. Phys.* 84, 4153 (1998).
12. L. Stollenwerk, S. Amiranashvili, J. P. Boeuf, and H. G. Purwins, *Phys. Rev. Lett.* 96, 255001 (2006).
13. F. Massines and G. Gouda *J. Phys. D Appl. Phys.* 31, 3411 (1998).
14. S. E. Babayan, J. Y. Jeong, V. J. Tu, J. Park, G. S. Selwyn, and R. F. Hicks, *Plasma Sources Sci. Technol.* 7, 286 (1998).
15. J. Y. Jeong, S. E. Babayan, V. J. Tu, J. Park, I. Henins, R. F. Hicks, and G. S. Selwyn, *Plasma Source Sci. Technol.* 7, 282 (1998).
16. J. Y. Jeong, S. E. Babayan, A. Schutz, V. J. Tu, J. Park, I. Henins, G. S. Selwyn, and R. F. Hicks, *J. Vac. Sci. Technol. A* 17, 2581 (1999).
17. J. Park, I. Henins, H. W. Herrmann, G. S. Selwyn, J. Y. Jeong, R. F. Hicks, D. Shim, and C. S. Chang, *Appl. Phys. Lett.* 76, 288 (2000).
18. S. E. Babayan, J. Y. Jeong, A. Schutze, V. J. Tu, M. Moravej, G. S. Selwyn, and R. F. Hicks, *Plasma Sources Sci. Technol.* 10, 573 (2001).
19. M. Moravej, X. Yang, G. R. Nowling, J. P. Chang, R. F. Hicks, and S. E. Babayan, *J. Appl. Phys.* 96, 7011 (2004).
20. L. Xu, P. Liu, R. J. Zhan, X. H. Wen, L. L. Ding, and M. Nagatsu, *Thin Solid Films* 506–507, 400 (2006).
21. Z. Hubicka, M. Cada, M. Sicha, A. Churpita, P. Pokorny, L. Soukup, and L. Jastrabik, *Plasma Sources Sci. Technol.* 11, 195 (2002).
22. V. Leveille and S. Coulombe, *Plasma Sources Sci. Technol.* 14, 467 (2005).
23. S. Yonson, S. Coulombe, V. Leveille, and R. L. Leask, *J. Phys. D Appl. Phys.* 39, 3508 (2006).
24. Q. Xiong, A. Nikiforov, X. Lu, and C. Leys, *J. Phys. D* 43, 415201 (2010).
25. M. Teschke, J. Kedzierski, E. G. Finantu-Dinu, D. Korzec, and J. Engemann, *IEEE Trans. Plasma Sci.* 33, 310 (2005).
26. Q. Li, J. T. Li, W. C. Zhu, X. M. Zhu, and Y. K. Pu, *Appl. Phys. Lett.* 95, 141502 (2009).
27. X. Lu, Z. Jiang, Q. Xiong, Z. Tang, X. Hu, and Y. Pan, *Appl. Phys. Lett.* 92, 081502 (2008).
28. J. L. Walsh and M. G. Kong, *Appl. Phys. Lett.* 93, 111501 (2008).
29. X. Lu, Z. Jiang, Q. Xiong, Z. Tang, and Y. Pan, *Appl. Phys. Lett.* 92, 151504 (2008).
30. M. Laroussi and X. Lu, *Appl. Phys. Lett.* 87, 112902 (2005).
31. A. Shashurin, M. N. Shneider, A. Dogariu, R. B. Miles, M. Keidar, *Appl. Phys. Lett.* 94, 231504 (2009).
32. E. Stoffels, I. E. Kieft, and R. E. J. Sladek, *J. Phys. D Appl. Phys.* 36, 2808 (2003).
33. X. Lu, Y. Cao, P. Yang, Q. Xiong, Z. Xiong, J. Hu, F. Zhou, W. Gong, Y. Xian, C. Zhou, Z. Tang, Z. Jiang, and Y. Pan, *IEEE Trans Plasma Sci.* 37, 668 (2009).
34. Y. C. Hong and H. S. Uhm, *Appl. Phys. Lett.* 89, 221504 (2006).
35. T. L. Ni, F. Ding, X. D. Zhu, X. H. Wen, and H. Y. Zhou, *Appl. Phys. Lett.* 92, 241503 (2008).
36. Y. C. Hong, H. S. Uhm, and W. J. Yi, *Appl. Phys. Lett.* 93, 051504 (2008).
37. Y. C. Hong and H. S. Uhm, *Phys. Plasmas* 14, 053503 (2007).
38. A H. Mohamed, J. F. Kolb, and K. H. Schoenbach, US Patent 7, 572, 998 B2 (2009).
39. Y. Hong, W. Kang, Y. Hong, W. Yi, and H. Uhm, *Phys. Plasmas* 16, 123502 (2009).
40. G. Fridman, *Plasma Chem. Plasma Process.* 26, 425 (2006).
41. S. Wu, X. Lu, Z. Xiong, and Y. Pan, *IEEE Trans. Plasma Sci.* 38, 3404 (2010).
42. Z. Machala, C. Laux, and C. Kruger, *IEEE Trans. Plasma Sci.* 33, 320 (2005).
43. X. Lu, Z. Xiong, F. Zhao, Y. Xian, Q. Xiong, W. Gong, C. Zou, Z. Jiang, and Y. Pan, *Appl. Phys. Lett.* 95, 181501 (2009).
44. S. Forster, C. Mohr, and W. Viol, *Surface Coat. Technol.* 200, 827 (2005).
45. S. Schneider, J. W. Lackmann, F. Narberhaus, J. E. Bandow, B. Denis, and J. Benedikt, *J. Phys. D Appl. Phys.* 44, 295201 (2011).

46. M. Qian, C. Ren, D. Wang, J. Zhang, and G. Wei, *J. Appl. Phys.* 107, 063303 (2010).
47. H. E. Porteanu, S. Kühn, and R. Gesche, *J. Appl. Phys.* 108, 013301 (2010).
48. T. Yuji, H. Kawano, S. Kanazawa, T. Ohkubo, and H. Akatsuka, *IEEE Trans. Plasma Sci.* 36, 976 (2008).
49. T.-C. Tsai and D. Staack, *Plasma Process. Polym.* 8, 523 (2011).
50. J. S. Sousa, K. Niemi, L. J. Cox, Q. T. Algwari, T. Gans, and D. O'Connell, *J. Appl. Phys.* 109, 123302 (2011).
51. G. Naidis, *Appl. Phys. Lett.* 98, 141501 (2011).
52. S. Reuter, K. Niemi, V. Schulz-von der Gathen, and H. F. Dobele, *Plasma Sources Sci. Technol.* 18, 015006 (2009).
53. K. Fricke, H. Steffen, T. von Woedtke, K. Schröder, and K.-D. Weltmann, *Plasma Process. Polym.* 8, 51 (2011).
54. J. S. Oh, O. T. Olabanji, C. Hale, R. Mariani, K. Kontis, and J. W. Bradley, *J. Phys. D Appl. Phys.* 44, 155206 (2011).
55. Z. Xiong, K. Takashiman, I. Adamovich, and M. Kushner, Simulation of high pressure ionization waves in straight and circuitous dielectric channels. *The 64th Gaseous Electronics Conference*, November 14–18, 2011, Salt Lake City, UT.
56. M. Kushner, Interaction of high pressure plasmas with their boundaries: channels, tubes, liquids and tissue. *The 30th International Conference on Phenomena in Ionized Gases*, August 28–September 2, 2011, Belfast, Northern Ireland.
57. G. Cho, H. Lim, J. H. Kim, D. J. Jin, G. C. Kwon, E. H. Choi, and H. S. Uhm, *IEEE Trans. Plasma Sci.* 39, 1234 (2011).
58. J. Dgheim, *Plasma Sources Sci. Technol.* 16, 211 (2007).
59. H. Kim, A. Brockhaus, and J. Engemann, *Appl. Phys. Lett.* 95, 211501 (2009).
60. J. Laimer, H. Reicher, and H. Stori, *Vacuum*, 84, 104 (2009).
61. J. Schafer, F. Sigeneger, R. Foest, D. Loffhagen, and K. D. Weltmann, *Eur. Phys. J. D*, 60, 531 (2010).
62. N. Georgescu, C. P. Lungu, A. R. Lupu, and M. Osiac, *IEEE Trans. Plasma Sci.* 38, 3156 (2010).
63. N. Bibinov, D. Dudek, P. Awakowicz, and J. Engemann, *J. Phys. D Appl. Phys.* 40, 7372 (2007).
64. A. V. Pipa and J. Ropcke, *IEEE Trans. Plasma Sci.* 37, 1000 (2009).
65. N. Knake, S. Reuter, K. Niemi, V. Schulz-von der Gathen, and J. Winter, *J. Phys. D Appl. Phys.* 41, 194006 (2008).
66. N. Knake, K. Niemi, S. Reuter, V. Schulz-von der Gathen, and J. Winter, *Appl. Phys. Lett.* 93, 131503 (2008).
67. A. V. Pipa, T. Bindemann, R. Foest, E. Kindel, J. Ropcke, and K. D. Weltmann, *J. Phys. D Appl. Phys.* 41, 194011 (2008).
68. Y. Hong, S. Yoo, and B. Lee, *J Electrostat.*, 69, 92 (2011).
69. Q. Y. Nie, C. S. Ren, D. Z. Wang, S. Z. Li, J. L. Zhang, and M. G. Kong, *Appl. Phys. Lett.* 90, 221504 (2007).
70. Q. Li, H. Takana, Y. K. Pu, and H. Nishiyama, *Appl. Phys. Lett.* 98, 241501 (2011).
71. D. Breden, K. Miki, and L. L. Raja, *Appl. Phys. Lett.* 99, 111501 (2011).
72. J. P. Lim, H. S. Uhm, and S. Z. Li, *Appl. Phys. Lett.* 90, 051504 (2007).
73. M. Laroussi and T. Akan, *Plasma Process. Polym.* 4, 777 (2007).
74. F. Iza, G. J. Kim, S. M. Lee, J. K. Lee, J. L. Walsh, Y. T. Zhang, and M. Kong, *Plasma Process. Polym.* 5, 322 (2008).
75. J. Jansky and A. Bourdon, *Appl. Phys. Lett.* 99, 161504 (2011).
76. T. Algwari and D. O'Connell, *Appl. Phys. Lett.* 99, 121501 (2011).
77. Z. Xiong, X. Lu, Q. Xiong, Y. Xian, C. Zou, J. Hu, W. Gong, J. Liu, F. Zou, Z. Jiang, and Y. Pan, *IEEE Trans. Plasma Sci.* 38, 1001 (2010).
78. Q. Xiong, X. Lu, J. Liu, Y. Xian, Z. Xiong, F. Zou, C. Zou, W. Gong, J. Hu, K. Chen, X. Pei, Z. Jiang, and Y. Pan, *J. Appl. Phys.* 106, 083302 (2009).
79. R. Stonies, S. Schermer, E. Voges, and J. A. C. Broekaert, *Plasma Sources Sci. Technol.* 13, 604 (2004).

80. Q. Xiong, X. Lu, Z. Jiang, Z. Tang, J. Hu, Z. Xiong, X. Hu, and Y. Pan, *IEEE Trans. Plasma Sci.* 36, 986 (2008).
81. X. Lu, Q. Xiong, Z. Tang, Z. Jiang, and Y. Pan, *IEEE Trans. Plasma Sci.* 36, 990 (2008).
82. J. D. Yan, C. F. Pau, S. R. Wylie, and M. T. C. Fang, *J. Phys. D Appl. Phys.* 35, 2594 (2002).
83. S. P. Kuo, O. Tarasenko, S. Popovic, and K. Levon, *IEEE Trans. Plasma Sci.* 34, 1275 (2006).
84. M. Laroussi, *Plasma Process. Polym.* 2, 391 (2005).
85. M. Laroussi, *IEEE Trans Plasma Sci.* 37, 714 (2009).
86. M. Kong, G. Kroesen, G. Morfill, T. Nosenko, T. Shimizu, J. Dijk, and J. Zimmermann, *New J. Phys.* 11, 115012 (2009).
87. S. Perni, X. Deng, G. Shama, and M. Kong, *IEEE Trans. Plasma Sci.* 34, 1297 (2006).
88. G. E. Morfill, M. Kong, and J. L. Zimmermann, *New J. Phys.* 11, 115011 (2009).
89. G. Fridman, G. Friedman, A. Gutsol, A. B. Shekhter, V. N. Vasilets, and A. Fridman, *Plasma Process. Polym.* 5, 503 (2008).
90. G. Fridman, A. Brooks, M. Galasubramanian, A. Fridman, A. Gutsol, V. Vasilets, H. Ayan, and G. Friedman, *Plasma Process. Polym.* 4, 370 (2007).
91. A. Shashurin, M. Keidar, S. Bronnikov, R. A. Jurjus, and M. A. Stepp, *Appl. Phys. Lett.* 93, 181501 (2008).
92. D. Mariotti, *Appl. Phys. Lett.* 92, 151505 (2008).
93. G. Kim, G. Kim, S. Park, S. Jeon, H. Seo, F. Iza, and J. Lee, *J. Phys. D Appl. Phys.* 42, 032005 (2009).
94. D. Dobrynin, G. Fridman, and G. Friedman, *New J. Phys.* 11, 115020 (2009).
95. R. E. J. Sladek, E. Stoffels, R. Walraven, P. J. A. Tielbeek, and R. A. Koolhoven, *IEEE Trans. Plasma Sci.* 32, 1540 (2004).
96. D. O'Connell, L. J. Cox, W. B. Hyland, S. J. McMahon, S. Reuter, W. G. Graham, T. Gans, and F. J. Currell, *Appl. Phys. Lett.* 98, 043701 (2011).
97. S. S. Yang, K. Kim, J. D. Choi, Y. C. Hong, G. Kim, E. J. Noh, and J. S. Lee, *Appl. Phys. Lett.* 98, 073701 (2011).
98. N. Abramzon, J. C. Joaquin, J. Bray, and G. Brelles-Marino, *IEEE Trans. Plasma Sci.* 34, 1304 (2006).
99. G. Daeschlein, S. Scholz, T. von Woedtke, M. Niggemeier, E. Kindel, R. Brandenburg, K. D. Weltmann, and M. Junger, *IEEE Trans. Plasma Sci.* 39, 815 (2011).
100. T. H. Chung, S. J. Kim, S. H. Bae, and S. H. Leem, *Appl. Phys. Lett.* 94, 141502 (2009).
101. Z. Xiong, X. Lu, A. Feng, Y. Pan, and K. Ostrikov, *Phys. Plasmas* 17, 123502 (2010).
102. X. Yan, F. Zou, S. Zhao, X. Lu, G. He, Z. Xiong, Q. Xiong, Q. Zhao, P. Deng, J. Huang, and G. Yang, *IEEE Trans Plasma Sci.* 38, 2451 (2010).
103. X. Zhou, Z. Xiong, Y. Cao, X. Lu, and D. Liu, *IEEE Trans Plasma Sci.* 38, 3370 (2010).
104. X. Yan, F. Zou, X. Lu, G. He, M. Shi, Q. Xiong, X. Gao, Z. Xiong, Y. Li, F. Ma, M. Yu, C. Wang, Y. Wang, and G. Yang, *Appl. Phys. Lett.* 95, 083702 (2009).
105. X. Lu, T. Ye, Y. Cao, Z. Sun, Q. Xiong, Z. Tang, Z. Xiong, J. Hu, Z. Jiang, and Y. Pan, *J. Appl. Phys.* 104, 053309 (2008).
106. Z. Xiong, T. Du, X. Lu, Y. Cao, and Y. Pan, *Appl. Phys. Lett.* 98, 221503 (2011).
107. J. P. Boeuf and L. C. Pitchford, *VI International Conference on Microplasmas*, April 3–6, 2011, Paris, France.
108. M. Laroussi and M. A. Akman, *AIP Adv.* 1, 032138, (2011).
109. X. Lu and M. Laroussi, *J. Appl. Phys.* 100, 063302 (2006).
110. G. A. Dawson and W. P. Winn, *Z. Phys.* 183, 159 (1965).
111. Y. Xian, X. Lu, Y. Cao, P. Yang, Q. Xiong, Z. Jiang, and Y. Pan, *IEEE Trans. Plasma Sci.* 37, 2068 (2009).
112. J. L. Walsh, F. Iza, N. B. Janson, V. J. Law, and M. G. Kong, *J. Phys. D* 43, 075201 (2010).
113. N. Mericam-Bourdet, M. Laroussi, A. Begum, and E. Karakas, *J. Phys. D* 42, 055207 (2009).
114. Y. Sakiyama, D. Graves, J. Jarrige, and M. Laroussi, *Appl. Phys. Lett.* 96, 041501 (2010).
115. G. V. Naidis, *J. Phys. D* 44, 215203 (2011).
116. W. Zhu, Q. Li, X. Zhu, and Y. Pu, *J. Phys. D* 42, 202002 (2009).

117. S. Wu, Q. Huang, Z. Wang, and X. Lu, *IEEE Trans. Plasma Sci.* 39, 2286 (2011).
118. X. Lu, Z. Jiang, Q. Xiong, Z. Tang, Z. Xiong, J. Hu, X. Hu, and Y. Pan, *IEEE Trans. Plasma Sci.* 36, 988 (2008).
119. E. Karakas and M. Laroussi, *J. Appl. Phys.* 108, 063305 (2010).
120. B. Sands, R. J. Leiweke, and B. Ganguly, *J. Phys. D* 43, 282001 (2010).
121. G. V. Naidis, *J. Phys. D* 43, 402001 (2010).
122. C. Jiang, M. T. Chen, and M. A. Gundersen, *J. Phys. D* 42, 232002 (2009).
123. Z. Xiong, X. Lu, Y. Xian, Z. Jiang, and Y. Pan, *J. Appl. Phys.* 108, 103303 (2010).
124. R. Ohyama, M. Sakamoto, and A. Nagai, *J. Phys. D Appl. Phys.* 42, 105203 (2009).
125. N. Jiang, A. Ji, and Z. Cao, *J. Appl. Phys.* 106, 013308 (2009).
126. N. Jiang, A. Ji, and Z. Cao, *J. Appl. Phys.* 108, 033302 (2010).
127. J. L. Walsh, J. J. Shi, and M. G. Kong, *Appl. Phys. Lett.* 88 171501 (2006).
128. J. L. Walsh, and M. G. Kong, *IEEE Trans. Plasma Sci.* 36, 954 (2008).
129. J. L. Walsh, and M. G. Kong, *Appl. Phys. Lett.* 91, 221502 (2007).
130. Q. Y. Nie, C. S. Ren, D. Z. Wang, S. Z. Li, J. L. Zhang, and M. G. Kong, *Appl. Phys. Lett.* 90, 221504 (2007).
131. Q. Y. Nie, Z. Cao, C. S. Ren, D. Z. Wang, and M. G. Kong, *New J. Phys.* 11, 115015 (2009).
132. Z. Cao, Q. Nie, D. L. Bayliss, J. L. Walsh, C. S. Ren, D. Z. Wang, and M. G. Kong, *Plasma Sources Sci. Technol.* 19, 025003 (2010).
133. Z. Cao, Q. Y. Nie, and M. G. Kong, *J. Phys. D Appl. Phys.* 42, 222003 (2009).
134. M. G. Kong, Q. Y. Nie, Z. Cao, C. S. Ren, and D. Z. Wang, *New J. Phys.* 11, 115015 (2009).
135. J. J. Shi, D. W. Liu, and M. G. Kong, *IEEE Trans. Plasma Sci.* 35, 137 (2007).
136. B. L. Sands, B. N. Ganguly, and K. Tachibana, *IEEE Trans. Plasma Sci.* 36, 956 (2008).
137. B. L. Sands, B. N. Ganguly, and K. Tachibana, *Appl. Phys. Lett.* 92, 151503 (2008).
138. K. Urabe, Y. Ito, K. Tachibana, and B. N. Ganguly, *Appl. Phys. Express*, 1, 066004 (2008).
139. K. Urabe, T. Morita, K. Tachibana, and B. N. Ganguly, *J. Phys. D Appl. Phys.* 43, 095201 (2010).
140. M. Laroussi, C. Tendero, X. Lu, S. Alla, and W. L. Hynes, *Plasma Process. Polym.* 3, 470 (2006).
141. M. Laroussi, W. Hynes, T. Akan, X. Lu, and C. Tendero, *IEEE Trans. Plasma Sci.* 36, 1298 (2008).
142. T. H. Chung, S. J. Kim, and S. H. Bae, *Phys. Plasmas*, 17, 053504 (2010).
143. L. W. Chen, P. Zhao, X. S. Shu, J. Shen, and Y. D. Meng, *Phys. Plasmas*, 17, 083502 (2010).
144. Q. Xiong, X. Lu, Y. Xian, J. Liu, C. Zou, Z. Xiong, W. Gong, K. Chen, X. Pei, F. Zou, J. Hu, Z. Jiang, and Y. Pan, *J. Appl. Phys.* 107, 073302 (2010).
145. Y. Xian, X. Lu, Z. Tang, Q. Xiong, W. Gong, D. Liu, Z. Jiang, Y. Pan, *J. Appl. Phys.* 107, 063308 (2010).
146. X. Lu, Q. Xiong, Z. Xiong, J. Hu, F. Zhou, W. Gong, Y. Xian, C. Zou, Z. Tang, Z. Jiang, and Y. Pan, *J. Appl. Phys.* 105, 043304 (2009).
147. X. Lu, Q. Xiong, Z. Xiong, J. Hu, F. Zhou, W. Gong, Y. Xian, C. Zhou, Z. Tang, Z. Jiang, and Y. Pan, *Thin Solid Films* 518, 967 (2009).
148. Y. Xian, X. Lu, J. Liu, S. Wu, D. Liu, and Y. Pan, *Plasma Sources Sci. Technol.* 21, 034013 (2012).
149. Y. Xian, X. Lu, S. Wu, P. Chu, and Y. Pan, *Appl. Phys. Lett.* 100, 123702 (2012).
150. J. Choi, N. Takano, K. Urabe, and K. Tachibana, *Plasma Sources Sci. Technol.* 18, 035013 (2009).
151. N. Knake, S. Reuter, K. Niemi, V. Schulz-von der Gathen, and J. Winter, *J. Phys. D* 41, 194006 (2008).
152. J. Jeong, J. Park, I. Henins, S. Babayan, V. Tu, G. Selwyn, G. Ding, and R. Hicks, *J. Phys. Chem.* A104, 8027–8032 (2000).
153. N. Srivastava and C. Wang, *IEEE Trans. Plasma Sci.* 39, 918 (2011).

8

Cavity Ringdown Spectroscopy of Plasma Species

Chuji Wang

8.1 Introduction

Cavity ringdown spectroscopy (CRDS) is a laser absorption spectral technique. Although several different forms of CRDS have been evolved from the original concept, the common features remain the same and exist in each form of CRDS: high detection sensitivity, high selectivity, absolute number density measurements, near real-time response, and immune-to-light source fluctuations. Over the past 10 years, CRDS has been extended from original spectroscopic studies to its applications in a wide variety of fields, such as environment, energy, medicine, materials, and sensing and diagnostics. One emerging application of CRDS is coupling CRDS with a plasma source, namely, plasma-cavity ringdown spectroscopy (P-CRDS), for plasma diagnostics and absolute number density measurements of plasma species in time and space. Along with latest advancements of nonthermal, low-temperature plasmas and their new applications in medicine, combustion, and material processing, P-CRDS, as a versatile diagnostic tool, begins to play an increasingly important role in the aforementioned fields. The objective of this chapter is to provide a basic guideline to those who are interested in the CRDS and P-CRDS techniques and have intent to develop and apply the techniques to their research. In addition to some detailed discussion of experimental configurations of a P-CRDS system, latest results of absolute number density measurements of more than 30 key plasma species, which have been measured by CRDS to date, including hydroxyl radical (OH), O_2 ($a^1\Delta_g$), O, N_2^+, $N_2(A)$, NH, SiH, and H_2O, are included as examples. Finally, some perspectives in future applications of P-CRDS are also presented.

8.2 Concept of CRDS

8.2.1 Introduction

In the original approach of CRDS, a laser pulse is trapped in a high-finesse optical cavity formed by two high-reflectivity mirrors. The many round trips of the laser pulse within the cavity effectively enhance the absorption path length in a sample gas contained in the cavity by several orders of magnitude. Therefore, detection sensitivity is increased correspondingly by the same orders of magnitude according to Beer's Law. The decay time for the light in the cavity rather than a change in intensity of the light is used to determine optical absorption. In this way, CRDS is immune to the adverse impact of the laser pulse intensity fluctuations. Following its introduction by O'Keefe and Deacon in 1988 [1], CRDS has been implemented for the spectroscopic study of numerous species, including molecules, free radicals, clusters, atoms, and ions in different experimental systems, such as chemical reaction cells, supersonic jets, high-temperature plasmas/ flames, and material processing chambers. Research and applications relevant to CRDS have extended the original implementation scheme to the use of novel optical cavities [2–6], different types of light sources [7–10], methods of cavity excitation and shutoff [11–17], and combinations with conventional instruments and analytical methods [18–27]. Although ringdown-based commercial instruments have been realized for detection of trace gases, thus far, mainly in semiconductor manufacturing and environmental monitoring [28–36], this technology has already shown much promise in a wide variety of application areas, such as combustion flame chemistry, plasma diagnostics [37–50], high-sensitivity elemental and isotopic measurements [51–54], breath gas analysis for medical diagnostics [55–58], and fiber ringdown techniques for remote sensing and sensor network [59–69]. CRDS has been applied in the spectral regions ranging from the ultraviolet to the infrared (IR) by using different laser sources. CRDS coupled with a frequency comb technique demonstrated the capability of measuring numerous chemical species simultaneously [70–72]. Using a pair of broadband prisms for a ringdown cavity, CRDS can detect multiple chemical species with absorption in different wavelength regions simultaneously in a single ringdown system. A single ringdown system formed by a pair of broadband prisms can detect multiple species concurrently in different wavelength regions [73].

To date, more than 10 review papers on CRDS have been published. The first historical overview of CRDS was published in a book edited by Busch and Busch [74]. This book contains many of the early publications in various aspects of CRDS. Berden et al. [75] have given a comprehensive review on experimental schemes and applications of CRDS. Scherer et al. [76] reviewed early publications in fundamental spectroscopic studies using CRDS. Miller and Winstead [77] have reviewed atomic and analytical applications of CRDS. Wagner et al. [78] have reviewed early CRDS studies in the IR spectral region. The application of CRDS for kinetics studies has been reviewed by Atkinson [79]. The technological innovations of cavity ringdown have been discussed in Vallance's review [80]. Paldus and Kachanov [81] have given a historical review of CRDS development. Recently, Mazurenka et al. [82] have reviewed the powerful application of diode lasers in CRDS and its variant, cavity-enhanced absorption spectroscopy (CEAS), also widely known as an integrated cavity output spectroscopy (ICOS). More recently, Wang has reviewed P-CRDS for elemental and isotopic measurements [83]. The most recent review of CRDS is given by Orr and He in 2011 [84]. Other reviews can be seen in the most recent CRDS book [85].

8.2.2 Theory and Operating Principle of CRDS

The number of experimental methods that are essentially based on the CRDS principle continues to grow. However, in its original form, CRDS is implemented by injecting and trapping a laser pulse in a stable optical cavity, which is formed by two highly reflective mirrors. The intensity of the light in the cavity decays exponentially with time at a rate determined by the round trip optical losses experienced by the laser pulse. This intensity decay is monitored using a photomultiplier tube or a photodiode placed behind the second mirror, and the ringdown time is determined by fitting the observed waveform to a single exponential function. A schematic diagram of the original CRDS concept is given in Figure 8.1.

For the simplest case, when the dominant losses are from mirror reflectivity and absorption of a sample gas in the cavity, the time constant (*ringdown time*) for the exponential decay is given approximately by [74]

$$\tau = \frac{L}{c\left[(1-R) + \alpha l_s\right]}, \tag{8.1}$$

where
L is the cavity length
R is the reflectivity of the cavity mirrors
α is the familiar Beer's Law absorption coefficient of a sample in the cavity
l_s is the length of the laser beam path through the sample
c is the speed of light. For a gas filling the entire cavity, $L = l_s$

In a constant concentration of the sample, as the laser wavelength is varied, the ringdown time will decrease upon tuning to an absorption wavelength of the sample (e.g., when α increases). Therefore, ringdown absorption spectra are obtained when the laser wavelength is scanned in a spectral range. Alternatively, when the laser wavelength is fixed on resonance with an absorption wavelength of the sample, variations in α caused by changes in the absorber concentration are reflected by changes in the ringdown time. In this way, the number density of the absorber is determined by measuring the different ringdown times.

A better understanding of the time behavior from Equation 8.1 can be gained from a closer examination of Figure 8.1, in which the results of the light intensity after the first few passes of a laser pulse

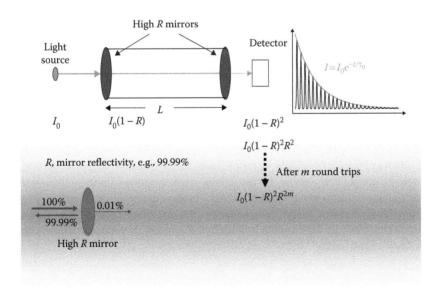

FIGURE 8.1 Concept of cavity ringdown spectroscopy. Effectively increases the absorption path length, for example, from 1 m to 10 km.

through an empty cavity are depicted. When a laser pulse is incident on the first cavity mirror, most of the energy, that is, 99.99%, is reflected away from the cavity. Now assuming that the fraction of the original pulse that enters the cavity has intensity I_0, after one pass through the cavity filled in a sample gas in the optical path length l_s, the intensity will be reduced according to Beer's Law to $I = I_0 e^{-\alpha l_s}$. Following reflection from the second cavity mirror, the intensity will be reduced by the finite mirror reflectivity to $I = I_0 Re^{-\alpha l_s}$. After m round trips through the cavity and substituting $R^{2m} = e^{2m \ln R}$, the intensity will be

$$I = I_0 R^{2m} e^{-2m\alpha l_s} = I_0 e^{-2m(-\ln R + \alpha l_s)}. \tag{8.2}$$

If the cavity is long enough that the pulse never overlaps itself (e.g., longer than one-half of the physical length of the laser pulse), a series of separated pulses decaying in time according to Equation 8.2 will be detected. In many cases, due to electronic response time or the use of shorter cavities, separated pulses are not detected, but rather a continuous exponential decay. The peak envelope of the separated pulses or the time constant for a continuous exponential decay can be found by converting from the discrete variable m to a continuous variable t such that $m = t/T_r$, where $T_r = 2L/c$ is the pulse round trip time in the cavity. In this case we have

$$I = I_0 e^{-tc(-\ln R + \alpha l_s)/L}. \tag{8.3}$$

This ringdown temporal behavior is illustrated in Figure 8.1. From Equation 8.3 we see that the characteristic time constant for the laser pulse decay is given by

$$\tau = \frac{L}{c(-\ln R + \alpha l_s)} \approx \frac{L}{c[(1-R) + \alpha l_s]} \tag{8.4}$$

as given in Equation 8.1. While this analysis does not provide a complete development of technique, it captures the essential elements of CRDS. Note that if the ringdown time is measured in the absence of a sample gas in the cavity, the reflectivity of the mirrors can be determined from the measured ringdown time, τ_0 (i.e., $\alpha = 0$ in Equation 8.4). Once the mirror reflectivity has been measured and a sample introduced in the cavity, CRDS provides an absolute measurement of the absorbance αl_s, from which the absolute number density n of the absorber is determined by

$$\text{Absorbance} = \alpha l_s = \sigma n l_s, \tag{8.5}$$

where σ is the absorption cross section of the absorber in the laser wavelength. This self-calibrating feature offers an advantageous benefit over other high-sensitivity laser spectroscopic techniques, such as laser-induced fluorescence (LIF) or resonantly enhanced multiphoton ionization (REMPI).

8.2.3 Detection Sensitivity of CRDS

The high sensitivity of the CRDS technique stems from the large number of passes that the light pulse makes through the sample in the cavity. For a 1 m long empty ringdown cavity with mirror reflectivity of 99.99%, the ringdown time is approximately 33.3 μs. This is equivalent to a 10 km path length traveled during one ringdown time. For analytical applications the detection sensitivity of CRDS is normally characterized by the detection limit, which is determined by a minimum detectable absorbance. From Equations 8.4 and 8.5, we have

$$\text{Absorbance} = \alpha l_s = \sigma n l_s = (1 - R)\frac{\Delta \tau}{\tau}, \tag{8.6}$$

where $\Delta \tau$ is the difference between τ_0 and τ, which are ringdown times with no absorber and with absorber present, respectively, or ringdown times when the laser wavelength is tuned off and on the absorption peak, respectively. Equation 8.6 indicates that for a given mirror reflectivity, the measured absorbance depends upon the quantity, $\Delta\tau/\tau$. When τ is replaced by τ_0 and $\Delta\tau$ replaced by σ_τ, (one standard deviation of the measured ringdown time), then the absorbance in Equation 8.6 represents the minimum detectable absorbance based on a one-σ_τ criterion. Namely,

$$A_{min} = (1-R)\frac{\sigma_\tau}{\tau_0}. \tag{8.7}$$

σ_τ/τ_0 is often referred to as the ringdown baseline stability or ringdown baseline noise, which is used to characterize a ringdown system's detection sensitivity limit. Clearly, Equation 8.7 shows that increased mirror reflectivity and a lower baseline noise will improve the detection sensitivity (a smaller A_{min}). In practice, σ_τ and τ_0 are the values determined after averaging a number of individual ringdown measurements, that is, 100 ringdown events, further improving the detection sensitivity. Alternatively, a detection sensitivity limit can also be characterized by explicitly including the assumption that the single decay measurement uncertainty is improved by averaging over N ringdown events [86],

$$A_{min} = (1-R)\frac{\sigma_\tau}{\tau\sqrt{N}}. \tag{8.8}$$

If the measuring time is restricted to one second, then N can be replaced by data collection frequency, f. Thus, the unit of A_{min} is $Hz^{-1/2}$. The A_{min} in Equation 8.8 is also called noise-equivalent minimum detectable absorbance.

Examining Equations 8.1 and 8.5, one can see that for a given species, the detection sensitivity in terms of the number density (n) is determined by a ringdown system's physical parameters, such as cavity length, sample path length, mirror reflectivity, and by the system electronic noise. Higher mirror reflectivity, longer sample path length, and lower baseline noise ($\sigma_\tau/\overline{\tau}$) will improve detection sensitivity. For a pulsed-CRDS system, the baseline noise is typically larger than 0.3%. For a continuous wave (CW)-CRDS system, however, the baseline noise can be much lower. The lowest baseline noise reported to date is 2.7×10^{-4} obtained in a near-infrared (NIR) CW-CRDS system with an average over 300 ringdown events [54]. Detection sensitivity on the order of 1×10^{-8} (absorbance) demonstrated in the early studies [87, 88] can now be readily achieved with the availability of ultra-high-reflectivity mirrors with reflectivity up to 99.9985%. In comparison, the experimental limit of $\sigma_\tau/\overline{\tau}$ in CRDS is close to the experimental detection limit of the fractional intensity $\Delta I/I$ in the single pass absorption spectroscopy (SPAS). However, the detection sensitivity limit of CRDS (A_{min}) is smaller than that of SPAS by a factor of $1/(1-R) \approx 10,000$. $1/(1-R)$ may be called ringdown cavity enhancement factor.

8.3 Principle of P-CRDS

P-CRDS combines a plasma source with the CRDS technique, in which the plasma source is a reservoir of the species of interest and located between two ringdown mirrors, and CRDS is used as a detector. P-CRDS is also characterized by the same advantages offered by CRDS: high sensitivity, absolute measurement of concentration, and fast response. P-CRDS follows the same measuring principle as CRDS. However, due to the additional optical losses resulting from the interaction of the laser beam with the plasma, Equation 8.4 is extended as

$$\tau^{plasma} = \frac{L}{c\left[(1-R) + \beta_{plasma}l_s + \beta_{air}(L-l_s) + \alpha l_s\right]}, \tag{8.9}$$

where β_{plasma} and β_{air} are the broadband scattering coefficients in the plasma zone and the nonplasma zone inside cavity, respectively, other parameters in Equation 8.9 have been defined previously. In practice, because the scattering coefficients vary slowly as a function of wavelength, such losses are simply incorporated into an effective mirror reflectivity R_{eff}. Thus, the ringdown time τ^p in Equation 8.9 can be expressed as

$$\tau^p = \frac{L}{c\left[\left(1-R_{eff}\right)+\alpha l_s\right]}. \tag{8.10a}$$

At a given wavelength, R_{eff} is determined by measuring τ_0^p, the ringdown time with no absorber present in the plasma or the laser wavelength is tuned off the absorption peak of the absorber. τ_0^p is often called ringdown time baseline of the P-CRD system, namely

$$\tau_0 = \frac{L}{c\left(1-R_{eff}\right)}. \tag{8.10b}$$

From Equation 8.10a and b, we have absorbance given by

$$\text{Absorbance} = \alpha l_s = \frac{L}{c}\left(\frac{1}{\tau^p}-\frac{1}{\tau_0^p}\right). \tag{8.11}$$

Equation 8.11 indicates that the P-CRDS technique simply measures two ringdown times to determine the absorbance, which relates to number density of the absorber of interest by Equation 8.5.

The density of absorbers can be position-dependent in a plasma and the absorption cross section of a given transition line is a function of frequency, a more appropriate expression for the absorbance is

$$\text{Absorbance} = \int_{-\infty}^{\infty}\int_0^{l_s}\sigma(v)n(r)\mathrm{d}r\mathrm{d}v, \tag{8.12}$$

where

r represents the position along the laser beam path through the plasma with a total laser path length of l_s
$n(r)$ is the concentration of absorbers at the plasma location r
v is the laser frequency
$\sigma(v)$ is the absorption cross section for the transition line at frequency v

Now let's discuss how to measure number density using Equations 8.11 and 8.12 in different cases.

Case 1. The number density of a plasma species is homogeneously distributed in the plasma and the line-width of the optical transition is significantly narrow. In this case, if the absorption cross section, $\sigma(v)$, in Equation 8.12 is approximated by the central absorption cross section, $\sigma(v_0)$, where v_0 is the laser wavelength locked on the absorption peak, then the number density is given by

$$n = \frac{L}{c\sigma(v_0)l_s}\left(\frac{1}{\tau^p(v_0)}-\frac{1}{\tau_0^p(v_0)}\right). \tag{8.13}$$

The measurement error mainly comes from the approximation of using the absorption at the central absorption cross section to replace the total absorption across the entire absorption lineshape. Typically, the measured number density can be over- or underestimated, depending on the absorption lineshape.

Case 2. The number density of a plasma species is homogeneously distributed in the plasma and the frequency distribution of the lineshape is considered. In this case, total absorbance integrated over the entire lineshape is counted for the absorbance given in Equation 8.12. Knowing that

$$S = \int_{-\infty}^{\infty}\sigma(v)\mathrm{d}v, \tag{8.14}$$

we have

$$n = \frac{A_{\text{total}}}{Sl_s},$$ (8.15)

where
S is the line intensity of an optical transition
A_{total} is the integrated absorbance over the entire experimental scan of an absorption lineshape

As long as the line intensity is known, n is more accurately determined than in Case 1.

Case 3. The number density of a plasma species is heterogeneously distributed in the plasma and the frequency distribution of the lineshape is considered. In this case, n in Equation 8.15 represents the spatially averaged and line-of-sight number density of the species, $\bar{n}(l_s)$, because A_{total} in Equation 8.15 is obtained from the spatially averaged and integrated lineshape. To obtain a spatially resolved number density, $n(r)$, we need to have a spatially resolved absorption cross section, $\sigma(v, r)$, and other techniques to convert the line-of-sight number density, $\bar{n}(l_s)$, to $n(r)$. For example, in a cylindrically symmetric plasma, if a laser beam path is aligned along with or parallel to the diameter of a circular cross section of the plasma, Abel inversion process (Equation 8.16) can be implemented to convert the measured line-of-sight density, $\bar{n}(l_s)$, to point measurements along the radii of the plasma section, $n(r)$ [89].

$$I(r) = -\frac{1}{\pi} \int \frac{I'(x)}{\sqrt{x^2 - r^2}} \, dx,$$ (8.16)

where $I'(x)$ is the derivative of $I(x)$. Here, $I(x)$ is the absorption signal intensity measured in the lateral position, x is the distance between the laser beam path and the diameter of the section of the plasma being probed, and $I(r)$ is the converted signal intensity at the radial location, r, along the radii of the plasma section, as shown in Figure 8.2. In practice, the circular cross section of the

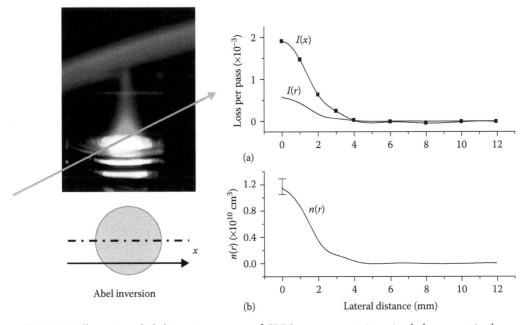

(a)

(b)

Abel inversion

Lateral distance (mm)

FIGURE 8.2 Illustration of Abel inversion process of CRDS measurements in a circularly symmetric plasma. (a) The line-of-sight CRDS measurement $I(x)$ along the laser beam path and the Abel inversion converted signal $I(r)$. (b) The number density $n(r)$ versus radius r.

plasma may be slightly asymmetric, thus numerical methods are often pursued to symmetrize the experimental data. Note that even using the Abel inversion process to obtain the spatially resolved number density in a cylindrically symmetric plasma, we still use a nonspatially resolved absorption cross section [$\sigma(v)$] to approximate the spatially resolved absorption cross section [$\sigma(v, r)$], which is not true in many situations. For the purpose of plasma diagnostics, such as measuring the distribution profile of electron density, gas kinetic temperature, and analyte/species density, the line-of-sight measurement may generate a large deviation [90–93]. However, for the application, in which the size of a plasma is very small [so small that not many data points of $I(x)$ can be obtained experimentally], the line-of-sight measurement is still applicable and quite practical.

As a result of the complicated linewidth broadening mechanisms under atmospheric plasma conditions, the lineshape of an absorption line is a Voigt profile and the absorption cross section can be expressed as [94,95]

$$\sigma_{ij}(v) = \frac{g_j}{g_i} \frac{\lambda^4}{4\pi^2 c} \frac{A_{ij}}{\Delta\lambda_D} V(a,0)(\pi \ln 2)^{\frac{1}{2}}, \tag{8.17}$$

where

g_j and g_i are the upper and lower state degeneracies, respectively
λ is the transition wavelength
c is the speed of light
A_{ij} is the spontaneous emission transition rate (also known as transition strength or transition Einstein A coefficient)

For most atomic/ionic transitions, these parameters are well documented in the literature. $\Delta\lambda_D$ is the Gaussian component of the broadened linewidth due to Doppler broadening and is determined by [94]

$$\Delta\lambda_D = 7.16 \times 10^{-7} \lambda_0 \sqrt{\frac{T}{M}}, \tag{8.18}$$

where

λ_0 is the central wavelength of the line
T is the plasma gas kinetic temperature (K)
M is the atomic mass (a.u.)

$V(a, 0)$ is the Voigt function determined by linewidth broadening mechanisms and the a parameter is determined by the ratio of Lorentzian and Gaussian components [96]. Linewidth broadening mechanisms in plasma, including Doppler broadening, Stark broadening, and collisional broadening, are complicated [96,97]. Which mechanism is dominant under a particular plasma condition is dependent on the plasma temperature(s), electron density, and location in the plasma. Detailed discussions on linewidth broadening mechanisms under plasma conditions can be found elsewhere [95–97]. For given experimental conditions, such as the plasma temperature and the lineshape broadening mechanism, the absorption cross sections can be calculated by using Equation 8.17. For most important plasma species, absorption cross sections of some well-known transition lines are published in the literature or spectroscopic database, that is, HITRAN database, *Journal of Physical Chemistry*, *Journal of Molecular Spectroscopy*, and *Journal of Quantum Spectroscopy Radiation Transfer*. Absorption cross section is temperature and pressure dependent; therefore, it determines the lineshape of a particular transition. The line intensity (the integration of the absorption cross section over the entire frequency range of a lineshape) is temperature dependent. Therefore, when the absorption-based CRDS is implemented to measure absolute number density, the plasma gas temperature is also needed to determine the transition line intensity if the intensity is temperature sensitive or the plasma temperature varies in a large range.

On the other hand, the P-CRDS technique can be used in plasma diagnostics to measure plasma gas temperature and electron density based on linewidth measurements. For example, fitting an experimental lineshape into a Voigt profile to generate Doppler broadening and Lorentzian broadening components, from which the gas kinetic temperature and electron density can be estimated [20].

8.4 Configurations of an Experimental P-CRDS System

A P-CRDS system consists of four major sections: (1) a ringdown cavity, (2) a laser source, (3) a plasma source, and (4) an electronic control and data acquisition system. Each of them is described below.

8.4.1 Ringdown Cavities

A ringdown cavity in its original form consists of two highly reflective mirrors with a mirror spacing that meets the stable cavity conditions. Nowadays, a ringdown cavity can have various forms, such as a cavity consisting of more than two high-reflectivity mirrors, a pair of super-polished prisms, a single piece of prism, a section of single-mode optical fiber with two identical fiber Bragg gratings inscribed on it, a single-mode fiber loop, and a cavity with a plasma or a flame in it. Although there are so many forms of ringdown cavities, the fundamental characteristics remain the same: laser light is trapped in a high-finesse cavity to experience many round trips to enhance the measurement sensitivity. Due to different geometries and/or different cavity configurations, experimentally observed ringdown times can range from nanoseconds to hundreds of microseconds. Berden et al. [98] demonstrated an extremely short ringdown cavity (3 cm). Le Grand and Le Floch reported a very long cavity (270 cm) [99]. A cavity used to demonstrate a compact acetone detection device was only 13 cm in length, which yielded a ringdown time of 700 ns in the deep ultraviolet (UV). For liquid detection, cavity lengths can even be on the order of micrometer. Conversely, using ultra-high-reflectivity mirrors in the IR, that is, $R = 99.9985\%$, a 43 cm cavity had a ringdown time of up to 100 μs, and a ringdown time of several hundred microseconds has been reported in a fiber loop ringdown system [100].

The effects of cavity design and implementation on CRDS was the subject of a lot of discussion in the early CRDS literature [101–105]. A ringdown cavity is a high-finesse etalon, and as such possesses a discrete mode structure dependent upon the design of the cavity. For a cavity constructed from two spherical surface mirrors of radii R_1 and R_2 separated by a distance L, the frequencies of the stable cavity modes are given by [106]

$$\nu = \frac{c}{2L}\left\{ n + (1 + m + q)\frac{\cos^{-1}\sqrt{(g_1 g_2)}}{\pi} \right\}, \tag{8.19}$$

where

$g_1 = 1 - \left(L/R_1 \right)$
$g_2 = 1 - \left(L/R_2 \right)$

The g_1 and g_2 parameters are dimensionless quantities that can be used to determine whether a given cavity design is stable. For a stable cavity, the condition $0 < g_1 g_2 < 1$ must hold. In Equation 8.19, m, n, and q are integers used to index the modes of the cavity. The n parameter represents the longitudinal mode number and is approximately of the order of $2L/\lambda$, and m and q represent the transverse mode number. For the lowest order mode, commonly referred to as TEM_{00}, m and q are zero. Here we see that for a cavity constructed from plane parallel mirrors where R_1 and R_2 are infinite, we recover the well-known frequency transmission of the Fabry–Perot type of etalon

$$\nu = \frac{nc}{2L}. \tag{8.20}$$

The frequency structure of the cavity modes can be visualized as consisting of TEM_{00} longitudinal modes whose frequency separation is given by Equation 8.20 as $c/2L$. The frequency space between these TEM_{00} modes will be filled to a greater or lesser degree by higher order transverse TEM_{lm} modes, depending upon the exact design of the cavity. When a cavity is subjected to excitation by a very narrow linewidth CW laser, not all laser frequencies are transmitted through the cavity [107]. Only those frequencies that correspond to stable modes of the cavity are transmitted. The behavior of such a cavity under pulsed excitation was also the subject of a lot of discussion. Arguments have been presented that the short coherence length of pulsed lasers in general precludes mode effects. Other concerns have been expressed that for the absorption features narrower than the mode spacing of the cavity, it is possible that absorption features could be missed (because only light at the mode frequencies rings down in the cavity) [108]. In practice, however, for pulsed laser excitation such effects have not been a concern. Cavities can be easily designed to possess a near continuum of transverse modes [109]. Typical pulsed laser linewidths are also wider than even the normal spacing between the TEM_{00} longitudinal modes, so that light from the pulsed laser is always effectively coupled into the cavity regardless of laser frequency or transverse mode structure. Finally, it has been noted that while mode effects may modulate the total intensity of the signal exiting the cavity, the signal usually never drops to zero [110]. Because the ringdown time, and not the intensity, is monitored in CRDS, a small modulation does not greatly hinder the ringdown measurement.

Often a mirror-based ringdown cavity can only be used for measurement of a very limited number of species due to the relatively narrow bandwidth of ultra-high-reflectivity mirrors. However, a broadband prism cavity, first introduced by Lehmann and Rabinowitz in 2000 [3], can have a spectral response ranging from the UV to IR. This type of prism cavity can potentially measure spectral fingerprints of various species in different spectral regions.

8.4.2 Laser Sources

In principle, there is no restriction on the laser sources to be used for a ringdown system. Both pulsed lasers and CW lasers with wavelengths ranging from the UV to the IR can be used, given availability of ringdown mirrors coated for the appropriate wavelength region. Commonly employed pulsed lasers include dye lasers pumped by Nd:YAG or excimer lasers, or widely tunable optical parametric oscillator (OPO) lasers. CW ring dye lasers pumped by Ar+ lasers and quantum cascade lasers (QCLs) have been also widely used for ringdown studies, as have mid-IR laser sources based on difference frequency generation (DFG) or Raman shifting techniques [111]. Inexpensive telecommunications diode lasers have been increasingly used in ringdown research and instrumentation due to low cost, small footprint, and high spectral resolution. Broadband laser sources have also been explored for ringdown studies. Engeln and Meijer [112] injected a broadband dye laser pulse into a ringdown cavity and used a Fourier transform spectrometer to separate the frequency components of the pulse exiting the cavity. Scherer [113] also utilized a broadband laser beam for the development of a ringdown spectrometer. Fiedler et al. [114] measured the C-H overtone of liquid benzene using incoherent broadband CEAS. Perhaps the most nonstandard laser system used for ringdown was reported by Engeln et al. [115,116], who implemented CRDS in the IR using a free electron laser system to obtain a spectrum of a thin film of C_{60}. Very recently, a broad bandwidth optical frequency comb has been coherently coupled to a ringdown cavity for real-time, quantitative measurements of trace concentrations, transition strengths and linewidths, and population redistributions for molecules, such as C_2H_2, O_2, H_2O, and NH_3 [70]. The same group has demonstrated an improved frequency comb-based CRDS system based on a mode-locked erbium-doped fiber laser source centered at 1.5 μm, resulting in an absorption sensitivity of 2×10^{-8} $cm^{-1} Hz^{-1/2}$ per detection channel for CO, NH_3, and C_2H_2 in a very large spectral bandwidth (1.45–1.65 μm) [71]. A palm-size, single-mode, single-wavelength, Q-switch diode laser has been very recently implemented for development of portable ringdown spectrometers [117].

A critical issue is the laser linewidth in selecting a laser source for an absorption spectroscopy method such as CRDS. If the laser used in CRDS has a large linewidth, two measurement problems may arise. First, hyperfine structures of an optical transition or an isotopic shift cannot be captured due to the lack of laser resolution. This is particularly true for atomic transitions and isotopic shift. Second, if the laser linewidth is too broad, accurate measurements for high-resolution spectroscopy and atomic absorption cannot be achieved (mostly, the overall absorption signal can be undermeasured). The effects of the laser linewidth on the accuracy of ringdown measurements have been experimentally observed and theoretically analyzed [118–120]. Even some pulsed lasers are suitable for moderate to high-resolution spectroscopy. For instance, a pulsed dye laser with double grating configuration can have a linewidth of 0.08 cm^{-1} at 590 nm (e.g., NarrowScan) and 0.03 cm^{-1} at 570 nm (e.g., Sirah). A widely tunable OPO system (e.g., the MOPO-HF) can have a linewidth of 0.075 cm^{-1} or narrower. In contrast, CW lasers such as external cavity diode lasers (ECDL) or distributed feedback (DFB) diode lasers typically have a laser linewidth of kilohertz to megahertz levels (1 cm^{-1} = 30 GHz). In general, a linewidth of 0.08 cm^{-1} is sufficient for scanning of rotational–vibrational lines of molecules or an atomic transition. However, a CW laser is preferred for very high-resolution CRDS studies.

Laser intensity in CRDS is not critical because CRDS does not measure absolute intensity change, rather CRDS measures the intensity decay rate. Experimentally, laser beam intensity can be as low as 1–2 μJ per pulse for a pulsed laser and tens of megawatts for a CW laser. In most cases, slightly higher laser intensity would give a better *S/N* ratio. Even though CRDS is insensitive to laser pulse-by-pulse fluctuations, the *S/N* ratio becomes affected when the fluctuations are greater than 10–20%. A typical power fluctuation rate for a new pulse laser system is smaller than 5%.

8.4.3 Coupling of a Plasma with a RD Cavity

A telescope system is often constructed to reform a laser pulse into a pseudo-Gaussian beam to be coupled into a ringdown cavity. Normally we let the Gaussian beam waist be located approximately in the center of the optical axis of the cavity. The average beam waist can be as small as 0.5 mm in diameter for a pulsed laser. For CW lasers, due to its high divergence of its beam, that is, a pig-tailed DFB diode laser, the beam can only be collimated for transmission up to 2–3 m in free space in a uniform shape. As long as the cross section of the collimated beam is smaller than 1 mm², a decent CW ringdown baseline can be obtained.

Figure 8.3 shows an example of the schematic of a P-CRDS system, in which the plasma jet is set perpendicular to the laser beam path and the position of the laser beam path in the plasma can be adjusted by tuning the plasma jet via a 3D high-precision mount with a spatial resolution of 0.25 mm. A pulsed-CRDS

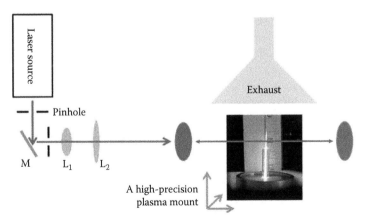

FIGURE 8.3 Coupling of a plasma source with a ringdown cavity. The plasma may be mounted in a high-precision mount for 3D adjustments.

system using an OPO or pulsed-Nd:YAG pumped dye laser is more robust than a diode laser CRDS when they are coupled with a plasma in terms of ringdown alignment. When a high-voltage discharge is coupled with a CRDS system, mostly the high-voltage discharge generates strong electromagnetic (EM) interferences with ringdown electronics. Effect of a plasma source on the ringdown baseline stability is mainly from gas flow turbulence of the plasma, not much from the plasma itself, that is, introduction of high temperature or plasma gases (ions, electrons, elements, other species). For a pulsed-CRDS plasma system, a ringdown baseline stability of 0.5% can be usually achieved with average over 30 ringdown events. Average over more than 500 ringdown events does not show obvious improvement of the baselines stability.

Laser beam propagation behavior in a ringdown cavity depends on the cavity geometry and the nature of the medium in the cavity. For example, different gases filling the cavity have different refractive indices, thus generating different scattering losses. Refractive index variations or the existence of a large refraction gradient can increase ringdown baseline noise, create beam-steering effects, or even prevent establishment of a ringdown event. For example, beam-steering effects have been observed in a ringdown cavity containing an atmospheric pressure plasma [121,122]. Severe beam deflections due to gas turbulence have been observed in plasma confined in tubing, preventing the establishment of a ringdown event. A careful design of the ringdown cavity geometry or an optimized plasma gas flow pattern can minimize beam-steering effects. Another adverse impact of a high-temperature, reactive plasma source on a CRDS system is degradation of mirror reflectivity. In most cases, for a given ringdown cavity, the ringdown baseline (τ_0^p) decreases significantly while the ringdown baseline stability may remain approximately unchanged. Consequently, the system's detection sensitivity decreases. Depending on experimental conditions and system configurations, we may use some protective means, that is, constant gas purge of ringdown mirrors, and anticorrosive coatings, to make the system more durable.

8.4.4 Electronics and Data Acquisition

For pulsed-CRDS, the bandwidth of the laser is typically several times larger than the longitudinal mode spacing of the ringdown cavity. Therefore, the light on each laser pulse is coupled into the cavity and no additional electronic parts are required to achieve cavity excitation. When a narrow linewidth diode laser is used as the light source, an additional electronic control is needed. The linewidth of a diode laser, that is, an ECDL or a DFB diode laser, is much narrower than the cavity mode spacing, implying that the laser must be tuned into resonance with a cavity mode so as to excite the cavity. Engeln et al. [123] first reported the use of an intensity modulated, narrow-linewidth CW ring dye laser for CRDS. This work utilized an unstablized ringdown cavity and depended upon random fluctuations (such as variations in the cavity length due to vibrations) to couple the laser into the cavity modes. To increase the coupling efficiency, a cavity with a dense mode structure was utilized. The phase shift of the intensity-modulated beam exiting the cavity was used to obtain a high-sensitivity spectrum. Later on Romanini et al. described a method of actively coupling narrow-linewidth CW lasers (such as ring dye and diode lasers). In these experiments, the length of the cavity was dithered by mounting a cavity mirror on a piezoelectric translator, as shown in Figure 8.4. By modulating the length of the cavity, the mode structure of the cavity is swept through the laser frequency. When the laser and cavity mode frequencies overlap, light is efficiently injected into the cavity mode, rapidly increasing the intensity of the light in the cavity (i.e., the cavity is excited). After these initial reports, several early studies using CW-CRDS were reported [124–130].

CW-CRDS has matured into a standard ringdown technique. The methods of achieving the cavity excitation can be simply classified into three dominant categories: (1) cavity length modulation to vary the cavity mode spacing, as shown in Figure 8.4; (2) current modulation to vary the laser frequency; and (3) a combination of the two methods. The first and the third methods have been almost abandoned due to the complexity and the high cost of the required instrumentation. Currently, most instrumentation

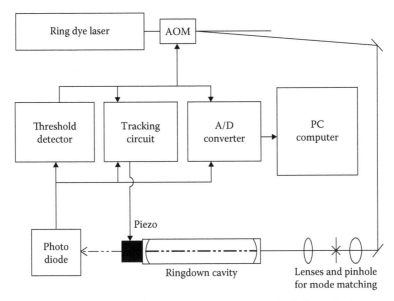

FIGURE 8.4 Cavity excitation and shutoff of a CW-CRDS system, in which the cavity length modulation and AMO methods are implemented. (Reprinted from Romanini, D., et al., *Chem. Phys. Lett.* 264, 316–22, 1997. Copyright 1997, with permission from Elsevier.)

and research using CW-CRDS implements the current modulation scheme. In CW-CRDS, once the cavity is excited, shutoff of the excitation is needed so as to observe the ringdown decay behavior. Typically, an acousto-optical modulator (AOM) is used to deflect the laser beam away from the optical axis of the ringdown cavity to shut off the cavity build-up, thus allowing the ringdown waveform to be subsequently recorded. This method is illustrated in Figure 8.4. The advantage of this method is that the sampling frequency can be as high as megahertz. However, this method requires an expensive AOM device and a trigger circuit to control the procedure. Another major disadvantage of this method includes a loss of laser power through the AOM. When a laser beam passes through the crystal of an AOM, losses in the observed laser power are typically 15–20%. This imposes a distinct disadvantage when a low-power DFB diode laser is used as the light source in a CW-CRDS system because lower signal intensity means lower *S/N* ratio. Although many CW-CRDS-based research efforts are still using an AOM, CRDS-based commercial instruments have avoided implementing this device [15,17,35].

An alternative approach to shut off the cavity excitation without using an AOM in CW-CRDS is controlling the laser current. Once the cavity is excited, the electrical signal from the detector is used to trigger a circuit, which drives the laser diode current below the threshold of the driving current. This procedure typically requires a specially designed electric circuit. This circuitry is a trivial engineering effort, but it is not readily available for research laboratories. Recently, a simple and effective method has been introduced to achieve the cavity excitation and shutoff using a standard off-the-shelf power combiner [17]. With this approach, the power combiner couples the output of a function generator with the output of a pulse generator to create a special voltage waveform to manipulate the diode laser current. The schematic is illustrated in Figure 8.5. This novel method eliminates the need to use special electronic modulation circuits to achieve current modulation. This method has been successfully used in several ringdown research studies and applications [131,132].

Ringdown signals from a PMT or photodiode detector are typically digitized using a digital oscilloscope, transient digitizer, or fast A/D converter interfaced to a computer. Ringdown software is typically in-house developed, but is also commercially available. To determine the ringdown time, the resulting data is fitted as a single exponential function. In practice, averaging of multiple waveforms is often employed to smooth out shot-to-shot variations in the ringdown signals and improve the signal-to-noise

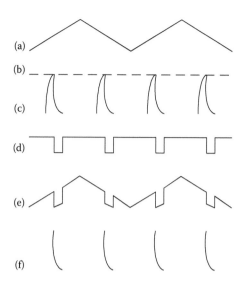

FIGURE 8.5 Cavity excitation and shutoff of CW-CRDS using a off the shelf frequency combiner. A schematic of the method of controlling the coupling of the CW laser to the ringdown cavity. (a) Sawtooth signal from the function generator; (b) trigger threshold; (c) energy buildup inside the cavity; (d) negative pulses from the pulse generator; (e) resultant modulation signal; and (f) ringing down of the signal.

ratio. Strictly speaking, such averaging is only valid if the waveforms are normalized in intensity. Alternatively, the ringdown time of each individual transient can be computed followed by averaging a number of the computed time constants. Some early ringdown experiments used a dual-channel boxcar integrator to compare two different regions of the ringdown signal and extract a quantity proportional to the ringdown time [133]. This approach yielded results comparable to data-fitting methods. However, the software method has the added advantage of providing a quick method of establishing if the ringdown waveform is a true single exponential decay. It is a simple matter to check the observed ringdown time in different regions of the transient to determine if the waveform is truly a single exponential function. In practice, for example, if the difference in the ringdown times obtained by fitting different ranges in the decay curve is less than about 1%, fitting a single exponential decay is justified. Cavity misalignments, linewidth effects, and index of refraction variations in the cavity can all lead to nonexponential or multiple exponential waveforms [134]. Scherer et al. have considered the effect of various data acquisition schemes in some detail and have concluded that while the temporal resolution of a digitizer system is important for determining the number of points sampled, the vertical resolution can be critical to the performance of a CRDS system.

8.5 Absolute Number Density Measurements of Plasma Species

Plasma species include atoms, ions, molecules, and free radicals. The plasma species can be populated in their electronic ground states and electronically excited states. Depending on plasma sources and their operating environments, typical atmospheric plasma and discharge species include OH, NO, N_2, N_2^+, NH, O_2, O, and H. When a plasma source interacts with materials or other media such as a combustion flame and semiconductor fabrication gases, many more plasma species can be generated. If additional gases or aerosols or chemical vapors are introduced into the plasma source, more complicated molecular fragments or elements and isotopes will also be generated. Determination of absolute number density of plasma species is of vital importance in understanding plasma gas kinetics, that is, in plasma-assisted combustion (PAC), in which absolute number densities of reactants and products play a critical role in helping understanding kinetic pathways. Knowing absolute number density of a plasma species of interest can help plasma source design to be tailored for a particular application, such as optimization and control processing dosage in plasma medicine or other plasma treatment processes. In plasma-based instrumentation, in which a plasma is used as an atomization source to generate an analyte of interest and analytic merit is often judged by detection sensitivity in terms of absolute number density of the analyte. In the field of electron–atom collision processes, determination of number density of atoms in a specific state can lead to determination of electron impact excitation cross section of a particular state of the atom [135]. In materials study, such as thrust material characterization and laser sputtering process, number density of elements is a key parameter to help understand properties of the materials. In plasma-assisted chemical vapor deposition (CVD), quantification of a reaction species can help improve material production yield and reduce fabrication costs through optimization and minimization of usage of precursors. There are many other fields and applications, in which we need to know absolute number density of species generated in plasma or in the medium between the plasma and the plasma-interacted targets.

Absolute measurements of plasma species pose several technological challenges in reality. (1) When the species of interest is a free radical or any reactive species, on-line and *in situ* measurement is necessary because sampling of plasma gas or plasma plume that contains the reactive species is not feasible because the lifetime of the reactive species is typically on orders of micro- or millisecond levels. (2) Measurement of species concentration must be nonintrusive to the plasma. For a small plasma source, a relatively large sampling probe may affect the plasma gas flow pattern or local EM field if the probe is metallic and inserted into the plasma. (3) High spatial resolution of a measuring technique is required so as to obtain a one-dimensional or two-dimensional profile in a small plasma, that is, a plasma needle, a plasma pencil, or a plasma jet array [136–138]. (4) Plasma species exist in a physically and chemically complicated matrix, techniques for quantification of plasma species must be highly selective and density measurement of a species with free of calibration would be highly desirable in many cases. Theoretically, many techniques, such as optical emission spectroscopy (OES), mass spectrometry (MS), chemical sensors, or laser-based techniques, can be implemented to determine number density of plasma species. Practically, however, laser-based methods, such as tunable diode laser absorption spectroscopy (TDLAS), LIF, REMPI, and CRDS are more attractive in terms of simplicity, high spatial resolution, high selectivity, nonintrusiveness, and field-deplorability. Due to the small size of a laser beam (less than 1 mm^2), either of the laser-based techniques has a high special resolution and capability of point-by-point or line-of-sight measurement. REMPI is a highly sensitive and highly selective laser technique for a lab-based plasma study though it requires a vacuum system. LIF, due to its high sensitivity and high special resolution, has been widely used in plasma measurements; advantages and limitations of LIF can be seen in Chapter 10. Both TDLAS and CRDS are absorption-based line-of-sight techniques. When species concentrations are not too low, such as higher than parts per million, for example, TDLAS is a convenient tool for density measurement of plasma species, details can be seen in Hanson's recent publications. Compared with TDLAS, CRDS is 10,000 times more sensitive if the same laser beam path length in the two methods is considered. If absorption cross section of an optical transition is known, both CRDS and TDLAS are self-calibration techniques.

Application of CRDS to absolute number density measurements in an atmospheric pressure plasma was first introduced by Miller and Winstead in 1997 [18]. Absolute number density of atomic lead (Pb) generated in an atmospheric pressure argon (Ar) inductively coupled plasma (ICP) was measured by pulsed-CRDS in 283 nm. This pioneering work extended the idea of the earlier publications in CRDS measurements of OH, CH, and mercury (Hg) in atmospheric flames reported by Meijer et al. in 1994 [107]. Since then, CRDS measurements of absolute number densities of N_2^+, trace metal elements, radioactive elements and isotopes, and OH radicals have been reported in different plasmas and different ringdown systems (pulsed- or CW-CRDS with different laser sources) [20,38,39,121,122]. Table 8.1 lists the species (atoms, ions, free radicals, and molecules), which have been measured by CRDS to date in plasma or plasma-interacted processes, such as ICP, microwave-induced plasma (MIP), microwave plasma torch (MPT), plasma jets, DC/AC discharges, dielectric barrier discharge (DBD), radio frequency (RF)

TABLE 8.1 The Plasma Species Which Have Been Measured by CRDS to Date

Species	Generation Environment	Wavelength (nm)
OH	ICP, MIP, MPT, plasma jets, DC discharge, DBD, PAC	308
O_2	Microwave discharge, ICP	253, 630, 1505, 1580
O	Microwave discharge	630
N_2^+	ICP	391, 505–611
N_2	Discharge	769–772
NH	Plasma jet	340
NH_2	Plasma jet	600, 224
Si	Thermal arc	251
SiH	Thermal arc	414
SiH_3	Thermal arc	200–260
CF	RF discharge	233
CF_2	RF discharge	253
SiF_2	RF discharge	220–300
C_2	DC arc jet	505–517, 231
CH	DC arc jet	387, 427, 431
CH_3	Thermal dissociation	1515, 216
Al, Mo, Fe, Ti, Mn, B	Ion beam sputtering	250–403
Pb	ICP, MPT	283
Hg	MPT	253, 405
Sr	ICP	688
$^{238}U(I)$, $^{238}U(II)$, $^{235}U(II)$	ICP	286, 358, 409
CH_3O_2	DC discharge	1350
C_6H_5	DC discharge	504–530
HO_2	DBD	237

discharges, thermal arc, ion beam sputtering, plasma etching, PAC flame, and plasma-enhanced CVD. Although combustion flames withhold the same common nature of high temperature as plasmas, the radicals and molecules which have been measured by CRDS in combustion flames only are not tabulated in Table 8.1. CRDS measurements of each of the species listed in Table 8.1 will be discussed in the following sections.

8.5.1 CRDS of OH in Plasmas and Discharges—A Detailed Example

The OH is one of the most extensively studied diatomic radicals, simply because of its abundant existence in many situations such as atmosphere [139,140], combustions [141,142], flames [143,144], shocks [145], and plasmas/discharges [146,147]. OH is frequently investigated for the thermal- and chemical diagnostics of combustion processes. High-temperature OH spectra are also of interest for the characterization of laser-induced air plasmas and laser-ignited combustible mixtures. In low-temperature plasma or cool plasma, OH radicals along with other reactive plasma species play an important role in plasma treatment of biomaterials [148–151]. For instance, in plasma sterilization process, the presence of OH radicals can compromise the function of membrane lipids because unsaturated fatty acids are susceptible to OH attacks. Abundant data on the effect of OH deactivation of different microorganisms are reported in plasma sterilization of air streams. Determination of OH radical concentration in many situations is very important. In low-temperature nonthermal PAC, absolute concentration of OH provides a critical test of PAC reaction kinetics models. Studies of the OH radical include fundamental spectroscopy, chemical dynamics/kinetics temperature diagnostics, and formation and loss mechanisms [152–155]. Spectroscopic constants and the pressure- and temperature-dependent absorption cross sections of OH radical have been well documented [155,156]. Using CRDS to directly measure absolute concentration of OH radical was first reported by Meijer et al. [107]. They measured the absolute density of the OH radical using pulsed-CRDS in heated plain air in the temperature range of 1000–1400 K. Later, a few studies about the absolute number density measurement of OH radicals in different combustion flames have been reported [157–159]. In a flame burner, the OH radical is typically generated by a gas mixture of H_2/CH_4/air that is flowed into the burner where a typical thermal temperature is less than 2000 K. OH concentrations in the different flames typically range from 1×10^{14} to 3×10^{16} molecule cm^{-3}. In plasmas such as ICP, the gas temperatures can be in the range of 2000–8000 K, depending on operating conditions. OH radical still exists in these high-temperature environments [20]. In low-temperature nonthermal plasmas where gas temperature is as low as room temperature, existence of OH radicals has also been observed.

CRDS measurement of OH in an atmospheric plasma was first reported in an atmospheric Ar ICP, in which CRDS of the OH $A^2\Sigma^+ \leftarrow X^2\Pi$ system were recorded [20,90]. Due to abundance of OH in the atmospheric plasma, the weak $S_{21}(1)$ rotational line in the OH (0–0) band was selected to measure OH number density profile along the axis of the ICP torch, which was powered at 200–300 W.

Figure 8.6 shows the part of CRDS of the OH $A^2\Sigma^+ \leftarrow X^2\Pi$ (0–0) band around 308 nm [90]. The location of the laser beam in the plasma (or called measured plasma height) was 2 mm above loaded coil (ALC) of the ICP torch. The experimental conditions were optimized for the best S/N ratio of the ringdown signal. The scanning step of the Nd:YAG pumped dye laser was 0.0006 nm. In Figure 8.6, each line can be assigned to a vibrational–rotational transition of the OH $A^2\Sigma^+ \leftarrow X^2\Pi$ (0–0) band with the help of LIFBASE [160]. As an example, the assignments of a few lines are listed on the top of the spectra. OH radical in plasma was so abundant that there was no way to avoid the absorption saturation for most of the strong lines in the P_1, P_2, Q_1, Q_2, R_1, R_2, Q_{21}, and R_{21} branches. Because of the strong absorption saturation of some transition lines, the spectral intensities in Figure 8.6 did not reflect the actual absorbance, only giving the relative intensity. Generally, the absorption saturation effect increased with the elevation of the measured plasma height (ALC), where higher OH density was present partially due to the stronger interaction of the plasma with humid atmosphere, which was favorable for OH generation via electron impact induced dissociation of H_2O. In the tail of plasma, scanning of a spectrum in a wide wavelength range was difficult due to the absorption saturation of strong lines even though the fitting window in the ringdown decay waveform was set much narrower and set at the beginning of the ringdown curve. Similar absorption saturation phenomenon was reported in CRDS of OH in flames.

The $S_{21}(1)$ line is a very weak transition, as shown in Figure 8.6. By measuring the total absorbance of this line at different plasma heights, a density profile of total OH at different plasma heights is obtained from the density population in the $S_{21}(1)$ line. The measured absorbance is obtained by the integration of absorbance over the whole lineshape. The absolute number density in the lower rotational energy level of the $S_{21}(1)$ line can be derived from ringdown measurements using

$$\text{Absorbance} = Snl_s \int_\nu \frac{L}{c}\left(\frac{1}{\tau^p(\nu)} - \frac{1}{\tau_0^p}\right), \tag{8.21}$$

where

n is the OH number density in the initial state of the $S_{21}(1)$ transition

τ_0^p is the plasma ringdown baseline obtained when the laser wavelength is tuned off the absorption region

σ_{ij} is the central absorption cross section of the transition and can be calculated by equation

$$\sigma_{ij}(\nu) = \frac{1}{8\pi c\nu_0^2}\frac{(2J'+1)}{(2J''+1)}A_{ij} \tag{8.22}$$

where ν_0, c, and A_{ij} are the central wavenumber of the transition, the speed of light, and the Einstein spontaneous emission coefficient, respectively. Here i and j represent the OH ground electronic state

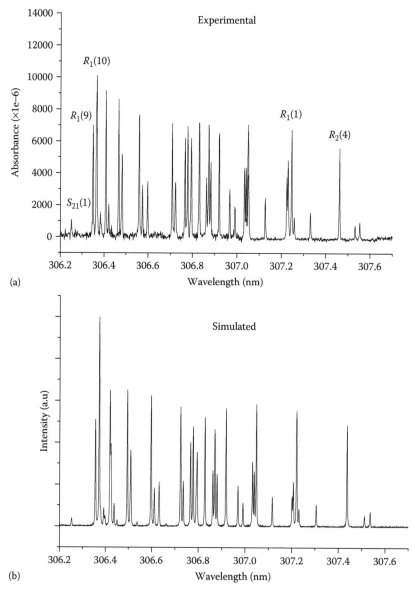

FIGURE 8.6 (a) Part of CRDS of the OH $A^2\Sigma^+ \leftarrow X^2\Pi$ (0–0) band recorded in the atmospheric pressure Ar ICP. Plasma power = 200 W; h (ALC) = 10 mm; x = 0 mm. (b) Part of the simulated absorption spectrum based on the Boltzmann distribution. The simulation conditions: resolution = 0.004 nm; lineshape = Voigt; Lorentzian component = 30%; $T_r = T_v$ = 2000 K; pressure = 760 Torr; 0.5% noise is added to the spectra. (From Wang, C., et al., *Appl. Spectrosc.*, 58, 734–40, 2004. Reprinted with permission from Society for Applied Spectroscopy.)

(X) and the first excited electronic state (A). For the $S_{21}(1)$ rotational line, all these parameters are documented in literature.

Figure 8.7 shows lineshapes of the $S_{21}(1)$ line of the OH A-X (0–0) band measured at different plasma heights [90]. Each increment in the spectra came from on average over 100 laser shots. The scanning step of the dye laser was 0.0003 nm. When the measured plasma height was higher than 25 mm ALC, a bigger baseline noise gave a worse *S/N* ratio spectrum. Although OH still existed in the height of 30 mm ALC, where the plasma was almost invisible, no good quality spectral scan was obtained beyond

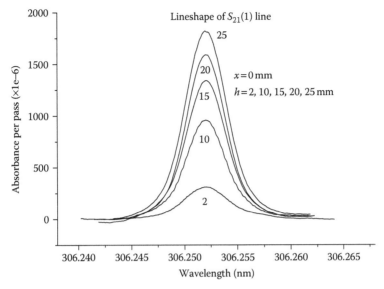

FIGURE 8.7 Measured ringdown spectral lineshapes of the $S_{21}(1)$ rotational line in the OH (0–0) band at different plasma heights. Plasma power = 200 W; $x = 0$ mm; h (ALC) = 2, 10, 15, 20, and 25 mm from the bottom to the top. (From Wang, C., et al., *Appl. Spectrosc.*, 58, 734–40, 2004. Reprinted with permission from Society for Applied Spectroscopy.)

this plasma height. Integrating the absorbance over each line profile given in Figure 8.7 yielded the number density through Equations 8.21 and 8.22. Notice that this density profile only gives the OH density populated on the lower energy level of the $S_{21}(1)$ line, rather than the density profile of total OH.

The total OH density at one plasma location is normally obtained from the intercept of a Boltzmann plot, in which the rotational temperature can be derived from the slope and used for calculation of total partition function. It should be noted that a Boltzmann plot is often obtained from scanning of each individual rotational line profile and the integrated absorbance of the rotational line; at least, five rotational lines need to be measured for a quality Boltzmann plot. To obtain a density distribution profile in the plasma, we need to repeatedly perform this procedure at different plasma locations. Additionally, this approach is not applicable to CRDS measurements of OH in the plasma where high-density OH is present. The scanning of CRDS of an entire band or an entire rotational branch is not achievable owing to the strong absorption saturation. Thus, it is unlikely to be able to obtain a Boltzmann plot from the scans of an entire rotational branch. An alternative approach is to use the measured density population in an individual vibrational–rotational line, that is, the $S_{21}(1)$ line, to calculate the total OH density with the thermal temperature derived from the recorded lineshape when the local thermal equilibrium (LTE) is justified [90].

In a nonequilibrium plasma, molecular population density of a vibrational–rotational state depends on the electron temperature, T_e, gas kinetic temperature, T_g, vibration temperature, T_v, and rotation temperature, T_r. In an LTE approximation, molecular population abides by the Boltzmann distribution. The relation of population distribution on an individual vibrational–rotational energy level with the total population is given by [140,155,156]

$$N_{V,J} = N_0 f_B(V, J) = N_0 \frac{1}{Q_e Q_V Q_r} \exp\left(- \frac{E_V(V)}{kT_V}\right)(2J + 1)\exp\left(- \frac{E(J)}{kT_r}\right), \tag{8.23}$$

where $N_{V,J}$ is the number density populated on the vibrational–rotational energy level (V, J), N_0 is the total number density, and $f_B(V, J)$ is the Boltzmann fraction; Q_e, Q_V, and Q_r are the electron, vibrational, and rotational partition functions, respectively. Q_e equals the electronic degeneracy, $g_e = (2 - \delta_{0,\Lambda})(2S + 1)$,

$2S + 1$ is the state spin multiplicity and $\delta_{o,\Lambda} = 1$ for Σ states and 0 for all others. For the OH electronic ground state $X^2\Pi$, $g_e = 4$. The vibrational partition function Q_V in the harmonic oscillator approximation is given by

$$Q_V = \frac{1}{1 - \exp\left(-1.4388\omega / T_V\right)}, \tag{8.24}$$

where

ω is the vibrational frequency
$\omega_{OH} = 3737.7941$ cm^{-1}
T_V is the vibrational temperature

The rotational partition function, Q_r, is calculated from the actual energy levels as follows:

$$Q_r = \sum_J (2J + 1)\exp\frac{-1.4388E(J)}{T_r}. \tag{8.25}$$

Goldman and Gillis [140] fitted a polynomial to the values of the rotational partition function $(200 \leq T_r \leq 6000 \text{ K})$,

$$Q_r\left(T_r\right) = \left(1.42 \times 10^{-6}\right)T_r^2 + 0.1485\,T_r - 4.1. \tag{8.26}$$

The rotational energy $E(J)$ in each line of the OH A-X (0–0) band is available in the literature [140]. The rotational energy of the ground state $J'' = 1.5$ of the $S_{21}(1)$ line in the OH (0–0) band is 0.056 cm^{-1}. Based on the measured population density in the $S_{21}(1)$ line by Equation 8.23, the total number density of OH radical at different plasma heights through the center of the plasma torch was determined to be from 0.27×10^{16} to 1.4×10^{16} molecule cm^{-3} with the highest density at the plasma tail [90]. Measured OH concentrations show a large density gradient at different plasma heights.

Several factors affect the measurement uncertainty of the OH density. First, the $S_{21}(1)$ line intensity is temperature dependent. However, variation of the thermal temperature only results in a small change of the total OH density. For example, in Equation 8.23, a change of the thermal temperature from 2000 K to 2800 K only generates 3.5% variation of the total OH density. Second, the measurement error of the OH density on the $S_{21}(1)$ line is affected by the error of the total absorbance integrated over the lineshape, 5%. Finally, the total absorption cross section of this line calculated by Equation 8.22 using Einstein coefficient has an error of 6%. Realistically, the overall measurement uncertainty of the total OH density will be less than 20%.

In the above example, OH radicals were measured using the rotational–vibrational line $S_{21}(1)$ and the OH concentrations in the plasma were high, on the order of 10^{15}–10^{16} molecule cm^{-3}. In this case, it is unnecessary to use CRDS to quantify the OH concentration; TDLAS would also be able to measure OH in those high concentrations. In many other cases, for instance, in low-density plasmas, or in some locations of the plasmas, where OH concentration is too low to be measured by single-pass laser absorption spectroscopy, the advantage of the high sensitivity of CRDS becomes obvious. For instance, in the downstream or far downstream of an atmospheric plasma jet, where OES become optically blind because there is no bright part of plasma plume in those regions [49,50]; and optical emissions are too weak to be detected. OH concentration in the far downstream of a microwave plasma jet is as low as 10^{12} molecule cm^{-3}, which is beyond the detection limit of typical SPAS.

Figure 8.8 shows the OES of an atmospheric Ar microwave plasma jet. It is clear that in the distance farther than $x = 10$ mm away from the plasma jet orifice, no OES was observed because it was too weak to be detected by the OES system. However, in the far downstream, $x > 12$ mm, CRDS of OH radicals were obtained and even at $x = 28$ mm, the existence of OH were still measurable by CRDS, as shown in Figure 8.9 [49]. This is a striking example that shows the advantage of the high sensitivity of CRDS

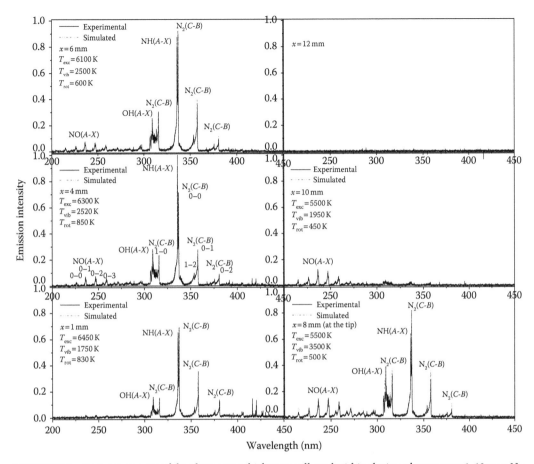

FIGURE 8.8 Emission spectra of the plasma jet, which were collected within the jet column at $x = 1$–12 mm. No OH emissions were detected at $x > 10$ mm. The spectral resolution = 0.07 nm. The emission intensities of the spectra collected at different locations were normalized.

and the results also indicate that using OES alone for plasma characterization would miss significant information in some cases.

In addition to the $S_{21}(1)$ line, numerous rotational lines of OH transition can also be used to measure OH number density. For instance, the $R_2(4)$ line in the (0–0) band of the OH system can also be used because this line has no overlap with other rotational lines. The temperature-dependent line intensities (cm per molecule) of the OH $R_2(4)$ line were calculated by [140]

$$S(T) = 3.721963 \times 10^{-20} \frac{T(\text{K})}{273.16} \frac{1}{8\pi c v^2} \left(\frac{N}{P} \right) \times \left(\frac{e^{-1.4388 E''/T}}{Q_{VR}} \right) A_{V''J''}^{V'J'} (2J' + 1) \left(1 - e^{\frac{-1.4388 v}{T}} \right) \tag{8.27}$$

where

    T is the temperature (K)
    v is the transition frequency of the OH $R_2(4)$ line of 32,517.473 cm^{-1}
    N is the total number density (molecule cm^{-3}) at pressure P (atm) and temperature T
    $A(V'J'/V''J'')$ is the Einstein coefficient (s^{-1})
    E'' is the lower state energy, that is, 429.458 cm^{-1}
    Q_{VR} is the vibrational–rotational partition function
    V and J are the vibrational and rotational quantum numbers

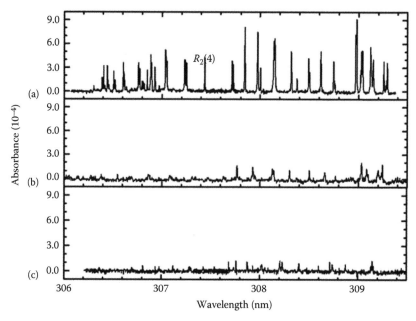

FIGURE 8.9 Existence of OH radicals in the jet flame zone was evidenced by CRDS of the OH $A^2\Sigma^+ \leftarrow X^2\Pi$ (0–0) band in the range of 306–309.5 nm. (a) Laser beam passing through the plasma jet at $x = 10$ mm (2 mm away from the jet tip), (b) $x = 18$ mm (10 mm away from the jet tip), (c) $x = 28$ mm (20 mm away from the jet tip—the far downstream part). (Reprinted with permission from Wang, C., et al., *Appl. Phys. Lett.*, 95, 051501-3, 2009. Copyright 2009, American Institute of Physics.)

For a known path length l_s cm, the absolute number density, n, of OH in the rotational energy level ($V'' = 0$, $J'' = 3.5$) was calculated by using Equations 8.14 and 8.15. The results of OH number densities along the jet axis at different locations x are tabulated in Table 8.2 [50]. In Table 8.2 the plasma gas temperatures were measured by a thermocouple. These temperatures were used to determine the temperature-dependent line intensities S of the OH $R_2(4)$ line. The lowest concentration of OH in the far downstream is only at sub-parts per million levels that are certainly not detectable by the SPAS because the laser path length in the plasma is only a few millimeters.

The information on the OH (the existence and number density in the far downstream), which cannot be accessed by OES or other techniques in this plasma region, is significant, that is, understanding of the

TABLE 8.2 Measured OH Number Densities in the Rotational–Vibrational Energy Level ($V'' = 0$, $J'' = 3.5$) of the Electronic Ground State at Different Locations along the He Plasma Jet Axis

x (mm)	Plasma Gas Temperature T(K)	Line Intensity $S(T)$ ($\times 10^{-17}$cm per molecule)	Path Length l_s(mm)	Integrated Absorbance ($\times 10^{-5}$ cm^{-1})	OH Number Density ($\times 10^{12}$ molecule cm^{-3})
2.0	606	1.221	1.0	1060.0	8688.52
4.0	570	1.222	1.2	741.0	5061.48
6.0	550	1.220	2.2	465.0	1732.49
8.0	530	1.215	4.5	263.0	480.35
10.0	510	1.210	6.2	142.0	189.31
12.0	500	1.206	8.0	79.1	82.40
16.0	480	1.196	11.0	9.6	7.32

Source: Srivastava, N. and C. Wang, *IEEE Trans. Plasma Sci.* 39, 918–24, 2011. © 2011 IEEE.

formation and loss mechanisms of the OH in the far downstream region. The OH radicals existing in the downstream and far downstream are relatively cool because the plasma gas temperature in those downstream regions is as low as 480 K and most OH radicals are populated in their electronic ground state (cool plasma species). However, their chemical reactivity is still as active as the (hot) excited state OH radicals.

Measurement of OH can also be achieved using CW-CRDS in the NIR spectral region [161,162]. OH overtone spectra are located around 1515 nm where OH absorption has minimal overlap with absorption of H_2O. Figure 8.10a shows CW-CRDS of the OH measured in an atmospheric AC air discharge [161]. The graph includes the spectra of H_2O and OH in the same spectral window at 1515 nm. Figure 8.10b shows the simulated spectra of the OH and atmospheric H_2O in the same spectral window. Both spectra are in good agreement.

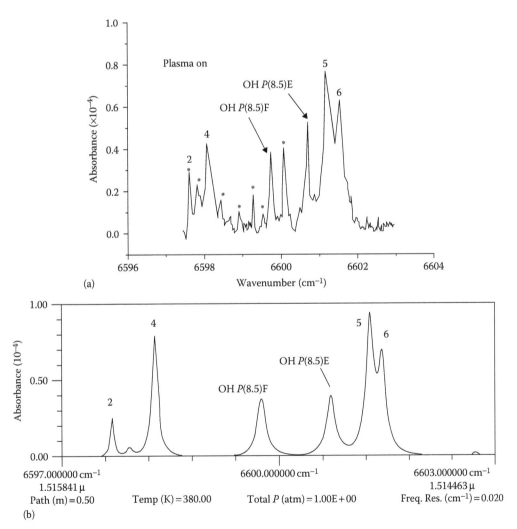

FIGURE 8.10 The experimentally measured (a) and simulated spectra (b) at wavelength near 1515 nm of the atmospheric water and OH radicals in the atmospheric AC discharge plasma. The laser path was 10 mm away from the axis of the electrodes. The plasma gas temperature at the laser path was determined to be 380 K. In the simulation, water and OH number densities were 1.9×10^{17} and 9.9×10^{14} molecule cm^{-3}, respectively. In Figure 10a, lines marked by an asterisk are attributed to transitions of high-temperature water. (With kind permission from Springer Science+Business Media: Srivastava, N., et al., *Eur. Phys. J. D*, 54, 77–86, 2009.)

TABLE 8.3 The Measured and Simulated Absorption at Wavelength near 1515 nm of Rotational–Vibrational Transitions of Atmospheric Water and OH Radicals in the Atmospheric AC Discharge Plasma ($T = 380$ K)

Line No.	Position (cm^{-1})	Assignment	Line strength (cm^{-1}) ×E-24	Absorbance (Exp.) ×1E-5	Absorbance* (Sim.) ×1E5
2	6597.926	(021-000)	N/A	N/A	N/A
4	6598.281	(120-000)	3.4456	4.2	4.3
OH P(8.5)F	6599.753	(2-0)	4.6136	0.16	0.15
OH P8.5E	6600.698	(2-0)	4.6545	0.15	0.15
5	6601.330	(120-000)	4.873	7.6	7.5
6	6601.406	(120-000)	3.228	6.3	5.1

Source: With kind permission from Springer Science+Business Media: Srivastava, N., et al., *Eur. Phys. J. D*, 54, 77–86, 2009.

Note: Both the measured and simulated absorbances of OH radicals were normalized to the 2-cm absorption path length.

*The simulated absorbance of water was normalized to the same experimental condition (2 cm path length in the plasma zone at 380 K and 40 cm path length in the nonplasma zone at 296 K).

Table 8.3 lists the transition lines measured by CW-CRDS at 1515 nm [161]. The assignments of the spectra were based on HITRAN database [163]. The OH $P(8.5)$F and $P(8.5)$E are spectral fingerprints which can be used to determine OH concentrations using CW-CRDS in the NIR.

The advantage of the CW-CRDS measurement of the OH generated in the atmospheric discharge is that absolute number densities of H_2O and OH radicals can be simultaneously measured; and the information can be used to determine the yield of OH generation from H_2O in humid AC discharge. Figure 8.11a shows CRDS of H_2O obtained in the lab atmosphere without discharge. This high S/N ratio CRDS of atmospheric water allows a comparison of the measured CRDS with the simulated ones using HITRAN database. Figure 8.11b shows the simulated spectra of H_2O at 1515 nm.

The ringdown measurements of the H_2O (120–000) band and the OH first overtone at 1515 nm in the atmospheric AC discharge at 380 K yielded OH concentration to be 1.1×10^{15} molecule cm^{-3}, corresponding to an OH formation yield in the atmospheric AC glow discharge of 4.8×10^{-3}. The system with a modest mirror reflectivity achieved a minimum detectable absorption coefficient of 8.89×10^{-9} cm^{-1}, which was equivalent to a 1σ detection limit of OH at 1515 nm to be 1.2×10^{13} molecule cm^{-3} [161].

OH radicals generated in DBD were also measured using CW-CRDS in the spectral region of 1515 nm [162]. More recently, OH radicals generated in different plasma jets using different plasmas gases or gas mixtures have been intensively studied using pulsed-CRDS technique [50,164,165]. To date, OH has been the most studied plasma species using CRDS technique, mainly due to its significance in plasma fundamentals and plasma applications.

8.5.2 CRDS of O_2 and O in Plasmas and Discharges

The excited state of O_2 molecules are among the most important reactive plasma species. Quantification of O_2 and O atoms are of vital importance in many applications, especially in plasma medicine [166–169]. O_2 has the electronic ground state $X^3\Sigma_g^-$, which is different from its two lowest excited states $a^1\Delta_g$ and $b^1\Sigma_g^+$ in terms of spin quantum number. According to the selection rules of electric dipole moment transition, transitions of $a^1\Delta_g - X^3\Sigma_g^-$, $b^1\Sigma_g^+ - X^3\Sigma_g^-$, and $b^1\Sigma_g^+ - a^1\Delta_g$ are forbidden [170]. Therefore, O_2 in the excited $a^1\Delta_g$ and $b^1\Sigma_g^+$ states have long lifetimes from tens of seconds to minutes in the absence of collision quenching. Spectroscopic measurements of these forbidden transitions require high sensitivity. Recently, spectroscopic constant measurements for the transition systems $a^1\Delta_g - X^3\Sigma_g^-$ and $b^1\Sigma_g^+ - X^3\Sigma_g^-$ have been carried out using Fourier-transform spectroscopy [171]. Several vibronic bands, such as (0–0),

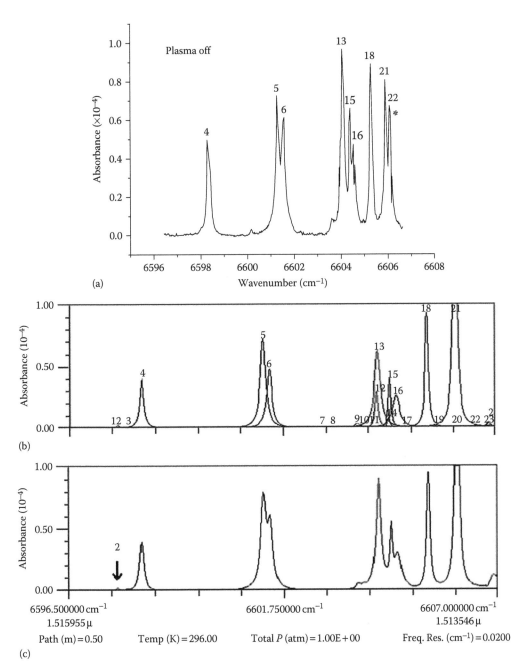

FIGURE 8.11 Comparison of the experimentally measured spectra at wavelength near 1515 nm of the atmospheric water with the simulated one. (a) CRDS of rotational–vibrational transitions of atmospheric water; (b) and (c) the simulated spectra based on HITRAN 96 database (path length = 50 cm, P = 1 atm, T = 296 K, lineshape = Voigt, resolution = 0.02 cm⁻¹, water concentration = 1.92×10^{17} molecule cm⁻³). The observed transition marked by an asterisk in (a) is not included in HITRAN 96 database. The transition marked by an arrow in (c) was not observed in the experimental spectra of (a). In (b), the simulated spectra are displayed by individual transition lines. In (c), the simulated spectra are displayed by composition of absorption. (With kind permission from Springer Science+Business Media: Srivastava, N., et al., *Eur. Phys. J. D*, 54, 77–86, 2009.)

(1–0), and (2–0) of the systems have been recorded. The $b^1\Sigma_g^+ - a^1\Delta_g$ transition is also called Noxon system, who first observed the emissions of the (0–0) band in the NIR (~1910 nm) [172].

Cavity ringdown measurement of O_2 was first reported by O'Keefe and Deacon in 1988 [1]. Historically, this was the first CRDS publication in which detection of the doubly forbidden transition $b^1\Sigma_g^+ - X^3\Sigma_g^-$ was recorded to demonstrate the high sensitivity of CRDS. The band origin of the (0–0) band (also called A band) of the transition is located in the vicinity of 763 nm, which can be covered by an Nd:YAG pumped dye laser or an OPO system. Figure 8.12 shows the CRDS of the $b^1\Sigma_g^+ - X^3\Sigma_g^-$ system. Many bands such as (1–0) and (2–0) of this system were measured by pulsed-CRDS in atmospheric and reduced pressure environments in the UV and red spectral regions. The line strengths of the rotational lines of the (0–0), (1–0), and (2–0) bands are well documented in HITRAN database.

The transition systems $a^1\Delta_g - X^3\Sigma_g^-$ and $b^1\Sigma_g^+ - a^1\Delta_g$ are located in the NIR [173,174]. With the advent of single-mode diode lasers, these two systems can be measured by CW-CRDS using a narrow linewidth (~kHz–MHz) DFB lasers at 1580 and 1505 nm, respectively. The former gives absolute number density of the $X^3\Sigma_g^-$ state; and the latter yields number density of the singlet $a^1\Delta_g$ state. Several rotational lines in the Q branch of the (0–1) hot band of the $a^1\Delta_g - X^3\Sigma_g^-$ system can be used as spectral fingerprints to quantify O_2 number density in the ground $X^3\Sigma_g^-$ state. Table 8.4 lists wavelength (cm^{-1}) and line intensities of five rotational lines in the wavelength region about 1581 nm [173]. More recently, the same system has been measured using CW-CRDS in the wavelength region of 1270 nm and O isotopes have also been studied [175].

Note that in plasma environments, implementation of CW-CRDS needs a special care because maintenance of quality CW-CRDS alignment is critical. The data shown in Table 8.4 were obtained in a steady low-pressure chamber. However, there is no issue in measurements of O_2 in the $X^3\Sigma_g^-$ state using a pulsed-CRDS system at atmospheric pressure.

To obtain absolute number density of oxygen in the singlet $a^1\Delta_g$ state, which is widely believed to play a crucial role in plasma treatments of biomaterials, $b^1\Sigma_g^+ \leftarrow a^1\Delta_g$ system may be chosen. The transition wavelength of this system is located in the telecommunications band and diode lasers at this wavelength can be readily affordable. Ringdown measurements of the (0–0) band of the Noxon $b^1\Sigma_g^+ \leftarrow a^1\Delta_g$ system have been

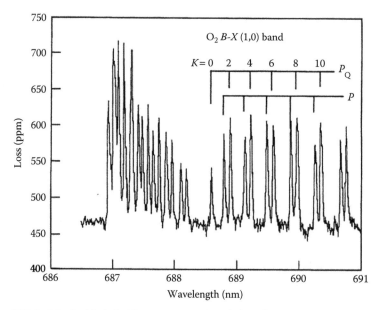

FIGURE 8.12 CRDS of the doubly forbidden transition of O_2 $b^1\Sigma_g^+ - X^3\Sigma_g^-$ system. (Reprinted with permission from O'Keefe A. and D. A. G. Deacon, *Rev. Sci. Instrum.*, 59, 2544–51, 1988. Copyright 1988, American Institute of Physics.)

TABLE 8.4 Wavenumbers and Intensities of the $^QQ(J)$ Lines of the $a^1\Delta_g(0) - X^3\Sigma_g^-(1)$ Band of O_2 Measured by CW-CRDS between 6325.2 and 6326.0 cm^{-1}.

Transition	Wavenumber (cm^{-1})	Line intensity k_{obs} (cm per molecule)	k_{obs}/k_{HITRAN}
$^QQ(13)$	6325.288	1.06×10^{-30}	0.136
$^QQ(11)$	6325.493	1.81×10^{-30}	0.194
$^QQ(9)$	6325.665	1.8×10^{-30}	0.175
$^QQ(5)$	6325.908	1.37×10^{-30}	0.160
$^QQ(3)$	6325.981	8.05×10^{-31}	0.146
Average			0.162 (23)

Source: Reprinted from Kassi, S., et al., *Chem. Phys. Lett.*, 409, 281–87, 2005. Copyright 2005, with permission from Elsevier.

Notes: Comparison with HITRAN values [CPL 409, 281–287 (2005)].

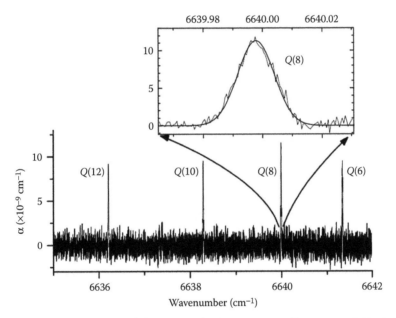

FIGURE 8.13 Part of the Q branch of $O_2\, b^1\Sigma_g^+(V=1) \leftarrow a^1\Delta_g(V=0)$ measured by CW-CRDS. The $O_2(a^1\Delta_g)$ density and the rotational temperature were $(4.6 \pm 1.0) \times 10^{15}$ cm^{-3} and 296 ± 5 K, respectively. (Reprinted from Földes, T., et al., *Chem. Phys. Lett.*, 467, 233–36, 2009. Copyright 2009, with permission from Elsevier.)

reported very recently [173]. Measurements of the $a^1\Delta_g$ state O_2 require excitation of O_2 from the electronic ground state to the singlet state. This can be readily achieved through electron impact excitation in many different plasmas or discharges. In an afterglow microwave discharge at gas (O_2) flow rate of 800 sccm, power of 800 W, and at 300 Pa, the number density of the singlet O_2 was about 6% of the total O_2 molecules. Figure 8.13 shows the CW-CRDS of part of the Q branch of the (1–0) band of the $b^1\Sigma_g^+ \leftarrow a^1\Delta_g$ system.

Oxygen atoms are readily generated in plasmas and discharges. O atoms also play an important role in many plasma applications. Ringdown measurements of O atoms can be achieved by using a pulsed laser in the wavelength region of 630 nm, which is in the same spectral window for the (2–0) band of the $b^1\Sigma_g^+ \leftarrow X^3\Sigma_g^-$ system. At 630 nm, absorption of O originates from the spin-forbidden transition $^1D_2 \leftarrow {}^3P_2$. Figure 8.14 shows the CRDS of O atoms generated in a microwave discharge of O_2 or Ar/NO$_2$ mixture [176].

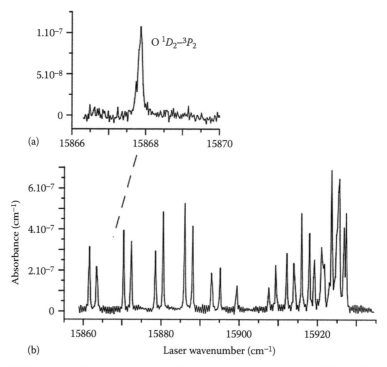

FIGURE 8.14 CRD spectra of (a) the oxygen $^1D_2 \leftarrow {}^3P_2$ atomic line, recorded with 1.7 Torr of O_2 flowing through the microwave discharge and (b) a portion of the $b^1\Sigma_g^+ - X^3\Sigma_g^-$ (2,0) band of O_2 (1/2 atm, microwave discharge off). An oscillating baseline has been subtracted from the upper trace. (Reprinted from Teslja, A. and P. J. Dagdigian, *Chem. Phys. Lett.*, 400, 374–78, 2004. Copyright 2004, with permission from Elsevier.)

The measured atomic O in the $^3P_2$ state was on the order of 10^{14} atom cm^{-3} [176]. Note that the laser linewidth used in this work was 0.15 cm^{-1}. A narrower linewidth of the laser beam would give a higher measurement accuracy for a given line intensity and laser beam path length.

8.5.3 CRDS of N_2^+ and N_2 in Plasmas and Discharges

Generation of N_2^+ needs electron energy of 15.577 eV or higher. As listed in Table 8.1, the (0–0) band of the first negative system of N_2^+ ($B^2\Sigma_u^+ - X^2\Sigma_g^+$) is located around 391 nm. This optically allowed strong band can be measured using pulsed-CRDS. Yalin et al. [121] measured the rotational–vibrational lines of $P(9)$–$P(17)$ of this band in an atmospheric DC discharge using CRDS with an OPO system as a light source. The laser pulse duration was ~7 ns and single pulse energy was ~1 mJ; and the laser linewidth was 0.14 cm^{-1}. The ringdown mirror reflectivity was approximately of 99.98%. The cavity parameter g as defined in Equation 8.19 was 0.875. Figure 8.15 shows CRDS of N_2^+ in the wavelength range of 390.5–391.5 nm. More than 25 rotational–vibrational lines are covered in that spectral region. Along with the CRDS, simulated spectra in the same spectral region are also given. Clearly, several lines can be used to quantify N_2^+ concentration. The discharge gas temperature was around 5600 K determined by Boltzmann plots of the rotational lines. The line strengths of N_2^+ rotational lines are documented in the literature [177,178].

The discharge was radically symmetric and the laser beam was aligned through the diameter of the circular cross section of the discharge and Abel inversion process was employed to determine the radically resolved number density of the N_2^+ in the plasma [121]. In the discharge radius of 4 mm, the number density of N_2^+ varied from 0.2×10^{12} to 3.9×10^{12} molecule cm^{-3} with a spatial resolution of 0.25 mm. The highest N_2^+ number density in the plasma was located in the center of the discharge.

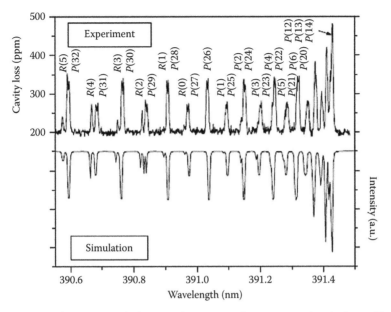

FIGURE 8.15 Cavity ringdown spectra of N_2^+ generated in an atmospheric nitrogen glow discharge. (From Yalin, A. P., et al., *Plasma Sources Sci. Technol.*, 11, 248–53, 2002. With permission.)

The same CRDS system was also used to study the temporally resolved variations of N_2^+. An additional narrow pulse of 1 μs was added to the base discharge, so that density of N_2^+ was varied. Change in the N_2^+ number density was monitored via ringdown measurements. This study demonstrates that CRDS can be implemented to quantify plasma species with not only a high spatial resolution (0.25 mm) but also a high temporal resolution above microsecond or more.

Because generation of N_2^+ ions requires relatively high electron temperature in plasma or discharge, not many CRDS measurements of N_2^+ number density have been reported to date. A much earlier study demonstrated another side of the ringdown measurements of N_2^+ in terms of different plasma generation environments and detection wavelengths [179]. The N_2^+ ions were generated in a discharge cooled by liquid nitrogen, so that rotational temperature of N_2^+ was only 150–200 K. Discharge gas pressure was 10^{-2} Torr. The overtone spectra of the N_2^+ $A^2\Pi_u - X^2\Sigma_g^+$ system were measured. In this study the pulsed laser was from an excimer-pumped dye laser with a linewidth of 0.15 cm^{-1}. A telescope system was used to reform the laser beam with a beam waist of 1 mm^2 before it entered the ringdown cavity. Figure 8.16 shows the rotationally resolved CRDS of N_2^+ in the wavelength range of 611.5–611.6 nm. Even with the third overtone, the spectra still has an *S/N* ratio better than 10.

The line intensity of this $A^2\Pi_u - X^2\Sigma_g^+$ system was published in the literature [180,181]. Due to the low-oscillator strengths of the (4–0) and (6–0) bands of the system (10^{-5} and 10^{-6}, respectively), the detection limits for these two bands using CRDS were 10^{14} molecule cm^{-3}. It should be emphasized that these CRDS measurements have utilized the high order of weak overtone, which is typically not accessible by other spectral techniques due to the low-spectral intensities. This is another example that shows the power of the high sensitivity of CRDS.

Nitrogen molecule (N_2) is a common atmospheric plasma/discharge species. N_2 has four low-lying electronic states, $X^1\Sigma_g^+$, $A^3\Sigma_u^+$, $B^3\Pi_g$, and $C^3\Pi_u$. The dissociative energy of N_2 is 12.139 eV. The electronic transition of the N_2 system $A^3\Sigma_u^+ \rightarrow X^1\Sigma_g^+$ is often called Vegard–Kaplan (VK) bands. Optical emissions of the VK bands are located in the spectral range of 200–460 nm and they are partially overlapped with emissions of the N_2 second positive group $C^3\Pi_u \rightarrow B^3\Pi_g$ in the spectral range of 360–390 nm. Furthermore, the VK bands are very weak due to the violation of spin transition selection rule ($\Delta S = 0$). Therefore,

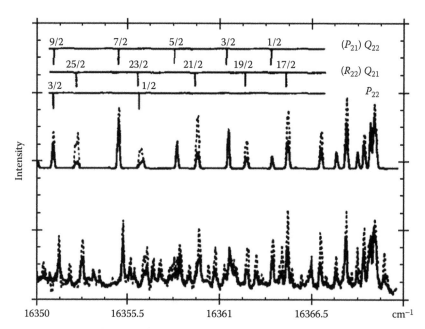

FIGURE 8.16 CRDS of the $A^2\Pi_u \leftarrow X^2\Sigma_g^+$ (4–0) transition of N_2^+ at 150–200 K (broken lines) and 350–400 K (solid lines). Bottom trace is the experimentally observed spectra. Top trace is the simulated spectra. (Reprinted from Kotterer, M., et al., *Chem. Phys. Lett.*, 259, 233–36, 1996. Copyright 1996, with permission from Elsevier.)

optical measurements of these two systems need to take care of the possible spectral overlap and the weak emissions of the VK bands. The N_2 first positive group $A^3\Sigma_u^+ - B^3\Pi_g$ is a preferred system that is used to measure the absolute number density of N_2. Ringdown measurements of N_2 in plasma have been reported only very recently [182]. In those studies, CRDS of the (2–0) band of the N_2 first positive group were recorded in the wavelength range of 769–772 nm, as shown in Figure 8.17. The measured CRDS match the simulated spectra very well. The excitation of the $A^3\Sigma_u^+$ state was achieved through electron impact excitation in the discharge. The discharge gas temperatures were determined through comparison of simulations with the measured emission spectra of the N_2 ($C{\rightarrow}B$) and ($B{\rightarrow}A$) systems in the spectral windows of 365–382 and 550–800 nm, respectively. In the nanosecond (ns)-pulsed discharges in air and in nitrogen with plasma gas temperature of 1600 ± 300 K and electron temperature of 12,000 K or higher, the N_2 number densities measured by CRDS were of the orders of 10^{14} and 10^{15} molecule cm^{-3}, respectively.

This first positive group can also be measured using TDLAS [183–185]. Due to the high sensitivity of CRDS, the VK bands may be measured using a pulsed-CRDS system. However, no publication has reported CRDS measurements of the VK bands to date.

8.5.4 CRDS of Important Radicals in Plasma-Assisted CVD and Material Growth

Free radicals, such as NH, NH_2, Si, SiH, SiH_3, CF, CF_2, SiF_2, C_2, CH, and CH_3 are important plasma species in plasma-assisted CVD. Knowing absolute concentrations of these radicals in plasma-assisted processes would provide crucial information on plasma operation, design, and optimization toward a high production yield with reduced costs. Most of the free radicals listed above are generated in discharge and afterglow discharge expansion in a vacuum-controlled reaction chamber in reduced pressure. Precursors of these radicals are typically semiconducting material fabrication gases, including CH_4, NH_3, CF_4, SiH_4, and SiF_4. Details of the generation conditions and their corresponding setup of the plasma devices can be read elsewhere. Here only brief spectral information on the transition

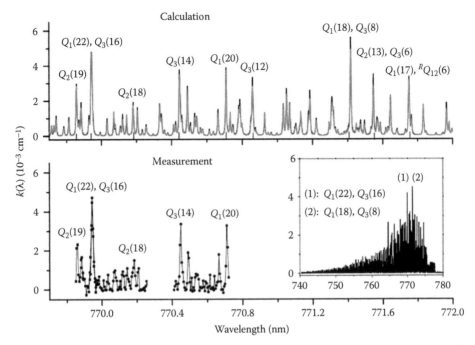

FIGURE 8.17 Absorption spectra of the (2 ← 0) vibrational band calculated with SPECAIR (red: Trot = 1600 K; black: Trot = 1000). Spectrum measured by CRDS in filamentary nitrogen discharge. The inset shows the entire 2 ← 0 band between 740 and 780 nm. (From Stancu, G. D., et al., *J. Phys. D Appl. Phys.*, 43, 124002-10, 2010. With permission.)

wavelengths of these radicals, which have been quantified by CRDS to date, are given for the practical purpose of the application of CRDS in quantification of plasma species.

NH and NH_2 are important radicals in the silicon nitride film growth and they can be generated in the downstream of an Ar plasma jet via injection of NH_3 gas that meets Ar plasma plume in the downstream. The discharge is usually created in the upstream at subatmospheric pressure. The discharge plume passes through a nozzle into a vacuum reaction chamber to form a supersonic jet, in which NH and NH_2 radicals are generated mainly through the initial reaction of Ar^+ with precursor NH_3, followed by dissociative recombination and ion–molecule reactions. An Nd:YAG pumped dye laser with frequency doubling at 340 nm can be used to measure NH and the direct output of a dye laser at 600 nm to measure NH_2. The 340 nm transition belongs to the (0–0) band of the NH $A^3\Pi \leftarrow X^3\Sigma^-$ system. At this wavelength region, the rotational line $P_{33}(9)$ at 339.6 nm has no overlap with other transition lines; therefore, it can be used for absolute number density measurements of NH. The line intensity is 8.3×10^{-21} cm^{-1} at the plasma gas kinetic temperature of 1750–1920 K. The measured NH density was around 2.5×10^{12} molecule cm^{-3} at the NH_3 flow rate of 3 sccs [186]. The NH_2 transitions belong to the $A^2A_1 \leftarrow X^2B_1$ system. One of the ^PQ rotational lines belonging to the (090–000) band can be used to measure NH_2 number density. The line intensity is on the order of 10^{-21} cm^{-1} at the plasma gas kinetic temperature of 1750–1920 K; this line intensity is strong enough for quantification of NH_2 at ppm levels.

Si, SiH, and SiH_3 are also important radicals in film growth and they can be generated in a thermal arc discharge of Ar–H_2 using SiH_4 as precursors [187,188]. Si atoms have been measured by CRDS in the thermal expansion of the Ar–H_2 discharge using the $4S^3P_{0,1,2} \leftarrow 3P^3P_{0,1,2}$ system which has a broad transition around 251 nm. The Si number density was measured to be of ~10^9 atom cm^{-3} in the plasma expansion. The SiH radical has been measured using the $A^2\Delta \leftarrow X^2\Pi$ electronic transition around 414 nm. The rotational lines of $Q_1(11.5)$, $Q_1(14.5)$, and $R_2(1.5)$ of the (0–0) band of the system around 413 nm have been used to quantify SiH number densities in the thermal plasma expansion. The rotational temperature in the plasma expansion was 1800 K. In the same experimental system SiH_3 was measured by CRDS using the $A^2A_1 \leftarrow X^2A_1$ system in the wavelength range of 200–260 nm. The measured number densities were on the order of 10^{12} molecule cm^{-3}.

CF, CF_2, and SiF_2 radicals play an important role in determining the etch rate, selectivity, and anisotropy of industrial plasma-etching processes. They can be readily generated via RF discharges or DC discharges of precursors of CF_4 or C_2F_6 in a low-pressure chamber. Although CF, CF_2, and SiF_2 have been extensively studied using LIF, absolute number density measurements of these radicals were first carried out by CRDS [40]. The (0–0) band of CF radical $A^2\Sigma^+ \leftarrow X^2\Pi$ system is located in 233 nm. Transitions of this band are partially overlapped with the weak overtone of CF_4 system. However, the rotational lines in the (1–0) and (2–0) bands of the CF system are clearly resolved, such as the P_{11}–P_{22} lines in the two bands, which can be used as spectral fingerprints for quantification of CF. The CF_2 radical has a long progression in the wavelength range of 237–300 nm. For example, the (050–000) band of CF_2 A–X system can be readily reached by frequency doubling of an Nd:YAG pumped dye laser at 253 nm. Molecular constants of CF_2 radicals are well documented. Number density of CF and CF_2 measured by CRDS can be as low as 10^{10} molecule cm^{-3} (tens of ppb) in reduced pressure. SiF_2 radical does not have resolved rotational structures though it has clear vibronic structures from the (000–000) band to the (050–000) band in the wavelength range of 220–300 nm. However, CRDS measurement of the integrated abortion from an entire vibrational band can still detect a number density of 10^{12} molecule cm^{-3} (sub-ppm) for SiF_2 radicals [40].

C_2, CH, and CH_3 are important radicals in the CVD processing. They can be generated by microwave discharges and DC arc discharges in the gas mixture of hydrocarbon–H_2 or hydrocarbon–Ar. CRDS measurements of C_2 have been reported by using the $d^3\Pi_g \leftarrow a^3\Pi_u$ Swan band system [41]. The P_1, P_2, P_3, R_1, R_2, and R_3 branches of the (0–0) band of the system are located in the spectral region of 505–517 nm and their rotational lines are clearly resolved. The electronic ground state of C_2 is $X^1\Sigma_g^+$. Number density of C_2 can also be measured through the $D^1\Sigma_u^+ \leftarrow X^1\Sigma_g^+$ system around 231 nm. CRDS measurements of the rotational lines in the R branches of the (0–0), (1–1), and (2–2) bands of the $D^1\Sigma_u^+ \leftarrow X^1\Sigma_g^+$ system in the spectral range around 231 nm have been reported. The measured number densities of C_2 in the $X^1\Sigma_g^+$ and $a^3\Pi_u$ states were on the order of 10^{12}–10^{13} molecule cm^{-3} in the 3.3% CH_4/H_2 mixture arc discharge at 6 kW. The concentration of C_2 in the X state is ~0.27 times higher than that of the C_2 in the a state. In the same experimental system, CRDS of CH have also been obtained in a narrow spectral range around 23,420–23,430 cm^{-1}, which spans a few strong and isolated rotational lines of the (0–0) band of the $A^2\Delta \leftarrow X^2\Pi$ transition. Figure 8.18 show the CRDS of C_2 and CH measured in the DC arc discharge jet [41]. The gas temperature of the radicals obtained via a Boltzmann plot was 3300 ± 200 K. Methyl radical CH_3 can be generated through direct thermal dissociation of CH_4 in the high temperature [189]. CH_3 can be measured by CRDS using the $\beta_1 \leftarrow X$ or $B^2A_1' \leftarrow X^2A_2''$ system at 216 nm. At 2300 K, a mixture of 0.5% CH_4 in H_2 slowly flowed through the reactor at total pressure of 20 Torr, the lowest CH_3 number density which could be measured was 3.0×10^{12} molecule cm^{-3} (25 ppm). Because the wavelength of the transition was in the deep UV, the laser source was a 248 nm ArF excimer laser that pumped dye with doubling of frequency.

8.5.5 CRDS of Sputtered Trace Atoms

CRDS has also been applied to measure trace atoms sputtered using ion beams. Ion beam sputtering plasma is created when an energetic ion beam bombards the surface of a metallic target. In the process, chemical species, such as small molecules, free radicals, ions, and neutral atoms are sputtered from the target and form sputtered plasma. Real-time, *in situ* monitoring sputter erosion is important in the operation of electric propulsion devices and the development of nuclear magnetic fusion. Recently, CRDS has been used to study sputter erosion of a metallic target [38,39]. In those studies, the ion beam source and the target were housed in a vacuum chamber with a background pressure of 10^{-6}–10^{-7} Torr. Energetic Ar ions bombarded a metallic target and elements were generated around the surface of the target.

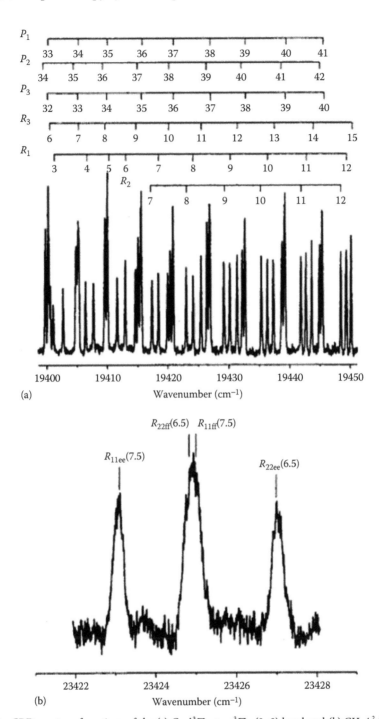

FIGURE 8.18 CRD spectra of portions of the (a) C_2 $d^3\Pi_g \leftarrow a^3\Pi_u$ (0–0) band and (b) CH $A^2\Delta \leftarrow X^2\Pi$ (0–0) band recorded in the DC arc jet under typical operating conditions of power (6 kW) and input gas composition (3.3% CH_4/H_2 ratio in excess Ar) at distances of $z = 20$ mm (C_2) and 10 mm (CH) from the substrate. The combs above the spectra indicate spectral line assignments. (Reprinted with permission from Wills, J. B., et al., *J. Appl. Phys.*, 92, 4213–22, 2002. Copyright 2002, American Institute of Physics.)

The UV laser beam (375–400 nm) from an OPO system was aligned several centimeters above the target surface. Laser parameters were a repetition rate of 10 Hz, a pulse duration of 7 ns, a pulse energy of 0.1 mJ, and a linewidth of 0.13 cm^{-1} at 385 nm. Ringdown spectra of optical transitions originating from the electronic ground state multiplet of elements were measured. Doppler linewidths of titanium (Ti) and molybdenum (Mo) atoms were of 7.5 ± 0.8 and 6.4 ± 0.4 pm, respectively. Those linewidth measurements gave translational temperatures for these two atoms of 34,000 and 51,000 K, respectively. CRDS was used to detect the densities of the sputtered elements, such as iron (Fe) and aluminum (Al). Given the system noise level of 2 ppm, the detection limits for Fe, Al, Mo, and Ti were obtained to be 7.5 × 10^6, 7.5 × 10^6, 7.5 × 10^6, and 8.7 × 10^6 atom cm^{-3}, respectively. These extremely high detection sensitivities (sub-ppt) are due to the large line intensities of the atomic transition in the UV.

Diode laser CW-CRDS has also been implemented to measure absolute number densities of atoms generated in ion beam etching processes [190]. In that study, the laser source had 30 GHz tenability with an output power of 10 mW with a linewidth <5 MHz. Manganese (Mn) and Boron (B) atoms have been measured *in situ* by CW-CRDS at 403 and 250 nm, respectively. The baseline noise of the CW-CRDS system was approximately 1%. At a fixed frequency, the minimum detectable absorbance was about 0.6 ppm with averaging over 1000 ringdown events.

8.5.6 CRDS of Environmentally Important Elements and Isotopes Generated in Atmospheric Plasmas

Real-time, *in situ* quantification of environmentally important trace heavy metals, radionuclides, and their isotopes are very important in analytical instrumentation and environmental monitoring (Figure 8.19). The current challenge is that almost all of the detection and quantification of the toxic trace elements are conducted based on the site sampling and back-to-lab analysis routine. In many applications, we need a field-deployable, highly sensitive, real-time, *in situ* instrument to serve the analytical purposes. Most trace toxic metals, such as Hg, Pb, strontium (Sr), and uranium (U), exist in the form of chemical compounds. Therefore, detection of those elements requires atomization of the elements from the existing compounds prior to quantification. Detections of these elements and isotopes were early

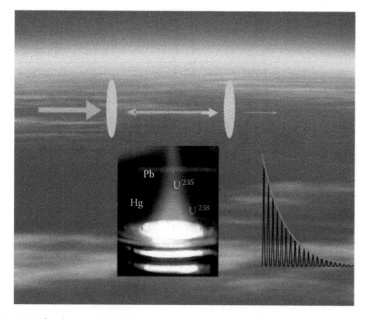

FIGURE 8.19   P-CRDS for elemental and isotopic measurements in an ICP.

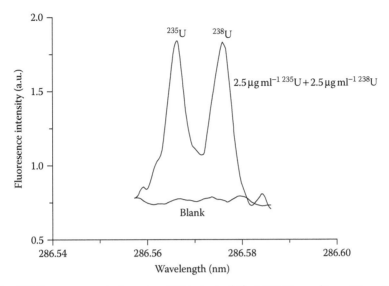

FIGURE 8.20 CRDS measurement of uranium (U) isotopic shift at 286.57 nm. (From Wang, C., et al., *Appl. Spectrosc.*, 57, 1167–72, 2003. Reprinted with permission from Society for Applied Spectroscopy.)

reported by using an atmospheric plasma as an atomization source and CRDS as a detector [18,20]. The standard analytical solutions containing the element compounds of interest in different concentrations were pumped into an ultrasonic nebulizer to generate chemical vapor to be injected into the plasma and the elements were generated through high-plasma temperature dissociation of the element-contained compounds. Because most elements have strong transitions in the UV–near visible spectral region, typically, a pulsed laser system, such as an Nd:YAG pumped dye with frequency doubling or an OPO system, is implemented as the laser source for CRDS measurements. The major constraint of the laser source in the atomic measurement is the requirement of a narrow linewidth in order to obtain an atomic line-shape scan, from which the total absorbance is used to determine absolute number density of the atom. To date, Hg, Pb, U, Sr, and their isotopes have been measured by CRDS coupled with different plasma sources, including low-power atmospheric ICP, MIP, and plasma jets [83]. The detection limits of these elements obtained range from ppb to ppt levels.

Figure 8.20 shows the CRDS measurement of a small isotopic shift (8.6 pm) of the uranium ^{238}U(II) and uranium ^{235}U(II) (radioactive) at 286.57 nm [21]. Many other analytical elements can also be measured using the P-CRDS technique.

The first diode laser CW-CRDS applied to measurements of plasma elements was reported in 2004 [51]. An ECDL CRDS system was developed to measure Sr generated in an atmospheric Ar ICP operated at power of 200–400 W. The laser diode output had a single wavelength at 688 nm with 30 GHz frequency tunability. More recently, a palm-size, single-mode diode laser has been demonstrated to develop a portable P-CRDS spectrometer via measuring elemental Hg in its metastable state. This gives a future direction for the development of the P-CRDS spectrometer toward a potable unit using a compact diode laser.

8.5.7 CRDS of Other Plasma Species (CH_3OO, C_6H_5, and HO_2) in Fundamental Spectroscopic Study

CRDS during the first few years after its introduction was almost exclusively contained in the fundamental spectroscopic studies of molecule, radicals, ions, and clusters existing or generated in different experimental systems, excluding a system containing a plasma source [74–76]. Fundamental studies, that is, measurements of new molecular constants, of some important free radicals generated through plasma-assisted processing have been reported only very recently. Some radicals are preferably studied using CRDS if they do not have high-fluorescence quantum yields or suffer from fluorescence quenching. One of the examples is the study of alkoxy radicals (i.e., CH_3O) and peroxy radicals (i.e., CH_3OO) generated in a DC discharge of a radical precursor in a vacuum chamber [191,192]. The generated radicals are cooled in a supersonic jet to very low rotational temperature (tens of K), so that spectral structures are simplified. Alkoxy and peroxy radicals play a key role in the formation of ozone, NO_x, OH, and HO_2 radicals in the troposphere. However, molecular constants and structures of the radicals, especially large peroxy radicals, are not available. Very recently, β-hydroxyethylperoxy (β-HEP) and their isotopologues generated in a narrow pulse DC discharge have been studied by pulsed-CRDS [193]. CRDS of the $\tilde{A}$–$\tilde{X}$ system of four different β-HEP isotopologues have been recorded in the NIR under supersonic jet-cooled condition. The pulsed NIR laser beam at 1350 nm was formed through DFG of the Nd:YAG 532 nm output with Ti:Sapphire at 7390 cm^{-1}. Figure 8.21 shows the first CRDS of the four different β-HEP isotopologues, and their new molecular constants have been determined.

In a similar way, phenyl radical (C_6H_5), generated in a DC discharge and cooled in a supersonic jet, has been studied recently using CRDS system, in which gas rotational temperature is only about 30 K [194]. C_6H_5 was generated through DC discharge of a C_6H_6–Ar mixture. Rotationally resolved spectra of the phenyl 1^2B_1–X^2A_1 transition in the wavelength range of 504–530 nm are reported. The lifetime of the excited state of the vibronic band 9_0^1 is only 96 ps due to its low quantum yield of 3.4×10^{-5} while the lifetime of the electronically excited state 1^2B_1 is 2.8 μs. UV CRDS of HO_2 radials generated in DBD has also been reported. In addition to the UV transition system, NIR transitions of HO_2 can also be measured by CRDS [195]. Except for the aforementioned radicals, few radicals generated via a plasma-assisted process have been reported for fundamental spectroscopic study to date.

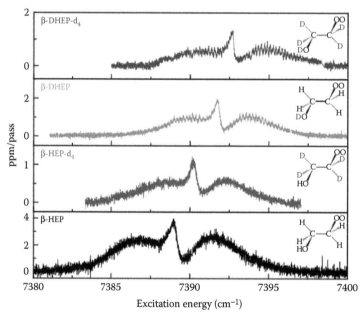

FIGURE 8.21 Experimental $\tilde{A}$–$\tilde{X}$ spectra of β-HEP isotopologues, demonstrating the differently resolved overall rotational contours. The site of deuteration is indicated in the upper right while the corresponding acronym for each isotopologue is shown in the upper left of each panel. (Reprinted with permission from Chen, M., et al., *J. Chem. Phys.*, 135, 184304–11, 2011. Copyright 2011, American Institute of Physics.)

TABLE 8.5 Selected Key Combustion Intermediates and Their Spectral Wavelengths

Wavelength (nm)	Species	Wavelength (nm)	Species
225–238	C_2H_3	656	H
242–258	O_2	659	H_2O
225–258	NO	688	O_2
298, 308	OH	730–755	H_2O_2
310–330	C_3H_3	762, 765	O_2
312	OH	776	N_2O
315, 334	NH	785, 797	NO_2
387	CN	812–819	H_2O
430–431	CH	1032	C_2H_2
436–442	C_2H_2O	1036	HCN
438–452	NO_2	1064	CO_2
444–445	C_2H_3	1270	O_2
469–573	HCN	1285	N_2O
505	N_2^+	1331	CH_4
562–582	NO_2	1509	OH
570	C_2H_2	1538	CO_2
615	HCO	1547	C_2H_2
622	CH_2	1591	CH_4
625–645	C_3H_8	1570	CO
627–635	N_2^+	1672	CO_2
628	O_2	3100–3170	CH_3
639	C_3H_4	3315	CH_4

Except for the common nature of the escalated temperature in plasma and in combustion, CRDS of combustion species are not included in the scope of discussion in this chapter. Combustion is one of the most sophisticate chemical processes, involving numerous radicals and intermediates; number of combustion intermediate species will be even larger when the combustion is assisted by a low-temperature nonthermal plasma source (PAC). Table 8.5 lists some important combustion species with their spectral fingerprints, which can be measured by CRDS. Some of the species as common species existing in both plasma and combustion flame have been discussed in previous sections. A lot of them, such as NO, CN, and HCN radicals, which have been measured by CRDS in flames [196–198], yet, have not been measured by CRDS in a plasma or discharge. For instance, no CRDS measurement of NO in plasma or discharge has been reported to date. Transitions of NO are located in the deep UV; measurements of NO need a deep UV laser source and encounter relatively low mirror reflectivity in the deep UV. Nevertheless, NO in plasma can be measured by CRDS. For CN and HCN radicals, they are not normally generated in a plasma or discharge though the laser wavelengths for measurements of CN and HCN are in the visible spectral region.

8.6 Coupling of P-CRDS with OES for Plasma Diagnostics

In addition to the absolute number density measurement of plasma species discussed previously, measurements of plasma gas temperature and electron density are also important for plasma diagnostics and application. CRDS can be used as a power tool for this purpose too [20,121,122]. For instance, high-resolution scans of a transition lineshape can lead to determination of plasma gas temperature and electron density by using lineshape-broadening effects. Gas temperature can often be estimated from Doppler broadening component of the lineshape. If Stark broadening is a dominant contribution to Lorentzian component of the broadened lineshape, electron density of the plasma under specifically known condition, that is, the range of the electron density, can be estimated. Although plasma gas temperature can be simply measured by a thermocouple, in many cases the large size of a thermal probe (compared with the size of a plasma itself) is intrusive. Furthermore, the temperature measurement using a thermocouple would suffer from low spatial resolution. Another popular means to estimate plasma temperatures (T_v, T_r) is to compare experimental emission spectra of some selected plasma species with their simulated spectra. Numerous examples for this method can be seen in the literature. The simulations of emission spectra of typical plasma species, such as OH, NO, and N_2 use SPECAIR or LIFBASE, which are freely available.

Electron density measurements can also be carried out through linewidth analysis using the hydrogen α-line and β-line in OES. Although the OES-based estimation of plasma temperatures and electron density methods are widely used, the CRDS-based lineshape broadening method can be a useful alternative to the existing methods. It should be noted that the linewidth broadening method is based on measurement of the ground-state plasma species; however, the OES method-based plasma temperature and electron density estimation are based on the excited state information of the plasma species. Note that in some regions of a plasma, that is, the far downstream of a plasma jet, OES are not detectable, the OES-based temperature and electron density estimation are not achievable. Therefore, when these two methods are combined and thorough plasma diagnostics can be expected in many cases. Especially, when a high-speed digital imaging is added to the spectroscopic system, a comprehensive diagnostics of plasma can be achieved. Figure 8.22 shows schematic of the experimental system which consists of CRDS, OES, and digital imaging for plasma diagnostics, in which the digital imaging gives information about the plasma shape, CRDS measurements offer absolute number density of species of interest

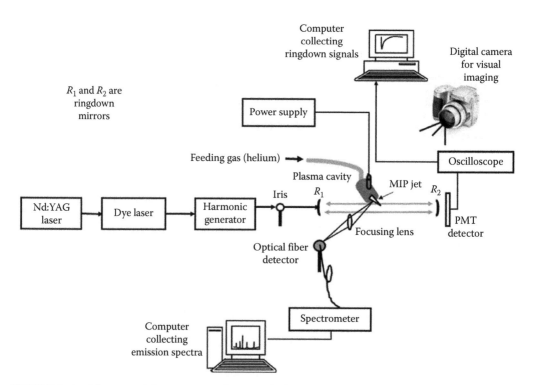

FIGURE 8.22 Schematic of the experimental setup, which consists of a microwave plasma source, a pulsed cavity ringdown spectroscopy system, an OES system, and a digital camera. (From Srivastava, N. and C. Wang, *IEEE Trans. Plasma Sci.*, 39, 918–24, 2011. © 2011 IEEE.)

as well as plasma gas temperature and electron density, and OES provides convenient identification of types of plasma species.

8.6.1 Gas Kinetic Temperature from Doppler Broadening

Estimation of plasma gas temperature using CRDS scans of a trace plasma species was earlier reported in the works of Wang et al. [20]. Recently, this approach has become a routine to obtain plasma gas temperature. The practical advantage of the method is to use a single CRDS scan of a lineshape of a transition to realize two purposes concurrently: number density measurements from the integrated absorbance over the entire lineshape and plasma gas temperature and electron density estimation from the linewidth-broadening analysis. As an example [20], measurements of plasma gas temperature in an atmospheric ICP using CRD scans of the transition of Pb at 283.3 nm are discussed below. The scanning range for the Pb line (283.3 nm) was 0.03 nm, and the scanning step was 0.0003 nm. At each data point, 100 laser shots were averaged. A 3–5-point smooth was applied to the data before plotting.

Figure 8.23 shows four typical lineshapes recorded through the diameter of plasma at four observation heights (2, 10, 18, and 30 mm ALC) in a low-power atmospheric Ar ICP. The measured absorbance was corrected using the background absorption from a blank (no Pb) solution and normalized to a 100 ng ml^{-1} (~100 ppb) concentration. At the wings of the lines, because the baseline noise is comparable with the weak off-line signal, the recorded lineshapes are somewhat affected by fluctuation noise (<5%). The recorded linewidth includes both instrument and physical broadening [199,200]. The former came from the laser linewidth (<0.65 pm). The latter included Doppler and collisional broadening contributions, which yielded Gaussian and Lorentzian lineshapes, respectively. Together the various broadening mechanisms result in a Voigt lineshape.

The commercially available software (PeakFit^{TM} 4.0) [201] can be used for the lineshape fitting to obtain Doppler and Lorentzian components. Gaussian and Lorentzian components are derived from

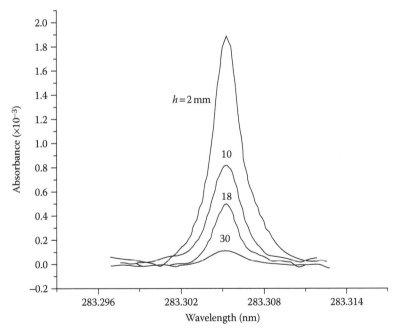

FIGURE 8.23 Measured lineshape of the Pb line (283.3 nm) versus height. ICP power = 200 W; lateral distance $x = 5.0$ mm; from top to bottom, $h = 2$, 10, 18, and 30 mm; Pb solution: 100, 100, 200, and 500 ng ml^{-1}. (From Wang, C., et al., *Appl. Spectrosc.*, 56, 386–97, 2002. Reprinted with permission from Society for Applied Spectroscopy.)

Voigt (Amplitude, Gaussian/Lorentzian width) fitting of the recorded line by the AutoFit Peaks III Gaussian Deconvolution Method, in which the Voigt function is defined as [201]

$$
y = \frac{a_0 \int\limits_{-\infty}^{\infty}\left(\exp\left(-t^2\right)\middle/ \frac{a_3^2}{2a_2^2}+\left(\frac{x-a_1}{\sqrt{2}a_2}-t\right)^2\right)}{\int\limits_{-\infty}^{\infty}\left(\exp\left(-t^2\right)\middle/ \frac{a_3^2}{2a_2^2}+t^2\right)},
\tag{8.28}
$$

where a_0, a_1, a_2, and a_3 are amplitude, central wave number, Gaussian width, and Lorentzian width, respectively. Prior to each fit, the measured wavelength needs to be converted to wavenumber units, the baseline subtracted, and the raw data smoothed. The smoothing levels are typically set at 20–30% for an overall ~98% noise reduction. After this data preparation, the AutoFit Peaks III Gaussian Deconvolution Method is used to fit each lineshape. A laser linewidth at 283.3 nm of 0.08 cm^{-1} was set as the instrument deconvolution FWHM for all fits. The amplitude rejection threshold was set at 5% so that the recorded data would fit to only one peak. In addition to these fixed parameters used in each fit, adjusting the Fourier domain truncation filter level to reduce noise produced by the deconvolution was necessary to obtain the best result as quantified by the coefficient of determination (r^2) and fit standard error. Typically, the adjusted range of the filter level is 80–86%. The general quality of the fits results in $r^2 > 0.97$ and fit standard errors <0.05.

The fitting results include the Lorentzian (w_L) and Gaussian components (w_G) [20]. The relative errors (standard error/average) of the Gaussian and Lorentzian components determined from lineshape fits are in the range of 12–20% for the various vertical positions. For instance, at 2 mm ALC, the CRDS signal is more stable (<5% variation), resulting in the best fitting error, 12%. At the tail of plasma, decreased stability (<10%) can result in a 20% fitting error. The plasma gas temperatures at different plasma locations were determined by w_G using Equation 8.18. The plasma temperatures estimated using this method was in the range of 2411–3099 K, which was in good agreement with those from other techniques [202]. The uncertainty of the gas kinetic temperature from the Gaussian component can be estimated from the fitting errors. For instance, at 2 mm ALC the fitted a_2 is 0.74 pm, corresponding to a gas kinetic temperature of 2755 K. However, the fitting error for the Gaussian component of the lineshape is 12%. Thus, a 12% uncertainty in the fitting yields a 25% uncertainty in the gas kinetic temperature.

8.6.2 Electron Density from Stark Broadening

It is more complicated to obtain electron density profiles from the Lorentzian component (w_L) of the lineshape shown in Figure 8.23. The collisional broadening responsible for the Lorentzian component may include resonance, Stark, and van der Waals components at atmospheric pressure plasma conditions, with Stark and van der Waals broadening being the most significant [203,204]. Although the electron density has previously been estimated by attributing the entire Lorentzian component (w_L) of an absorption line to Stark broadening (caused by interaction with charged particles), more recent work revealed significant differences between electron densities derived from Thomson scattering and Voigt lineshape analysis using diode laser absorption data [204]. This difference was especially obvious at the plasma center and edge. In the later study, Regt et al. demonstrated that higher neutral particle densities at the center and edge (due to the lower heavy particle temperature) resulted in van der Waals broadening being dominant in these areas. In the intermediate plasma region, where higher gas temperatures result in lower neutral particle densities, Stark broadening was a significantly larger contributor to the total Lorentzian width. Using the Lorentzian component to calculate the electron density may overestimate the electron density by a factor of 3–4 over the Thomson scattering method. Thus, although not highly accurate, a reasonable upper bound for the electron density can be readily determined in the intermediate region by attributing the total

Lorentzian component to Stark broadening. With this assumption, an upper bound on the electron density can be derived from the theoretical formula [203]

$$w_S \approx 2\left[1 + 1.75 \times 10^{-4}\, n_e^{1/4}\alpha\left(1 - 0.068\, n_e^{1/6} T_e^{-1/2}\right)\right]10^{-16}\, wn_e, \tag{8.29}$$

where
  w_S (Å) is the theoretical width (FWHM) due to Stark broadening
  n_e is the electron density
  T_e is the electron temperature
  w is the electron impact parameter (Å)
  α is the ion broadening parameter

Note that application of Equation 8.29 is restricted by two conditions: (1) The ratio of the mean distance between ions, ρ_m, and the Debye radius, ρ_{Debye}, accounting for shielding by electrons is less than 1, and (2) only the quasi-static approximation is used in the Stark broadening calculation. These two considerations are justified in some plasmas conditions, including the ICP, that is, $T_e \sim 7500$, $n_e \sim 10^{15}$ cm^{-3}. Stark broadening parameters (w, α) can be chosen from various tabulated values with consideration of transition wavelength, estimated ranges of electron density and electron temperature. In the calculation using Equation 8.29, w, α, and T_e were set to 0.01 (Å), 0.082, and 7500 K, respectively [203]. Note that the Stark width is very insensitive to the electron temperature. For example, a 1000 K change in T_e produces less than 10% change in the calculated electron density. The electron densities estimated using Equation 8.29, based on the Lorentzian components are shown in Figure 8.24. As noted above, the actual electron density is lower than the estimate from Equation 8.29, suggesting a maximum electron density of approximately 10^{15} electron cm^{-3}. This value is in line with typical literature values in the range of 5×10^{14} to 5×10^{15} electron cm^{-3}. The Voigt parameter a obtained from the fitting is in the range of 0.6–1.1. The a values are observed to decrease slightly with an increase of observation height. The gas temperatures, and thus the gas density, are seen to remain reasonably constant at lower heights in this

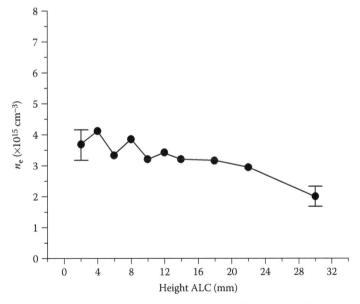

FIGURE 8.24 Vertical profile of electron density as measured through the diameter of the torch. ICP power = 200 W. The two error bars show the maximum and minimum errors. (From Wang, C., et al., *Appl. Spectrosc.*, 56, 386–97, 2002. Reprinted with permission from Society for Applied Spectroscopy.)

atmospheric pressure plasma. However, a steady decrease in electron density with height is noted in this region. The reduction in the Lorentzian component due to the corresponding decrease in Stark broadening causes the *a* parameter to decrease. Higher in the plasma, where the gas temperature dropped considerably, thus increasing the gas density, an increase in van der Waals broadening and decrease in Doppler broadening would cause the *a* parameter to increase again.

8.7 Perspectives of P-CRDS in Future Applications

In addition to its high sensitivity, fast response, and ease of setup, the unique advantage of CRDS over its counterparts, such as LIF, REMPI, and Raman, is the capability of absolute number density measurements. Just because of this feature alone, CRDS has huge potential in many fields, from environment monitoring, special gas industry, medicine, material, and sensing. Of the plasma-related fields, the merging arena of plasma medicine, PAC, and plasma-assisted material processing are the most important and active fields of research and application; however, potential of CRDS in these fields has not been fully reached.

One of the current challenges in the new field of plasma medicine is to understand mechanisms of the plasma interactions with medical subjects, such as tissues, bacteria, DNA, wounding surfaces, liquids, and medical tools [205]. Numerous evidences are showing inarguable effects of low-temperature nonthermal plasma treatments and a few key radicals and plasma agents have been identified as the dominant plasma agents which play a key role in the plasma treatment of biomaterials. For instance, it is widely accepted that O_2, O, NO, and OH all play a role in the plasma treatments. However, a detailed picture of reaction kinetics for plasma treatments still does not exist yet. This scientific gap hinders the further advancement of the plasma medicine to a new height. Only complete understanding of the nonthermal plasma treatment reaction mechanisms and kinetic pathways at the molecular and bimolecular levels, can make a qualitative jump happen in the new field of plasma medicine. All chemical reactions are related by reactants and products, some of them or most of them are reaction intermediates whose lifetimes are short due to high reactivity. Therefore, knowing the number densities of the reaction intermediates is like clearing the bottleneck step toward understanding reaction mechanisms of plasma medicine.

The role of CRDS in plasma biomedical treatments can be classified into two groups. One is the CRDS-based plasma diagnostics which will lead to an optimal plasma design and plasma operation in terms of plasma-generated radical sources. The other is on-line monitoring of the plasma treatment processing, in which adjustment of a plasma species dosage in terms of number density per volume can be guided by near real-time measurements of the plasma species *in situ*. Currently, the study of plasma medicine is facilitated by using OES, which offers a convenient way to show both presence and type of plasma species; and even relative concentrations of the excited state plasma species can be estimated. In most cases, gas temperature of a nonthermal plasma used in the plasma medicine is very low, even close to room temperature, which is significantly lower than electronic excitation temperature of a plasma species. Consequently, when the OES is not able to detect emissions of the plasma species, no spectroscopic information can be obtained from the OES measurements. For instance, due to short lifetimes of OH at atmospheric pressure, it is often assumed that OH-assisted plasma treatments can only be executed using the bright part of a plasma jet. However, recently high-sensitivity ringdown measurements of OH shows that OH radicals can exist in a far distance from the bright part of the atmospheric plasma jet. This observation implies that the dark part of the plasma jet can be also utilized for OH-assisted plasma treatments. Therefore, CRDS measurements of the ground state of plasma species, which are most likely as chemically reactive as the excited states of the species, would offer an alternative yet an essential approach leading to a better understanding and control of plasma treatments.

Another increasingly interesting area of plasma interactions with media is PAC, which is an energetically prospective and highly promising technology to enhance combustion performance [206–208]. Similar to plasma medicine, many rock-solid evidences have shown the effects of the plasma enhancement of combustion in terms of shortening of ignition delay time (IDT), reduction of pollutant emissions, improvement of fuel efficiency, and enhancement of combustion flame holding.

To date, various types of low-temperature nonthermal plasmas [209,210], including silent discharge, corona discharge, DBD, ns-pulsed discharge, RF discharge, microwave discharge, MPT, fast ionization wave, and DC glow discharge, have been implemented in PAC technology. The entire combustion process may be categorized into three major stages: (1) pre-ignition, (2) ignition, and (3) post-ignition combustion processes. In each stage, PAC has been demonstrated to be advantageous in each of the aforementioned areas over conventional combustion (without assistance of a plasma). For instance, combustion with plasma pretreated fuel mixtures significantly increases combustion efficiency and reduces NO emissions [211]. Theoretical modeling shows that IDT can be reduced from tens of milliseconds in a typical thermal ignition to a few hundreds of nanoseconds in a plasma-assisted ignition under extreme conditions [212]. A high-voltage, repetitively pulsed, nanosecond DBD can result in combustion flame stabilization with up to 50% higher coflow speeds; a single electrode corona discharge between a platinum electrode and the flame base has been demonstrated to maintain the flame at a 20% higher coflow speed [213]. An MPT can increase flame volume significantly and has high-energy transfer efficiency [214]. Numerous examples can be further listed to show this promising PAC technology [215,216].

Despite the significant strides made in the technological developments of PAC during the past decade, several fundamentally important attributes of PAC have not been thoroughly investigated. For example, questions, such as how plasma enhances the combustion performance and what are the reaction pathways and detailed reaction mechanisms in PACs, are not clearly answered.

Understanding the combustion processes relies on a better description of the coupling between chemistry, fluid mechanics, and heat/mass transfer. However, understanding PAC also requires plasma physics because the development of PAC technologies is determined in many aspects by a degree of insight into the processes occurring in the plasma. For example, in PAC, the plasma provides sources of reactive species, which rely on the impact of energetic electrons from the plasma to induce dissociation, ionization, and vibrational and electronic excitation of the parent fuel molecules. All combustion processes rely on the breakdown of the fuel being burned into free radicals, ions, and excited neutral species, which initiate the combustion process (a chain of chemical reactions). However, current understanding of the roles of reactive species (radicals, ions, and excited neutral species) in PACs is still in debate, and the rate constants and kinetic mechanisms of the reactions involving radicals as well as other reactive species remain little known, to a large extent, due to the challenge of obtaining experimental data of the absolute number densities of reactants/products of the reactions pre-, during, and post-ignition. This scientific gap, in part, results in an incomplete understanding of PAC for further technological developments and applications. The quantitative information regarding radicals and other reactive species that physically and chemically link the plasma source with the entire combustion processes may offer a unique bottleneck channel toward better understanding of PAC. In this regard, CRDS measurements of plasma and combustion intermediates will help advance PAC technology.

Finally, plasma-assisted material processing has long been an interesting field for plasma applications. In the nonthermal plasma-enhanced CVD [217,218], plasma injects electrical energy into a molecular system (a feedstock gas) to initiate chemical reaction through electron impact dissociation, followed by fragmentation, ionization, and a chain of chemical reactions among the precursor molecules, fragments, free radicals, neutral excited species, ions, and energetic electrons. In a nonthermal plasma, T_e is typically of 2–5 eV, which is sufficient for dissociation of feedstock gas molecules; plasma gas kinetic temperature can be as low as room temperature. Highly energetic electrons collide with relatively cool neutral gases with little waste of enthalpy. Additionally, the low plasma gas temperature has no adverse effect of high-temperature thermal CVD on the processing environment. For example, in the deposition of silicon nitride for the insulating capping, nonthermal plasma-enhanced CVD can be performed at temperature near 300°C while nonplasma CVD would require temperature of 900°C, at which an aluminum-made device will be melt.

Nonthermal plasmas used in plasma-enhanced CVD include RF discharge, microwave discharge, and DC or AC discharge [219,220]. The presence of chemically reactive species at low-gas kinetics temperature is the most significant characteristic of low-temperature nonthermal plasmas. The common chemical species present in atmospheric pressure plasmas are reactive nitrogen species, such as N_2, N_2^+, NH, and CN; and reactive oxygen species, such as O, O_2, O_3, NO, and OH. Other radicals include CH, N, Ar, He, and H, depending on the plasma gases to be used. At reduced pressure, presence of trace organic compounds and trace solvents will result in generation of OH [221,222]. In many low-temperature nonthermal plasma applications, such as plasma activation of CVD, reactive oxygen species are considered to play a critical role. For instance, in the plasma-enhanced metal organic CVD of ZnO, the effect of the carrier gas on deposition rate was proven to be related to the occurrence of OH radicals; and OH and O radicals did promote the film growth [223]. Knowing absolute concentrations of reactive plasma species, including OH and O, in time and in space, is a key step toward understanding of the formation mechanism of the plasma species. The determination of OH radical concentrations will help optimize and control plasma operation and adjust the treatment dosage of the species. In addition, metal organic CVD processes create abundant metal elements, such as Y, Gd, Ba, and Cu, depending on precursors [224]. Knowledge of absolute concentrations of these elements will help optimize the ratio of plasma gas/precursor vapor; thus, the efficiency of precursor use can be maximized. This is another promising field where CRDS begins to play its role.

8.8 Summary

CRDS is a versatile laser spectroscopic technique. Its unique advantages in plasma applications are high sensitivity, absolute number density measurement with self-calibration, and high temporal resolution. The line-of-sight density measurement using CRDS can be converted to point measurement using Abel inversion process to achieve high spatial resolution. In addition to the absolute number density measurements of plasma species, CRDS can also be implemented for plasma diagnostics through high-resolution scanning of a lineshape, from which linewidth broadening components are used to determine plasma gas kinetic temperature and plasma electron density. When CRDS is employed with OES and high-resolution (time and spatial) imaging, thorough plasma diagnostics can be achieved. In special cases, where OES becomes blind due to weak or no plasma emissions in some locations of plasma, CRDS can still serve as a powerful tool to obtain information on absolute number density as well as identity of plasma species. To date, more than 30 species including free radicals, atoms, ions, and isotopes, which are generated in plasmas and discharged, have been measured by CRDS. With growth of plasma application fields, especially, the emerging fields of plasma medicine, PAC, and plasma-enhanced material processing, CRDS will play an increasingly important role for years to come.

Acknowledgments

This work is supported by the National Science Foundation through the grant CBET-1066486. The author thanks Zhennan Wang and Wei Wu for assistance in the preparation of the manuscript.

References

1. O'Keefe A. and D. A. G. Deacon. 1988. Cavity ring-down optical spectrometer for absorption measurements using pulsed laser sources. *Rev. Sci. Instrum.* 59:2544–51.
2. Pipino, A. C. R., J. W. Hudgens, and R. E. Hule. 1997. Evanescent cavity ring-down spectroscopy with a total-internal-reflection minicavity. *Rev. Sci. Instrum.* 68:2978–89.
3. Lehmann K. K. and P. Rabinowitz. 2000. High-finesse optical resonator for cavity ring-down spectroscopy based upon Brewster's angle prism retroreflectors. US Patent No. 6,097,555.

4. Gupta, M., H. Jiao, and A. O'Keefe. 2002. Cavity-enhanced spectroscopy in optical fibers. *Opt. Lett.* 27:1878–80.

5. Tarsa, P. B., P. Rabinowitz, and K. K. Lehmann. 2002. Passive optical fiber resonator for cavity ring-down spectroscopy. *Paper Presented at the 224 ACS National Meeting.* Boston, MA, August 18–22.

6. Brown, R. S., L. Kozin, Z. Tong, R. D. Oleschuk, and H.-P. Loock. 2002. Fiber-loop ring-down spectroscopy. *J. Chem. Phys.* 117:10444–47.

7. Scherer, J. J., D. Voelkel, D. J. Rakestraw, J. B. Paul, C. P. Collier, R. J. Saykally, and A. O'Keefe. 1995. Infrared cavity ringdown laser absorption spectroscopy (IR-CLAS). *Chem. Phys. Lett.* 245:273–80.

8. Hallock, A. J., E. S. F. Berman, and R. N. Zare. 2003. Use of broadband, continuous-wave diode lasers in cavity ring-down spectroscopy for liquid samples. *Appl. Spectrosc.* 57:571–73.

9. Paldus, B. A., C. C. Harb, T. G. Spence, et al. 2000. Cavity ringdown spectroscopy using mid-infrared quantum-cascade lasers. *Opt. Lett.* 25:666–68.

10. Paul, J. B., J. J. Scherer, A. O'Keefe, et al. 2002. Infrared cavity ringdown and integrated cavity output spectroscopy for trace species monitoring. *Proc. SPIE.* 4577:1–11.

11. Romanini, D., J. Gambogi and K. K. Lehmann. 1995. Cavity ring down spectroscopy with CW diode laser excitation. *Proceedings of the 50th International Symposium on Molecular Spectroscopy,* ed. T. A. Miller. Columbus, OH: Department of Chemistry, Ohio State University, p. 284.

12. Romanini, D., A. A. Kachanov, and F. Stoeckel. 1997. Cavity ring-down spectroscopy: Broad band absolute absorption measurements. *Chem. Phys. Lett.* 270:546–50.

13. He, Y. and B. J. Orr. 2002. Rapid-swept, continuous-wave cavity ringdown spectroscopy with optical heterodyne detection: Single- and multi-wavelength sensing of gases. *Appl. Phys. B* 75:267–80.

14. Zare, R. N., J. Martin, and B. A. Paldus. 1998. Deflecting light into resonant cavities for spectroscopy. US Patent No. 5,815,277.

15. Augustine, R., C. R. Krusen, C. Wang, and W. B. Yan. 2007. System and method for controlling a light source for cavity ring-down spectroscopy. US Patent No. 7,277,177 B2.

16. Koirala, S. P. 2003. Plasma cavity ringdown spectroscopy—A powerful technique for elemental measurements and plasma diagnostics. MS thesis, Mississippi State University.

17. Wang, C., S. T. Scherrer, and C. B. Winstead. 2001. A simple method and device for control of cavity energy buildup and shutoff in cw-cavity ringdown spectroscopy: application for ringdown measurements of atmospheric CH_4, CO_2, and H_2O at 1.65 µm. The US DOE-DIAL 40395-13 Report, "Instrumentation Development, Measurement and Performance Evaluation of Environmental Technologies" by Diagnostic Instrumentation and Analysis Laboratory, Mississippi State University. October 2001.

18. Miller, G. P. and C. B. Winstead. 1997. Inductively coupled plasma cavity ringdown spectrometry. *J. Anal. At. Spectrom.* 12:907–12.

19. Winstead, C. B., F. J. Mazzotti, J. Mierzwa, and G. P. Miller. 1999. Preliminary results for electrothermal atomization-cavity ringdown spectroscopy (ETA-CRDS). *Anal. Commun.* 36:277–79.

20. Wang, C., F. J. Mazzotti, G. P. Miller, and C. B. Winstead. 2002. Cavity ringdown spectroscopy for diagnostic and analytical measurements in an inductively coupled plasma. *Appl. Spectrosc.* 56:386–97.

21. Wang, C., F. J. Mazzotti, G. P. Miller, and C. B. Winstead. 2003. Isotopic measurements of uranium using inductively coupled plasma cavity ringdown spectroscopy. *Appl. Spectrosc.* 57:1167–72.

22. Wang, C., F. J. Mazzotti, J. Mierzwa, G. P. Miller, Y. Duan, and C. B. Winstead. 2002. Cavity ringdown spectroscopy in atmospheric pressure plasma applications. *Paper Presented at the 53rd Pittsburgh Conference.* New Orleans, LA, March 17–22.

23. Duan, Y., C. Wang, and C. B. Winstead. 2003. Exploration of microwave plasma source cavity ring-down spectroscopy for elemental measurements. *Anal. Chem.* 75:2105–11.

24. Bechtel, K. L., R. N. Zare, A. A. Kachanov, S. S. Sanders, and B. A. Paldus. 2005. Moving beyond traditional UV-visible absorption detection: Cavity ring-down spectroscopy for HPLC. *Anal. Chem.* 77:1177–82.

25. Snyder, K. L. and R. N. Zare. 2003. Cavity ring-down spectroscopy as a detector for liquid chromatography. *Anal. Chem.* 75:3086–91.

26. van der Sneppen, L., A. E. Wiskerke, F. Ariese, C. Gooijer, and W. Ubachs. 2006. Improving the sensitivity of HPLC absorption detection by cavity ring-down spectroscopy in a liquid-only cavity. *Anal. Chim. Acta* 558:2–6.

27. van der Sneppen, L., A. E. Wiskerke, F. Ariese, and C. Gooijer. 2006. Cavity ring-down spectroscopy for detection in liquid chromatography: Extension to tunable sources and ultraviolet wavelengths. *Appl. Spectrosc.* 60:931–35.

28. Dudek, J. B., P. B. Tarsa, A. Velasquez, M. Wladyslawski, P. Rabinowitz, and K. K. Lehmann. 2003. Trace moisture detection using continuous-wave cavity ring-down spectroscopy. *Anal. Chem.* 75:4599–605.

29. Vasudev, R., A. Usachev, and W. R. Dunsford. 1999. Detection of toxic compounds by cavity ring-down spectroscopy. *Environ. Sci. Technol.* 33:1936–39.

30. Mazurenka, M. I., B. L. Fawcett, J. M. F. Elks, D. E. Shallcross, and A. J. Orr-Ewing. 2003. 410-nm diode laser cavity ring-down spectroscopy for trace detection of NO_2. *Chem. Phys. Lett.* 367:1–9.

31. Fawcett, B. L., A. M. Parkes, D. E. Shallcross, and A. J. Orr-Ewing. 2002. Trace detection of methane using continuous wave cavity ring down spectroscopy at 1.65 μm. *Phys. Chem. Chem. Phys.* 4:5960–65.

32. Awtry, A. R. and J. H. Miller. 2002. Development of a CW-laser-based cavity-ringdown sensor aboard a spacecraft for trace air constituents. *Appl. Phys. B* 75:255–60.

33. Bakhirkin, Y. A., A. A. Kosterev, R. F. Curl, et al. 2006. Sub-ppbv nitric oxide concentration measurements using CW thermoelectrically cooled quantum cascade laser-based integrated cavity output spectroscopy. *Appl. Phys. B* 82:149–54.

34. Kosterev, A. A., A. L. Malinovsky, F. K. Tittel, et al. 2001. Cavity ringdown spectroscopic detection of nitric oxide with a continuous-wave quantum-cascade laser. *Appl. Opt.* 40:5522–29.

35. Wang, C., N. Srivastava, B. A. Jones, and R. B. Reese. 2008. A novel multiple species ringdown spectrometer for *in situ* measurements of methane, carbon dioxide, and carbon isotope. *Appl. Phys. B* 92:259–70.

36. Crosson, E. R. 2008. A cavity ring-down analyzer for measuring atmospheric levels of methane, carbon dioxide, and water vapor. *Appl. Phys. B* 92:403–08.

37. Schwabedissen, A., A. Brockhaus, A. Georg, and J. Engemann. 2001. Determination of the gas-phase Si atom density in radio frequency discharges by means of cavity ring-down spectroscopy. *J. Phys. D Appl. Phys.* 34:1116–21.

38. Yalin, A. P., V. Surla, M. Butweiller, and J. D. Williams. 2005. Detection of sputtered metals with cavity ring-down spectroscopy. *Appl. Opt.* 44:6496–505.

39. Surla, V., P. J. Wilbur, M. Johnson, J. D. Williams, and A. P. Yalin. 2004. Sputter erosion measurements of titanium and molybdenum by cavity ringdown spectroscopy. *Rev. Sci. Instrum.* 75:3025–30.

40. Booth, J. P., G. Cunge, L. Biennier, D. Romanini, and A. Kachanov. 2000. Ultraviolet cavity ring-down spectroscopy of free radicals in etching plasmas. *Chem. Phys. Lett.* 317:631–36.

41. Wills, J. B., J. A. Smith, W. E. Boxford, J. M. F. Elks, M. N. R. Ashfold, and A. J. Orr-Ewing. 2002. Measurements of C_2 and CH concentrations and temperatures in a DC arc jet using cavity ring-down spectroscopy. *J. Appl. Phys.* 92:4213–22.

42. Wahl, E. H., T. G. Owano, C. H. Kruger, P. Zalicki, Y. Ma, and R. N. Zare. 1996. Measurement of absolute CH_3 concentration in a hot-filament reactor using cavity ring-down spectroscopy. *Diam. Relat. Mater.* 5:373–77.

43. Evertsen, R., R. L. Stolk, and J. J. ter Meulen. 2000. Investigations of cavity ring down spectroscopy applied to the detection of CH in atmospheric flames. *Combust. Sci. Technol.* 157:341–42.

44. Zalicki, P., Y. Ma, R. N. Zare, E. H. Wahl, T. G. Owano, and C. H. Kruger. 1995. Measurement of the methyl radical concentration profile in a hot-filament reactor. *Appl. Phys. Lett.* 67:144–46.

45. Kessels, W. M. M., A. Leroux, M. G. H. Boogaarts, J. P. M. Hoefhagels, M. C. M. van de Sanden, and D. C. Schram. 2001. Cavity ring down detection of SiH_3 in a remote SiH_4 plasma and comparison with model calculations and mass spectrometry. *J. Vac. Sci. Technol. A* 19:467–76.

46. Smith, J. A., J. B. Wills, H. S. Moores, et al. 2002. Effects of NH_3 and N_2 additions to hot filament activated CH_4/H_2 gas mixtures. *J. Appl. Phys.* 92:672–81.

47. Bakowski, B., G. Hancock, R. Peverall, G. A. D. Ritchie, and L. J. Thornton. 2004. Characterization of an inductively coupled N_2 plasma using sensitive diode laser spectroscopy. *J. Phys. D Appl. Phys.* 37:2064–72.

48. Wang, C., N. Srivastava, S. Scherrer, P. R. Jang, T. S. Dibble, and Y. Duan. 2009. Optical diagnostics of a low power–low gas flow rates atmospheric-pressure argon plasma created by a microwave plasma torch. *Plasma Sources Sci. Technol.* 18:025030–41.

49. Wang, C., N. Srivastava, and T. S. Dibble. 2009. Observation and quantification of OH radicals in the far downstream part of an atmospheric microwave plasma jet using cavity ringdown spectroscopy. *Appl. Phys. Lett.* 95:051501–3.

50. Srivastava, N. and C. Wang. 2011. Determination of OH radicals in an atmospheric pressure helium microwave plasma jet. *IEEE Trans. Plasma Sci.* 39:918–24.

51. Wang, C., S. P. Koirala, S. T. Scherrer, Y. Duan, and C. B. Winstead. 2004. Diode laser microwave induced plasma cavity ringdown spectrometer: Performance and perspective. *Rev. Sci. Instrum.* 75:1305–13.

52. Wang, C., S. T. Scherrer, Y. Duan, and C. B. Winstead. 2005. Cavity ringdown measurements of mercury and its hyperfine structures at 254 nm in an atmospheric microwave plasma: Spectral interference and analytical performance. *J. Anal. At. Spectrom.* 20:638–44.

53. Duan, Y., C. Wang, S. T. Scherrer, and C. B. Winstead. 2005. Development of alternative plasma sources for cavity ring-down measurements of mercury. *Anal. Chem.* 77:4883–89.

54. Crosson, E. R., K. N. Ricci, B. A. Richman, et al. 2002. Stable isotope ratios using cavity ring-down spectroscopy: Determination of $^{13}C/^{12}C$ for carbon dioxide in human breath. *Anal. Chem.* 74:2003–07.

55. Wang, C., S. T. Scherrer, and D. Hossain. 2004. Measurements of cavity ringdown spectroscopy of acetone in the ultraviolet and near-infrared spectral regions: Potential for development of a breath analyzer. *Appl. Spectrosc.* 58:784–91.

56. Mürtz, M., D. Halmer, M. Horstjann, S. Thelen, and P. Hering. 2006. Ultra sensitive trace gas detection for biomedical applications. *Spectrochim. Acta A Mol. Biomol. Spectrosc.* 63:963–69.

57. Wang, C. and A. B. Surampudi. 2008. An acetone breath analyzer using cavity ringdown spectroscopy: An initial test with human subjects under various situations. *Meas. Sci. Technol.* 19:105604–10.

58. Wang, C., A. Mbi, and M. Shepherd. 2010. A study on breath acetone in diabetic patients using a cavity ringdown breath analyzer: Exploring correlations of breath acetone with blood glucose and glycohemoglobin A1C. *IEEE Sens. J.* 10:54–63.

59. Wang, C. and S. T. Scherrer. 2004. Fiber ringdown pressure sensors. *Opt. Lett.* 29:352–54.

60. Wang, C. and S. T. Scherrer. 2004. Fiber loop ringdown for physical sensor development: Pressure sensor. *Appl. Opt.* 43:6458–64.

61. Tarsa, P. B., A. D. Wist, P. Rabinowitz, and K. K. Lehmann. 2004. Single-cell detection by cavity ringdown spectroscopy. *Appl. Phys. Lett.* 85:4523–25.

62. Tong, Z., A. Wright, T. McCormick, R. Li, R. D. Oleschuk, and H.-P. Loock. 2004. Phase-shift fiber-loop ring-down spectroscopy. *Anal. Chem.* 76:6594–99.

63. Tarsa, P. B., P. Rabinowitz, and K. K. Lehmann. 2004. Evanescent field absorption in a passive optical fiber resonator using continuous-wave cavity ring-down spectroscopy. *Chem. Phys. Lett.* 383:297–303.

64. Wang, C. 2005. Fiber ringdown temperature sensors. *Opt. Eng.* 44:030503–2.

65. Loock, H.-P. 2006. Ring-down absorption spectroscopy for analytical microdevices. *Trends Anal. Chem.* 25:655–64.

66. Wang, C. and A. Mbi. 2006. Optical superposition in double fiber loop ringdown. *Proc. SPIE.* 6377:637702–8.

67. Wang, C. 2009. Fiber loop ringdown—A time-domain sensing technique for multi-function fiber optic sensor platforms: Current status and design perspectives. *Sensors* 9:7595–621.

68. Waechter, H., J. Litman, A. H. Cheung, J. A. Barnes, and H.-P. Loock. 2010. Chemical sensing using fiber cavity ring-down spectroscopy. *Sensors* 10:1716–42.

69. Wang, C. and C. Herath. 2010. High-sensitivity fiber-loop ringdown evanescent-field index sensors using single-mode fiber. *Opt. Lett.* 35:1629–31.

70. Thorpe, M. J., K. D. Moll, R. J. Jones, B. Safdi, and J. Ye. 2006. Broadband cavity ringdown spectroscopy for sensitive and rapid molecular detection. *Science* 311:1595–99.

71. Thorpe, M. J., D. D. Hudson, K. D. Moll, J. Lasri, and J. Ye. 2007. Cavity-ringdown molecular spectroscopy based on an optical frequency comb at 1.45-1.65 µm. *Opt. Lett.* 32:307–09.

72. Pipino, A. C. R., J. W. Hudgens, and R. E. Huie. 1997. Evanescent wave cavity ring-down spectroscopy for probing surface processes. *Chem. Phys. Lett.* 280:104–12.

73. Lehmann, K. K., P. S. Johnston, and P. Rabinowitz. 2009. Brewster angle prism retroreflectors for cavity enhanced spectroscopy. *Appl. Opt.* 48:2966–78.

74. Busch, K. W. and M. A. Busch. 1999. *ACS Symposium Series 720: Cavity-Ringdown Spectroscopy—An Ultratrace-Absorption Measurement Technique.* Oxford: Oxford University Press.

75. Berden, G., R. Peeters, and G. Meijer. 2000. Cavity ring-down spectroscopy: Experimental schemes and applications. *Int. Rev. Phys. Chem.* 19:565–607.

76. Scherer, J. J., J. B. Paul, A. O'Keefe, and R. J. Saykally. 1997. Cavity ringdown laser absorption spectroscopy: History, development, and application to pulsed molecular beams. *Chem. Rev.* 97:25–51.

77. Miller, G. P. and C. B. Winstead. 2000. Cavity ringdown laser absorption spectroscopy. In *Encyclopedia of Analytical Chemistry: Instrumentation and Applications*, ed. R. A. Meyers. Chichester: John Wiley & Sons Ltd., pp. 10734–50.

78. Wagner, D. R., G. P. Miller, and C. B. Winstead. 2002. Infrared cavity ringdown spectroscopy. In *Handbook of Vibrational Spectroscopy*, ed. J. M. Chalmers and P. R. Griffiths. Chichester: John Wiley & Sons Ltd., pp. 866–80.

79. Atkinson, D. B. 2003. Solving chemical problems of environmental importance using cavity ring-down spectroscopy. *Analyst* 128:117–25.

80. Vallance, C. 2005. Innovations in cavity ringdown spectroscopy. *New J. Chem.* 97:867–74.

81. Paldus, B. A. and A. A. Kachanov. 2005. An historical overview of cavity-enhanced methods. *Can. J. Phys.* 83:975–99.

82. Mazurenka, M., A. J. Orr-Ewing, R. Peverall, and G. A. D. Ritchie. 2005. Cavity ring-down and cavity enhanced spectroscopy using diode lasers. *Annu. Rep. Prog. Chem. Sect. C Phys. Chem.* 101:100–42.

83. Wang, C. 2007. Plasma-cavity ringdown spectroscopy (P-CRDS) for elemental and isotopic measurements. *J. Anal. At. Spectrom.* 22:1347–63.

84. Orr, B. J. and Y. He. 2011. Rapidly swept continuous-wave cavity-ringdown spectroscopy. *Chem. Phys. Lett.* 512:1–20.

85. Berden, G. and R. Engeln. 2009. *Cavity Ring-Down Spectroscopy: Techniques and Applications.* West Sussex: Wiley-Blackwell Publishing Ltd.

86. van Zee, R. D., J. T. Hodges, and J. P. Looney. 1999. Pulsed, single-mode cavity ringdown spectroscopy. *Appl. Opt.* 38:3951–60.

87. Romanini, D. and K. K. Lehmann. 1995. Cavity ringdown overtone spectroscopy of HCN, H^{13}CN and HC^{15}N. *Chem. Phys.* 102:633–42.

88. Romanini, D., A. A. Kachanov, N. Sadeghi, and F. Stoeckel. 1997. CW cavity ring down spectroscopy. *Chem. Phys. Lett.* 264:316–22.

89. Yang, P. and R. M. Barnes. 1989. Modification of Abel inversion technique for line width calculation. *Spectrochim. Acta B At. Spectrosc.* 44:561–70.

90. Wang, C., F. J. Mazzotti, S. P. Koirala, C. B. Winstead, and G. P. Miller. 2004. Measurements of OH radicals in a low-power atmospheric inductively coupled plasma by cavity ringdown spectroscopy. *Appl. Spectrosc.* 58:734–40.

91. Baer, D. S. and R. K. Hanson. 1992. Tunable diode laser absorption diagnostics for atmospheric pressure plasmas. *J. Quant. Spectrosc. Radiat. Transf.* 47:455–75.

92. van der Mullen, J. A. M. 1989. On the atomic state distribution function in inductively coupled plasmas—I. Thermodynamic equilibrium considered on the elementary level. *Spectrochim. Acta B At. Spectrosc.* 44:1067–80.

93. Laux, C. O., T. G. Spence, C. H. Kruger, and R. N. Zare. 2003. Optical diagnostics of atmospheric pressure air plasmas. *Plasma Sources Sci. Technol.* 12:125–38.

94. Siegman, A. E. 1986. Lasers. Mill Valley, CA: University Science Books.

95. Kirkbright, G. F. and M. Sargent. 1974. *Atomic Absorption and Fluorescence Spectroscopy*. London: Academic Press.

96. Breene, R. G. 1961. *The Shift and Shape of Spectral Lines*. Oxford: Pergamon Press.

97. Eckert, H. V. 1977. *Physics and Technology of Low-Temperature Plasma*, English edition. Ames, IA: Iowa State University Press.

98. Berden, G., R. Engeln, P. C. M. Christianen, J. C. Maan, and G. Meijer. 1998. Cavity ring down spectroscopy on the oxygen A band in magnetic fields up to 20 T. *Phys. Rev. A At. Mol. Opt. Phys.* 58:3114–23.

99. Le Grand, Y. and A. Le Floch. 1990. Sensitive dichroism measurements using eigenstate decay times. *Appl. Opt.* 29:1244–46.

100. Stewart, G., P. Shields, and B. Culshaw. 2004. Development of fibre laser systems for ring-down and intracavity gas spectroscopy in the near-IR. *Meas. Sci. Technol.* 15:1621–28.

101. Scherer, J. J., D. Voelkel, D. J. Rakestraw, et al. 1995. Infrared cavity ringdown laser absorption spectroscopy (IR-CRLAS). *Chem. Phys. Lett.* 245:273–80.

102. Martin, J., B. A. Paldus, P. Zalicki, et al. 1996. Cavity ring-down spectroscopy with fourier-transform-limited light pulses. *Chem. Phys. Lett.* 258:63–70.

103. Hodges, J. T., J. P. Looney, and R. D. van Zee. 1996. Laser bandwidth effects in quantitative cavity ring-down spectroscopy. *Appl. Opt.* 35:4112–16.

104. Lehmann, K. K. and D. Romanini. 1996. The superposition principle and cavity ring-down spectroscopy. *J. Chem. Phys.* 105:10263–77.

105. Hodges, J. T., J. P. Looney, and R. D. van Zee. 1996. Response of a ring-down cavity to an arbitrary excitation. *J. Chem. Phys.* 105:10278–88.

106. O'Shea, D. C., W. R. Callen, and W. T. Rhodes. 1977. *Introduction to Lasers and Their Applications*. Reading, MA: Addison-Wesley Publishing Company.

107. Meijer, G., M. G. H. Boogaarts, R. T. Jongma, D. H. Parker, and A. M. Wodtke. 1994. Coherent cavity ring down spectroscopy. *Chem. Phys. Lett.* 217:112–16.

108. O'Keefe A. and O. Lee. 1989. Trace gas analysis by pulsed laser absorption spectroscopy. *Am. Lab.* 21:19–24.

109. Koplow, J. P., D. A. V. Kliner, and L. Goldberg. 1998. Development of a narrow-band, tunable, frequency-quadrupled diode laser for UV absorption spectroscopy. *Appl. Opt.* 37:3954–60.

110. Pearson, J., A. J. Orr-Ewing, M. N. R. Ashfold, and R. N. Dixon. 1997. Spectroscopy and predissociation dynamics of the Ã^{1}A" state of HNO. *J. Chem. Phys.* 106:5850–73.

111. Paul, J. B., C. P. Collier, R. J. Saykally, J. J. Scherer, and A. O'Keefe. 1997. Direct measurement of water cluster concentrations by infrared cavity ringdown laser absorption spectroscopy. *J. Phys. Chem.* 101:5211–14.

112. Engeln, R. and G. Meijer. 1996. A fourier transform cavity ring down spectrometer. *Rev. Sci. Instrum.* 67:2708–13.

113. Scherer, J. J. 1998. Ringdown spectral photography. *Chem. Phys. Lett.* 292:143–53.

114. Fiedler, S. E., A. Hese, and A. R. Ruth. 2005. Incoherent broad-band cavity-enhanced absorption spectroscopy of liquids. *Rev. Sci. Instrum.* 76:023107–13.

115. Engeln, R., E. van den Berg, G. Meijer, L. Lin, G. M. H. Knippels, and A. F. G. van der Meer. 1997. Cavity ring down spectroscopy with a free-electron laser. *Chem. Phys. Lett.* 269:293–97.

116. Engeln, R., G. von Helden, A. J. A. van Roij, and G. Meijer. 1999. Cavity ring down spectroscopy on solid C_{60}. *J. Chem. Phys.* 110:2732–33.

117. Wang, C., P. Sahay, and S. T. Scherrer. 2012. Electron impact excitation-cavity ringdown absorption spectroscopy of elemental mercury at 405 nm. *J. Anal. At. Spectrom.* 27:284–92.

118. Jongma, R. T., M. G. H. Boogaarts, I. Holleman, and G. Meijer. 1995. Trace gas detection with cavity ring down spectroscopy. *Rev. Sci. Instrum.* 66:2821–28.

119. Lock, J. A. and J. T. Hodges. 1996. Far-field scattering of an axisymmetric laser beam of arbitrary profile by an on-axis spherical particle. *Appl. Opt.* 35:4283–90.

120. Yalin, A. P. and R. N. Zare. 2002. Effect of laser lineshape on the quantitative analysis of cavity ring-down signals. *Laser Phys.* 12:1065–72.

121. Yalin, A. P., C. O. Laux, C. H. Kruger, and R. N. Zare. 2002. Spatial profiles of N_2^+ concentration in an atmospheric pressure nitrogen glow discharge. *Plasma Sources Sci. Technol.* 11:248–53.

122. Yalin, A. P., R. N. Zare, C. O. Laux, and C. H. Kruger. 2002. Temporally resolved cavity ring-down spectroscopy in a pulsed nitrogen plasma. *Appl. Phys. Lett.* 81:1408–10.

123. Engeln, R., G. von Helden, G. Berden, and G. Meijer. 1996. Phase shift cavity ring down absorpion spectroscopy. *Chem. Phys. Lett.* 262:105–09.

124. He, Y., M. Hippler, and M. Quack. 1998. High-resolution cavity ring-down absorption spectroscopy of nitrous oxide and chloroform using a near-infrared CW diode laser. *Chem. Phys. Lett.* 289:527–34.

125. Romanini, D., P. Dupré, and R. Jost. 1999. Non-linear effects by continuous wave cavity ringdown spectroscopy in jet-cooled NO_2. *Vib. Spectrosc.* 19:93–106.

126. Paldus, B. A., J. S. Harris Jr., J. Martin, X. Kie, and R. N. Zare. 1997. Laser diode cavity ring-down spectroscopy using acousto-optic modulator stabilization. *J. Appl. Phys.* 82:3199–204.

127. Paldus, B. A., C. C. Harb, T. G. Spence, et al. 1998. Cavity-locked ring-down spectroscopy. *J. Appl. Phys.* 83:3991–97.

128. Levenson, M. D., B. A. Paldus, T. G. Spence, C. C. Harb, J. S. Harris Jr., and R. N. Zare. 1998. Optical heterodyne detection in cavity ring-down spectroscopy. *Chem. Phys. Lett.* 290:335–40.

129. Schulz, K. J. and W. R. Simpson. 1998. Frequency-matched cavity ring-down spectroscopy. *Chem. Phys. Lett.* 297:523–29.

130. Hahn, J. W., Y. S. Yoo, J. Y. Lee, J. W. Kim, and H. W. Lee. 1999. Cavity ringdown spectroscopy with a continuous-wave laser: Calculation of coupling efficiency and a new spectrometer design. *Appl. Opt.* 38:1859–66.

131. Cias, P., C. Wang, and T. S. Dibble. 2007. Absorption cross-sections of the C-H overtone of volatile organic compounds: 2 Methyl-1,3-Butadiene (Isoprene), 1,3-Butadiene, and 2,3-Dimethyl-1,3-Butadiene. *Appl. Spectrosc.* 61:230–36.

132. Wang, C., N. Srivastava, J. Cambre, B. A. Jones, and R. B. Reese. 2007. Development of a portable ringdown spectrometer for greenhouse gases and carbon isotope. *Paper Presented at the 34th FACSS Annual Meeting.* Memphis, TN, October 14–18.

133. Romanini, D. and K. K. Lehmann. 1993. Ring-down cavity absorption spectroscopy of the very weak HCN overtone bands with six, seven, and eight stretching quanta. *J. Chem. Phys.* 99:6287–301.

134. Naus, H., I. H. M. van Stokkum, W. Hogervorst, and W. Ubachs. 2001. Quantitative analysis of decay transients applied to a multimode pulsed cavity ringdown experiment. *Appl. Opt.* 40:4416–26.

135. Wang, C., P. Sahay, and S. T. Scherrer. 2011. A new optical method of measuring electron impact excitation cross section of atoms: Cross section of the metastable 6s6p $^3P_0$ level of Hg. *Phys. Lett. A* 375:2366–70.

136. Stoffels, E., A. J. Flikweert, W. W. Stoffels, and G. M. W. Kroesen. 2002. Plasma needle: A non-destructive atmospheric plasma source for fine surface treatment of (bio)materials. *Plasma Sources Sci. Technol.* 11:383–88.

137. Laroussi, M. and X. Lu. 2005. Room-temperature atmospheric pressure plasma plume for biomedical application. *Appl. Phys. Lett.* 87:113902–03.

138. Pei, X., Z. Wang, Q. Huang, S. Wu, and X. Lu. 2011. Dynamics of a plasma jet array. *IEEE Trans. Plasma Sci.* 39:2276–77.

139. Dorn, H.-P., R. Neuroth, and A. Hofzumahaus. 1995. Investigation of OH absorption cross sections of rotational transitions in the $A^2\Sigma^+$, $\upsilon' = 0 \leftarrow X^2\Pi$, $\upsilon'' = 0$ band under atmospheric conditions: Implications for tropospheric long-path absorption measurements. *J. Geophys. Res.* 100:7397–409.

140. Goldman, A. and J. R. Gillis. 1981. Spectral line parameters for the $A^2\Sigma$-$X^2\Pi(0,0)$ band of OH for atmospheric and high temperatures. *J. Quant. Spectrosc. Radiat. Transf.* 25:111–35.

141. Gardiner, W. C. 1984. *Combustion Chemistry.* New York: Springer-Verlag.

142. Crosley, D. R. 1989. Semiquantitative laser-induced fluorescence in flames. *Combust. Flame* 78:153–67.

143. Linteris, G. T., V. D. Knyazev, and V. I. Babushok. 2002. Inhibition of premixed methane flames by manganese and tin compounds. *Combust. Flame* 129:221–38.

144. Rea, E. C., A. Y. Chang, and R. K. Hanson. 1987. Shock-tube study of pressure broadening of the $A^2\Sigma^+$-$X^2\Pi(0,0)$ band of OH by Ar and N_2. *J. Quant. Spectrosc. Radiat. Transf.* 37:117–27.

145. Laux, C. O., C. H. Kruger, and R. N. Zare. 2001. Diagnostics of atmospheric pressure air plasmas. *Paper Presented at Arbeitsgemeinschaft Plasma Physik (APP) Spring Meeting.* Bad Honnef, Germany, February 18–21.

146. Shin, D. N., C. W. Park, and J. W. Hahn. 2000. Detection of OH($A^2\Sigma^+$) and O(^{1}D) emission spectrum generated in a pulsed corona plasma. *Bull. Korean Chem. Soc.* 21:228–32.

147. Laroussi, M. 2009. Low-temperature plasmas for medicine? *IEEE Trans. Plasma Sci.* 37:714–25.

148. Lu, X., Z. Xiong, F. Zhao, et al. 2009. A simple atmospheric pressure room-temperature air plasma needle device for biomedical application. *Appl. Phys. Lett.* 95:181501–3.

149. Kong, M.G., G. Kroesen, G. Morfill, et al. 2009. Plasma medicine: An introductory review. *New J. Phys.* 11:115012–35.

150. Stoffels, E., Y. Sakiyama, and D. B. Graves. 2008. Cold atmospheric plasma: Charged species and their interactions with cells and tissues. *IEEE Trans. Plasma Sci.* 36:1441–57.

151. Lu, X., T. Ye, Y. Cao, et al. 2008. The roles of the various plasma agents in the inactivation of bacteria. *J. Appl. Phys.* 104:053309–5.

152. Spaanjaars, J. J. L., J. J. ter Meulen, and G. Meijer. 1997. Relative predissociation rates of OH ($A^2\Sigma^+$, $\upsilon' = 3$) from combined cavity ring down–laser-induced fluorescence measurements. *J. Chem. Phys.* 107:2242–48.

153. Mcllroy, A. 1999. Laser studies of small radicals in rich methane flames: OH, HCO, and $^1CH_2$. *Isr. J. Chem.* 39:55–62.

154. Greet, P. A., W. J. R. French, G. B. Burns, P. F. B. Williams, R. P. Lowe, and K. Finlayson. 1998. OH(6-2) spectra and rotational temperature measurements at Davis, Antarctica. *Ann. Geophys.* 16:77–89.

155. Luque, J. and D. R. Crosley. 1998. Transition probabilities in the $A^2\Sigma^+$-$X^2\Pi$ electronic system of OH. *J. Chem. Phys.* 109:439–48.

156. Carlone, C. and F. W. Dalby. 1969. Spectrum of the hydroxyl radical. *Can. J. Phys.* 47:1945–57.

157. Cheskis, S., I. Derzy, V. A. Lozovsky, A. Kachanov, and D. Romanini. 1998. Cavity ring-down spectroscopy of OH radicals in low pressure flame. *Appl. Phys. B* 66:377–81.

158. Lozovsky, V. A., I. Derzy, and S. Cheskis. 1998. Nonequilibrium concentrations of the vibrationally excited OH radical in a methane flame measured by cavity ring-down spectroscopy. *Chem. Phys. Lett.* 284:407–11.

159. Mercier, X., E. Therssen, J. F. Pauwels, and P. Desgroux. 1999. Cavity ring-down measurements of OH radical in atmospheric premixed and diffusion flames: A comparison with laser-induced fluorescence and direct laser absorption. *Chem. Phys. Lett.* 299:75–83.

160. Luque, J. 2003. LIFBASE, Windows Version 1.9.130.

161. Srivastava, N., C. Wang, and T. S. Dibble. 2009. A study of OH radicals in an atmospheric AC discharge plasma using near infrared diode laser cavity ringdown spectroscopy combined with optical emission spectroscopy. *Eur. Phys. J. D* 54:77–86.

162. Liu, Z., X. Yang, A. Zhu, G. Zhao, and Y. Xu. 2008. Determination of the OH radical in atmospheric pressure dielectric barrier discharge plasmas using near infrared cavity ring-down spectroscopy. *Eur. Phys. J. D* 48:365–73.

163. www.hitran.com

164. Srivastava, N. and C. Wang. 2011. Effects of water addition on OH radical generation and plasma properties in an atmospheric argon microwave plasma jet. *J. Appl. Phys.* 110:053304–9.

165. Wang C. and N. Srivastava. 2010. Cavity ringdown spectroscopic measurements of OH number densities in atmospheric microwave plasma jets operating with different plasma gases (Ar, Ar/N$_2$, and Ar/O$_2$). *Eur. Phys. J. D* 60:465–67.

166. Deng, X., J. Shi, and M. Kong. 2006. Physical mechanisms of inactivation of *Bacillus subtilis* spores using cold atmospheric plasmas. *IEEE Trans. Plasma Sci.* 34:1310–16.

167. Laroussi, M., J. P. Richardson, and F. C. Dobbs. 2002. Effect of nonequilibrium atmospheric pressure plasmas on the heterotrophic pathways of bacteria and on their cell morphology. *Appl. Phys. Lett.* 81:772–74.

168. Fridman, G., G. Friedman, A. Gutsol, A. B. Shekhter, V. N. Vasilets, and A. Fridman. 2008. Applied plasma medicine. *Plasma Process. Polym.* 5:503–33.

169. Xiong, Z., Y. Cao, X. Lu, and T. Du. 2011. Plasmas in tooth root canal. *IEEE Trans. Plasma Sci.* 39:2968–69.

170. Herzberg, G. and K. P. Huber. 1950. *Molecular Spectra and Molecular Structure*, Vols. III and VI. New York: Van Nostrand Reinhold.

171. Cheah, S.-L., Y.-P. Lee, and J. F. Ogilvie. 2000. Wavenumbers, strengths, widths and shifts with pressure of lines in four bands of gaseous $^{16}O_2$ in the systems a$^1\Delta_g$-X^3 Σ_g^- and b^1 Σ_g^+-X^3 Σ_g^-. *J. Quant. Spectrosc. Radiat. Transf.* 64:467–82.

172. Noxon, J. F. 1961. Observation of the (b^1 Σ_g^+-a$^1\Delta_g$) transition in O$_2$. *Can. J. Phys.* 39:1110–19.

173. Kassi, S., D. Romanini, A. Campargue, and B. Bussery-Honvault. 2005. Very high sensitivity CW-cavity ring down spectroscopy: Application to the a$^1\Delta_g$(0)-X^3 Σ_g^-(1) O$_2$ band near 1.58 µm. *Chem. Phys. Lett.* 409:281–87.

174. Földes, T., P. Čermák, M. Macko, P. Veis, and P. Macko. 2009. Cavity ring-down spectroscopy of singlet oxygen generated in microwave plasma. *Chem. Phys. Lett.* 467:233–36.

175. Leshchishina, O., S. Kassi, I. E. Gordon, L. S. Rothman, L. Wang, and A. Campargue. 2010. High sensitivity CRDS of the a$^1\Delta_g$-X^3 Σ_g^- band of oxygen near 1.27 µm: Extended observations, quadrupole transitions, hot bands and minor isotopologues. *J. Quant. Spectrosc. Radiat. Transf.* 111:2236–45.

176. Teslja, A. and P. J. Dagdigian. 2004. Determination of oxygen atom concentrations by cavity ring-down spectroscopy. *Chem. Phys. Lett.* 400:374–78.

177. Michaud, F., F. Roux, S. P. Davis, A. Nguyen, and C. O. Laux. 2000. High-resolution fourier spectrometry of the $^{14}N_2^+$ ion. *J. Mol. Spectrosc.* 203:1–8.

178. Laux, C. O., R. J. Gessman, C. H. Kruger, F. Roux, F. Michaud, and S. P. Davis. 2001. Rotational temperature measurements in air and nitrogen plasmas using the first negative system of N$_2^+$. *J. Quant. Spectrosc. Radiat. Transf.* 68:473–82.

179. Kotterer, M., J. Conceicao, and J. P. Maier. 1996. Cavity ringdown spectroscopy of molecular ions: A$^2\Pi_u$←X^2 Σ_g^+ (6-0) transition of N$_2^+$. *Chem. Phys. Lett.* 259:233–36.

180. Lofthus, A. and P. H. Krupenie. 1977. The spectrum of molecular nitrogen. *J. Phys. Chem. Ref. Data* 6:113–307.

181. Green, M. E. and C. M. Western. 1996. A deperturbation analysis of the B $^3\Sigma_u^-$ (v' = 0-6) and the B''$^3\Pi_u$(v' = 2-12) states of S$_2$. *J. Chem. Phys.* 104:848–64.

182. Stancu, G. D., M. Janda, F. Kaddouri, D. A. Lacoste, and C. O. Laux. 2010. Time-resolved CRDS measurements of the N$_2$(A$^3\Sigma_u^+$) density produced by nanosecond discharges in atmospheric pressure nitrogen and air. *J. Phys. Chem.* 114:201–08.

183. Stancu, G. D., F. Kaddouri, D. A. Lacoste, and C. O. Laux. 2010. Atmospheric pressure plasma diagnostics by OES, CRDS and TALIF. *J. Phys. D Appl. Phys.* 43:124002–10.

184. Scripter, C., E. Augustyniak, and J. Borysow. 1993. Direct spectroscopic detection of $A'^5\Sigma_g^+$ state of N_2. *Chem. Phys. Lett.* 201:194–98.

185. Augustyniak, E. and J. Borysow. 1994. Kinetics of the ($A^3\Sigma_u^+$, v = 0) state of N_2 in the near afterglow of a nitrogen pulsed discharge. *J. Phys. D Appl. Phys.* 27:652–60.

186. van den Oever, P. J., J. H. van Helden, C. C. H. Lamers, et al. 2005. Density and production of NH and NH_2 in an Ar-NH_3 expanding plasma jet. *J. Appl. Phys.* 98:093301–10.

187. Hoefnagels, J. P. M., A. A. E. Stevens, M. G. H. Boogaarts, W. M. M. Kessels, and M. C. M. van de Sanden. 2002. Time-resolved cavity ring-down spectroscopic study of the gas phase and surface loss rates of Si and SiH_3 plasma radicals. *Chem. Phys. Lett.* 360:189–93.

188. Kessels, W. M. M., J. P. M. Hoefnagels, M. G. H. Boogaarts, D. C. Schram, and M. C. M. van de Sanden. 2001. Cavity ring down study of the densities and kinetics of Si and SiH in a remote Ar-H_2-SiH_4 plasma. *J. Appl. Phys.* 89:2065–73.

189. Zalicki, P., Y. Ma, R. N. Zare, et al. 1995. Methyl radical measurement by cavity ring-down spectroscopy. *Chem. Phys. Lett.* 234:269–74.

190. Yamamoto, N., A. P. Yalin, L. Tao, T. B. Smith, A. D. Gallimore, and Y. Arakawa. 2009. Development of real-time boron nitride erosion monitoring system for hall thrusters by cavity ring-down spectroscopy. *Trans. Space Technol. Jpn.* 7:Pb_1–Pb_6.

191. Pushkarsky, M. B., S. J. Zalyubovsky, and T. A. Miller. 2000. Detection and characterization of alkyl peroxy radicals using cavity ringdown spectroscopy. *J. Chem. Phys.* 112:10695–98.

192. Just, G. M. P., P. Rupper, T. A. Miller, and W. Leo Meerts. 2009. High-resolution cavity ringdown spectroscopy of the jet-cooled ethyl peroxy radical $C_2H_5O_2$. *J. Chem. Phys.* 131:184303–11.

193. Chen, M., G. M. P. Just, T. Codd, and T. A. Miller. 2011. Spectroscopic studies of the $\tilde{A}$ – $\tilde{X}$ electronic spectrum of the β-hydroxyethylperoxy radical: Structure and dynamics. *J. Chem. Phys.* 135:184304–11.

194. Freel, K., J. Park, M. C. Lin, and M. C. Heaven. 2011. Cavity ring-down spectroscopy of the phenyl radical in a pulsed discharge supersonic jet expansion. *Chem. Phys. Lett.* 507:216–20.

195. Liu, Z., Y. Xu, X. Yang, A. Zhu, G. Zhao, and W. Wang. 2008. Determination of the HO_2 radical in dielectric barrier discharge plasmas using near-infrared cavity ring-down spectroscopy. *J. Phys. D Appl. Phys.* 41:045203–7.

196. Pillier, L., C. Moreau, X. Mercier, J. F. Pauwels, and P. Desgroux. 2002. Quantification of stable minor species in confined flames by cavity ring-down spectroscopy: Application to NO. *Appl. Phys. B* 74:427–34.

197. Mercier, X., L. Pillier, J. F. Pauwels, and P. Desgroux. 2001. Quantitative measurement of CN radical in a low-pressure methane/air flame by cavity ring-down spectroscopy. *C. R. Acad. Sci. Paris* 2:965–72.

198. Miller, J. H., A. R. Awtry, M. E. Moses, A. D. Jewell, and E. L. Wilson. 2002. Measurements of hydrogen cyanide and its chemical production rate in a laminar methane/air, non-premixed flame using CW cavity ringdown spectroscopy. *Proc. Combust. Inst.* 29:2203–09.

199. Boumans, P. W. J. M. and J. J. A. M. Vrakking. 1986. The widths and shapes of about 350 prominent lines of 65 elements emitted by an inductively coupled plasma. *Spectrochim. Acta B At. Spectrosc.* 41:1235–75.

200. Mitchell, A. C. G. and M. W. Zemansky. 1934. *Resonance Radiation and Excited Atoms*. Cambridge: Cambridge University Press.

201. *PeakFitTM 4.0 for Windows User's Manual*, 1997. Chicago, IL: SPSS Inc.

202. Kornblum, G. R. and L. de Galan. 1977. Spatial distribution of the temperature and the number densities of electrons and atomic and ionic species in an inductively coupled RF argon plasma. *Spectrochim. Acta B At. Spectrosc.* 32:71–96.

203. Griem, H. R. 1964. *Plasma Spectroscopy*. New York: McGraw-Hill Inc.

204. de Regt, J. M., R. D. Tas, and J. A. M. van der Mullen. 1996. A diode laser absorption study on a 100 MHz argon ICP. *J. Phys. D Appl. Phys.* 29:2404–12.

205. Starikovskaia, S. M. 2006. Plasma assisted ignition and combustion. *J. Phys. D Appl. Phys.* 39:R265–R299.
206. Starikovskii, A. Y., N. B. Anikin, I. N. Kosarev, E. I. Mintoussov, S. M. Starikovskaia, and V. P. Zhukov. 2006. Plasma-assisted combustion. *Pure Appl. Chem.* 78:1265–98.
207. Plasma 2010 Committee, Plasma Science Committee, Board on Physics and Astronomy, Division of Engineering and Physical Sciences, National Research Council of the National Academies. 2007. *Plasma Science, Advancing Knowledge in the National Interest.* Washington, DC: The National Academies Press.
208. Matveev, I. B. and L. A. Rosocha. 2008. Third special issue on plasma-assisted combustion. *IEEE Trans. Plasma Sci.* 36:2885–86.
209. Ju, Y. 2006. Plasma assisted combustion. *Paper Presented at the Northern Section Annual Meeting of Japanese Society of Aeronautics and Astronautics.* Sendai, Japan, March 9.
210. Rosocha, L. A., D. M. Coates, D. Platts, and S. Stange. 2004. Plasma-enhanced combustion of propane using a silent discharge. *Phys. Plasmas* 11:2950–56.
211. Campbell, C. S. and F. N. Egolfopoulos. 2005. Kinetic paths to radical-induced ignition of methane/air mixtures. *Combust. Sci. Technol.* 177:2275–98.
212. Kim, W., H. Do, M. G. Mungal, and M. A. Cappelli. 2006. Flame stabilization enhancement and NO_x production using ultra short repetitively pulsed plasma discharges. *Paper Presented at the 44th AIAA Aerospace Sciences Meeting and Exhibit.* Reno, NV, January 9–12.
213. Tan, W. and T. A. Grotjohn. 1995. Modeling the electromagnetic field and plasma discharge in a microwave plasma diamond deposition reactor. *Diam. Relat. Mater.* 4:1145–54.
214. Stockman, E. S., S. H. Zaidi, R. B. Miles, C. D. Carter, and M. D. Ryan. 2010. Measurements of combustion properties in a microwave enhanced flame. *Combust. Flame* 156:1453–61.
215. Kim, W., M. G. Mungal, and M. A. Cappelli. 2010. The role of *in situ* reforming in plasma enhanced ultra lean premixed methane/air flames. *Combust. Flame* 157:374–83.
216. Ombrello, T., S. H. Won, Y. Ju, and S. Williams. 2010. Flame propagation enhancement by plasma excitation of oxygen. Part II: Effects of $O_2(a^1\Delta_g)$. *Combust. Flame* 157:1916–28.
217. Jones, A. C. and M. L. Hitchman. 2009. *Chemical Vapour Deposition: Precursors, Processes and Applications.* Cambridge, UK: Royal Society of Chemistry.
218. Li, Y., D. Mann, M. Rolandi, et al. 2004. Preferential growth of semiconducting single-walled carbon nanotubes by a plasma enhanced CVD method. *Nano Lett.* 4:317–21.
219. Kodas, T. T. and M. J. Hampden-Smith. 2007. *The Chemistry of Metal CVD.* Weinheim, Germany: Wiley-VCH Verlag GmbH.
220. Park, W. I., D. H. Kim, S. -W. Jung, and Gyu-Chul Yi. 2002. Metalorganic vapor-phase epitaxial growth of vertically well-aligned ZnO nanorods. *Appl. Phys. Lett.* 80:4232–34.
221. Rossnagel, S. M., J. J. Cuomo, and W. D. Westwood. 1990. *Handbook of Plasma Processing Technology: Fundamental, Etching, Deposition and Surface Interactions.* New York: Noyes Publications.
222. Nakamura, A., S. Shigemori, Y. Shimizu, T. Aoki, and J. Temmyo. 2004. Hydroxyl-radical-assisted growth of ZnO films by remote plasma metal-organic chemical vapor deposition. *Jpn. J. Appl. Phys.* 43:7672–76.
223. Chen, Y., V. Selvamanickam, Y. Zhang, et al. 2009. Enhanced flux pinning by $BaZrO_3$ and $(Gd,Y)_2O_3$ nanostructures in metal organic chemical vapor deposited GdYBCO high temperature superconductor tapes. *Appl. Phys. Lett.* 94:062513–5.
224. Selvamanickam, V., A. Guevara, Y. Zhang, et al. 2010. Enhanced and uniform in-field performance in long (Gd,Y)-Ba-Cu-O tapes with zirconium doping fabricated by metal–organic chemical vapor deposition. *Supercond. Sci. Technol.* 23:014014–6.

9

Laser-Induced Fluorescence Methods for Transient Species Detection in High-Pressure Discharges

Santolo De
Benedictis and
Giorgio Dilecce

9.1 Introduction

Atmospheric pressure (AP) discharges are generally highly transient and spatially inhomogeneous media. The measurement and temporal monitoring of transient species requires a sensitive, *in situ* technique with high temporal and spatial resolution. Laser-induced fluorescence (LIF) has all these requisites. It is a solution to those cases in which simple emission spectroscopy is not applicable because the species can be nonemitting or, even if emitting electronic states exist, the emission cannot be correlated to the corresponding ground state. This is often the case in atmospheric pressure discharges (Dilecce et al., 2010a). LIF spectroscopy is discussed in textbooks on laser spectroscopy (Demtröder, 2003) and laser chemistry (Telle et al., 2007). Diagnostic methods based on LIF spectroscopy have been reviewed for combustion (Smith and Crosley, 2002) and for low-pressure plasmas (Freegarde and Hancock, 1997; Amorim et al., 2000; Barnat, 2011). In this chapter the LIF technique is described, from its basic principles to sophisticated techniques such as two-photon absorption LIF and optical–optical double resonance (OODR) LIF. The treatment will be restricted to LIF with pulsed lasers only due to its time-resolution capability. Peculiarities of LIF application at atmospheric pressure will be addressed due to the presence of fast collision processes, including electronic quenching (Q), vibrational (VET) and rotational (RET) energy transfers, and predissociation, that have a great influence on LIF measurements. Diagnostics aspects related to various applications of atmospheric pressure discharges will be described, with a focus on LIF on OH and of OODR-LIF on the nitrogen triplet metastable $N_2(A^3\Sigma_u^+)$, which are good model cases for highlighting the characteristics of the technique.

9.2 Basics of LIF Spectroscopy

9.2.1 Principles and Mathematical Expressions

LIF is based on a two-step process. In the first step molecules or atoms in the electronic initial state $|i>$ are pumped (excited) to an upper electronic state $|j>$ by absorbing one laser photon with energy $h\nu_{ij}$. In the second step $|j>$ undergoes spontaneous radiative decay to a lower state $|k>$, and the corresponding $h\nu_{jk}$ photon is the spectroscopic observable called the fluorescence signal. A conceptual scheme of the LIF process is depicted in Figure 9.1. The states $|i>$, $|j>$, and $|m>$ represent atomic levels and molecular rovibronic levels for atoms and molecules, respectively. Depopulation of level $|j>$ is determined by the sum of radiative transitions and by collision processes whose final state is not defined here. A more general scheme allows for collisional population of another state $|m>$ from $|j>$, with observation of fluorescence from $|m>$. The LIF scheme shown in Figure 9.1 can be described by a set of coupled rate equations* for the densities N of the three levels:

$$\frac{dN_i}{dt} = -B_{ij}\rho_\nu(t)\left(N_i - \frac{g_i}{g_j}N_j\right),$$

$$\frac{dN_j}{dt} = B_{ij}\rho_\nu(t)\left(N_i - \frac{g_i}{g_j}N_j\right) - \left(A_j + Q_j\right)N_j, \tag{9.1}$$

$$\frac{dN_m}{dt} = R_{jm}N_j - \left(A_m + Q_m\right)N_m.$$

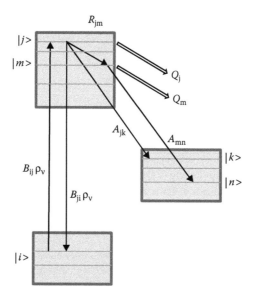

FIGURE 9.1   Scheme of LIF process and quantities involved in its mathematical description.

---

* The rate equations approach is valid if coherence effects can be neglected. In most LIF experiments this is true because the coherence time of non-single-mode pulsed laser (<0.1 ns) is shorter than the few nanoseconds pumping time (Avan and Cohen-Tannoudji, 1977). In addition, due to the inhomogeneous spatial profile and the pulse-to-pulse intensity fluctuations of the laser beam, coherence effects are masked by the averaging over a large number of shots (Altkorn and Zare, 1984). When, instead, coherent effects are important, a more general formulation based on the density matrix formalism has to be used (Demtröder, 2003).

The interaction of the two-level $|i>$ and $|j>$ system with the resonant laser field comprises an absorption process, with rate $R_{ij}(t) = B_{ij}\rho_v$, and a stimulated emission process, with rate $B_{ji}\rho_v$ and $B_{ji} = (g_i/g_j)B_{ij}$, in which g_i and g_j are the multiplicities of states $|i>$ and $|j>$. The absorption coefficient B_{ij} m^3 J^{-1} s^{-2}, multiplied by the spectral radiant energy density of the laser field ρ_v J m^{-3} s, gives an absorption rate. Dealing with pulsed lasers, the spectral radiant energy density is a function of time. Equation 9.1 is also valid in the case of a cw laser, in which ρ_v is constant in time and a stationary solution is looked for. The total radiative rates of levels $|j>$ and $|m>$ are $A_j = \sum_k A_{jk}$ and $A_m = \sum_n A_{mn}$, Q_j and Q_m are the collision quenching rates, and R_{jm} is the $|j> \rightarrow |m>$ collision transfer rate. The observable LIF signal is then the *direct* LIF or the *collisional* LIF:

$$S^d(t) = CA_{jk}N_j(t) \quad \text{or} \quad S^c(t) = CA_{mn}N_m(t) \tag{9.2}$$

or the sum of the two. The constant C depends on the measurement apparatus and will be detailed later. Formula $B_{ij}\rho_v$ for the absorption rate is valid only in the case of constant ρ_v in the spectral interval of the absorption line. In a real case one has to calculate an overlap integral as follows. Defining

$$\rho_v(t,v) = \rho_L(t)\varepsilon(v - v_L) \quad \text{with} \quad \int_{-\infty}^{\infty} \varepsilon(v - v_L)dv = 1,$$

$$B_{ij}(v) = B_{ij}b(v - v_{ij}) \quad \text{with} \quad \int_{-\infty}^{\infty} b(v - v_{ij})dv = 1, \tag{9.3}$$

where $b(v - v_{ij})$ is the normalized absorbing line profile given by Doppler and pressure broadening. Predissociation broadening may also be important. $\varepsilon(v - v_L)$ is the normalized laser line shape that, if not known, can be determined by measuring the LIF excitation spectra around an isolated line whose line shape is known. The absorption rate is then given by

$$R_{ij}(t) = B_{ij}\rho_L(t)\int_{-\infty}^{\infty} \varepsilon(v - v_L)b(v - v_{ij})dv = B_{ij}\rho_L(t)\psi, \tag{9.4}$$

where ψ is the overlap integral of the laser and of the absorption line profiles and can be written symbolically as the convolution $\varepsilon \otimes b$. The quantity $\rho_L(t)$ for a pulsed laser can be calculated as $\rho_L(t) = EF(t)/cS$, with c the velocity of light, S the laser beam section area, E the pulse energy, and $F(t)$ the normalized time profile of the laser pulse. It can be measured by a fast photomultiplier and a digitizing oscilloscope. The absorption is often described in terms of a cross section σ_{ij}, and the spectral radiant energy density is substituted, for the calculation of the absorption rate, by the spectral photon flux φ_v:

$$\sigma_{ij} = \frac{hv_{ij}}{c}B_{ij} \quad \text{and} \quad \phi_v = \frac{c}{hv_{ij}}\rho_v. \tag{9.5}$$

It is customary in the literature to describe the laser field by the quantity $I_v = c\rho_v$. The corresponding spectrally integrated quantities

$$I_L = \int_{-\infty}^{\infty} I_v dv \quad \text{and} \quad \Phi_L = \int_{-\infty}^{\infty} \phi_v dv \tag{9.6}$$

are referred to as *intensity* and *photon flux*. In case of low absorption rate, the depletion of $|i>$ state population is negligible, so that the first two equations of the Equation 9.1 can be decoupled, with a constant N_i in the second equation. The latter can then be solved as [for $N_j(0) = 0$]

$$N_j(t) = N_i B_{ij}\psi e^{-t/\tau}\int_0^t \rho_L(t')e^{t'/\tau}dt', \tag{9.7}$$

where $\tau = 1/(A_j + Q_j)$ is the effective lifetime of $|j>$. This is the linear LIF case where linearity is intended with respect to variations of the absorption rate, that is, variations of the laser pulse energy. The extreme opposite case is that of fully saturated LIF, which is achieved when, on increasing the laser pulse energy, the limits of absorption and stimulated emission (ASE) are reached. Full saturation is never achieved, being characteristic of a perfectly isolated two-level system, because of strong spatial inhomogeneities of real-life pulsed laser beams.

9.2.2 Experimental Arrangements and Instrumentation

9.2.2.1 Geometry of the Experiment

The typical geometry of a LIF experiment envisages the optical collection of fluorescence at right angle with respect to the laser beam direction. It is shown schematically in Figure 9.2. This geometry minimizes the collection of scattered laser light in principle. The collection is schematized by a cone with the aperture (solid angle) Ω of the optical collection system. Fluorescence light is not emitted isotropically in all directions. Given a linearly polarized laser light, its absorption produces an alignment of the transition dipoles along the laser electric field direction, such that the fluorescence from an aligned population of dipoles depends on the angle θ between the direction of observation and the laser electric field (z-axis). The measured signal can be formally written as

$$S^d(t) \propto A_{jk} N_j(t) \int_\Omega K(\theta) d\omega \quad \text{with} \quad \int_{4\pi} K(\theta) d\omega = 1, \tag{9.8}$$

where $K(\theta)$ is the result of a *nascent* angular distribution that evolves in time during the fluorescence interval. The nascent distribution is peculiar to each atomic/molecular system under observation. It is a blend of angular distributions of linearly ($\Delta M = 0$) and circularly ($\Delta M = \pm 1$) polarized emitted photons (M is the projection of the angular momentum along the laser electric field, z-axis). Accordingly, the fluorescence light is also polarized. A complete treatment of this issue can be found in the works of Greene and Zare (1983) and Altkorn and Zare (1984). The time evolution of $K(\theta)$ is determined by collisions, and, in the molecular case, rotation, that quickly bring the nascent dipoles' alignment to an isotropic distribution. In De Benedictis et al. (1997) a full treatment of the atomic $He(4p^3P_2 \rightarrow 2^3S_1)$ case, taking into account M-changing collisions, was given after the observation made in the works of De Benedictis and Dilecce (1995) that in the same system fluorescence isotropization took place in a characteristic time of about 5 ns Torr^{-1}. It is then clear that at pressures above a few Torr the fluorescence can be safely considered

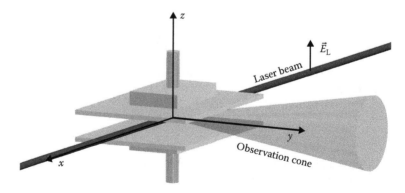

FIGURE 9.2 Typical geometry of a LIF experiment in a discharge. In the figure a dielectric barrier discharge is crossed longitudinally by the laser beam, and fluorescence is collected at a right angle. The laser light is vertically polarized along the z-axis. The light collection cone has the aperture of the optical collection–detection system (see text).

isotropic, and it will be considered as such in the remainder of the text. With a constant $K(\theta)$ the integral in Equation 9.8 is equal to $\Omega/4\pi$. Such a fraction is normally quite low, of the order of some 10^{-3}, Ω being limited by a number of factors such as optical windows, if the discharge is enclosed into a vacuum vessel, or, at atmospheric pressure, by the small interelectrode distance (1–5 mm). Another important geometrical parameter is the sampled volume V_s, that is, the fraction of discharge volume from which fluorescence photons can reach the detector. Each detection system has an entrance surface (the input slit of a spectrograph/monochromator, the surface of an optic fiber, and so on), and its image is transported by the optical collection system onto the region crossed by the laser beam. V_s is then determined by the intersection of this image with the laser beam. Proper alignment of the optical collection with the laser beam is important if V_s has to be determined with sufficient accuracy.

9.2.2.2 Lasers

Pulsed, tuneable lasers are the primary choice for LIF in gas discharges. The very high power of pulsed lasers is fundamental for achieving a fluorescence signal that overcomes plasma-induced emission (PIE), and the intrinsic time resolution of pulsed LIF is extremely useful for investigating the kinetics of the species by monitoring its evolution in time-varying discharges. The most common tuneable pulsed laser system is the dye laser pumped by the second harmonic (at 532 nm) of a solid-state Nd–YAG laser or by an excimer laser. Dye lasers roughly cover the range 500–950 nm when pumped at 532 nm or even about 350 nm with an excimer laser pump at pulse energies of the order of some tens of millijoules. Lower wavelengths down to about 200 nm can be obtained by a nonlinear conversion system mostly based on nonlinear crystals like BBO or LBO. Configurations like second and third harmonic generation (SHG, THG), mixing of the dye beam with the pump laser beam, or Raman shift can be realized. Common dye lasers have a nominal bandwidth (Bw) of about 0.2 cm^{-1}, for example, about 7.2×10^{-3} nm at 600 nm. If this laser beam is converted to 300 nm by SHG, the Bw is doubled. Nonlinear conversion is made at the expense of pulse energy. Each stage has efficiencies of the order of 10–30%, so that after SHG one can obtain a pulse energy of a few millijoules. Further stages reduce the energy requirement accordingly. Another tuneable pulsed laser type is the optical parametric oscillator (OPO) pumped again by an Nd–YAG. It features a very large tuneability range, 410–2200 nm, so that by a simple SHG it can reach 205 nm. It is very versatile because the wavelength change is fast (compared to the dye where the dye solution must be changed), but it normally has a larger Bw, typically 5 cm^{-1}. More sophisticated OPO configurations can reach Bw of the order of those of a dye laser. Finally the titanium–sapphire laser has a shorter tuneability range, about 700–900 nm, when pumped by an Nd–YAG. All these lasers have the common feature of being pumped by a Q-switched laser, and the temporal behavior of their laser pulse is dictated by that of the pump laser. The pulse duration is about 4–10 ns, while the pulse repetition rate is normally 10 Hz, sometimes 20 Hz, and, in a few cases, when not much energy is required, it can reach a value of 50 Hz. Q-switched lasers have large shot-by-shot energy fluctuations. One further advantage of running a LIF experiment in the linear regime is that the average of many shots of the LIF signal can be calculated from Equation 9.1 with the corresponding average laser energy.

9.2.2.3 Detection Systems

The detection of fluorescence photons goes through three stages: a spectral filter, a detector, and a device for the measurement of the detector outcome. Spectral selection is necessary to suppress (reduce) scattered laser radiation and to reduce the detected PIE to that of the same electronic transition as the fluorescence one. Interference filters or a monochromator can be used to this end. Fluorescence light is transported into the monochromator by a lens optical system or by an optical fiber. The spectral selection and optical collection system is characterized by its transmission (T) and Bw. T comprises many terms. Each (clean) transmitting surface on the optical path (windows, lenses) has reflection losses of about 4%. The monochromator has a wavelength and polarization-dependent grating efficiency. Optical fibers and interferences filters have also wavelength-dependent transmission figures.

The Bw determines in which fraction the spectral fluorescence is detected (especially in the case of molecules, single-fluorescence bands, or even more than one band, can be detected fully or partly depending on Bw). The main detector type used in LIF is the photomultiplier tube (PMT), being very sensitive and fast. PMTs are characterized by the quantum efficiency η of the photocathode, the gain G (of the order of 10^6–10^7), and their saturation characteristics. This is a very important issue. For fast light pulse measurements the parameter is the pulse saturation due to the space charge in the last dynode space. It is important to check the linearity of the PMT against the light pulse height and to work within the linear range. If the fluorescence light is larger, it is convenient to reduce it by putting neutral filters in the optical path, leaving the geometry and Bw of the system unchanged. The signal from a photomultiplier is a current. It is converted to voltage by the potential drop across the input resistance of the measurement device. Early devices employed for pulsed LIF signal measurements were gated integrators with a boxcar averager. These devices measure the integral of the PMT signal in a predefined time interval T_G called gate. Nowadays digitizing oscilloscopes are a better choice because they can recover the full $S(t)$ signal. This is advantageous because $S(t)$ contains information on the collision quenching rate Q_j. The integrated signal can anyway be calculated numerically, and this is what is normally done when a large number of measurements is done, and it is impractical to fit model-calculated $S(t)$ to measured ones. Finally, modern intensified charge-coupled device (ICCD) detectors that are gateable for time-resolved measurements down to nanosecond resolution, coupled to monochromators, are very useful for the measurement of LIF fluorescence spectra, which is important especially in molecular cases. Such devices measure the accumulation of charge (photoelectrons) within the gate interval. With sufficient fluorescence light it is also possible by ICCD spectrographs to capture the temporal evolution of fluorescence spectra.

9.2.2.4 Measured Signal

The constant C in Equation 9.2, which defines the effective measured signal by a photomultiplier and oscilloscope system, can be written as

$$C = V_s \frac{\Omega}{4\pi} T \eta e G R, \tag{9.9}$$

where

e is the electron charge
R is the input resistance of the oscilloscope, normally 50 Ω

Each of these parameters is affected by errors and/or it is known only by its nominal value. The measured signal is in most cases the integral over a defined time gate T_G. Restricting the analysis to direct fluorescence in the linear case, using Equation 9.7:

$$J = C \int_0^{T_G} A_{jk} N_j(t) \mathrm{d}t = CA_{jk} N_i B_{ij} \psi \int_0^{T_G} \left[e^{-t/\tau} \int_0^t \rho_L(t') e^{t'/\tau} \mathrm{d}t' \right] \mathrm{d}t \tag{9.10}$$

with $T_G \gg \tau$, that is, with a measurement gate that contains the whole fluorescence pulse, the integration limit T_G can be set to ∞, and after elementary mathematics one obtains

$$J = CN_i B_{ij} \psi \overline{\rho_L} \frac{A_{jk}}{A_j + Q_j} \quad \text{with} \quad \overline{\rho_L} = \int_0^\infty \rho_L(t) \mathrm{d}t. \tag{9.11}$$

The quantity $Y = A_{jk}/A_j + Q_j$ is the so-called quantum yield, the fraction of N_j that can be observed by detecting the fluorescence, the other part being lost by radiationless collision quenching. Note that the importance and physical meaning of the quantum yield are the same in the nonlinear LIF case too. Fluorescence measurement by ICCD is of the gate integrated type, so that in this case the constant ηeGR in Equation 9.9 must be substituted by the corresponding gain and quantum efficiencies of the ICCD, and the fluorescence spectrum must be integrated.

9.2.3 High-Pressure Peculiarities

Fast collision processes at high pressure have a number of consequences of both fundamental and practical nature. The quantum yield can be drastically lowered by collision quenching. With typical radiative rates of the order of 10^7 s^{-1} and typical Q rate coefficients of the order of 10^{-11} cm^3 s^{-1}, it is common to find Y values, and accordingly LIF signals, that at atmospheric pressure are 2 orders of magnitude lower than at low pressure. VET collisions in some instances are fast so as to make it preferable to observe the collision fluorescence rather than the direct one. Q and VET rates are strongly dependent on the collision partner. The gas mixture composition, as well as the collision rate constants of the main mixture components, must be accurately known to determine the total quenching rate in the mixture. When Q cannot be a priori determined, the measurement of the fluorescence pulse and of its decay after the laser pulse permits an empirical determination of Q. At high pressure, however, this method is rarely available because the fluorescence pulse decay practically coincides with the laser pulse temporal profile, due to quenching rates that determine lifetimes of $|j\rangle$ in the nanosecond timescale. When LIF is applied to inhomogeneous media like flames, in which the gas composition is a strong and unknown function of the spatial position, a way out of the problem of unknown quenching rates is given by the so-called predissociative LIF (LIPF). If $|j\rangle$ is a predissociative level, its quenching rate is dominated by predissociation that is much faster than collision quenching, making the LIF outcome independent of the gas composition, at the expense, of course, of sensitivity. For example, the CH (B, $v = 1$, $N = 9$) state, used in the works of Luque et al. (2002) as predissociative $|j\rangle$, has a lifetime of 75 ± 10 ps, that is, more than 1 order of magnitude larger than collision quenching in any flame mixture.

Equation 9.1 becomes incomplete at high pressure because how many $|j\rangle$ collision quenching events terminate to $|i\rangle$ is normally unknown. In addition, in the molecular case, $|i\rangle$ is a single or a small group of rotational levels, and the RET collisions may contribute to refill $|i\rangle$ from adjacent rotational levels even within the short laser pulse duration. This phenomenon is negligible in the linear regime, in which the depletion of the lower level by laser absorption is small (Luque et al., 2002; Cathey et al., 2008). The linear regime range must be determined case by case. As a rule of thumb, in case of diatomic radicals like OH, CH, and CN at atmospheric pressure, deviations from the linear regime start at laser pulse energies of the order of 10 μJ.

RET collisions in the upper $|j\rangle$ level quickly redistribute excitation in the whole rotational manifold. Fluorescence spectra at high pressure often show thermal rotational distributions even if only one rotational line has been pumped, unless thermalization is prevented by collision quenching. Generally, then, fluorescence must be detected by a spectral Bw large enough to include a whole vibronic band, and the collision–radiative rate coefficients to use are those relevant to the whole band. When rotational thermalization is not achieved, the fluorescence spectrum must be measured and rotational state-specific rate constants used when available.

Interelectrode spacings in AP discharges are small, of the order of a few millimeters, and comparable to the laser beam section. Strong diffused laser light superimposed on the true LIF signal is unavoidable. LIF excitation–detection schemes with different bands, allowing spectral separation of the LIF signal from the scattered laser, are then compulsory. When maximum sensitivity is required, however, another type of indirect laser scattered light comes into play. The hardware materials of the discharge apparatus irradiated by spatial "wings" of the laser beam produce a broadband fluorescence, at wavelengths larger than the laser one, that cannot be completely filtered out, unless the LIF detection wavelength is lower than the laser one, which is rarely the case. The spatial wings of the laser beam are produced by diffusion from all the beam manipulation optics and from the discharge vessel windows.

9.2.4 Internal States' Distributions

The internal state population factor is essential for quantitative LIF density measurement. It can be determined by the measurement of LIF excitation spectra, that is, spectra obtained by scanning the laser wavelength in an appropriate spectral range with fixed spectral and time interval (gate) observation of the fluorescence. In this way the determination of the rotational distribution or temperature (T_{rot}) is also routinely carried out. The gas temperature (T_{gas}) is a fundamental parameter to qualify most temperature-sensitive applications of atmospheric pressure nonthermal plasmas, and at atmospheric pressure T_{rot} of ground-state species is equal to T_{gas}. The rotational temperature can be derived from LIF rotational excitation spectra using a Boltzmann plot representation once a sufficient number of isolated spectral lines of different quantum number are detected. This is frequently the case under low-pressure conditions at high temperatures of thousands of degrees. Note that most collision processes (depolarization, RET, VET, and Q) are rotational level dependent and, if not properly considered, may lead to systematic bias in the evaluated T_{rot}, as seen in LIF thermometry of low-pressure flames by molecular radicals such as OH, CH, NH, CN, and NO (Rensberger et al., 1989; Luque and Crosley, 1999b). In the CH LIF thermometry of flames (Luque and Crosley, 1999b), rotational Boltzmann plots obtained by changing the observation time gate width and delay with respect to the laser pulse gave T_{rot} differences within 20%. The most dependable values were obtained by minimizing the impact of collision relaxation using short gate duration at the prompt of a laser shot or by considering fast predissociative rotational levels when available. In these cases the fluorescence comes out mainly from rotational levels close to the pumped ones. This approach is, however, detrimental for the quantum yield. In atmospheric pressure nonthermal plasmas collisions may restrict the fluorescence duration to the timescale of the laser pulse width, preventing any choice of the detection gate width. In addition, the gas temperature is close to the ambient one; thus, the number of isolated lines are too few for a dependable Boltzmann plot. The rotational temperature must therefore be inferred from fits of partially resolved excitation spectra by numerical spectra simulation codes. One such simulation code for OH, NO, CH, and CN radicals is LIFBASE (Luque and Crosley, 1999a). LIF excitation spectra can also address the measurement of vibrational distributions, as will be seen in a forthcoming paragraph for the case of CN radical.

9.2.5 Two-Photon LIF

The single-photon absorption LIF scheme is possible if an optically allowed electronic transition is available for the $|i\rangle$ state to be probed, with a wavelength within the range of tuneable lasers. The possibility of absorption can be extended by multiphoton absorption, in particular two-photon absorption, in which the final level $|j\rangle$ is reached through an intermediate level $|int\rangle$ in a two-step process. $|int\rangle$ can be a virtual level or a real stationary electronic state. In the first case the LIF technique takes the name of TALIF and in the second case OODR-LIF. States with equal parity can be connected by two-photon absorption, that is, a further extension of the single-photon scheme in which only opposite parity is allowed.

9.2.5.1 TALIF

The two photons for a TALIF scheme can be of different wavelengths, but only cases with two equal photons of the same laser beam have been applied in gas discharges research for measuring atomic ground-state species, in particular of nonmetallic atoms, given that for metallic atoms single-photon schemes are readily available. The treatment is then restricted to the single laser case, with $\nu_{ij} = 2\nu_L$. The rate equation for TALIF is the same as that for a single photon, in which the absorption rate is calculated as follows:

$$R(t) = \sigma^{(2)}\psi(\Delta\nu)G^{(2)}\left(\frac{I_L(t)}{h\nu_L}\right)^2 \tag{9.12}$$

where $\sigma^{(2)}$ is the two-photon absorption cross section (in cm^4), $\psi(\Delta\nu = 2\nu_L - \nu_{ij}) = b(2\nu - \nu_{ij}) \otimes \varepsilon(\nu - \nu_L) \otimes \varepsilon(\nu - \nu_L)$. The absorption rate is proportional to the square of the laser field (intensity). The effect of fluctuations of the laser intensity on the LIF signal cannot be simply averaged as in the linear case. Even if the time profile of the laser beam is measured shot by shot, fluctuations occurring in a timescale shorter than the response time of the measurement (and longer than the coherence time) are averaged in the measurement of $F(t)$. To deal with this problem the absorption rate is multiplied by the second-order correlation function, $G^{(2)} = \langle f^2(t) \rangle / \langle f(t) \rangle^2 = \langle f^2(t) \rangle / F^2(t)$, where $f(t)$ is the *true* laser pulse profile (i.e., the profile that would be measured by a sufficiently fast detector as to capture all its time variations). For multimode lasers like those used in pulsed LIF experiments, $G^{(2)}$ tends to 2, the value for chaotic light.

C, Cl, F, H, N, O, S, I, and Xe atoms have been detected by TALIF with laser wavelengths in the 205–305 nm range (except for fluorine at 170 nm). The known cross sections are low, of the order of 10^{-35} cm^4. All of them obtained by theoretical calculations (Omidvar, 1980a,b; Saxon and Eichler, 1986) except the cross section of O atoms for which an experimental measurement is available (Bamford et al., 1986). The calculation is based on the summation of out-of-resonance absorption by all possible electronic intermediate states. The low cross-sectional value requires high laser intensities, and laser wavelengths in the UV are produced by nonlinear conversion methods that greatly reduce the available pulse energy. It is thus necessary to focus the laser beam to achieve sufficient intensities, taking advantage of the dependence of $R(t)$ on I_L^2. The high intensity of focused beams opens the way to two more processes: three-photon ionization, due to absorption of a third photon by the $|j\rangle$ state, and amplified spontaneous emission. ASE occurs when a sufficient population inversion is created between the $|j\rangle$ and $|k\rangle$ states and is a threshold process. Both processes depend on the $|j\rangle$ state density and act as a depletion mechanism of $|j\rangle$. They introduce a strong nonlinearity in the LIF signal as a function of I_L^2, such that increasing I_L starts to be disadvantageous after a certain value (Pezé et al., 1993). These processes could be important at low pressure [see Amorim et al. (1994) for a complete description]. At high pressure fast collision quenching prevents a sufficiently high $|j\rangle$ population from being ionized and ASE from playing a significant role.

9.2.5.2 OODR-LIF

When $|int\rangle$ is a resonant electronic state the two-photon absorption rate considerably increases, together with the instrumental complexity tied to the production of two laser beams at different wavelengths that must be synchronized in the nanosecond timescale. OODR-LIF belongs to the *pump-and-probe* class of experiments for investigating the rovibronic relaxation dynamics of the intermediate $|int\rangle$ state (Demtröder, 2003). Its use for analytical purposes may appear a useless complication because a single-photon scheme is available with $|int\rangle$ as the final state. The possibility to pump a different state by an OODR absorption may, nevertheless, be useful at a high pressure for selecting $|j\rangle$ states with better quantum yield than $|int\rangle$. To the best of our knowledge the only case reported in literature is that of OODR-LIF detection of the $N_2(A^3\Sigma_u^+)$ metastable state (Dilecce et al., 2007) in which the quantum yield of $|j\rangle$ is about 10^2 times larger than that of $|int\rangle$, as will be detailed later. The mathematical description of the OODR-LIF process is not straightforward in the molecular case because it requires not just the addition of another equation for the $|int\rangle$ state population to Equation 9.1. The whole rotational manifold of $|int\rangle$ must be described because RET collisions are fast enough to modify the LIF outcome deeply. The overpopulation of $|int\rangle$ rotational levels produced by absorption of the first photon is in fact quickly redistributed within the manifold well within the laser pulse duration. Such a complete modeling was not addressed in the works of Dilecce et al. (2007), where it was found more convenient to rely on a calibration technique for the absolute $N_2(A^3\Sigma_u^+)$ density measurement.

9.3 Calibration

The final calculation of absolute density values from a LIF experiment is often full of uncertainties, so that calibration techniques are advisable in the best case and mandatory in the worst one. The wide variety of calibration methods, which are often specific to the particular discharge system under investigation, can be roughly grouped into three classes. In the first class LIF scattering, with unknown scatterer density, is compared to a well-defined different scattering process with known scatterer density in the same geometrical and instrumental arrangement. The ratio of LIF and known scattering outcomes gets rid of the constant C and is proportional to the ratio of the scatterer density. Rayleigh scattering and TALIF on noble gases are the two available examples. The second class includes all local (i.e., in the same point as that of LIF measurement) methods for the measurement of the unknown radical density by other less sensitive and flexible techniques in selected favorable conditions. Once calibrated, absolute values can be extrapolated to all other conditions by LIF relative measurements. In the third class local or nonlocal sources of known radical density are employed.

9.3.1 Rayleigh Scattering

In this method (Kunze, 1968) the LIF signal is compared to the Rayleigh scattering signal from a reference gas, typically a noble gas (Döbele et al., 2005) or N_2 (Krames et al., 2001; Nemschokmichal et al., 2011). The Rayleigh signal is given by

$$S_R(t) = V_s T \eta e G R N_R I_L(t) \int_\Omega \frac{\partial \sigma_R}{\partial \omega} d\omega \qquad (9.13)$$

with the same constants and laser intensity as for LIF and N_R as the scatterer density. The differential scattering cross section $\partial \sigma_R / \partial \omega$ depends on the symmetry properties of the gas, which lead to a polarization pattern of the scattered radiation (Born and Wolf, 1999; Eberhard, 2010). For a spherically symmetric scatterer, such as a noble gas, the differential cross section has a dipole radiation pattern and depends on θ only:

$$\frac{\partial \sigma_R}{\partial \theta} = \frac{1}{4\pi} \frac{3}{2} \sigma_R^s \sin^2(\theta). \qquad (9.14)$$

The total cross section σ_R^s can be calculated from the polarizability α or from the refractive index n, using the Lorentz–Lorentz equation:

$$\sigma_R^s = \frac{8\pi^3 \alpha^2}{3\varepsilon_0^2 \lambda^4} = \frac{24\pi^3 \nu^4}{c^4 N_R^2} \left(\frac{n^2(\lambda)-1}{n^2(\lambda)+2} \right)^2 \cong \frac{32\pi^3}{3\lambda^4} \left(\frac{n(\lambda)-1}{N_R} \right)^2 \qquad (9.15)$$

with the approximation valid for $n \simeq 1$. For nonspherical symmetric scatterers like diatomic molecules, the total cross section is written in the form of the symmetric cross section of Equation 9.15 multiplied by the King's correction factor $F(\lambda)$ (Naus and Ubachs, 2000; Miles et al., 2001) as $\sigma_R = \sigma_R^s F(\lambda)$. King's correction factor, the refractive index, and the Rayleigh cross section for standard air in the wavelength range 0.2–1.0 µm are reported in Miles et al. (2001). The linear dependence of the Rayleigh signal on laser intensity limits the calibration to the linear LIF regime. However, LIF measurements carried out under weak–intermediate saturation can be normalized to the linear regime and then calibrated (van Lessen et al., 1998; Nemschokmichal et al., 2011). Rayleigh scattering applied to real discharge devices is affected by spurious radiation scattering by discharge electrodes and vacuum vessel walls or discharge hardware (Krames et al., 2001). In large-volume geometries, as in low-pressure discharges, spurious radiation can be well accounted for (Nemschokmichal et al., 2011), while in the small geometries of atmospheric pressure discharges this task is more difficult (Nemschokmichal and Meichsner, 2013a,b).

9.3.2 TALIF on Noble Gases

This method, introduced in Goehlich et al. (1998), is based on the comparative measurement of TALIF on a noble gas—Xe, Kr, and Ar—that has a two-photon resonance close to that of the atomic species under investigation. Because the laser wavelengths are very close, the laser beam characteristics can be assumed

TABLE 9.1 Transitions of the TALIF Process for N, O, and H Atoms, with Those of the Noble Gases Chosen for Calibration, Kr for N and H, Xe, for O

	Absorption	λ_L (nm)	$\sigma^{(2)}$ (10^{-35} cm^4)	Emission	λ_E (nm)
N	$2p^{34}S_{3/2} \rightarrow 3p^4S_{3/2}$	206.7	1.324^a	$\rightarrow^4S_{1/2,3/2,5/2}$	742–747
H	$1s^2S_{1/2} \rightarrow 3d^4D_{3/2,5/2}$	205.1		$\rightarrow 2p^2P_{1/2,3/2}$	656.5
Kr	$4p^{61}S_0 \rightarrow 5p'[3/2]_2$	204.2	$0.67 \times \sigma^{(2)}(N)^b$	$\rightarrow 5s'[3/2]_1$	826.5
			$0.62 \times \sigma^{(2)}(H)^b$		
O	$2p^{43}P_{2,1,0} \rightarrow 3p^3P_{2,1,0}$	225.6	2.66 ± 0.80^c	$\rightarrow 3s^3S$	844.9
			1.319 ± 0.2^d		
Xe	$5p^{61}S_0 \rightarrow 6p'[3/2]_2$	224.3	$1.9 \times \sigma^{(2)}(O)^e$	$\rightarrow 6s'[1/2]_2$	834.6

Source: ^{a}Omidvar, K., *Phys. Rev. A*, 30, 2805, 1980a; ^{b}Niemi, K. et al., *J. Phys. D Appl. Phys.*, 34, 2330–2335, 2001; ^{c}Bamford, D. J. et al., *Phys. Rev. A*, 34, 185–198, 1986; ^{d}Saxon, R. P. and J. Eichler, *Phys. Rev. A*, 34, 199–206, 1986; ^{e}Niemi, K. et al., *Plasma Sources Sci. Technol.*, 14, 375–386, 2005.

to be identical and, from a practical point of view, when using a dye laser, the dye solution must not be changed. In Table 9.1 the excitation (absorption)–detection (emission) transitions for TALIF on N and H atoms, calibrated with Kr, and on O atoms, calibrated with Xe, are reported (Niemi et al., 2005). Strictly speaking, the application of this method does not require knowledge of the absolute cross sections of both the unknown atom and the noble gas, but it is enough to know their ratio. These ratios for Kr and Xe and N, H, and O were determined by Niemi et al. (2001, 2005) in a flow tube reactor in which the concentration of N, H, and O atoms was measured by titration methods. They are given in Table 9.1 and have estimated uncertainties of about 50% (Niemi et al., 2001). At high pressure the ratio of quantum yields of the two emission transition must also be known. A comprehensive compilation of the relevant collision rate coefficients and emission rates can be found in Niemi et al. (2001, 2005). It is advisable to apply the method to the linear TALIF regime to avoid partial saturation, ASE, and three-photon ionization. Strictly speaking, the constant C is completely ruled out only if λ_E is the same for both emission transitions, which is likely the case for the O–Xe pair only. A correction factor taking into account the relative spectral response of the whole optical collection—spectrometer chain at the two emission wavelengths must be applied. This applies to the Rayleigh scattering calibration procedure as well.

9.3.3 Sources of Known Radical Density

Microwave discharges in a flow tube reactor have been employed for producing steady-state flow of O, N, and H atoms for TALIF calibration (Bittner et al., 1991; Döbele et al., 2005). The atom flow is quantified by titration methods (Clyne and Nip, 1979) that are based on fast chemical reactions of radicals with a suitable titration gas injected downstream of the discharge. Titration is based on the extinction point of the atomic radical loss curve measured by TALIF as a function of increased titration gas flow injected into the main afterglow stream. At this end point the radical flow is quantitatively correlated to that of the titrating gas (Tserepi et al., 1997; Niemi et al., 2005). The method requires a laminar flow condition and a proper titration gas injection and monitoring point that guarantee both complete mixing of gases and complete development of titration kinetics. The following well-known reactions are used for quantifying the flow and local density of O, N, and H atoms, with NO_2 and NO as titrating gases:

$$H + NO_2 \rightarrow OH + NO \quad k = 1.3 \times 10^{-10} \, cm^3 s^{-1}, \tag{9.16}$$

$$O + NO_2 \rightarrow NO + O_2 \quad k = 9.5 \times 10^{-12} \, cm^3 s^{-1}, \tag{9.17}$$

$$N + NO \rightarrow N_2 + O \quad k = 2.0 \times 10^{-11} \, cm^3 s^{-1}, \tag{9.18}$$

with the respective commonly accepted rate constants (Döbele et al., 2005). Equation 9.16 is also employed for the calibration of OH LIF measurements (Stevens et al., 1994). Flow tube calibrated sources typically operate at a pressure of a few Torr. Their use for TALIF calibration at higher pressure requires that quantum yields be duly taken into account.

Local generation of known radical quantities can be achieved by photochemical reactions, and in particular by laser photodissociation, provided the relevant cross sections are known. For example, in the works of Stevens et al. (1994) photodissociation of water by light at 185 nm from a mercury lamp was used to produce OH. In the works of Ono et al. (1992) and Marro and Graham (2008) TALIF on Cl atoms was calibrated by generating Cl by photodissociation of CCl_4 by the same TALIF laser at 233.3 nm.

9.3.4 Auxiliary Measurement Techniques

Absorption spectroscopy provides a mean for directly measuring the concentration of radicals. Absorption is a line-of-sight measurement and therefore does not have the same spatial resolution as LIF does. Absorption laser techniques based on cavity-enhanced absorption spectroscopy (CEAS) and cavity ring-down spectroscopy (CRDS) have a sensitivity comparable to that of LIF. A combination of such absorption methods and LIF on the same electronic transitions, and by employing the same laser, has provided a powerful apparatus for quantitative, time-, and space-resolved measurements, largely adopted in combustion research (Dreyer et al., 2001; McIlroy and Jeffries, 2002). Single-pass absorption with conventional light sources is less sensitive, and its lack of time resolution is a challenge in detecting transient species. Calibration conditions must be chosen in which the temporal variation of the radical density is monitored by LIF and its average value calibrated by absorption. Recently UV LED sources have been utilized for broadband absorption spectroscopy (BBAS). A pulsed LED has been used in a simple time-resolved BBAS measurement with 100 μs time resolution for the calibration of LIF on OH in a pulsed dielectric-barrier discharge (DBD) (Dilecce et al., 2012).

Known energy transfers between electronically excited states active in a reference condition can be used for *in situ* calibration. In Dilecce et al. (2007) the problem was the calibration of OODR-LIF on a $N_2(A)$ triplet metastable in a DBD at atmospheric pressure. The reference condition was the post-discharge of the DBD with N_2 + NO trace gas feed, in which it was possible to show that $N_2(A)$ is the precursor of both NO-γ and N_2 second positive system (SPS) emissions thanks to the following energy transfers:

$$N_2(A^3\Sigma_u^+) + NO(X^2\Pi) \rightarrow NO(A^2\Sigma^+) + N_2(X^1\Sigma_g^+), \tag{9.19}$$

$$N_2(A^3\Sigma_u^+) + N_2(A^3\Sigma_u^+) \rightarrow N_2(C^3\Pi_u) + N_2(X^1\Sigma_g^+), \tag{9.20}$$

whose rate constants are well known. The measured ratio of SPS to NO-γ emissions is then, in the reference condition, proportional to the density ratio $[N_2(A)]/[NO]$ through a known proportionality constant. Direct LIF measurement of [NO] is the last step to achieve $[N_2(A)]$. Similarly in the works of Dilecce et al. (2002, 2004) OH LIF was calibrated in a low-pressure radio frequency (RF) discharge after the observation that, in addition to Equation (9.19), another energy transfer involving $N_2(A)$ was the dominant precursor of the OH 3064 Å system:

$$N_2(A^3\Sigma_u^+) + OH(X^2\Sigma) \rightarrow OH(A^2\Sigma^+) + N_2(X^1\Sigma_g^+), \tag{9.21}$$

such that the ratio of OH and NO-γ emission is proportional to the ratio [OH]/[NO]. The simultaneous [NO] LIF measurement closed the procedure.

The calibration strategy can be specific to the device and condition under observation. In the works of Dilecce et al. (2000), the O TALIF in the expansion region of a plasma jet was calibrated by measuring the collision quenching of the $O(3p^3P_2)$ laser-excited state in a pure O_2 system with high dissociation degree.

Because the collision rate constant by O is significantly lower than that by O_2, the quenching rate could be correlated to the O density.

9.4 LIF Application to Active Species in Atmospheric Nonthermal Plasmas

9.4.1 OH

The measurement of OH in an atmospheric pressure DBD is a model case of the application of LIF to AP discharges. Due to research in combustion and atmospheric chemistry, the $OH(A^2\Sigma^+,v)$ vibronic states have been mostly investigated. Q and VET rate coefficients for the main air constituents, including water, and for some hydrocarbons are known (Wysong et al., 1990), together with their dependence on gas temperature (Tamura et al., 1998) and rotational level (Copeland et al., 1985; Cattolica and Mataga, 1991; Hemming et al., 2001), and their routes following the formation of van der Waals complexes with the quencher species (Lester et al., 1997). Collision rate coefficients for the main noble gases and air constituents at $T = 300$ K are reported in Table 9.2. The interest in the OH radical in AP discharges is also considerable. Many applications of AP discharges, like waste gas treatment, plasma-assisted combustion, plasma medicine, and, in general, plasmas in contact with liquids or discharges in liquid media, deal with gas mixtures with some degree of humidity. Large amounts of OH are formed and can play an important role in the plasma chemistry of the mentioned applications. Intense efforts have been recently devoted to this issue: LIF on OH in a pulsed corona discharge (Ono and Oda, 2002, 2003; Kanazawa et al., 2007; Nakagawa et al., 2011), in a DBD (Sankaranarayanan et al., 2000; Magne et al.,

TABLE 9.2 Rate Coefficients for $OH(A^2\Sigma^+, v = 0,1)$ Total Collision Quenching (Q + VET) and $1 \to 0$ Vibrational Relaxation at $T = 300$ K by He, Ar, N_2, O_2, and H_2O (10^{-11} cm^3 s^{-1})

	v = 0	v = 1	1 → 0	References
He		0.004 ± 0.0015^a	0.002^a	Dilecce and De Benedictis (2011)
			0.13 ± 0.04	Lengel and Crosley (1978)
Ar	$\leq 0.03^a$	0.3^a	0.27^a	Dilecce and De Benedictis (2011)
			0.41 ± 0.03	Williams and Crosley (1996)
			0.32 ± 0.04	Lengel and Crosley (1978)
N_2	2.8 ± 1.2^a	23.6 ± 1.5^a	23.3 ± 2.2^a	Williams and Crosley (1996)
	2.7 ± 1.0			Copeland et al. (1985)
		24.1 ± 1.3	21.1 ± 2.1	Burris et al. (1988)
		19 ± 1.5	24 ± 3	Copeland et al. (1988)
			14	German (1976)
H_2O	68.0 ± 6.1^a			Wysong et al. (1990)
	69.1 ± 5.0			Bailey et al. (1999)
		66 ± 4^a	≤ 12	Copeland et al. (1988)
			7.3 ± 0.5^a	Williams and Crosley (1996)
O_2	9.6 ± 1.2	20.6 ± 1.5	2.1 ± 0.2	Williams and Crosley (1996)
	9.2 ± 1.5			Copeland et al. (1985)
		20.2 ± 1.6	3.3 ± 0.7	Burris et al. (1988)
		17	1.5	German (1976)
Radiation	s^{-1}	s^{-1}		
A_{tot}	1.458×10^7	1.329×10^7		German (1975)
$A_{(0,0)}$	1.451×10^7			LIFBASE database
$A_{(1,1)}$		8.678×10^6		LIFBASE database

Source: [a] Values recommended in Dilecce, G. and S. De Benedictis, *Plasma Phys. Control. Fusion*, 53, 124006, 2011.
Notes: Radiative rates are also reported.

2007), in a pulsed DBD (Dilecce and De Benedictis, 2011; Dilecce et al., 2011), in a pulsed discharge over a liquid surface (Kanazawa et al., 2011), in a pin-to-pin single filament discharge (Verreycken et al., 2012), in a nanosecond H_2–air plasma for plasma-assisted combustion at 50–100 Torr (Choi et al., 2011). The excitation–detection scheme makes use of the transitions of the 3064 Å system of OH:

$$OH(X^2\Pi, v = 0) + h\nu_L \rightarrow OH(A^2\Sigma^+, v' = 1) \rightarrow OH(X^2\Pi, v'' = 1) + h\nu_F, \qquad (9.22)$$

which combines good absorption and emission coefficients with the requirement of sufficient spectral separation between the laser and detection wavelengths. Excitation of the $v' = 3$ level by absorption of laser light at 248 nm can also be used (Ono and Oda, 2002, 2003), which is useful at high pressure because the $v' = 3$ level is predissociative. Due to large spectral spacing of the rotational lines and to the absence of strong band heads, absorption on single rovibronic transition is possible and advantageous both for signal magnitude and for a good definition of the absorption process.

Q and VET rate coefficients are strongly dependent on the colliding gas, as shown in Table 9.2. An example of what can happen to the fluorescence features when the discharge gas mixture changes is shown in Figure 9.3, where numerical solutions of Equation 9.1 in Ar–H_2O mixtures at atmospheric pressure are shown (Dilecce and De Benedictis, 2011). At low water vapor content the electronic quenching Q is low and VET is fast, so the collision fluorescence from level $v' = 0$ prevails. With an increase in the H_2O partial pressure Q increases much faster than VET, so that less vibrational relaxation collisions are possible within the lifetime of $v' = 1$, and the direct fluorescence turns larger than the collision one. These calculations were found to be in good agreement in the works of Dilecce and De Benedictis (2011) with time-resolved measurements by a photomultiplier of $S(t)$ and with spectrally resolved measurements by a spectrograph equipped with an ICCD of $J_{LIF}(\lambda)$. Detection limits are also dependent on the gas mixture through the

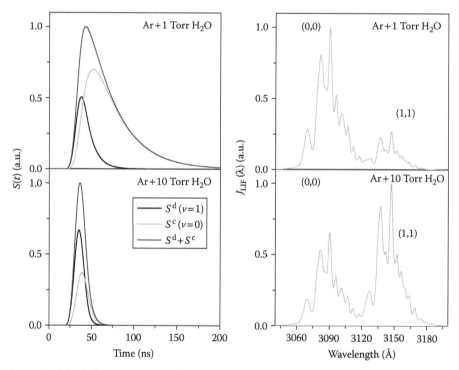

FIGURE 9.3 Model calculations of OH LIF fluorescence pulses in Ar–H_2O mixtures and of the spectrally resolved fluorescence, $J_{LIF}(\lambda)$. The latter is calculated by LIFBASE spectra of the (0,0) and (1,1) emission bands, weighted by time-integrated $S^c(t)$ and $S^d(t)$, respectively.

influence on Q. A minimum detectable [OH] density of 3×10^8 cm^{-3} was found with the apparatus of Dilecce et al. (2011) in He plus saturated water vapor mixture. In air plus saturated water vapor the limit was estimated to increase to about $0.5–1 \times 10^{10}$ cm^{-3}.

LIF can be an invasive diagnostic. In particular for the OH case two photochemical processes may interfere with the measurement:

$$O_3 + h\nu_L \rightarrow O(^1D) + O_2 \tag{9.23a}$$

$$O(^1D) + H_2O \rightarrow 2OH \tag{9.23b}$$

$$H_2O_2 + h\nu_L \rightarrow 2OH \tag{9.24}$$

In the Ono and Oda (2003) process Equation 9.23a and b was accounted for by subtracting its contribution from the total LIF signal after its evaluation with a known amount of ozone. In the works of Dilecce et al. (2011), Equation 9.24 was ruled out by taking advantage of the fact that OH produced by H_2O_2 photodissociation is rotationally hot (Gericke et al., 1986). The measurement of LIF excitation spectra of OH photo fragments showed that even at atmospheric pressure the memory of hot nascent nonthermal rotational distribution was preserved. In the discharge no trace of such a distribution was found, leading to the conclusion that OH produced by a process associated with Equation 9.24 was negligible compared to that produced by the discharge.

9.4.2 $N_2(A^3\Sigma_u^+)$

The $N_2(A)$ triplet metastable is a long-lived species characterized by the weak spin-forbidden Vegard–Kaplan emission. Its detection at low and at high pressure by LIF is usually based on a one-photon scheme of transitions of the $N_2(B^3\Pi_g \rightarrow A^3\Sigma_u^+)$ first positive system (FPS). Due to many FPS bands, many excitation–detection schemes can be adopted, with the possibility of detecting the whole $\nu = 0–9$ vibrational manifold of $N_2(A)$ (De Benedictis and Dilecce, 1997). Whatever the scheme, FPS ro-vibronic bands have a complex triplet fine structure, formed by 27 branches and multiple band heads (Herzberg, 1950; Geisen et al., 1987). In Figure 9.4 the contribution of various ro-vibronic branches to the excitation LIF spectrum of the (3,0) band, with detection on the (3,1) band, is shown. High sensitivity is achieved when the laser is tuned close to the strong ro-vibronic band heads, like that of P_{11} in the figure, with multiple and undetermined rotational lines absorption. Single line excitation is possible in a region where rotational lines are sufficiently spaced, but at the expense of lower sensitivity. In the works of De Benedictis et al. (1998) single (8,4) band $P_{12}(9)$ line excitation and single (8,5) band $Q_{11}(8)$ line detection allowed the evaluation of absolute $N_2(A)$ density in an RF discharge at low pressure. Band-head LIF excitation was applied in an AP point-to-plane corona discharge by Teramoto et al. (2009) in various gas mixtures with nitrogen. A limit of the one-photon scheme is the low quantum yield at atmospheric pressure. Due to low emission probabilities of FPS bands, of the order of 10^4 s^{-1} (Loftus and Krupenie, 1977), and high Q rates, with rate coefficients of the order of 10^{-11} cm^3 s^{-1} (Piper, 1992), the quantum yield can be of the order of $10^{-3}–10^{-4}$. In the works of Ono et al. (2009c) the strong FPS emission from a pulsed positive corona discharge was overwhelming, in spite of the large $N_2(A)$ density of 10^{13} cm^{-3}, such that it was necessary to make the LIF measurement at 4 μs in the postdischarge to reduce the discharge emission. To overcome this problem an OODR-LIF scheme was proposed in the works of Dilecce et al. (2005):

$$N_2(A^3\Sigma_u^+, \nu'' = 0) + h\nu_{L1} \rightarrow N_2(B^3\Pi_g, \nu' = 3) + h\nu_{L2} \rightarrow$$
$$\rightarrow N_2(C^3\Pi_u, \nu = 2) \rightarrow N_2(B^3\Pi_g, \nu' = 1) + h\nu_F \tag{9.25}$$

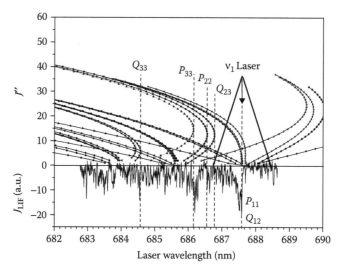

FIGURE 9.4 Branch contributions to excitation LIF spectrum of (3,0) band of FPS with detection on (3,1) band. J'' is the lower rotational number (Hund's case a). The detection window of (3,1) seen on the excitation spectrum is shown. Only the most important turning branches are labeled. The arrow indicates the laser position used for the OODR experiment. (From Dilecce, G., et al., *Plasma Sources Sci. Technol.*, 16, 511–522, 2007. With permission of IOP Publishing, Ltd.)

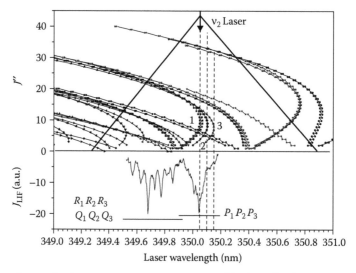

FIGURE 9.5 Branch contributions to excitation LIF spectrum of (2,3) SPS band with detection on (2,1) band. The detection window of (2,1) band seen on the excitation spectrum is shown. Labels 1, 2, and 3 indicate the strong turning branches P_{11}, P_{22}, and P_{33}. (From Dilecce G. et al., *Plasma Sources Sci. Technol.*, 16, 511–522, 2007. With permission of IOP Publishing, Ltd.)

that, due to the much larger emission rates of the $N_2(C^3\Pi_u \rightarrow B^3\Pi_g)$ SPS (Loftus and Krupenie, 1977), yields a LIF signal that is 2 orders of magnitude larger. The LIF excitation spectrum of the second step, measured when the first laser is in the position indicated by the arrow in Figure 9.4, is shown in Figure 9.5. The spectral detection window collects fluorescence from almost all the strongest branches. This scheme was applied to a N_2 diffuse DBD in the works of Dilecce et al. (2007), with peak densities of about 1.5×10^{13} cm^{-3}, and a time evolution in the discharge cycles, in agreement with model calculations (Massines et al., 2005). The minimum detectable density was about 10^{10} cm^{-3}. With small O_2 admixtures it was possible to correlate the diffuse-to-filamentary transition to the increase of $N_2(A)$ loss rate. In Ambrico et al. (2008, 2009) and Simek et al. (2010) it was also applied to the space afterglow of a surface DBD.

9.4.3 N and O

N and O atoms have been investigated by TALIF in AP nonthermal plasma sources due to their relevance to materials sciences, plasma medicine, pollution control, ozone generation, and plasma-assisted combustion. Density measurements at atmospheric pressure have focused on pure N_2, O_2, or mixtures with He/Ar; air-like, including NO_x. $N(2p^{34}S_{3/2})$ and $O(2p^{43}P_2)$, ground-state sublevels are monitored by applying the calibrated TALIF scheme outlined before, and the total ground state population is reconstructed according to the Boltzmann factor of the sublevels. Peak N densities of 3×10^{14} cm^{-3} were measured in a N_2 DBD in the Townsend regime (Es-Sebbar et al., 2009), and similar densities were found in the filamentary regime, in spite of the higher expected densities in the filaments. Although LIF is space-resolved, the filaments occur randomly in space and time in the probed volume, so that the many shots-averaged TALIF outcome correctly describes the average density in the volume but not the density inside each microdischarge. Spatially and temporally resolved relative measurements, synchronized to the filament, were conducted in a pin-to-plate DBD in Lukas et al. (2001), showing an N density restricted to a 200 µm thick channel within about 100 µs of the filament ignition. In the filament of a pin-to-plate corona discharge in N_2, a N density of the order of 10^{16} cm^{-3} was measured (Ono et al., 2009b). In the same pin-to-plate corona discharge in a N_2–O_2 mixture, relative O atom density measurements were made (Ono et al., 2005, 2009a). In these papers it was pointed out that part of the TALIF signal was produced by O_2 and O_3 photodissociation by the intense focused 226 nm laser beam. The O_2 photodissociation contribution was measured without discharge. O_3 is formed in the discharge, and the contribution of its photodissociation to the TALIF signal was measured in late postdischarge, when O density was reasonably negligible. Time-resolved O atom density measurements were reported in the works of Stancu et al. (2010) in a pin-to-pin nanosecond repetitive pulse (NRP) discharge at atmospheric pressure for flame ignition and stabilization. High density values were found ranging from 4×10^{14} cm^{-3} to 2×10^{16} cm^{-3} when the discharge changes from diffuse to filamentary regime on raising the applied voltage. An RF-excited atmospheric pressure plasma jet (APPJ) was investigated in the works of Niemi et al. (2005). Two-dimensional maps of He + 0.5% O_2 mixture effluent from the coaxial discharge were recovered, and the density was found to decrease from 2×10^{15} cm^{-3} at the nozzle exit to about 10^{13} cm^{-3} 10 cm downstream. Extension of TALIF investigations to the discharge volume was carried out in a micro-APPJ specifically designed for direct optical access (Knake et al., 2008; Knake and Shultz-von der Gaten, 2010). The spaced-resolved exploration of the discharge channel revealed high atom density of about 10^{13} cm^{-3} with He + 0.5% O_2 gas feed. Finally, O atom densities measurements in a N_2 DBD with O_2 and N_2O additives have been reported in the works of Massines et al. (2008) together with N and NO measurements by LIF. Densities of the order of 10^{14} cm^{-3} were found when O_2 and N_2O were added in quantities of some hundreds of parts per million.

9.4.4 CH and CN

Although CH might have a role in hydrocarbon reforming (Chen et al., 2008) or VOC abatement (Kim, 2004), its detection in atmospheric nonthermal plasmas has not been explored. The only such application to the best of our knowledge is that of the works of Dilecce et al. (2010a, 2010b), in which relative LIF measurements were ancillary to the determination of processes giving rise to CH Gerö bands emission [CH ($A^2\Delta \rightarrow X^2\Pi$)]. The state of the art on CH LIF detection/measurement has come to light from combustion research. Almost all possible transitions involving the three electronic states $A^2\Delta$, $B^2\Sigma^-$, and $C^2\Sigma^+$ have been employed (Hirano et al., 1992; Luque et al., 2002; Moreau et al., 2003). The C state is predissociative and the B state is predissociative for $v = 0$, $N \geq 14$ and $v = 1$, $N \geq 7$. Quenching rate constants by various collision partners have been measured for *A* state in the works of Tamura et al. (1998), while *B* state collision processes have been characterized in a flame environment in the works of Luque et al. (2000). Densities as low as 10^{11}–10^{12} cm^{-3} have been detected in atmospheric pressure flames (Moreau et al., 2003), and planar LIF has been demonstrated in the works of Tsujishita et al. (1993). All the possible $(A,B,C \rightarrow X)$ electronic transitions are strongly diagonal. The best combination of diagonal/off-diagonal absorption/emission coefficients is given by the scheme:

$$CH(X^2\Pi, v') + h\nu_L \rightarrow CH(B^2\Sigma^-, v) \rightarrow CH(X^2\Pi, v'') + h\nu_F, \tag{9.26}$$

with $v' = 0$ and $v = v'' = 1$, adopted from the works of Dilecce et al. (2010b) for a He–CH$_4$ DBD, with absorption of the $Q_1(2)$ rotational line at $\lambda_L = 363.798$ nm and observation centered at about 405 nm with 10 nm Bw. The measurement is difficult because of the low quantum yield and the low CH density and highlights all the difficulties of LIF application to AP discharges. It was found that the CH reactivity restricts its presence to the discharge phase, with fast density decay in the postdischarge phase, an upper limit of 10^{11} cm^{-3} being estimated.

The cyano radical, CN, is abundant in discharges containing hydrocarbons and nitrogen, and its violet system emission, $CN(B^2\Sigma^+ \rightarrow X^2\Sigma^+)$, characterizes the emission spectra of these discharges in conditions in which both gas-phase and gas-surface reactions play a role (Dilecce et al., 2009). After reacting in N$_2$+CH$_4$ mixtures, a brownish deposit on the dielectric electrode surface is formed. In presence of the deposit, a strong CN emission was still observed even with pure N$_2$ gas feed. In this study LIF was employed for relative, time-resolved measurement to look for correlations between the CN ground state and the emission, that is, to validate or negate the presence of CN(B) state excitation processes involving the CN ground state. The excitation–detection scheme employed transitions of the violet system bands:

$$CN(X^2\Sigma^+, v') + h\nu_L \rightarrow CN(B^2\Sigma^+, v) \rightarrow CN(X^2\Sigma^+, v'') + h\nu_F \tag{9.27}$$

with $v - v' = 0$ and $v - v'' = -1$ for spectral separation of fluorescence and laser radiation. Laser excitation scans in the whole spectral range of the $\Delta v = 0$ sequence revealed the features of the ground-state vibrational distribution. As shown in Figure 9.6, vibrational excitation up to level $v' = 9$ was observed in pure N$_2$ gas feed, while small addition of CH$_4$ to the gas mixture turned to a lower but still suprathermal vibrational distribution. LIF spectra then, in conditions of low vibrational relaxation, were able to reveal exothermic CN formation mechanisms that were likely identified as $C + N + M \rightarrow CN(A, B) + M$ plus radiative cascade in the case of pure N$_2$ + surface deposit, replaced, on addition of methane, by $CH + N \rightarrow CN(A, B)$ plus radiative cascade.

9.5 Concluding Remarks

The big advantage of LIF is its unique combination of sensitivity, time, and space resolution, while its major imperfection is that in most cases it requires a calibration technique for reliable absolute density measurements. Universal recipes, as always, do not exist, while each atomic/molecular case has its own peculiarities. A LIF measurement is, in principle, and from an analytical point of view, an absorption process, but with a different observable with respect to absorption techniques. Such a different observable, the fluorescence, is the reason of the following LIF advantages compared to absorption: higher sensitivity (except for cavity-enhanced absorption); higher spatial resolution, from the geometrical crossing of absorption and observation; and double selectivity—fluorescence spectral selectivity in addition to laser absorption selectivity. The disadvantage of a different observable is its strong dependence on collisional processes in the upper electronic level, which at high pressure reduces the sensitivity and complicates the interpretation of the outcomes. Such a pitfall can sometimes be a boon since, if the collisional processes are quantitatively well known, the analysis of electronic quenching and vibrational relaxation can provide information on the gas composition when it is unknown. We finally point out that LIF has an added

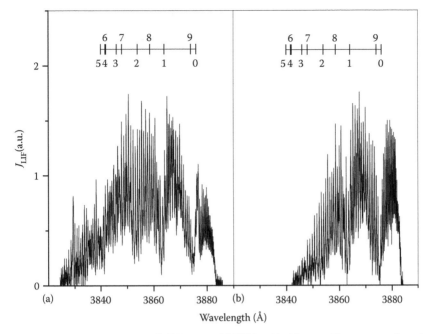

FIGURE 9.6 LIF excitation spectrum of CN in an AP-DBD with (a) pure N_2 gas + surface deposit and (b) N_2 + 0.2% CH_4. The band origins of the $\Delta v = 0$ bands are indicated, with v labeling. (Data from Dilecce, G. et al., *Plasma Sources Sci. Technol.*, 18, 015010, 2009. With permission of IOP Publishing, Ltd.)

value as a mean for investigation of collisional processes in electronic states with selective excitation. These studies can be very useful both in fundamental collision processes research and in fields where collision–radiative models are required, like in the analysis of emission spectra of discharges and of upper atmosphere plasmas.

References

Altkorn, R. and R. Zare (1984). Effects of saturation on laser-induced fluorescence measurements of population and polarization. *Ann. Rev. Phys. Chem.* 35, 265–289.

Ambrico, P. F., M. Simek, G. Dilecce, and S. De Benedictis (2008). On the measurement of $N_2(A^3\Sigma_u^+)$ metastable in N_2 surface dielectric barrier discharge at atmospheric pressure. *Plasma Chem. Plasma Process.* 28, 299–316.

Ambrico, P. F., M. Simek, G. Dilecce, and S. De Benedictis (2009). $N_2(A^3\Sigma_u^+)$ time evolution in N_2 atmospheric pressure surface dielectric barrier discharge driven by ac voltage under modulated regime. *Appl. Phys. Lett.* 94, 231503.

Amorim, J., G. Baravian, and J. Jolly (2000). Laser-induced resonance fluorescence as a diagnostic technique in non-thermal equilibrium plasmas. *J. Phys. D Appl. Phys.* 33, R51–R65.

Amorim, J., G. Baravian, M. Touzeau, and J. Jolly (1994). Two-photon laser induced fluorescence and amplified spontaneous emission atom concentration measurements in O_2 and H_2 discharges. *J. Appl. Phys.* 76, 1487–1493.

Avan, P. and C. Cohen-Tannoudji (1977). Two-level atom saturated by a fluctuating resonant laser beam. Calculation of the fluorescence spectrum. *J. Phys. B At. Mol. Opt. Phys.* 10, 155.

Bailey, A. E., D. E. Heard, D. A. Henderson, and P. H. Paul (1999). Collisional quenching of OH($A^2\Sigma^+$, $v' = 0$) by H_2O between 211 and 294 K and the development of a unified model for quenching. *Chem. Phys. Lett.* 302, 132–138.

Bamford, D. J., L. E. Jusinski, and W. K. Bischel (1986). Absolute two-photon absorption and three-photon ionization cross sections for atomic oxygen. *Phys. Rev. A* 34, 185–198.

Barnat, E. V. (2011). Multi-dimensional optical and laser-based diagnostics of low-temperature ionized plasma discharges. *Plasma Sources Sci. Technol.* 20, 053001.

Bittner, J., A. Lawitzki, U. Meier, and K. Kohse-Höinghaus (1991). Nitrogen atom detection in low-pressure flames by two-photon laser-excited fluorescence. *Appl. Phys. B-Lasers O.* 52, 108–116.

Born, M. and E. Wolf (1999). *Principles of Optics.* Cambridge University Press, Cambridge.

Burris, J., J. J. Butler, T. J. McGee, and W. S. Heaps (1988). Collisional deactivation rates for $A^2\Sigma^+(v' = 1)$ state of OH. *Chem. Phys.* 124, 251–258.

Cathey, C., J. Cain, H. Wang, M. A. Gundersen, C. Carter, and M. Ryan (2008). OH production by transient plasma and mechanism of flame ignition and propagation in quiescent methane–air mixtures. *Combust. Flame* 154, 715–727.

Cattolica, R. J. and T. G. Mataga (1991). Rotational-level-dependent quenching of OH($A^2\Sigma$, $v' = 1$) by collisions with H_2O in a low-pressure flame. *Chem. Phys. Lett.* 182, 623–631.

Chen, H. L., H. M. Lee, S. H. Chen, Y. Chao, and M. B. Chang (2008). Review of plasma catalysis on hydrocarbon reforming for hydrogen production–Interaction, integration, and prospects. *Appl. Catal. B Environ.* 85, 1–9.

Choi, I., Z. Yin, I. V. Adamovich, and W. R. Lempert (2011). Hydroxyl radical kinetics in repetitively pulsed hydrogen-air nanosecond plasmas. *IEEE Trans. Plasma Sci.* 39, 3288–3299.

Clyne, M. A. A. and W. S. Nip (1979). Generation and measurement of Atom and radical concentration in flow system. In D. W. Setser (Ed.), *Reactive Intermediates in the Gas Phase: Generation and Monitoring.* Academic Press, New York. pp. 1–57.

Copeland, R. A., M. J. Dyer, and D. R. Crosley (1985). Rotational-level-dependent quenching of $A^2\Sigma^+$ OH and OD. *J. Chem. Phys.* 82, 4022–4032.

Copeland, R. A., M. L. Wise, and D. R. C. and (1988). Vibrational energy transfer and quenching of hydroxyl ($A^2\Sigma^+$, $v' = 1$). *J. Phys. Chem.* 92, 5710–5715.

De Benedictis, S. and G. Dilecce (1995). Laser-induced fluorescence measurements of He(2^3S) decay in He-N_2/O_2 pulsed RF discharges: Penning ionization. *J. Phys. D Appl. Phys.* 28, 2067–2076.

De Benedictis, S. and G. Dilecce (1997). Rate constants for deactivation of N_2(A, v = 2 7) by O, O_2, and NO. *J. Chem. Phys.* 107, 6219–6229.

De Benedictis, S., G. Dilecce, and V. L. Lepore (1997). Laser-induced fluorescence and polarized light in the collision regime: the He(2s 3 S 4p 3 P) case. *J. Phys. B At. Mol. Opt. Phys.* 30, 5367–5379.

De Benedictis, S., G. Dilecce, and M. Simek (1998). LIF measurement of $N_2(A^3\Sigma_u^+, u = 4)$ population density in a pulsed rf discharge. *J. Phys. D Appl. Phys.* 31, 1197–1205.

Demtröder, W. (2003). *Laser Spectroscopy.* Springer-Verlag, Berlin/Heidelberg/New York.

Dilecce, G., P. F. Ambrico, and S. De Benedictis (2004). An ambient air RF low-pressure pulsed discharge as an OH source for LIF calibration. *Plasma Sources Sci. Technol.* 13, 237–244.

Dilecce, G., P. F. Ambrico, and S. De Benedictis (2005). Optical–optical double resonance LIF detection of $N_2(A^3\Sigma_u^+)$ in high pressure gas discharges. *Plasma Sources Sci. Technol.* 14, 561–565.

Dilecce, G., P. F. Ambrico, and S. De Benedictis (2007). $N_2(A^3\Sigma_u^+)$ density measurement in a dielectric barrier discharge in N_2 and N_2 with small O_2 admixtures. *Plasma Sources Sci. Technol.* 16, 511–522.

Dilecce, G., P. F. Ambrico, and S. De Benedictis (2010a). Optical diagnostics in dielectric barrier discharges at atmospheric pressure. *Pure Appl. Chem.* 82, 1201–1207.

Dilecce, G., P. F. Ambrico, and S. De Benedictis (2010b). CH spectroscopic observables in He–CH_4 and N_2–CH_4 atmospheric pressure dielectric barrier discharges. *J. Phys. D Appl. Phys.* 43 (12), 124004.

Dilecce, G., P. F. Ambrico, G. Scarduelli, P. Tosi, and S. De Benedictis (2009). CN($B^2\Sigma^+$) formation and emission in a N_2–CH_4 atmospheric pressure dielectric barrier discharge. *Plasma Sources Sci. Technol.* 18, 015010.

Dilecce, G., P. F. Ambrico, M. Simek, and S. De Benedictis (2002). New laser-induced fluorescence scheme for simultaneous OH and NO by a single laser set-up. *Appl. Phys. B* 75, 131–135.

Dilecce, G., P. F. Ambrico, M. Simek, and S. De Benedictis (2011). LIF diagnostics of hydroxyl radical in atmospheric pressure He–H_2O dielectric barrier discharges. *Chem. Phys.* doi:10.1016/j.chemphys.2011.03.012.

Dilecce, G., P. F. Ambrico, M. Simek, and S. De Benedictis (2012). OH density measurement by time-resolved broad band absorption spectroscopy in a Ar–H_2O dielectric barrier discharge. *J. Phys. D Appl. Phys* 45, 125203.

Dilecce, G. and S. De Benedictis (2011). Laser induced fluorescence detection of OH in He/Ar–H_2O DBD. *Plasma Phys. Controlled Fusion* 53, 124006.

Dilecce, G., M. Vigliotti, and S. De Benedictis (2000). A TALIF calibration method for quantitative oxygen atom density measurement in plasma jets. *J. Phys. D Appl. Phys.* 33, L53–L56.

Döbele, H. F., T. Mosbach, K. Niemi, and V. Shultz-von der Gaten (2005). Laser-induced fluorescence measurements of absolute atomic densities: concepts and limitations. *Plasma Sources Sci. Technol.* 14, S31–S41.

Dreyer, C. B., S. M. Spuler, and M. Linne (2001). Calibration of laser induced fluorescence of the OH radical by cavity ringdown spectroscopy in premixed atmospheric pressure flames. *Combust. Sci. Technol.* 171, 163–190.

Eberhard, W. L. (2010). Correct equations and common approximations for calculating Rayleigh scatter in pure gases and mixtures and evaluation of differences. *Appl. Opt.* 49, 1116–1130.

Es-Sebbar, E., C. Sarra-Bournet, N. Naud, F. O. Massines, and N. Gherardi (2009). Absolute nitrogen atom density measurements by two-photon laser-induced fluorescence spectroscopy in atmospheric pressure dielectric barrier discharges of pure nitrogen. *J. Appl. Phys.* 106, 073302.

Freegarde, T. G. M. and G. Hancock (1997). A guide to laser-induced fluorescence diagnostics in plasmas. *J. Phys. IV Colloque C4* 7, 15–29.

Geisen, H., D. Neuschäfer, and C. Ottinger (1987). Hyperfine structure of $N_2(B^3\Pi_g$ and $A^3 \Sigma_u^+)$ from LIF measurements on a beam of metastable N_2 molecules. *Z. Phys. D* 4, 263–290.

Gericke, K.-H., S. Klee, and F. J. Comes (1986). Dynamics of H_2O_2 photodissociation: OH product state and momentum distribution characterized by sub-Doppler and polarization spectroscopy. *J. Chem. Phys.* 85 (8), 4463–4479.

German, K. R. (1975). Radiative and predissociative lifetimes of the $v' = 0,1$, and 2 levels of the $A^2\Sigma^+$ state of OH and OD. *J. Chem. Phys.* 63, 5252.

German, K. R. (1976). Collision and quenching cross sections in the $A^2\Sigma^+$ state of OH and OD. *J. Chem. Phys.* 64, 4065–4068.

Goehlich, A., T. Kawetzki, and H. F. Döbele (1998). On absolute calibration with xenon of laser diagnostic methods based on two-photon absorption. *J. Chem. Phys.* 108, 9362–9370.

Greene, C. and R. Zare (1983). Determination of product population and alignment using laser-induced fluorescence. *J. Chem. Phys.* 78, 6741.

Hemming, B. L., D. R. Crosley, J. E. Harrington, and V. Sick (2001). Collisional quenching of high rotational levels in $A^2\Sigma^+$ OH. *J. Chem. Phys.* 115, 3099–3104.

Herzberg, G. (1950). *Molecular Spectra and Molecular Structure of Diatomic Molecules*. Van Nostrand Company Inc., Princeton, NJ.

Hirano, A., M. Ipponmatsu, and M. Tsujishita (1992). Two-dimensional digital imaging of the CH distribution in a natural gas/oxygen flame at atmospheric pressure and detection of A-state emission by means of C-state excitation. *Opt. Lett.* 17, 303–304.

Kanazawa, S., H. Kawano, S. Watanabe, T. Furuki, S. Akamine, R. Ichiki, T. Ohkubo, M. Kocik, and J. Mizeraczyk (2011). Observation of OH radicals produced by pulsed discharges on the surface of a liquid. *Plasma Sources Sci. Technol.* 20, 034010.

Kanazawa, S., H. Tanaka, A. Kajiwara, T. Ohkubo, Y. Nomoto, M. Kocik, J. Mizeraczyk, and J.-S. Chang (2007). LIF imaging of OH radicals in DC positive streamer coronas. *Thin Solid Films* 515 (2), 4266–4271.

Kim, H.-H. (2004). Nonthermal plasma processing for air-pollution control: A historical review, current issues, and future prospects. *Plasma Process. Polym.* 1, 91–110.

Knake, N., K. Niemi, S. Reuter, V. Shultz-von der Gaten, and J. Winter (2008). Absolute atomic oxygen density profiles in the discharge core of a microscale atmospheric pressure plasma jet. *Appl. Phys. Lett.* 93, 131503.

Knake, N. and V. Shultz-von der Gaten (2010). Investigations of the spatiotemporal build-up of atomic oxygen inside the micro-scaled atmospheric pressure plasma jet. *Eur. Phys. J. D* 60, 645–652.

Krames, B., T. Glenewinkel-Meyer, and J. Meichsner (2001). In situ determination of absolute number densities of nitrogen molecule triplet states in an rf-plasma sheath. *J. Appl. Phys.* 89, 3115.

Kunze, H. J. (1968). The laser as tool for plasma diagnostics. In W. Lochte-Holtgreven (Ed.), *Plasma Diagnostic Techniques*. North-Holland publishers, Amsterdam. pp. 550–616.

Lengel, R. K. and D. R. Crosley (1978). Energy transfer in $A^2\Sigma^+$ OH. Vibrational. *J. Chem. Phys.* 68, 5309–5324.

Lester, M. I., R. A. Loomis, and R. L. Schwartz (1997). Electronic Quenching of $OH(A^2\Sigma^+, v' = 0, 1)$ in Complexes with Hydrogen and Nitrogen. *J. Phys. Chem. A* 101, 9195–9206.

Loftus, A. and P. H. Krupenie (1977). The spectrum of molecular nitrogen. *J. Phys. Chem. Ref. Data* 6, 113.

Lukas, C., M. Spaan, V. Shultz-von der Gaten, M. Thomson, R. Wegst, H. F. Döbele, and M. Neiger (2001). Dielectric barrier discharges with steep voltage rise: Mapping of atomic nitrogen in single filaments measured by laser-induced fluorescence spectroscopy. *Plasma Sources Sci. Technol.* 10, 18171–18175.

Luque, J. and D. Crosley (1999a). International report P99-009. Technical report, SRI—free download at http://www.sri.com/psd/lifbase/.

Luque, J. and D. R. Crosley (1999b). Radiative, collisional, and predissociative effects in CH laser-induced-fluorescence flame thermometry. *Appl. Opt.* 38, 1423–1433.

Luque, J., R. J. H. Klein-Douwel, J. B. Jeffries, and D. R. Crosley (2000). Collisional processes near the CH $B^2\Sigma^-$ $v' = 0, 1$ predissociation limit in laser-induced fluorescence flame diagnostics. *Appl. Phys. B* 71, 85–94.

Luque, J., R. J. H. Klein-Douwel, J. B. Jeffries, G. P. Smith, and D. R. Crosley (2002). Quantitative laser-induced fluorescence of CH in atmospheric pressure flames. *Appl. Phys. B* 75, 779–790.

Magne, L., S. Pasquiers, N. Blin-Simiand, and C. Postel (2007). Production and reactivity of the hydroxyl radical in homogeneous high pressure plasmas of atmospheric gases containing traces of light olefins. *J. Phys. D Appl. Phys.* 40, 3112–3127.

Marro, F. G. and W. G. Graham (2008). Atomic chlorine two photon LIF characterization in a Cl_2 ICP plasma. *Plasma Sources Sci. Technol.* 17, 015007.

Massines, F., E. Es-Sebbar, N. Gherardi, D. Naudé, D. Tsyganov, P. Ségur, and S. Pancheshnyi (2008). Comparison of Townsend dielectric barrier discharge in N_2, N_2–O_2 and N_2–N_2O behaviour and density of radicals. *35th EPS Conference on Plasma Phys. Hersonissos,* 9–13 June 2008. Vol.32D, p. 2.169.

Massines, F., N. Gherardi, N. Naudé, and P. Ségur (2005). Glow and Townsend dielectric barrier discharge in various atmosphere. *Plasma Phys. Control. Fusion* 47, B577–B588.

McIlroy, A. and J. B. Jeffries (2002). Cavity ringdown spectroscopy for concentration measurements. In K. Kohse-Höinghaus and J. B. Jeffries (Eds.), *Applied Combustion Diagnostics*. Taylor & Francis, New York.

Miles, R., W. Lempert, and J. Forkey (2001). Laser Rayleigh scattering. *Meas. Sci. Technol.* 12, R33.

Moreau, C., E. T. P. Desgroux, J. Pauwels, A. Chapput, and M. Barj (2003). Quantitative measurements of the CH radical in sooting diffusion flames at atmospheric pressure. *Appl. Phys. B* 76, 597–602.

Nakagawa, Y., R. Ono, and T. Oda (2011). Density and temperature measurement of OH radicals in atmospheric-pressure pulsed corona discharge in humid air. *J. Appl. Phys.* 110, 073304.

Naus, H. and W. Ubachs (2000). Experimental verification of Rayleigh scattering cross sections. *Opt. Lett.* 25 (5), 347–349.

Nemschokmichal, S., F. Bernhardt, B. Krames, and J. Meichsner (2011). Laser-induced fluorescence spectroscopy of and absolute density calibration by Rayleigh scattering in capacitively coupled rf discharges. *J. Phys. D Appl. Phys.* 44, 205201.

Nemschokmichal, S. and J. Meichsner (2013a). $N_2(A^3\Sigma_u^+)$ metastable density in nitrogen barrier discharges: I. LIF diagnostics and absolute calibration by Rayleigh scattering. *Plasma Sources Sci. Technol.* 22, 015005 (8 pp).

Nemschokmichal, S. and J. Meichsner (2013b). $N_2(A^3\Sigma_u^+)$ metastable density in nitrogen barrier discharges: II. Spatio-temporal behaviour in filamentary mode. *Plasma Sources Sci. Technol.* 22, 015006 (8 pp).

Niemi, K., V. Shultz-von der Gaten, and H. F. Döbele (2001). Absolute calibration of atomic density measurements by laser-induced fluorescence spectroscopy with two-photon excitation. *J. Phys. D Appl. Phys.* 34, 2330–2335.

Niemi, K., V. Shultz-von der Gaten, and H. F. Döbele (2005). Absolute atomic oxygen density measurements by two-photon absorption laser-induced fluorescence spectroscopy in an RF-excited atmospheric pressure plasma jet. *Plasma Sources Sci. Technol.* 14, 375–386.

Omidvar, K. (1980a). Erratum: Two-photon excitation cross section in light and intermediate atoms in frozen-core LS-coupling approximation. *Phys. Rev. A* 30, 2805.

Omidvar, K. (1980b). Two-photon excitation cross section in light and intermediate atoms in frozen-core LS-coupling approximation. *Phys. Rev. A* 22 (4), 1576–1587.

Ono, K., T. Oomori, M. Tudi, and K. Namba (1992). Measurements of the Cl atom concentration in radiofrequency and microwave plasmas by two-photon laser induced fluorescence: Relation to the etching of Si. *J. Vac. Sci. Technol. A* 10, 1071–1079.

Ono, R. and T. Oda (2002). Dynamics and density estimation of hydroxyl radicals in a pulsed corona discharge. *J. Phys. D Appl. Phys.* 35, 2133–2138.

Ono, R. and T. Oda (2003). Dynamics of ozone and OH radicals generated by pulsed corona discharge in humid-air flow reactor measured by laser spectroscopy. *J. Appl. Phys.* 93 (10), 5876–5882.

Ono, R., K. Takezawa, and T. Oda (2009a). Two-photon absorption laser-induced fluorescence of atomic oxygen in the afterglow of pulsed positive corona discharge. *J. Appl. Phys.* 106, 043302.

Ono, R., Y. Teramoto, and T. Oda (2009b). Measurement of atomic nitrogen in N_2 pulsed positive corona discharge using two-photon absorption laser-induced fluorescence. *Jap. J. Appl. Phys.* 48, 122302.

Ono, R., C. Tobaru, Y. Teramoto, and T. Oda (2009c). Laser-induced fluorescence of $N_2(A^3\Sigma_u^+)$ metastable in N_2 pulsed positive corona discharge. *Plasma Sources Sci. Technol.* 18, 025006.

Ono, R., Y. Yamashita, K. Takezawa, and T. Oda (2005). Behaviour of atomic oxygen in a pulsed dielectric barrier discharge measured by laser-induced fluorescence. *J. Phys. D Appl. Phys.* 38, 2812–2816.

Pezé, P., A. Paillous, J. Siffre, and B. Dubreuil (1993). Quantitative measurements of oxygen atom density using LIF. *J. Phys. D Appl. Phys.* 26, 1622–1629.

Piper, L. G. (1992). Energy transfer studies on $N_2(X^1\Sigma_g^+, v)$ and $N_2(B^3\Pi_g)$. *J. Chem. Phys.* 97, 270–275.

Rensberger, K. J., J. B. Jeffries, R. A. Copeland, K. Kohse-Höinghaus, M. L. Wise, and D. R. Crosley (1989). Laser-induced fluorescence determination of temperatures in low pressure flames. *Appl. Opt.* 28, 3556–3566.

Sankaranarayanan, R., B. Pashaie, and S. K. Dhalib (2000). Laser-induced fluorescence of OH radicals in a dielectric barrier discharge. *Appl. Phys. Lett.* 77 (19), 2970–2972.

Saxon, R. P. and J. Eichler (1986). Theoretical calculation of two-photon absorption cross sections in atomic oxygen. *Phys. Rev. A* 34, 199–206.

Simek, M., P. F. Ambrico, S. D. Benedictis, G. Dilecce, V. Prukner, and J. Schmidt (2010). $N_2(A^3\Sigma_u^+)$ behaviour in a N_2–NO surface dielectric barrier discharge in the modulated ac regime at atmospheric pressure. *J. Phys. D Appl. Phys.* 43 (12), 124003.

Smith, K. C. and D. R. Crosley (2002). Detection of minor species with laser techniques. In K. Kohse-Höinghaus and J. B. Jeffries (Eds.), *Applied Combustion Diagnostics*. Taylor & Francis, New York.

Stancu, G. D., F. Kaddouri, D. A. Lacoste, and C. O. Laux (2010). Atmospheric pressure plasma diagnostics by OES, CRDS and TALIF. *J. Phys. D Appl. Phys.* 43, 124002.

Stevens, P. S., J. H. Mather, and W. H. Brune (1994). Measurement of tropospheric OH and HO_2 by laser induced fluorescence. *J. Geophys. Res.* 99, 3543–3557.

Tamura, M., P. A. Berg, J. E. Harrington, J. Luque, J. B. Jeffries, G. P. Smith, and D. R. Crosley (1998). Collisional quenching of CH(A), OH(A), and NO(A) in low pressure hydrocarbon flames. *Combust. Flame* 114, 502–514.

Telle, H. H., A. G. Ureña, and R. Donovan (2007). *Laser Chemistry: Spectroscopy, Dynamics and Applications.* John Wiley & Sons Ltd.

Teramoto, Y., R. Ono, and T. Oda (2009). Measurement of $N_2(A^3\Sigma_u^+)$ metastable in N_2 pulsed positive corona discharge with trace amounts of additives. *J. Phys. D Appl. Phys.* 42, 235205.

Tserepi, A., E. Wurzberg, and T. A. Miller (1997). Two-photon-excited stimulated emission from atomic oxygen in rf plasmas: Detection and estimation of its threshold. *Chem. Phys. Lett.* 265, 297–302.

Tsujishita, M., M. Ipponmatsu, and A. Hirano (1993). Visualization of the CH molecule by exciting $C^2\Sigma^+(v = 1)$ state in turbulent flames by planar laser induced fluorescence. *Jpn. J. Appl. Phys.* 32, 5564–5569.

van Lessen, M., R. Schnabel, and M. Kock (1998). Population densities of Fe I and Fe II levels in an atomic beam from partially saturated LIF signals. *J. Phys. B At. Mol. Opt. Phys.* 31, 1931–1946.

Verreycken, T., R. M. van der Horst, A. H. F. M. Baede, E. M. V. Veldhuizen, and P. J. Bruggeman (2012). Time and spatially resolved LIF of OH in a plasma filament in atmospheric pressure $He–H_2O$. *J. Phys. D Appl. Phys.* 45, 045205.

Williams, L. R. and D. R. Crosley (1996). Collisional vibrational energy transfer of OH (A). *J. Chem. Phys.* 104, 6507–6514.

Wysong, I. J., J. B. Jeffries, and D. R. Crosley (1990). Quenching of $A^2\Sigma^+$ OH at 300 K by several colliders. *J. Chem. Phys.* 92, 5218–5222.

III

Applications

285

10

Plasma Technology in Silicon Photovoltaics

Shaoqing Xiao,
Shuyan Xu, and
Haiping Zhou

10.1 Introduction

Photovoltaic (PV) materials have commanded a great deal of attention in recent years because PV power appears to be one of the major technologies for tackling current greenhouse or global warming effects. Among the PV materials, silicon (Si), as the second most abundant element on earth, has attracted tremendous interest for applications in solar cells. Thus, Si-based solar cells have been reported everywhere, from electronic archives to proceedings of symposium, not to mention numerous monographs and edited books [1–10]. In general, Si-based solar cells can be divided into two categories: bulk silicon technology and thin-film silicon technology. Nowadays, the PV market is largely dominated by bulk silicon technology with roughly 90% including monocrystalline Si (c-Si) and polycrystalline silicon (poly-Si), which reach industrial conversion efficiencies of around 19% and 17%, respectively. The cost of bulk silicon technology has reduced significantly as the conversion efficiency has increased drastically by 11% in the past two years with the introduction and application of more sophisticated technologies. However, thin-film silicon technology constitutes one of the most promising ways for manufacturing low-cost PV solar cells and modules even if industrial cells show current record conversion efficiencies of only around 13% [11]. Indeed, thin-film technology implies a substantial potential for cost reduction due to thinner layer usage, cheap deposition processes, compatibility with large area and mass production, as well as a large choice of rigid or flexible substrates (glass, metal, plastic, etc.).

In the past twenty years, a range of plasma sources and plasma facilities, such as capacitively coupled radio-frequency (RF) plasma-enhanced chemical vapor deposition (PECVD) [9,10,12,13], very high frequency (VHF) PECVD [5,7,8,14–16], microwave plasma (MP) CVD [17–21], and inductively coupled plasma (ICP) CVD [22–24] have been successfully applied for the fabrication of PV devices including bulk Si solar cells and thin-film Si solar cells. With the growing consensus that anthropogenic climate change is now inevitable, plasma technology is set to play a key role in the development of a cleaner and more environmentally conscious world, especially in the development of PV power as a

287

clean, sustainable energy source [25]. Plasma sources have various applications in bulk Si solar cells, for example, in surface passivation by means of silicon nitride (SiN_x:H) films [24,26–28] and in plasma etching processes for removal of phosphorus silicate glass (PSG) or parasitic emitters [22,29], and for wafer cleaning [30–33] as well as masked and mask-free surface texturization [34–36]. For thin-film Si solar cells and "heterojunction with intrinsic thin-layer" (HIT) solar cells, however, almost all the fabrication processes including the formation of the p–n junction, interface passivation, and the surface texturization for light trapping are implemented in plasma facilities [4–16]. As such, plasma-assisted fabrication of Si-based PV materials and complex assemblies is a topic that is increasing in importance, both for fundamental research and for existing and potential industrial applications. The desire to produce Si-based PV materials with consistent performance and reproducibility and at an acceptable cost is stimulating a growing number of major research and research infrastructure programs and a rapidly increasing number of publications in the field of plasma fabrication in Si-based PV devices. Therefore, large-scale utilization of Si-based PV materials prepared in plasma facilities has become an ever-increasing industry in the world.

It is the superior performance and extensive application of plasma technology in Si-based PV devices that motivates us to present an overview on this topic, and our purpose is to highlight the main characteristics and competitive advantages of plasma-aided fabrication approaches, methods, and techniques in PV applications. In this chapter, we explain the choice of plasmas and processes as well as the most commonly used technologies for manufacturing Si-based solar cells both in research and in industry, shed light on some of the most important points and subjects raised in individual contributions, and also furnish a brief survey of relevant research efforts in a broader context. In addition, the most significant milestones achieved in each area of plasma-based synthesis of PV devices are pinpointed and the main directions, opportunities, and challenges for future fundamental and applied research in this undoubtedly hot research area are identified.

This chapter is organized as follows. Because the fabrication of various thin films has been the most extensive application of plasmas owing to their unique ability to dissociate and activate complex gaseous molecular matter, we first introduce SiH_4/H_2 discharges for the synthesis of Si thin films including hydrogenated amorphous silicon (a-Si:H) and microcrystalline silicon (μc-Si:H) as well as the diagnostic tools employed during the discharge to better understand the formation mechanism. We then present the development and current status of thin-film solar cells commonly prepared by PECVD based on extensive and profound research in this field. Third, we briefly introduce plasma-assisted fabrication techniques for thin dielectric or semiconducting layers (SiN) for surface passivation and antireflection. Fourth, we discuss the interaction between hydrogen (nondeposited gas) plasma and a Si surface, with the main focus on the reactions of atomic hydrogen with Si. Fifth, we elaborate on the plasma cleaning of Si surfaces, which is necessary for fabricating high-efficiency Si-wafer-based solar cells or HIT solar cells. Sixth, we discuss plasma etching for the formation of large arrays of vertically aligned nanostructures (NSs), namely, surface texturization. In the next section we mainly focus on p-to-n-type conductivity conversion (PNTCC) based on the plasma etching effect. PNTCC can simultaneously result in the formation of a p–n junction and thus a single-junction solar cell. Finally, we highlight some of the most important features and competitive advantages of plasma-based fabrication techniques and identify major future challenges and directions in solar cell research. This chapter ends with a short summary in the last section.

10.2 Plasma Technologies for Si Thin Films and Solar Cells

10.2.1 SiH_4/H_2 Discharge

Thin films of a-Si:H and μc-Si:H are commonly synthesized using SiH_4 and H_2 precursors from a variety of plasma sources, such as PECVD [37–39], VHF-PECVD [40–42], and ICPCVD [43–46]. It has long been clear that film properties and semiconductor quality, such as the morphological, structural, optical, and electrical properties, are strongly dependent on the plasma properties and processing conditions including substrate temperature, RF power, deposition pressure, hydrogen-dilution ratio, excitation frequency, and electrode configuration. In the interest of controlling such plasma processing methodologies to optimize production cost and film properties, a thorough understanding of the physical and chemical properties of plasmas is necessary [47,48], which requires knowledge of chemistry, thermodynamics, gas transport, heat transfer, and film growth kinetics. Although the SiH_4/H_2 discharge has been widely studied by many groups [49–53], knowledge of the radical chemistry and physics of the plasma involved is still incomplete. Indeed, a US National Research Council Board report on plasma processing [54] stated that "plasma process control remains largely rudimentary and is performed predominantly by trial and error," an expensive procedure that limits growth and innovation in the plasma industry. The same report thus concluded that "a clear research imperative in the next decade will therefore be to increase our knowledge of the chemical and physical interactions in such plasmas of electrons, ions and radicals with neutral species."

By their nature plasmas are dominated by electron and ion interactions; accordingly, in understanding any plasma it is crucial to have an authoritative database for such interactions between electron and atom/molecule including the processes of excitation, ionization, and dissociation. In SiH_4/H_2 discharge, electron collisions with silane produce H and SiH_n, namely, H, SiH_2, SiH_3, SiH_1, and Si, in order of importance. All but SiH_3 are reactive with silane. The main reactions that are expected to contribute to the consumption of SiH_4 in SiH_4/H_2 discharge are listed as follows [49]:

$$e^- + SiH_4 \rightarrow SiH_n + (4-n)H \ (n = 0-3), \tag{10.1}$$

$$H + SiH_4 \rightarrow H_2 + SiH_3, \tag{10.2}$$

$$SiH_2 + SiH_4 \rightarrow Si_2H_6, \ \text{plus some} \ \ Si_2H_4 + H_2 \text{ and } Si_2H_2 + 2H_2, \tag{10.3}$$

$$SiH_1 + SiH_4 \rightarrow Si_2H_5, \ \text{plus some} \ \ Si_2H_3 + H_2, \tag{10.4}$$

$$Si + SiH_4 \rightarrow Si_2H_4 \ \text{ and } \ Si_2H_2 + H_2. \tag{10.5}$$

Specifically, electron collisions with both SiH_4 and H_2 can produce a large amount of atomic hydrogen H. Atomic hydrogen is important because it is formed in nearly all electron impact collisions. The most abundant dissociation product, H, reacts primarily with silane to yield SiH_3 at high silane pressures but reaches surfaces at low silane pressures. SiH_3 radicals are assumed to be responsible for film growth; they are so-called growth precursors. The SiH_2 radicals react quickly with silane and are assumed to be precursors to higher silanes, that is, Si_nH_{2n+2}. Some of the resultant activated $Si_2H_6^*$ dissociates into $Si_2H_4 + H_2$ or $Si_2H_2 + 2H_2$ before being stabilized as Si_2H_6 (stable disilane) by additional gas collisions. Disilane is well favored at high pressure and is observed as the primary reaction product at normal operating pressures [55]. SiH is a minor component. Si_2H_5 can also be produced by $H + Si_2H_6$ reactions, and, as Si_2H_5 reacts slowly if at all with silane, it contributes to film growth, as has been observed at high silane pressure [56]. Despite the minor production of Si, some of the Si_2H_4 resulting from the $Si + SiH_4$ and $SiH_2 + SiH_4$ reactions further reacts with SiH_4 to produce a small amount of Si_3H_8 (stable trisilane).

We have reviewed the database for electron-induced chemical reactions but what role do ions play in SiH_4/H_2 discharges? Ionization of SiH_4, Si_2H_6, and H_2 can produce a large number of different positive ions [57,58], as shown below. For SiH_4, Si_2H_6, and H_2, these ions are SiH_2^+, $Si_2H_4^+$, and H_2^+, respectively.

$$SiH_4 + e \rightarrow SiH_2^+ + 2H + 2e^-, \tag{10.6}$$

$$Si_2H_6 + e \rightarrow Si_2H_4^+ + 2H + 2e^-, \qquad\qquad (10.7)$$

$$H_2 + e \rightarrow H_2^+ + 2e. \qquad\qquad (10.8)$$

Clearly, the ions play a critical role in establishing and maintaining the physical conditions of the plasma, but they also drive much of the chemistry through ion–molecule reactions:

$$SiH_2^+ + SiH_4 \rightarrow SiH_3^+ + SiH_3, \qquad\qquad (10.9)$$

$$H_2^+ + H_2 \rightarrow H_3^+ + H. \qquad\qquad (10.10)$$

Indeed, such ion–molecule reactions may produce new chemical species and dominate neutral chemistry. It should also be noted that ions and electrons may be removed simultaneously from the plasma by (dissociative) electron recombination:

$$Si_2H_4^+ + e \rightarrow Si_2H_4 \quad or \quad Si_2H_2 + H_2. \qquad\qquad (10.11)$$

In addition to the generation and reaction of radicals, the plasma–surface interaction of the neutral particles has a crucial impact on the film properties. The plasma–wall interaction of the neutral particles is described by a so-called sticking model [59,60]. In this model only the radicals react and absorb on the walls forming a layer (or more often multilayers), while nonradical neutrals (H_2, SiH_4, and Si_nH_{2n+2}) are reflected into the discharge. The probability of such absorption is highly variable depending upon the surface material, the surface temperature, and, of course, the gas itself [61–63]. The plasma–wall interaction may provide a limiting factor in solar cell processing because the surface conditions change throughout the manufacturing process and changes may occur from "batch to batch," restricting the reproducibility achievable on the "production line." One "simple" solution is to coat the surfaces of the reaction chamber with a known amount of material at the start of each processing cycle. Thus whilst the actual surface chemistry may remain unknown, by starting each cycle with the same surface conditions and running the reactor under identical working conditions, the surface chemistry may be reproduced in each cycle, leading to a greater homogeneity in the final products. The plasma–surface interaction can result in a significant reduction of the surface diffusion activation energy through the transfer of momentum from the impinging species to the growth surface, a higher surface temperature through ion bombardment, high growth rates through the creation of the adsorption sites via the fragmentation and rearrangement of the chemical bonds on the surface by the impinging particles, and some other effects [64–67]. Other plasma-related effects (such as electric field- and polarization-related effects) can significantly increase the rates of dissociation of silane molecules in the gas phase, promote the fast transport of the reactive species to the growth surface, and further reduce surface diffusion activation barriers [64–69].

To investigate the flux of radicals and ions responsible for film growth in a SiH_4/H_2 discharge, many plasma diagnostics and modeling such as optical emission spectroscopy (OES) [70–75], mass spectroscopy (MS) [74,75], laser absorption spectroscopy (LAS) [76], UV spectroscopy [77], and Langmuir probe measurements of plasma density and electron temperature [64,78,79] have been employed. Figure 10.1 [80] shows the steady-state number densities of chemical species including radical and ionic species in SiH_4 and SiH_4/H_2 plasmas used for preparing device-grade µc-Si:H and a-Si:H. Steady-state number densities of chemical species are basically determined by the balance between their generation rate and their annihilation rate. Therefore, highly reactive species such as SiH_2, SiH, and Si with SiH_4 and H_2 (short lifetime species) take a much smaller value than SiH_3, which shows no reactivity with SiH_4 and H_2 (long lifetime species) in steady-state plasma, although the generation rates of those species are not so different. It is clearly seen in Figure 10.1 that the SiH_3 radical is the dominant chemical species for the growth of both µc-Si:H and a-Si:H, although the density ratio of short lifetime species to SiH_3 varies depending on the plasma-generation condition. Steady-state density of atomic hydrogen (H) varies widely in the plasma, as shown in Figure 10.1. This is mainly due to the change in the hydrogen-dilution ratio R (H_2/SiH_4) in the starting-source-gas materials, that is, the density of atomic hydrogen increases with increasing R. Based on the fact that μc-Si:H is formed with increasing R at constant electron density in the plasma and constant substrate (surface) temperature, it has been suggested that atomic hydrogen plays an important role in μc-Si:H growth [81], although SiH_3 is the dominant film precursor for both μc-Si:H and a-Si:H growth [82].

MS is a simple technique that can provide a great deal of information about the excited neutrals and ionized species in the plasma. A typical example of Quadrupole MS (QMS) spectra for the deposition of μc-Si p/i/n core structure in a continuous process using a low-frequency (460 kHz), low-pressure, thermal nonequilibrium ICP deposition system from the reactive silane precursor gas diluted with hydrogen is shown in Figure 10.2. The details of the plasma reactor and the setup of QMS can be found

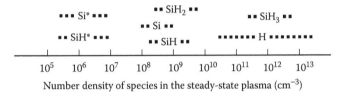

FIGURE 10.1 Number density of chemical species in the realistic steady-state plasmas measured or predicted by various diagnostic techniques. (Reproduced from *J. Non Cryst. Solids*, 338–340, Matsuda, A., Microcrystalline silicon: Growth and device application, 1–12, Copyright 2004, with permission from Elsevier.)

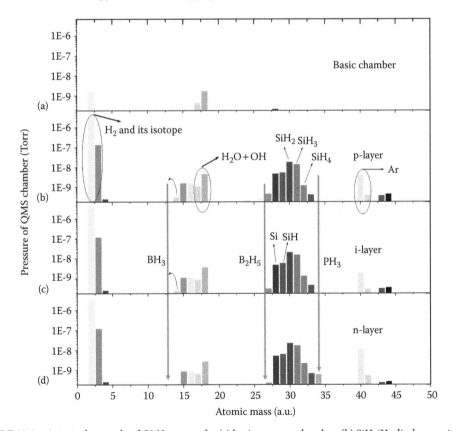

FIGURE 10.2 A typical example of QMS spectra for (a) basic vacuum chamber, (b) SiH_4/H_2 discharge with B_2H_6 doped (p-type), (c) SiH_4/H_2 discharge (intrinsic), and (d) SiH_4/H_2 discharge with PH_3 doped (n-type) during the deposition of μc-Si p/i/n core structure in a continuous process using a low-frequency (460 kHz), low-pressure, thermal nonequilibrium ICP deposition system from the reactive silane precursor gas diluted with hydrogen.

in Xiao et al.'s report [46]. The peaks at 28, 29, 30, 31, and 32 amu, which correspond to Si, SiH$_1$, SiH$_2$, SiH$_3$, and SiH$_4$, respectively, hardly change with the doping condition from the p layer through the intrinsic layer to the n layer. It is clear that SiH$_3$ and SiH$_2$ are the major components, while SiH$_1$ and Si are the minor components. Furthermore, the insignificant amount of SiH$_4$ in comparison with the total SiH$_x$ ($0 \leq x \leq 3$) implies that the inletting silane gas can be completely dissociated. The significant hydrogen flux in the chamber leads to very high surface hydrogen coverage, which increases the surface diffusion length and thus promotes the growth of µc-Si. The clear variation of BH$_3$ (or B$_2$H$_5$) and PH$_3$ from Figure 10.2b–d indicates the doping transition from the p layer through the intrinsic layer to the n layer.

10.2.2 Si Thin-Film Solar Cells

After acquiring an understanding of the physics and chemistry of plasma-synthesized Si thin films, we now review Si thin-film solar cells based on plasma-aided fabrication technologies. Si thin-film solar cells are either amorphous or microcrystalline. Perhaps the most important feature of a-Si and µc-Si material is that a wide range of temperatures (from room temperature to 400°C) can be used for deposition. Room-temperature deposition allows the use of a variety of substrates (e.g., glass, metal, and plastic), and, in particular, the possibility of using low-cost plastic polyethylene terephthalate (PET), which could be a significant advantage in reducing the cost of the modules. Recent work on a-Si:H and µc-Si:H single-junction thin-film solar cells in the laboratory yielded a record conversion efficiency of 10.2% [83] and 10.3% [6], respectively. The weakness of present commercially produced a-Si:H thin film solar modules is their relatively low conversion efficiency, in the order of 6.5% [84], and the poor long-term stability under light soaking, called the Staebler–Wronski effect [85]. The conventional way to improve its stability is advantageously combining a-Si:H with µc-Si:H to form "micromorph" tandem solar cells. A "micromorph" tandem cell consists of a top (through which light enters first) a-Si:H cell, deposited on a bottom µc-Si:H cell, as shown in Figure 10.3a. Compared to a-Si, µc-Si absorbs light within a wider spectral range. This difference in the range of absorption of a-Si:H and µc-Si:H is attributed to different bandgap energy values: the bandgap of a-Si:H is equal to 1.75 eV,

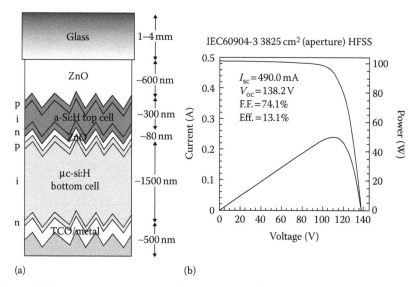

(a)   (b)

FIGURE 10.3 (a) Schematic device structure of the "micromorph" tandem solar cells; (b) a typical *I–V* performance of the "micromorph" tandem solar cells confirmed by AIST. (From Yamamoto, K. et al., *Prog. Photovolt. Res. Appl.*, 13, 489, 2005.)

while that of μc-Si:H is lower at 1.1 eV (similar to crystalline silicon) [14,86]. Nevertheless, within its range of absorption, the absolute value of the absorption coefficient of a-Si:H (quasi-direct gap) is higher than that of μc-Si:H (indirect gap). Therefore, such a combination of materials takes advantage of a large part of the solar spectrum (as compared to single-junction cells), and the tandem solar cell conversion efficiency is consequently increased, with up to 13.4% (initial efficiency) for minimodules (910 × 455 mm²), as demonstrated by Kaneka Corp. [87]. A typical performance of the "micromorph" tandem solar cells confirmed by National Institute of Advanced Industrial Science and Technology (AIST) is shown in Figure 10.3b [88].

For low-cost manufacturing of plasma-synthesized silicon thin-film solar cells, high-rate deposition must be achieved without any reduction in the cell's efficiency. In conventional capacitively coupled RF-PECVD processes [89,90], however, high-rate deposition of device-grade μc-Si:H or a-Si:H films has generally been difficult because of such problems as ion bombardment and powder formation that are often encountered. Bombardment of the growth surface by high-energy ions accelerated in the plasma sheath often results in excessive defect formation, and hence significant deterioration of the film quality [89,90]. The solar cell performance is dominated by a recombination of photocarriers and therefore is sensitive to the density of defects in the film. Thereby how to simultaneously achieve high growth rates and obtain high-quality thin films becomes an important issue for the current plasma techniques in the field of Si-based thin film solar cells. Recently, low-pressure, thermal nonequilibrium, high-density ICPs have been demonstrated to be an effective and versatile process environment to resolve this issue [43–46,91–94]. Kosku et al. [91] reported a remarkably high growth rate of ~6 nm s^{-1} for μc-Si:H films with VHF-ICP. In our previous studies [45] on low-frequency ICP processes, a very high growth rate of up to 2.4 nm s^{-1} for μc-Si:H films with a high crystallinity was achieved. In comparison with conventional capacitively coupled RF plasma sources, the ICPs operating in the electromagnetic mode have several prominent merits: (1) high densities of the plasma species; in particular, the electron number density n_0 in the pressure range of a few pascals can reach 10^{12}–10^{13} cm^{-3} and is up to 2–3 orders of magnitude higher than that of the capacitively coupled plasma discharges [95–98]; (2) low plasma sheath potentials (several or tens of volts) near the chamber wall or deposition substrate, which minimizes ion bombardment on the deposited films; (3) low electron temperatures (a few electron volts) in a broad range of discharge conditions; and (4) excellent uniformity of the plasma parameters in the radical and axial directions [95,99]. This high value of electron number density achieved in the ICP-based process can lead to a large amount of SiH$_x$ (x = 1, 2, 3) radicals and atomic hydrogen flux through the intense inelastic interactions between electrons and silane molecules via the processes of dissociation, excitation, and ionization of the reactive silane precursor gas. A substantial amount of SiH$_x$ (x = 1, 2, 3) radicals on the growing surface of the film account for the high growth rate [100]. On the other hand, strong fluxes of atomic hydrogen from the plasma can lead to a higher surface coverage by bonded hydrogen, a higher etching rate to remove the disordered material, and a local heating through hydrogen recombination on the growing surface.

In improving the deposition rate, the high-density and low-temperature MP has also been developed to synthesize a-Si and μc-Si [17–20]. The use of microwaves for plasma excitation has advantages over that of the generally used RF or VHF exciting plasma for the deposition of silicon films [17–20]. Microwaves can produce high-density plasma (~10^{12} cm^{-3}), which enables high-speed deposition without accompanying difficulties in enlarging process chambers such as VHF equipment where standing waves are generated. However, the μc-Si:H films prepared by the MP technique usually contain high void fraction, which is hardly suitable for a device-quality material. This is caused by the high dissociation of SiH$_4$ and hydrogen in the plasma, which is determined by the high-energy part of electron energy distribution (EED) above 10 eV and the chemical reactivity of SiH$_4$ [101–103]. Therefore to suppress the excess dissociation of source gas and to improve the rigidity of the Si network, a new source material is required in place of SiH$_4$ for fabricating device-grade μc-Si films with less volume fraction of void for Si thin-film solar cells.

10.2.3 SiN Thin Films for Surface Passivation

After elaborating on the plasma-aided synthesis of Si thin films and Si thin-film solar cells, we briefly introduce the plasma-assisted fabrication of thin dielectric or semiconducting layers (SiN) for surface passivation. The use of PECVD-deposited or MP CVD-deposited SiN technology in the fabrication of bulk silicon solar cells is becoming pervasive. The expansion of this low-temperature technology is attributed to a substantial improvement in solar cell efficiency resulting from the deposition of SiN in conjunction with fire-through techniques for the metallization step. This enhancement can be ascribed to three driving forces. First, a large amount of hydrogen originating from plasma gas dissociation and incorporated in the SiN film can be driven into the solar cell during the metallization step, leading to an excellent bulk passivation for c-Si or poly-Si solar cells. An efficiency improvement of 1–1.5% (absolute), which can be due to hydrogenation of defects in the volume of the c-Si or poly-Si wafers, has been observed [104,105]. Second, another driving force is the surface passivation effect. SiN has been proven to provide very low surface recombination velocities both on phosphorus-diffused regions [106] and on p-type and n-type wafers [107–110]. Third, another advantage is the antireflective (ARC) properties of the nitride layer, which reduces light reflection significantly. SiN film deposited by PECVD or MP CVD is therefore a common method to combine bulk and surface passivation with ARC to improve the electrical properties of solar cells. Nevertheless, the degree of enhancement depends on the deposition and annealing process parameters.

For the PV application, SiN is usually made from a gas mixture of SiH_4 and NH_3. Silane is a source of silicon and hydrogen. Ammonia, in addition to being a source of nitrogen, aids in depositing SiN with a high ratio of incorporated hydrogen [111]. In SiH_4/NH_3 discharge, electron collisions with NH_3 produce H and all of the NH_n: NH_2, NH, and N. The dissociation of NH_3 is quite similar to that of SiH_4. The reactions between the nitrogen- and hydrogen-containing species in the plasma result in an amorphous solid deposit commonly denoted as a-SiN_x:H or simply SiN. These films contain up to 40% hydrogen [112] bonded to either Si or nitrogen. Their optical properties (absorption, reflection, and refractive index) depend greatly on the concentration and chemical distribution of hydrogen, Si, and nitrogen in the film, that is, on the deposition conditions.

Recently, the group led by Professor Xu Shuyan at Plasma Sources and Application Centre, NIE, Nanyang Technological University, Singapore, developed an innovative and proven plasma-atomized deposition (PAD) technique to synthesize SiN_x films with superior surface passivation and antireflection layers [113]. The PAD process is based on low-energy, high-rate production of atomic radicals in the ICP. In this process, the solid material (Si target) is first locally sputtered and then fully atomized/ionized by high-density plasmas. The atoms/radicals undergo a reaction to form SiN_x passivation coatings. By applying PAD technology to industrial diffusion solar cells, Professor Xu's group achieved an energy conversion efficiency of 18.43% (see Figure 10.4), which was tested at the Solar Energy Research Institute of Singapore for 5″, p-type, industrial-grade silicon wafer-based solar cells. This performance is already comparable/superior to that of industrial solar cells passivated by standard PECVD-SiN_x or MP CVD-SiN_x. This novel process is environmentally friendly and much less expensive compared to the commonly used PECVD and microwave CVD processes because it uses only Si solid and N_2/H_2 gaseous precursors. The PAD process can also eliminate the use of explosive and flammable gas, namely, SiH_4 and NH_3. This novel green plasma technology can replace the existing standard processes and can also be introduced in established production cycles to fabricate solar cells with higher efficiency.

10.3 Plasma Etching in Si-Based PV Applications

10.3.1 Hydrogen Interaction

The above section mainly focused on the plasma-aided fabrication of Si or SiN thin films for PV applications. Now we turn our attention to the plasma etching effect. Hydrogen plasma (or atoms) is well known to modify the morphology, structure, and properties of silicon materials, such as

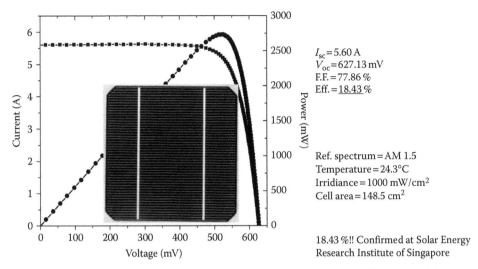

$I_{sc} = 5.60\,\text{A}$
$V_{oc} = 627.13\,\text{mV}$
F.F. = 77.86 %
Eff. = 18.43 %

Ref. spectrum = AM 1.5
Temperature = 24.3°C
Irridiance = 1000 mW/cm²
Cell area = 148.5 cm²

18.43 %!! Confirmed at Solar Energy
Research Institute of Singapore

FIGURE 10.4 A typical performance of the crystalline Si wafer-based solar cells passivated by PAD–SiN. (From Xiao, S. Q. et al., unpublished.)

crystalline silicon (c-Si), a-Si:H, and μc-Si:H, including phase changes [114]. The interaction of hydrogen plasma with the c-Si surface gives rise to diverse phenomena [115–118], including wafer cleaning (namely, native oxide and contaminant removal) [30–33,119], passivation of defects and impurities, activation of dopants [120], creation of donor-like states (namely, the formation of PNTCC) [121–125], surface texturization [34–36], as well as etching and hydrogen incorporation into the near-surface layers [31], and, thus, has been the subject of very active research for a long time.

These phenomena are caused by the interaction of atomic hydrogen with silicon surfaces, which rearranges and removes Si atoms according to reactions of adsorption (e.g., passivation of dangling bonds), abstraction, insertion of hydrogen, and etching of silicon (see Figure 10.5) [126]. These reactions can be schematized as follows (where the –Si represents a dangling bond):

$$H + (-Si) \rightarrow H-Si \,(\text{adsorption}), \tag{10.12}$$

$$H + H-Si \rightarrow H-H_{(g)} + (-Si)\,(\text{abstraction}), \tag{10.13}$$

$$H + Si-Si \rightarrow H-Si + (-Si)\,(\text{insertion}), \tag{10.14}$$

$$H + Si-SiH_3 \rightarrow SiH_{4(g)} + (-Si)\,(\text{etching}). \tag{10.15}$$

During exposure of Si surfaces to hydrogen plasma, more than one of these reactions can occur simultaneously. Also shown in Figure 10.5 are rate expressions. Each reaction rate depends on the surface coverage or local concentration of the reacting species, which are –Si, H–Si, and Si–Si in Equations 10.12, 10.13, and 10.14, respectively. These reactions may have different rates in the "bulk" compared to in the surface due to differences in the rate coefficient or the local concentration (coverage). The rate coefficient of Equation 10.12 is the "sticking probability," S, and is generally assumed to be near unity. Schulze and Henzler measured S on the Si(111)-(7 × 7) surface and obtained a value of 1 within 30% [127]. The rate term of Equation 10.13 is given in Figure 10.5 assuming a reaction with monohydride species. We write the rate term of Equation 10.14 as $k_I F_H\{Si-Si\}$, where $\{Si-Si\}$ represents the coverage or bulk density of strained Si–Si bonds and F_H the hydrogen atom flux. The rate coefficients k_I will be larger for strained Si–Si bonds compared to unstrained bonds. For simplicity, we write one example rate expression for

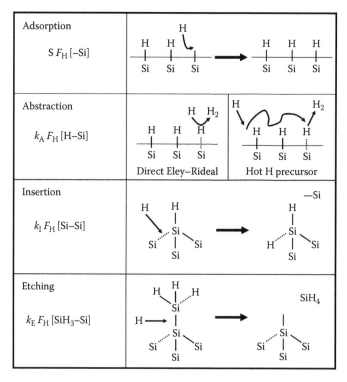

FIGURE 10.5 Schematic illustration of four principal reaction pathways for the interaction of atomic hydrogen with Si surface: adsorption (sticking), abstraction, insertion, and etching. Also shown are characteristic expressions for the reaction rates of each process. (Reprinted with permission from Chiang, C. M. et al., *J. Phys. Chem. B*, 101, 9537, 1997. Copyright 1997 American Chemical Society.)

Equation 10.15 as $k_E F_H \{Si-SiH_3\}$. If there are other surface precursors to etching in addition to $Si-SiH_3$, a similar rate expression should be written for each.

The relative contribution of these processes depends on various factors, such as the crystallographic orientation [115], doping [128,129], and the temperature [130,131] of silicon as well as on hydrogen exposure and coverage [129]. The kinetics of the Si surface modification arises from a competition among all the above processes and factors. Therefore, there is a strong technological and fundamental interest in understanding and controlling Si surface morphology and properties induced by the interaction with hydrogen plasma. The interaction of atomic hydrogen with Si surface has so far been extensively studied using a variety of spectroscopic tools such as multiple internal reflection Fourier transform infrared absorption spectroscopy (MIR-FTIR) [132–134], scanning tunneling microscope (STM) [115,135], temperature-programmed desorption (TPD) spectrometry [136,137], and mass spectrometry [138]. A comprehensive description of the work until 1995 is provided in the review by Waltenburg and Yates [139]. The dynamics of the interaction of hydrogen with Si surfaces with respect to adsorption and desorption was reviewed by Kolasinski [140]. More recently, the hydrogen surface chemistry of Si surfaces was dealt with in the paper by Oura and coworkers [116].

10.3.2 Plasma Cleaning

It was reported that cleaning of Si substrate surfaces is necessary for fabricating high-quality Si epitaxial films [141], high-efficiency bulk Si solar cells, and HIT solar cells. To this aim different wet cleaning methods have been studied, but they are all difficult to control and produce dangerous residuals, so recently attention was focused on dry cleaning treatment, which is one of the most favorable methods for precleaning Si surfaces and mass producing solar cells. Hydrogen plasma cleaning is considered to be an effective way of removing native oxide and hydrocarbon contaminants on the Si surface at low temperatures [142–144]. Therefore, a number of investigations have been carried out on the structural and chemical properties of Si surfaces that are treated with hydrogen plasma. Auger electron spectroscopy (AES) [145] and FTIR studies [146,147] revealed that hydrogen plasma treatment of Si surfaces considerably reduces the level of contaminant species such as oxygen and hydrocarbons. Reflection high-energy electron diffraction (RHEED) [145] and low-energy electron diffraction (LEED) studies [148] showed that hydrogen plasma cleaning produces Si surfaces with (1×1)-, (2×1)-, and (3×1)-reconstructed structures, depending on the process pressure and the substrate temperature. Furthermore, residual gas analysis (RGA) experiments and atomic force microscopy (AFM) studies showed that during hydrogen plasma treatment at low substrate temperature, desorption of SiH_x fragments occurs, leading to a rough surface [148,149]. These studies suggest that hydrogen plasma treatment rearranges and removes the surface Si atoms. Moreover, transmission electron microscopy (TEM) studies revealed that during etching of Si surfaces, hydrogen atoms diffuse into the bulk Si crystal to produce platelet defects that are predominantly oriented along {111} crystallographic planes and that etching of the Si(100) surface preferentially occurs at portions where {111} platelet defects intersect the surface [143,150]. Shinohara et al. [151] have used infrared absorption spectroscopy (IRAS) in the multiple internal reflection geometry to investigate the interaction of hydrogen-terminated Si(100), (110), and (111) surfaces with hydrogen plasma at room temperature. IRAS data show that at initial stages of H-plasma treatment, surface hydride species (SiH_n, $n = 1,2,3$) are removed from the surface. A long-term H-plasma treatment of Si(100) and (110) surfaces reproduces monohydride species and creates hydrogen-terminated Si vacancies at subsurface regions, that is, near the surface. On Si(111), no hydride species are reproduced even after long-term H-plasma treatment. It is suggested that monohydride is more stable against attack of hydrogen radicals than higher hydride species SiH_2 and SiH_3. The formation of Si vacancies depends on the crystallographic orientation of the Si surface: the formation of Si vacancies is more favored on Si(110) than on Si(100), and Si vacancy does not occur on Si(111).

Apart from hydrogen plasma, Ar [152] and CF_4/O_2 [30] plasmas have been used to clean Si surfaces. However, for Ar plasma, the substrate is easily damaged by ion bombardment. The substrate damage is reported to be generated mostly by heavy ion (Ar) bombardment. These effects can degrade the quality of the grown layers and the interfaces between them. The ion-induced damage must be annealed dynamically at considerably high temperatures (~800°C) to ensure defect-free epitaxy. Hydrogen plasma cleaning, on the other hand, is chemical etching by lighter hydrogen atoms and ions, and it has been used as an excellent substitute for Ar plasma cleaning. In addition to reduced substrate damage, the silicon surface can be passivated by hydrogen as a result of hydrogen plasma cleaning. This passivation suppresses the recontamination of the cleaned surface and reduces the recombination velocity of the carriers at the interface, which is quite beneficial for the synthesis of high-quality Si epitaxial films as well as high-efficiency bulk Si solar cells and HIT solar cells. In comparison with hydrogen plasma, CF_4/O_2 treatment represents a good choice in the cleaning process and passivation of the textured p-type crystalline silicon surface [30]. However, the use of CF_4 may produce toxic fluorides, and therefore this technique is not environmentally friendly.

10.3.3 Plasma Texturing

Among the plasma etching effects, surface texturization is of great interest in industrial applications for Si solar cells. Surface texturing needs to be improved to increase the solar cell short-circuit current and hence the solar cell conversion efficiency due to the enhanced absorption properties of the silicon surface. At present, wet chemicals are routinely used by the PV industry for texturing as well as for cleaning the Si surface. Potassium hydroxide (KOH) [153,154], sodium hydroxide (NaOH) [155,156], and tetramethyl ammonium hydroxide [$(CH_3)_4NOH$] [157] are currently used to texture the Si surface.

Hydrofluoric acid (HF) combined with DI water is a standard solution used to remove native SiO_2, which is present on Si wafers due to ambient oxidation [158].

A major limitation of the current Si wafer-based solar cells is the high cost per watt of energy produced. Moreover, many hazardous and corrosive chemicals and gases are involved in wet etching processes. This leads to adverse environmental impacts and an enormous cost of waste disposal. The plasma dry etching process offers excellent promise in drastically reducing wafer consumption and the use of chemical etching (or cleaning) agents as well as in eliminating handling of hazardous acids and solvents. The greatest advantage of dry plasma etching processing is the full control over all process parameters, which may ultimately lead to clean, fully automated, perfectly reliable, highly reproducible, in-line integrated, high-throughput processes where wafers are automatically transported through the production line. In addition to ease of automation, plasma dry etching results in isotropic or anisotropic etch profiles for directional etching without using the crystal orientation of silicon with a better process control and cleanliness. Furthermore, it is not merely a matter of industrial and environmental sustainability: thinner wafers (<180 μm) pose serious problems of yield and require different and independent conditioning of the front (minimum reflectance) and rear (least surface damage) sides. Plasma dry etching technology offers a series of new avenues to overcome these bottlenecks and increases the flexibility of the process: (1) high-yield processing on ultrathin wafers; (2) etching/texturing independent of initial surface preparation; (3) controllable degree of roughness; (4) easily adjustable etching rates, suitable for micrometer-thick layers (e.g., epitaxial); and (5) independent treatment and conditioning of the front and rear surface (single side emitter removal, different front/rear structures).

In general, large arrays of vertically aligned Si NSs, such as nanowires, nanorods, nanocones, and nanotips [159–165], are always fabricated by a plasma etching process to achieve surface texturization [34–36,166,167], which can be used to create topographically enhanced light-trapping PV cells, similar to the black silicon solar cells [168–173]. Bai et al. [159] reported the high-density aligned arrays made of one-dimensional (1D) silicon NSs, including nanocone, nanorod, and nanowire, which are fabricated by plasma etching in a hot-filament CVD apparatus using a gas mixture of hydrogen, nitrogen, and methane. The authors attributed the fabrication of aligned silicon NS arrays to plasma etching together with redeposition of silicon. Hsu et al. [162] proposed and reported a one-step, self-masked dry etching technique for fabricating uniform and high-density nanotip arrays over a large area in a cost-effective manner, using a Seki high-density electron cyclotron resonance (ECR) plasma reactor. The nanotip formation mechanism is related to both the SiC-nanosized clusters formed from the reaction of $SiH_4 + CH_4$ plasma [174] and the etching effect of an $Ar + H_2$ plasma. The SiC nanoclusters uniformly distributed over the substrate surface act as nanomasks against etching because of their higher hardness and chemical inertness. The interspace between every two SiC nanoclusters can be etched by an $Ar + H_2$ plasma to form high-density and high-aspect-ratio nanotip arrays. The work published by Yang et al. [163] discusses a well-aligned silicon nanograss that was fabricated on a silicon wafer by hydrogen plasma etching alone, that is, without the use of a mask or a catalyst. A selective etching mechanism in which the hydrogen ion flux is accelerated toward the substrate by the low-frequency bias and the partial native oxide is sputtered off has been proposed. In a recent work [166], a dry and lithography-free texturing process for crystalline silicon wafers using SF_6/O_2 plasma in an ICP system was investigated. Upon ICP etching in SF_6/O_2, pillar-shaped NSs are formed on the Si surface and the reflectivity in a broad spectral range is greatly suppressed. High-aspect-ratio vertically aligned Si NSs with a low reflectance have been obtained. Similarly, Moreno et al. [167] have studied a dry and free mask texturing process of crystalline silicon wafers using SF_6/O_2 plasmas in a reactive ion etching (RIE) system. At optimized processing parameters, an average reflectance value as low as 6% can be realized. It is believed that two opposite effects take place on the SF_6/O_2 plasma: an etching process due to fluorine radicals, which are very efficient for silicon etching, and a redeposition process due to residual SiO_xF_y radicals, which produce a masking effect [175]. Micromasks enhance the texturing of the c-Si surface and therefore reduce light reflectance, besides increasing light trapping.

Apart from these fundamental researches, Yoo et al. [176] have fabricated a large-area (156 × 156 mm) poly-Si solar cell by maskless surface texturing using a SF_6/O_2 RIE. By optimizing the processing steps, they accomplished texturing by reducing silicon loss by half of that in a wet texturing process as well as achieved conversion efficiency, open-circuit voltage, short-circuit current density, and fill factor as high as 16.1%, 619 mV, 33.5 mA cm^{-2}, and 77.7%, respectively. Almost at the same time, Lee et al. [177] developed damage-free RIE texturing for polycrystalline Si solar cells by using SF$_6$/Cl$_2$/O$_2$ gas mixtures. This improved texturing process can produce a better surface with less surface recombination velocity due to the removal of RIE-induced damage and high control of nanocone shape. Applying this damage-free etching technique to polycrystalline Si solar cells, they achieved a consistent absolute increase in short-circuit current gain of ~1.3 mA cm^{-2}, an increase in cell efficiency gain of 0.7%, where the gas ratio and plasma power density were keys to mitigate the plasma-induced damage during the RIE processing while maintaining decent surface reflectance. This shows that the surface damage induced during RIE processing can be removed by *in situ* processing, which means that there is no need for additional wet processing.

Most plasma dry etching techniques listed above use toxic, corrosive gases such as CF$_4$, C$_2$F$_6$, NF$_3$, and SF$_6$. These gases are not environmentally friendly and cause green house effects. In addition, these processes use relatively high temperature, which influences the optical properties of the wafer surface. In our previous studies [178,179], we reported on a novel and green technique for fabricating large arrays of vertically aligned Si NSs via "self-organized" maskless etching in low-temperature Ar + H$_2$ plasmas. The unique feature of these works is in the use of externally generated high-density ICP, as summarized in the above sections. Figure 10.6a and b show two typical images in sharp contrast prepared at different

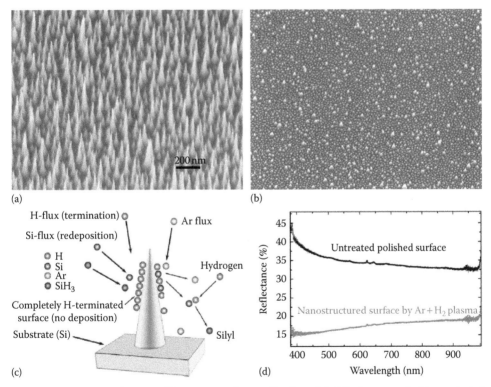

FIGURE 10.6 (a) A typical image of large arrays of vertically aligned Si NSs (nanocones); (b) a typical photo of roundish nanodots (no NSs can be observed); (c) schematic of the main elementary processes on a Si surface exposed to low-temperature Ar + H$_2$ plasma. H termination impedes silicon material redeposition; concurrent Ar$^+$ ion bombardment improves conditions for Si attachment. Silyl radicals contribute to the growth through redeposition mediated by their interaction with the H-terminated surfaces; (d) a typical comparison of the reflectance for the sample before and after hydrogen plasma etching. (Reprinted with permission from Xu, S. et al., *Appl. Phys. Lett.*, 95, 111505, 2009. Copyright 2009 American Institute of Physics.)

experimental conditions. Vertical nanocones were achieved in Figure 10.6a, whereas in Figure 10.6b only roundish nanodots were obtained. We systematically and carefully investigated the effect of a variety of experimental parameters, such as the gas composition (the H_2:Ar ratio), substrate temperature, etching time, and discharge power on the formation of vertical single-crystalline silicon NSs [179]. It has been shown that the H_2:Ar ratio, substrate temperature, and discharge power significantly influence the surface morphology. When the H_2 content is either too low or too high, only semispherical nanodots can be produced on the surface. When the substrate temperature is too low (200°C), the NT array is very nonuniform, with a large amount of short NSs present on the surface. At the optimum temperature of 300°C, NT arrays of the highest mean aspect ratio (up to 4.0) can be produced, albeit with a high degree of nonuniformity. A further increase in the substrate temperature leads to the formation of highly uniform arrays with a decreased mean aspect ratio. Power below 1.0 kW is insufficient to produce nanocones; instead, a high-density nanodot array can be formed. Based on these experimental results, we have developed an NS growth scenario and a simple model that provide the main observations [179]. In the present case of Ar + H_2, as shown in Figure 10.6c, argon serves as the sputtering environment. The intense flux of the energetic Ar ions removes the hydrogen atoms from the surface and thus helps in keeping this surface open (unterminated), while hydrogen terminates the surface (and hence reduces the surface energy) and acts as a reactant. The process of nanocone formation is based on at least two parallel chemical–physics processes, namely, the etching of the surface with the reactive plasma accompanied by simultaneous redeposition of Si etched away from the surface. A combination of these processes with thermokinetically driven dynamic surface reconstruction eventually results in NS size and shape, besides impacting the array macrocharacteristics such as NS density and size (aspect ratio). The detailed NS growth scenario and interpretation of these results can be found elsewhere [179]. Large arrays of vertically aligned Si NSs enhance light absorption, which is also called the black silicon effect. As shown in Figure 10.6d, low reflectance was realized after hydrogen and argon plasma etching in the wavelength region from 400 to 1000 nm, demonstrating the excellent output of this technique, which is environmentally friendly, using minimum chemicals and environmentally clean gases. The technique introduces fine surface texturing on the silicon substrate, because of which light gets trapped on the surface thereby improving the short-circuit current in the bulk silicon solar cell. Similarly, Dhamrin et al. [180] have also developed a new etching technique based on remote hydrogen plasma for crystalline silicon wafers. An excellent optical property of the etched surfaces is found where low surface reflectance below 2% is realized at the wavelength regions of 500–900 nm. In addition, the technique is applied to chemically textured crystalline silicon wafers, demonstrating the technique's ability to further enhance already textured surfaces.

10.3.4 Plasma-Induced PNTCC

Plasma etching can not only clean Si substrate surfaces and produce large arrays of vertically aligned Si NSs used for light trapping in solar cells but also lead to the creation of donor-like states, which implies PNTCC, and simultaneously the formation of the p–n junction. In practice, it is well known that the formation of p–n junction is realized by high-temperature phosphorus diffusion in silicon wafers in the c-Si or poly-Si solar cell industry. However, this technique usually involves energy-consuming thermal diffusion-based processes and toxic chemicals/gases. In recent years, various methods other than traditional diffusion have been employed to create PNTCC on p-type Si substrates. One technique is the use of high-energy (>100 eV) ion beam (Ar+, H+, or e–) bombardment and implantation as reported [181–184]. The other is RF [121–124] and direct current [125] hydrogen plasma immersion, followed by long-time annealing at a temperature below 550°C.

The origin of PNTCC is still a matter of debate and the key locates at the origin of excess donors. Up to now, there have been several accepted points of view on this topic: (1) substitutional boron atoms (acceptor) can be transferred into the electrically inactive interstitial position by the generated interstitial Si atoms, which is known as the kickout mechanism [185]. However, this can only explain the loss of an acceptor but not the occurrence of a donor. (2) The weakly bonded interstitial hydrogen, which reacts with the impurities, as described by Wolkenstein's theory, can also behave as a donor [186]. (3) Helium-like double-oxygen-related thermal donors (OTDs) in oxygen-rich (O_i density ~10^{18} cm^{-3}) silicon are attributed to the presence of larger oxygen clusters (O_2, O_3, O_4, etc.) [187]. (4) The single shallow thermal donor (STD) [188] incorporates a hydrogen atom to form hydrogen-related STD, that is, STDH. In the case of OTD and STDH, hydrogen atoms are believed to play a crucial role. At low temperature, theoretical investigations [189] have revealed that an activation energy of 0.84 eV is needed for hydrogen to diffuse into silicon. At high temperatures, however, hydrogen diffusion occurs with a significantly lower activation energy of about 0.4 eV [123]. When hydrogen diffuses deeply into the silicon wafer it reduces the potential barrier for O_i migration, and then several O_i atoms migrate through the wafer until they reach an appropriate local site in the Si lattice where an OTD is created. Therefore, hydrogen plays a catalytic role that enhances the formation of OTD. Indeed, Newman et al. found that at 450°C formulation of OTD could be enhanced by a possible factor of 5–10 in the presence of hydrogen [190].

In particular, we present an example of one-step formation of a p–n junction by means of PNTCC on a p-type substrate exposed to Ar + H_2 plasma in the low-frequency (460 kHz) ICP system [121,191]. The heating and Ar + H_2 plasma treatment are simultaneous without any additional annealing process. As the p-type surface is converted to an n-type surface, large arrays of vertically aligned Si NSs can also be formed on the surface at an appropriate range of experimental conditions, as discussed in the above sections. A schematic of the experimental setup is shown in Figure 10.7a, and a representative photo of the discharge can be seen in Figure 10.7b. A scheme of the process and a representative SEM image of the Si NS arrays created on the surface of a boron-doped p-type Si wafer are shown in Figure 10.7c and d. A schematic of the nanoarray-based solar cell is shown in Figure 10.7e. This cell (typical size 2 × 2 cm², with an Al finger grid as a front electrode and an Al metal pad as a back electrode) only required As-processed Si wafers with a nanostructured top surface, an Al back reflector, and metal electrodes.

Let us discuss the details of p–n junction formation. In this process, both hydrogen and argon play important yet different roles. During plasma treatment, the flat wafer surface is intensively heated up by high-density electrons, and sputtered by heavy argon ions impinging on the Si. Reactive hydrogen ions and radicals also roughen the Si surface by hydrogen-assisted etching through the formation of volatile Si-containing molecules (SiH$_n$, $n = 1, 2, 3$) of varying complexity. The process consists of two

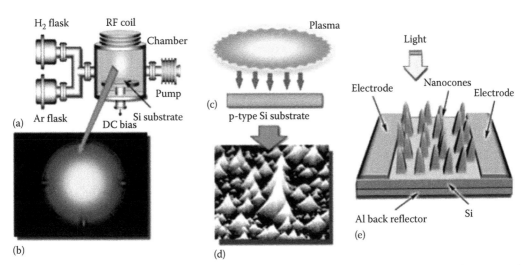

FIGURE 10.7 (a) Schematic of the experimental setup for one-step formation of p–n junction by means of PNTCC; (b) photograph of the Ar + H_2 plasma discharge; (c) scheme of the PNTCC process; (d) representative SEM image of a Si NS array; and (e) scheme of the nanoarray-based solar cell.

stages. At the stage of NS formation, the hydrogen/argon ratio determines the resulting microscopic morphology of the surface. Hydrogen radicals provide very strong heating of the silicon surface due to the exothermic recombination, that is, release of the dissociation energy [192]. Eventually, a hierarchical two-level array of NS is formed, and the surface is covered with very small and dense nanocones [179,180].

At the next stage, the p–n junction is formed under the nanostructured surface. The first process is the diffusion of boron atoms from the bulk silicon to the surface (heated up by the ion flux), extraction of boron atoms from Si by atomic hydrogen, and their removal from the surface in the form of volatile boron–hydrogen compounds B_xH_y. The presence of B_xH_y was detected by QMS in our plasma diagnostic experiments. As a result, a boron-depleted subsurface layer is formed. Second, after partial removal of boron-related acceptor sites, an n-type layer is formed due to the hydrogen-assisted formation of OTDs and STDs (note that thermal donors are able to compensate the boron doping in Si [193]). Besides, Si–H–Si centers that act as donors can be formed in Si [194]. Moreover, energetic H atoms from the plasma can form positively charged bond-center sites; this in turn provides passivation of the shallow acceptors by annihilating the free holes by electrons from hydrogen atoms [195]. This eventually leads to PNTCC, and an effective gradient p–n junction is formed.

Figure 10.8a displays the dark *I–V* curve of the plasma-generated p–n junction, which exhibits an extraordinary rectifying characteristic (rectifying ratio up to 10^3) comparable to the common p–n junction formed by diffusion. It is worth noting that the resultant p–n junction presents an open-circuit voltage (V_{oc}) of 520 mV, a short-circuit current (J_{sc}) of 28 mA cm^{-2}, and an efficiency of 10% measured under one sun illumination, as shown in Figure 10.8b. Such evident PV signal indicates that the resultant junction is of real potential to convert solar radiation into electrical energy for practical use.

Thus, our experiments have demonstrated that a simple and environmentally friendly plasma treatment results in a simultaneous formation of the two essential features that enabled the effective operation of the solar cell prototype. First, an effective p–n junction was formed in the subsurface; second, an effective light-trapping two-level hierarchical array of vertically aligned Si nanocones capable of effectively increasing the light absorption of the wafer surface was grown in the same process. Such a fabrication technique would be a powerful candidate for low-cost PV applications due to low energy consumption, dry etching instead of chemical etching, and no usage of toxic gas such as phosphor and boron as compared to the industrial high-energy diffusion bulk silicon solar cells.

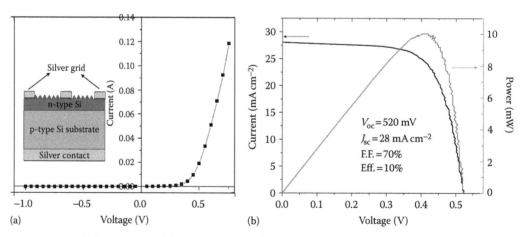

FIGURE 10.8 (a) Dark *I–V* and (b) photo *I–V* curves for a typical p–n junction solar cell by means of PNTCC on p-type substrate exposed to Ar + H$_2$ plasma in the low-frequency ICP system. The inset shows the schematic structure of the PNTCC solar cells.

10.4 Competitive Advantage of Plasma Technology in PV Devices

A particular advantage of plasma technology in PV applications, especially in the field of Si-based solar cells, is the deterministic control of the plasma species under the influence of the electromagnetic field. This gives the possibility of isotropic or anisotropic etching of Si substrates, as well as precise and effective control of photonic and electronic properties of the materials by careful engineering of the thin-film composition and microstructure by appropriate, *in situ* control of the plasma process parameters. By combining plasma diagnostics with process and reactor modeling, one can easily assess the effects of process variables on performance and eventually optimize the growth process to meet the device requirement. This high degree of control is particularly desirable for fabricating large-area thin-film Si solar cells. It is also promising for applications in heterojunction structures and light emitting diodes. Indeed, the plasma-based synthesis of μc-Si/a-Si and related materials and their integration in solar cell prototypes has made a significant impact on the development of this multidisciplinary research field and has been widely adopted internationally. These plasma-aided fabrication processes have also attracted significant interest from established and emerging industries. Among the plasma-synthesized Si thin-film solar cells, the stacking solar cells including the "micromorph" tandem solar cell and the triple-junction solar cell will be the platform for novel PV devices that are highly efficient, low-cost substitutes for existing solar cells (mostly based on bulk crystalline silicon) and as such will open new horizons for highly efficient green power generation.

Apart from precise and effective control of Si thin-film fabrication, plasma technology is advantageous for dry plasma processing, including the removal of PSG or parasitic emitters, wafer cleaning, masked or mask-free surface texturization, and the direct formation of the p–n junction by means of PNTCC. Recently, increasing chemical waste disposal costs and environmental and water [196] concerns are being given a lot of importance. In addition, because of the demand for integrated solutions in process equipment, the PV industry has been showing an increasing interest in dry etching processes as well. One prominent advantage of dry plasma processing is the full control over all process parameters, which allows for good reliability and reproducibility, especially in industrial production [35]. For the heterostructure devices such as HIT solar cells, one of the most important steps in the fabrication of amorphous crystalline heterojunction is wafer cleaning before amorphous layer deposition in order to prevent the introduction of defects. To this end different wet etching methods have been proposed, but they are all difficult to control and produce dangerous residuals. However, the plasma dry etching cleaning procedures may solve this problem and be easier to transfer to solar cell mass production.

For bulk Si solar cells, the first step is wafer cleaning by using hazardous acids and solvents such as HF and HNO_3 followed by wet etching by using alkaline solutions with isopropyl alcohol (IPA) addition to achieve surface texturing. In contrast, the plasma dry etching processes can realize both wafer cleaning and surface texturization in one simultaneous step. As a result, most wet chemistry can be replaced by only a simple dry etching process, leading to the reduction of normal cost of chemical etching agents and chemical waste disposal. Actually, PSG formed during emitter diffusion is normally removed by plasma dry etching in solar cell processing. It is necessary to mention again the plasma-assisted fabrication of SiN films for c-Si or poly-Si solar cells. SiN has been proven to provide excellent passivation and antireflection properties and therefore is a common method for improving the solar cell efficiency. The PAD technology for the synthesis of high-quality SiN films is promising to replace the existing standard processes and can be introduced into the established production cycles to fabricate solar cells with higher efficiency.

Finally we emphasize on the simultaneous formation of large arrays of Si NSs and PNTCC on p-type substrates using the plasma etching processes. These simple and cost-effective processes are of great importance in plasma fabrication for solar cells because they can achieve both surface texturing and p–n junction in one continuous and fast process. Moreover, the formation of the p–n junction is realized without using any toxic gas such as phosphor and boron, which contribute to low-cost PV applications in comparison with the industrial high-energy diffusion solar cells.

10.5 Outlook

Due to the overwhelming variety of applications in PV research and industry for plasma technology, it is not an easy task to prescribe a single plasma technology for solar cells in the near future. However, we can highlight some of the most important challenges, unresolved issues, and opportunities in plasma-aided fabrication of PV materials that arise from the topics reviewed in this chapter.

Silicon thin-film modules based on plasma-synthesized a-Si:H and μc-Si:H tandem cells are one of the most promising future technologies for providing cost-effective and efficient PV electricity. Necessary prerequisites for cost-effective mass production of thin-film solar cells incorporating μc-Si films are the demonstration of high deposition rates and scalability to large areas. As reviewed in Section 10.2.2, low-pressure, thermal nonequilibrium, high-density ICPs have been demonstrated as an effective and versatile process environment to resolve the deposition rate issue. However, the reported efficiency of μc-Si:H solar cells prepared by ICPs still remains low, and achieving high-efficiency μc-Si:H solar cells is a very challenging task. Another promising route to obtain high deposition rates for μc-Si:H is the use of VHF (>60 MHz) in combination with high deposition pressures in PECVD reactors. However, problems concerning the process upscaling toward square meter size like standing waves, skin effect, or voltage uniformity may hinder homogeneous film growth in the VHF regime (see References [13,197] and references therein). At the high-pressure and high-power condition applied for high-rate μc-Si:H deposition, significant powder formation occurs, and dust is visible in the deposition chamber already after only a few runs. An important and related issue is the development of plasma etching cleaning processes that significantly shorten the maintenance cycle and thus increase effective operation time. Moreover, effective light trapping is essential to obtain high cell efficiencies at small absorber layer thickness. Usual methods such as surface texturing used in bulk silicon solar cells for light trapping cannot be applied to thin-film Si cells. A promising way found recently [198,199] is to use the rapidly emerging field of plasmonics and nanoplasmonics in particular to enhance the optical absorption of the photoactive layer. In recent studies published by Akimov et al. [200,201], the plasmonic effect of silver nanoparticles deposited on the top surface of the thin-film a-Si:H solar cell on light trapping inside the photoactive layer was studied. It has been shown that by excitation of higher-order surface plasmon resonances in larger nanoparticles, one can increase the energy transmitted into the a-Si layer as well as decrease the optical absorption related to metallic nanoparticles. Thus, the overall broadband optical absorption in the photoactive layer can be significantly improved.

Large arrays of vertically aligned Si NSs in combination with PNTCC simultaneously produced by the plasma dry etching processes are of great potential in PV applications. For bulk silicon solar cells, one of the primary costs is the starting silicon wafer, which requires extensive purification to maintain reasonable performance [202–204]. Therefore, reducing the required silicon's quality and quantity will help drive large-scale implementation of silicon PVs. Using solar cells with nanostructured p–n junctions may solve both of these problems simultaneously by orthogonalizing the direction of light absorption and charge separation while allowing for improved light scattering and trapping [205]. Furthermore, such NSs easily remind one of third-generation solar cells. NSs of semiconductor materials exhibit quantization effects when the electronic particles of these materials are confined by potential barriers to very small regions of space. Quantum confinement of electronic particles (negative electrons and positive holes) in NSs produces unique optical and electronic properties that have the potential to enhance the power conversion efficiency of solar cells for PV production at lower cost. These approaches and applications are labeled third-generation solar photon conversion. Prominent among these unique properties is the efficient formation of more than one electron–hole pair (called excitons in nanocrystals) from a single absorbed photon. Therefore, plasma dry etching processes will play an important role in third-generation solar cells in the near future. Accordingly, the detailed understanding and control of plasma-synthesized NSs as well as the technological status of NSs for third-generation PV cells need to be further investigated

10.6 Conclusion

The interest in plasma technology for PV applications has gained much attention in the last decade. This chapter presents some typical examples of the rapid progress in the field. We emphasize that plasma technology has been widely applied in Si-based solar cells, and this field is rapidly expanding. Plasma technology not only is limited to the synthesis of Si-based (including SiN) thin films as well as Si-based thin-film solar cells, but also covers plasma etching techniques such as the removal of PSG or parasitic emitters, wafer cleaning, masked or mask-free surface texturization, and the direct formation of the p–n junction by means of PNTCC. The quest for low-cost and high-efficiency solar cells is still on. Many highly qualified researchers and engineers in both academia and industry seek solutions to improve cell efficiency and reduce cost. All in all, this is a vigorous industry that will continue to grow in the next few decades.

References

1. D. M. Chapin, C. S. Fuller and G. L. Pearson, *J. Appl. Phys.* 25 (1954): 676.
2. M. A. Green, K. Emery, D. L. King et al., *Prog. Photovoltaics Res. Appl.* 9 (2001): 49.
3. W. Schmidt and B. Woesten, 2000 *Proceedings of the 16th European PV Solar Energy Conference* (Glasgow), 1–5 May, James & James (Science Publishers) Ltd. p. 1083.
4. D. E. Carlson and C. R. Wronski, *Appl. Phys. Lett.* 28 (1976): 671.
5. J. Meier, R. Fluckiger, H. Keppner et al., *Appl. Phys. Lett.* 65 (1994): 860.
6. Y. Mai, S. Klein, R. Carius et al., *Appl. Phys. Lett.* 87 (2005): 073503.
7. K. Yamamoto, A. Nakajima, M. Yoshimi et al., 2006 *Proceedings of the 2006 IEEE 4th World Conference on Photovoltaic Energy Conversion* (Hawaii), Vol. 1–2, 7–12 May, IEEE, pp. 1489–1492.
8. J. Meier, E. Vallat-Sauvain, S. Dubail et al., *Sol. Energy Mater. Sol. Cells* 66 (2001): 73.
9. M. Taguchi, M. Tanaka, T. Matsuyama et al., 1990 *Technical Digest of the International PVSEC-5* (Kyoto), 26–30 November, General Chairman of International PVSEC-5, pp. 689–692.
10. M. Taguchi, Y. Tsunomura, H. Inoue et al., 2009 *Proceedings of the 24th European Photovoltaic Solar Energy Conference* (Hamburg), 21–25 September, WIP-Renewable Energies, pp. 1690–1693.
11. A. Nakajima, M. Yoshimi, T. Sawada et al., 2004 *Proceedings of 19th European Photovoltaic Solar Energy Conference* (Paris), 7–11 June, WIP-Munich and ETA-Fluence, pp. 1567–1570.
12. J. Muller, O. Kluth, S. Wieder et al., *Sol. Energy Mater. Sol. Cells* 66 (2001): 275.
13. T. Roschek, T. Repmann, J. Muller et al., *J. Vac. Sci. Technol. A* 20 (2002): 492.
14. O. Vetterl, F. Finger, R. Carius et al., *Sol. Energy Mater. Sol. Cells* 62 (2000): 97.
15. J. Meier, H. Keppner, S. Dubail et al., *Mater. Res. Soc. Symp. Proc.* 507 (1998): 139–144.
16. S. Y. Myong, K. Sriprapha, Y. Yashiki et al., *Sol. Energy Mater. Sol. Cells* 92 (2008): 639.
17. H. Shirai, T. Arai, and H. Ueyama, *Jpn. J. Appl. Phys.* 37 (1998): L1078–L1081.
18. H. Inoue, K. Tanaka, Y. Sano et al., *Jpn. J. Appl. Phys.* 50 (2011): 036502.
19. W. J. Soppe, C. Devilee, M. Geusebroek et al., *Thin Solid Films* 515 (2007): 7490–7494.
20. H. Jia, J. K. Saha, N. Ohse et al., *Eur. Phys. J. Appl. Phys.* 33 (2006): 153–159.
21. J. K. Saha, N. Ohse, K. Hamada et al., *Sol. Energy Mater. Sol. Cells* 94 (2010): 524–530.
22. M. Hofmann, J. Rentsch, and R. Preu, *Eur. Phys. J. Appl. Phys.* 52 (2010): 11101.
23. J. W. Leem, Y. M. Song, Y. T. Lee et al., *Appl. Phys. B-Lasers Opt.* 100 (2010): 891.
24. I. O. Parm, K. Kim, D. G. Lim et al., *Sol. Energy Mater. Sol. Cells* 74 (2002): 97.
25. G. Federici, P. Andrew, P. Barabaschi et al., *J. Nucl. Mater.* 313–316 (2003): 11–22.
26. C. Leguijt, P. Lolgen, J. A. Eikelboom et al., *Sol. Energy Mater. Sol. Cells* 40 (1996): 297.
27. T. Lauinger, J. Moschner, A. G. Aberle et al., *J. Vac. Sci. Technol. A* 16 (1998): 530.
28. M. Xu, S. Xu, J. W. Chai et al., *Appl. Phys. Lett.* 89 (2006): 251904.
29. W. A. Nositschka, O. Voigt, A. Kenanoglu et al., *Prog. Photovolt. Res. Appl.* 11 (2003): 445.
30. M. Tucci, E. Salurso, F. Roca et al., *Thin Solid Films* 403–404 (2002): 307.

31. C. Forster, F. Schnabel, P. Weih et al., *Thin Solid Films* 455–456 (2004): 695.

32. Y. Momonoi, K. Yokogawa, and M. Izawa, *J. Vac. Sci. Technol. B* 22 (2004): 268.

33. A. Strass, W. Hansch, P. Bieringer et al., *Surf. Coat. Technol.* 97 (1997): 158.

34. S. H. Zaidi, D. S. Ruby, and J. M. Gee, *IEEE Tran. Electron Devices* 48 (2001): 1200.

35. J. Yoo, K. Kim, M. Thamilselvan et al., *J. Phys. D Appl. Phys.* 41 (2008): 125205.

36. H. Nagayoshi, K. Konno, S. Nishimura et al., *Jpn. J. Appl. Phys.* 44 (2005): 7839.

37. T. Arai and H. Shirai, *J. Appl. Phys.* 80 (1996): 4976.

38. Y. Fukuda, Y. Sakuma, C. Fukai et al., *Thin Solid Films* 386 (2001): 256.

39. C. Das, A. Dasgupta, S. C. Saha et al., *J. Appl. Phys.* 91 (2002): 9401.

40. C. Droz, E. V. Sauvain, J. Bailat et al., *Sol. Energy Mater. Sol. Cells* 81 (2004): 61.

41. C. Niikura, M. Kondo, and A. Matsuda, *J. Non Cryst. Solids* 338–340 (2004): 42.

42. A. H. M. Smets, T. Matsui, and M. Kondo, *J. Appl. Phys.* 104 (2008): 034508.

43. N. Kosku and S. Miyazaki, *J. Non Cryst. Solids* 352 (2006): 911.

44. J. Li, J. Wang, M. Yin et al., *J. Appl. Phys.* 103 (2008): 043505.

45. Q. Cheng, S. Xu, S. Huang et al., *Cryst. Growth Des.* 9 (2009): 2863.

46. S. Q. Xiao, S. Xu, D. Y. Wei et al., *J. Appl. Phys.* 108 (2010): 113520.

47. T. Makabe and Z. Petrovic, 2006 *Plasma Electronics* (London: Taylor & Francis).

48. T. J. M. Boyd and J. J. Sanderson, 2008 *The Physics of Plasmas* (Cambridge: Cambridge University Press).

49. P. Horvath and A. Gallagher, *J. Appl. Phys.* 105 (2009): 013304.

50. E. Amanatides, A. Hammad, E. Katsia et al., *J. Appl. Phys.* 97 (2005): 073303.

51. S. Sriraman, S. Agarwal, E. S. Aydil et al., *Nature* 418 (2002): 62.

52. S. Klein, F. Finger, R. Carius et al., *J. Appl. Phys.* 98 (2005): 024905.

53. I. B. Denysenko, K. Ostrikov, S. Xu et al., *J. Appl. Phys.* 94 (2003): 6097.

54. 1996 *Database Needs for Modeling and Simulation of Plasma Processing* (Washington, DC: National Research Council) ISBN-10: 0-309-05591-1.

55. J. R. Doyle, D. A. Doughty, and A. Gallagher, *J. Appl. Phys.* 71 (1992): 4771.

56. P. Kae-Nune, J. Perrin, J. Guillon et al., *Plasma Sources Sci. Technol.* 4 (1995): 250.

57. E. Krishnakumar and S. K. Srivastava, *Contrib. Plasma Phys.* 35 (1995): 395.

58. H. Tawara and T. Kato, *At. Data Nucl. Data Tables* 36 (1987): 167.

59. J. Perrin, Y. Takeda, N. Hirano et al., *Surf. Sci.* 210 (1989): 114.

60. A. Matsuda, K. Nomoto, Y. Takeuchi et al., *Surf. Sci.* 227 (1990): 50.

61. S. Morisset and A. Allouche, *J. Chem. Phys.* 129 (2008): 024509.

62. W. P. Leroy, S. Mahieu, R. Persoons et al., *Plasma Process. Polym.* 32 (2009): 401.

63. G. Kokkoris, A. Goodyear, M. Cooke et al., *J. Phys. D Appl. Phys.* 41 (2008): 195211.

64. P. P. Rukevych, K. Ostrikov, S. Xu et al., *J. Appl. Phys.* 96 (2004): 4421.

65. K. Ostrikov, *Rev. Mod. Phys.* 77 (2005): 489.

66. I. Levchenko, A. E. Rider, and K. Ostrikov, *Appl. Phys. Lett.* 90 (2007): 193110.

67. K. Ostrikov, *Vacuum* 83 (2008): 4.

68. K. Ostrikov, I. Levchenko, and S. Xu, *Pure Appl. Chem.* 89 (2008): 1909.

69. A. D. Arulsamy and K. Ostrikov, *Phys. Lett. A* 373 (2009): 2267.

70. K. N. Ostrikov, S. Xu and A. B. M. Shafiul Azam, *J. Vac. Sci. Technol. A* 20 (2002): 251.

71. A. Hammad, E. Amanatides, D. E. Rapakoulias et al., *J. Phys. IV Colloq* 11 (2001): Pr3-779.

72. L. Feitknecht, J. Meier, P. Torres et al., *Sol. Energy Mater. Sol. Cells* 74 (2002): 539.

73. Y. T. Gao, X. D. Zhang, Y. Zhao et al., *Acta Phys. Sin.* 55 (2006): 1497.

74. M. Heintze, R. Zedlitz, and G. H. Bauer, *J. Phys. D Appl. Phys.* 26 (1993): 1781.

75. M. N. van den Donker, B. Rech, W. M. M. Kessels et al., *New J. Phys.* 9 (2007): 280.

76. A. Matsuda, *Plasma Phys. Control. Fusion* 39 (1997): 431.

77. A. F. I. Morral, P. R. I. Cabarrocas, and C. Clerc, *Phys. Rev. B* 69 (2004): 125307.

78. R. Huang, X. Y. Lin, Y. P. Yu et al., *J. Phys. D Appl. Phys.* 39 (2006): 4423.

Plasma Technology in Silicon Photovoltaics

79. K. N. Ostrikov, I. B. Denysenko, E. L. Tsakadze et al., *J. Appl. Phys.* 92 (2002): 4935.
80. A. Matsuda, *J. Non Cryst. Solids* 338–340 (2004): 1–12.
81. A. Matsuda, *J. Non Cryst. Solids* 59–60 (1983): 767.
82. A. Natsuda and T. Goto, *Mater. Res. Soc. Symp. Proc.* 164 (1990): 3.
83. S. Bauer, B. Schroeder, W. Herbst et al., 1998 *Proceedings of the 2nd WCPVSEC* (Vienna), 6–10 July, European Commission, p. 363.
84. J. Meier, J. Spitznagel, U. Kroll et al., 2003 *Proceedings of the 3rd World Conference on Photovoltaic Energy Conversion* (Osaka), Vol. A-C, 11–18 May, IEEE, pp. 2801–2805.
85. D. L. Staebler and C. R. Wronski, *Appl. Phys. Lett.* 31 (1977): 292.
86. A. V. Shah, J. Meier, E. Vallat-Sauvain et al., *Sol. Energy Mater. Sol. Cells* 78 (2003): 469.
87. H. Takatsuka, Y. Yamauchi, Y. Takeuchi et al., 2006 *Proceedings of the 4th World Conference on Photovoltaic Energy Conversion* (Hawaii), Vol. 1–2, 7–12 May, IEEE, pp. 2028–2033.
88. K. Yamamoto, A. Nakajima, M. Yoshimi et al., *Prog. Photovolt. Res. Appl.* 13 (2005): 489.
89. Y. Mai, S. Klein, R. Carius et al., *J. Appl. Phys.* 97 (2005): 114913.
90. M. Kondo, M. Fukawa, L. Guo et al., *J. Non Cryst. Solids* 266–269 (2000): 84.
91. N. Kosku and S. Miyazaki, *Thin Solid Films* 511–512 (2006): 265.
92. N. Kosku, F. Kurisu, M. Takegoshi et al., *Thin Solid Films* 435 (2003): 39.
93. Q. J. Cheng, S. Xu, and K. Ostrikov, *Nanotechnology* 20 (2009): 215606.
94. Q. J. Cheng, S. Xu, and K. Ostrikov, *J. Mater. Chem.* 19 (2009): 5134.
95. S. Xu, K. N. Ostrikov, Y. Li et al., *Phys. Plasmas* 8 (2001): 2549.
96. K. N. Ostrikov, S. Xu, and M. Y. Yu, *J. Appl. Phys.* 88 (2000): 2268.
97. Z. L. Tsakadze, L. Levchenko, K. Ostrikov et al., *Carbon* 45 (2007): 2022.
98. I. B. Denysenko, S. Xu, J. D. Long et al., *J. Appl. Phys.* 95 (2004): 2713.
99. B. Y. Moon, J. H. Youn, S. H. Won et al., *Sol. Energy Mater. Sol. Cells* 69 (2001): 139.
100. M. Goto, H. Toyoda, M. Kitagawa et al., *Jpn. J. Appl. Phys.* 35 (1996): L1009.
101. E. Amanatides, D. Mataras, D. E. Rapakoulias et al., *Jpn. J. Appl. Phys.* 90 (2001): 5799–5807.
102. L. Sansonnens, A. A. Howling, and C. Hollenstein, *Plasma Sources Sci. Technol.* 7 (1998): 114–118.
103. B. Strahm, A. A. Howling, L. Sansonnens et al., *J. Vac. Sci. Technol.* 25 (4) (2007): 1198–1202.
104. W. J. Soppe, B. G. Duijvelaar, and S. E. A. Schiermeier, 2000 *Proceedings of the 16th European PVSEC* (Glasgow), 1–5 May, James & James (Science Publishers) Ltd., pp. 1420–1423.
105. F. Duerinckx and J. Szlufcik, *Sol. Energy Mater. Sol. Cells* 72 (2002): 231.
106. M. J. J. Kerr, A. Cuevas, and J. H. Bultman, *J. Appl. Phys.* 89 (2001): 3821.
107. M. J. J. Kerr and A. Cuevas, *Semiconduct. Sci. Tech.* 17 (2002): 166.
108. J. Schmidt and A. G. Aberle, *J. Appl. Phys.* 81 (1997): 6186.
109. H. Mackel and R. Ludemann, *J. Appl. Phys.* 92 (2002): 2602.
110. J. D. Moschner, J. Henze, J. Schmidt et al., *Prog. Photovolt.* 12 (2004): 21.
111. H. O. Pierson, 1996 *Handbook of Refractory Carbides and Nitrides* (Westwood, NJ: Noyes Publications) p. 290, chapter 15.
112. A. Luque and S. Hegedus, 2003 *Handbook of Photovoltaic Science and Engineering* (New York: Wiley) p. 284, chapter 7.
113. S. Q. Xiao, S. Xu, S. Y. Huang et al., A novel plasma-atomized deposition technique for fabricating silicon nitride films with superior surface passivation and antireflection performances (unpublished).
114. O. Vetterl, P. Hapke, L. Houben et al., *J. Appl. Phys.* 85 (1999): 2991.
115. J. J. Boland, *Adv. Phys.* 42 (1993): 129.
116. K. Oura, V. G. Lifshits, A. A. Saranin et al., *Surf. Sci. Rep.* 35 (1999): 1.
117. A. Dinger, C. Lutterloh, and J. Kuppers, *J. Chem. Phys.* 114 (2001): 5338.
118. M. Niwano, M. Terashi, and J. Kuge, *Surf. Sci.* 420 (1999): 6.
119. Z. Suet, D. J. Paul, J. Zhang et al., *Appl. Phys. Lett.* 90 (2007): 203501.
120. A. Vengurlekar, S. Ashok, C. E. Kalnas et al., *Appl. Phys. Lett.* 85 (2004): 4052.
121. H. P. Zhou, L. X. Xu, S. Xu et al., *J. Phys. D Appl. Phys.* 43 (2010): 505402.

122. E. Simoen, Y. L. Huang, Y. Ma et al., *J. Electrochem. Soc.* 156 (2009): H434.
123. Y. L. Huang, Y. Ma, R. Job et al., *J. Appl. Phys.* 96 (2004): 7080.
124. E. Simoen, C. Claeys, J. M. Raff et al., *Mater. Sci. Eng. B* 134 (2006): 189.
125. A. G. Ulyashin, Y. A. Bunmay, R. Job et al., *Appl. Phys. A: Mater. Sci. Process.* 66 (1998): 399.
126. C. M. Chiang, S. M. Gates, S. S. Lee et al., *J. Phys. Chem. B* 101 (1997): 9537.
127. G. Schulze and M. Henzler, *Surf. Sci.* 124 (1983): 336.
128. J. I. Pankove, C. W. Magee, and R. O. Wance, *Appl. Phys. Lett.* 47 (1985): 748.
129. N. H. Nickel, G. B. Anderson, N. M. Johnson et al., *Physica B* 273–274 (1999): 212.
130. N. H. Nickel, W. B. Jackson, and J. Walker, *Phys. Rev. B* 53 (1996): 7750.
131. S. Veprek and F. A. Sarott, *Plasma Chem. Plasma Process.* 2 (1982): 233.
132. Y. J. Chabal and K. Raghavachari, *Phys. Rev. Lett.* 54 (1985): 1055.
133. M. K. Weldon, V. E. Marsico, Y. J. Chabal et al., *J. Vac. Sci. Technol. B* 15 (1997): 7978.
134. S. S. Lee, M. Kong, S. F. Bent et al., *Phys. Chem.* 100 (1996): 20015.
135. J. J. Boland, *Surf. Sci.* 261 (1992): 17.
136. W. Widdra, S. I. Yi, R. Maboudian et al., *Phys. Rev. Lett.* 74 (1995): 2074.
137. S. Sinniah, M. G. Sherman, L. B. Lewis et al., *J. Chem. Phys.* 92 (1990): 5700.
138. S. M. Gates, R. R. Kunz, and C. M. Greenlief, *Surf. Sci.* 207 (1989): 364.
139. H. N. Waltenburg and J. T. Yates, *Chem. Rev.* 95 (1995): 1589.
140. K. W. Kolasinski, *J. Mod. Phys. B* 9 (1995): 2753.
141. T. J. Donahue and R. Reif, *J. Appl. Phys.* 57 (1985): 2757.
142. H. W. Kim, Z. H. Zhou, and R. Reif, *Thin Solid Films* 302 (1997): 169.
143. H. S. Tae, S. J. Park, S. H. Hwang et al., *J. Vac. Sci. Technol. B* 13 (1995): 908.
144. K. Choi, S. Gosh, J. Lim et al., *Appl. Surf. Sci.* 206 (2003): 355.
145. B. Anthony, T. Hsu, R. Qian et al., *J. Electron. Mater.* 19 (1990): 1027.
146. E. S. Aydil, R. A. Gottscho and Y. J. Chabal, *Pure Appl. Chem.* 66 (1994): 1381.
147. Z. H. Zhou, E. S. Aydil, R. A. Gottscho et al., *J. Electrochem. Soc.* 140 (1993): 3316.
148. R. J. Carter, T. P. Schneider, J. S. Montgomery et al., *J. Electrochem. Soc.* 141 (1994): 3136.
149. J. S. Montgomery, T. P. Schneider, R. J. Carter et al., *Appl. Phys. Lett.* 67 (1995): 2194.
150. K. H. Hwang, E. Yoon, K. W. Whang et al., *Appl. Phys. Lett.* 67 (1995): 3590.
151. M. Shinohara, T. Kuwano, Y. Akama et al., *J. Vac. Sci. Technol. A* 21 (2003): 25.
152. T. R. Yew and R. Rief, *J. Appl. Phys.* 68 (1990): 4681.
153. D. L. King and M. E. Buck, 1991 *Proceedings of the 22nd IEEE Photovoltaic Specialist Conference* (Las Vegas), 7–11 October, *IEEE*, pp. 303–308.
154. J. M. Kim and Y. K. Kim, *Sol. Energy Mater. Sol. Cells* 81 (2004): 239.
155. E. Vazsonyi, K. De Clercq, R. Einhaus et al., *Sol. Energy Mater. Sol. Cells* 57 (1999): 179.
156. M. Edwards, S. Bowden, U. Das et al., *Sol. Energy Mater. Sol. Cells* 92 (2008): 1373.
157. D. Iencinella, E. Centurioni, R. Rizzoli et al., *Sol. Energy Mater. Sol. Cells* 87 (2005): 725.
158. G. W. Trucks, K. Raghavachari, G. S. Higashi et al., *Phys. Rev. Lett.* 65 (1990): 504.
159. X. D. Bai, Z. Xu, S. Liu et al., *Sci. Technol. Adv. Mater.* 6 (2005): 804.
160. K. Seeger and R. E. Palmer, *Appl. Phys. Lett.* 74 (1999): 1627.
161. T. Tada, T. Kanayama, K. Koga et al., *J. Phys. D Appl. Phys.* 31 (1998): L21.
162. C. H. Hsu, H. C. Lo, C. F. Chen et al., *Nano Lett.* 4 (2004): 471.
163. M. C. Yang, J. Shieh, C. C. Hsu et al., *Electrochem. Solid State Lett.* 8 (2005): C131.
164. K. Ostrikov and S. Xu, 2007 *Plasma-Aided Nanofabrication: From Plasma Sources to Nanoassembly* (Weinheim: Wiely-VCH) ISBN 978-3-527-40633-3.
165. K. Ostrikov and A. B. Murphy, *J. Phys. D Appl. Phys.* 40 (2007): 2223.
166. S. H. Rsu, C. Yang, W. J. Yoo et al., *J. Korean Phys. Soc.* 54 (2009): 1016.
167. M. Moreno, D. Daineka, and P. R. I. Cabarrocas, *Sol. Energy Mater. Sol. Cells* 94 (2010): 733.
168. J. S. Yoo, I. O. Parm, U. Gangopadhyay et al., *Sol. Energy Mater. Sol. Cells* 90 (2006): 3085.
169. L. L. Ma, Y. C. Zhou, N. Jiang et al., *Appl. Phys. Lett.* 88 (2006): 171907.

170. S. Koynov, M. S. Brandt, and M. Stutzmann, *Appl. Phys. Lett.* 88 (2006): 203107.
171. Y. Jia, J. Wei, K. Wang et al., *Adv. Matter.* 20 (2008): 4594.
172. H. C. Yuan, V. E. Yost, M. R. Page et al., *Appl. Phys. Lett.* 95 (2009): 123501.
173. E. Garnett and P. Yang, *Nano Lett.* 10 (2010): 1082.
174. W. H. Lee, J. C. Lin, C. Lee et al., *Diamond Relat. Mater.* 10 (2001): 2075.
175. H. Jansen, M. De Boer, H. Wensink et al., *J. Microelectron* 32 (2001): 769.
176. J. Yoo, G. Yu and J. Yi, *Sol. Energy Mater. Sol. Cells* 95 (2011): 2.
177. K. Lee, M. H. Ha, J. H. Kim et al., *Sol. Energy Mater. Sol. Cells* 95 (2011): 66.
178. S. Xu, I. Levchenko, S. Y. Huang et al., *Appl. Phys. Lett.* 95 (2009): 111505.
179. I. Levchenko, S. Y. Huang, K. Ostrikov et al., *Nanotechnology* 21 (2010): 025605.
180. M. Dhamrin, N. H. Ghazali, M. S. Jeon et al., 2006 *Conference Record of the 2006 IEEE 4th World Conference on Photovoltaic Energy Conversion* (Hawaii), Vol. 1–2, 7–12 May, IEEE, pp. 1395–1398.
181. O. Zinchuk, N. Drozdov, A. Fedotov et al., *Vacuum* 83 (2009): s99.
182. V. P. Popov, I. E. Tyschenko, L. N. Safronov et al., *Thin Solid Films* 403–404 (2002): 500.
183. D. Barakel, A. Ulyashin, I. Perichaud et al., *Sol. Energy Mater. Sol. Cells* 72 (2002): 285.
184. T. Yamaguchi, S. J. Taylor, S. Watanabe et al., *Appl. Phys. Lett.* 72 (1998): 1226.
185. A. Buzynin, A. Luk'yanov, V. Osiko et al., *Nucl. Instrum. Methods Phys. Res. B* 186 (2002): 366.
186. T. Wolkenstein, 1973 *Physico-Chimie de la Surface des Semiconducteurs* (Moscow: MIR).
187. W. Götz, G. Pensl, and W. Zulehner, *Phys. Rev. B* 46 (1992): 4312.
188. R. C. Newman, M. J. Ashwin, R. E. Pritchard et al., *Phys. Status Solidi* 210 (1998): 519.
189. P. De'ak, L. L. Snyder, J. L. Lindstrom et al., *Phys. Lett. A* 126 (1988): 427.
190. R. C. Newman, J. H. Tucker, A. R. Brown et al., *J. Appl. Phys.* 70 (1991): 3061.
191. S. Xu, S. Y. Huang, I. Levchenko et al., *Adv. Energy Mater.* 1 (2011): 373.
192. M. Wolter, I. Levchenko, H. Kersten et al., *Appl. Phys. Lett.* 96 (2010): 133105.
193. M. Bruzzi, D. Menichelli, M. Scaringella et al., *J. Appl. Phys.* 99 (2006): 093706
194. A. Janotti and C. G. Van de Walle, *Nature* 6 (2007): 44.
195. C. G. Van de Walle, P. J. H. Denteneer, Y. Bar-Yam et al., *Phys. Rev. B* 39 (1989): 10791.
196. G. Agostinelli, H. F. W. Dekkers, S. de Wolf et al., 2004 *Proceedings of the 19th EU PVSEC* (Paris), 7–11 June, WIP-Munich and ETA-Fluence, p. 132.
197. B. Rech, T. Roschek, T. Repmann et al., *Thin Solid Films* 427 (2003): 157.
198. D. Derkacs, S. H. Lim, P. Matheu et al., *Appl. Phys. Lett.* 89 (2006): 093103.
199. M. D. Yang, Y. K. Liu, J. L. Shen et al., *Opt. Express* 16 (2008): 15754.
200. Y. A. Akimov, K. Ostrikov, and E. P. Li, *Plasmonics* 4 (2009): 107.
201. Y. A. Akimov, W. S. Koh, and K. Ostrikov, *Opt. Express* 17 (2009): 10195.
202. M. Fawer-Wasswer, 2004 *Sarasin Sustainable Investment Report* (Basel), November, Sarasin.
203. M. A. Green, *Sol. Energy* 76 (2004): 3.
204. E. A. Alsema, *Prog. Photovolt.* 8 (2000): 17.
205. B. M. Kayes, H. A. Atwater, and N. S. Lewis, *J. Appl. Phys.* 97 (2005): 114302.

11

Environmental Applications of Plasmas

Keping Yan

11.1 Overview of Nonthermal Plasma for Pollution Control

With regard to environmental applications of nonthermal plasmas (NTPs), the most popular techniques used to be electrostatic precipitation (ESP) and ozone generation. Both of them have a history of over a century. In NTPs, usually induced by high electric field or magnetic field, free electrons, ions, and radicals can be generated through electron-impact dissociation and the ionization of molecules. The electrostatic precipitator and the ozone generator are based on ion and O radical generation, respectively. Figure 11.1 gives a summary of the development of NTPs focused on environmental applications. The initial literature can be traced back to 1796, when a German research group discovered that hydrocarbons could be converted into oil-like products if exposed to an electrical discharge. The famous ozone generator and ESP were invented in 1857 and 1908, respectively. After that, the application of NTP in DeSO$_2$ and DeNO$_x$ of flue gas became concerns (Matteson et al. 1972; Masuda 1988; Clements et al. 1989). In recent decades, the NTP technique was introduced in other fields, such as for abatement of volatile organic compounds (VOCs) (Toshiaki 1997; Urashima and Chang 2000; Oda 2003), vehicle exhaust gas treatment (Higashi et al. 1992; Yao et al. 2001; Chae 2003), water

311

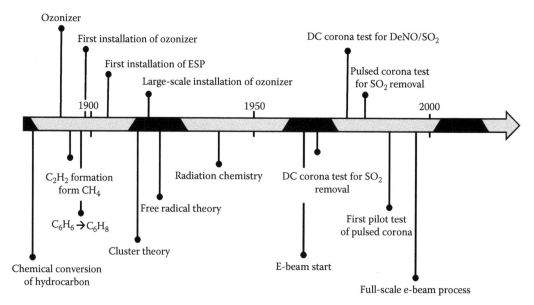

FIGURE 11.1 The history of NTP application in environment-related fields. (From Kim, H.-H.: Nonthermal plasma processing for air pollution control: A historical review, current issues, and future prospects. *Plasma Processes and Polymers.* 2004. 1(2): 91–110. Copyright Wiley-VCH Verlag GmbH & Co. KGaA. Reproduced with permission.)

treatment (Lukes et al. 2005; Locke et al. 2006), and sterilization (Lerouge et al. 2000; Ching et al. 2001; Huang et al. 2009). NTP was also combined with other techniques for pollution control, especially the hybrid NTP/wet process system (Yan et al. 2006) and the hybrid plasma catalysis system (Yan et al. 1998; Durme et al. 2008).

NTP applications in various environment-related fields, such as the chemical reactions for pollution control with NTP, have been investigated. Some examples of investigated chemical reactions are hydrocarbon conversion; cracking of gaseous hydrocarbons; polymerization; the formation and decomposition of nitrogen oxides; NH_3 synthesis; NO_x and CO oxidation; hydrogen production; electrical discharge-enhanced combustion; OH, N, and H_2O_2 generation; hydrogen sulfide decomposition; SO_2 oxidation by O_3; SO_2 to $(NH_4)_2SO_4$ salt conversion with NH_3 injection; and effects of electrical discharge on chemical catalysis. There are also many types of NTP systems that have been developed for environmental applications, roughly divided into three groups: pulsed streamer corona (PSC) energized by high-voltage pulse sources, dielectric barrier discharge (DBD) energized by AC power sources or pulsed power sources, and packed-bed corona (PBC) energized by either AC or pulsed power sources. There is another way to generate an NTP with microwaves that is energized by a radio frequency power source. Up to now, most commercially available corona plasma systems are based on DBD technology.

11.1.1 Electrostatic Precipitation

In 1883, the British physicist Sir Oliver Lodge came up with the idea of ESP by suggesting in an article in *Nature* that ESP could be used to clean polluted air consisting of fog and smoke. In the early 1900s, Dr. F.G. Cottrell in California, USA, developed the first applicable ESP on an industrial scale, which was described in his first ESP patent (US Patent No. 895,729 in 1908). In the following ten years, several ESPs were installed for applications such as for removal of cement kiln dust, lead smelter fumes, tar, and pulp and paper alkali salts. Harry J. White made a great contribution to promote scientific understanding and industrial progress of ESP. In 1963, he wrote a very famous book *Industrial Electrostatic*

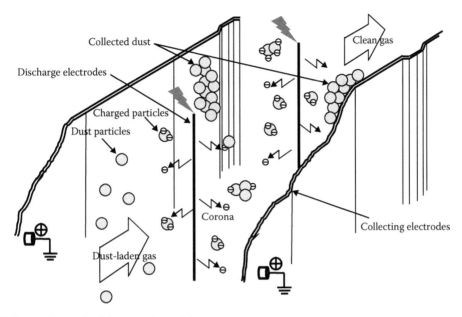

FIGURE 11.2 Principle of electrostatic precipitator.

Precipitation. After that Oglesby and Nichols reviewed almost all aspects related to the technique, such as historical development, theory of ESP, development of equipment, applications, and economics in their book *Electrostatic Precipitation.*

As shown in Figure 11.2, wire electrodes (also called discharging electrodes) are placed between two grounded plate electrodes (also called the collecting electrodes) to produce a localized high nonuniform electric field. In this region, free electrons are accelerated to high velocities when a sufficiently high voltage (larger than corona onset voltage) is applied to the wire electrode. If the energy of these electrons is high enough, the ionization process will be repeated for many times so that large quantities of free electrons and positive ions are formed within the corona region. Dust particles can be charged by these generated electrons and ions, and dragged toward the plate electrode by the Coulomb force. As long as the charged particles arrive at the plate electrode, the electric charge runs away through the ground connection while the particles remain. When a dust layer is formed on the collecting electrode and its thickness reaches a predetermined level, a rapping hammer or an intensive acoustic apparatus is used to remove the layer to a hopper located below. In summary, ESP works in three steps: particle charging by generated electrons and ions; particle transportation under an electric field; and particle removal from the collecting electrode by mechanical force.

For ESP fields, the famous Deutsch equation describes the basic law about the relationship between dust collecting efficiency and treatment time or collecting area for a given gas flow. It was discovered experimentally based on field tests by Anderson in 1919, and first derived through theoretical considerations by Deutsch in 1922. So it is also called the Deutsch–Anderson equation:

$$\eta = 1 - e^{(-\omega A/Q)}, \tag{11.1}$$

where

  η is the collecting efficiency
  ω the charged particle migration velocity, which means the moving velocity in the direction perpendicular to the collecting electrode (cm/s)
  A the collecting plate area (m²)
  Q the gas flow rate (m³/s)

The migration velocity ω is in principle a constant for given conditions (particle size and shape, electric field strength, gas composition and temperature, etc.). The theoretical derivation details of ω can be found in chapter 5 of Oglesby and Nichols's book *Electrostatic Precipitation*, but it is usually calculated from the measured collecting efficiency η, measured or calculated gas flow Q, and actual collection area A, because the theoretical result always deviates significantly from the real condition. It is observed empirically that ω is a constant within a wide efficiency range. Here, A/Q is also defined as the specific collection electrode area, which indicates the collecting area required per cubic meter per second of gas. It is always used to compare (normalize) one ESP to another. In ESP operation, there are some other factors that influence its performance, for instance, dust resistivity, reentrainment, back corona, and sparking.

Different compositions of dust result in different resistivities. In principle, Si, Al, and Ca often increase the resistivity, while Na and Fe do the opposite. A difference of up to eight times can be achieved for various fly ashes from coal burning, which ranges from 10^4 to 10^{12} $\Omega \cdot$cm. It is a general perception that ESP works most efficiently around 10^9–10^{11} $\Omega \cdot$cm. Reentrainment may occur at low-value resistivity as electrostatic charge is drained off too quickly, while high-value resistivity will slow down this phenomenon and generate back corona. Additionally, charging becomes difficult for high-resistivity dust, especially for fine particles. Several models have been built to predict resistivity in terms of composition, water concentration, and other factors (Arrondel et al. 2006; Li et al. 2012). It can also be measured in the laboratory with a standardized method recommended by IEEE, or *in situ* with equipment installed in the flue gas stream at a specific point. Dust resistivity is so important that almost every ESP company has its own database as a business secret.

As mentioned before, high dust resistivity will cause back corona. Figure 11.3 illustrates that a back corona current through a dust layer multiplied with the apparent resistivity of the dust generates a voltage (electrical field) drop through the dust layer. When the field strength breaks a critical point, a breakthrough of the gas in the dust layer happens. The mechanical force generated by local sparking makes the dust at the layer surface rebound into gas flow. Typical current–voltage (*I/V*) curves under back corona conditions are shown in Figure 11.4. More back corona results in lower corona current in the ESP.

In addition to dust resistivity and back corona, rapping, direct erosion, or scouring by aerodynamic force and sparking can also induce dust at the collection layer surface reentering into the gas flow, which is so-called reentrainment. The amount of reentrained dust is extremely difficult to measure, but it can be reduced by mechanical or electrical methods, for instance, off-flow rapping, gas distribution optimizing with a straightener or splitter, and spark rate control by using intelligent electric circuit.

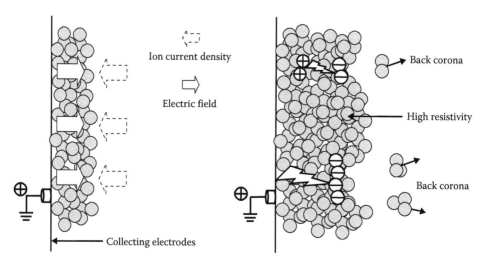

FIGURE 11.3 Schematic of back corona process.

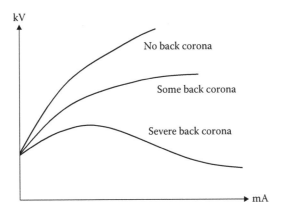

FIGURE 11.4 Typical *I/V* curves under back corona condition.

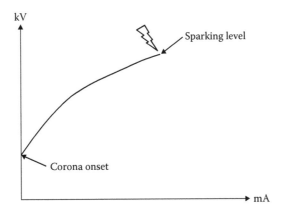

FIGURE 11.5 Normal *I/V* curve of ESP between corona onset and sparking level.

Unlike back corona, here sparking means electrical breakdown between the discharging electrode and the collecting electrode. This not only induces reentrainment, but also limits the effective electrical power input into the precipitator. Figure 11.5 shows the normal *I/V* curve (*I–V* relation) of ESP. At the sparking level, when an arc channel is formed, the voltage between the electrodes even drops to zero, while the current increases rapidly because the impedance of the arc channel is extremely small. Different gas composition, gas temperature, dust load, and dust buildup on electrodes result in different sparking levels.

In the past 30–40 years, emission limits have been reduced from several hundred milligrams per normal cubic meter to less than 25 mg/Nm³; thus fine particle (including PM2.5 particles) control is becoming more and more urgent. To meet the low emission limits, several state-of-the-art techniques have been included in the ESP design and operation. Computational fluid dynamics software is used to simulate gas flow field in an ESP in order to optimize gas flow quality; more efficient power sources, for instance, three-phase DC source, switch mode DC source, and DC superposed pulse source, are used to make particles charging more efficient; electrical and chemical agglomeration methods are applied to enlarge fine particles before their removal (Zhu et al. 2010). In the modern ESP industry, there are also some research activities under way. Combining ESP and the fabric filter (FF is good at fine particle removal, but its pressure drop is large) is one of them. Its basic idea is that if the particle is charged in an upstream ESP, the pressure drop at FF bags will decrease because of the repulsive force between particles with the homogeneous charge. Hybrid dry ESP–wet ESP is another promising technique. The reentrainment from upstream dry ESP can be removed in the following wet ESP. J.S. Chang proposed an integrated electrostatic gas cleaning system that could remove particle and gaseous pollutants simultaneously. In this system, ESP consists of different techniques in each field for the oxidation and agglomeration of fine particles and the effective collection of secondary particles generated from the gas phase by NTP processes. The NTP reactor also should be integrated for effective reductions or oxidations of gaseous pollutants (Chang 2003).

11.1.2 Ozone Generation

In 1857, Siemens invented an ozone generation method, which could generate ozone from atmospheric pressure oxygen or air. In his invention, the discharge was initiated in an annular gap between two coaxial glass tubes by applying a sufficiently high alternating voltage. Because glass walls function as dielectric barriers, the discharge is called DBD (sometimes called "silent discharge"). Since then, DBD ozone generators are being widely used for industrial applications, such as drinking water treatment and odor emission control.

DBD ozone generators convert O_2 to O_3 with the O radical generated by electrical discharges in two steps. First, O_2 molecules are dissociated by accelerated electron impact. If nitrogen is present or the electrical discharge is initiated in air, the O radical can also be generated by reactions with N atoms or excited N_2 molecules. Second, ozone is formed by three-body reaction involving O and O_2:

$$O + O_2 + M \rightarrow O_3^* + M \rightarrow O_3 + M, \tag{11.2}$$

where M is a third collision partner: O_2, O_3, and O, or, in the case of air, also N_2. O_3^* stands for a transient excited state in which the ozone molecule is initially formed after the reaction of an O atom with an O_2 molecule.

In air discharges, nitrogen ions N^+, N_2^+, nitrogen atoms, and excited atomic and molecular species $[N_2(A^3\Sigma_u^+)$ and $N_2(B^3\Pi_g)]$ involved in the reaction can produce additional oxygen atoms for ozone formation. Details of the reaction path are described in the following equations. Almost half of the ozone formed in air discharges results from these indirect processes. Except for ozone, a variety of nitrogen oxide species (NO, N_2O, NO_2, NO_3, and N_2O_5) are generated. As a result, ozone formation in air takes longer (about 100 μs) than in pure oxygen (about 10 μs).

$$N + O_2 \rightarrow NO + O, \tag{11.3}$$

$$N + NO \rightarrow N_2 + O, \tag{11.4}$$

$$N_2(A) + O_2 \rightarrow N_2O + O, \tag{11.5}$$

$$N_2(A, B) + O_2 \rightarrow N_2 + 2O. \tag{11.6}$$

Because of the accompanying ozone formation reaction, some undesired side reactions consume the O atom imposing upper limits on the O atom concentration or the degree of dissociation. As described in the following equations, if the O atom concentration is too high, recombination to O_2 becomes dominant:

$$O + O + M \rightarrow O_2 + M, \tag{11.7}$$

$$O + O_3 + M \rightarrow 2O_2 + M, \tag{11.8}$$

$$O + O_3^* + M \rightarrow 2O_2 + M. \tag{11.9}$$

Besides side reactions, another important impact factor should be considered: reaction temperature. High temperature results in rising ozone decomposition velocity; so in the industrial DBD ozone generator, the recycled cooling system is essential to take reaction heat away.

To optimize ozone generation, several researchers have investigated the DBD process to optimize the discharge condition with simulating models or plasma diagnostic tools with respect to discharging gap

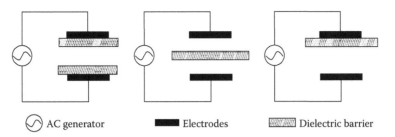

FIGURE 11.6 DBD discharge configurations. (With kind permission from Springer Science+Business Media: *Plasma Chemistry and Plasma Processing*. Dielectric-barrier discharges: Their history, discharge physics, and industrial applications. 23, 2003, 1–46, Kogelschatz, U.)

width, dielectric properties, operating pressure, voltage wave form, and driving frequency. Important contributions to the fundamental understanding and industrial applications of DBD were made by Kogelschatzs et al. at the ABB company.

Figure 11.6 gives the three typical DBD configurations, which describe the spatial relationship between electrodes and dielectric barrier. In traditional ozone generators, cylindrical Pyrex (Duran) glass tubes (sometimes ceramic tubes are also used) are mounted inside stainless-steel tubes with a diameter about 20–50 mm to form annular discharge gaps about 0.5–1 mm in width. Conductive coatings (aluminum films) and a brush-shaped electrode inside the glass tubes are connected to the high-voltage supply operated in line frequency or derived from a motor generator. High-performance ozone generators now use nonglass dielectrics derived from a high-frequency switch source (0.5–5 kHz), which generates a square-wave current or some other special wave form. As higher operating frequencies can deliver the desired power density at much lower operating voltages, from typically 20 kV in the past to <5 kV now, less electrical stress on the dielectrics makes the discharge gap more narrow.

DBD ozone production ranges from a few milligrams per hour to several kilograms per hour. Large ozone generators use several hundred discharge tubes. Figure 11.7 shows a large ozone generator produced by Ozonia Ltd. that has a capacity of 60 kg/h.

11.1.3 NTP-Induced Gas Cleaning

In addition to dust removal and ozone generation, NTP-induced gas cleaning research and industrial practices include NO_x reduction and oxidation, Hg oxidation, flue gas preprocessing, plasma-induced soot combustion, decomposition of dilute VOCs, diesel exhaust treatment, and tar cracking. Sections 11.4–11.6 will discuss these subjects in detail.

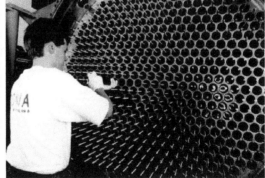

FIGURE 11.7 Large ozone generator producing 60 kg ozone per hour. (From Kogelschatz, U. et al., *Pure Appl. Chem.*, 71, 1823, 1999. © IUPAC.)

Generally speaking, the main function of NTP in pollution control is to convert one kind of compound into another one, which may be easily handled. The most distinctive characteristic of NTP as a chemical process is its ability to induce various chemical reactions at atmospheric pressure and room temperature. Energetic electrons in NTPs produce active species—free radicals and ions—as well as additional electrons through electron impact dissociation, excitation, and ionization of background gas molecules. These active species oxidize, reduce, or decompose the pollutant molecules. For example, NO, NO_2, and SO_2 in flue gas are insoluble, and active species produced by NTPs can oxidize them into soluble compounds, such as HNO_2, HNO_3, NO_3, N_2O_5, and SO_3. For Hg reduction in flue gas, a Hg atom should be oxidized by an active species before emission because it is highly toxic. For low-concentration VOC abatement, tar cracking, and plasma-induced soot combustion, NTP mainly decomposes a giant organic molecule into a small one. Details of active species generation and global reactive kinetics will be discussed in the next section.

Table 11.1 lists some compounds that were investigated for emission control with NTP techniques, where "$\sqrt{}$" means the NTP generation mode has been adopted. Most of the research and development on these subjects is still under investigation in order to improve the efficiency, reduce the cost, and promote industrial applications. Although different kinds of NTP generation modes have been investigated, commercially available NTP systems are mainly based on DBD technology, and most investigations on mobile exhaust gas treatment, soot, and NO_x removal are operated by DBD. For large-volume gas cleaning, pulsed corona plasma techniques are considered to be more suitable than DBD and PBC, not only because of high-pressure drop across DBD and PBC reactors, but also because of the risk of insulator failure.

A large-scale demonstration of NTP technology was initiated by ENEL (Italy) at the Marghera power plant in the late 1980s, and since then many other tests have been conducted in China, Korea, and Japan. H.-H. Kim has summarized all large-scale demonstrations of NTP technology, including e-beam for flue gas cleaning (Table 11.2).

11.1.4 NTP-Induced Water Cleaning

In the last century, mechanical, chemical, and biological treatment methods have been developed and implemented all over the world to protect water sources and produce clean water for people's daily life. To conform to stringent drinking water standards proves difficult where low concentrations of persistent organic pollutants (POPs) are present. The potential toxicity of these pollutants creates even more difficulties to design and implement a proper cleaning process system. To tackle this problem, the so-called advanced oxidation technologies (AOTs) has been developed, which aim to produce *in situ* strong oxidizers to degrade POPs and other tough pollutants. Normal AOTs include ozone–UV oxidation, hydrogen peroxide–UV oxidation, Fenton oxidation, photocatalytic oxidation, wet oxidation, radiolysis, ultrasonic irradiation, and electrical discharge.

The electrical discharge for water cleaning applications may be created directly in the liquid but also in the gas phase with a different electrode configuration. The localized plasma induced by a discharge can emit intensive ultraviolet radiation and generate shockwaves and active species. So it is a multiple mode-of-action approach for water treatment. Locke et al. (2006) gave a comprehensive review on this topic including different electrohydraulic discharge modes, different reactor types, and basic physical principles of electrohydraulic discharge and their chemical and biological effects. Details of this subject will be discussed in Section 11.7.

11.1.5 NTP-Induced Surface Disinfection

In NTPs, energetic electrons collide with the background gas, causing an enhanced level of dissociation, excitation, and ionization, while ions and neutrals remain relatively cold and do not cause any thermal damage to articles they come in contact with. Based on this characteristic, treatment of some

TABLE 11.1 List of Compounds Recently Investigated for Emission Control with Electrical Discharge NTPs

Compounds	Formula	PSC	DBD	PBC
Ammonia	NH_3	√	√	
Benzene	C_6H_6			√
Butane	C_4H_{10}	√		
Carbon dioxide	CO_2	√	√	√
Carbon monoxide	CO	√	√	
Carbon tetrachloride	CCl_4		√	√
Chlorobenzene	C_6H_5Cl		√	
Cyclohexane	C_6H_{12}			√
(*E*)-1,2-Dichloroethylene	$ClHC=CHCl$	√		
Dioxin and furans		√		
Dichlorodifluoromethane	(CFC-12) CCl_2F_2		√	
Dichloromethane	CH_2Cl_2	√		
Ethene	$CH_2=CH_2$	√		
Ethanol	CH_3CH_2OH			√
Fluorosilicate	SiF_4	√		
Formaldehyde (methanol)	HCHO		√	
Hexane	C_6H_{14}			√
Hg vapor	Hg	√		
Hydrogen chloride	HCl	√		
Hydrogen cyanide	HCN		√	
Hydrogen sulfide	H_2S	√	√	√
Odor		√	√	
Methane	CH_4	√	√	√
Methanol	CH_3OH	√		
Methyl acetate	CH_3COOCH_3	√		
Methylene chloride	CH_2Cl_2	√		
Methyl ethyl ketone	$CH_3COC_2H_5$		√	
Naphthalene	$C_{10}H_8$	√		
Nitrogen fluorides	NF_3			√
Nitrogen oxides	N_2O, NO, NO_2	√	√	√
Pentane	C_5H_{12}	√		
Perchloroethylene	$Cl_2C=CCl_2$	√		
Henantrene	$C_{14}H_{10}$	√		
Phenol	C_6H_6O	√		
Propane	$CH_3CH_2CH_3$	√		
Styrene	C_8H_8	√	√	
Sulfur dioxide	SO_2	√	√	√
Tar		√		
Trichloroethylene (TCE)	$CHCl=CCl_2$	√	√	
Toluene	$C_6H_5CH_3$	√	√	√
Trichlorotrifluoroethane (CFC-113)	$CCl_2F–CClF_2$		√	√
Xylenes	C_8H_{10}		√	

Source: Yan, K., Corona plasma generation. Ph.D. Dissertation, TU/e, Eindhoven, 2001.

Notes: PSC: Pulsed streamer corona, DBD: Dielectric barrier discharge, PBC: Packed-bed corona.

TABLE 11.2 Large-Scale Demonstration of NTP Technology for Flue Gas Cleaning

Institution	Year[a]	Reactor	Power supply	Flow rate (Nm³/h)	Temperature (°C)	Target Gases	Additive (ppm)	SIE (Wh/Nm³)	DRE (%)
ENEL, Marghera Power station	1990	Wire–plate (plasma + ESP)	Pulse	1,000	<100	NO_x (430–550 ppm) SO_2 (360–550 ppm)	NH_3 $Ca(OH)_2$	~15	NO_x (60) SO_2 (80)
EPRI–LANL (mobile system)	1996	DBD	AC	18	RT	VOCs mixtures (TCE, TCA, PCE, toluene)	–	18	>90
Fujisawa pilot plant	1996	Corona + EB	DC	1,200	<65	NO (60 ppm) SO_2 (230 ppm)	NH_3 (S.R.)	NI + 3kGy	NO_x (85) SO_2 (99)
Hitachi (diesel engine)	1998	Cylindrical DBD	AC	115		NO_x (500 ppm)	HC	~5	60
Korea Cottrell	1998	Wire–plate	Pulse (RSG, 99 kV, 300 Hz)	10,000	70–150	NO_x SO_2	NH_3 (S.R.)	–	NO_x (50) SO_2 (89)
KIMM (oil-fired)	1998	Wire–plate	Pulse (MPC)	2,000	100	NO (157 ppm) SO_2 (315 ppm)	NH_3 (800) C_2H_4 (550)	–	NO_x (85) SO_2 (95)
Dalian University	1998	Wire–plate	Pulse (RSG)	1,400	70	SO_2 (2,000 ppm)	NH_3 (S.R.)	3.9	SO_2 (80)
TUT (incinerator)	2000	Wet ESP (rod–cylinder)	DC	1,000	<80	Dioxin (73 ng–TEQ/Nm³)	–	<0.2	90
Masuda Research Inc.	2001	PPCP (wire–cylinder)	DC-bias pulse	6,700		Dioxin (Dust <5 mg/Nm³)	Propane (100)	~6	~92.4
RIST	2001	Wire–plate (plasma + ESP)	Pulse (MPC, 200 kV, 40 kW)	5,000	150	NO_x (73 ppm) SO_2 (150 ppm)	C_3H_6 (48)	4	NO_x (60) SO_2 (90)
Doosan Heavy Industry	2003	Retrofitted ESP	Pulse (MPC, 120 kV, 160 kW, 300 Hz)	35,000	120	NO_x (300 ppm) SO_2 (300 ppm)	NH_3 (S.R.) C_2H_4 (600)	9	NO_x (70) SO_2 (95)
Takuma Co.[b] (Waste water plant)	2003	Surface corona with catalyst	AC (16 kW)	51,000	RT	Odor	–	0.2	99
Pohang University	2003	PPCP (wire–plate)	Pulse (MPC)	42,000	170	NO_x (120 ppm) SO_2 (150 ppm)	NH_3 (370) C_3H_6 (55)	1.4	NO_x (70) SO_2 (99)

Source: Kim, H.-H.: Nonthermal plasma processing for air-pollution control: A historical review, current issues, and future prospects. *Plasma Processes and Polymers.* 2004. 1(2). 91–110. Copyright Wiley-VCH Verlag GmbH & Co. KGaA. Reproduced with permission.
[a] based on the publication year
[b] full-scale plant

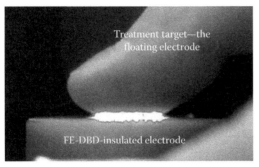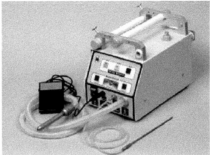

FIGURE 11.8 NTP-assisted surface treatment and available commercial instrument. (From Fridman, G. et al.: Applied Plasma Medicine. *Plasma Processes and Polymers.* 2008. 5. 503–533. Copyright Wiley-VCH Verlag GmbH & Co. KGaA. Reproduced with permission.)

heat-sensitive materials such as cells and tissues with NTP become possible. This subject has a very wide biological and medical application potential, and has attracted many researchers and engineers in the past decade. Up to now, most research is on the relationship between NTP characteristics and its functional efficiency. Limited work has been done on the cellular level to explain the effects of NTPs. Professor Laroussi from Old Dominion University and Professor Fridman from Drexel University have significantly contributed toward promoting the research and application practices in this field. Several commercial instruments have been developed for tooth surface treatment, wound healing, blood coagulation, anaphylactic rhinitis treatment, and so on (Figure 11.8).

11.2 Corona Discharge Modes and Chemical Reactivity

It is well known that for a point–plate electrode arrangement in air, a positive DC electrical discharge changes from onset streamer to Hermstein glow, prebreakdown streamer, and then to spark breakdown when the applied voltage is increased. The discharge patterns depend on electrode geometry, gas flow rate, and compositions. The glow and streamer coronas were distinguished by means of optical and electrical measurements. For streamer corona, both current and voltage waveforms show pulsed characteristics and streamers may cross the electrode gap. Within the pulse duration, the voltage across the gap drops several hundred volts. The peak streamer current may reach up to a few hundred milliamperes. When the Hermstein glow occurs, the discharge light emission can only be observed near the tip of the high-voltage electrode under a very stable corona current. The corona current may also show a high-frequency oscillation with small amplitudes. When a spark breakdown occurs, the voltage drops to almost zero with a large current pulse short circuit. A large amount of NO_x can be produced when spark breakdown occurs. For the purpose of gas cleaning, spark breakdown should be avoided (Yan et al. 1999a).

With regard to basic relationships between electrical discharge modes and chemical reactions, Akerlof and Wills (1951) gave a very comprehensive bibliography on chemical reactions in electrical discharges. Yan et al. systematically investigated the NO to NO_2 conversion for the purpose of gas cleaning and concluded that the glow discharge induces negligible chemical reactions in comparison to a streamer corona, while a prebreakdown streamer is less efficient than an onset streamer. Details are described in Figure 11.9.

Figure 11.10 shows a typical discharge photograph in a wire–plate reactor. The diameter of the high-voltage wire is 140 μm, and the plate-to-plate gap is 50 mm. Because of effects of gas composition on the corona discharge pattern and the sensitivity of the DC streamer corona to electrode misarrangements, the transition from streamer corona to glow discharge occurs. As a result, glow and streamers (prebreakdown streamer) are often generated simultaneously, and both are less efficient for pollution control. As such, it is impossible to generate a large-volume streamer (onset streamer) corona with a positive DC source. The effective method to solve this problem, especially restrict the

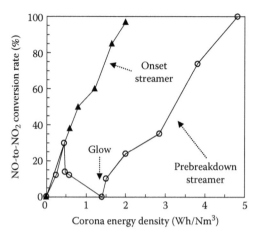

FIGURE 11.9 Comparison of NO-to-NO$_2$ conversion with a point–plate-type corona reactor under different discharge modes. (From Yan, K. Corona plasma generation, Ph.D. Dissertation, TU/e, Eindhoven, 2001.)

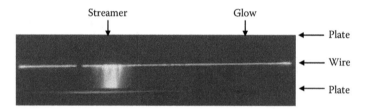

FIGURE 11.10 Positive corona discharge stationary photograph in a wire–plate corona reactor. (From Yan, K. Corona plasma generation, Ph.D. Dissertation, TU/e, Eindhoven, 2001.)

spark breakdown, is a short-pulsed energization system, whose pulse width should be shorter than the streamer transition time. There are two kinds of power sources available that can realize this: one is a DC with superimposed high frequency AC (10–60 kHz) and the other is a pulsed power source (Yan et al. 1999c, 2001c).

Considering the electrical and optical measurement results of PSC in a wire–cylinder reactor, which is shown in Figure 11.11, streamers can be divided into two phases: primary streamer and secondary streamer. On increasing the applied voltage, primary streamers can cross the gap, and then the secondary streamers start to develop after the primary streamer approaches the cylinder. Corona energy is mainly transferred into streamer channels with primary streamer propagation and secondary streamer development, and most radicals are generated during this period. Winand et al. confirmed this by photographing the streamer propagation process with high-speed ICCD. The results are shown in Figure 11.12.

The elementary process of radical formation and reaction in NTPs can be broadly divided into a primary process and a secondary process based on the timescale of streamer propagation (Kim 2004). As shown in Figure 11.13, the primary process includes ionization, excitation, dissociation, light emission, and charge transfer. Because the electric field strength at the streamer head can reach 150–200 kV/cm and streamer propagation velocity can be as fast as several hundred kilometers per second, the typical timescale of the primary process is around 10^{-8} s. The possibility of low-energy electrons attaching themselves to molecules and forming a negative ion exists (called attachment, equals electron thermalization in Figure 11.13). Molecules like O$_2$ attach easily, while N$_2$ and noble gases do not attach at all. This strongly influences the characteristics of the discharge in gas mixtures because it limits the mobility of negative charge carriers. The efficiency of the primary process is highly dependent on energization methods

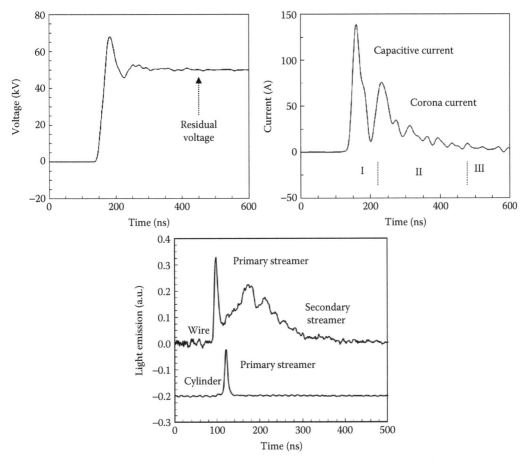

FIGURE 11.11 Typical voltage, current, and light emission waveforms for streamer corona plasma generation in air in a wire–cylinder reactor. The length and inner and outer diameters of the wire–cylinder reactor are 800, 3, and 155 mm, respectively. Two quartz fibers are used to focus two small regions near the corona wire and the cylinder, respectively. The two photomultipliers are sensitive in the range of 185–930 nm and have a rise time of 2 ns. (From Yan, K. Corona plasma generation, Ph.D. Dissertation, TU/e, Eindhoven, 2001.)

and their parameters, such as pulse, DC/pulse, AC, AC/pulse or DC, voltage rise time, and frequency. The secondary process is the subsequent chemical reactions involving the products of primary processes (electrons, radicals, ions, and excited molecules). Here radical reactions play a key role in pollutant control, for example, the oxidation and reduction of NO_x and SO_x in flue gas. Some additional radical species and reactive molecules (O_3, HO_2, and H_2O_2) are also formed by radical–neutral recombination in the secondary processes. The details of radical generation and global chemical kinetics of radical reactions will be discussed afterward.

11.2.1 Initial Radical Generation

There are three main pathways of radical generation in corona discharge: formation due to reaction with ions of plasma, formation in reaction with excited molecules, and formation through other radical–neutral reactions (Bityurin et al. 2000).

In the first pathway, radical formation can occur in ion recombination reactions, charge transfer reactions, or ion neutralization reactions. For example, in the presence of water vapor, the OH radical is

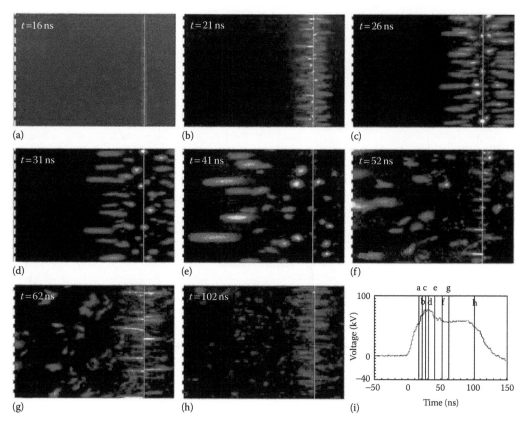

FIGURE 11.12 Time-resolved 5 ns gate time ICCD pictures. White line: reactor wire. Dotted line: reactor wall. The time the picture is taken, relative to the moment the voltage on the reactor begins to increase, is shown in the top-left corner of the pictures. Picture size: ~7 × 5 cm². Positive polarity. Pulse voltage 74 kV, rise rate 2.7 kV ns⁻¹. Pulse width: 110 ns. Airflow rate: 30 Nm³h⁻¹. (From Winands, G.J.J. et al., *J. Phys. D Appl. Phys.*, 41, 234–244, 2008. With permission.)

mainly generated by charge transfer reactions, which can be categorized into two mechanisms. One is OH radical production of positive ions exchange at water molecules:

$$N_2^+ + H_2O = H_2O^+ + N_2, \tag{11.10}$$

$$H_2O^+ + H_2O = H_3O^+ + OH. \tag{11.11}$$

If the ionization potential of an ion is higher than that of water molecule, a third particle (marked as "M") participates in the reaction:

$$A^+ + H_2O + M = A^+(H_2O) + M. \tag{11.12}$$

The dissociative reactions of water cluster ions also generate OH radicals:

$$A^+(H_2O) + H_2O = H_3O^+ + OH + A, \tag{11.13}$$

$$A^+(H_2O) + H_2O = H_3O^+(OH) + A, \tag{11.14}$$

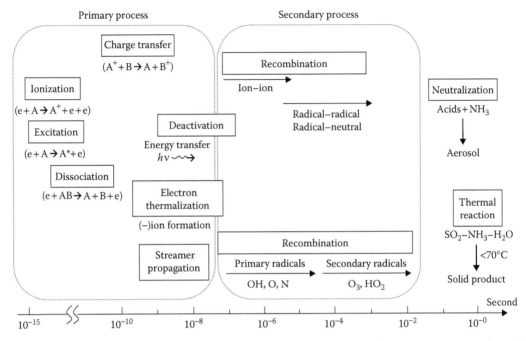

FIGURE 11.13 Timescale events of elementary processes in the NTP process. (From Kim, H.-H.: Nonthermal plasma processing for air-pollution control: A historical review, current issues, and future prospects. *Plasma Processes and Polymers.* 2004. 1. 91–110. Copyright Wiley-VCH Verlag GmbH & Co. KGaA. Reproduced with permission.)

The other mechanism for OH radical generation is atomic hydrogen formation in reactions of dissociative ion recombination:

$$M^- + H_3O^+(H_2O)_n = M + H + (H_2O)_{n+1}, \tag{11.15}$$

$$H + HO_2 = 2OH, \tag{11.16}$$

$$H + O_3 = OH + O_2. \tag{11.17}$$

It should be pointed out that in heterogeneous conditions, the main pathway for OH radical production in water aerosols in DBD is

$$O_3(aq) + OH^-(aq) = OH(aq) + O_3^-. \tag{11.18}$$

In the second pathway, excited molecules play an important role in the radical generation process. One of the most well-known excited molecules in corona plasma in air mixture is $O(^1\Delta)$, which is formed in oxygen and nitrogen reactions:

$$N_2^* + O_2 = N_2 + O(^1\Delta) + O, \tag{11.19}$$

$$O_2^* + M = M + O(^1\Delta) + O. \tag{11.20}$$

Then, $O(^1\Delta)$ abstracts one atomic hydrogen from a water molecule to generate two OH radicals:

$$O(^1\Delta) + H_2O = OH + OH. \tag{11.21}$$

The rate coefficient of this reaction is large compared with $O(^1\Delta)$ quenching reactions. Therefore, small water vapor concentration can generate large quantities of OH radicals.

In the third pathway, active particles are often generated in the intermediate stage of radical–neutral reactions, for example, peroxyl radicals (HO_2) play the role of an intermediate active particle in the chain mechanism for transforming the OH radical in hydrocarbon decomposition processes:

$$OH + CO = CO_2 + H, \tag{11.22}$$

$$H + O_2 = HO_2, \tag{11.23}$$

$$HO_2 + NO = NO_2 + OH. \tag{11.24}$$

Here, peroxyl radicals are often produced through other radical reaction or ion neutralization:

$$OH + O_3 = HO_2 + O_2, \tag{11.25}$$

$$OH + H_2O_2 = HO_2 + H_2O, \tag{11.26}$$

$$H_3O^+ + O_2^- = HO_2 + H_2O. \tag{11.27}$$

In Equation 11.24, the produced OH radicals react with the chain-carrying agent to reform the peroxyl radical and restart the chain reactions again. In hydrocarbon decomposition processes, ethene, propene, ethanol, and dimethyl are well-known chain-carrying agents.

11.2.2 Global Chemical Kinetics and Multiple Processing

The main feature of pollutant degradation mechanisms with NTP is change of the energetic threshold of the reaction and/or the provision of a chain character for the processes. Here, a simplified global chemical kinetic model was proposed to discuss global chemical kinetics of NTP processing and evaluate it. The model consists of one radical formation process and three reactions: radical production R_1, pollutant removal reaction R_2, radical linear termination reaction R_3, and radical nonlinear termination reaction R_4 (Yan et al. 2001a).

$$\text{Radical production process, } k_1 \quad M + \text{corona} \rightarrow R, \tag{R$_1$}$$

$$\text{Pollutant removal reaction, } k_2 \quad X + R \rightarrow A, \tag{R$_2$}$$

$$\text{Radical linear termination reaction, } k_3 \quad R + M \rightarrow B, \tag{R$_3$}$$

$$\text{Radical non linear termination reaction, } k_4 \quad R + R \rightarrow C, \tag{R$_4$}$$

where
   R is the radical generated by corona discharge; X is the pollutant; A, B and C are all the by-products; M is the bulk gas compounds
   k_2, k_3, and k_4 are the reaction rate constants
   k_1 is the initial radical production efficiency or radical yield in terms of corona power density, which is defined as the ratio of average corona plasma power over reactor volume

Multiple radical propagation and branching reactions are simplified by a one-step pollutant removal process (R_2). Linear (R_3) plus nonlinear (R_4) radical termination reactions simplify various types of termination processes. Moreover, we assume that the concentrations are uniformly distributed inside the reactor, and the radicals are uniformly generated during the residence time t_{res}, which is defined as the ratio of the reactor volume over the gas flow rate. The initial radical production is directly proportional to the corona power, which is dependent on corona discharge patterns and gas compositions.

As mentioned before, the Hermstein glow discharge produces negligible initial radicals in comparison to streamer corona. A prebreakdown streamer is less efficient than an onset streamer, and a secondary streamer is less efficient in comparison to a primary streamer. Radicals are mainly produced during primary streamer propagation. Optimizing the streamer energy transfer increases the initial radical production efficiency k_1, which can be calculated with the energy costs of radical production. For N radical production in N_2, the energy cost for producing each N radical is about 170 eV, while for O radical production in oxygen-rich gases, the cost is about 50 eV.

According to the model and the assumption above, we obtain the following kinetic equations for the radical R and pollutant X:

$$\frac{d[R]}{dt} = k_1 \cdot P - k_2 \cdot [X] \cdot [R] - k_3 \cdot [M] \cdot [R] - 2k_4 \cdot [R]^2, \quad (11.28)$$

$$\frac{d[X]}{dt} = -k_2 \cdot [X] \cdot [R], \quad (11.29)$$

where

[R], [X], and [M] are the radical R, the pollutant X, and the bulk gas M concentrations, respectively
P is the corona plasma power density in watts per cubic meter

According to Equation 11.28, the radical lifetime τ_r and radical concentration [R] can be approximated by Equations 11.30 and 11.31, respectively.

$$\frac{1}{\tau_r} \approx k_2 \cdot [X]_{in} + k_3 \cdot [M] + 2\sqrt{2k_1 k_4 P}, \quad (11.30)$$

$$[R] \approx k_1 \cdot P \cdot \tau_r \cdot \left[1 - \exp\left(-\frac{t}{\tau_r} \right) \right], \quad (11.31)$$

where [X] refers to the X initial concentration.

Because of the short lifetime of the radicals, their concentration can be estimated by $k_1 \cdot P \cdot \tau_r$ (see Equation 11.31) or by stationary-state approximation. Thus, Equation 11.28 becomes

$$k_1 \cdot P - k_2 \cdot [X] \cdot [R] - k_3 \cdot [M] \cdot [R] - 2k_4 \cdot [R]^2 = 0 \quad (11.32)$$

or

$$[R] = \frac{-(k_2 \cdot [X] + k_3 \cdot [M]) + \sqrt{(k_2 \cdot [X] + k_3 \cdot [M])^2 + 8k_1 k_4 P}}{4k_4}, \quad (11.33)$$

where [X] and [R] refer to average concentration after stationary-state approximation. Substituting Equation 11.33 into Equation 11.29, and integrating it with time, a general solution can be obtained in terms of the corona power density, the residence time, and/or the corona energy density.

11.2.2.1 Insignificant Radical Terminations

A small corona energy density usually corresponds to a low radical concentration. In this case, we assume that the two radical termination reactions R_3 and R_4 do not play a very significant role or that the radical termination reactions R_3 and R_4 are very slow in comparison with reaction R_2. Thus, we have the following relations:

$$k_2 \cdot [X] \cdot [R] \gg k_3 \cdot [M] \cdot [R],$$

$$k_2 \cdot [X] \cdot [R] \gg 2k_4 \cdot [R]^2. \tag{11.34}$$

According to Equations 11.33 and 11.29, we have:

$$[R] = \frac{k_1 \cdot P}{k_2 \cdot [X]}, \tag{11.35}$$

$$\frac{d[X]}{dt} = -k_1 \cdot P. \tag{11.36}$$

Integrating Equation 11.36 up to the residence time t_{res}, we obtain:

$$[X]_{in} - [X]_{out} = k_1 \cdot P \cdot t_{res} \tag{11.37}$$

or

$$[X]_{in} - [X]_{out} = k_1 \cdot E, \tag{11.38}$$

where

$$E = P \cdot t_{res} \tag{11.39}$$

$[X]_{out}$ and E are the final pollutant concentration and the corona energy density, respectively

According to Equation 11.38, the amount of the removed pollutant is directly proportional to the corona energy density. Thus, the energy yield or the so-called G value is a constant, which is defined as the removed molecules per 100 eV corona energy.

11.2.2.2 Significant Linear Radical Termination

At larger corona energy density, the radical concentration increases. As a result, radical termination can become significant, which results in lower energy efficiency. In the following discussions, we assume that the radical linear termination reaction R_3 becomes the dominant process to determine the radical concentration inside a reactor. Then we have the following relations:

$$k_2 \cdot [X] \cdot [R] \ll k_3 \cdot [M] \cdot [R],$$

$$k_3 \cdot [M] \cdot [R] \gg 2k_4 \cdot [R]^2. \tag{11.40}$$

According to Equations 11.33 and 11.29, we have:

$$[R] = \frac{k_1 \cdot P}{k_3 \cdot [M]}, \tag{11.41}$$

$$\frac{d[X]}{dt} = -\frac{k_2 \cdot k_1 \cdot P}{k_3 \cdot [M]} \cdot [X]. \tag{11.42}$$

Integrating Equation 11.42 up to the residence time t_{res}, we obtain:

$$\frac{[X]_{out}}{[X]_{in}} = \exp\left(-\frac{E}{\beta}\right), \tag{11.43}$$

where

$$\beta = \frac{k_3 \cdot [M]}{k_2 \cdot k_1}.$$ (11.44)

The energy efficiency can be evaluated by the energy cost per removed molecule in terms of eV per molecule, the energy yield (*G* value) in molecules per 100 eV, or by the β value in J/L (or Wh/Nm³). The smaller the energy cost in terms of eV per molecule, the larger the *G* value, the smaller the β value, and the higher the removal efficiency. Rosocha and Korzekwa (1999) derived Equation 11.43 with a complicated chemical kinetic model under the assumption that the degree of removal is low. According to Equation 11.43, the energy cost for each removed molecule increases and the energy yield decreases for higher corona energy density.

With regard to the influences of the gas temperature, according to Equations 11.43 and 11.44, and the Arrhenius temperature dependence of the reaction rate k_2 and k_3, the *G* value $G(-X)$ (or the energy yield) and the β value can be approximated by:

$$\ln\left[G(-X)\right] = A_1 - \frac{B_1}{T} \quad \text{or} \quad \log\left[G(-X)\right] = A_2 - \frac{B_2}{T},$$ (11.45)

$$\ln(\beta) = C_1 + \frac{D_1}{T} \quad \text{or} \quad \ln(\beta) = C_2 + \frac{D_2}{T},$$ (11.46)

where
A_1, B_1, A_2, B_2, C_1, D_1, C_2, and D_2 are coefficients
T is the gas temperature

11.2.2.3 Significant Nonlinear Radical Termination

As discussed above, in this case, we assume that the radical nonlinear termination reaction R_4 becomes the dominant process to determine the radical concentration. Then, we have the following relations:

$$k_2 \cdot [X] \cdot [R] \ll 2k_4 \cdot [R]^2,$$ (11.47)

$$k_3 \cdot [M] \cdot [R] \ll 2k_4 \cdot [R]^2.$$

According to Equations 11.33 and 11.29, we have:

$$[R] = \sqrt{\frac{k_1 \cdot P}{2k_4}},$$ (11.48)

$$\frac{d[X]}{dt} = -k_2 \cdot \sqrt{\frac{k_1 \cdot P}{2k_4}} \cdot [X].$$ (11.49)

Integrating Equation 11.42 up to the residence time t_{res}, we obtain:

$$\frac{[X]_{out}}{[X]_{in}} = \exp\left(-k_2 \cdot \sqrt{\frac{k_1}{2k_4}} \cdot \sqrt{E} \cdot \sqrt{t_{res}}\right).$$ (11.50)

The removal rate depends on the square root of the corona energy density. The first-order approximation of Equation 11.50 can be expressed as:

$$\frac{[X]_{in} - [X]_{out}}{[X]_{in}} = \alpha \cdot \sqrt{E} \cdot \sqrt{t_{res}}, \tag{11.51}$$

where

$$\alpha = k_2 \cdot \sqrt{\frac{k_1}{2k_4}} \tag{11.52}$$

Because the energy yield $G(-X)$ is directly proportional to $\frac{[X]_{in} - [X]_{out}}{E}$, according to Equation 11.51, the energy efficiency increases by increasing the corona residence time and the initial concentration, and it decreases by increasing the corona energy density. The energy yield is inversely proportional to the square root of the corona energy density.

11.2.2.4 Multiple Processing

For large-scale industrial applications, it is very common to use multiple chemical reactors. Figure 11.14 shows the schematic diagram of in-parallel and in-series processing with two identical reactors. The two reactors can be connected in parallel or in series. We also suppose that the total corona power P is equally distributed within the two reactors. For in-parallel processing, the total gas flow rate F is also equally divided into two parts.

According to Equations 11.38, 11.43, and 11.50, one can easily confirm that for either insignificant or significant radical termination processing, the in-series and in-parallel processing would give identical pollution removal efficiency for a given corona energy density. Moreover, for both insignificant and significant linear radical termination chemical kinetics, the reactor number does not affect the removal efficiency. In other words, the residence time does not affect the processing. However, when nonlinear radical termination reactions play a significant role as expressed by Equations 11.50 and/or 11.51, multiple treatments always give a better removal efficiency. Table 11.3 provides a comparison between

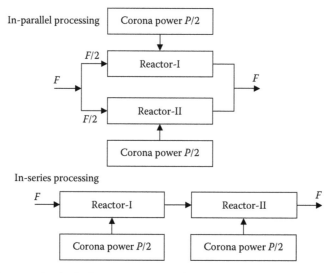

FIGURE 11.14 Schematic of multiple plasma treatments with two reactors. (With kind permission from Springer Science+Business Media: *Plasma Chemistry and Plasma Processing*, From chemical kinetics to streamer corona reactor and voltage pulse generator, 21, 2001a, 107–137, Yan, K. et al.)

TABLE 11.3 Improvements with Multiple Treatments

Reactors	Two reactors	One reactor
Total corona power (W)	P	P
Total gas flow rate (m³/s or m³/h)	F	F
Total reactor volume (m³)	V	$V/2$
Corona energy density (J/L or Wh/Nm³)	$E = P/F$	$E = P/F$
Corona residence time (s)	t_{res}	$t_{res}/2$
Removal rate	$\dfrac{[X]_{2-out}}{[X]_{in}} = \exp\left(-\alpha \cdot \sqrt{E} \cdot \sqrt{t_{res}}\right)$	$\dfrac{[X]_{1-out}}{[X]_{in}} = \exp\left(-\alpha \cdot \dfrac{1}{\sqrt{2}} \cdot \sqrt{E} \cdot \sqrt{t_{res}}\right)$
Comparison	$\dfrac{[X]_{2-out}}{[X]_{1-out}} = \exp\left[-\alpha \cdot \left(1 - \dfrac{1}{\sqrt{2}}\right) \cdot \sqrt{E} \cdot \sqrt{t_{res}}\right] < 1$	

Source: With kind permission from Springer Science+Business Media: *Plasma Chemistry and Plasma Processing*, From chemical kinetics to streamer corona reactor and voltage pulse generator, 21, 2001b, 107–137, Yan, K., et al.

single-reactor processing and two-reactor processing. The values of $[X]_{1-out}$ and $[X]_{2-out}$ are the final concentrations with one and two reactors, respectively. Increasing the residence time t_{res} always improves the efficiency and/or reduces the energy consumption.

The real corona plasma chemical reactions and the pollutant distribution inside a reactor are much more complicated in comparison to the above model. Both linear and nonlinear radical termination reactions may play significant roles. More radicals are also involved. Chain branching and propagation may also affect the treatment. The analyses above only gives very basic relations among chemical kinetics, pollution removal, and corona reactor design. Table 11.4 summarizes the above discussions.

11.3 Power Source and Streamer Corona Plasma Reactor

With regard to various applications, it is impossible to conclude which type of corona plasma is the best for all kinds of purposes. It is generally considered that because of high-pressure drop across DBD and PBC reactors and the risk of insulator failure, these reactors are less attractive for a large-volume gas cleaning. Pulsed corona plasma techniques are considered to be suitable for large-volume gas cleaning. Here, for large volume of streamer corona plasma generation, two kinds of driving modes—pulsed power mode and DC/pulse mode—are introduced. A noble nanosecond pulsed power system proposed by K. Yan will be described in detail.

11.3.1 Energization Methods and Circuit Topology

As mentioned in Section 11.2, glow discharge is not efficient for the application of $DeNO_x$ and $DeSO_2$. It is also the same to arc discharge because of the great heat loss in gas heating and production of NO in a much higher level than originally present in the flue gas. The most efficient NTP mode is streamer corona. The transition from corona to arc can be avoided by applying voltage pulses of sufficiently short duration.

A streamer corona plasma may last a few tens of nanoseconds; thus, to generate a corona plasma of 10 kW average at 1000 pps, 50 ns pulse duration, and 100 kV peak voltage, the output impedance and the peak power of the high-voltage pulse generator should be around 50 Ω and 200 MW, respectively. At the moment, solid-state switches are still expensive, and magnetic compression techniques do not have high enough energy efficiency for these kinds of high-voltage pulse power and corona plasma applications. Because of high hold-off voltage, large possible currents, and small forward voltage drop, high-pressure triggered gas-filled spark-gap switch is one of the most cost-effective switches for 1–10 kW corona plasma applications, provided the lifetime and the repetition rate can be improved. Moreover, to reliably

TABLE 11.4   Chemical Kinetics and Corona Plasma Processing

1	Insignificant radical terminations	
	Removal rate	$[X]_{in} - [X]_{out} = \alpha \cdot E$
	Reactors	In-parallel and in-series processing lead to identical results.
	Energy yield or the G value	1. Constant; 2. May not significantly depend on the gas temperature.
	Residence time	No significant effects.
	Removal efficiency	Depends on the corona energy density.
Comment		The most efficient plasma processing.
2	Significant linear radical termination	
	Removal rate	$\dfrac{[X]_{out}}{[X]_{in}} = \exp\left(-\dfrac{E}{\beta}\right)$
	Reactors	1. In-parallel and in-series processing give identical results; 2. Increasing reactor number cannot improve the efficiency.
	Energy yield or the G value	1. Decreases by increasing the corona energy density; 2. May significantly depend on the gas temperature.
	Residence time	No significant effects.
	Removal efficiency	1. Depends on the corona energy density; 2. Less efficient than the insignificant radical termination processing.
Comment		Can use hybrid plasma reactors to increase the efficiency.
3	Significant nonlinear radical termination	
	Removal rate	$\dfrac{[X]_{out}}{[X]_{in}} = \exp\left(-\alpha \cdot \sqrt{E} \cdot \sqrt{t_{res}}\right)$ or $\dfrac{[X]_{in} - [X]_{out}}{[X]_{in}} = \alpha \cdot \sqrt{E} \cdot \sqrt{t_{res}}$
	Reactors	1. Increasing the reactor number improves the efficiency; 2. Multiple treatments increase the total efficiency; 3. In-parallel may give a better efficiency.
	Energy yield or the G value	1. Decreases by increasing the corona energy density; 2. Increases with increasing the residence time; 3. May significantly depend on the gas temperature.
	Residence time	Increasing the residence time improves the efficiency.
	Removal efficiency	1. Depends on the corona energy density and the residence time; 2. Less efficient than the insignificant radical termination processing.
Comment		Can use hybrid plasma reactors to increase the efficiency.

Source: With kind permission from Springer Science+Business Media: *Plasma Chemistry and Plasma Processing,* From chemical kinetics to streamer corona reactor and voltage pulse generator, 21, 2001a, 107–137, Yan, K., et al.

produce corona plasmas, the high-voltage pulse generator should be immune to a spark breakdown inside the reactor or any other kind of short circuit. The triggered spark-gap switches indeed allow a very robust design of the corona circuitry (Heesch and Laan 2000a; Yan et al. 2001b). With regard to voltage-rise time, pulse duration, repetition rate, peak voltage, peak current, average power, and costs, Table 11.5 lists the suitable switches in terms of the pulse duration for pulsed power applications.

Generally speaking, pulsed power is generated by a sequence of energy conversion steps as indicated in Figure 11.15: from main AC power to high-voltage pulse power generating corona plasma in four wire–cylinder reactors in parallel. With regard to technical developments, the system can be roughly divided into six parts as indicated by Parts I–VI, respectively. Part I is AC–DC–pulse conversion mainly including a resonant charging circuit and a pulse transformer generating pulse wave of the order of a

TABLE 11.5 Comparison of Various Kinds of Switches for Pulsed Power Applications

Switches	Pulse duration			
	<100 ns	100 ns–0.5 μs	0.5–5 μs	>5 μs
Solid state			Suitable	Most suitable
Magnetic compression		Suitable	Most suitable	
Spark-gap	Most suitable			

Source: Yan, K., Corona plasma generation, Ph.D. Dissertation, TU/e, Eindhoven, 2001.

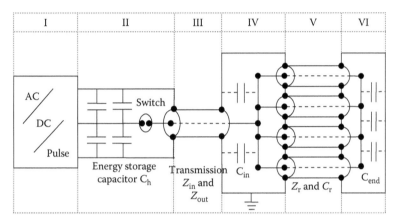

FIGURE 11.15 A schematic diagram of high-voltage pulse generator and four parallel wire–cylinder reactors. (From Yan, K., Corona plasma generation, Ph.D. Dissertation, TU/e, Eindhoven, 2001.)

microsecond. The greatest technical concerns of this part are energy conversion efficiency, charging and discharging rates, pulse repetition rate, energy per pulse, stability of the output, and electromagnetic compatibility of the system; Part II contains energy storage capacitor C_h and spark-gap switch to compress the timescale of output from Part I into a nanosecond. The rise time of the voltage pulse mainly depends on the switch-on time and/or the switch inductance; Parts III–V constitute pulsed power transmission, which contains transmission line transformer (TLT). TLT is mainly constructed with transmission cables and magnetic cores. The most critical issues for constructing a TLT are the peak power and the energy losses caused by the secondary mode current and the skin effect; Part VI is the load. The matching between load and transmission line is very important for improving the energy efficiency of the whole system, especially for industrial practice, but it is always ignored in a lot of literature and techniques. Details of the matching will be introduced in the next section.

According to the general idea indicated in Figure 11.15, a novel circuit of pulsed power source based on multiple gap switch and TLT transmission line was proposed by Yan in 2001. The main electrical circuit with a resistive load is shown in Figure 11.16.

The pulse transformer T_R separates the low and high voltage parts of the circuit. The low-voltage part mainly consists of a main filter, a set of rectifiers, three air-core inductors L_1, L_2, and L_3, three thyristors Th_1, Th_2, and Th_3, two energy storage capacitors C_0 and C_L, and primary windings of the pulse transformer T_R. The three thyristors are switched consecutively in order to charge the high-voltage capacitor C_h. As indicated in Figure 11.16, three RC snubbers and one silicon-surge voltage suppressor S are used for avoiding overvoltage on these thyristors.

The high-voltage part mainly consists of secondary windings of the pulse transformer T_R, two high-voltage diodes D_1 and D_2, two damping resistors R_4 and R_5, an air-core inductor L_4, a high-voltage capacitor C_h, a triggered spark-gap switch, an LCR trigger circuit, and a TLT transmission line. The free diode D_2 together with the damping resistor R_5 is designed to protect the high-voltage diode D_1 when the main spark-gap switch prefires. An RC snubber is used to protect the diode D_1 during normal operation.

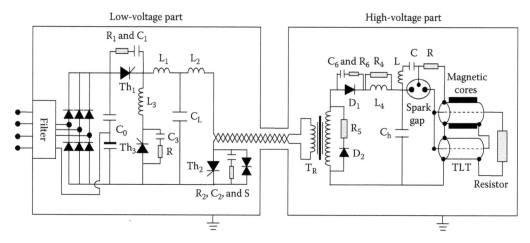

FIGURE 11.16 A schematic diagram of high-voltage pulse generator and four parallel wire–cylinder reactors. (From Yan, K., Corona plasma generation, Ph.D. Dissertation, TU/e, Eindhoven, 2001.)

A four-step process generates the high-voltage pulse. In the first step, the low-voltage capacitor C_L is resonantly charged via the energy storage capacitor C_0, the thyristor Th_1, and the inductor L_1, where $C_0 \gg C_L$. During the second step, the high-voltage capacitor C_h is resonantly charged via C_L, L_2, T_R, Th_2, D_1, and L_4. Then the stored energy in C_h is transferred to the load via the trigged spark-gap switch and the TLT. Before the low-voltage capacitor C_L is recharged again, the third thyristor Th_3 is used to rectify the voltage polarity on C_L via Th_3, L_3, and L_1. After the main spark gap breaks down, the voltage on the capacitor C_h decreases with a time constant of $Z_{in}C_h$, where Z_{in} is the input impedance of the TLT. The capacitor C_h and the input impedance of the TLT determine the high-voltage pulse duration. The switching voltage (or the maximum charging voltage on the capacitor C_h) of the spark-gap switch and the voltage gain of the TLT determine the output peak voltage. Typical output waveform of this circuit with 50 Ω resistive load is shown in Figure 11.17.

One of the critical issues for corona plasma applications is the total energy consumption of the system. Generally speaking, the total energy consumption per unit reaction products depends on both electrical energy conversion efficiency of the corona plasma system and chemical effectiveness of the processing. Here, the electrical energy conversion efficiency of the high-voltage pulse generator will be analyzed below.

With regard to the electric circuit in Figure 11.16, the total energy conversion efficiency η of the main electric circuit can be divided into two separate energy conversion efficiencies. One is the energy conversion efficiency η_c for the AC–DC–pulse converter charging the high-voltage capacitor C_h. The other one is the efficiency η_t transferring the stored energy in the capacitor to a load, which is about 96% with a matched resistive load. The total energy conversion efficiency can be evaluated by the following equation:

$$\eta = \eta_c \cdot \eta_t. \tag{11.53}$$

For the first energy conversion efficiency η_c, the main energy losses are caused by the three RC snubbers, the R&L snubber, the pulse transformer T_R, the three thyristors, and the three air-core inductors. For the second energy efficiency η_t, the main energy losses are caused by the spark-gap switch, the TLT transmission line, the stray capacitance of the R&L snubber, the LCR trigger circuit, the RC snubber (for the high-voltage diode), and the matching between the load and the generator (see Figure 11.16). Details of these energy loss are analyzed in the works of Yan (2001).

Based on the research and development of the high-voltage pulse power generator described above and the investigations on DC-superimposed pulsed corona energization, a hybrid pulsed power source

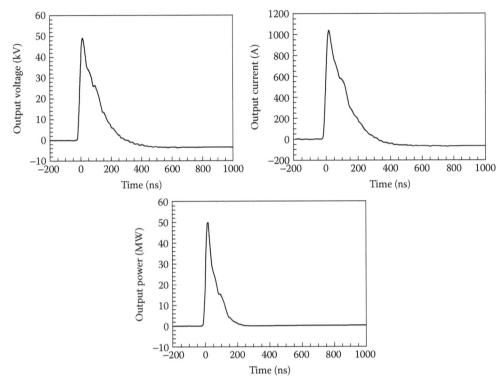

FIGURE 11.17 Typical output voltage, current, and power waveforms of the electric circuit described in Figure 11.16 with a 50 Ω resistive load. The energy conversion efficiency is around 96%. (From Yan, K., Corona plasma generation, Ph.D. Dissertation, TU/e, Eindhoven, 2001.)

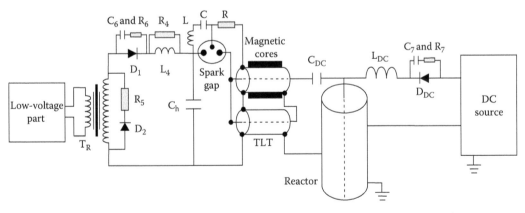

FIGURE 11.18 A schematic diagram of the HPPS and a wire–cylinder-type corona reactor. (From Yan, K., Corona plasma generation, Ph.D. Dissertation, TU/e, Eindhoven, 2001.)

(HPPS) and a wire–cylinder type reactor, shown in Figure 11.18, are designed for corona plasma applications. The HPPS consists of two parts: a high-voltage pulse generator as shown in Figure 11.16, and a DC charging unit. These two parts are coupled together with a coupling capacitor C_{DC}. The corona reactor is energized by two separate sources. Before applying the high-voltage pulse, the reactor is resonantly charged to V_{DC} by the DC source via the high-voltage diode D_{DC}, the inductor L_{DC}, the coupling capacitor C_{DC}, and the TLT transmission line (see Figure 11.18). By firing the main spark-gap switch, the energies stored in the coupling capacitor C_{DC} and in the high-voltage capacitor C_h are transferred to the reactor simultaneously. After pulsed energization, the reactor and the capacitor C_{DC} are resonantly charged again by the DC source in order to prepare for the next high-voltage pulse.

The DC resonant charging inductor L_{DC} is a 1-m-long helical transmission line. It is designed to have the following characteristics: (1) during the pulsed energization, its characteristic impedance for the fast pulses is much larger than the output impedance of the TLT; (2) When a spark breakdown occurs inside the reactor, the inductor can prevent large current flow from the DC source; and (3) during the interpulse duration, the coupling capacitor C_{DC} and the reactor in parallel can be resonantly charged to a given voltage V_{DC}. In order to transfer the pulsed energy stored in the high-voltage capacitor C_h to the reactor, the two energy storage capacitors C_{DC} and C_h, the maximum charging voltages V_{max} and V_{DC} are designed to obey the following inequality:

$$V_{DC} \geq \frac{1}{2} \cdot \frac{C_h}{C_{DC}} \cdot V_{max} \tag{11.54}$$

If the left- and right-hand sides in the above relation become equal, the voltage across the capacitor C_{DC} would decrease to zero after the pulsed energization. In this case, the total charge injected by the TLT equals the initial charge on the capacitor C_{DC}.

With regard to DBD, except for the AC driving mode, it can also be operated under the pulsed power mode. As discussed in Section 11.1.2, the traditional ozone generator based on the DBD technique is derived from a high-voltage supply operated in linear frequency or a motor generator. High-performance ozone generator systems now use a high-frequency switching source (0.5–5 kHz), which generates a square-wave current or some other special wave form. Sometimes, a high-frequency short-pulsed power source is also used. As higher operating frequencies can deliver the desired power density at much lower operating voltages, from typically 20 kV in the past to <5 kV now, less electrical stress on the dielectrics makes the discharge gap more narrow.

The PBC is an interesting combination of the DBD and the sliding surface discharge. In this system, a high AC voltage (about 15–30 kV) is normally applied to a packed bed of dielectric pellets and creates a non-equilibrium plasma in the void between the pellets. A typical scheme for organizing a PBC and its discharge picture are shown in Figure 11.19. The discharge chamber, shown in the figure, consists of coaxial cylinders with an inner metal electrode and an outer tube made of glass. The dielectric pellets are placed in the annular gap. A metal foil or screen in contact with the outside surface of the tube serves as the ground electrode.

The inner electrode is connected to a high-voltage AC power supply operated at a fixed frequency of 50/60 Hz or at variable frequencies. K. Yan designed an efficient high-frequency AC power supply with MOSFET switch as shown in Figure 11.20. To optimize the matching between the power generator and the discharge reactor, the pulse width of the MOSFET units is designed to match the resonance frequency of the load and the transformer. Switching operation is close to the zero-current state. Several generators with output power in the order of 2 kW have been fabricated. The pulse frequency and peak voltage are usually below 40 kHz and 30 kV, respectively. Energy conversion efficiency of larger than 85% from the main power source line to the corona reactor could be obtained.

11.3.2 Matching between Power Source and Reactor

Matching between a power source and a corona plasma reactor becomes more important when scaling up the system. A good matching can not only increase the total energy efficiency but it can also lengthen the switch lifetime and reduce investment and operational costs. Here the matching methods for a high-voltage pulse generator, as was described in Section 11.3.1 for a corona plasma reactor, will be discussed.

Optimal corona plasma energization and/or matching between a generator and a reactor can be achieved by considering the following three distinct energization stages: before, during, and after corona plasma generation (see current waveform in Figure 11.11; I–III refer to the three energization stages). When a positive voltage pulse is applied to a small-size (short-length) corona reactor, a displacement

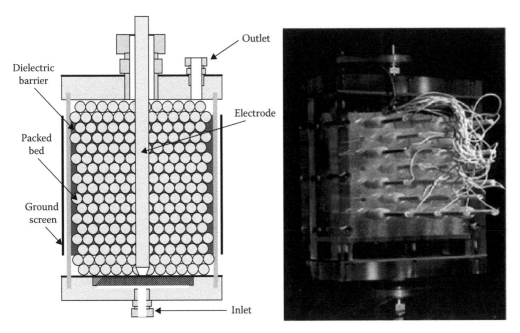

FIGURE 11.19 Packed-bed corona: scheme and picture. (From Fridman, A. et al., *J. Phys. D Appl. Phys.*, 38, R1, 2005. With permission.)

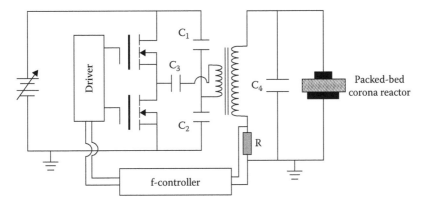

FIGURE 11.20 Schematic of AC high-voltage power supply. (Reprinted from *Journal of Electrostatics*, 44, Yan, K. et al., Corona induced non-thermal plasmas: Fundamental study and industrial applications, 17–39, Copyright 1998, with permission from Elsevier.)

(or capacitive) current flows through the reactor before the applied voltage pulse reaches the corona inception voltage. During this period, the corona reactor can be described as a capacitive load. When the voltage exceeds the inception voltage, streamers propagate from the wire to the cylinder. As a result, energy is transferred from the generator to the plasma inside the reactor. At this time, a resistive model can be used to evaluate the energy transfer efficiency. After corona plasma quenching (the corona current drops to almost zero), the voltage on the reactor is called the residual voltage.

An example is shown in Figure 11.21, which includes a schematic diagram of the high-voltage discharge circuit and a wire–plate corona plasma reactor. The high-voltage capacitor C_h is connected to a two-stage TLT via a triggered spark-gap switch S. The output of the two-stage TLT is connected in series, so the output impedance $Z_{out} = 2 \times Z_0$, where Z_0 is the characteristic impedance of single transmission cable. Two capacitors C_{in} and C_{end} are installed at both ends of the reactor to improve the matching

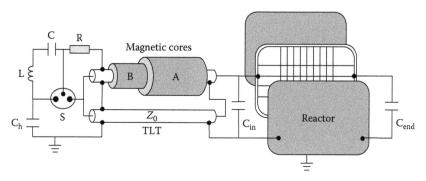

FIGURE 11.21 A schematic diagram of the high-voltage circuit and a wire–plate corona reactor. (From Yan, K., Corona plasma generation, Ph.D. Dissertation, TU/e, Eindhoven, 2001.)

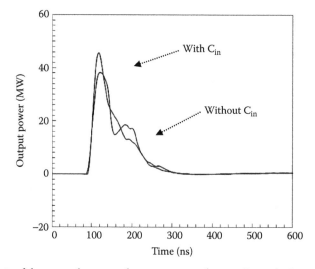

FIGURE 11.22 Effects of the system layout on the output power for a 50 Ω matched resistive load and the TLT shown in Figure 11.21. (From Yan, K., Corona plasma generation, Ph.D. Dissertation, TU/e, Eindhoven, 2001.)

before the applied voltage reaches the inception voltage. For example, Figure 11.22 shows the effects of matching capacitor C_{in}. The peak output power is increased, and the voltage pulse reflection is avoided by compensating for the stray inductance of the system layout.

According to the analysis above, the simple models and equivalent circuits described in Figure 11.23 are used to evaluate the energy transfer efficiency. Before corona streamer is generated, the matching between cable-type voltage-pulse generator and corona reactor can be simplified as a capacitor connecting to a pulsed excited cable (left of Figure 11.23). The capacitance C_r of a corona reactor, the output impedance Z_{out} of a voltage pulse generator, and the time τ for the voltage pulse to reach the inception value should follow such a relationship:

$$\tau = 2 \cdot Z_{out} \cdot C_r \quad \text{or} \quad \lambda_1 = \frac{\tau}{Z_{out} \cdot C_r} = 2, \tag{11.55}$$

where λ_1 is defined as a time constant ratio. The condition $\lambda_1 = 2$ is the first criterion for optimization of the output impedance and the size of the corona reactor. The maximum energy transfer efficiency is around 97%. The voltage on the reactor V_c at $t = \tau$ is about 1.14 V_0. For corona plasma generation, it is most desirable that the corona inception voltage equals V_c ($V_c = 1.14 \ V_0$ at $t = \tau$), when the output current from the source reaches its maximum. The time constant τ is then equal to the time to reach the

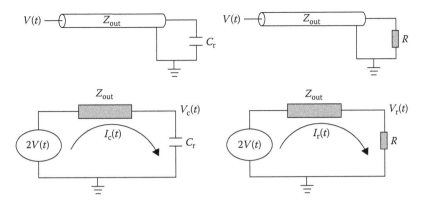

FIGURE 11.23 Simplified models and equivalent circuits for matching corona reactor to cable-type voltage-pulse generator before and during streamer is generated. (From Yan, K., Corona plasma generation, Ph.D. Dissertation, TU/e, Eindhoven, 2001.)

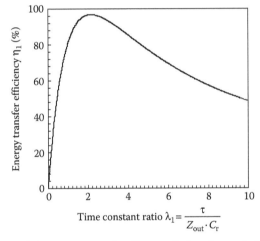

FIGURE 11.24 Dependence of the energy transfer efficiency before the applied voltage reaches the corona inception voltage on the time constant ratio. (From Yan, K., Corona plasma generation, Ph.D. Dissertation, TU/e, Eindhoven, 2001.)

inception voltage. The relationship between time constant ratio λ_1 and energy transfer efficiency η_1 is indicated in Figure 11.24. Details of the calculation process can be found in the works of Yan (2001).

During the process of streamer propagation, the capacitive load can be substituted by a resistive load, which is shown in the right of Figure 11.23. The equivalent impedance R_r of the corona reactor, which is defined as the ratio of the peak voltage over the peak current (Heesch et al. 1999; Yan et al. 2001a), and the output impedance Z_{out} should follow the relation below:

$$R_r = Z_{out} \quad \text{or} \quad \lambda_2 = \frac{R_r}{Z_{out}} = 1, \tag{11.56}$$

where λ_2 is defined as the impedance ratio. The energy transfer efficiency in this process can be derived as follows:

$$\eta_2 = \frac{\int I_r(t) \cdot V_r(t) \cdot dt}{\int V(t)/Z_{out} \cdot V(t) \cdot dt} = \frac{4R_r \cdot Z_{out}}{\left(R_r + Z_{out}\right)^2} = \frac{4\lambda_2}{\left(1+\lambda_2\right)^2} \tag{11.57}$$

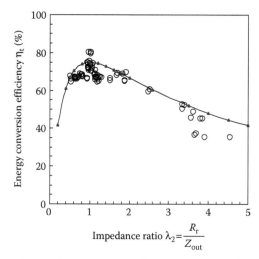

FIGURE 11.25 Dependence of the total energy transfer efficiency on the impedance ratio. The points are experimental results of total energy transfer efficiency under the condition of $Z_{out} = 200\ \Omega$. The line is 75% of the calculation result of energy transfer efficiency in Equation 11.57. (From Yan, K., Corona plasma generation, Ph.D. Dissertation, TU/e, Eindhoven, 2001.)

A maximum energy transfer efficiency of 100% can be obtained by setting $R_r = Z_{out}$. In this case, the voltage pulse on the load is identical to the input voltage pulse. It can also be confirmed that the energy transfer efficiency does not depend on the voltage pulse waveform, but only on the impedance ratio. Details of the calculation process can also be obtained from the works of Yan (2001).

Figure 11.25 gives the dependence of the total energy transfer efficiency η_t on the impedance ratio λ_2. Here, the total energy transfer efficiency η_t can be calculated by using Equation 11.58. The calculation results fit the experimental data well. After the corona streamer quenches, the reactor equivalent impedance becomes much larger than the output impedance of the generator, and the energy transfer is blocked. The remaining energy in the circuit in this period is mainly dissipated in the spark-gap switch and in the TLT cables.

$$\eta_t = \frac{\int I_r(t) \cdot V_r(t) \cdot \mathrm{d}t}{\frac{1}{2} C_h V_{max}^2}. \tag{11.58}$$

11.4 Flue Gas Cleaning

In addition to dust removal, $DeNO_x$ and $DeSO_x$ are the most popular topics in flue gas cleaning. NO_x and SO_2 in the flue gas from coal-fired power plants and in exhaust gas from automobiles are the main sources of acid rain. The traditional industrial treatments for NO_x and SO_2 are selective catalytic reduction (SCR) and flue gas desulfurization (FGD), which use NH_3 together with catalysts to convert NO into N_2 and use lime or other alkaline material (CaO, NaOH, etc.) to convert SO_2 into gypsum, respectively. These complex methods need huge investments and have a common problem of end products treatment.

The idea of NO_x and SO_2 removal with NTP came from the fact that corona discharge in ESP made a serious corrosion problem. The cause was found to be the conversion of SO_2 into sulfuric acid, which was induced by the generated ozone. Several investigations were conducted from the 1970s to the 1980s (Matteson et al. 1972; Masuda and Hirano 1981). The first large-scale demonstration of NTP technology in flue gas cleaning was started by ENEL (Italy) at the Marghera power plant in the late 1980s (Dinelli et al. 1990). The results were 60% NO_x and 80% SO_2 removal, with initial concentrations

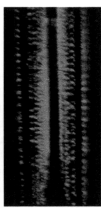

FIGURE 11.26 Outward appearance of the 12,000 Nm³/h streamer plasma FGD system in Guangdong Province, China, and the streamer corona plasma in the reactor. (Reprinted from *Chemical Engineering Journal*, 116, Yan, K. et al., A semi-wet technological process for flue gas desulfurization by corona discharges at an industrial scale, 139–147, Copyright 2006, with permission from Elsevier.)

TABLE 11.6 The Parameters of the 12,000 Nm³/h Streamer Plasma FGD System in Guangdong Province, China

Parameters	Value
Gas volume	12,000 Nm³/h
DC voltage	48.4 kV
AC peak–peak voltage	4 kV
Energy cost per unit volume of gas	0.28 Wh/Nm³
Mole oxidation energy cost	22.1 Wh/mol
Desulfurization rate	96.2%

Source: Yan, K. et al., *Chemical Engineering Journal,* 116, 139–147, 2006.

of 400 and 530 ppm respectively. The energy consumption is around 12–14 Wh/Nm³. Figure 11.26 shows a DC/AC-driving semiwet streamer plasma FGD system with parameters given in Table 11.6.

11.4.1 NO_x Reduction and Oxidation

Plasma-induced NO_x removal mainly converts relatively insoluble species, such as NO and NO_2, to very soluble compounds, such as HNO_2, HNO_3, NO_3, and N_2O_5. Most experimental investigations focused on the oxidation of NO to NO_2 with different gaseous mixtures and energization systems. It was found that NO-to-NO_2 conversion significantly depends on gaseous mixtures. O_2 could greatly suppress the reduction process and enhance the oxidation process. If O_2 concentration is higher than 3.6% (Yan et al. 1999b), the reduction process-induced NO removal can be negligible, which means NO is completely converted into NO_2. While, if O_2 concentration is less than 1%, the reduction process is dominant (Tas 1995). Under higher corona-specific energy density (>30 Wh/Nm³), the amount of produced NO_2 could be larger than the removed NO because of $N(^2D)$-induced NO formation and NO-to-NO_2 conversion. Another critical problem in NO oxidation is that the NO_2 concentration should be controlled at a relatively low level because the reverse reaction $NO_2 + O \rightarrow NO + O_2$ would decrease oxidation efficiency. As a result, much more NO_2 can be reduced to NO under large corona energy density. The reverse reaction will greatly increase the energy costs for NO removal by NTP because the energy required for this process is 1 order of magnitude higher (≥100 eV/NO) than that for the oxidation process (50 eV/NO in the gas-phase reaction (Tas et al. 1997). Details of NO oxidation and the reduction pathway can be found in the works of Yan et al. (1999b).

By adding H_2O or wet ammonia (NH_3), heterogeneous reactions increase the NO_x removal efficiency as the relative humidity on the surface of solid media increases (Chae and Dessiaterik 2000). In the liquid phase, the energy cost per NO oxidation is around 10 eV, which is lower than in gas (Choulakova et al. 1997). The produced NO_2 per unit NO removed can be greatly decreased because of NO_2 oxidation by OH radicals. Approximately 100% of NO_2 can be removed in humidified air to form nitric acid. With injection of NH_3, the produced NO_2 from NO oxidation can also be decreased. Nitric acid in aerosols reacts with NH_3 to form NH_4NO_3, which is highly useful as a fertilizer because of its high nitrogen content and less negative side effects. Even without H_2O, NO and NH_3 can form NH_4NO_3 under corona discharge. The mean diameter of the white particle is around 100 nm (Diato 1997).

With regard to the energization method for NTP-induced NO oxidation and reduction, DC/AC power source and pulsed power sources are most investigated. The DC/AC method means DC with superimposed high-frequency AC (10–60 kHz). In comparison to a corona discharge with a DC power source, DC/AC-energized or pulsed corona plasma are much less sensitive to gas composition and electrode misarrangement, provided the AC peak–peak voltage is higher than 1.0 kV (Yan 2001).

11.4.2 SO_2 Removal

It is well known that SO_2 is less soluble in water than SO_3. On exposure to sunlight, SO_2 is slowly oxidized by active species O_3 or H_2O_2 in the atmosphere. Once SO_3 is formed, it attracts water molecules because of its strong hygroscopic nature. If the moisture in air is large enough, these H_2SO_4 clusters will grow into droplets that end up as acid rain. Because the concentration of immediates O_3 or H_2O_2 in natural air is very low, the timescale of SO_2 oxidation is too long (hours or days) for industrial applications. Radicals generated by NTP can accelerate the oxidation process. From the 1960s to the 1970s, several researchers investigated direct oxidation and removal of SO_2 under corona discharge conditions (Paul 1964; Palumbo and Fraas 1971; Matteson et al. 1972). Several conclusions have been made: that SO_2 oxidation is a zero-order reaction, and its rate-determining step is the formation of atomic oxygen by electrical discharge; that pulsed corona discharge is more effective than DC corona discharge, and the conversion efficiency decreases in the absence of water vapor.

Another pathway for SO_2 conversion is to use wet ammonia (NH_3) to absorb SO_2 in flue gas first, then oxidize sulfite into sulfate. The traditional method for industrial FGD is to use a giant air-blasting system to pump air into the solution to oxidize ammonium sulfite. This method has a drawback that the required initial concentration is usually small (<1 mol/L). If the solution is diluted for oxidation, the energy consumption for late-stage crystallization rises steeply. Since the 1970s, investigations on plasma-assisted FGD have been carried out all over the world. It has been anticipated that plasma-assisted FGD will enhance the efficiency and eliminate NH_3, SO_3, and fine particle or aerosol emission simultaneously. Plasma-induced radicals, such as OH, O, and H, play a key role in sulfite oxidation.

SO_2 and NH_3 react spontaneously and can form various substances depending on different reaction conditions. If the water concentration approaches those of SO_2 and NH_3, the main products are NH_3SO_2 and $(NH_3)_2SO_2$. These products form sulfite-containing aerosols in the same way as H_2SO_4. In the plasma-assisted system, aerosol formation with diameter from 10 nm to 1 μm is enhanced due to the presence of UV light (Christensen et al. 1994). Even in clean air, under DC corona discharge and pulsed corona discharge conditions, the amount of aerosol generated with average diameter of 10 nm/s is 3×10^2 and 3×10^6 per cm^3, respectively (Paris et al. 1996). In the presence of aerosols, heterogeneous reactions are much more complex than gas-phase reactions.

Besides, ozone and hydrogen peroxide are often used in sulfite oxidation. Some researchers have also investigated the use of catalysts like Fe(III), Mn(III), and NO. It has been found that the combined removal of NO and SO_2 is much more efficient than separate removal. The mechanism is not understood as yet, and aerosol effects should also be taken into account. Some researchers established a model to describe the process (Veldhuizen et al. 1997), but it was not enough for understanding giant heterogeneous reaction processes.

11.4.3 Hg Oxidation

Because the concentration of mercury (Hg) in coal is quite small (0.02–0.25 ppm, an average of 0.09 ppm), uncontrolled emissions from a typical 500-MW coal-fired plant would be less than 250 pounds per year, or less than 1 pound per day (Change and Offen 1995). Nonetheless, Hg emission should be controlled for its high toxicity. Mercury is present mainly in the vapor phase in utility power plant flue gas. It is difficult for existing emissions control devices, such as ESP, FF, and scrubbers, to capture Hg vapor. Therefore, methods are needed for converting Hg vapor to the solid phase, capturing Hg on solid sorbent, or converting Hg to a soluble form. The key step of these conversion processes is Hg oxidation, because water solubility of elemental mercury (Hg^0) is very low; however, oxidized mercury (Hg^{2+}) compounds are easy to be captured by wet scrubbing. Main Hg oxidation reactions with O_3 and OH radical are as follows (Calvert and Lindberg 2005):

$$Hg^0 + O_3 \rightarrow HgO + O_2, \tag{11.59}$$

$$Hg^0 + OH \rightarrow HgOH, \tag{11.60}$$

$$HgOH + O_2 \rightarrow HgO + HO_2 \quad \text{or} \quad HgOH + X \rightarrow XHgOH, \tag{11.61}$$

where X can be OH, HO, RO_2, RO, NO, or NO_2.

For the NTP technique, active species play a key role in Hg oxidation. The pioneering work by Masuda et al., has demonstrated the combined removal of NO_x, SO_2, and trace elements Hg using intense pulsed corona discharges with a pulse frequency of 50 Hz. The Hg removal test made at an incineration plant indicated that Hg vapor of 0.5 mg/Nm^3 (0.06 ppm) could be 100% converted into solid particulates (oxides and chlorides) within 2 s of gas residence time through a wide range of gas temperatures (30–300°C). Mercury removal was enhanced by the addition of HCl or Cl_2 (Masuda 1988). Liang et al. found that the collection efficiency depended not only on the pulse voltage, pulse frequency, and residence time, but also on the initial concentration of Hg in the gas flow (Liang et al. 1998, 2002).

Similar results were also obtained by Xu et al. where the Hg^0 oxidation efficiency depended on the radicals (OH, HO_2, and O) and the active species (O_3, H_2O_2, etc.) produced by the pulsed corona discharge. With increasing pulse peak voltage, pulse frequency, electrode number, and residence time, Hg^0 oxidation efficiency could be improved, while the opposite occurs with increasing initial Hg^0 concentrations (Xu et al. 2009).

Apart from pulsed corona discharge and DC corona discharge, the application of the DBD reactor system has also been investigated. Different additions also impact the Hg^0 oxidation efficiency. The addition of H_2O to the gas mixture of HCl in N_2 accelerates the oxidation of Hg^0, although no appreciable effect of H_2O alone on the oxidation of Hg^0 has been observed (Jeong and Jurng 2007). The presence of NO decreased Hg^0 removal efficiency compared to Hg^0-only case due to the competition for ozone and O radical between Hg^0 and NO (Ko et al. 2008). However, Chen et al. got the opposite result wherein the presence of NO_x enhanced Hg oxidation in the DBD reactor (Chen et al. 2006).

Hg removal with NTP is a relative new subject and a lot of problems are still there. Much more basic research and experiments need to be carried out before practical industrial applications.

11.4.4 Flue Gas Preprocessing and Fine Particle Collection

As discussed in Section 11.1.1, ESP does not perform so well in fine particle (PM2.5) collection as in large particle collection. One of the most common techniques to solve this problem is flue gas preprocessing. The diameter of particles are enlarged before they enter into ESP using methods of bipolar charging and gas flow-induced agglomerator. For example, Zukeran et al. (2000) reported the agglomeration phenomena by using AC corona for particle charging and DC electric field for particle agglomeration. Hautanen et al. (1995) proposed that Brownian effects play a significant role in the agglomeration process.

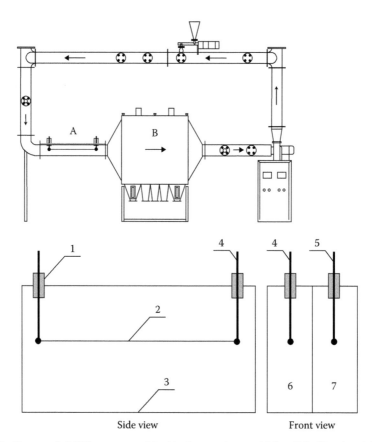

Side view

Front view

FIGURE 11.27 The recycled ESP system and its bipolar precharger (right side). (Reprinted from *Journal of Electrostatics*, 68, Zhu, J.B. et al., Electrostatic precipitation of fine particles with a bipolar pre-charger, 174–178, Copyright 2010, with permission from Elsevier.)

One example of a bipolar precharger in an ESP system can be seen in Figure 11.27. On the left side, a bipolar precharger (marked as "A") is fixed before ESP (marked as "B") inlet. The details of the bipolar precharger are shown on the right side: (1) insulator, (2) corona wire, (3) duct, (4) positive high-voltage connector, (5) negative high-voltage connector, (6) positive corona charger, and (7) negative corona charger. The length of two identical corona wire is 1000 mm. The height and width of each corona charger are 150 and 75 mm, respectively.

In this system, Zhu et al. reported that by using the bipolar precharger, the collection efficiency can be significantly improved for particles of all sizes. The results are shown in Figure 11.28.

Up to now, there has been limited research on the hydrodynamic characteristics of fine particles with diameter less than 1 μm in the precipitator and their dependence on collection efficiency. One technique was proposed recently, which combined particle image velocimetry (PIV) and electrical low pressure impactor (ELPI). PIV is used to test 2D or 3D hydrodynamic properties without any disturbance in the flow field, while ELPI can test the fine particle collection efficiency on site in a wide diameter range.

11.4.5 Plasma-Induced Soot Combustion

The research and industrial application of plasma-induced soot combustion mainly focuses on the treatment of diesel engine exhaust, which is usually combined with NO_x and SO_2 removal (Clements et al. 1989; Higashi et al. 1992; Chae 2003). Figure 11.29 gives a typical result of ultrafine particle removal from diesel exhaust gas by using an NTP reactor.

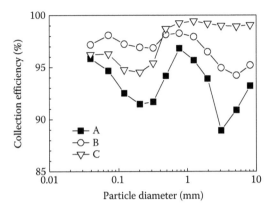

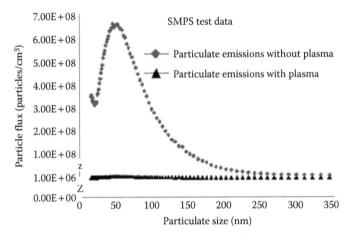

FIGURE 11.28 Effects of the bipolar precharger on the grade collection efficiency under an initial mass concentration of 50 mg/Nm³ and gas flow rate of 725 Nm³/h. The corresponding corona currents for the precharger are 9.0 mA at +5.0 kV, 105 mA at −13.0 kV, and 136 mA at −15.0 kV, respectively. Test conditions: A, without using pre-charger; B, voltage pairs on the precharger (+5.0 kV, −13.0 kV); C, voltage pairs (+5.0 kV, −15.0 kV). (Reprinted from *Journal of Electrostatics*, 68, Zhu, J.B. et al., Electrostatic precipitation of fine particles with a bipolar pre-charger, 174–178, Copyright 2010, with permission from Elsevier.)

FIGURE 11.29 Ultrafine particle testing with and without an NTP reactor, by using scanning mobility particle sizer (SMPS). (Reprinted from *SAE paper*, Thomas, S.E. et al., Non thermal plasma aftertreatment of particulate—Theoretical limits and impact on reactor design, 2000-01-1926, Copyright 2000, with permission from Elsevier.)

High collection efficiency of soot in diesel engine exhaust gas can be achieved. For example, Higashi et al. reported that 100% of the soot was removed within 1 min when over 7 kV was applied on the NTP reactor (Higashi et al. 1992). More than 90% particle removal efficiency was achieved by using a single-stage NTP–chemical hybrid process (Kuroki et al. 2002). A direct example of NTP-induced soot removal can be obtained from Okubo et al. (2008). In this chapter, the NTP is induced inside the honeycomb path of the DPF (ceramic diesel particulate filter, widely used for the collection of carbon soot from diesel engine) oxidizes NO to NO_2, then the generated NO_2 together with activated oxygen species incinerate the deposited carbon soot simultaneously. Figure 11.30 shows the photographs of DPF before and after NTP treatment.

The main conclusions regarding the interaction between NO_x, SO_2 oxidation, and soot removal are as follows: removal efficiency of SO_2 can be improved by the presence of fly-ash particles in the gas stream, and a reduction in the ash resistivity by the conversion products of SO_2 and NO_x results in ash collection improvement (Clements et al. 1989). But the reduction of NO_2 to NO occurs during the process of carbon

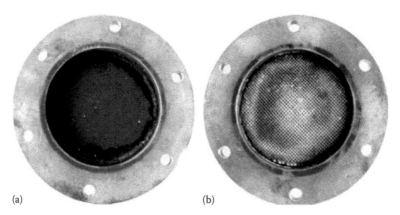

(a)   (b)

FIGURE 11.30   Photographs of DPF (a) before and (b) after NTP treatment. (With kind permission from Springer Science+Business Media: *Plasma Chemistry and Plasma Processing*, Total diesel emission control technology using ozone injection and plasma desorption, 28, 2008, 173–187, Okubo, M. et al.)

soot oxidation. According to Rajesh and Mark (2002), pulse corona discharge induced an initial gas-phase oxidation of NO to NO_2 with soot, followed by heterogeneous conversion of NO_2 to NO on the soot surface.

11.4.6 Integrated Flue Gas Cleaning System and Industrial Demonstration

The integrated flue gas cleaning system should have a capability of simultaneously removing fine particles, acid gases such as SO_2, NO_x, and the poisonous element Hg. The conventional integrated flue gas cleaning systems used in power plants burn three different kinds of fuels, which have been listed in Figure 11.31. ESP is combined with traditional $DeNO_x$ and $DeSO_2$ methods, for example, selective catalytic reactors and wet scrubbers. In a normal coal-fired power plant, the dust concentration in the flue gas determines the position of ESP in the integrated system. For high dust concentration, a wet $DeNO_x$ system is set before ESP to premove dust so as to reduce the load of ESP, while for low dust concentration, ESP can be set before the $DeNO_x$ and $DeSO_2$ system.

Figure 11.32 illustrates an integrated coal-fired flue gas cleaning system. A dry-type plasma reactor is set before ESP for flue gas conditioning and the preoxidation of acid pollutants. In this section, fine particles are precharged and agglomerated into bigger ones, and NO and SO_2 are oxidized to NO_2 and SO_3 and poisonous heavy metals such as Hg. After the dust is collected in the ESP, acid pollutants and the heavy metal in the oxidized state are absorbed in an ammonia absorber. The wet-type plasma reactor is used for posttreatment of the absorbing agent, in which sulfite is oxidized to sulfate for product recycling; ammonia leakage and SO_3-containing acid fog are controlled simultaneously. Figure 11.33 shows an integrated system, in which an FF is set before a dry-type plasma reactor to substitute the ESP.

K. Yan provided the design for the power supply in a plasma reactor for removal of acid pollutants, as shown in Figure 11.34. As discussed in Section 11.2, for a given reactor size, the maximum energy transfer for primary streamer propagation is mainly dependent on the voltage rise rate and pulse duration. However, the total energy transfer depends on the stored energy, the voltage level, and the size of reactor. For realizing an energy-efficient process, the stored energy and the maximum transferred energy should be matched when designing the power supply and the plasma reactor.

11.5 VOCs and Odor Emission Abatement

For most of the commercial pollution control techniques, such as catalytic and thermal incineration, adsorption, and absorption, the lower the concentration, the higher the cost per unit of pollutant treatment. For NTP techniques, the lower the concentration, the lower the cost. In recent decades, the abatement of low-concentration VOCs is one of the fastest-growing industrial applications of NTPs.

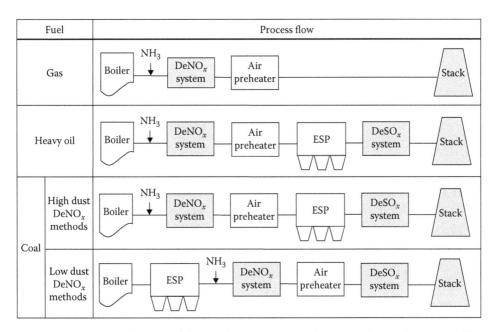

FIGURE 11.31 Conventional integrated flue gas cleaning system used in power plants, where wet scrubbers are used as the DeSO$_x$ system and a selective catalytic reactor is used as the DeNO$_x$ system. (Reprinted from *Journal of Electrostatics*, 57, Chang, J.S., Next generation integrated electrostatic gas cleaning systems, 273–291, Copyright 2003, with permission from Elsevier.)

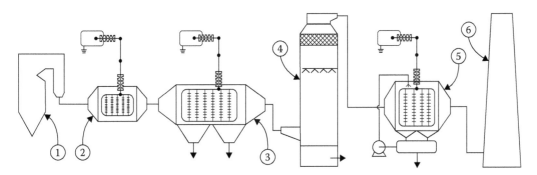

FIGURE 11.32 The schematic diagram of a demonstration of integrated coal-fired flue gas cleaning system. (1) the boiler, (2) the dry-type plasma reactor, (3) the ESP, (4) the ammonia absorber, (5) the wet-type plasma reactor, (6) the stack.

The mechanism for decomposing plasma-induced VOCs can be divided into two pathways: high-energy electron impact with VOCs and molecular and radical reaction. Which reaction type is dominant depends on background gas composition, the molecular structure of VOCs, and the discharge mode. For example, for decomposing halogenated hydrocarbons with NTP, high-energy electron impact is preferred because it has a strong electron adsorption ability (Mok et al. 2008). Although hydrocarbons have relatively strong chemical reactivity, the decomposition process is mainly initiated by radical or excited atomic N and excited atomic O (Nair 2004; Zhang et al. 2011). Recently, Schiorlin et al. confirmed that ionic reactions are responsible for the initial stages of toluene decomposition induced by +DC corona (Schiorlin et al. 2009). The initiation stage might be in the timescale of picoseconds to nanoseconds. After that, the reaction at the molecular level continues with a timescale of nanoseconds

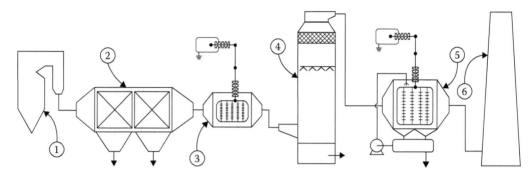

FIGURE 11.33 The schematic diagram of another demonstration of integrated coal-fired flue gas cleaning system. (1) the boiler, (2) the FF, (3) the dry-type plasma reactor, (4) the ammonia absorber, (5) the wet-type plasma reactor, (6) the stack.

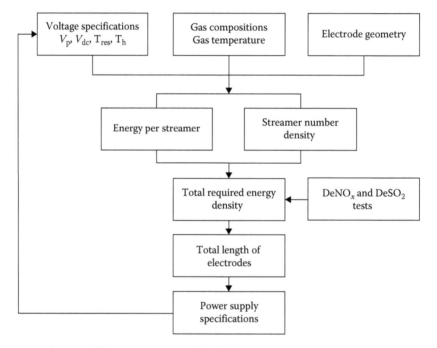

FIGURE 11.34 Schematic of designing power supply and plasma reactor. (Reproduced from Yan, K. et al., *Journal of Electrostatics*, 44, 17–39, 1998.)

to microseconds. This process leads to two different types of termination depending on the reaction conditions. Under conditions of high temperature, high energy density, and low initial concentration of VOCs, the oxidation reaction is dominant and the final products are water and CO_2, while if the initiation concentration is relatively high and temperature and energy density are low, a chain reaction between the active sites of intermediates occurs leading to the formation of large quantities of aerosol or polymer film (Yan et al. 2001a) (Table 11.7).

With regard to the energy density needed for abatement of low-concentration VOC emission in air, it was found that when the corona energy density is around a few Wh/Nm³, the global kinetics shows insignificant radical terminations, as indicated in Figure 11.35, for styrene removal. While if corona energy density becomes larger than 2.0 Wh/Nm³, the energy yield decreases very rapidly because of significant radical terminations. The typical energy cost per single styrene molecular removal is about 90 eV.

TABLE 11.7 Typical Mechanism for Removal of Plasma-Induced VOCs

VOCs	Molecular formula	Discharge type	Experimental conditions	Initial reaction
Halogenated hydrocarbon	$C_2H_2F_2$	Packed bed reactor derived by AC source	$C_0 = 200–1000$ ppm Dry air, 150–250°C	High-energy electron impact
	CCl_4	Pulsed corona	$C_0 = 100$ ppm Dry air, 25–300°C	High-energy electron impact
	CF_2Br_2	+DC corona	$C_0 = 100$ ppm Wet air, room temperature	Positive ions
Alcohols	CH_3OH	Pulsed corona	$C_0 = 100$ ppm Dry air, room temperature	Positive ions, radicals, atoms
Hydrocarbon	C_5H_{12}	+DC corona	$C_0 = 100$ ppm Wet air, room temperature	Positive ions
	C_5H_{12}	−DC corona	$C_0 = 100$ ppm Wet air, room temperature	Radicals, atoms
	C_7H_8	Pulsed corona	$C_0 = 500$ ppm Wet air, room temperature	Radicals, atoms

Source: Zhang, X.M., Removal of volatile organic compounds by non-thermal plasma technology, Ph.D. Dissertation, Zhejiang University, Hangzhou, China, 2011. With permission.

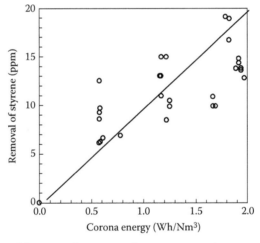

FIGURE 11.35 Dependence of the removed styrene on the corona energy density in air. The length and inner and outer diameters of the wire–cylinder reactor are 3000, 3, and 160 mm, respectively. The initial styrene concentration is around 20–30 ppm. The gas flow rate and the average corona power are up to 1000 Nm³/h and 1.2 kW, respectively. (From Yan, K., Corona plasma generation, Ph.D. Dissertation, TU/e, Eindhoven, 2001.)

Figure 11.36 shows a pilot plant of an emission abatement system for mobile VOCs developed by Drexel Plasma Institute. It is operated in the pulsed corona streamer mode and derived from a nanosecond pulsed power source.

11.5.1 Plasma-Induced VOCs to Aerosol Conversion

As discussed before, aerosol is one of the by-products of VOC decomposition, and cannot be avoided. Several researchers have tested the amount of aerosol generated and its diameter distribution (Yamamoto and Jang 1999; Parissi et al. 2000; Kim et al. 2005). Their common opinion is that the diameter of the aerosol is mainly less than 1 μm. As mentioned in Section 11.4.2, even in clean air, fine particles can

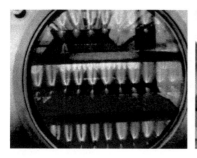

FIGURE 11.36 Large-volume, atmospheric pressure pulsed discharge plasma for treating VOC emissions. Left picture: discharge operation at 4 kW. Right picture: overview of the discharge system-pilot plant. (From Fridman, A. et al., *J. Phys. D Appl. Phys.*, 38, R1, 2005. With permission.)

be generated under conditions of corona discharge. The amount of aerosol generated is around $10^4/cm^3$ with a diameter normally less than 100 nm. When the discharge occurs in a flue gas containing 110 ppm styrene, aerosol generation rises to $10^6/cm^3$ (Zhang 2011). These fine particles containing organic substances are very harmful for our respiratory system (Saiyasitpanich et al. 2006).

Nucleation plays a key role in the aerosol formation process. Borra investigated aerosol generation under corona streamer discharge, and came up with two hypotheses for the nucleation mechanism. One is nucleation of metallic vapors, which are produced in the discharging gap either through thermal effects (sublimation of the metallic grid) or through atomization of the cathode by electrical sputtering (ionic bombardment) with ejection of metallic atoms; the other is gaseous species with low saturation vapor pressure (Borra et al. 1998). Under discharge conditions in clean air, products such as O_3, NO_2, N_2O_5, and HNO_3 have lower saturation vapor pressure than O_2, N_2, and CO_2, while under the condition of VOC abatement with NTP, the decomposition pathway and intermediates formed both in the gas phase and in the solid phase are so complex that there is still no effective tool to trace this process directly, while the aerosol products can be analyzed *in situ* and collected by an ESP system. Figure 11.37 shows the result of number concentration of aerosol with respect to particle diameter

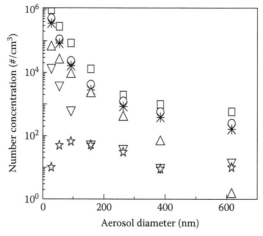

FIGURE 11.37 Number concentration of aerosol with respect to particle diameter in the experiment of streamer corona-induced benzene and toluene decomposition. The experiment is operated under the following conditions: initial concentration of benzene and toluene are 100–110, 160–170, and 130–140 ppm, respectively. □ styrene (RH ≈ 42%); ○ toluene (RH ≈ 42%); ✳ benzene (RH ≈ 42%); ▽ air discharge (RH = 32%); △ air discharge (RH = 87%); ☆ without discharge. (From Zhang, X.M., Removal of volatile organic compounds by non-thermal plasma technology, Ph.D. Dissertation, Zhejiang University, Hangzhou, China, 2011. With permission.)

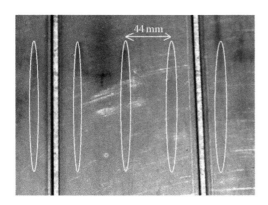 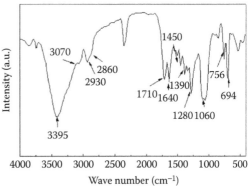

FIGURE 11.38 Photograph of aerosol deposit on ground plate in the recycled ESP system and the FTIR spectrum of the aerosol deposit. (From Zhang, X.M., Removal of volatile organic compounds by non-thermal plasma technology, Ph.D. Dissertation, Zhejiang University, Hangzhou, China, 2011. With permission.)

by using Dekati-produced ELPI. These aerosols are formed during streamer corona-induced benzene and toluene decomposition and collected in the recycled ESP described in Figure 11.27. A photograph of aerosol deposit on the ground plate is shown in Figure 11.38. The area in deep color corresponds to the discharging electrode, whose gap distance is about 44 mm. The FTRI spectrum is quite similar to the styrene polymer film produced by Luo with NTP (2003).

11.5.2 Plasma-Initiated Catalysis Intensification

The combination of NTP and chemical catalysis is an efficient way to increase energy efficiency and optimize by-product distribution for VOC decomposition. It has attracted a lot of research interests in the past decades. Yamamoto and Mizuno have done a lot of work on this subject since 1990s. Recently, J. van Durme provided a comprehensive review in Table 11.8, in which IPC and PPC refer to in-plasma catalysis and postplasma catalysis, respectively. IPC is also defined as plasma-driven catalysis (PDC), which introduces the catalyst in the plasma zone, while PPC is defined as plasma-enhanced (assisted) catalysis (PEC), which places the catalyst after the plasma zone. The schematic of these two kinds of plasma catalysis system can be seen in Figure 11.39.

In a PDC system, the catalyst material can be inserted into the reactor in several ways: as coating on the reactor wall or electrodes, as a packed bed (granulates, coated fibers, and pellets), or as a layer of catalyst material (powder, pellet, granulates, and coated fiber) (Figure 11.40).

The catalyst increases the reaction rate by offering an alternative reaction route with a lower overall activation energy. In the presence of catalyst, the gas-phase reaction is substituted by the surface reaction, which is usually called the heterogeneous catalysis reaction. Introducing catalysts into the plasma discharge may change the type of discharge or induce a shift in the distribution of the accelerated electrons. These processes again influence the production of excited and short-living active species or the formation of stable active species. The active species and high-energy electrons trigger physical changes in the catalyst and consequently affect surface adsorption, which is the first step in a heterogeneous catalysis reaction.

11.6 Fuel Gas Cleaning

11.6.1 Plasma-Induced Tar Cracking

Biomass gasification is an efficient way to convert organic or fossil-based carbonaceous materials into carbon monoxide, hydrogen, and carbon dioxide by treating the material at high temperature, without combustion, but with a controlled amount of oxygen and/or steam. The gas mixture produced is called syngas (synthesis gas or synthetic gas), and it is a fuel. The power derived from

TABLE 11.8   Overview of Recently Published Papers on Plasma Catalysis

Plasma type	VOC	Flow rate (mL/min)	Concentration range (ppm)	Catalysis	Position	Maximum removal efficiency	Energy cost (g/kWh)
DBD	Toluene	315	240	MnO_2/Al_2O_3 MnO/AC Fe_2O_3/MnO_2	IPC IPC IPC	55	11
Corona	n-Heptane			TiO_2	IPC	50.4	
DBD	Toluene Ammonia	667×10^3	10–50	CuO/MnO_2	PPC	50	
DBD	Toluene	100–500	50	$MnO_2/Al/Ni$	IPC	>95	1
DBD	Benzene	250	300–380	TiO_2 MnO_2	IPC IPC	16	3 4
Pulsed corona	Toluene Benzene Hexane Methane	100	300	AlO_2 Silica gel	IPC IPC	>95 75 25 5	156
DBD	SF_6 NF_3 CF_4 C_2F_6	600	300	$CuO/ZnO/MgO/Al_2O_3$	IPC IPC IPC IPC	>99 >99 66 83	
DBD	Benzene Toluene o,m,p-Xylene Formic acid	4,000	75–110	Ag/TiO_2	IPC IPC IPC IPC	96	11
DBD	Toluene	500	250 100	MnO_x/CoO_x	IPC	>99 >99	2 1

Reactor	Compound	Concentration		Catalyst		Conversion	Ref
Coil type (AC)	Benzene	4,000–10,000	200	Ag/TiO$_2$	IPC	>99	10
				Ni/TiO$_2$	IPC	>99	
				Ag/Al$_2$O$_3$	IPC	>99	
				Pt/Al$_2$O$_3$	IPC	>99	
				Pd/Al$_2$O$_3$	IPC	>99	
				Ferrierite	IPC	>99	
				Ag/H-Y	IPC	>99	
DBD	Formaldehyde	605	140	Ag/CeO$_2$	IPC	92	6
+DC corona	TCE	1,500	100	Ti/O$_2$	IPC	85	3
Streamer	Toluene	133×10^3	45	CuOMnO$_2$/Al$_2$O$_3$	PPC	96	33
+DC Corona	Toluene	10,000	0.5	CuOMnO$_2$/Al$_2$O$_3$	PPC	>99	
Coil type (AC)	Toluene	500	200	Zeolites	IPC (discontinue)		
Dielectric pellet-bed reactor	CFC-12	1,000	500	TiO$_2$	IPC	27	36
DBD	Isopropanol	500	250	MnO$_x$/CoO$_x$	IPC	>100	9
DBD	Trichloroethylene	510	430	Au/SBA-15	PPC	>99	
DBD	Trichloroethylene	500	250	MnO$_2$	PPC	97	
DBD	Dichloromethane	1,000	500	γ-Al$_2$O$_3$	IPC/PPC	51/43	
				α-Al$_2$O$_3$	PPC	34	
				TiO$_2$	PPC	39	
				HZSM-5	PPC	41	
				NaZSM-5	PPC	38	
				NaA	PPC	37	
				NaX	PPC	40	

Source: With kind permission from Springer Science+Business Media: *Applied Catalysis B: Environmental*, Combining non-thermal plasma with heterogeneous catalysis in waste gas treatment: A review, 78, 2008, 324–333, van Durme, J. et al.

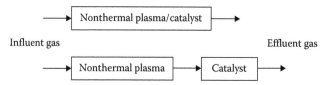

FIGURE 11.39   Schematic overview of two plasma catalyst hybrid configurations: PDC (up) and PEC (down).

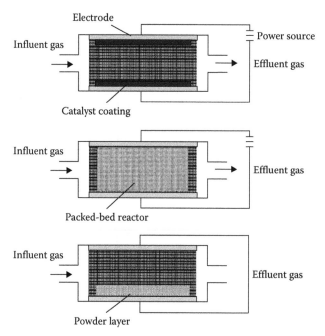

FIGURE 11.40   The most common catalyst insertion methods for PDC system configuration. (Reprinted from *Applied Catalysis B: Environmental*, 78, van Durme, J. et al., Combining non-thermal plasma with heterogeneous catalysis in waste gas treatment: A review, 324–333, Copyright 2008, with permission from Elsevier.)

gasification of biomass and combustion of the resultant gas is considered to be a source of renewable energy. However, one of the problems associated with biomass gasification is the production of tar. Thermal efficiency is high as the gas exit temperatures are relatively low. However, this means that tar production is significant at typical operation temperatures. Tar is a complex mixture of polyaromatic hydrocarbons. It imposes serious limitations in the use of product gas due to fouling of downstream process equipment, so product gas must be extensively cleaned before use. The typical limit for tar concentration is less than 100 mg/Nm³ for an internal combustion engine (Hasler and Nussbaumer 1999).

Several kinds of methods have been applied in tar removal, for example, mechanism collection, thermal cracking, catalyst cracking, and corona plasma cracking. Cracking means decomposing tar molecules into lighter ones. The research team group at Electrical Engineering Department, TU/e, studied the tar cracking process with pulsed corona plasma and conducted an industrial demonstration, which is shown in Figure 11.41. It was manufactured by the Biomass Technology Group, Enschede, the Netherlands. The power source driving the tar removal reactor is similar to the one shown in Figure 11.18. Figure 11.42 indicates that the particle removal efficiency of the pulsed corona system varies from 72% to 95%, and increases with energy density. The particle generation and collection process is similar to aerosol formation and collection in VOCs decomposition with ESP techniques. Neeft (2002) and Paasen and Rabou (2004) also confirmed the validity of ESP capturing tar aerosols.

FIGURE 11.41 Overview of the "hybrid" pulsed corona system and the BTG gasifier (left) and pulsed corona reactor with pulsed source and piping to the gasifier (right). (From Nair, S.A. et al., *Ind. Eng. Chem. Res.*, 43, 1649–1658, 2004. With permission.)

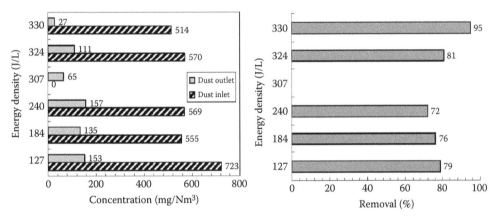

FIGURE 11.42 The test results of particle removal efficiency via energy density of the system in Figure 11.41. (From Nair, S.A. et al., *Ind. Eng. Chem. Res.*, 43, 1649–1658, 2004. With permission.)

Figure 11.43 provides the data for tar removal efficiency, including for both heavy tar and light tar. Tar removal at energy density of around 200 J/L indicates conversion of heavy tar fractions into light tar. The total tar removal efficiency can reach 43% at the energy density level of 330 J/L. The heavy tar cracking efficiency with this system was also studied by Heesch et al. at two energy density levels of 148 and 161 J/L, respectively. Part of the heavy tar was decomposed into light tar under both energy density levels (Heesch and Laan 2000). By comparing these test results, it can be concluded that heavy tar conversion dominates the removal process under the energy density of 300 J/L, while above 300 J/L input energy is effectively utilized for light tar removal.

11.6.2 Plasma-Assisted Gasification

Except for biomass, gasification can also begin with material that would otherwise have been disposed of such as biodegradable waste. A plasma-assisted waste gasification system converts waste stream reaction residues into a clean synthesis gas (syngas). The "plasma" here refers to arc plasma that connects the two electrodes in a graphite furnace and heats the feedstock (solid waste) to a very high temperature.

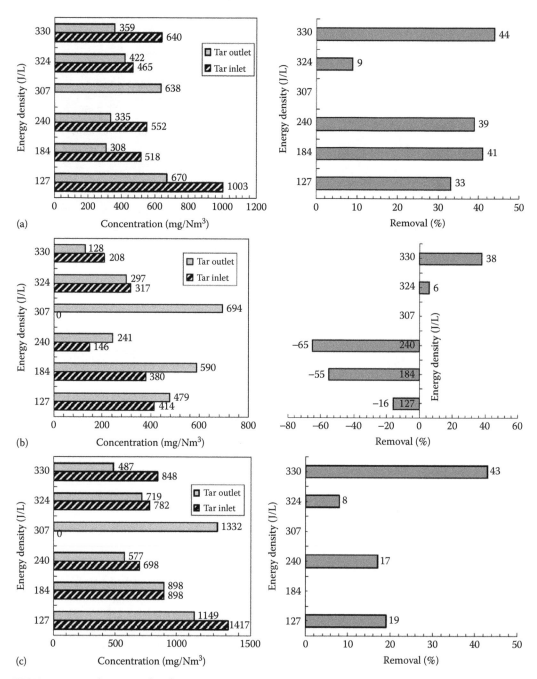

FIGURE 11.43 The test results of tar removal efficiency ((a) heavy tar, (b) light tar, and (c) total tar) via energy density of the system in Figure 11.41. (From Nair, S.A. et al., *Ind. Eng. Chem. Res.*, 43, 1649–1658, 2004. With permission.)

Thus plasma-assisted gasification uses the external heat from an arc tunnel to gasify the waste, resulting in very little combustion. Almost all of the carbon is converted to fuel gas through the following simplified reactions (Schuster et al. 2001):

$$R1: C(s) + CO_2 = 2CO \qquad \text{Boudouard equilibrium,} \qquad (11.62)$$

$$\text{R2: } C(s) + 2H_2 = CH_4 \qquad \text{Hydrogenating gasification,} \qquad (11.63)$$

$$\text{R3: } C(s) + H_2O = CO + H_2 \qquad \text{Heterogeneous water gas shift reaction,} \qquad (11.64)$$

$$\text{R4: } CH_4 + H_2O = CO + 3H_2 \qquad \text{Methane decomposition,} \qquad (11.65)$$

$$\text{R5: } CO + H_2O = CO_2 + H_2 \qquad \text{Water gas shift reaction.} \qquad (11.66)$$

Based on these reactions, several thermodynamic equilibrium models have been built to analyze the gasification process and evaluate different effect factors, such as moisture content, air flow rate, and operating temperature. According to simulation results, moisture content (up to 25%) is essential for the gasification chemical reactions as it increases the hydrogen concentration. An increase in the air flow rate is disadvantageous for the dilution of the produced gas with nitrogen introduced by air. Low gasification temperature results in high tar concentration in the syngas, and high gasification temperature leads to high carbon monoxide concentration and a small decrease in hydrogen concentration at temperatures over 1073 K. While the total concentration of these two main synthesis gas components (H_2 and CO) increases (Mountouris et al. 2006).

Figure 11.44 shows that a plasma-assisted waste gasification system normally comprises four parts: a waste pretreatment system (also called "feedstock preparation system"), a plasma furnace reactor, a gas cleaning system, and the energy recovery system. The waste pretreatment system is used to make the waste conform to the inlet requirements of the plasma reactor, especially adjust the moisture content in waste. Two graphite electrodes extend into the plasma furnace reactor. The reactor inside can be divided into three zones: a bottom zone for melting waste stream reaction residues and forming a slag pool; a middle zone for converting the waste stream into the syngas; and a top zone having at least one plasma arc torch for controlling the temperature and composition of the syngas. Before the syngas enters the energy recovery system, it should go through the gas cleaning system, in which elimination of acid gases (HCl, SO$_x$), tar removal, heavy metal removal, suspended particulate collection, and moisture conditioning are achieved. The energy recovery system can be a hydrogen- or methanol-burned electrical generator.

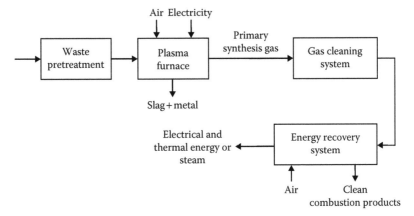

FIGURE 11.44 Block diagram of plasma-assisted gasification system. (Reprinted from *Energy Conversion and Management*, 47, Mountouris, A. et al., Solid waste plasma gasification: Equilibrium model development and exergy analysis, 1723–1737, Copyright 2006, with permission from Elsevier.)

11.7 Water Cleaning

11.7.1 Overview of Plasma System and Aqueous Reaction Engineering

Just like gas cleaning with NTP, pollutants in water, such as organic compounds or microorganisms, can be decomposed by the physical and chemical effects of a discharge-induced NTP, for example, pyrolysis reaction, photolysis reaction, primary and secondary molecular, ionic, or radical reactions. The localized regions of high temperature and pressure of an NTP form intensive shock and acoustic waves, which have been used for inactivating microorganisms. Thus, electrical discharge-based water cleaning is a multiple-mode approach. The dominant factor varies for different discharging modes. Based on the reactor and electrode configuration described in Figure 11.45, two types of discharging

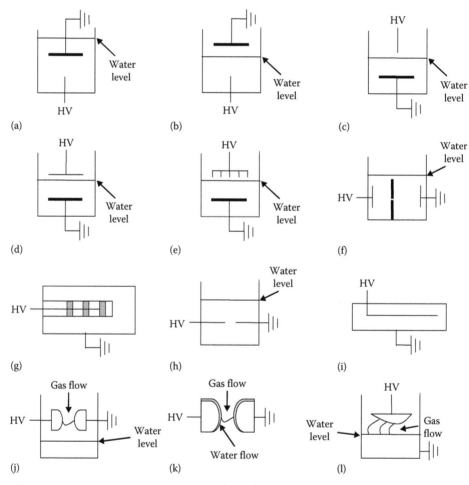

FIGURE 11.45 Schematic of reactor and electrode configurations: (a) point-to-plane liquid-phase corona reactor; (b) point-to-plane with air gap liquid phase corona reactor; (c) single point-to-plane glow discharge reactor; (d) plane-to-plane glow discharge reactor; (e) multiple point-to-plane glow discharge reactor; (f) pinhole reactor; (g) ring electrode reactor; (h) point-to-point arc reactor; (i) wire-to-cylinder reactor; (j) gas-phase gliding arc over water surface; (k) gliding arc with water film; and (l) gliding arc to water surface. (Reprinted from *Industrial and Engineering Chemical Research*, 45, Locke, B.R. et al., Electrohydraulic discharge and nonthermal plasma for water treatment, 882–905, Copyright 2006, with permission from Elsevier.)

FIGURE 11.46 Streamer corona discharge (left) in reactor A and pulsed arc discharge (right) in reactor E.

modes can be categorized: liquid phase and liquid/gas-phase electrical discharges. In the reactor A, B, F, G, H, and I, discharging electrodes are submerged in water. Streamer corona plasma or arc plasma can be generated at the tip of the discharging electrode or between two electrodes driven by a DC or pulsed power source. A typical streamer corona discharge in water is shown in Figure 11.46 (left). While in the reactor C, D, E and J, K, L, both electrodes are in the gas phase, or the discharging electrode is in the gas phase and the water body acts like the grounded electrode. For example, in reactor E the discharge between the discharging electrode and the water surface is as seen in Figure 11.46 (right). Details of the reactor configuration and the driving power source can be found in Locke et al. (2006).

The mechanism of gas-phase discharge has been studied extensively, while the process of discharge in water is still a gray box. Depending on the relationship between the applied voltage and water conductivity, at least four initiation mechanisms can be included: (1) bubble, (2) microexplosive, (3) ionization, and (4) electrothermal. The first mechanism is suitable for nondegassed water and pulse duration ranging from units of hundreds of microseconds; the second one is likely to happen with a small tip radius (from units to tens of microns), small interelectrode distance (submillimeter), and nanosecond voltage pulse duration; the third one is likely to happen when the field strength is sufficiently high for ionization of water molecules (normally field strength should be larger than 10^7 V/cm and pulse duration needs to be shorter than 10^{-8} s); the last one can be realized in large values of product of water conductivity into pulse duration (Ushakov 2007). For most discharges in water, especially when water is not used as the dielectric medium, the electrothermal process can explain most experimental results. The electrothermal process occurs as follows: when high-voltage conduction current flows under the effect of the electric field, water in the near-electrode regions with maximum field strength is heated up to boiling point, and as long as vapor-gas cavities are formed, electron avalanche and ionization processes such as gas-phase discharge take place. The electrothermal process has been confirmed in the works of Lisitsyn et al. (1999) through energy balance analysis and streamer propagation velocity measurement. So the chemical reaction induced by electrical discharge in water is still primarily a gas-phase process, and the active species are mainly formed in the vapor-gas cavities.

By measuring the emission intensity at the corresponding wavelength, the generation of different active species can be determined. For example, the typical emission wavelength of OH is 309 nm. Figure 11.47 provides the waveforms of input power and OH intensity of arc discharge in water. It can be concluded that the OH radical is mainly formed in the arc tunnel, while the formation in the prebreakdown period can be neglected. Comparing the emission spectra of arc discharge and streamer corona discharge in Figure 11.48, the emission intensity of arc discharge is much stronger than the streamer corona discharge under the same electrical energy. It indicates that the generation volume of radicals is much bigger in the arc tunnel than in the streamer tunnel. Water conductivity also has an effect on the emission spectrum characteristics. As water conductivity increases from 1 μs/cm, the emission intensity

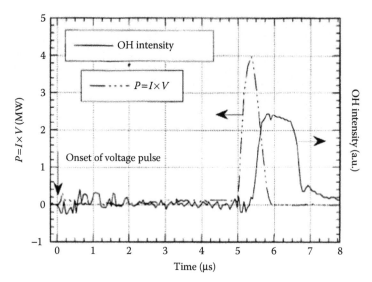

FIGURE 11.47 Time characteristics of OH radical emission and input power for the arc discharge in distilled water. (Reprinted from *Journal of Electrostatics*, 43, Sun, B. et al., Non-uniform pulse discharge-induced radical production in distilled water, 115–126, Copyright 1998, with permission from Elsevier.)

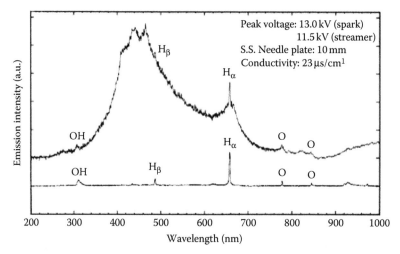

FIGURE 11.48 Emission spectra from arc discharge (top curve) and from streamer corona discharge (bottom curve) in distilled water with needle–plate electrodes. (Reprinted from *Journal of Electrostatics*, 43, Sun, B. et al., Non-uniform pulse discharge-induced radical production in distilled water, 115–126, Copyright 1998, with permission from Elsevier.)

at 309 nm first increases with increasing current, then decreases for a relatively low electric field in a high-conductivity liquid, and reaches the maximum around tens of μs/cm.

Besides active species, the intensive UV radiation from an electrical discharge in water also contributes to pollutant degradation, especially microorganism inactivation. For pulsed arc discharge in water or at the water surface, the power of UV radiation can reach the level of GW, and the intensive UV radiation plays an important role in sterilization or disinfection (Ching et al. 2001; Zheng et al. 2012). This can also be confirmed indirectly by adding a UV absorbant. Figure 11.49 shows *Escherichia coli* disinfection efficiency under the conditions of arc discharge at the water surface

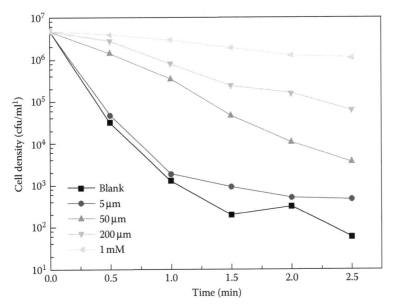

FIGURE 11.49 *E. coli* disinfection efficiency under arc discharge with addition of different concentrations of uridine.

with the addition of uridine. Uridine can absorb UV radiation to decrease its damage on DNA. The results in Figure 11.49 show that with increasing uridine concentration in water, the disinfection efficiency decreases rapidly. When the uridine concentration reaches 1 mmol/L, some *E. coli* are killed after 2.5 min of treatment.

References

Akerlof, G.C. and Wills, E. 1951. *Bibliography of Chemical Reactions in Electric Discharges*. Washington, DC: Office of Technical Services, Department of Commerce.

Arrondel, V., Bacchiega, G., Caraman, N. et al. 2006. A friendly tool to assist plant operators and design engineers to control fly ash emissions: ORCHIDEE. In *Proceedings of 10th International Conference on Electrostatic Precipitation*, June 25–29, Australia, pp. 1–14.

Bityurin V.A., Potapkin B.V. and Deminsky M.A. 2000. Chemical activity of discharges. In *Electrical Discharges for Environmental Purposes, Fundamentals and Applications*, ed. E.M. van Veldhuizen, 49–117. New York: NOVA Science Publishers Incorporation.

Borra, J.P., Goldman, A., Goldman, M. et al. 1998. Electrical discharge regimes and aerosol production in point-to-plane DC high-pressure cold plasmas: Aerosol production by electrical discharges. *Journal of Aerosol Science*, 29(5–6): 661–674.

Calvert, J.G. and Lindberg, S.E. 2005. Mechanism of mercury removal by O$_3$ and OH in the atmosphere. *Atmospheric Environment*, 39: 3355–3367.

Chae, J.O. 2003. Non-thermal plasma for diesel exhaust treatment. *Journal of Electrostatics,* 57(3–4): 251–262.

Chae, J.O. and Dessiaterik, Y. 2000. *Non-Thermal Plasma Technology for Pollution Control*. Inchon: Inha University.

Chang, J.S. 2003. Next generation integrated electrostatic gas cleaning systems. *Journal of Electrostatics*, 57(3–4): 273–291.

Change, R. and Offen, G.R. 1995. Mercury emission control technologies: An EPRl synopsis. *Power Engineering*, 99(11): 51–57.

Chen, Z., Mannava, D.P. and Mathur, V.K. 2006. Mercury oxidization in dielectric barrier discharge plasma system. *Industrial and Engineering Chemistry Research*, 45(17): 6050–6055.

Ching, W.K., Colussi, A.J., Sun, H.J. et al. 2001. *Escherichia coli* disinfection by electrohydraulic discharges. *Environmental Science and Technology*, 35(20): 4139–4144.

Choulakova, E.V., Denisenko, V.P., Fridman, A.A. et al. 1997. Mechanisms of NO_x oxidation stimulated by electron beam. In *Proceedings of 13th International Symposium on Plasma Chemistry*, August 18–22, Beijing, China, p. 2096.

Christensen, P.S., Wedel, S. and Livbjerg, H. 1994. The kinetics of the photolytic production of aerosols from SO_2 and NH_3 in humid air. *Chemical Engineering Science*, 49: 460–465.

Clements, J.S., Mizuno, A., Finney, W.C. et al. 1989. Combined removal of SO_2, NO_x, and fly ash from simulated flue gas using pulsed streamer corona. *IEEE Transactions on Industry Applications*, 25(1): 62–69.

Diato, S. 1997. Particle growth in the corona discharge with ammonia addition for the removal of nitrogen oxides. In *13th International Symposium on Plasma Chemistry*, August 18–22, Beijing, China, p. 1799.

Dinelli, G., Civitano, L. and Rea, M. 1990. Industrial experiments on pulse corona simultaneous removal of NO_x and SO_2 from flue gas. *IEEE Transactions on Industry Applications*, 26:535.

Fridman, A., Chirokov, A. and Gutsol, A. 2005. Non-thermal atmospheric pressure discharges. *Journal of Physics D: Applied Physics*, 38(2): R1.

Fridman, G., Gutsol, A., Shekhter, A.B. et al. 2008. Applied plasma medicine. *Plasma Processes and Polymers*, 5: 503–533.

Hasler, P. and Nussbaumer T. 1999. Gas cleaning requirements for IC engine applications fixed bed biomass gasification. *Biomass and Bioenergy*, 16: 385–395.

Hautanen, J., Watanabe, T., Tsuchida, Y. et al. 1995. Brownian agglomeration of bipolarly charged aerosol particles. *Journal of Aerosol Science*, 26(s1): s21–s22.

Higashi, M., Uchida, S., Suzuki, N. et al. 1992. Soot elimination and NO_x and SO_x reduction in diesel-engine exhaust by a combination of discharge plasma and oil dynamics. *IEEE Transactions on Plasma Science*, 20(1): 1–12.

Huang, Y.F., Yan, H., Li, S.R. et al. 2009. Multi-electrode electrohydraulic discharge for sterilization and disinfection. In *2009 IEEE Pulsed Power Conference*, June 28–July 2, Washington, DC, pp. 883–888.

Jeong, J. and Jurng, J. 2007. Removal of gaseous elemental mercury by dielectric barrier discharge. *Chemosphere*, 68(10): 2007–2010.

Kim, H.-H. 2004. Nonthermal plasma processing for air-pollution control: A historical review, current issues, and future prospects. *Plasma Processes and Polymers*, 1(2): 91–110.

Kim, H.-H., Kobara, H., Ogata, A. et al. 2005. Comparative assessment of different non-thermal plasma reactors on energy efficiency and aerosol formation from the decomposition of gas-phase benzene. *IEEE Transactions on Industrial Applications*, 41(1): 206–214.

Ko, K.B., Byun, Y., Cho, M., et al. 2008. Influence of HCl on oxidation of gaseous elemental mercury by dielectric barrier discharge process. *Chemosphere*, 71(9): 1674–1682.

Kogelschatz, U. 2003. Dielectric-barrier discharges: Their history, discharge physics, and industrial applications. *Plasma Chemistry and Plasma Processing*, 23(1): 1–46.

Kuroki, T., Takahashi, M., Okubo, M. et al. 2002. Single-stage plasma-chemical process for particulates, NO_x and SO_x simultaneous removal. *IEEE Transactions on Industry Applications*, 38(5): 1204–1209.

Lerouge, S., Fozza, A.C., Wertheimer, M.R. et al. 2000. Sterilization by low-pressure plasma: The role of vacuum-ultraviolet radiation. *Plasmas and Polymers*, 5(1): 31–46.

Li, X.Y., Zhang, X.M., Zhu, J.B. et al. 2012. Sensitivity analysis on the maximum ash resistivity in terms of its compositions and gaseous water concentration. *Journal of Electrostatics*, 70(1): 83–90.

Liang, X., Jayaram, S., Looy, P. et al. 1998. Pulse corona application for the collection of mercury from flue gas. In *Conference Record of the 1998 IEEE International Symposium on Electrical Insulation*, June 7–10, Arlington, VA, pp. 521–524.

Liang, X., Looy, P.C., Jayaram, S. et al. 2002. Mercury and other trace elements removal characteristics of DC and pulse-energized electrostatic precipitator. *IEEE Transactions on Industry Applications*, 38(1): 69–76.

Lisitsyn, I.V., Nomlyama, H., Katsuki, S. et al. 1999. Thermal processes in a streamer discharge in water. *IEEE Transactions on Dielectrics and Electrical Insulation*, 6(3): 351–356.

Locke, B.R., Sato, M., Sunka, P. et al. 2006. Electrohydraulic discharge and nonthermal plasma for water treatment. *Industrial and Engineering Chemical Research*, 45: 882–905.

Lukes, P., Clupek, M., Sunka, P. et al. 2005. Degradation of phenol by underwater pulsed corona discharge in combination with TiO_2 photocatalysis. *Research on Chemical Intermediates*, 31: 285–294.

Luo, H.L., Sheng, J. and Wan, Y.Z. 2003. Plasma polymerization of styrene with carbon dioxide under glow discharge conditions. *Applied Surface Science*, 253(12): 5203–5205.

Masuda, S. 1988. Pulse corona induced plasma chemical process: A horizon of new plasma chemical technologies. *Pure and Applied Chemistry*, 60(5): 727–731.

Masuda, S. and Hirano, M. 1981. Enhancement of electron beam denitration process by means of electric field. *Radiation Physics and Chemistry*, 25: 223.

Matteson, M.J., Stringer H.L. and Busbee W.L. 1972. Corona discharge oxidation of sulfur dioxide. *Environmental Science and Technology*, 6(10): 895–901.

Mok, Y.S., Demidyuk, V. and Whitehead, J.C. 2008. Decomposition of hydrofluorocarbons in a dielectric-packed plasma reactor. *Journal of Physical Chemistry A*, 112(29): 6386–6591.

Mountouris, A., Voutsas, E. and Tassios, D. 2006. Solid waste plasma gasification: Equilibrium model development and exergy analysis. *Energy Conversion and Management*, 47(13–14): 1723–1737.

Nair, S.A. 2004. Corona plasma for tar removal. Ph.D. Dissertation. Eindhoven: TU/e.

Nair, S.A., Yan, K., Pemen, A.J.M. et al. 2004. Tar removal from biomass-derived fuel gas by pulsed corona discharges. A chemical kinetic study. *Industrial and Engineering Chemistry Research*, 43(7): 1649–1658.

Neeft, J.P.A. 2002. Physical removal of tar aerosol from biomass producer gases by ESP and RPS. In *Proceedings of the European Conference on Biomass for Energy, Industry and Climate Protection*, June 17–21, Amsterdam, The Netherlands, pp. 31–40.

Oda, T. 2003. Non-thermal plasma processing for environmental protection: Decomposition of dilute VOCs in air. *Journal of Electrostatics*, 57(3–4): 293–311.

Okubo, M., Arita, N., Kuroki, T. et al. 2008. Total diesel emission control technology using ozone injection and plasma desorption. *Plasma Chemistry and Plasma Processing*, 28(2): 173–187.

Paasen, S.V.B and Rabou. 2004. Tar removal with wet ESP: Parametric study. In *Proceedings of the 2nd World Conference and Technology Exhibition on Biomass for Energy, Industry and Climate Protection*, May 10–14, Rome, Italy, pp. 205–210.

Palumbo, F.J. and Fraas, F. 1971. Removal of sulfur from stack gases by an electric discharge. *Journal of the Air Pollution*, 21(3): 143–144.

Paris, P., Laan, M., Mirme, A. et al. 1996. Corona discharge as a generator of aerosols. In *Proceedings of HAKONE V*, September 2–4, Brno, pp. 185–189.

Parissi, L., Odic, E., Goldman, A. et al. 2000. Temperature effects on plasma chemical reactions application to VOC removal from flue gases by dielectric barrier discharges. In *Electrical Discharges for Environmental Purposes, Fundamentals and Applications*, ed. E.M. van Veldhuizen, 279–314. New York: NOVA Science Publishers Incorporation.

Paul, E.M. and Thomas T.M. 1964. A point-to-plane electrostatic precipitator for particle size sampling. *American Industrial Hygiene Association Journal*, 25(1): 8–14.

Rajesh, D. and Mark, J.K. 2002. Repetitively pulsed plasma remediation of NO_x in soot laden exhaust using dielectric barrier discharges. *Journal of Physics D: Applied Physics*, 35(22): 2954–2968.

Rosocha, L.A. and Korzekwa, R.A. 1999. Advanced oxidation and reduction processes in the gas phase using non-thermal plasmas. *Journal of Advanced Oxidation Technologies*, 4: 247–264.

Saiyasitpanich, P., Keener, T.C. and Lu, M. 2006. Collection of ultra-fine diesel particulate matter (DPM) in cylindrical single-stage wet electrostatic precipitators. *Environmental Science and Technology*, 40(24): 7890–7895.

Schiorlin, M., Marotta, E., Rea, M. et al. 2009. Comparison of toluene removal in air at atmospheric conditions by different corona discharges. *Environmental Science and Technology*, 43(24): 9386–9392.

Schuster, G., Löffler, G., Weigl, K. et al. 2001. Biomass steam gasification—An extensive parametric modeling study. *Bioresource Technology*, 77(1): 71–79.

Sun, B., Sato, M., Harano, A. et al. 1998. Non-uniform pulse discharge-induced radical production in distilled water. *Journal of Electrostatics*, 43:115–126.

Tas, M.A. 1995. Plasma induced catalysis: A feasibility study and fundamentals. Ph.D. Dissertation. Eindhoven: TU/e.

Tas, M.A., van Hardeveld, R. and van Veldhuizen, E.M. 1997. Reactions of NO in a positive streamer corona plasma. *Plasma Chemistry and Plamsa Processing*, 17: 371.

Thomas, S.E., Martin, A.R., Raybone, D. et al. 2000. Non thermal plasma aftertreatment of particulate—Theoretical limits and impact on reactor design. *SAE Paper*. 2000-01-1926.

Toshiaki, Y. 1997. VOC decomposition by nonthermal plasma processing—A new approach. *Journal of Electrostatics*, 42(1–2): 227–238.

Urashima, K. and Chang, J.S. 2000. Removal of VOC from air streams and industrial flue gases. *IEEE Transactions on Dielectrics and Electrical Insulation*, 7 (5): 602–614.

Ushakov, V.Y. 2007. *Impulse Breakdown of Liquids*. Berlin: Springer.

van Durme, J., Dewulf, J., Leys, C. et al. 2008. Combining non-thermal plasma with heterogeneous catalysis in waste gas treatment: A review. *Applied Catalysis B: Environmental*, 78(3–4): 324–333.

van Heesch, E.J.M. and van der Laan, P.C.T. 2000a. Power sources for electrical discharges. In *Electrical Discharges for Environmental Purposes, Fundamentals and Application*, ed. E.M. van Veldhuizen, 118–155. New York: NOVA Science Publishers Incorporation.

van Heesch, E.J.M., Pemen, A.J.M., Yan, K. et al. 2000b. Pulsed corona tar cracker. *IEEE Transactions on Plasma Science*, 28(5): 1571–1575.

van Heesch, E.J.M., Yan, K., Pemen, A.J.M. et al. 1999. Matching between a pulsed power source and corona reactor for producing non-thermal plasma. *14th International Symposium on Plasma Chemistry*, 1063–1068.

van Veldhuizen, E.M., Rutgers, W.R., Bityurin, V.A. et al. 1997. Multi-phase modeling of flue gas cleaning in pulse corona reactor. In *Proceedings of XII International Conference on Gas Discharge and Their Applications*, September 8–12, Greifswald, Germany, pp. 401–406.

Winands, G.J.J., Liu, Z., Pemen, A.J.M. et al. 2008. Analysis of streamer properties in air as function of pulse and reactor parameters by ICCD photography. *Journal of Physics D: Applied Physics*, 41(23): 234–244.

Xu, F., Luo, Z.Y., Cao, W. et al. 2009. Simultaneous oxidation of NO, SO_2 and Hg^0 from flue gas by pulsed corona discharge. *Journal of Environmental Sciences*, 21(3): 328–332.

Yamamoto, T. and Jang, B.W.L. 1999. Aerosol generation and decomposition of CFC-113 by the ferroelectric plasma reactor. *IEEE Transactions on Industrial Applications*, 35(4): 736–742.

Yan, K. 2001. Corona plasma generation. Ph.D. Dissertation. Eindhoven: TU/e.

Yan, K., Hui, H.X., Cui, M. et al. 1998. Corona induced non-thermal plasmas: Fundamental study and industrial applications. *Journal of Electrostatics*, 44(1–2): 17–39.

Yan, K., Kanazawa, S., Ohkubo, T. et al. 1999a. Streamer-glow-spark transition in a corona radical shower system during NO_x removal operation. *Transactions of the Institute of Electrical Engineers of Japan. A*, 119-A: 938–944.

Yan, K., Kanazawa, S., Ohkubo, T. et al. 1999b. Oxidation and reduction processes during NO_x removal with corona-induced nonthermal plasma. *Plasma Chemistry and Plasma Processing*, 19(3): 421–443.

Yan, K., Li, R., Zhu, T. et al. 2006. A semi-wet technological process for flue gas desulfurization by corona discharges at an industrial scale. *Chemical Engineering Journal*, 116: 139–147.

Yan, K., van Heesch, E.J.M., Pemen, A.J.M. et al. 2001a. From chemical kinetics to streamer corona reactor and voltage pulse generator. *Plasma Chemistry and Plasma Processing*, 21: 107–137.

Yan, K., van Heesch, E.J.M., Pemen, A.J.M. et al. 2001b. A 10 kW high-voltage pulse generator for corona plasma generation. *Review of Scientific Instruments*, 72: 2443–2447.

Yan, K., Yamamoto, T., Kanazawa, S. et al. 1999c. Control of flow stabilized positive corona discharge modes and NO removal characteristics in dry air by CO_2 injections. *Journal of Electrostatics*, 46: 207–219.

Yan, K., Yamamoto, T., Kanazawa, S. et al. 2001c. NO removal characteristics of a corona radical shower system under DC and AC/DC superimposed operations. *IEEE Transactions on Industry Applications*. 37(5): 1499–1504.

Yao, S., Suzuki, E. and Nakayama, A. 2001. Oxidation of activated carbon and methane using a high-frequency pulsed plasma. *Journal of Hazardous Materials*, 83(3): 237–242.

Zhang, X.M. 2011. Removal of volatile organic compounds by non-thermal plasma technology. Ph.D. Dissertation. Hangzhou, China: Zhejiang University.

Zhang, X.M., Zhu, J.B., Li, X. et al. 2011. Characteristics of styrene removal with an AC/DC streamer corona plasma system. *IEEE Transactions on Plasma Science*, 39(6): 1482–1488.

Zheng, C., Xu, Y.Z., Huang, H.M. et al. 2012. Water disinfection by pulsed atmospheric air plasma along water surface. *AIChE Journal*, DOI: 10.1002/aic.13929.

Zhu, J.B., Zhang, X.M., Chen, W.L. et al. 2010. Electrostatic precipitation of fine particles with a bipolar pre-charger. *Journal of Electrostatics*, 68(2): 174–178.

Zukeran, A., Ikeda, Y., Ehara, Y. et al. 2000. Agglomeration of particles by ac corona discharge. *Electrical Engineering in Japan*, 130(1): 30–36.

12

Assessment of Potential Applications of Plasma with Liquid Water

Peter J. Bruggeman
and Bruce R. Locke

12.1 Introduction

The analysis and development of electrical discharge plasma formed either inside a liquid water phase or in a gas phase contacting the liquid is of increasing importance due to the production of highly reactive radicals (e.g., ·OH) and molecules (e.g., H_2O_2), the generation of strong UV emission, and the production of shockwaves [1,2].

Such plasmas have been, and are continuing to be, investigated for medical, environmental, chemical, material functionalization, and synthesis applications [3–5]. Two very successful medical applications are the plasma scalpel [6], which uses the plasma for tissue ablation, and the spark discharge to produce focused shockwaves for lithotripsy [7]. In these two medical applications the energy efficiency is not a critical factor because the total cost is not dominated by power requirements, and other factors such as control, safety, and medical considerations are critical. In contrast, in applications that process large volumes of fluids or other materials the energy cost may be of utmost importance. For example, disinfection of liquids for medical and food processing and the destruction of toxic or harmful organic compounds in liquids for pollution control require high efficiency (as well as high chemical conversion and yield) with minimal operating costs [3–5]. In addition, for the production of commodity (large amounts of relatively inexpensive) chemicals, such as H_2 and H_2O_2, the energy costs should also be reduced as much as possible while achieving production rates and efficiencies as high as possible with performance that matches or exceeds competitive technologies. Other environmental factors such as emissions of global warming compounds should also be considered. In contrast, for high-value (and/or small quantity) products, such as nanoparticles [8], the energy costs may not be significant.

In this chapter, we will discuss four representative cases of applications of plasma with liquid water where the energy efficiency is crucial. These cases include the following:

1. H_2O_2 production
2. H_2 production
3. Destruction of organic chemical species
4. Microbial disinfection

We review published literature concerning the conversion, removal efficiencies, or inactivation amounts, and, where possible, make comparisons with competing technologies. We do not focus in detail on chemical reaction mechanisms, which have been discussed in other recent reviews [3–5]. While preliminary conclusions will be drawn about the competitiveness of the various techniques for the above applications, it is important to note that the data obtained for these comparisons come from a wide range of different studies conducted, in some cases, under very different operating conditions. Therefore, plasma processes that have efficiencies that compare to conventional methods by less than 1 order of magnitude warrant more detailed analysis and study, while processes that are more than 1 or 2 orders of magnitude less efficient than the best available technology, including other plasma processes, may not warrant further development and study, at least for volume processing applications. Detailed economic analysis that includes all operating, capital, and equipment costs are needed, and, to our knowledge, no such detailed analysis has been reported in the literature for plasma with liquids. Depending upon the system, in some cases capital costs, or other factors, can outweigh the operating cost. Some reports on economic analysis of gas-phase plasma, for example, can be referred to the works of Bromberg et al. [9]. In addition, commercial feasibility requires assessment of the potential for system scale-up and analysis of factors such as durability and maintenance-free long-term performance. These issues are outside the scope of this review, and in any case limited work has been reported on these topics.

Typical measures of performance of electrically driven chemical processes are the energy yields given in either grams per kilowatt-hour or electronvolt per molecule, and the electrical energy per order (EEO) of magnitude reduction is often used for many chemical oxidation or degradation processes (particularly those spanning a number of orders of magnitude in degradation) [10] and the similar D value for microbiological inactivation (see following sections). In this review, we will focus on energy yields for the formation of H_2 and H_2O_2 and chemical degradation. We will focus on EEO values for selected results of chemical destruction and D values for biological inactivation given that biological inactivation generally spans many orders of magnitude and most papers, at best, report only 1–2 orders of magnitude degradation. While most studies report energy delivered by the plasma, it is usually not possible to determine the efficiency of the power supply from a published source; clearly, more work is needed to develop high-efficiency power supplies and to match the plasma to the power supply [11].

Because water has a relatively high heat capacity (4.2 J g^{-1} K^{-1}) and a high heat of evaporation, that is, 2257 kJ kg^{-1} at 100°C [12], it is very important that the energy input be efficiently coupled to the plasma and that the plasma energy not be used excessively for heating and evaporating water. These thermal effects cannot be completely eliminated because plasma temperatures generally exceed the boiling point of water, and certainly one of the essential features of a plasma contacting a liquid is the initiation of reactions between the plasma-generated species and the compounds from the liquid, including of course H_2O. Thus, analysis of the energy and mass transfer as well as chemical reactions at the *interface* between the plasma and the liquid are very important and require further study [13,14]. In addition, the energy coupled to the plasma should be directed toward dissociation and production of active radicals and other species rather than to gas heating. The defining feature of a nonthermal plasma is the high electron temperature relative to the gas temperature whereby electron collisions with ambient temperature gas species control the subsequent chemistry, while in thermal plasma all species are near thermodynamic equilibrium [15,16]. Arcs and sparks are thermal plasmas, while gas-phase dielectric barrier and pulsed corona are examples of nonthermal plasmas. The reactor geometry and the plasma generation methods vary widely and can strongly affect the performance. See Figure 12.1 for a few reactor and

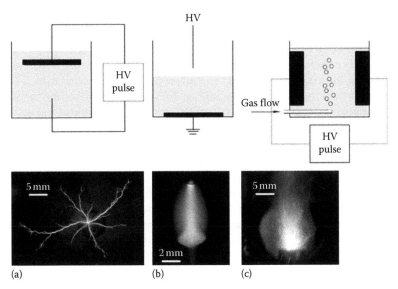

FIGURE 12.1 An example of reactor geometry and image of (a) a corona-like discharge in water, (b) a glow discharge in the gas phase above a water electrode, and (c) a discharge generated in a bubble. (Reproduced with permission from Samukawa, S., et al., *J. Phys. D Appl. Phys.*, 45, 2012. © IOP.)

electrode geometry examples, and for a more detailed survey of reactor geometries refer to the works of Bruggeman and Leys [1], Locke et al. [2], and Locke et al. [3].

In Section 12.2, different types of discharges in and in contact with liquids and their plasma parameters are summarized. The synthesis of H_2O_2 and H_2 from water are thereafter reviewed, followed by an overview of organic chemical removal efficiencies and analysis of bacterial inactivation. Finally, issues to improve energy efficiency are discussed, and suggestions for future research are made.

12.2 Classification of Discharges and Corresponding Plasma Properties

Three general groups of discharges with liquids can be classified based upon the plasma-phase distribution [1,2]:

1. Plasma formed directly in the liquid phase with a streamer-like discharge or arc/spark discharges
2. Gas-phase plasma that contacts with a liquid reservoir where the liquid can function as an electrode (in some cases this type of discharge is combined with the previous type)
3. Multiphase plasma formed by spraying liquids into the gas phase or the inverse case where gases are bubbled through the liquid. This group includes discharges such as diaphragm and capillary discharges, which are created in vapor voids caused by local Ohmic heating.

Schematic diagrams of each of these three major types of discharges are shown in Figure 12.1. There is an extremely wide diversity of combinations of different electrodes, power supplies, gas–liquid contacting methods, and reactor volumes that have been studied [1–3]. Because every group in the world working in the field of plasmas with liquids has its own plasma reactor, there are no standard reactors used by all. While these differences demonstrate the creativity in this field and the desire to continue to improve the systems, the lack of a specific control and the lack of consistently reported experimental conditions make it very difficult to make comparisons across all of the published work. There is thus a continuing need for all researchers to report as much detail about the characteristics of their plasma, the properties of the power supplies and reactors, and the solution conditions. For more details the reader is referred to the reviews of Locke and Bruggeman [1,2].

Plasmas belonging to one of the three specific groups mentioned above can have very large differences in gas temperature and electron density and temperature, and these factors can lead to orders of magnitude differences in radical densities, UV emissions, and the occurrence of shockwaves. The reactivity of the plasma strongly depends on input power, excitation voltage (DC, AC, MW, RF, and pulsed DC), the gas composition (where gas is used), and the liquid properties (e.g., conductivity, pH, and additives such as salts, ions, particles from the electrodes, and catalysts). Some typical plasma parameters for various plasma processes are summarized in Table 12.1. For example, in the case of corona-like discharges in water there is a large range of temperatures and densities. This range is often related to the excitation voltage properties (e.g., the pulse width) and the amount of deposited energy per pulse. Indeed, in corona discharge by DC excitation, electron densities of about 10^{21} m^{-3} and gas temperatures of 1600 K have been found [17], while in microsecond pulsed discharges electron densities up to 10^{25} m^{-3} and gas temperatures of about 5000 K have been found [18]. Remarkably, the H_2O_2 production rate in the liquid phase of these two extreme cases is identical and on the order of 1 g kWh^{-1}. This may be due to the limitations of the chemical transfer and reactions at the plasma–liquid interface [14]. It has been suggested, based upon gas-phase plasma studies [19,20], that radical recombination and quenching at the steep boundary transition regions between the plasma and liquid (or surrounding environment) can control the rates of H_2O_2 production, and therefore, the efficiencies; in such a case the properties (e.g., temperature) inside the plasma are not critical as long as sufficient dissociation of water occurs.

Nevertheless, it is very important when describing applications to characterize the plasma by power measurements, gas temperatures, and electron density measurements. Gas temperatures can often be obtained from rotational spectra of OH(A–X) or N_2(C–B), although the nonequilibrium rotational distributions of the OH(A) often have to be considered in the temperature determination [17,21]. Electron densities can be easily obtained by optical emission spectroscopy in most water-containing plasmas due to their relatively large electron density (mostly $>10^{20}$ m^{-3}) and the presence of hydrogen Balmer lines in the emission spectrum, which allow accurate Stark broadening measurements even with moderate resolution spectrometers; see Figure 12.2 for an example of H_α line broadening as a function of solution conductivity for a streamer-like discharge directly in water [7,17]. Note that the large range in electron temperature of the different discharges shown in Table 12.1 is often due to the fact that T_e is difficult to measure in high-pressure plasmas.

TABLE 12.1   Overview of the Typical Gas Temperature (T_g), Electron Temperature (T_e), and Electron Density (n_e) Range in Which the Main Plasma Types Are Operating

Plasma	T_g (K)	T_e (eV)	n_e (m^{-3})	Radical Densities	UV	Shock Waves
Corona-like in liquid water	1,500–7,000	1–10	10^{21}–10^{25}	+	+	+
Capillary/diaphragm in liquid water	500–3,000	2–10	10^{20}–10^{21}	+	+	(+)
Diffuse glow-like	300–1,000	1–4	10^{16}–10^{19}	+	+	−
Filamentary DBD-like	300–500	2–5	10^{20}–10^{21}	+	++	−
Pulsed corona (gas phase)	300–500	2–10	10^{20}–10^{21}	+	+	−
Spark*	500–5,000	1–3	10^{20}–10^{24}	++	++	+++
MW	500–5,000	1–3	10^{20}–10^{22}	++	++	−
Arcs	3,000–20,000	~1	10^{23}–10^{25}	+++	++++	++++
Plasma jets (cold)	300–600	1–10	10^{17}–10^{21}	+	+	−
Gliding arc	2,500–10,000	1–2	10^{17}–10^{19} (averaged)	++	++	−

Notes: 1 eV equals approximately 11,600 K.

Deviations from the above shown typical parameters exist. For each discharge the relative importance of the production of radicals, UV, and shock waves are presented. (For more details the reader is referred to the works of Bruggeman, P. and C. Leys, *J. Phys. D Appl. Phys.*, 42, 1–28, 2009 as well as Chapter 2 of this book.)

*As the spark is transient it bridges the gap of plasma properties between the streamer and the arc. When not stated explicitly the discharge is assumed to be in the gas phase. For both MW and arcs the given properties are also valid in liquids.

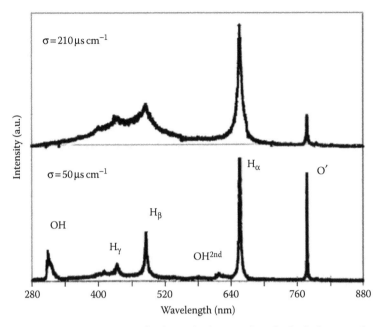

FIGURE 12.2 Typical emission spectrum of a direct discharge in liquid. The hydrogen Balmer lines and the OH(A–X) emission is clearly visible. The spectrum is presented for two different solution conductivities. The line broadening of the hydrogen Balmer lines significantly increases with increasing solution conductivities, which can be correlated to an increase in electron density (after correction for the pressure broadening). (Reproduced with permission from Sunka, P., et al., *Plasma Sources Sci. Technol.*, 8, 258–265, 1999. © IOP.)

Table 12.1 also shows the relative importance of radical concentrations, UV emission, and shockwave production. Because radical densities often scale with the electron density, both higher gas temperature and electron density can lead to large radical (e.g., ·OH) densities due to thermal dissociation and electron dissociation reactions:

$$H_2O \rightarrow \cdot OH + \cdot H, \qquad\qquad R1$$

$$H_2O + e^- \rightarrow \cdot OH + \cdot H + e^-. \qquad\qquad R2$$

The electron-induced dissociation rate of water is strongly dependent on the electron temperature, which can cause the ratio of ·OH to electron density, n_{OH}/n_e, in plasmas to vary significantly. For example, average ·OH density in an radio frequency (RF)-excited cold diffuse glow discharge at atmospheric pressure in a He–H_2O mixture is about 10^{20} m^{-3}, while the electron density is on the order of 10^{17} m^{-3} [22]. In a plasma filament, produced by a nanosecond voltage pulse, which has a peak electron density between 10^{22} and 10^{23} m^{-3}, the ·OH density is of the same order of magnitude [23]. Although the radical density is the highest for the high-density plasma, the n_{OH}/n_e ratio is 3 orders of magnitude larger for the low-density plasma. In the latter case the maximum ·OH density occurs in the early afterglow when the electron temperature is low. It has been shown by Verreycken et al. [23] that the ·OH production can be explained by recombination reactions such as the following:

$$H_2O^+ + H_2O \rightarrow H_3O^+ + \cdot OH, \qquad\qquad R3$$

$$H_3O^+ + e^- \rightarrow \cdot OH + 2H. \qquad\qquad R4$$

In the RF glow discharge the discharge can be considered to be continuous and the electron temperature will be 2–4 eV. In such a discharge electron-induced dissociation reactions of water are the dominant process for ·OH production [21,22].

It is thus very important to consider these different plasma properties in addition to the transport of radicals to the "treatment zone" or inversely the transport of target species to the plasma region. For example, in the case of H_2O_2 the high reactivity of such short-lived radicals as ·OH tends to limit its range of influence and ultimately determines the H_2O_2 production rate in the liquid phase as mentioned.

All plasmas containing water will have emission in the UV due to the strong band of OH(A–X) at 309 nm (see Figure 12.2). Normally the higher the electron density, the more broadband UV is produced by the plasma due to Brehmstrahlung or H_2 broadband UV emission. Plasmas that have particularly strong UV emission are spark and arc discharges. The UV emission is normally strongly related to the working gas. As is well known, typical low-pressure mercury gas discharge lamps are a rich source of UV emission at 254 nm. Also, dielectric barrier discharges (DBDs) are used to produce efficiently UV in noble gases due to the formation of excimers [24]. Excimers such as Ar_2^* yield broadband emission in vacuum UV around 125 nm. Often for applications when Ar is used, an admixture of water and/or air is present, which strongly quenches excimer formation [25]. This quenching strongly reduces excimer emission in cold plasma jets. In water the transparency of vacuum UV is limited, and when significant amounts of impurities are present UV transparency is also significantly reduced.

Only discharges formed directly in the liquid or in bubbles can produce shockwaves. The discharges that have the highest energy density (arcs and sparks) can produce the strongest shockwaves [7,26]. Streamer-like discharges produce shock fronts [18], while discharges produced in bubbles in the liquid such as in capillary or diaphragm-like discharges can produce jets and/or weaker shockwaves at the time of implosion of the bubble [27,28]. Gas-phase plasmas normally do not induce any shockwaves in the liquid, but they can induce liquid mixing. Three common situations where gas-phase plasma affects the liquid motion are (1) cold plasma jets or plumes impinging directly on the liquid surface, (2) ionic wind formed by, for example, DC corona discharges above a liquid surface, and (3) plasma–water interface dynamics and interfacial instabilities such as the formation of Taylor cones [29,30].

12.3　Chemical Synthesis of H_2O_2 and H_2 from Water

Analysis of the thermal equilibrium dissociation of water into H_2, H, O_2, O, and OH at 1 bar as a function of the temperature (Figure 12.3) shows a significant amount of water dissociation for $T > 2500$ K and that the maximum mole fraction of H_2 is about 20% at a temperature of 3400 K. As shown in Table 12.1, these high gas temperatures can easily be obtained in some plasma reactors. Equilibrium conversion can be shifted by removal of reaction products, and thus radical quenching and phase distribution can perhaps enhance formation of a desired species even in a plasma in or near equilibrium. Under nonequilibrium conditions in nonthermal plasmas, the kinetics of radical quenching and phase distribution can affect process efficiency [20]. In the case of H_2O_2, which is normally formed by ·OH recombination, high temperature leads to its thermal dissociation and a reduced net rate of formation. As shown in Reactions R5 and R6, two competing reactions affect the ·OH:

$$·OH + ·OH + M \rightarrow H_2O_2 + M, \hspace{4cm} R5$$

$$OH + ·OH \rightarrow H_2O + O. \hspace{4cm} R6$$

The branching ratio between these reactions depends on the gas temperature, which can strongly determine the efficiency of H_2O_2 production independently of the efficiency to produce ·OH. The second reaction causes ·OH to recombine to water, leading to a direct loss of energy. The reactions of H follow a similar pattern:

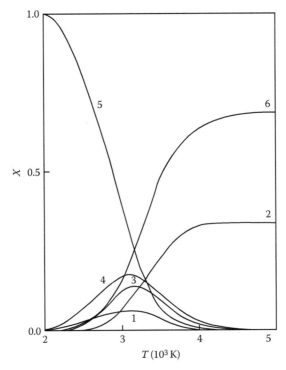

FIGURE 12.3 Equilibrium composition of water vapor as a function of temperature (1) O_2, (2) O, (3) OH, (4) H_2 (5) H_2O, and (6) H. (Reproduced with permission from Fridman, A., *Plasma Chemistry*, Cambridge University Press, Cambridge, 2008. © Cambridge University Press.)

$$H + H + M \rightarrow H_2 + M, \qquad \text{R7}$$

$$H + OH + M \rightarrow H_2O + M. \qquad \text{R8}$$

While the argument is somewhat simplified because there are generally many more reactions in the plasma, clearly efficient H_2O_2 and H_2 production require not only efficient dissociation of water (R1 and R2) but also controlled recombination toward the desired reaction products (R5 through R8). More details of the reactions can be found in the literature [14,20,21,31–34]. The efficiency of H_2O_2 and H_2 production by plasmas interacting with water are separately discussed in the next two subsections.

12.3.1 H_2O_2 Production

Worldwide, 4 metric tons/year of H_2O_2 is produced (US$3 billion) for a wide range of applications in chemical production [e.g., as a bleaching agent in pulp and paper manufacturing (60%)], disinfection and cleaning [in food, pharmaceutical, medical settings, and electronics (5%–10%)], and pollution control [as an oxidant in pollution treatment (–12%)]. The prevailing industrial method for making hydrogen peroxide is based upon the anthraquinone process, which utilizes hydrogen (H_2) from natural gas (CH_4) to hydrogenate anthraquinone followed by oxidation with oxygen from air. The resulting product requires large-scale countercurrent extraction and purification steps. This process is economical only at the very large scale (>1000 metric tons/year).

While the overall thermodynamics of the H_2/O_2 reaction

$$H_2 + O_2 \rightarrow H_2O_2 \qquad \text{R9}$$

to produce H_2O_2 via anthraquinone are favorable (exothermic), natural gas and H_2 production can add 35% to the total cost, and generate significant quantities of CO_2 with an inherent adverse environmental impact. While the advertised shipping costs for industrial-grade commodity H_2O_2 are quite modest and benefit from high-volume production, the shipping costs for smaller quantities of high-purity H_2O_2 can be relatively high. Environmental impact costs are rarely, if ever, considered in the shipping price (as well as in the overall manufacturing and storage costs). However, 7.4% of greenhouse gas emissions within the chemical industry are directly attributable to transportation. In industries that demand high-purity H_2O_2 (such as in food, medical, and electronics industries) the purification costs can also be quite high. The variation in prices of H_2O_2 in small amounts can be large, but it is relatively cheap. The price ranges from 1 g kWh^{-1} for retail amounts of higher purity to 130 g kWh^{-1} for commodity-scale high concentrations with low purity assuming \$0.1 kWh^{-1} [32]. A technology for smaller-scale local production that does not rely on natural gas would generate significant interest. The development of highly efficient small-scale (<500 metric tons/year) processes to produce H_2O_2 from water and air or just water can potentially lead to significant savings in terms of reducing the carbon footprint/environmental impact through improved manufacturing and reduced shipping and storage costs (refrigeration is required for storage) by providing onsite and on-demand sources of this chemical. The global H_2O_2 market is growing at over 8% annually in part due to the fact that it is considered a "green" chemical in its application, and therefore much more attention should be made to produce H_2O_2 in sustainable ways.

In contrast to the conventional method discussed above, most plasma processes aim to produce H_2O_2 directly from water. The thermodynamic limit of the endothermic reaction to produce H_2O_2 from water is 400 g kWh^{-1}:

$$2H_2O \rightarrow H_2O_2 + H_2, \quad \Delta H = 3.2 \text{ eV molecule}^{-1}\left(400\,\text{g}\,\text{kWh}^{-1}\right). \qquad \text{R10}$$

Table 12.2 summarizes the literature on the methods to form H_2O_2 by water-containing plasma and more details are given in the works of Locke and Shih [32]. Discharges directly in the liquid phase tend to be less efficient than those that are formed in a gas phase that contacts the liquid (e.g., over water, in bubbles, or with water spray). For example, discharges in bubbles inside liquid water tend to more efficiently produce H_2O_2 compared to direct liquid spark, capillary, diaphragm, and even corona discharges. The energy needed to form a plasma directly in the liquid is much higher than that in a gas phase and typically in liquids much of the energy is expended in mechanical (pressure–volume expansion and shockwaves) and thermal (heating) processes [14]. There are many other factors, including multiple chemical reactions with additives or materials sputtered from the electrodes, that affect H_2O_2, and we refer to the recent review for more details [32].

As shown in Table 12.2, the spread in data is rather small within each discharge type except for gliding arc discharges and gas-phase corona and DBD above water. In the case of the gliding arc discharge with water droplets sprayed into the plasma, it was found that low-power pulsed gliding arc discharges produce H_2O_2 significantly more efficiently than their low-frequency and high power AC or DC counterparts. This is due to (1) the steep decrease in gas temperature from the dissociation zone in the arc toward the downstream recombination zone where H_2O_2 is produced, (2) the low power that prevents excessive heating of the gas and liquid, and (3) the insufficient energy input to vaporize the water droplet. Large temporal and spatial gradients allow freezing the chemistry during recombination and reducing the residence time of H_2O_2 in the hot plasma zone, thus reducing the thermal or radical dissociation of the formed H_2O_2 as a consequence. The (liquid-phase) water droplets can serve as collectors for the highly soluble H_2O_2, thereby protecting them from degradation by the (gas-phase) plasma. An additional advantage with a low-power gliding arc is that the water droplets injected into the discharge can increase the amount of water entering the plasma zone significantly compared to the saturation vapor pressure at room temperature.

TABLE 12.2 Summary of the Efficiency and Production Rate of Reported H_2O_2 Production

	Input	Generation Rate (g h^{-1})	Energy Efficiency (g kWh^{-1})
Spark	Liquid water	~8.6 10^{-2}	0.43–0.55
Pulsed corona	Liquid water	0.01–0.21	0.96–3.64
Capillary/diaphragm discharge	Liquid water	0.02–0.36	0.1–1
Contact glow discharge electrolysis	Liquid water	0.03–0.64	0.8–1.6
RF/MW discharge	Liquid water	~0.1	0.46–0.64
Discharges in bubbles	Air/Ar/O_2 in liquid H_2O	2.3×10^{-3}–26	0.4–8.4
Hybrid discharges	Water + air	2.7×10^{-3}–6.6×10^{-2}	0.37–1
Gas phase corona discharges	Air/Ar + water surface	5.7×10^{-5}–2.5×10^{-2}	0.13–5
DBD above liquid	Air/water surface	2.5×10^{-4}–0.12	0.04–2.7
MW	Steam	48	24
DBD	Humid gas	1.8×10^{-3}–1.6×10^{-2}	1.14–1.7
	H_2–O_2 mixtures (note different chemistry)	0.15–2.8	12.5–80
Gliding arc	Water droplets (in Ar)	0.02–0.14	0.57–80
Electron beam			8.9
Ultrasound	Dissolved gas + liquid water	1.2–9.8×10^{-3}	0.01–0.12
Vacuum UV	Vapor or liquid water		13–33
Electrolysis			112.4–227.3

Source: Locke, B. R. and K.-Y. Shih, *Plasma Sources Sci. Technol.*, 20, 034006, 2011.

The basic formation steps of H_2O_2 production in the gas phase are given by R1, R2, and R5. This pathway for H_2O_2 (R1, R2, and R5) resembles the reaction mechanism of ozone in DBD reactors very closely:

$$O_2 + e^- \rightarrow O + O + e^-, \hspace{3cm} \text{R11}$$

$$O_2 + O + M \rightarrow O_3 + M. \hspace{3cm} \text{R12}$$

Both reaction sets (for O_3 and H_2O_2) are favored at high electron temperature, low gas temperatures, and high pressures. Ozone generation in DBD reactors is extremely efficient, and it is a well-established commercialized technology [24]. It can be noted that after 100 years of optimizing gas-phase ozone production, commercial energy yields are about 12% of thermodynamic efficiency, but lab-scale demonstrations have reached 44% [32]. In the case of H_2O_2, as shown in Table 12.2, laboratory demonstrations have reached 20% of the thermodynamic efficiency. A key difference with commercial ozone generation compared to H_2O_2 generation is that there are no viable competitive technologies for ozone production, while there are several for H_2O_2. Nevertheless, H_2O_2 from pure water could be of significant environmental benefit, particularly for small-scale applications.

Alkaline electrolysis produces H_2O_2 from water (H_2O) and O_2 through an endothermic reaction, and the laboratory energy yields can be quite high, but this process requires a highly alkaline solution. In such a system the feed gas must not contain any acid gases (e.g., CO_2 present in air), which can lower the pH. There are other electrochemical methods that produce H_2O_2 over a range of pH values, but these processes have not been widely commercialized, nor have they penetrated the market for commodity H_2O_2. The production of H_2O_2 (and H_2) directly from H_2O (in the absence of O_2) has a larger ΔH than from that from H_2O/O_2, but it is advantageous because of the simultaneous production of H_2 and there is no requirement for pure oxygen.

While the conventional anthraquinone production method relies on Reaction R9, other catalytic and plasma processes have also sought to utilize this reaction. In plasma without liquids, Zhou et al. [35,36]

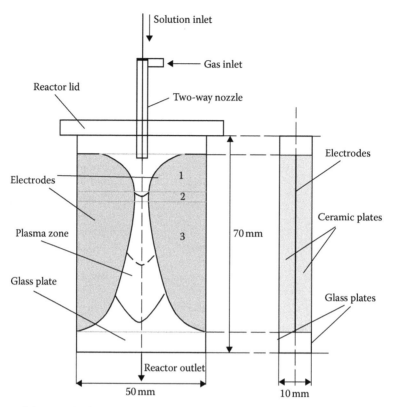

FIGURE 12.4 Gliding arc discharge reactor with water spray. Carrier gas is injected with a liquid through a nozzle at the top of the reactor and the resulting aerosol flows through between the electrodes where the plasma is formed. (Reproduced with permission from Burlica, R. and B. R. Locke, *IEEE Trans. Ind. Appl.*, 44, 482–489, 2008. © 2008 IEEE.)

have shown that with a nonexplosive mixture of H_2 and O_2, energy yields of up to 80 g kWh^{-1} (based on electrical discharge energy) can be reached. It should be noted that plasma energy is used to drive an exothermic reaction and that the feed components H_2 and O_2 have to be supplied. We will discuss H_2 production in Section 12.3.2, but any comparison of Reactions R9 with R10 require consideration of the cost of the feed components and value of the products. One should thus be cautious in comparing energy yields for different reaction processes.

In summary, the highest efficiency reactor for producing H_2O_2 from liquid water utilized low-power pulses with water droplets sprayed into the reactor (see Figure 12.4). The system requires clean water and a carrier gas (Ar that could be recycled). Further development of processes that can efficiently derive H_2O_2 from H_2O and avoid significant purification or extraction steps, that can operate over a wide range of scales, that reduce the emissions of global warming compounds from transportation and storage requirements, and that have efficiencies comparable to or better than current manufacturing and distribution costs may lead to wider utilization of onsite (plasma) production methods for H_2O_2.

12.3.2 H_2 Production

The major large-scale current use of H_2 is in chemical synthesis (e.g., ammonia, methanol); however, the potential to use H_2 as a fuel in mobile and stationary applications is stimulating the development of highly efficient and environmentally benign H_2 production processes with much emphasis on fuel cell applications. The standard method to produce H_2 is by the steam methane reforming of natural gas:

$$CH_4 + H_2O \rightarrow CO + 3H_2. \hspace{3cm} \text{R13}$$

This reaction is complemented by the reversible water shift gas reaction:

$$CO + H_2O \leftrightarrow CO_2 + H_2. \hspace{3cm} \text{R14}$$

Brown estimated [37], based upon the enthalpy input needed to form H_2, the theoretical energy yields for methane (95 g kWh^{-1} or 0.98 eV molecule^{-1}), gasoline, diesel, and jet fuel (80 g kWh^{-1}), ethanol (120 g kWh^{-1}), and methanol (128 g kWh^{-1}); however, when accounting for the lower heating value (LHV) of the input compounds (e.g., 208 kJ mol^{-1} of useable H_2 for CH_4), energy costs were approximately 25 g kWh^{-1} for all of these feed compounds. For realistic estimates it is always important to account for composition and temperature effects and the reversible nature of the water gas shift reaction; see the report by Brown for a more detailed discussion of the thermodynamic considerations needed to make these estimates [37]. Steam reforming is generally a large-scale mature process, although much current work is under way to develop small-scale and portable units for fuel cell applications. The major disadvantages of steam reforming for pure H_2 formation (to be used in energy conversion) are that it requires organic feed compounds, and it therefore produces CO_2. However, this is of course an advantage in the production of synthesis gas (mixture of H_2 and CO), which is very important industrially for the synthesis of larger organic compounds. The commercial cost of H_2 (assuming \$0.1 kWh^{-1} at \$5 kg^{-1}) is approximately 20 g kWh^{-1} and targets for developing the hydrogen economy are (\$2–4 kg^{-1}) 25–50 g kWh^{-1} [38]. (Of course, these costs are based on market value and not on a single process step.)

In addition to using steam reforming, synthesis gas can be produced by partial oxidation with oxygen and steam or CO_2 reforming with or without added oxygen [38,39]. A variety of gas-phase plasma processes from thermal to nonthermal plasma with and without added catalysts have been studied to convert hydrocarbons and other organic materials to H_2 [9,39–42]. It is not always clear in the reported studies whether the energy value of the feed compounds have been considered. Nevertheless, values of H_2 formation from methane reforming are on the order of 10 g kWh^{-1}. Dry methane in microwave plasma gave 17 g kWh^{-1} [41], while methane in air gave a similar value of 22 g kWh^{-1} (corrected to 13 g kWh^{-1} with LHV) [42]. Because there are many competitive methods, some commercialized, for H_2 production, plasma processes will need to demonstrate clear advantages in terms of operation and efficiency [38].

The thermodynamic limit of producing H_2 from water vapor is

$$H_2O(g) \rightarrow H_2(g) + 0.5\,O_2(g), \quad \Delta H = 2.6 \text{ eV molecule}^{-1} \left(28.7 \text{ g kWh}^{-1}\right). \hspace{1cm} \text{R15}$$

Electrolysis, thermolysis (heat supplied by solar or nuclear energy), and plasma with liquid water are potential methods to produce H_2 with water as the hydrogen source rather than hydrocarbons [38]. Electrolysis consumes, like electrical discharge plasma, electrical energy, although it has two additional disadvantages: the use of expensive catalysts and electrolytic baths. Nevertheless, commercial electrolysis plants for H_2 production have been reported [38]. The energy efficiency is approximately 70%–80% of the thermodynamic limit of the following reaction:

$$H_2O(l) \rightarrow H_2(g) + 0.5O_2(g), \quad \Delta H = 2.95 \text{ eV molecule}^{-1}, \hspace{1.5cm} \text{R16}$$

giving 20 g kWh^{-1}. Note that the reaction enthalpy is larger compared to the production of H_2 from H_2O in the vapor phase because of the water enthalpy change from liquid to vapor.

H_2 can also be formed from CO_2 and H_2O using concentrated solar radiation, which heats up the system to highly elevated temperatures. The energy efficiencies compared to the reaction enthalpies of CO_2 and H_2O dissociation are very low (0.4% for 80% of the produced fuel) [43], although this estimate includes the energy cost necessary to separate air from the purge gas, and the energy input is abundantly available solar energy. Utilization of solar cells with an efficiency of 10% to produce electricity to generate the plasma is competitive to efficiently produce H_2 by plasmas from H_2O when the energy efficiency of the plasma process is better than 5% (1.4 g kWh^{-1}). For plasma to compete with electrolysis (and only considering energy efficiency and not environmental impact or other production costs) the efficiency of the plasma process should be about 70% (20 g kWh^{-1}). For more information on different H_2 production technologies, see the review by Holladay et al. [38].

The summary of the efficiencies of H_2 production in different water-containing plasmas, shown in Table 12.3, demonstrates a large variation in H_2 production energy yields. The most energy-efficient plasmas are an MW discharge at reduced pressure, pulsed gliding arc with water spray, and microdischarge in porous ceramics. Direct discharges in the liquid are not recommended for H_2 production from H_2O. Note that all three of these relatively efficient plasmas have an elevated gas temperature, which is necessary for large amounts of water dissociation. However, thermal arcs cannot compete with the above-mentioned discharges because of large heat losses from the arc to the surroundings. The radical recombination is also crucial; this is especially the case for the pulsed gliding arc discharge for which the quenching speed (drop in gas temperature per second) is large. Boudesocque [50] calculated that for an arc discharge the hydrogen recovery efficiency (i.e., the amount of hydrogen produced at room temperature compared to the amount of hydrogen in the core of the arc) increases from 1.6% up to 37% if the quenching rate increases from 10^6 to 10^8 K s^{-1}. Thus large spatial and temporal gradients are required to reduce the loss of atomic H in the core of the arc toward unwanted reactions that, for example, produce water (Reaction R8). This loss reaction leads to lower densities of the H_2 in the exhaust. It is thus beneficial to guide the recombination of ·H toward H_2 (Reaction R7). The complete use of radicals and atoms to form additional H_2 (and O_2) (which occurs during ideal quenching) can increase the efficiency by almost a factor of 2 [20].

From theoretical considerations one can deduce that nonequilibrium dissociation stimulated by vibrational excitation and dissociative attachment could be a more energy-efficient path for water dissociation compared to thermal dissociation. However, specifically for water, which has a large vibrational–translational (V–T) energy transfer rate, rather high ionization degrees (10^{-4}) are necessary to ensure that the vibrational excitation is faster than the V–T relaxation [20]. The dissociation is thus restricted by the necessary large ionization degree and the back reaction of ·OH to water (R8). Both limitations can be addressed by adding CO_2 because CO_2 has a lower V–T relaxation rate that

TABLE 12.3 Overview of Energy Efficiencies and Concentrations of H_2 Production from Water by Different Plasma Reactors as Reported in Literature

Plasma	Input	Concentration of H_2 (%)	Energy Cost (g kWh^{-1})	Reference
MW plasma	H_2O vapor	9	10	[20]
AC gliding arc	Water spray in N_2 and Ar	1.36	1.3	[44]
Pulsed gliding arc	Water spray in Ar	0.04	13	[45]
Pulsed corona in liquid water	Liquid water	0.4	0.25	[46]
Packed bed	2% H_2O in Ar	0.04	0.12	[47]
Sliding discharge	H_2O vapor	60	1.2	[48]
Microdischarge in porous ceramics	H_2O vapor (preheated)	0.9	15	[49]
Arc submerged in liquid H_2O	Graphite electrode	55	0.83	[50]
Steam arc jet (thermal)	H_2O	0.4	0.13	[50]

allows operation at lower electron densities while maintaining high vibrational excitation, and CO reacts with ·OH producing CO_2 and ·H, which can significantly reduce the ·OH density and the recombination to water. It can be concluded that the energy efficiency of water dissociation can be close to 60% or 17 g kWh^{-1} H$_2$ [20]. This value is close to the efficiency of H$_2$ production by electrolysis. More details about the CO_2–H_2O chemistry can be found in the works of Fridman [20].

12.4 Water Purification and Treatment

The growing world population and the increase in production, consumption, and utilization of compounds that can pollute the environment and affect human health are driving the development of more efficient, effective, and safe water purification and wastewater treatment technologies. The fields of wastewater and water purification are quite large and we will not review them here, but we note that a number of conventional technologies based upon biological (activated sludge for municipal waste), physical/mechanical (filtration, sedimentation, and adsorption), and chemical (chlorine, ozone) methods are commonly used [51]. The choice of a method, or more commonly combination of methods, depends upon the characteristics of the water to be treated (i.e., the type and amounts of contaminants, temperature, pH, conductivity, and suspended solids) as well as the goals of the treatment (e.g., for drinking water, industrial use, or for discharge to the environment). Advanced water and wastewater treatment methods to inactivate microorganisms and to degrade toxic or hazardous compounds include the advanced oxidation/reduction technologies (AOTs; sometimes denoted as AOPs for advanced oxidation processes in the literature). AOTs generate highly reactive radicals of which ·OH is the most important. An example of AOTs is treatment based upon ozone, hydrogen peroxide, and/or UV [52]. Plasma processes are widely used and commercially successful to produce ozone and generate UV in water treatment applications. In such cases the plasma does not contact the wastewater, but *indirectly* supplies either ozone or UV for water treatment. In contrast, *direct* application of thermal and nonthermal plasma methods to water are often classified as AOTs because of the formation of radicals and UV emissions inside or in contact with the liquid phase. Plasmas in or in contact with liquids could be beneficial because they have the ability to produce oxidizing species (including ozone and OH radicals), UV, and, in the case of intense plasmas in liquids, shockwaves. Table 12.4 [53] shows a qualitative comparison of various water treatment methods for selected target pollutants including microorganisms, algae, urine components, volatile organic compounds (VOCs), and inorganics. Clearly, no single method is effective for all types of pollutants, and generally most applications require a combination of

TABLE 12.4 Qualitative Comparison of Various Water Treatment Methods for Selected Target Pollutants

Target Pollutants	UV–C	UV Photocatalyst	Ozone	Electron Beam	γ-Ray	Glow Discharge	Barrier Discharge	Pulsed Corona	Pulsed Arc	Sand Gravel	Cl/ClO₂
Microorganisms	◉	Δ	◉	O	◉	O	O	◉	◉	Δ	O
Oxidation power	X	◉	◉	◉	◉	◉	◉	◉	◉	X	◉
Algae destruction	O	X	Δ	X	X	X	X	Δ	◉	Δ	X
Urine components destruction	X	O	◉	◉	◉	Δ	Δ	◉	◉	X	O
VOCs destruction	X	O	O	◉	O	Δ	Δ	◉	O	X	X
Removal of inorganics	X	Δ	Δ	Δ	Δ	Δ	Δ	O	O	O	X

Source: Chang, J.-S., *Sci. Tech. Adv. Mater.*, 2, 571–576, 2001, © IOP.

Notes: ◉, good; O, adequate; Δ, partial; X, none.

methods [54,55]. To determine economic feasibility, energy costs must be considered and comparison with conventional and alternative technologies made. In the next two sections we consider the destruction of organic compounds and inactivation of bacteria by plasma processes with liquid water in order to assess the potential of such processes in water treatment.

12.5 Organic Compound Degradation

A wide range of organic compounds from complex organic dyes and antibiotics to phenols, chlorinated, and other compounds have been degraded by electrical discharge plasmas in or in contact with liquid water. Here we seek to summarize results that report energy cost. Table 12.5 summarizes some key results given in a recent detailed review on the decolorization of organic dyes for a variety of electrical discharge reactor types [56]. Clearly, energy yields for 50% decolorization are much higher in the case of gas-phase plasma contacting liquid surfaces (water spray and falling films) and the best cases can reach several hundred grams per kilowatt-hour (g/kWh). Gas discharges require lower initiation electric fields than liquids, and typically energy costs are higher to produce plasma in the liquid [typically Joules per pulse (J/pulse) and higher in the liquid]. It should be noted, however, that while corona-like millijoules per pulse (mJ/pulse) discharge have been developed for liquids [57,58], the energy efficiencies for liquid-phase reactions in these very low-energy discharges have not been determined. Discharges over thin water films or with water droplets lead to high rates of mass transfer of reactive species from the plasma into the liquid and relatively low energy costs to generate the plasma. It should be noted that these chemical species (dyes) are not volatile so all decolorization reactions must take place in the liquid or at the liquid–gas interface. Interpretation of the results is complicated by the effects of the gas composition and, although much more detailed analysis of reaction mechanisms and economic feasibility is required to assess these results regarding the role of ozone formed in the gas phase to compare such results with direct ozonolysis methods, we can make some general comparisons based on data from the AOT literature. For example, dye decolorization by direct ozone can be over 250 g kWh^{-1} (95% reduction of acid orange) [59] and UV/H$_2$O$_2$ has been reported to be in the range of 33 g kWh^{-1} (93% decolorization of acid blue dye) [60] to 220 g kWh^{-1} (90% azo dye decolorization) [61]. Electrochemical methods have been reported to give energy yields of 62 g kWh^{-1} (reactive blue dye) [62]. Clearly, there is a wide range of energy yields for dye decolorization, but it is also true that the best plasma processes (i.e., plasma in gas with thin liquid films or droplets) may be very competitive with the other AOTs. The key issue for future

TABLE 12.5 Summary of Decolorization of Various Dyes Using Different Electrical Discharge Methods

Method	Energy Yield (g kWh^{-1})
Direct discharge in water (PC, RF, MW, GD)	0.02–0.08
Discharge with bubbles in water (PCD, MW, DD)	0.16–1.3
Discharge over thin films	
PCD, DBD	1.5–9.4
PCD	294–566
Discharge in water spray	
Pulsed glide arc	13
PCD	622

Source: Mededovic, S. and B. Locke, *J. Phys. D Appl. Phys.*, 42, 049801, 2009.

Notes: PC, pulsed corona; RF, radio frequency; MW, microwave; GD, glow discharge; DD, diaphragm discharge. Energy yields were all given at 50% decolorization conversion of the initial compound.

work is to determine where the *direct* plasma process differs from the *indirect* (conventional ozone generation) plasma process.

In comparison with the degradation of other chemical species it should be noted that the molecular mass of many of the dyes used for the data in Table 12.5 can be quite large (to several hundred amus), and therefore the energy yields in moles per Joule (mol/J) should be considered for comparison. For example, the highest efficiency reported in Table 12.5, 622 g kWh^{-1}, was found in a water spray pulsed corona discharge reactor for the dye indigo carmine [63]. Indigo carmine has a molecular mass of 466, which gives 4×10^{-7} mol J^{-1}. This value indicates higher efficiency than for any other data for all compounds and reactor types shown in Tables 12.5 through 12.9, where the best values are of the order 1×10^{-8} mol J^{-1} for falling film, corona discharge over water, and gas bubbling reactors. Clearly, g/kWh can be used when comparing the same compound across different reactors, but mol/J should be considered when comparing different compounds and the relative role of reactive species.

Phenol is perhaps the most commonly studied organic compound in such electrical discharges with liquids (and indeed in AOTs in general), and this is due to its high solubility in water, low volatility, known reaction pathways with ·OH and ozone, and representative chemical structure. Table 12.6 summarizes results on phenol destruction in direct water discharges. Clearly, chemical yields are relatively low, but they can be significantly enhanced through the addition of iron salts (leading to the Fenton reaction from H$_2$O$_2$ formed in the liquid [64]), bubbling oxygen through the discharge electrode [65,66] (making gas–liquid discharges to form ozone), or by adding catalysts (activated carbon [67,68], TiO$_2$ [69], and zeolites [70–72]).

Table 12.7 shows energy yields for various other compounds, including chloro- and nitrophenols, atrazine, and TNT by direct liquid water discharges. Without additives, energy yields are approximately 10^{-10}–10^{-9} mol J^{-1} across many compounds and reactor types. Again, efficiency can be improved with the additives mentioned above.

For discharges over water Table 12.8 shows, depending upon the compound and the gas composition, that energy yields can reach 10^{-8} mol J^{-1}, and some very complex compounds such as some antibiotics can be degraded. The EEO for one antibiotic, oxacillin, was 3.7 kWh m^{-3} per order of magnitude [73], which compares very well with ozone treatment of ciprofloxacin at 3.2 kWh m^{-3} per order of magnitude [74]. It can be noted that for ciprofloxacin the EEO (in kWh m^{-3} per order) varied from 16.3 for direct UV, to 5.2 for UV/O$_3$, to 5.2 UV/H$_2$O$_2$/O$_3$, to 3.2 O$_3$, and 2.2 H$_2$O$_2$/O$_3$ [74]. One of the most efficient plasma reactors, the gas–liquid falling film reactor, shown in Figure 12.5, was used to degrade antibiotics

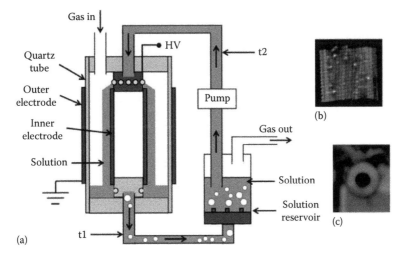

FIGURE 12.5 Falling liquid film reactor utilizing a dielectric barrier discharge in an annular configuration where the liquid falls along the inner electrode and gas flows over the liquid surface. (a) Experimental setup; (b) side view of plasma reactor; (c) top view of plasma reactor. (Reproduced with permission from Magureanu, M., et al., *Water Res.*, 44, 3445–3453, 2010. © 2010 IEEE.)

and other compounds [73,75–77] and a similar configuration has been used by other researchers [78]. Oxygen carrier gases are typically preferred due to high reactivity of organics with ozone, the formation of OH radicals from ozone and hydrogen peroxide, and even possible direct reactions of produced oxygen radicals with the organic pollutant. Air typically leads to nitrogen oxides (which can affect pH and conductivity as well as participate in various reactions), and these and other reactive nitrogen species (RNS) can be very important for sterilization [3–5,79–81].

The combination of a direct discharge in water with a gas-phase discharge over water where the ground electrode is in the gas phase, termed a series reactor, as shown in Table 12.9, leads to moderate improvements in efficiency compared to the case of direct discharge in the liquid phase. For example, phenol energy yield reached 10^{-9} mol J^{-1} with further improvement to 10^{-8} mol J^{-1} upon the addition of activated carbon, which showed catalytic activity in addition to its adsorption properties.

There is only a limited amount of data, as shown in Table 12.10, for other organic compounds in the water spray reactors. In the case of dichlorophenol, efficiency reached 10^{-8} mol J^{-1}, which is about a factor of 2 lower than the best dye decoloration mentioned above. It is recommended that more work be conducted to compare spray reactors with falling film reactors to assess efficiency and relative performance.

Plasma processes should be compared to the competitive advanced oxidation methods including direct ozone treatment, Fenton's methods (addition of H_2O_2 and iron salts), direct UV, and various other combinations of ozone, H_2O_2, UV, and catalysts [52]. Another emerging technique is the electro-Fenton process whereby H_2O_2 is generated by electrolysis directly in the liquid containing the contaminant [82]. Reported energy costs for phenol oxidation by direct ozone fall in the range of 2–120 g kWh^{-1} (10^{-8}–10^{-6} mol J^{-1}), while those by the Fenton method are approximately 10–25 g kWh^{-1} (10^{-7} mol J^{-1}) [83,84]. As shown in Table 12.8, the best case for corona discharge in the gas phase with a water film is 10^{-8} mol J^{-1}. Clearly, for primary phenol oxidation, ozonation is the best choice; phenol, chlorophenol, nitrophenol, and related compounds have relatively high reactivity with ozone [52], but this also depends upon pH. In the case of mineralization, the cost for ozone treatment was 3–30 g kWh^{-1} [based on chemical oxygen demand (COD)] [84], while TOC (total organic carbon) removal for phenol in a gliding arc plasma with water spray was in the range of 1–5 g kWh^{-1} [85]. It can be noted, however, that the gas–liquid falling film reactor led to 24% TOC removal with energy cost of 15 g kWh^{-1} for nitrophenol (unsubstituted phenol was not tested) (M. Magureanu, pers. comm.).

Not all compounds are subject to direct ozone attack and compounds such as benzene, nitrobenzene, trichloroethylene, and chloroform have very low reaction rates with ozone [52]. The ·OH is a less selective oxidant that rapidly reacts near the diffusion limit with many organic compounds. In order to produce ·OH in conventional AOTs various combinations of ozone, hydrogen peroxide, and UV light can be used [86–88]. Because plasma produced in or over liquid water is known to produce ozone (with oxygen supply), hydrogen peroxide, and UV, general aspects of the chemical reaction mechanisms could be very similar, but specific pathways may differ due to different amounts of these species, for example, ozone and H_2O_2, and UV generated in various plasma processes.

Atrazine is a common, hard to degrade pesticide that has been studied in electrical discharge reactors as well as in electrochemical reactors. Electrochemical degradation was found to be 0.3 g kWh^{-1} [89], which is comparable to the pulsed arc (0.15 g kWh^{-1}), while corona streamer-like discharges directly in water gas 0.025 g kWh^{-1} without additives and up to 2.3 g kWh^{-1} with the Fenton reaction.

In addition to the initial degradation of the target compound, care must be taken to consider the formation of by-products, which can be toxic or hazardous, and the need for complete oxidation or mineralization. Both issues are common to all AOTs. The second issue is closely related to the concentration of the target pollutant. Nonthermal plasmas, chemical oxidation, and biological oxidation processes can be used when the COD is typically less than 10 g l^{-1}, while thermal plasmas, supercritical oxidation and incineration should be used when the COD is larger than this value and thus a high concentration of waste is present [52]. (Note that COD used in this reference is a measure of TOC, excluding some aromatics that are not oxidized, and it includes chemically reduced inorganic substances that are oxidized in the test [51].) Generally, AOT processes preferentially function in the relatively low-concentration waste where radical scavenging and losses are reduced; this is probably the domain where nonthermal plasma technology is best suited. High-concentration waste may be more suitable for thermal plasma, which has similarities to supercritical oxidation and incineration. High degrees of mineralization may reduce problems associated with toxic by-products, but it is not always necessary to use the more expensive (or energy intensive) process for complete destruction of the target pollutant; in some cases the AOT, or plasma, process can break down a target pollutant into components that are easily degradable by biological treatment.

Only limited data on TOC removal efficiencies for complex organics in plasma processes has been reported. For example, TOC removal in the gas–liquid falling film reactor of the antibiotics oxacillin and amoxicillin are approximately 0.5 g kWh^{-1} at 20% reduction. Examples of TOC reduction in a competitive AOT, electro-Fenton, are for enrofloxacin, 7 g kWh^{-1}, and acid yellow dye, 7–14 g kWh^{-1} [90]. We recommend that more work be undertaken to determine mineralization costs for plasma and competitive AOTs for the same compounds under similar conditions of concentration and other solution properties. Another major issue is to test these processes using real wastewater. Many of the AOTs, particularly O_3, UV, and H_2O_2 (Fenton), are currently successfully used to treat a wide range of contaminated waters [91,92]. The best competitive plasma processes should also be tested under similar conditions.

TABLE 12.6 Phenol Degradation in Various Electrical Discharges Formed Directly in the Liquid Phase

Discharge Type	Electrode Configuration	P (W)	EY (mol J^{-1})	EY (g kWh^{-1})	EEO	Reference	Notes
Streamer-like	Needle–plate	60	8.87E–10	3.0E–01	8.2E+02	[93]	No gas
Streamer-like	Needle–plate	60	3.55E–09	1.2E+00	2.5E+02	[93]	O_2 bubbles
Streamer-like	Needle–plate	60	8.27E–09	2.8E+00	5.1E+01	[93]	AC
Streamer-like	Needle–plate	60	1.30E–10	4.4E–02	1.3E+03	[94]	No additives
Streamer-like	Needle–plate	60	1.40E–10	4.7E–02	1.1E+03	[94]	No additives
Streamer-like	Needle–plate	60	2.00E–10	6.8E–02	8.5E+02	[94]	No additives
Streamer-like	Needle–plate	109	3.00E–10	1.0E–01	1.7E+03	[95]	No additives
Streamer-like	Needle–plate	109	3.50E–10	1.2E–01	1.4E+03	[95]	TiO_2
Streamer-like	Needle–plate	200	1.60E–10	5.4E–02	1.7E+03	[96]	No additives
Streamer-like	Needle–plate	200	8.30E–10	2.8E–01	9.2E+01	[96]	With $FeCl_2$
Streamer-like	Needle–plate		1.04E–09	3.5E–01	3.7E+02	[97]	No additives
Spark	Needle–plate		2.28E–09	7.7E–01	9.2E+01	[97]	No additives
Spark	Needle–plate		6.91E–09	2.3E+00	2.3E+01	[97]	H_2O_2 added
Streamer-like	Needle–plate	2.2	1.10E–09	3.7E–01	1.5E+02	[98]	No additives
Streamer-like	Needle–plate	2.2	4.30E–09	1.5E+00	5.9E+01	[98]	O_3 added
Streamer-like	Needle–plate	2.2	4.50E–09	1.5E+00	5.9E+01	[98]	Silica gel
Streamer-like	Needle–plate	2.2	5.70E–09	1.9E+00	1.3E+01	[98]	O_3 and silica gel
Streamer-like	Needle(7)–plate	26.9	1.46E–08	4.9E+00	1.2E+02	[99]	O_2 gas bubbles
Streamer-like	Needle(7)–plate	26.9	2.32E–08	7.9E+00	6.8E+01	[99]	O_2 gas bubbles TiO_2
DC glow	Needle–plate	140	5.32E–10	1.8E–01	1.1E+03	[100]	Na_2SO_4
DC glow	Needle–plate	54	2.63E–09	8.9E–01	1.3E+02	[100]	$Na_2SO_4 + Fe^{2+}$

Notes: P, power; EY, energy yield; EEO, electrical energy per order of magnitude [(kWh m^{-3})/order of magnitude].

TABLE 12.7   Summary of Degradation of Various Organic Compounds by Direct Discharge in Water

Chemical	MW	Discharge Type	Electrode Configuration	P (W)	EY (mol J^{-1})	EY (g kWh^{-1})	Reference	Notes
4-Chlorophenol	129	DC diaphragm glow	Plate–plate	45	1.0E–09	4.8E–01	[102]	No additives, strong pH dependence between pH 2 and 9
2-Chlorophenol	129	Streamer-like	Needle–plate	54	3.5E–09	1.6E+00	[95]	w/ferrous
2-Chlorophenol	129	Streamer-like	Needle–plate	54	1.9E–10	8.8E–02	[95]	No additives
4-Chlorophenol	129	Streamer-like	Needle–plate	35	1.3E–10	6.0E–02	[102]	No additive
4-Nitrophenol	129	Streamer-like	Needle–plate	35	5.5E–09	2.5E+00	[102]	w/FeCl$_2$
Atrazine	216	Streamer-like	Needle–plate	60	3.2E–11	2.5E–02	[103]	NiCr electrode
Atrazine	216	Streamer-like	Needle–plate	60	2.0E–09	1.6E+00	[103]	FeSO$_4$, NiCr
Atrazine	216	Streamer-like	Needle–plate	60	3.0E–09	2.3E+00	[103]	FeSO$_4$, NiCr
Atrazine	216	Pulsed arc	Rod to rod		1.6E–10	1.2E–01	[104]	Best case
Hydroquinone	110	Pulsed arc	Rod to rod		6.4E–10	2.5E–01	[104]	
4-Chlorophenol	129	Pulsed arc	Rod to rod		4.6E–10	2.1E–01	[105]	
4-Chlorophenol	129	Pulsed arc	Rod to rod		5.4E–10	2.5E–01	[105]	
3,4-Dichloroaniline	161	Pulsed arc	Rod to rod		4.0E–10	2.3E–01	[105]	
2,4,6-Trinitrotoluene	227	Pulsed arc	Rod to rod		1.0E–10	8.5E–02	[105]	
2,4,6-Trinitrotoluene	227	Pulsed arc	Rod to rod		5.0E–10	4.1E–01	[105]	w/O$_3$ added
Bisphenol A	228	DC glow discharge		50	2.7E–10	2.2E–01	[106]	Na$_2$SO$_4$
Bisphenol A	228	DC glow discharge		50	1.9E–09	1.6E+00	[106]	NaCl
Bisphenol A	228	DC glow discharge		50	9.1E–09	7.5E+00	[106]	Na$_2$SO$_4$, 0.33 mMFe + 3

Notes: P, power; EY, energy yield; MW, molecular mass.

TABLE 12.8   Summary of Degradation of Various Organic Compounds by Discharge in Gas Phase over the Liquid Water

Chemical	MW	Discharge Type	Electrode Configuration	P (W)	EY (mol J^{-1})	EY (g kWh^{-1})	Reference	Notes
Phenol	94	Pulsed corona	Over water		1.38E–08	4.7E+00	[107]	G range: 0.13–0.25
Phenol	94	Pulsed corona	Over water		1.72E–08	5.8E+00	[107]	G range: 0.10–0.23
Phenol	94	Pulsed corona	Multipin over water	1	1.85E–08	6.3E+00	[108]	In air
Phenol	94	Pulsed corona	Multipin over water	1	6.90E–08	2.3E+01	[108]	In oxygen
Phenol	94		Falling film	8	2.66E–09	9.0E–01	[109]	O$_2$ carrier
Atrazine	216	Pulsed corona			7.67E–10	6.0E–01	[110]	
Penta-chlorophenol	266	Glow discharge	Over water		3.13E–12	3.0E–03	[111]	50 Torr, Argon
Oxacillin	398	DBD	Falling film	2	1.88E–08	2.7E+01	[73]	
Amoxicillin	349	DBD	Falling film	2	8.36E–08	1.1E+02	[73]	
Ampicillin	333	DBD	Falling film	2	2.42E–08	2.9E+01	[73]	
Pentoxifylline	278	DBD	Falling film	1.2	2.00E–08	2.0E+01	[75]	Tap water
Phenol	94	DC	Over water	8.75	1.68E–09	5.7E–01	[112]	O$_2$:N$_2$ 20:80, negative polarity better

(continued)

TABLE 12.8 (continued)

Chemical	MW	Discharge Type	Electrode Configuration	P (W)	EY (mol J^{-1})	EY (g kWh^{-1})	Reference	Notes
Sulfonol	50	DBD	Falling film	5	6.74E–10	1.2E–01	[113]	O$_2$ carrier, 8 s residence time
Phenol	94	DBD	Falling film	5	4.26E–10	1.4E–01	[113]	O$_2$ carrier, 8 s residence time
Sodium lauryl sulfate	288	DBD	Falling film	95 mW cm^{-3}, 256 s	1.43E–08	1.5E+01	[114]	O$_2$ carrier
Sodium lauryl sulfate	288	DBD	Falling film		2.89E–10	3.0E–01	[115]	O$_2$ carrier, liquid flow 2E–4 l s^{-1}
Phenol	94	DBD	Falling film		3.25E–09	1.1E+00	[115]	O$_2$ carrier, liquid flow 2E–4 l s^{-1}
Sulfonol	50	DBD	Falling film		1.50E–09	2.7E–01	[115]	O$_2$ carrier, liquid flow 2E–4 l s^{-1}
Acetic acid	60	DBD	Falling film	0.93 W cm^{-3}	1.24E–08	2.7E+00	[116]	O$_2$ carrier
Acetic acid	60	DBD	Falling film	1			[117]	Ar carrier
Acetic acid	60	DBD	Falling film	1			[117]	Ne carrier

Notes: P, power; EY, energy yield; MW, molecular mass; DBD, dielectric barrier discharge; DC, direct current.

TABLE 12.9 Summary of Degradation of Various Organic Compounds by the Hybrid Series Reactor with Combined Gas and Liquid Discharge

Chemical	EY (mol J^{-1})	EY (g kWh^{-1})	Reference	Notes	TOC	Conductivity (μs cm^{-1})	Gas	pH
Phenol	8.9E–10	3.0E–01	[118]	SS, 5 cm gap	0.2	150	Stagnant air	5
Phenol	2.1E–09	7.0E–01	[118]	SS	0.3	150	O$_2$ flowing 150 SCCM	5
Phenol	1.1E–08	3.7E+00	[118]	SS	2.9	150	O$_2$ flowing 150 SCCM + activated carbon	5
Phenol	4.1E–09	1.4E+00	[118]	SS	0.5	150	FeSO$_4$	5
Phenol	6.5E–09	2.2E+00	[118]	SS	0.6	150	FeSO$_4$ + O$_2$	5
Phenol	1.2E–08	4.0E+00	[118]	SS	3.2	150	FeSO$_4$ + O$_2$ + AC	5
Phenol	4.7E–09	1.6E+00	[118]	RVC	1.6	150	O$_2$ flowing	5
Phenol	5.0E–09	1.7E+00	[118]	RVC, 8 cm gap			FeSO$_4$ + O$_2$	5
Phenol	4.4E–10	1.5E–01	[119]	RVC		100	Ar	5.1
Phenol	5.6E–10	1.9E–01	[119]	RVC		100	O$_2$	5.1
Phenol	5.2E–10	1.8E–01	[94]	RVC			Ar	3.6
Phenol	4.4E–10	1.5E–01	[94]	RVC			Ar	5.1
Phenol	4.8E–10	1.6E–01	[94]	RVC			Ar	10.6
Phenol	3.9E–10	1.3E–01	[94]	RVC			O$_2$	3.6
Phenol	5.6E–10	1.9E–01	[94]	RVC			O$_2$	5.1
Phenol	1.4E–09	4.6E–01	[94]	RVC			O$_2$	10.6
Catechol	4.9E–10	1.9E–01	[119]	RVC		100	Ar	5.1
Catechol	7.3E–10	2.9E–01	[119]	RVC		100	O$_2$	5.1
Resorcinol	4.9E–10	1.9E–01	[119]	RVC		100	Ar	5.1
Resorcinol	5.7E–10	2.3E–01	[119]	RVC		100	O$_2$	5.1
Hydroquinone	4.5E–10	1.8E–01	[119]	RVC		100	Ar	5.1
Hydroquinone	9.3E–10	3.7E–01	[119]	RVC		100	O$_2$	5.1
2-Chlorophenol	5.5E–10	2.5E–01	[119]	RVC		100	Ar	5.1

(continued)

TABLE 12.9 (continued)

Chemical	EY (mol J^{-1})	EY (g kWh^{-1})	Reference	Notes	TOC	Conductivity (µs cm^{-1})		Gas	pH
2-Chlorophenol	6.2E–10	2.9E–01	[119]	RVC		100	O$_2$		5.1
3-Chlorophenol	6.4E–10	3.0E–01	[119]	RVC		100	Ar		5.1
3-Chlorophenol	6.8E–10	3.1E–01	[119]	RVC		100	O$_2$		5.1
4-Chlorophenol	7.2E–10	3.3E–01	[119]	RVC		100	Ar		5.1
4-Chlorophenol	6.4E–10	3.0E–01	[119]	RVC		100	O$_2$		5.1
2-Nitrophenol	6.3E–10	3.2E–01	[119]	RVC		100	Ar		5.1
2-Nitrophenol	5.4E–10	2.7E–01	[119]	RVC		100	O$_2$		5.1
3-Nitrophenol	5.1E–10	2.6E–01	[119]	RVC		100	Ar		5.1
3-Nitrophenol	5.6E–10	2.8E–01	[119]	RVC		100	O$_2$		5.1
4-Nitrophenol	5.2E–10	2.6E–01	[119]	RVC		100	Ar		5.1
4-Nitrophenol	3.8E–10	1.9E–01	[119]	RVC		100	O$_2$		5.1

Notes: SS, stainless-steel ground electrode; RVC, reticulated vitreous carbon electrode; P = 60 W; EY, energy yield; MW, molecular mass; 5 cm gap unless noted otherwise.

TABLE 12.10 Summary of Degradation of Various Organic Compounds by the Water Spray Reactor

Chemical	MW	Discharge Type	P (W)	EY (mol J^{-1})	EY (g kWh^{-1})	Reference	TOC	Conductivity (µs cm^{-1})	pH
Phenol	94	Gliding arc	234			[85]	4.6 g kWh^{-1}	256	7.8
Phenol	94	Gliding arc	234			[85]	1.4 g kWh^{-1}	256	7.8
Dichlorophenol	163	Pulsed corona		1.70E–08	1.0E+01	[120]			

Notes: EY, energy yield; MW, molecular mass; TOC, total organic carbon.

12.6 Disinfection

Many methods for disinfecting liquids and surfaces, based upon chemical and/or physical, mechanical, and thermal phenomena (e.g., high pressure, high temperature, UV, gamma irradiation, and chemical oxidation by ozone or chlorine), have been developed and studied. Due to limitations related to cost, user acceptance, formation of residues in the liquids or on surfaces, damage to the underlying surfaces, and acquired microbial resistance, there is a continuing need to develop more cost-effective and efficient sterilization methods. Plasma processes may offer an alternative to these conventional methods. However, detailed analysis of the mechanisms, efficiencies, and energy costs is required. Because reviews of aspects related to the mechanisms of plasma interaction with biological cells have been reported [4,121–124], the present work seeks to assess energy efficiency of plasma disinfection of liquids. As in the case with the degradation of the chemical compounds discussed in the last section, many factors can affect the performance. These factors include those related to the materials to be treated (e.g., microbial type, solution composition, temperature, scale of treatment required, and degree of sterilization required) and the characteristics of the plasma and the power supply. It is not possible to consider all of these factors in the present review and indeed there is insufficient data from the literature for a complete technical and economic analysis. Therefore, we will focus on consideration of the energy yield and the D value in Joules per milliliter per order inactivation of bacteria by a range of plasma processes and compare these results to direct UV and ozone treatment.

To place plasma technology in context we start with an order of magnitude cost estimate of using a nonoptimized treatment by an UV mercury lamp for *Escherichia coli* inactivation in water. As per Soloshenko et al. [125], the cost in UV flux energy to reduce the bacteria content by a factor of 10 for

254 nm emission is approximately 0.5 mJ cm^{-2} for a 3 ml volume (diameter of 32 and 3 mm thickness). A Philips TUV PLS 9 W lamp has an UV output (mainly 254 nm) efficiency of about 25%. By using this lamp for the above sample at a distance of 4 cm, an irradiance of 9 mW cm^{-2} is achieved [126]. This means that one can reduce the *E. coli* in the above-mentioned volume by consuming an electrical energy of 0.5 J ml^{-1}.

A similar estimate can be obtained for ozonation of drinking water. It is generally accepted that when a residual ozone level of 0.4 mg l^{-1} is maintained for 4 min, drinking water is disinfected [127]. Assuming a production efficiency of 100 g kWh^{-1} for ozone (optimized systems can obtain efficiencies over 200 g kWh^{-1} [128]), this yields an energy cost of 1.4×10^{-2} J ml^{-1}. Of course this value will strongly depend on the efficiency with which ozone is mixed with the water and how fast it is consumed in the water. Even if only 10% of the ozone transports into the liquid, an energy efficiency of about 0.1 J ml^{-1} is obtained.

Note that the above are estimates that do not consider the most efficient transfer of UV or O_3 to the water. It is thus clear that before plasma processes can be considered to have potential for practical disinfection one needs to reach the order of magnitude of energy efficiencies estimated above for UV and ozone treatment. Of course not only energy efficiency is of importance. UV and ozone treatment also have practical disadvantages and physical limitations. The UV penetration in polluted water containing ions and organic compounds can be relatively small, which allows treatment of only thin films of polluted liquids. Ozone is remotely produced and injected in the water. Direct plasma treatment produces O_3 *in situ* without transport losses and can combine both UV and ozone and also directly generate ·OH, which has a significantly higher reactivity.

TABLE 12.11 Overview of Reported *D* Values (Energy Necessary for 1 Log Reduction per ml Liquid) for *E. coli* Inactivation in Water in Literature

Plasma	*D* value* (J ml^{-1})	Liquid Conductivity (µs cm^{-1})	Initial Bacterial Density (CFU ml^{-1})	Reference
Surface discharge	0.3	0.1	10^6	[129]
Pulsed corona in water	3	1.6		[130]
Low frequency AC in air	23		10^5–10^6	[131]
Pulsed arc in water	18.7		10^7	[132]
DBD in air (bubbling)	0.29			[133]
Pulsed corona in water	33.3	0.365	10^4–10^5	[134]
Pulsed arc in water	2.1	Drinking water	10^5–10^6	[135]
Pulsed corona in water	45	0.1	10^6–10^7	[136]
Capillary discharge in water	5.4	0.9 NaCl in H_2O	10^7	[137]
Corona in water	18	0.2	10^5	[138]
PEF	<5	13	10^5	[139]
Streamers in air bubbles	13		10^5–10^6	[140]
PEF	40	Low	10^8–10^9	[141]
Spark arc	0.6		10^6	[142]
Spark arc	1		4×10^4	[143]
Pulsed corona in air	0.1	0.9	10^7–10^8	[144]
Low voltage 10 µs pulsed discharge in liquid	158	0.9 NaCl in H_2O	2.5×10^5	[145]
Surface streamers	8.6	Tap	10^7	[146]
Spark discharges in water	0.1–0.4	0.2	10^4–10^6	[147]
Packed-bed air bubble discharge	9	0.91–15.7	10^6	[148]

Notes: *The data in this table are obtained from graphs in publications by assuming initially exponential deactivation as function of treatment time.

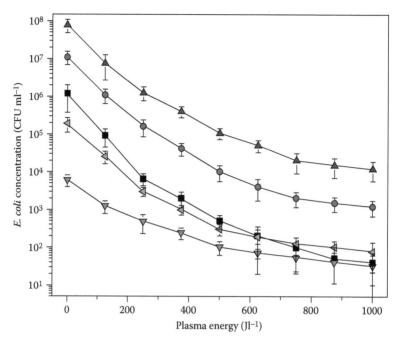

FIGURE 12.6 Survival plot for different initial bacterial concentrations for a spark discharge in the liquid phase. (Reprinted with permission from Yang, Y., et al., *Rev. Sci. Instrum.*, 82, 2011. Copyright 2011, American Institute of Physics.)

A typical bacterial inactivation curve that is observed in bacterial inactivation in liquids is shown in Figure 12.6. The graph has the typical dependence of the colony-forming units (CFU) per milliliter that survive the plasma treatment as a function of the dissipated energy. The reason why CFU are used is due to the analysis methods, which is often based on plating diluted samples of the treated bacteria samples and letting them multiply. Each surviving bacterium will multiply and form a colony. Basically, counting the colonies gives an accurate value of the amount of surviving bacteria just after the plasma treatment. However, this method does not always allow differentiation between inactivated (nonreproducing bacteria) and killed bacteria. This is why we refer to bacteria inactivation rather than bacteria kill in this work. Figure 12.6 also clearly shows that the efficiency of bacterial inactivation can depend on the bacteria concentration in the liquid. Some studies therefore use units of amount of energy per inactivated bacterium (e.g., Reference [135]). In this work we choose to follow the approach shown in Figure 12.6.

Instead of presenting CFU per milliliter as a function of dissipated energy, it is more common in the literature to use a treatment time axis and if no information on the plasma power or energy per pulse (with pulse frequency) is given, it is impossible to deduce energy efficiency. Browsing through hundreds of papers we found that data from a very limited number of papers allowed calculation of the energy per volume, which is necessary to reduce the bacterial concentration by 1 order of magnitude (D value). The D value is essentially the same as the EEO value used in AOT.

An overview of the D values for *E. coli* treatment by direct plasma treatment of water is given in Table 12.11. Additionally, the liquid conductivity and the initial bacterial concentration before treatment are indicated, at least when reported in the references cited.

The large range of energy efficiencies for *E. coli* inactivation for different plasmas is immediately apparent. The best energy efficiencies are obtained for spark discharges and gaseous discharges including DBD, pulsed streamers in the air gap above water, or surface discharges. All discharges produced in the liquid that involve evaporating water are less energy efficient except for spark/arc discharges. In view of the fact that the heat capacity and latent heat of water are very large this is of course to be expected.

It is believed that in the case of spark/arc discharges strong UV emission and radical chemistry rather than the shockwaves or thermal effects are the cause of bacteria inactivation [149]. The UV emission and the radical density are significant in arc discharges compared to capillary and pulsed corona discharges. When the bacterial inactivation is due to chemistry (such as in capillary discharges), it is more efficient to produce the plasma in the gas phase and transport the reactive species to the liquid phase for the treatment. The molecular transport in these cases is a key issue to ensure efficiencies and it is clear that surface discharges [150] and the DBD with gas bubbling [133] are most efficient in this case.

The same type of discharge can have significantly different energy efficiency for bacterial inactivation depending upon the pulse characteristics. This effect is analogous to the known changes in energy efficiency of radical production or chemical conversion in gas-phase corona discharges depending on the excitation pulse width [151,152]. However, efficient transport of chemically active species from the gas phase to the liquid phase and the related dispersion (and size reduction) of bubbles in the liquid phase are very important for efficient bacterial inactivation. This could, at least partially, explain the fact that very similar discharges such as reported by de Wever et al. [133], Marsili et al. [140], and Zhang et al. [148], which all produce filamentary discharges in bubbles, have D values of 0.29, 13, and 9 J ml^{-1}, respectively.

The liquid conductivity is not an important factor when the discharge is produced in the gas phase but it has a significant effect in liquid-phase discharges. Particularly for corona-like discharges in a liquid the discharge parameters significantly change with increasing liquid conductivity [153]. This is due to the fact that plasma initiation in a highly conductive medium is more difficult and more electrical current is lost through Ohmic conduction of the medium.

TABLE 12.12 Overview of Reported D Values and Maximum Reported Inactivation for Different Bacteria by Plasma Inactivation in Water

Plasma	Bacterium	D Value[a] (J ml^{-1})	Max Log Reduction[a]	Reference
Pulsed corona in water	*Escherichia coli* (GN)	3	4	[130]
	Bacillus subtilis (GP)	17	4	
	B. subtilis sp.	No effect	–	
Low-frequency AC in air	*E. coli* (GN)	23	5	[131]
	Staphylococcus aureus (GN)	34	5	
	Yeast	31	3	
DBD in air (bubbling)	*E. coli* (GN)	0.29	7	[133]
	E. hirae (GP)	0.29	7	
	B. subtilis sp.	–	0.5	
	Arthrobacter sp. (GP)	0.41	7	
	Streptococcus thermophilus (GP)	0.76	5.6	
	Marginella gilvae B (GP)	9	1	
Pulsed arc in water	*E. coli* (GN)	~2.1	6	[135]
	Staphyloccus (GP)	Similar	6	
Streamers in air bubbles	*B. subtilis* (GP)	4	6	[154]
	B. subtilis (sp.) (initial 10^4)	2.5	6	
	B. subtilis (sp.) (initial 10^7)	No effect	–	
Pulsed corona in water	*Enterococcus faecalis* (GP)	60	1	[138]
	E. coli (GN)	15	5	
PEF	*Listeria monocytogenes* (GP)	–	2	[139]
	Candida albicans (fungus)	–	2	
	S. aureus (GP)	–	4.5	
	Yersinia enterocolitica (GN)	–	6	
	E. coli (GN)	–	8	

(continued)

TABLE 12.12 (continued)

Plasma	Bacterium	D Value[a] (J ml^{-1})	Max Log Reduction[a]	Reference
Streamers in air bubbles	E. coli (GN)	13.0	5	[140]
	S. aureus (GP)	13.0	5	
	Salmonella enteritidis (GN)	11.8	5	
	B. cereus (GP)	4.1	5	
Pulsed corona in water	Saccharomyces cerevisiae (yeast)	72	1.4	[155]
Air cold plasma jet	B. cereus (sp.)	96–252	1	[156]
PEF	Pseudomonas fluorescens (GN)	85	3	[157]
	B. cereus (sp.)	500	1	

[a] The maximum log reduction is considered for the same treatment time of each experiment individually. It does not mean that it is the absolute minimum that can be obtained in some cases. It can only be used to yield information on the relative difficulty to inactivate different bacteria within the same experiment.

In addition to the effect of different plasma conditions on bacterial inactivation it is important to note that bacterial inactivation by plasmas also strongly depends on the morphology and structure of the bacterium. This is of importance because several kinds of bacteria can be present in practical water treatment applications. Table 12.12 gives an overview of the plasma inactivation of different bacteria types for the same plasma conditions. It is immediately clear that large variations in inactivation efficiencies can occur for identical treatment conditions. Kong et al. [158] state in their review paper that the cold atmospheric plasma treatment is more effective against Gram-negative compared to Gram-positive bacteria. This is attributed to the difference in cell wall structure. In line with this statement, it can be deduced from Table 12.11 that E. coli (GN) is often the easiest to inactive. However, even between two different Gram-negative bacteria, a significant different response to plasma treatment occurs, and it is hard to predict at present the trends in every plasma reactor without knowledge of the key inactivation mechanisms of each bacterium. Spores, which are dormant encapsulated bacteria, require, as expected, significantly more energy to inactivate compared to bacteria in the vegetative state. This is particularly important when disinfection of liquids is required.

An important difference between bacteria inactivation and chemical degradation is that for the latter mainly reactive oxygen species (ROS) (such as O_3, $\cdot O$, and $\cdot OH$) species are considered to be important, while in the case of bacteria inactivation NO-related chemistry and RNS in the liquid phase are considered to play a major role in inactivation [4,79–81]. It must be concluded that biological systems are extremely complex. Mutations within one bacterium type can even lead to a change in how this bacterium reacts to plasma inactivation and metabolism of surviving cells after plasma treatment has been shown to change [159]. This makes predicting trends even more difficult compared to dealing with chemical compounds that are chemically much better defined.

12.7 Conclusions

A wide range of electrical discharge reactors from direct discharge in the liquid phase to gas-phase discharge over a liquid, bubbles through the liquid, and water spray have been studied for applications in chemical synthesis (e.g., H_2 and H_2O_2 generation), chemical degradation (e.g., dyes, phenols, antibiotics), and microbial inactivation (e.g., E. coli). In the case of molecular synthesis it is important that water dissociation is efficient and that not too much energy is wasted in heating. Additionally, rapid thermal quenching is necessary to favor the desired reaction products. Reduced pressure microwave and pulsed gliding arc discharges were found to be the most efficient for H_2 generation, while DBD and pulsed gliding arc discharges with water spray are the most efficient for H_2O_2 generation. In the pulsed gliding arc water spray reactor low power with short pulses leads to highest efficiency. Direct discharges in water are less efficient than gas-phase discharges in contact with the liquid phase. Plasma generation of H_2 and H_2O_2 from clean water has some significant advantages over other methods, but while energy yields are comparable to existing processes, further work is needed to develop reactors of appropriate scale for applications.

The degradation of chemical species followed a similar trend as seen in the production of H_2O_2 whereby gas-phase discharges in contact with the liquid (as a film, aerosol, or bubble) could be important. Liquid-phase chemistry is complicated by the effects of solution pH, conductivity, additives, and gas composition (when present). Nevertheless, oxygen is the best gas to use for chemical degradation, leading to ROS, while air leads to the formation of RNS, which may be of equal or more importance in bacterial inactivation. The underlying chemistry of the plasma interactions with water is fundamentally similar to the major pathways in other AOTs that utilize O_3, H_2O_2, and/or UV. The best plasma processes, for example, the falling water film and water spray reactors, can be comparable in energy efficiency to conventional AOTs, for example, ozone and the Fenton process, but more work is needed to establish where plasma has true advantages.

Bacterial inactivation efficiency depends on the efficient production of radicals and UV. In some cases where chemical reactions dominate the inactivation mechanisms, efficient transport of the radicals, particularly at the plasma–liquid interface, is important. Both spark discharges in liquids and DBD or pulsed corona discharges in the gas phase are most efficient for bacterial inactivation.

Acknowledgments

Peter J. Bruggeman thanks Richard van de Sanden, Daan C. Schram, and Koen van Gils for helpful discussions. Bruce R. Locke acknowledges partial support by the National Science Foundation (CBET-0932481) and Florida State University.

References

1. Bruggeman, P. and C. Leys. 2009. Non-thermal plasmas in and in contact with liquids. *Journal of Physics D: Applied Physics* 42:1–28.
2. Locke, B. R., P. Sunka, M. Sato, M. Hoffmann, and J. S. Chang. 2006. Electrohydraulic discharge and non thermal plasma for water treatment. *Industrial & Engineering Chemistry Research* 45:882–905.
3. Locke, B. R., P. Lukes, and J. L. Brisset. 2012. Elementary chemical and physical phenomena in electrical discharge plasma in gas–liquid environments and in liquids. In *Plasma Chemistry and Catalysis in Gases and Liquids*. M. M. V. I. Parvulescu, P. Lukes, editors. Wiley-VCH Verlag GmbH & Co. KGaA, Weinheim, pp. 185–241.
4. Lukes, P., J. L. Brisset, and B. R. Locke. 2012. Biological effects of electrical discharge plasma in water and in gas–liquid environments. In *Plasma Chemistry and Catalysis in Gases and Liquids*. M. M. V. I. Parvulescu, P. Lukes, editors. Wiley-VCH Verlag GmbH & Co. KGaA, Weinheim, pp. 309–352.
5. Lukes, P., B. R. Locke, and J. L. Brisset. 2012. Aqueous-phase chemistry of electrical discharge plasma in water and in gas–liquid environments. In *Plasma Chemistry and Catalysis in Gases and Liquids*. M. M. V. I. Parvulescu, P. Lukes, editors. Wiley-VCH Verlag GmbH & Co. KGaA, Weinheim, pp. 243–308.
6. Woloszko, J., K. R. Stalder, and I. G. Brown. 2002. Plasma characteristics of repetitively-pulsed electrical discharges in saline solutions used for surgical procedures. *IEEE Transactions on Plasma Science* 30:1376.
7. Sunka, P., V. Babicky, M. Clupek, M. Fuciman, P. Lukes, M. Simek, J. Benes, B. Locke, and Z. Majcherova. 2004. Potential applications of pulse electrical discharges in water. *Acta Physica Slovaca* 54:135–145.
8. Hieda, J., N. Saito, and O. Takai. 2008. Exotic shapes of gold nanoparticles synthesized using plasma in aqueous solution. *Journal of Vacuum Science & Technology A* 26:854–856.

9. Bromberg, L., D. R. Cohn, A. Rabinovich, N. Alexeev, A. Samokhin, R. Ramprasad, and S. Tamhankar. 2000. System optimization and cost analysis of plasma catalytic reforming of natural gas. *International Journal of Hydrogen Energy* 25:1157–1161.

10. Bolton, J. R., K. G. Bircher, W. Tumas, and C. A. Tolman. 2001. Figures-of-merit for the technical development and application of advanced oxidation technologies for both electric- and solar-driven systems—(IUPAC Technical Report). *Pure and Applied Chemistry* 73:627–637.

11. Winands, G. J. J., Z. Liu, E. J. M. van Heesch, A. J. M. Pemen, and K. Yan. 2008. Matching a pulsed-power modulator to a streamer plasma reactor. *IEEE Transactions on Plasma Science* 36:243–252.

12. Lide, D. 2012. *CRC Handbook of Chemistry and Physics*. CRC Press, Boca Raton.

13. Samukawa, S., M. Hori, S. Rauf, K. Tachibana, P. Bruggeman, G. Kroesen, J. C. Whitehead, et al., 2012. The 2012 plasma roadmap (part: Plasmas in and in contact with liquids: A retrospective and an outlook, P. Bruggeman). *Journal of Physics D: Applied Physics* 45:253001.

14. Locke, B. R. and S. M. Thagard. 2012. Analysis and review of chemical reactions and transport processes in pulsed electrical dishcarge plasma formed directly in liquid water. *Plasma Chemistry and Plasma Processing* 32:875–917.

15. Lieberman, M. A. and A. J. Lichtenberg. 1994. *Principles of Plasma Discharges and Materials Processing*. John Wiley & Sons, Inc., New York.

16. Fridman, A. and L. A. Kennedy. 2004. *Plasma Physics and Engineering*. Taylor and Francis, New York.

17. Bruggeman, P., D. Schram, M. A. Gonzalez, R. Rego, M. G. Kong, and C. Leys. 2009. Characterization of a direct dc-excited discharge in water by optical emission spectroscopy. *Plasma Sources Science and Technology* 18:025017.

18. An, W., K. Baumung, and H. Bluhm. 2007. Underwater streamer propagation analyzed from detailed measurements of pressure release. *Journal of Applied Physics* 101:053302.

19. Locke, B. R. and S. Mededovic-Thagard. 2009. Analysis of chemical reactions in gliding arc reactors with water spray. *IEEE Transactions on Plasma Science* 37:494–501.

20. Fridman, A. 2008. *Plasma Chemistry*. Cambridge University Press, Cambridge.

21. Bruggeman, P. and D. C. Schram. 2010. On OH production in water containing atmospheric pressure plasmas. *Plasma Sources Science and Technology* 19:045025.

22. Bruggeman, P., G. Cunge, and N. Sadeghi. 2012. Absolute OH density measurements by broadband UV absorption in diffuse atmospheric-pressure He–H_2O RF glow discharges. *Plasma Sources Science and Technology* 21:035019.

23. Verreycken, T., R. M. van der Horst, A. Baede, E. M. van Veldhuizen, and P. J. Bruggeman. 2012. Time and spatially resolved LIF of OH in a plasma filament in atmospheric pressure He–H_2O. *Journal of Physics D: Applied Physics* 45:045205.

24. Kogelschatz, U. 2003. Dielectric-barrier discharges: their history, discharge physics, and industrial applications. *Plasma Chemistry and Plasma Processing* 23:1–45.

25. Brandenburg, R., H. Lange, T. von Woedtke, M. Stieber, E. Kindel, J. Ehlbeck, and K. D. Weltmann. 2009. Antimicrobial effects of UV and VUV radiation of nonthermal plasma jets. *IEEE Transactions on Plasma Science* 37:877–883.

26. Delius, M., E. Hoffmann, G. Steinbeck, and P. Conzen. 1994. Biological effects of short-waves induction of arrhythmia in piglet hearts. *Ultrasound in Medicine and Biology* 20:279–285.

27. Schaper, L., C. P. Kelsey, P. Ceccato, A. Rousseau, K. R. Stalder, and W. G. Graham. 2011. Pre- to post-discharge behavior in saline solution. *IEEE Transactions on Plasma Science* 39:2670–2671.

28. Tachibana, K., Y. Takekata, Y. Mizumoto, H. Motomura, and M. Jinno. 2011. Analysis of a pulsed discharge within single bubbles in water under synchronized conditions. *Plasma Sources Science and Technology* 20:034005.

29. Bruggeman, P., L. Graham, J. Degroote, J. Vierendeels, and C. Leys. 2007. Water surface deformation in strong electrical fields and its influence on electrical breakdown in a metal pin-water electrode system. *Journal of Physics D: Applied Physics* 40:4779–4786.

30. Bruggeman, P., J. Van Slycken, J. Degroote, J. Vierendeels, P. Verleysen, and C. Leys. 2008. DC electrical breakdown in a metal pin water electrode system. *IEEE Transactions on Plasma Science* 36:1138–1139.

31. Liu, D. X., P. Bruggeman, F. Iza, M. Z. Rong, and M. G. Kong. 2010. Global model of low-temperature atmospheric-pressure He + H_2O plasmas. *Plasma Sources Science and Technology* 19:025018.

32. Locke, B. R. and K.-Y. Shih. 2011. Review of the methods to form hydrogen peroxide in electrical discharge plasma with liquid water. *Plasma Sources Science and Technology* 20:034006.

33. Mededovic, S. and B. Locke. 2009. Primary chemical reactions in pulsed electrical discharge channels in water (vol 40, pg 7734, 2007). *Journal of Physics D: Applied Physics* 42:049801.

34. Mededovic, S. and B. Locke. 2007. Primary chemical reactions in pulsed electrical discharge channels in water. *Journal of Physics D: Applied Physics* 40:7734–7746.

35. Zhou, J. C., H. C. Guo, X. S. Wang, M. X. Guo, J. L. Zhao, L. X. Chen, and W. M. Gong. 2005. Direct and continuous synthesis of concentrated hydrogen peroxide by the gaseous reaction of H-2/O-2 non-equilibrium plasma. *Chemical Communications* 1631–1633.

36. Zhao, J. L., J. C. Zhou, J. Su, H. C. Guo, X. S. Wang, and W. M. Gong. 2007. Propene epoxidation with in-site H_2O_2 produced by H_2/O_2 non-equilibrium plasma. *AIChE Journal* 53:3204–3209.

37. Brown, L. F. 2001. A comparative study of fuels for on-board hydrogen production for fuel-cell-powered automobiles. *International Journal of Hydrogen Energy* 26:381–397.

38. Holladay, J. D., J. Hu, D. L. King, and Y. Wang. 2009. An overview of hydrogen production technologies. *Catalysis Today* 139:244–260.

39. Cormier, J. M. and I. Rusu. 2001. Syngas production via methane steam reforming with oxygen: Plasma reactors versus chemical reactors. *Journal of Physics D: Applied Physics* 34:2798–2803.

40. Chao, Y., C. T. Huang, H. M. Lee, and M. B. Chang. 2008. Hydrogen production via partial oxidation of methane with plasma-assisted catalysis. *International Journal of Hydrogen Energy* 33:664–671.

41. Jasinski, M., M. Dors, H. Nowakowska, G. V. Nichipor, and J. Mizeraczyk. 2011. Production of hydrogen via conversion of hydrocarbons using a microwave plasma. *Journal of Physics D: Applied Physics* 44:194002.

42. Luche, J., O. Aubry, A. Khacef, and J. M. Cormier. 2009. Syngas production from methane oxidation using a non-thermal plasma: Experiments and kinetic modeling. *Chemical Engineering Journal* 149:35–41.

43. Chueh, W. C., C. Falter, M. Abbott, D. Scipio, P. Furler, S. M. Haile, and A. Steinfeld. 2010. High-flux solar-driven thermochemical dissociation of CO_2 and H_2O using nonstoichiometric ceria. *Science* 330:1797–1801.

44. Porter, D., M. Poplin, F. Holzer, W. Finney, and B. Locke. 2009. Formation of hydrogen peroxide, hydrogen, and oxygen in gliding arc electrical discharge reactors with water spray. *IEEE Transactions on Industry Applications* 45:623–629.

45. Burlica, R., K. Y. Shih, and B. R. Locke. 2010. Formation of H_2 and H_2O_2 in a water-spray gliding arc nonthermal plasma reactor. *Industrial & Engineering Chemistry Research* 49:6342–6349.

46. Kirkpatrick, M. and B. Locke. 2005. Hydrogen, oxygen, and hydrogen peroxide formation in aqueous phase pulsed corona electrical discharge. *Industrial & Engineering Chemistry Research* 44:4243–4248.

47. Kabashima, H., H. Einaga, and S. Futamura. 2003. Hydrogen generation from water, methane, and methanol with nonthermal plasma. *IEEE Transactions on Industry Applications* 39:340–345.

48. Malik, M. A. and K. H. Schoenbach. 2012. New approach for sustaining energetic, efficient and scalable non-equilibrium plasma in water vapours at atmospheric pressure. *Journal of Physics D: Applied Physics* 45:132001.

49. Koo, I. G., M. Y. Choi, J. H. Kim, J. H. Cho, and W. M. Lee. 2008. Microdischarge in porous ceramics with atmospheric pressure high temperature H_2O/SO_2 gas mixture and its application for hydrogen production. *Japanese Journal of Applied Physics* 47:4705–4709.

50. Boudesocque, N., C. Vandensteendam, C. Lafon, and C. Girold. 2006. Hydrogen production by thermal water splitting using a therma plasma. In *16th World Hydrogen Energy Conference, WHEC 16*, June 13–16, Lyon, France.

51. Eckenfelder, W. W. 1989. *Industrial Water Pollution Control*. McGraw Hill, New York.

52. Tarr, M. A., editor. 2003. *Chemical Degradation Methods for Wastes and Pollutants, Environmental and Industrial Applications*. Marcel Dekker, New York.

53. Chang, J.-S. 2001. Recent development of plasma pollution control technology: A critical review. *Science and Technology of Advanced Materials* 2:571–576.

54. Scott, J. P. and D. F. Ollis. 1995. Integration of chemical and biological oxidation processes for water treatment: review and recommendations. *Environmental Progress* 14:88–103.

55. Scott, J. P. and D. F. Ollis. 1996. Engineering models of combined chemical and biological processes. *Journal of Environmental Engineering* 122:1110–1114.

56. Malik, M. A. 2010. Water purification by plasmas: Which reactors are most energy efficient? *Plasma Chemistry and Plasma Processing* 30:21–31.

57. Starikovskiy, A., Y. Yang, Y. I. Cho, and A. Fridman. 2011. Nonequilibrium liquid plasma generation. *IEEE Transactions on Plasma Science* 39:2668–2669.

58. Starikovskiy, A., Y. Yang, Y. I. Cho, and A. Fridman. 2011. Non-equilibrium plasma in liquid water: Dynamics of generation and quenching. *Plasma Sources Science and Technology* 20:024003.

59. Shu, H. Y. and M. C. Chang. 2005. Decolorization effects of six azo dyes by O_3, UV/O_3 and UV/H_2O_2 processes. *Dyes and Pigments* 65:25–31.

60. Kasiri, M. B. and A. R. Khataee. 2012. Removal of organic dyes by UV/H_2O_2 process: Modelling and optimization. *Environmental Technology* 33:1417–1425.

61. Aleboyeh, A., M. E. Olya, and H. Aleboyeh. 2008. Electrical energy determination for an azo dye decolorization and mineralization by UVM/H_2O_2 advanced oxidation process. *Chemical Engineering Journal* 137:518–524.

62. Neti, N. R. and R. Misra. 2012. Efficient degradation of Reactive Blue 4 in carbon bed electrochemical reactor. *Chemical Engineering Journal* 184:23–32.

63. Minamitani, Y., S. Shoji, Y. Ohba, and Y. Higashiyama. 2008. Decomposition of dye in water solution by pulsed power discharge in a water droplet spray. *IEEE Transactions on Plasma Science* 36:2586–2591.

64. Grymonpre, D., A. Sharma, W. Finney, and B. Locke. 2001. The role of Fenton's reaction in aqueous phase pulsed streamer corona reactors. *Chemical Engineering Journal* 82:189–207.

65. Clements, J. S., M. Sato, and R. H. Davis. 1987. Preliminary investigation of prebreakdown phenomena and chemical reactions using a pulsed high-voltage discharge in water. *IEEE Transactions on Industry Applications* 23:224–235.

66. Sharma, A. K., B. R. Locke, P. Arce, and W. C. Finney. 1993. A preliminary study of pulsed streamer corona discharge for the degradation of phenol in aqueous solutions. *Hazardous Waste and Hazardous Materials* 10:209–219.

67. Grymonpre, D., W. Finney, R. Clark, and B. Locke. 2003. Suspended activated carbon particles and ozone formation in aqueous-phase pulsed corona discharge reactors. *Industrial & Engineering Chemistry Research* 42:5117–5134.

68. Lu, N., J. Li, X. X. Wang, T. C. Wang, and Y. Wu. 2012. Application of double-dielectric barrier discharge plasma for removal of pentachlorophenol from wastewater coupling with activated carbon adsorption and simultaneous regeneration. *Plasma Chemistry and Plasma Processing* 32:109–121.

69. Lukes, P., M. Clupek, P. Sunka, F. Peterka, T. Sano, N. Negishi, S. Matsuzawa, and K. Takeuchi. 2005. Degradation of phenol by underwater pulsed corona discharge in combination with TiO_2 photocatalysis. *Research on Chemical Intermediates* 31:285–294.

70. Peternel, I., H. Kusic, N. Koprivanac, and B. Locke. 2006. The roles of ozone and zeolite on reactive dye degradation in electrical discharge reactors. *Environmental Technology* 27:545–557.

71. Kusic, H., N. Koprivanac, I. Peternel, and B. Locke. 2005. Hybrid gas/liquid electrical discharge reactors with zeolites for colored wastewater degradation. *Journal of Advanced Oxidation Technologies* 8:172–181.

72. Kusic, H., N. Koprivanac, and B. Locke. 2005. Decomposition of phenol by hybrid gas/liquid electrical discharge reactors with zeolite catalysts. *Journal of Hazardous Materials* 125:190–200.

73. Magureanu, M., D. Piroi, N. B. Mandache, V. David, A. Medvedovici, C. Bradu, and V. I. Parvulescu. 2011. Degradation of antibiotics in water by non-thermal plasma treatment. *Water Research* 45:3407–3416.

74. Lester, Y., D. Avisar, I. Gozlan, and H. Mamane. 2011. Removal of pharmaceuticals using combination of $UV/H_2O_2/O$-3 advanced oxidation process. *Water Science and Technology* 64:2230–2238.

75. Magureanu, M., D. Piroi, N. B. Mandache, V. David, A. Medvedovici, and V. I. Parvulescu. 2010. Degradation of pharmaceutical compound pentoxifylline in water by non-thermal plasma treatment. *Water Research* 44:3445–3453.

76. Magureanu, M., D. Piroi, N. B. Mandache, and V. Parvulescu. 2008. Decomposition of methylene blue in water using a dielectric barrier discharge: Optimization of the operating parameters. *Journal of Applied Physics* 104:103306.

77. Magureanu, M., N. B. Mandache, and V. I. Parvulescu. 2007. Degradation of organic dyes in water by electrical discharges. *Plasma Chemistry and Plasma Processing* 27:589–598.

78. Li, J., T. C. Wang, N. Lu, D. D. Zhang, Y. Wu, T. W. Wang, and M. Sato. 2011. Degradation of dyes by active species injected from a gas phase surface discharge. *Plasma Sources Science and Technology* 20:034019.

79. Graves, D. B. 2012. The emerging role of reactive oxygen and nitrogen species in redox biology and some implications for plasma applications to medicine and biology. *Journal of Physics D: Applied Physics* 45:263001.

80. Brisset, J.-L. and E. Hnatiuc. 2012. Peroxynitrite: A re-examination of the chemical properties of non-thermal discharges burning in air over aqueous solutions. *Plasma Chemistry and Plasma Processing* 32:655–674.

81. Brisset, J.-L., B. Benstaali, D. Moussa, J. Fanmoe, and E. Njoyim-Tamungang. 2011. Acidity control of plasma-chemical oxidation: applications to dye removal, urban waste abatement and microbial inactivation. *Plasma Sources Science and Technology* 20:034021.

82. Sires, I. and E. Brillas. 2012. Remediation of water pollution caused by pharmaceutical residues based on electrochemical separation and degradation technologies: A review. *Environment International* 40:212–229.

83. Esplugas, S., J. Gimenez, S. Contreras, E. Pascual, and M. Rodriguez. 2002. Comparison of different advanced oxidation processes for phenol degradation. *Water Research* 36:1034–1042.

84. Krichevskaya, M., D. Klauson, E. Portjanskaja, and S. Preis. 2011. The cost evaluation of advanced oxidation processes in laboratory and pilot-scale experiments. *Ozone: Science & Engineering* 33:211–223.

85. Yan, J. H., C. M. Du, X. D. Li, X. D. Sun, M. J. Ni, K. F. Cen, and B. Cheron. 2005. Plasma chemical degradation of phenol in solution by gas–liquid gliding arc discharge. *Plasma Sources Science and Technology* 14:637–644.

86. Hoigne, J. and H. Bader. 1976. The Role of hydroxyl radical reactions in ozonation processes in aqueous solutions. *Water Research* 10:377.

87. Hoigne, J. 1998. Chemistry of aqueous ozone and transformation of pollutants by ozonation and advanced oxidation. In *The Handbook of Environmental Chemistry Vol. 5 Part C Quality and Treatment of Drinking Water II.* J. Hrubec, editor. Springer-Verlag, Berlin Heidelberg, pp. 83–141.

88. Hoigne, J. 1988. The chemistry of ozone in water. In *Process Technologies for Water Treatment.* S. Stucki, editor. Plenum Press, New York, pp. 121–143.

89. Malpass, G. R. P., D. W. Miwa, S. A. S. Machado, P. Olivi, and A. J. Motheo. 2006. Oxidation of the pesticide atrazine at DSA (R) electrodes. *Journal of Hazardous Materials* 137:565–572.

90. Ruiz, E. J., C. Arias, E. Brillas, A. Hernandez-Ramirez, and J. M. Peralta-Hernandez. 2011. Mineralization of acid yellow 36 azo dye by electro-Fenton and solar photoelectro-Fenton processes with a boron-doped diamond anode. *Chemosphere* 82:495–501.

91. Rosal, R., A. Rodriguez, J. A. Perdigon-Melon, M. Mezcua, M. D. Hernando, P. Leton, E. Garcia-Calvo, A. Aguera, and A. R. Fernandez-Alba. 2008. Removal of pharmaceuticals and kinetics of mineralization by O(3)/H(2)O(2) in a biotreated municipal wastewater. *Water Research* 42:3719–3728.

92. Lester, Y., H. Mamane, and D. Avisar. 2012. Enhanced removal of micropollutants from groundwater, using pH modification coupled with photolysis. *Water, Air, & Soil Pollution* 223:1639–1647.

93. Grymonpre, D. R., W. C. Finney, R. J. Clark, and B. R. Locke. 2003. Suspended activated carbon particles and ozone formation in aqueous phase pulsed corona discharge reactors. *Industrial & Engineering Chemistry Research* 42:5117–5134.

94. Lukes, P. and B. R. Locke. 2005. Degradation of phenol in a hybrid series gas–liquid electrical discharge reactor. *Journal of Physics D: Applied Physics* 38:4074–4081.

95. Lukes, P., M. Clupek, V. Babicky, P. Sunka, G. Winterova, and V. Janda. 2003. Non-thermal plasma induced decomposition of 2-chlorophenol in water. *Acta Physica Slovaca* 53:423–428.

96. Sunka, P., V. Babicky, M. Clupek, P. Lukes, M. Simek, J. Schmidt, and M. Cernak. 1999. Generation of chemically active species by electrical discharges in water. *Plasma Sources Science and Technology* 8:258–265.

97. Sun, B., M. Sato, and J. S. Clements. 2000. Oxidative processes occurring when pulsed high voltage discharges degrade phenol in aqueous solution. *Environmental Science & Technology* 34:509–513.

98. Malik, M. A. 2003. Synergistic effect of plasmacatalyst and ozone in a pulsed corona discharge reactor on the decomposition of organic pollutants in water. *Plasma Sources Science and Technology* 12:S26–S32.

99. Wang, H. J., J. Li, X. Quan, and Y. Wu. 2008. Enhanced generation of oxidative species and phenol degradation in a discharge plasma system coupled with TiO_2 photocatalysis. *Applied Catalysis B: Environmental* 83:72–77.

100. Gao, J. Z., Y. J. Liu, W. Yang, L. M. Pu, J. Yu, and Q. F. Lu. 2003. Oxidative degradation of phenol in aqueous electrolyte induced by plasma from a direct glow discharge. *Plasma Sources Science and Technology* 12:533–538.

101. Wang, L. 2009. 4-Chlorophenol degradation and hydrogen peroxide formation induced by DC diaphragm glow discharge in an aqueous solution. *Plasma Chemistry and Plasma Processing* 29:241–250.

102. Dang, T. H., A. Denat, O. Lesaint, and G. Teissedre. 2008. Degradation of organic molecules by streamer discharges in water: coupled electrical and chemical measurements. *Plasma Sources Science and Technology* 17.

103. Mededovic, S. and B. R. Locke. 2007. Atrazine degradation using pulsed electrical discharge in water. *Industrial & Engineering Chemistry Research* 46:2702–2709.

104. Karpel vel Leitner, N., G. Syoen, H. Romat, K. Urashima, and J.-S. Chang. 2005. Generation of active entities by the pulsed arc electrohydraulic discharge system and application to removal of atrazine. *Water Research* 39:4705–4714.

105. Willberg, D. M., P. S. Lang, R. H. Hochemer, A. Kratel, and M. R. Hoffmann. 1996. Degradation of 4-chlorophenol, 3,4-dichloroanilin, and 2,4,6-trinitrotoluene in an electrohydraulic discharge. *Environmental Science & Technology* 30:2526–2534.

106. Wang, L., X. Z. Jiang, and Y. J. Liu. 2008. Degradation of bisphenol A and formation of hydrogen peroxide induced by glow discharge plasma in aqueous solutions. *Journal of Hazardous Materials* 154:1106–1114.

107. Hoeben, W. F. L. M., E. M. van Veldhuizen, W. R. Rutgers, C. A. M. G. Cramers, and G. M. W. Kroesen. 2000. The degradation of aqueous phenol solutions by pulsed positive corona discharges. *Plasma Sources Science and Technology* 9:361–369.

108. Hoeben, W. F. L. M., E. M. van Veldhuizen, W. R. Rutgers, and G. M. W. Kroesen. 1999. Gas phase corona discharges for oxidation of phenol in an aqueous solution. *Journal of Physics D: Applied Physics* 32:L133–L137.

109. Ognier, S., C. Fourmond, S. Bereza, and S. Cavadias. 2009. Treatment of polluted water by gas–liquid discharge plasma reactor: role of ozone and active species. *High Temperature Material Processes* 13:439–452.
110. Hoeben, W. F. L. M., E. M. van Veldhuizen, H. A. Classens, and W. R. Rutgers. 1997. Degradation of phenol and atrazine in water by pulsed corona discharges. In *13th International Symposium on Plasma Chemistry*, August 18–22, Beijing University Press, Beijing, p. 1843.
111. Sharma, A. K., G. B. Josephson, D. M. Camaioni, and S. C. Goheen. 2000. Destruction of pentachlorophenol using glow discharge plasma process. *Environmental Science & Technology* 34:2267–2272.
112. Sano, N., T. Kawashima, J. Fujikawa, T. Fugimoto, T. Kitai, and T. Kanki. 2002. Decomposition of organic compounds in water by direct contact of gas corona discharge: Influence of discharge conditions. *Industrial & Engineering Chemistry Research* 41:5906–5911.
113. Bobkova, E. S., V. I. Grinevich, N. A. Ivantsova, and V. V. Rybkin. 2012. A study of sulfonol decomposition in water solutions under the action of dielectric barrier discharge in the presence of different heterogeneous catalysts. *Plasma Chemistry and Plasma Processing* 32:97–107.
114. Bubnov, A. G., V. I. Grinevich, N. A. Kuvykin, and O. N. Maslova. 2004. The kinetics of plasma-induced degradation of organic pollutants in sewage water. *High Energy Chemistry* 38:41–45.
115. Bobkova, E. S., V. I. Grinevich, N. A. Ivantsova, and V. V. Rybkin. 2012. Influence of various solid catalysts on the destruction kinetics of sodium lauryl sulfate in aqueous solutions by DBD. *Plasma Chemistry and Plasma Processing* 32:703–714.
116. Bobkova, E. S., A. A. Isakina, V. I. Grinevich, and V. V. Rybkin. 2012. Decomposition of aqueous solution of acetic acid under the action of atmospheric-pressure dielectric barrier discharge in oxygen. *Russian Journal of Applied Chemistry* 85:71–75.
117. Katayama, H., H. Honma, N. Nakagawara, and K. Yasuoka. 2009. Decomposition of persistent organics in water using a gas–liquid two-phase flow plasma reactor. *IEEE Transactions on Plasma Science* 37:897–904.
118. Grymonpre, D. R., W. C. Finney, R. J. Clark, and B. R. Locke. 2004. Hybrid gas–liquid electrical discharge reactors for organic compound degradation. *Industrial & Engineering Chemistry Research* 43:1975–1989.
119. Lukes, P. and B. R. Locke. 2005. Degradation of substituted phenols in hybrid gas–liquid electrical discharge reactor. *Industrial & Engineering Chemistry Research* 44:2921–2930.
120. Yee, D. C., S. Chauhan, E. Yankelevich, V. Bystritski, and T. K. Wood. 1998. Degradation of perchloroethylene and dicholorophenol by pulsed-electric discharge and bioremediation. *Biotechnology and Bioengineering* 59:438–444.
121. Laroussi, M. 1996. Sterilization of contaminated matter with an atmospheric pressure plasma. *IEEE Transactions on Plasma Science* 24:1188–1191.
122. Laroussi, M. 2000. Biological decontamination by nonthermal plasma. *IEEE Transactions on Plasma Science* 28:184–188.
123. Laroussi, M. 2002. Nonthermal decontamination of biological media by atmospheric-pressure plasma: Review, analysis and prospects. *IEEE Transactions on Plasma Science* 30:1409–1415.
124. Laroussi, M. 2005. Low temperature plasma-based sterilization: Overview and state-of-the-art. *Plasma Processes and Polymers* 2:391–400.
125. Soloshenko, I. A., V. Y. Bazhenov, V. A. Khomich, V. V. Tsiolko, and N. G. Potapchenko. 2006. Comparative research of efficiency of water decontamination by UV radiation of cold hollow cathode discharge plasma versus that of low- and medium-pressure mercury lamps. *IEEE Transactions on Plasma Science* 34:1365–1369.
126. www.philips.com/uvpurification (accessed on March 2013).
127. http://ozonia.com/ozone.php (accessed on March 2013).
128. Kogelschatz, Y. and B. Eliasson. 1995. Chapter 26, Ozone generation and applications. In *Handbook of Electrostatic Processes*. J. S. Chang, A. J. Kelly, J. M. Crowley, editors. Marcel Dekker, Inc., New York, pp. 581–605.

129. Anpilov, A., E. Barkhudarov, N. Christofi, V. Kopev, I. Kossyi, and M. Taktakishvili. 2004. The effectiveness of a multi-spark electric discharge system in the destruction of microorganisms in domestic and industrial wastewaters. *Journal of Water and Health* 2:267–277.

130. Abou-Ghazala, A., S. Katsuki, K. H. Schoenbach, F. C. Dobbs, and K. R. Moreira. 2002. Bacterial decontamination of water by means of pulsed-corona discharges. *IEEE Transactions on Plasma Science* 30:1449–1453.

131. Chen, C. W., H. M. Lee, and M. B. Chang. 2008. Inactivation of aquatic microorganisms by low-frequency AC discharges. *IEEE Transactions on Plasma Science* 36:215–219.

132. Ching, W. K., A. J. Colussi, H. J. Sun, K. H. Nealson, and M. R. Hoffmann. 2001. *Escherichia coli* disinfection by electrohydraulic discharges. *Environmental Science & Technology* 35:4139–4144.

133. de Wever, H., H. Elslander, W. Boenne, L. Diels, D. Zander, M. de Roeck, and R. Rego. 2007. Disinfection and cleaning of water by a novel plasma treatment concept. In *3rd International Congress on Cold Atmospheric Pressure Plasmas: Sources and Applications*, July 10–13, Ghent, Belgium.

134. Dors, M., E. Metel, J. Mizeraczyk, and E. Marotta. 2008. Coli bacteria inactivation by pulsed corona discharge in water. *International Journal of Plasma Environmental Science & Technology* 2:34–37.

135. Efremov, N. M., B. Y. Adamiak, V. I. Blochin, S. J. Dadshev, K. I. Dmitriev, V. N. Semjonov, V. F. Levashov, and V. F. Jusbashev. 2000. Experimental investigation of the action of pulsed electrical discharges in liquids on biological objects. *IEEE Transactions on Plasma Science* 28:224–228.

136. Fudamoto, T., T. Namihira, S. Katsuki, H. Akiyama, T. Imakubo, and T. Majima. 2008. Sterilization of *E. coli* by underwater pulsed streamer discharges in a continuous flow system. *Electrical Engineering in Japan* 164:1–7.

137. Hong, Y. C., H. J. Park, B. J. Lee, W.-S. Kang, and H. S. Uhm. 2010. Plasma formation using a capillary discharge in water and its application to the sterilization of *E. coli*. *Physics of Plasmas* 17:053502.

138. Lukes, P., M. Clupek, V. Babicky, and T. Vykouk. 2007. Bacterial inactivation by pulsed corona discharge in water. In *16th IEEE International Pulsed Power Conference*, June 17–22, Albuquerque, NM, pp. 320–323.

139. Mazurek, B., P. Lubicki, and Z. Staroniewicz. 1995. Effect of short HV pulses on bacteria and fungi. *IEEE Transactions on Dielectrics and Electrical Insulation* 2:418–425.

140. Marsili, L., S. Espie, J. G. Anderson, and S. J. MacGregor. 2002. Plasma inactivation of food-related microorganisms in liquids. *Radiation Physics and Chemistry* 65:507–513.

141. Ohshima, T., K. Sato, H. Terauchi, and M. Sato. 1997. Physical and chemical modifications of high-voltage pulse sterilization. *Journal of Electrostatics* 42:159–166.

142. Rutberg, P. G. 2002. Some plasma environmental technologies developed in Russia. *Plasma Sources Science and Technology* 11:A159–A165.

143. Rutberg, P. G., V. A. Kolikov, V. E. Kurochkin, L. K. Panina, and A. P. Rutberg. 2007. Electric discharges and the prolonged microbial resistance of water. *IEEE Transactions on Plasma Science* 35:1111–1118.

144. Satoh, K., S. J. MacGregor, J. G. Anderson, G. A. Woolsey, and R. A. Fouracre. 2007. Pulsed-plasma disinfection of water containing *Escherichia coli*. *Japanese Journal of Applied Physics Part 1–Regular Papers Brief Communications & Review Papers* 46:1137–1141.

145. Sakiyama, Y., T. Tomai, M. Miyano, and D. B. Graves. 2009. Disinfection of *E. coli* by nonthermal microplasma electrolysis in normal saline solution. *Applied Physics Letters* 94:161501.

146. Shmelev, V. M., N. V. Evtyukhin, and D. O. Che. 1996. Water sterilization by pulse surface discharge. *Chemical Physics Reports* 15:463–468.

147. Yang, Y., H. Kim, A. Starikovskiy, Y. I. Cho, and A. Fridman. 2011. Note: An underwater multi-channel plasma array for water sterilization. *Review of Scientific Instruments* 82:096103.

148. Zhang, R. B., L. M. Wang, Y. Wu, Z. C. Guan, and Z. D. Jia. 2006. Bacterial decontamination of water by bipolar pulsed discharge in a gas–liquid–solid three-phase discharge reactor. *IEEE Transactions on Plasma Science* 34:1370–1374.

149. Edebo, L. and I. Selin. 1968. The effect of the pressure shock wave and some electrical quantities in the microbicidal effect of transient electric arcs in aqueous systems. *Journal of General Microbiology* 50:253.

150. Anpilov, A. M., E. M. Barkhudarov, N. Christofi, V. A. Kop'ev, I. A. Kossyi, M. I. Taktakishvili, and Y. Zadiraka. 2002. Pulsed high voltage electric discharge disinfection of microbially contaminated liquids. *Letters in Applied Microbiology* 35:90–94.

151. van Heesch, E. J. M., G. J. J. Winands, and A. J. M. Pemen. 2008. Evaluation of pulsed streamer corona experiments to determine the O* radical yield. *Journal of Physics D: Applied Physics* 41:234015.

152. Hackam, R. and H. Akiyama. 2000. Air pollution control by electrical discharges. *IEEE Transactions on Dielectrics and Electrical Insulation* 7:654–683.

153. Gupta, S. B. and H. Bluhm. 2008. The potential of pulsed underwater streamer discharges as a disinfection technique. *IEEE Transactions on Plasma Science* 36:1621–1632.

154. Kadowaki, K., T. Sone, T. Kamikozawa, H. Takasu, and S. Suzuki. 2009. Effect of water-surface discharge on the inactivation of *Bacillus subtilis* due to protein lysis and DNA damage. *Bioscience, Biotechnology, and Biochemistry* 73:1978–1983.

155. Sato, M., T. Ohgiyama, and J. S. Clements. 1996. Formation of chemical species and their effects on microorganisms using a pulsed high-voltage discharge in water. *IEEE Transactions on Industry Applications* 32:106–112.

156. Sun, P., H. Wu, N. Bai, H. R. Wang, H. Feng, W. Zhu, J. Zhang, and J. Fang. 2012. Inactivation of *Bacillus subtilis* sproes in water by a direct-current, cold atmospheric-pressure air plasma microjet. *Plasma Processes and Polymers* 9:157–164.

157. van Heesch, E. J. M., A. J. M. Pemen, P. A. H. J. Huijbrechts, P. C. T. van der Laan, K. J. Ptasinski, G. J. Zanstra, and P. de Jong. 2000. A fast pulsed power source applied to treatment of conducting liquids and air. *IEEE Transactions on Plasma Science* 28:137–142.

158. Kong, M. G., G. Kroesen, G. Morfill, T. Nosenko, T. Shimizu, J. van Dijk, and J. L. Zimmermann. 2009. Plasma medicine: An introductory review. *New Journal of Physics* 11:115012.

159. Laroussi, M., O. Minayeva, F. C. Dobbs, and J. Woods. 2006. Spores survivability after exposure to low-temperature plasmas. *IEEE Transactions on Plasma Science* 34:1253–1256.

160. Burlica, R. and B. R. Locke. 2008. Pulsed plasma gliding arc discharges with water spray. *IEEE Transactions on Industry Applications* 44:482–489.

13

Plasma-Assisted Surface Modification of Polymeric Biomaterials

Nathalie De Geyter,
Peter Dubruel,
Rino Morent, and
Christophe Leys

13.1 Introduction on Surface Engineering of Biomaterials

It is obvious that the interactions at the surface of a material in contact with its biological surroundings are of significant interest, in particular for artificial implants in contact with proteins and cells [1,2]. The first important feature is the enhancement in biocompatibility and the biointegration of the implant. Biomaterials are defined as substances other than foods or drugs contained in therapeutic or diagnostic systems [3,4]. They were used in ancient Egypt culture, and examples of replacing missing human tissue with biomaterials such as artificial eyes of glass or wooden teeth are known. However, nowadays implants are much more sophisticated and the demands much more stringent: only materials with limited risks with regard to rejections and complications are allowed to be implanted [5].

Therefore, a lot of research is oriented toward the improvement of the biocompatibility and biointegration of the implant by modifying the implant surface [6,7].

In some applications such as stents, contact lenses, and catheters, it is crucial to avoid the adsorption of unspecific proteins [8]. Protein adsorption is a complex process depending mainly on the surface energy: the more hydrophilic a surface, the less proteins adsorb on the surface [2]. Other factors that influence protein adsorption are the electrostatic interaction between proteins and surfaces, the protein concentration, and the structural properties of proteins [9].

For certain applications, the biomaterial surface should have antibacterial properties [10]. It is known that there is a competition between cells and bacteria to grow on implants. With an antibacterial surface, the growth of bacteria is diminished or avoided in favor of desired cell growth. The leakage of toxic compounds from the biomaterial into the body should be avoided and the implant surface should be built to prevent release of toxic compounds into the biological environment or the human body.

At first research was oriented toward biostable implants, but in the past few decades biodegradable implants have become more and more prominent with the boom in tissue engineering. In tissue

401

engineering biological substitutes are developed to restore, improve, or maintain original tissue action [11–14]. In the past decade a shift from biostable to biodegradable implants has taken place. The field of tissue engineering focuses on the development of biological substitutes that restore, maintain, or improve tissue function [11–14]. The three most important goals of a tissue engineering scaffold are [15]

- Enclosing a space that will shape the regenerating tissue.
- Providing the temporary function in a defect while the tissue regenerates.
- Promoting ingrowth of tissue.

As the original scaffold is only a temporary substitute, the degradation products of the surface and bulk material have to be biocompatible to avoid inflammation when the scaffold degrades. Thus, the tissue engineering scaffold should have a surface that enhances cell attachment, proliferation, and differentiation. However, the necessary functional groups to promote cell growth are absent on the surface of most of the implants created from typical synthetic biomaterials. Therefore surface modification is of crucial importance to prepare successful tissue engineering scaffolds.

13.2 Surface Modification Techniques

Although most research is oriented toward the modification of only the surface of a biomaterial, some work has also been done on bulk modification. To achieve surfaces with desired properties, one could think about producing materials that contain the same specifications throughout the whole material (bulk and surface). This so-called bulk modification is a strategy whereby cell-signaling peptides are incorporated into the biomaterials, and the resulting recognition sites are not only present on the surface but also in the bulk of the materials [16]. Such an approach is sometimes used for applications with injectable biomimetic materials such as hydrogels and microspheres. Bulk incorporation of functional groups in biomaterials offers opportunities to fine-tune degradation rates. However, this chapter will focus on pure surface modification, which differs from bulk modification in that a polymeric material is fabricated from a bulk material and only the surface is subsequently altered.

Because the literature available in this field is significant, it is not the intent of this chapter to be all-embracing and to cover all the research done on this subject during the past few decades. It is also clear that not all the described techniques will be of the same interest for biomaterials. Therefore, this section will discuss the most common surface modification strategies for biomaterials at an introductory level. It can also be understood that a lot of research combines methods or uses methods that are not always straightforward to categorize. Nonthermal plasma treatment will be discussed only briefly in this section because Section 13.2 is devoted to this technique.

13.2.1 Ultraviolet Treatment and Photografting

A method for modifying surfaces of polymers is ultraviolet (UV) treatment, which has been extensively studied for surface graft polymerization of polymers in the presence of a photoinitiator. Extensive reviews on this subject can be found in References [17–19]. The functionalities introduced at the surface depend on the chemical structure of the grafted polymer. Three main operating modes of UV treatment can be distinguished: (1) VUV, that is, the sample is kept under vacuum conditions; (2) the sample is submerged in an inert gas such as argon; or (3) the sample is covered with a monomer solution. Often, the photoinitiator is either precoated on the substrate or available in the solution. The most commonly used photoinitiator is benzophenone (BPO), which is excited by UV light and which can, when it relaxes, abstract hydrogen atoms from the polymer film, thus creating free radicals at the surface. BPO was also used as a photoinitiator in Goda et al.'s study [20] to increase the protein resistance of polydimethylsiloxane. A phosphorylcholine group functional methacrylate monomer was graft polymerized on the surface, and by controlling the graft density the resistance to protein adsorption could be tuned.

The development of polymers with amine side chains was reported in the literature. Grafting of 2-aminoethyl methacrylate (AEMA) was described by Yu et al. [21]. The improvement in antifouling properties on polypropylene (PP) membranes by photoinduced graft polymerization of acrylamide (AAm) was reported by the same group in another paper [22].

The photografting process is a complex process. The interaction between molar ratio, monomer concentration, photoinitiator, irradiation time, UV, and solvent determines the final success of the grafting process [23]. These complex interactions are not fully understood, and therefore no predictive model is available as yet. Therefore, for each particular combination of monomer and surface, an optimization study is needed. Also, the selection of the applied solvent has a great influence on the result and the extent of grafting, as demonstrated by Edlund et al. [24]. However, the use of solvents is one of the major drawbacks of the above-mentioned case studies. To avoid the use of solvents, a solvent-free grafting technique was developed: biodegradable polymers were subjected to the vapor phase of a monomer, while the grafting reaction was induced by photoinitiation [24–29].

13.2.2 Wet-Chemical Methods for Surface Modification

A major class of methods includes the wet-chemical methods. This involves the reaction between a chemical compound in solution and a surface [15]. In general, these wet-chemical methods are very useful, but some drawbacks have to be mentioned. The most important disadvantage is that the reactions are in most cases nonspecific, and therefore they simultaneously introduce several functional groups. Also, the degree of surface modification may not be repeatable between polymers of different molecular weights, crystallinity, or tacticity [30]. Surface etching occurs, and irregular surface etching has already been reported [31]. Moreover, as this surface degradation affects the outer shell of the material, mechanical performance can be lowered [32] or faster degradation can occur.

In this section, three typical examples will be shortly described: (1) hydrolysis, (2) aminolysis, and (3) hydrogen peroxide treatment (which is also considered as a wet-chemical method).

13.2.2.1 Hydrolysis

When a polyester is hydrolyzed via auto-catalytic cleavage of main-chain ester bonds, hydroxyl and carboxyl end groups are formed [33], and in most cases surface roughness and surface hydrophilicity increases [34]. This results in increased cell attachment and cell spreading. Typically, subsequent immobilization of natural proteins tends to enhance cell adhesion and cell viability [32,35].

13.2.2.2 Aminolysis

Similar to hydrolysis, aminolysis is as well a surface degradation reaction. The surface roughness also tends to increase accompanied by an increase in the wettability. Therefore, it is a challenging task to evaluate to what extent the increased wettability can be attributed to increased roughness and/or to what extent the change in wettability is due to the introduction of amine groups on the surface.

13.2.2.3 Hydrogen Peroxide Treatment

Hydrogen peroxide decomposes to hydroxyl radicals. These radicals are more reactive than other oxidative chemicals in the presence of UV light [36]. By immersing samples in hydrogen peroxide solution and by simultaneous irradiation with UV light, hydroperoxide groups are formed on the polymer surface [37]. This is called photooxidation. In the next step, the sample is immersed in a monomer solution and exposed a second time to UV light to initiate graft polymerization [37–40].

13.2.3 Ozone Treatment

A third strategy consists in the exposure of a biomaterial to ozone in order to oxidize the surface. This technique can be used as an ozone-alone treatment; however, when combined with UV radiation the process is substantially accelerated. Ozone can be used as such, but it was found that a combination of ozone/UV irradiation significantly fastened the process. Following ozone treatment, subsequent grafting of monomers on the oxidized surface is possible. Liu et al. studied the grafting of different hydrophilic polymers such as poly(ethylene oxide) (PEO), chitosan, and poly(vinyl alcohol) (PVA) to an ultrafiltration membrane composed of polyethersulfone by ozone/UV. The modified films demonstrated increased hydrophilicity, reduced protein adsorption, and increased roughness [41].

Again, as was the case for the previously described wet-chemical methods, ozone treatment is a nonspecific technique resulting in different functional groups simultaneously incorporated at the surface. A major drawback of this technique is the degradation of the polymers. Ozone tends to enhance this polymer degradation [42].

13.2.4 High-Energy Radiation

In industry, the most commonly used radiation types include γ-radiation and e-beam radiation [43]. Ion beams are used to produce ion implantation in the outer surface layer or to deposit coatings. For application with polymers, several ions have been used such as ions from hydrogen and helium atoms as well as ions of gold or uranium [43]. The implantation of ions at the surface does not immediately introduce functional groups onto the surface. However, the surface chemistry and thus the surface properties are altered. It should be noted that for polymer treatment high-energy irradiation can cause additional chemical effects. Recombination of free radicals leads to cross-linking, while chain cleaving leads to polymer degradation [24,44]. In most cases both processes occur simultaneously. Radiation at higher energy is not considered a viable route for the modification of most biodegradable polymers [45,46].

In the past 20 years, the interaction between biomaterials and γ-radiation has generated significant interest [47] because high-energy photons can be employed to modify the surface. Radical sites and free molecules on the surface are created by these photons and can lead to propagation or termination reactions such as recombination. Comparable to plasma treatment, two approaches can be distinguished for γ-ray irradiation: postirradiation and syn-irradiation. Gamma-ray irradiation can initiate the polymerization and grafting of a monomer to a surface [48–51] (syn-irradiation), or it can produce peroxides and hydroperoxides on a surface, which can be subsequently used as initiating sites for graft polymerization [52] (postirradiation). Finally, γ-ray irradiation in water can lead to hydroxylation of the surface [50].

13.2.5 Self-Assembly

Self-assembly is the process whereby a disordered system of components forms and organizes a structure or pattern because of specific, local interactions among the components themselves, without external direction [15]. This means that certain molecules are able to interact with the surface to form a self-assembled monolayer (SAM) on the surface. Different functionalities can be introduced onto the surface determined by the chemical structure of the molecules of which the SAM layer is built up. A typical example is the interaction between gold and *n*-alkane thiols. The SAMs on gold have also been engineered to enable dynamic control of interfacial activity, such as wettability and chemical reactivity of the surface. The applicability of SAMs is somewhat limited because they are typically formed on gold and silver. To overcome this limitation, block-copolymers of an amphiphilic nature are being investigated to form self-assembled layers onto polymers [53–55]. These layers are formed by hydrophobic interactions.

13.2.6 Nonthermal Plasma Treatment

As this chapter focuses on nonthermal plasma treatment in the next part, the major techniques will be outlined very briefly here.

Different functionalities can be directly or indirectly introduced by nonthermal plasma treatment on inert surfaces. Direct modification includes the reactive ammonia (NH_3) plasma, which is known to introduce, among others, amines, where air and oxygen plasmas introduce a mixture of mainly COOH and OH functionalities. Argon plasmas are typically used to introduce free radicals. These free radicals can interact with the atmosphere to produce oxygen-containing functionalities. Indirect modification is based on the grafting of polymers containing the desired functionalities onto the surface. Both strategies and recent advancements in the field will be extensively discussed in Section 13.3.

The use of a plasma in a surface modification technique has some major advantages. The major advantage is that the technique is solvent free, and therefore the use of hazardous solvents is avoided. Second, normal nonthermal plasma treatment will not affect the mechanical properties of the polymer, while some other techniques will significantly influence the mechanical properties of the outer layer of the implant. As mentioned above, a wet-chemical treatment of a surface can cause (partial) degradation and scissions of the polymer at the surface. This results in a decrease in the mechanical strength and faster degradation.

Another advantage is that any plasma is able to uniformly alter the surface chemistry of a surface, regardless of the geometry. The technique can be used on complex objects such as for 3D components for tissue engineering or artificial organs [56], nanoparticles, and films [57].

13.3 Nonthermal Plasma Surface Modification

In Section 13.2, some surface modification technologies and their possible advantages and drawbacks were discussed. Plasma-based strategies for the surface modification of (biodegradable) polymers, and biomaterials in general, have already demonstrated great potential, and this chapter will focus on these plasma-based approaches. This book deals with low-temperature (or nonthermal or cold or nonequilibrium) plasmas and its applications. It is quite obvious that the high temperatures used in thermal plasmas damage polymers and practically all biopolymer surface modification applications employ nonthermal or cold plasmas.

To begin with, the interaction between plasmas and materials is discussed in general. The different approaches for surface modification that evolve from this interaction and the research undertaken in this area, focusing on materials, monomers, and plasma types, will be described.

13.3.1 Plasma–Surface Interactions

A nonthermal plasma is an extremely reactive environment in which several interactions between a polymeric surface and a plasma are possible. According to the interaction result, one can distinguish between three main types of plasma reactions: plasma polymerization, plasma activation, and plasma etching (or ablation). Plasma polymerization occurs when an organic monomer is converted into reactive fragments. These reactive fragments can polymerize and form deposits on the substrate, resulting in a coating of the surface. The polymerization of the monomer can be initiated in two ways: the surface is covered with a liquid monomer or the discharge gas contains the monomer in the vapor phase. In both situations, substrate and coating are bombarded with ions from the plasma, resulting in plasma etching. However, not all gases lead to reactive intermediates that are able to polymerize. This is the case for different so-called inert gases such as He, Ar, O_2, N_2, and air. These gases do not deposit a polymerized coating but create or substitute functional groups, or create radicals, on the surface. This is a typical plasma activation strategy.

It is clear that the process vapor, the substrates, and the process conditions determine which process—deposition, substitution, or etching—is dominant. This process was defined by Yasuda et al. [58] as competitive ablation polymerization (CAP) (see Figure 13.1).

These different interactions result in different plasma-based surface modification strategies that will be extensively discussed in the following paragraphs.

13.3.2 Plasma Activation

In the case of plasma activation, the discharge is generated in an inert gas such as Ar, He, O_2, N_2, NH_3, and CF_4, and not in a monomer in the vapor phase (see Figure 13.2). This results in the introduction of chemical functionalities at the surface or in the creation of free radicals. The latter can be used for

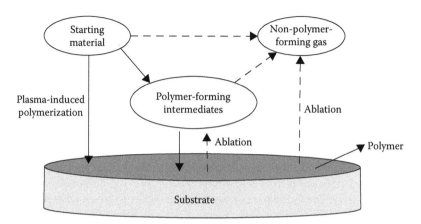

FIGURE 13.1 A schematic representation of competitive ablation polymerization. (Adapted from Yasuda, H., *Plasma Polymerization*, Academic Press, Orlando, FL, 1985.)

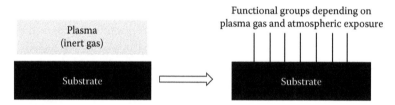

FIGURE 13.2 A schematic representation of plasma activation in inert gases.

cross-linking or surface grafting. In the following step, the introduced functionalities can bind polymers or other molecules to the surface to achieve the desired surface properties. Very often, plasma activation is mainly used to increase the hydrophilicity of the surface.

Plasma activation of a surface has different effects. Typically, surface wettability is increased due to the incorporation of functional groups as is roughness; moreover, chain scission and cross-linking can occur. Some of these changes are linked to each other in the sense that the final surface properties are a complex interplay between the original polymer material, the used gas, the processing parameters, and the storage time (aging) and storage conditions [59–61].

Plasma treatment in gases containing oxygen leads to the introduction of oxygen-containing functional groups such as carboxylic acid groups, hydroxyl groups, and peroxide (due to postplasma reactions). Carboxylic acid groups may also be introduced by discharges generated in CO_2 or CO plasmas. Using CO_2 as gas in which the plasma is generated will also result in additional functional groups such as hydroxyls, ketones, aldehydes, and esters [62]. Nitrogen, ammonia, and N_2/H_2 plasmas incorporate primary, secondary, and tertiary amines, as well as amides. Polymers treated in a pure Ar or He plasma will not immediately lead to the incorporation of new functionalities on the polymer surface, but free radicals will be created on the surface [63]. These free radicals can react with oxygen from the atmosphere [64]. However, it should be noted that it is sometimes very difficult to work in pure Ar or pure He. The presence of only a small number of oxygen impurities can result in the incorporation of oxygen functionalities during treatment. It is clear that plasma activation is not a selective technique because it does not result in a unique functionality.

Aging effects, that is, effects of postplasma rearrangements (surface adaptation) and reactions (postplasma oxidation), should not be underestimated [59–61]. Siow et al. [62] presented an excellent overview of studies on aging effects. They suggested that the surface chemistry should be characterized at the time of biological testing to establish the correlation between chemical surface composition and biological response [62]. Sharma et al. [65] demonstrated that oxygen levels on the polycarbonate (PC) surface continued to increase until 72 h after He plasma treatment. The wettability of the PC surface dropped initially from 93°C to 30°C but recovered up to 67°C due to aging effects. This dynamic behavior of plasma-treated surfaces during aging was also studied by Murakami et al., [66,67] and they linked it to a reorientation of the introduced polar groups. A surface has the tendency to minimize the interfacial energy. If an O_2 plasma-treated poly(dimethyl siloxane) (PDMS) surface is aged in air, the surface returns to a low-energy state. This phenomenon can be explained by a reorientation of the polar groups toward the interface, minimizing the interfacial free energy [15,68]. However, when the sample is stored in water, or in an aqueous phase, the polar groups are still present on the surface, minimizing the interfacial free energy by expanding the polar groups toward the polar solvent. Also, temperature is a critical factor responsible for aging behavior of plasma-treated polymers. A rapid change of the contact angle at higher temperatures supports the assumption that the changes are caused by polymer chain motion, reorienting the polar groups into the bulk [59,69]. Another effect is postplasma oxidation, that is, a reaction between the remaining radicals and the atmospheric oxygen.

Wan et al. [70] reported that the modification of poly-L-lactide (PLLA) substrates with NH_3 plasma generates surfaces to which cells adhere.

Ligands with a biological activity can be immobilized on the activated surface. Ho et al. showed that the peptide Arg–Gly–Asp–Ser (RGDS) was immobilized using Ar plasma treatment on PLLA porous scaffolds [71]. They noted that such modification can make PLLA scaffolds more suitable for the culture of osteoblast-like cells and for generating bone-like tissues. NH_3 plasma treatment and subsequent immobilization of collagen improved cell affinity of a polylactide (PLA) film toward fibroblasts as evidenced by Zhao et al. [72]. Compared to the films with collagen adsorbed on the surface or just NH_3 plasma-activated polymer films, the covalently immobilized collagen films were superior with regard to cell attachment and proliferation.

Masson et al. [73] successfully immobilized hyaluronic acid on several polymers that were modified by Ar and NH_3 plasma activation. They showed that steric effects were involved in the reactivity of the hyaluronic acid toward the groups on the surface.

For mimicking the natural extracellular matrix (ECM), the surface chemistry and presence of ligands should be considered. The ECM is a fibrous environment consisting of different proteins such as collagen and fibronectin. There is growing interest in novel technologies such as electrospinning (ESP) that are able to mimic the structural properties of the ECM by creating micro- and nanofibers [15]. Therefore, ESP–polycaprolactone (PCL) fibers were coated with gelatin [74]. PCL fibers were treated with air plasmas to introduce carboxylic groups, which were then covalently coupled with gelatin. The authors have shown that endothelial cells spread better and proliferate on the gelatin-grafted PCL. They reported that endothelial cells were oriented along gelatin-grafted aligned nanofibers, as opposed to nongrafted aligned nanofibers.

Gugala et al. [75] showed the influence of different plasma treatments on the attachment, growth, and activity of rat osteoblasts cultured on PLA. They observed that treatment with an NH_3 plasma was more efficacious than with O_2 or SO_2/H_2 plasmas.

The adhesion and growth of mouse fibroblast cells was greatly enhanced by the anchorage of basic fibroblast growth factor on CO_2-plasma activated poly(lactic-*co*-glycolic acid) (PLGA) films [76].

Although Ar plasma treatment is successful in many applications, Woodfield et al. observed that it was not suitable for surface modification of poly(ethylene oxide)terephthalate-*co*-poly(butylene)terephthalate (PEOT/PBT) block copolymers. The adhesion of chondrocyte was improved but the ability of these substrates to maintain the chondrocyte phenotype was reversed [77].

13.3.3 Plasma Postirradiation Grafting

The use of He and Ar plasma is known to introduce mainly radicals on the surface. If those free radicals are subjected to the atmosphere or O_2, peroxides and hydroperoxides are formed. Those formed functionalities can then be used to initiate a polymerization reaction (see Figure 13.3). This is the

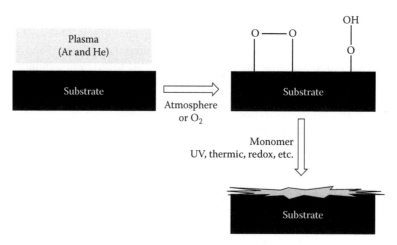

FIGURE 13.3   A schematic representation of plasma postirradiation grafting.

so-called postirradiation grafting technique. Note that in this case the monomer is not subjected to the plasma. This implies that the polymer that will be grafted has a similar composition as polymers obtained by conventional polymerization. In literature, postirradiation grafting is often referred to as "plasma-induced graft (co)polymerization." It leads to covalent bonding of selective functionalities onto the surface.

It is clear that the number of (hydro)peroxides present at the surface will have a great effect on the grafting density because every peroxide group is a potential initiating site. Mostly these (hydro)peroxides are determined by spectroscopic reaction with 1,1-diphenyl-2-picrylhydrazyl (DPPH) [52,56,78–80]. Also, 2,2,6,6-tetramethyl-1-piperidinyloxy (TEMPO) has been used for demonstrating the presence of free radicals on silicon surfaces [81].

Müller and Oehr [82] compared plasma activation, plasma polymerization, and plasma grafting for the introduction of primary amino groups on microfiltration membranes. They observed that the introduction of primary amine groups is maximal with the plasma postgrafting method. According to the authors, this phenomenon was due to monomer fragmentation in the plasma. When the monomer is subjected to plasma several side reactions can occur. In that case, a more heterogeneous formation of several types of amino compounds can occur due to hydrogen abstraction from the amino group in the plasma. In general, the major advantage of the postgrafting approach is the ability to avoid these side reactions and thus obtain a high specificity.

As for plasma-treated polymers, aging is equally important for postirradiation grafted substrates [62]. The effect of a 30 days' storage period on the grafting density of acrylic acid (AA) on poly(ethylene terephthalate) (PET) films was analyzed by Gupta et al. [83]. They noted significant changes in surface chemistry at ambient temperature. However, if the surface was functionalized with collagen, the effect of aging was far less important.

Kim et al. [84] reported the immobilization of heparin and insulin on a PET surface grafted with AA to obtain a better blood compatibility. Coagulation tests revealed that PET grafted with PEO demonstrated only a slight improvement, while the PET surfaces with PEO and immobilized heparin or insulin showed a significant improvement. This illustrates once more that the immobilization of biomacromolecules is crucial for certain applications.

To enhance cell adhesion and proliferation in tissue engineering, the immobilization of collagen on a postirradiation grafted substrate has been extensively studied [32,85–89]. The coating of collagen, in general, was also studied in order to reduce the foreign body reaction [7,90] and the development of tumors [91]. These collagen-immobilized surfaces were tested with a wide range of cells such as human umbilical vein endothelial cells (HUVECs) [89], human dermal fibroblasts (HDFs) [85,88], and different types of smooth muscle cells [32,87]. In all cases, the collagen-immobilized surfaces got a favorable response from cells.

Postirradiation grafting is not limited to the immobilization of collagen alone. Sano et al. grafted AAm to high-density polyethylene (HDPE) films and subsequent modification led to carboxyl groups or amino groups for immobilization of protein A, which could be successfully bound to both functionalities.

13.3.4 Plasma Syn-Irradiation Grafting

Plasma syn-irradiation grafting is a typical plasma-induced polymerization strategy and includes the adsorption of a monomer to the substrate surface, which is then subjected to a plasma (see Figure 13.4). Radicals are created in the adsorbed monomer layer and a substrate surface, resulting in a cross-linked polymer top layer.

In literature, Sasai et al. [92] reported the grafting of vinylmethylether-maleic anhydride (VEMA) onto low-density polyethylene (LDPE). In the first step, VEMA was coated on the LDPE surface layer using a mixed solvent cyclohexanon/p-xylene. Afterward, the substrate was dried with the VEMA on its surface. This surface was then exposed to an Ar plasma to covalently graft the VEMA to the LDPE surface. The authors applied this technique to create COOH groups on the surface, which were subsequently used to immobilize an atom transfer radical polymerization (ATRP) initiator [92].

Combinations of plasma activation and plasma syn-irradiation grafting can also be found in the literature. Ding et al. reported the immobilization of chitosan on PLLA [93]. In the first step, they exposed the pristine substrates to plasma in an inert gas. Afterward, they exposed the plasma-activated surfaces to the atmosphere to create peroxides. Then chitosan was adsorbed to the surface, and after drying the substrate was exposed to the plasma a second time. The substrates demonstrated poor cell adhesion, but proliferation was as in glass substrates [93].

13.3.5 Plasma Polymerization or Plasma Deposition

The last approach discussed in this chapter is plasma polymerization or plasma deposition. A monomer in the vapor phase is introduced in the plasma in which it is converted into reactive fragments (see Figure 13.5). These reactive fragments can recombine to polymers in the gas phase (so-called plasma-state polymerization). The polymers formed in a plasma will not necessarily have a structure and composition that is comparable to polymers achieved by conventional polymerization techniques. It is noteworthy that plasma polymerization occurs with many monomers in the vapor phase, even if they do not contain unsaturated bonds or cyclic structures. These polymers can be deposited on the substrate, thus creating a plasma-deposited polymer coating on the surface [15].

Plasma deposition differs from plasma grafting in that it coats the substrate rather than the covalently binding species to the modified polymer surface [94]. Using the same monomer, the plasma-polymerized polymers possess different chemical and physical properties compared to polymers obtained using more conventional synthesis methods.

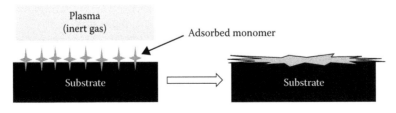

FIGURE 13.4 A schematic representation of syn-irradiation grafting.

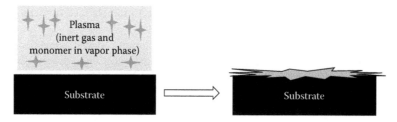

FIGURE 13.5 A schematic representation of plasma polymerization.

A wide variety of substrates and monomers have been used in applications. A correlation between the plasma chemistry and the character of the deposited film has been reported [94,95]. First of all, in plasmas the monomer can react in different ways resulting in different fragments. For example, a dissociation–ionization reaction will lead to a different reactive fragment than a protonation reaction, followed by a dehydrogenation reaction. Different chemical pathways will depend on the different combinations of plasma conditions such as power, pressure, and the duty factor. Therefore, to achieve homogeneous coatings with a retention of functional groups and to avoid cross-linking reactions, low discharge power and low-duty cycle have to be employed [96–99].

When the plasma is pulsed, pulsing frequency will determine the plasma chemistry and consecutively the deposited film [100]. Choukourov et al. showed that the concentration of free amines was higher at lower pulse duration (<0.5 ms), suggesting that this microsecond range is more preferable for technological processes. Kuhn et al. [97] also observed this for hydroxyl-containing coatings.

One can assume that for each monomer an optimization study should be performed, varying both the applied plasma operating conditions and the gases. In addition, possible side reactions should be taken into account [97]. Little fundamental theoretical knowledge or a general predictive model is available. In general, one could state that "softer" plasma reaction conditions lead to fewer side reactions [15].

Stability and aging are also a major concern for plasma-polymerized thin layers [101]. Drews et al. published an interesting study on the hydrolysis and stability of thin pulsed plasma maleic anhydride coatings. They observed that shorter treatment time gave rise to thinner films that were more stable than thicker films [102]. The hydrolysis and stability of the coatings strongly depended on the plasma power. With increasing plasma power, coatings became more stable. This can be explained by a higher cross-linking at higher plasma power, thus resulting in better stability. Also, the number of available carboxylic acid groups on the surface increased with decreasing power. This is consistent with the proposed dependence of the cross-linking as a function of the plasma power.

Taking into account the need of 3D scaffolds in tissue engineering, the work of Barry et al. [94] on the modification of PLA scaffolds is worth mentioning in which they compared plasma grafting with plasma polymerization by using allylamine. X-ray photoelectron spectroscopy (XPS) measurements of the scaffolds at set positions showed that the grafting resulted in a more homogenous N concentration through the scaffold, while plasma polymerization led to a higher N concentration on the outer and inner surface than the grafting procedure. The authors reported that deposition-treated scaffolds showed higher metabolic activity than the grafted scaffolds and that cells were found in the center of the deposited scaffolds but not in the grafted scaffolds.

Ju et al. treated a 3D porous PLLA scaffold with an AA plasma and observed improved cell adhesion and proliferation of chondrocytes as compared to unmodified scaffolds [103].

In literature, one can find data on the adsorption of plasmid DNA on different surfaces. First, allylamine is plasma polymerized and then grafting of PEO is performed. Hook et al. [104] reported that DNA preferably binds to the allylamine plasma-deposited coating and that both hydrophobic and electrostatic interactions contribute to the binding.

Apart from these and other adsorption studies of biomolecules, bioactive proteins were also covalently immobilized on plasma-polymerized films [105]. Fibroblasts were cultured on modified samples [94,106–108].

All papers reported positive effects such as fast adhesion and high cell activity. It was observed that fibroblasts tend to attach preferentially to the more hydrophilic regions achieved by plasma polymerization of allylamine or isopropyl alcohol [106,108].

13.4 Trends and Conclusions

Desmet et al. [15] evaluated the number of papers they discussed in their review on surface engineering of polymers and stated that it is a very fast developing field. The number of publications on biodegradable polymers has substantially increased in recent decades. Surface modification of biodegradable polymers is of interest for many research groups, and its use has risen in the past 10 years. However, this does not mean that there is a shift from biostable to biodegradable polymers because one can find in literature a significant number of recent publications dealing with surface engineering of biostable polymers. Research on tissue engineering and regenerative medicine has created a strong demand for biodegradable polymers with highly advanced tailored surfaces, while, at the same time, there is still a lot of research necessary to improve existing biostable implants.

A major aim of plasma-based strategies is to prevent the use of hazardous solvents in the process. The use of such hazardous organic solvents might pose problems with regard to cell viability. However, this could be considered a general trend in surface engineering. There has been significant research on less common materials such as plasma sources, gas mixtures, monomers, and biomaterials. In future, this broader research will enable the scientific community to develop scaffolds with desired mechanical and chemical specifications.

Also worth mentioning is the need for a better understanding of the aging effects of plasma-treated surfaces. The knowledge of the actual composition of the surface at the time of immobilization is desirable, and therefore better storage procedures should be developed and described.

Up till now, most research has been conducted on surface engineering of 2D substrates. Some (but only few) studies have demonstrated the applicability of plasma techniques for surface modification of the interior of 3D porous scaffolds. Nevertheless, given the specific needs of tissue engineering, surface modification of those 3D porous scaffolds will become increasingly important.

To conclude, one can state that a combination of a patient-specific, highly porous, 100% interconnected scaffold adapted with state-of-the-art plasma surface modification technology can and will greatly improve the success of modern tissue engineering.

References

1. Chilkoti, A., and J. A. Hubbell. 2005. Biointerface science. *MRS Bulletin* 30:175–176.
2. Nath, N., J. Hyun, H. Ma, and A. Chilkoti. 2004. Surface engineering strategies for control of protein and cell interactions. *Surface Science* 570:98–110.
3. Langer, R., and D. A. Tirrell. 2004. Designing materials for biology and medicine. *Nature* 428:487–492.
4. Peppas, N. A., and R. Langer. 1994. New challenges in biomaterials. *Science* 263:1715–1720.
5. Chen, V. J., and P. X. Ma. 2004. Nano-fibrous poly(L-lactic acid) scaffolds with interconnected spherical macropores. *Biomaterials* 25:2065–2073.
6. Thevenot, P., W. J. Hu, and L. P. Tang. 2008. Surface chemistry influences implant biocompatibility. *Current Topics in Medicinal Chemistry* 8:270–280.
7. van Wachem, P. B., M. Hendricks, E. H. Blaauw, F. Dijk, M. Verhoeven, P. T. Cahalan, and M. J. A. van Luyn. 2002. (Electron) microscopic observations on tissue integration of collagen-immobilized polyurethane. *Biomaterials* 23:1401–1409.
8. Chu, P. K., J. Y. Chen, L. P. Wang, and N. Huang. 2002. Plasma-surface modification of biomaterials. *Materials Science and Engineering R-Reports* 36:143–206.
9. Kondo, A., and H. Fukuda. 1998. Effects of adsorption conditions on kinetics of protein adsorption and conformational changes at ultrafine silica particles. *Journal of Colloid and Interface Science* 198:34–41.

10. Chen, K. S., Y. A. Ku, H. R. Lin, T. R. Yan, D. C. Sheu, and T. M. Chen. 2006. Surface grafting polymerization of *N*-vinyl-2-pyrrolidone onto a poly(ethylene terephthalate) nonwoven by plasma pretreatment and its antibacterial activities. *Journal of Applied Polymer Science* 100:803–809.

11. Bonassar, L. J., and C. A. Vacanti. 1998. Tissue engineering: The first decade and beyond. *Journal of Cellular Biochemistry* 30–31:297–303.

12. Langer, R., and J. P. Vacanti. 1993. Tissue engineering. *Science* 260:920–926.

13. Griffith, L. G., and G. Naughton. 2002. Tissue engineering—Current challenges and expanding opportunities. *Science* 295:1009–1014.

14. Mironov, V., G. Prestwich, and G. Forgacs. 2007. Bioprinting living structures. *Journal of Materials Chemistry* 17:2054–2060.

15. Desmet, T., R. Morent, N. De Geyter, C. Leys, E. Schacht, and P. Dubruel. 2009. Nonthermal plasma technology as a versatile strategy for polymeric biomaterials surface modification: A review. *Biomacromolecules* 10:2351–2378.

16. Shin, H., S. Jo, and A. G. Mikos. 2003. Biomimetic materials for tissue engineering. *Biomaterials* 24:4353–4364.

17. Uyama, Y., K. Kato, and Y. Ikada. 1998. Surface modification of polymers by grafting. In *Grafting/Characterization Techniques/Kinetic Modeling*, eds. H. Galina, Y. Ikada, K. Kato, R. Kitamaru, J. Lechowicz, Y. Uyama, and C. Wu. Springer-Verlag Berlin, Berlin 33. pp. 1–39.

18. Kato, K., E. Uchida, E. T. Kang, Y. Uyama, and Y. Ikada. 2003. Polymer surface with graft chains. *Progress in Polymer Science* 28:209–259.

19. Deng, J. P., L. F. Wang, L. Y. Liu, and W. T. Yang. 2009. Developments and new applications of UV-induced surface graft polymerizations. *Progress in Polymer Science* 34:156–193.

20. Goda, T., R. Matsuno, T. Konno, M. Takai, and K. Ishihara. 2008. Photografting of 2-methacryloyloxyethyl phosphorylcholine from polydimethylsiloxane: Tunable protein repellency and lubrication property. *Colloids and Surfaces B—Biointerfaces* 63:64–72.

21. Yu, H. Y., Z. K. Xu, H. Lei, M. X. Hu, and Q. Yang. 2007. Photoinduced graft polymerization of acrylamide on polypropylene microporous membranes for the improvement of antifouling characteristics in a submerged membrane-bioreactor. *Separation and Purification Technology* 53:119–125.

22. Yu, H. Y., J. M. He, L. Q. Liu, X. C. He, J. S. Gu, and X. W. Wei. 2007. Photoinduced graft polymerization to improve antifouling characteristics of an SMBR. *Journal of Membrane Science* 302:235–242.

23. Xing, C. M., J. P. Deng, and W. T. Yang. 2002. Surface photografting polymerization of binary monomers maleic anhydride and *n*-butyl vinyl ether on polypropylene film I. Effects of principal factors. *Polymer Journal* 34:801–808.

24. Edlund, U., M. Kallrot, and A. C. Albertsson. 2005. Single-step covalent functionalization of polylactide surfaces. *Journal of the American Chemical Society* 127:8865–8871.

25. Kallrot, M., U. Edlund, and A. C. Albertsson. 2007. Covalent grafting of poly(L-lactide) to tune the in vitro degradation rate. *Biomacromolecules* 8:2492–2496.

26. Kallrot, M., U. Edlund, and A. C. Albertsson. 2006. Surface functionalization of degradable polymers by covalent grafting. *Biomaterials* 27:1788–1796.

27. Wirsen, A., H. Sun, and A. C. Albertsson. 2005. Solvent-free vapor-phase photografting of acrylamide onto poly(ethylene terephthalate). *Biomacromolecules* 6:2697–2702.

28. Wirsen, A., H. Sun, and A. C. Albertsson. 2005. Solvent free vapour phase photografting of acrylamide onto poly(methyl methacrylate). *Polymer* 46:4554–4561.

29. Wirsen, A., H. Sun, L. Emilsson, and A. C. Albertsson. 2005. Solvent free vapor phase photografting of maleic anhydride onto poly(ethylene terephthalate) and surface coupling of fluorinated probes, PEG, and an RGD-peptide. *Biomacromolecules* 6:2281–2289.

30. Goddard, J. M., and J. H. Hotchkiss. 2007. Polymer surface modification for the attachment of bioactive compounds. *Progress in Polymer Science* 32:698–725.

31. Desai, S. M., and R. P. Singh. 2004. Surface modification of polyethylene. *Advances in Polymer Science* 169:234–291.
32. Chong, M. S. K., C. N. Lee, and S. H. Teoh. 2007. Characterization of smooth muscle cells on poly(epsilon-caprolactone) films. *Materials Science and Engineering C—Biomimetic and Supramolecular Systems* 27:309–312.
33. Jiao, Y. P., and F. Z. Cui. 2007. Surface modification of polyester biomaterials for tissue engineering. *Biomedical Materials* 2:R24–R37.
34. Cao, Y., W. Liu, G. Zhou, and L. Cui. 2007. Tissue engineering and tissue repair in immunocompetent animals: Tissue construction and repair. *Handchirurgie Mikrochirurgie Plastische Chirurgie* 39:156–160.
35. Choong, C. S. N., D. W. Hutmacher, and J. T. Triffitt. 2006. Co-culture of bone marrow fibroblasts and endothelial cells on modified polycaprolactone substrates for enhanced potentials in bone tissue engineering. *Tissue Engineering* 12:2521–2531.
36. Perincek, S. D., K. Duran, A. E. Korlu, and M. I. Bahtiyari. 2007. Ultraviolet technology. *Tekstil Ve Konfeksiyon* 17:219–223.
37. Guan, J. J., C. Y. Gao, L. X. Feng, and J. C. Shen. 2001. Surface modification of polyurethane for promotion of cell adhesion and growth 1: Surface photo-grafting with *N,N*-dimethylaminoethyl methacrylate and cytocompatibility of the modified surface. *Journal of Materials Science—Materials in Medicine* 12:447–452.
38. Zhu, Y. B., C. Y. Gao, J. J. Guan, and J. C. Shen. 2004. Promoting the cytocompatibility of polyurethane scaffolds via surface photo-grafting polymerization of acrylamide. *Journal of Materials Science—Materials in Medicine* 15:283–289.
39. Zhu, Y. B., C. Y. Gao, X. Y. Liu, and J. C. Shen. 2002. Surface modification of polycaprolactone membrane via aminolysis and biomacromolecule immobilization for promoting cytocompatibility of human endothelial cells. *Biomacromolecules* 3:1312–1319.
40. Guan, J. J., G. Y. Gao, L. X. Feng, and J. C. Sheng. 2000. Surface photo-grafting of polyurethane with 2-hydroxyethyl acrylate for promotion of human endothelial cell adhesion and growth. *Journal of Biomaterials Science—Polymer Edition* 11:523–536.
41. Liu, S. X., J. T. Kim, and S. Kim. 2008. Effect of polymer surface modification on polymer–protein interaction via hydrophilic polymer grafting. *Journal of Food Science* 73:E143–E150.
42. Singh, B., and N. Sharma. 2008. Mechanistic implications of plastic degradation. *Polymer Degradation and Stability* 93:561–584.
43. Clough, R. L. 2001. High-energy radiation and polymers: A review of commercial processes and emerging applications. *Nuclear Instruments and Methods in Physics Research Section B—Beam Interactions with Materials and Atoms* 185:8–33.
44. Sodergard, A. 2004. Perspectives on modification of aliphatic polyesters by radiation processing. *Journal of Bioactive and Compatible Polymers* 19:511–525.
45. Gupta, M. C., and V. G. Deshmukh. 1983. Radiation effects on poly(lactic acid). *Polymer* 24:827–830.
46. Loo, S. C. J., C. P. Ooi, and Y. C. F. Boey. 2004. Radiation effects on poly(lactide-co-glycolide) (PLGA) and poly(L-lactide) (PLLA). *Polymer Degradation and Stability* 83:259–265.
47. Amato, I., G. Ciapettia, S. Pagani, G. Marletta, C. Satriano, N. Baldini, and D. Granchi. 2007. Expression of cell adhesion receptors in human osteoblasts cultured on biofunctionalized poly-(epsilon-caprolactone) surfaces. *Biomaterials* 28:3668–3678.
48. Yang, Y., M. C. Porte, P. Marmey, A. J. El Haj, J. Amedee, and C. Baquey. 2003. Covalent bonding of collagen on poly(L-lactic acid) by gamma irradiation. *Nuclear Instruments and Methods in Physics Research Section B—Beam Interactions with Materials and Atoms* 207:165–174.
49. Cho, E. H., S. G. Lee, and J. K. Kim. 2005. Surface modification of UHMWPE with gamma-ray radiation for improving interfacial bonding strength with bone cement (II). *Current Applied Physics* 5:475–479.

50. Shojaei, A., R. Fathi, and N. Sheikh. 2007. Adhesion modification of polyethylenes for metallization using radiation-induced grafting of vinyl monomers. *Surface and Coatings Technology* 201:7519–7529.
51. Shin, Y. M., K. S. Kim, Y. M. Lim, Y. C. Nho, and H. Shin. 2008. Modulation of spreading, proliferation, and differentiation of human mesenchymal stem cells on gelatin-immobilized poly(L-lactide-co-epsilon-caprolactone) substrates. *Biomacromolecules* 9:1772–1781.
52. Shim, J. K., H. S. Na, Y. M. Lee, H. Huh, and Y. C. Nho. 2001. Surface modification of polypropylene membranes by gamma-ray induced graft copolymerization and their solute permeation characteristics. *Journal of Membrane Science* 190:215–226.
53. Kubies, D., L. Machova, E. Brynda, J. Lukas, and F. Rypacek. 2003. Functionalized surfaces of polylactide modified by Langmuir–Blodgett films of amphiphilic block copolymers. *Journal of Materials Science—Materials in Medicine* 14:143–149.
54. Popelka, S., L. Machova, and F. Rypacek. 2007. Adsorption of poly(ethylene oxide)-block-polylactide copolymers on polylactide as studied by ATR–FTIR spectroscopy. *Journal of Colloid and Interface Science* 308:291–299.
55. Murphy, K. A., J. M. Eisenhauer, and D. A. Savin. 2008. Synthesis, self-assembly and adsorption of PEO-PLA block copolymers onto colloidal polystyrene. *Journal of Polymer Science Part B—Polymer Physics* 46:244–252.
56. Lee, S. D., G. H. Hsiue, C. Y. Kao, and P. C. T. Chang. 1996. Artificial cornea: Surface modification of silicone rubber membrane by graft polymerization of pHEMA via glow discharge. *Biomaterials* 17:587–595.
57. Ryu, G. H., W. S. Yang, H. W. Roh, I. S. Lee, J. K. Kim, G. H. Lee, D. H. Lee, B. J. Park, M. S. Lee, and J. C. Park. 2005. Plasma surface modification of poly(D,L-lactic-co-glycolic acid)(65/35) film for tissue engineering. *Surface and Coatings Technology* 193:60–64.
58. Yasuda, H. 1985. *Plasma Polymerization*. Academic Press, Orlando, Fl.
59. De Geyter, N., R. Morent, and C. Leys. 2008. Influence of ambient conditions on the ageing behaviour of plasma-treated PET surfaces. *Nuclear Instruments and Methods in Physics Research Section B—Beam Interactions with Materials and Atoms* 266:3086–3090.
60. Morent, R., N. De Geyter, C. Leys, L. Gengembre, and E. Payen. 2007. Study of the ageing behaviour of polymer films treated with a dielectric barrier discharge in air, helium and argon at medium pressure. *Surface and Coatings Technology* 201:7847–7854.
61. Morent, R., N. De Geyter, M. Trentesaux, L. Gengembre, P. Dubruel, C. Leys, and E. Payen. 2010. Influence of discharge atmosphere on the ageing behaviour of plasma-treated polylactic acid. *Plasma Chemistry and Plasma Processing* 30:525–536.
62. Siow, K. S., L. Britcher, S. Kumar, and H. J. Griesser. 2006. Plasma methods for the generation of chemically reactive surfaces for biomolecule immobilization and cell colonization—A review. *Plasma Processes and Polymers* 3:392–418.
63. Grace, J. M., and L. J. Gerenser. 2003. Plasma treatment of polymers. *Journal of Dispersion Science and Technology* 24:305–341.
64. France, R. M., and R. D. Short. 1997. Plasma treatment of polymers—Effects of energy transfer from an argon plasma on the surface chemistry of poly(styrene), low density poly(ethylene), poly(propylene) and poly(ethylene terephthalate). *Journal of the Chemical Society—Faraday Transactions* 93:3173–3178.
65. Sharma, R., E. Holcomb, S. Trigwell, and M. Mazumder. 2007. Stability of atmospheric-pressure plasma induced changes on polycarbonate surfaces. *Journal of Electrostatics* 65:269–273.
66. Murakami, T., S. Kuroda, and Z. Osawa. 1998. Dynamics of polymeric solid surfaces treated with oxygen plasma: Effect of aging media after plasma treatment. *Journal of Colloid and Interface Science* 202:37–44.
67. Murakami, T., S. Kuroda, and Z. Osawa. 1998. Dynamics of polymeric solid surfaces treated by oxygen plasma: Plasma-induced increases in surface molecular mobility of polystyrene. *Journal of Colloid and Interface Science* 200:192–194.

68. Morra, M., E. Occhiello, R. Marola, F. Garbassi, P. Humphrey, and D. Johnson. 1990. On the aging of oxygen plasma-treated polydimethylsiloxane surfaces. *Journal of Colloid and Interface Science* 137:11–24.

69. Chan, C. M., T. M. Ko, and H. Hiraoka. 1996. Polymer surface modification by plasmas and photons. *Surface Science Reports* 24:3–54.

70. Wan, Y. Q., J. Yang, J. L. Yang, J. Z. Bei, and S. G. Wang. 2003. Cell adhesion on gaseous plasma modified poly-(L-lactide) surface under shear stress field. *Biomaterials* 24:3757–3764.

71. Ho, M. H., L. T. Hou, C. Y. Tu, H. J. Hsieh, J. Y. Lai, W. J. Chen, and D. M. Wang. 2006. Promotion of cell affinity of porous PLLA scaffolds by immobilization of RGD peptides via plasma treatment. *Macromolecular Bioscience* 6:90–98.

72. Zhao, J. H., J. Wang, M. Tu, B. H. Luo, and C. R. Zhou. 2006. Improving the cell affinity of a poly(D, L-lactide) film modified by grafting collagen via a plasma technique. *Biomedical Materials* 1:247–252.

73. Mason, M., K. P. Vercruysse, K. R. Kirker, R. Frisch, D. M. Marecak, C. D. Prestwich, and W. G. Pitt. 2000. Attachment of hyaluronic acid to polypropylene, polystyrene, and polytetrafluoroethylene. *Biomaterials* 21:31–36.

74. Ma, Z. W., W. He, T. Yong, and S. Ramakrishna. 2005. Grafting of gelatin on electrospun poly(caprolactone) nanofibers to improve endothelial cell spreading and proliferation and to control cell orientation. *Tissue Engineering* 11:1149–1158.

75. Gugala, Z., and S. Gogolewski. 2006. Attachment, growth, and activity of rat osteoblasts on polylactide membranes treated with various lowtemperature radiofrequency plasmas. *Journal of Biomedical Materials Research Part A* 76A:288–299.

76. Shen, H., X. X. Hu, J. Z. Bei, and S. G. Wang. 2008. The immobilization of basic fibroblast growth factor on plasma-treated poly(lactide-co-glycolide). *Biomaterials* 29:2388–2399.

77. Woodfield, T. B. F., S. Miot, I. Martin, C. A. van Blitterswijk, and J. Riesle. 2006. The regulation of expanded human nasal chondrocyte re-differentiation capacity by substrate composition and gas plasma surface modification. *Biomaterials* 27:1043–1053.

78. Suzuki, M., A. Kishida, H. Iwata, and Y. Ikada. 1986. Graft-copolymerization of acrylamide onto a polyethylene surface pretreated with a glow-discharge. *Macromolecules* 19:1804–1808.

79. Lee, S. D., G. H. Hsiue, and C. Y. Kao. 1996. Preparation and characterization of a homobifunctional silicone rubber membrane grafted with acrylic acid via plasma-induced graft copolymerization. *Journal of Polymer Science Part A—Polymer Chemistry* 34:141–148.

80. Huang, C. Y., W. L. Lu, and Y. C. Feng. 2003. Effect of plasma treatment on the AAc grafting percentage of high-density polyethylene. *Surface and Coatings Technology* 167:1–10.

81. Lewis, G. T., G. R. Nowling, R. F. Hicks, and Y. Cohen. 2007. Inorganic surface nanostructuring by atmospheric pressure plasma-induced graft polymerization. *Langmuir* 23:10756–10764.

82. Muller, M., and C. Oehr. 1999. Plasma aminofunctionalisation of PVDF microfiltration membranes: Comparison of the in plasma modifications with a grafting method using ESCA and an amino-selective fluorescent probe. *Surface and Coatings Technology* 116:802–807.

83. Gupta, B., J. Hilborn, C. Plummer, I. Bisson, and P. Frey. 2002. Thermal crosslinking of collagen immobilized on poly(acrylic acid) grafted poly(ethylene terephthalate) films. *Journal of Applied Polymer Science* 85:1874–1880.

84. Kim, Y. J., I. K. Kang, M. W. Huh, and S. C. Yoon. 2000. Surface characterization and in vitro blood compatibility of poly(ethylene terephthalate) immobilized with insulin and/or heparin using plasma glow discharge. *Biomaterials* 21:121–130.

85. Duan, Y., Z. Wang, W. Yan, S. Wang, S. Zhang, and J. Jia. 2007. Preparation of collagen-coated electrospun nanofibers by remote plasma treatment and their biological properties. *Journal of Biomaterials Science—Polymer Edition* 18:1153–1164.

86. Lee, S. D., G. H. Hsiue, P. C. T. Chang, and C. Y. Kao. 1996. Plasma-induced grafted polymerization of acrylic acid and subsequent grafting of collagen onto polymer film as biomaterials. *Biomaterials* 17:1599–1608.

87. Gupta, B., J. G. Hilborn, I. Bisson, and P. Frey. 2001. Plasma-induced craft polymerization of acrylic acid onto poly(ethylene terephthalate) films. *Journal of Applied Polymer Science* 81:2993–3001.

88. Cheng, Z. Y., and S. H. Teoh. 2004. Surface modification of ultra thin poly (epsilon-caprolactone) films using acrylic acid and collagen. *Biomaterials* 25:1991–2001.

89. Foo, H. L., A. Taniguchi, H. Yu, T. Okano, and S. H. Teoh. 2007. Catalytic surface modification of roll-milled poly(epsilon-caprolactone) biaxially stretched to ultra-thin dimension. *Materials Science and Engineering C—Biomimetic and Supramolecular Systems* 27:299–303.

90. Kinoshita, Y., T. Kuzuhara, M. Kirigakubo, M. Kobayashi, K. Shimura, and Y. Ikada. 1993. Soft tissue reaction to collagen-immobilized porous polyethylene—Subcutaneous implantation in rats for 20 wk. *Biomaterials* 14:209–215.

91. Kinoshita, Y., T. Kuzuhara, M. Kirigakubo, M. Kobayashi, K. Shimura, and Y. Ikada. 1993. Reduction in tumor-formation on porous polyethylene by collagen immobilization. *Biomaterials* 14:546–550.

92. Sasai, Y., M. Oikawa, S. I. Kondo, and M. Kuzuya. 2007. Surface engineering of polymer sheet by plasma techniques and atom transfer radical polymerization for covalent immobilization of biomolecules. *Journal of Photopolymer Science and Technology* 20:197–200.

93. Ding, Z., J. N. Chen, S. Y. Gao, J. B. Chang, J. F. Zhang, and E. T. Kang. 2004. Immobilization of chitosan onto poly-L-lactic acid film surface by plasma graft polymerization to control the morphology of fibroblast and liver cells. *Biomaterials* 25:1059–1067.

94. Barry, J. J. A., M. M. C. G. Silva, K. M. Shakesheff, S. M. Howdle, and M. R. Alexander. 2005. Using plasma deposits to promote cell population of the porous interior of three-dimensional poly(D, L-lactic acid) tissue-engineering scaffolds. *Advanced Functional Materials* 15:1134–1140.

95. Guerin, D. C., D. D. Hinshelwood, S. Monolache, F. S. Denes, and V. A. Shamamian. 2002. Plasma polymerization of thin films: Correlations between plasma chemistry and thin film character. *Langmuir* 18:4118–4123.

96. Cho, D. L., P. M. Claesson, C. G. Golander, and K. S. Johansson. 1990. Structure and surface-properties of plasma polymerized acrylic-acid layers. *Abstracts of Papers of the American Chemical Society* 199:28-PMSE.

97. Kuhn, G., I. Retzko, A. Lippitz, W. Unger, and J. Friedrich. 2001. Homofunctionalized polymer surfaces formed by selective plasma processes. *Surface and Coatings Technology* 142:494–500.

98. Chua, C. K., K. F. Leong, C. M. Cheah, and S. W. Chua. 2003. Development of a tissue engineering scaffold structure library for rapid prototyping. Part 2: Parametric library and assembly program. *International Journal of Advanced Manufacturing Technology* 21:302–312.

99. Chu, L. Q., X. N. Zou, W. Knoll, and R. Forch. 2008. Thermosensitive surfaces fabricated by plasma polymerization of N,N-diethylacrylamide. *Surface and Coatings Technology* 202:2047–2051.

100. Choukourov, A., H. Biederman, D. Slavinska, M. Trchova, and A. Hollander. 2003. The influence of pulse parameters on film composition during pulsed plasma polymerization of diaminocyclohexane. *Surface and Coatings Technology* 174:863–866.

101. Morent, R., N. De Geyter, M. Trentesaux, L. Gengembre, P. Dubruel, C. Leys, and E. Payen. 2010. Stability study of polyacrylic acid films plasma-polymerized on polypropylene substrates at medium pressure. *Applied Surface Science* 257:372–380.

102. Drews, J., H. Launay, C. M. Hansen, K. West, S. Hvilsted, P. Kingshott, and K. Almdal. 2008. Hydrolysis and stability of thin pulsed plasma polymerised maleic anhydride coatings. *Applied Surface Science* 254:4720–4725.

103. Ju, Y. M., K. Park, J. S. Son, J. J. Kim, J. W. Rhie, and D. K. Han. 2008. Beneficial effect of hydrophilized porous polymer scaffolds in tissue-engineered cartilage formation. *Journal of Biomedical Materials Research Part B—Applied Biomaterials* 85B:252–260.

104. Hook, A. L., H. Thissen, J. Quinton, and N. H. Voelcker. 2008. Comparison of the binding mode of plasmid DNA to allylamine plasma polymer and poly(ethylene glycol) surfaces. *Surface Science* 602:1883–1891.

105. Puleo, D. A., R. A. Kissling, and M. S. Sheu. 2002. A technique to immobilize bioactive proteins, including bone morphogenetic protein-4 (BMP-4), on titanium alloy. *Biomaterials* 23:2079–2087.
106. Zelzer, M., R. Majani, J. W. Bradley, F. Rose, M. C. Davies, and M. R. Alexander. 2008. Investigation of cell-surface interactions using chemical gradients formed from plasma polymers. *Biomaterials* 29:172–184.
107. Ren, T. B., T. Weigel, T. Groth, and A. Lendlein. 2008. Microwave plasma surface modification of silicone elastomer with Allylamine for improvement of biocompatibility. *Journal of Biomedical Materials Research Part A* 86A:209–219.
108. Mitchell, S. A., M. R. Davidson, N. Emmison, and R. H. Bradley. 2004. Isopropyl alcohol plasma modification of polystyrene surfaces to influence cell attachment behaviour. *Surface Science* 561:110–120.

14

Emerging Applications of Plasmas in Medicine: Fashion versus Efficacy

Chunqi Jiang

14.1 Introduction

In recent years, plasma applications in medicine have emerged to be a new and fascinating field. In particular, nonthermal atmospheric pressure (NTAP) plasmas have made possible treatments of heat-sensitive biomaterials and living tissue. The electrons in these plasmas, which have much higher energies than the heavier ions and neutral species, collide with background atoms and molecules causing enhanced levels of dissociation, excitation, and ionization. Reactive neutral species and charged particles, generated by these collisions, interact with the materials under treatment, while the bulk gas remains near room temperature. Among the potential plasma agents for bacterial deactivation, reactive chemical species including reactive oxygen species (ROS) and reactive nitrogen species (NO, NO_2) were considered to play crucial roles in the NTAP plasma-induced antimicrobial process (Laroussi 2005; Lu et al. 2008b). Charged particles such as O_2^- as well as electric fields may also importantly contribute to the antimicrobial process (Babaeva et al. 2012). In comparison, heat and ultraviolet (UV) radiation were not considered to play significant roles for bacterial inactivation by the NTAP plasmas (Laroussi and Leipold 2004).

Recent extensive research and development of NTAP plasma sources have enabled their applications in rapid, contact-free sterilization of medical devices and biomaterials, as well as suggested their highly promising potential in dental treatment, wound healing, tissue engineering, and cancer treatment. As more and more researchers are interested in these emerging application potentials, questions arise: Are these promised/suggested application potentials real? Is it a superficial or scientifically based trend that drives the interest/motivation? Have the efficacy of the well-known applications been completely proved? This chapter intends to answer these questions by providing an overview of the low-temperature plasma sources for biomedical applications. A brief introduction to NTAP plasma sources for biomedical applications is first given. A review of the diverse biomedical applications with emphasis in areas of biomaterial and instrument sterilization, dental treatment, and wound care is provided followed by a discussion of the technical challenges that must be overcome for commercialization of the NTAP plasma-based technologies. Nevertheless, this chapter by no means serves as a complete record of all the research activities in the fields. For further material on these and additional topics, the reader is referred to other reviews by Laroussi and Akan (2007), Moreau et al. (2008), Kong et al. (2009), Lloyd et al. (2010), and Weltmann et al. (2010).

14.2 NTAP Plasmas

To date, there are many different types of NTAPs that have been applied for biomedical applications. According to the range of driving frequencies and electrode configurations, the NTAP discharges can be categorized as the following.

14.2.1 Corona Discharges

Corona discharges occur when a sharp nonuniform electric field forms near one or both electrodes and the localized field near the electrode(s) is much stronger than in the rest of the gap (Razer 1997). Typical configurations of corona discharges are point-to-plane, point-to-point, and concentric rod-to-cylinder. For modern industrial applications, such as surface treatment and cleansing of gas and liquid exhaust streams, arrays of multiple points or large scales of concentric wire to cylinder are proposed and employed (Akishev et al. 1993; Vercammen et al. 1997). However, the application of continuous corona discharge is limited by very low currents to prevent arc formation and hence a low rate of treatment of materials and exhaust streams. Pulsed corona discharges allow higher application voltage and peak power than continuous corona and are therefore more attractive in environmental and biomedical applications with their enhanced chemistry and increased energy efficiency.

14.2.2 Microhollow Cathode Discharges

Microhollow cathode discharges (MHCDs) (Stark and Schoenbach 1999) are nonthermal glow discharges capable of stable DC operation at atmospheric pressure. Perforated metal–dielectric–metal configurations and the submillimeter-range characteristic dimensions (i.e., the thickness and the aperture diameter) of the dielectric material enable high-pressure breakdown at relatively low voltages. These MHCD plasmas may serve as the cathode or the electron source for a larger volume glow discharge, which allows more efficient processing and treatment in environmental and biomedical applications (Jiang et al. 2005; Becker et al. 2006; Kolb et al. 2008).

14.2.3 Dielectric Barrier Discharges

In a dielectric barrier discharge (DBD), dielectric layers are used to cover at least one of the electrodes. DBDs, also known as silent discharges, can be operated at a relatively wide frequency range, that is, from 50 Hz up to radio frequency (RF), and pulsed modes. The typical operation frequencies for DBDs are between 50 Hz and 500 kHz. By applying a sinusoidal voltage of sufficient amplitude to the electrodes, a large number of microdischarges occur due to charge accumulation on the dielectric. The dielectric serves both as a discharge "current stopper" by limiting the amount of charge transported by a single microdischarge and as a discharge "transporter" by distributing the microdischarges over the entire electrode surface (Eliasson and Kogelschatz 1991). Although the DBD was originally considered filamentary, relatively large volume, diffused atmospheric pressure glow discharges can also be generated in a pulsed DBD in helium or in a mixture of helium and small traces of other gases (Donohoe and Wydeven 1979; Kanazawa et al. 1988). Therefore diffuse DBDs are used in biomedical applications (Laroussi 1996, 2002). In addition, one atmosphere uniform glow discharges in air generated by a water-cooled DBD reactor powered with RF were also reported for room-temperature sterilization of surfaces and fabrics (Kelly-Wintenberg et al. 1998).

14.2.4 RF Discharges

RF discharges are obtained when the gas is subjected to an oscillating electromagnetic field generated by inductive or capacitive coupling of the power. In an inductively coupled RF discharge, a high-frequency (typically in the range of 0.1–100 MHz) current is passed through a solenoid coil that surrounds a dielectric tube filled with the gas to be studied. The most common industrial frequency is 13.56 MHz. Pulsed discharges are produced if a sufficiently strong current pulse is fed into the coil. High-pressure, inductively coupled RF plasmas have relatively high gas temperature (>7000 K) and have been typically used for material heating as equilibrium plasmas (Razer 1997). In capacitively coupled RF discharges, the systems of two electrodes behave as capacitors to the high-frequency voltage. The two electrodes may be bare and in direct contact with the discharge plasma or may be insulated by dielectric layers. Geometry of the electrodes may be as simple as two parallel plates or may have more complex configurations such as coaxial wire to cylinders.

14.2.5 Microwave Discharges

Microwave-induced plasmas are mostly produced in waveguide structures or resonant cavities where the electromagnetic fields were excited at microwave frequencies between 300 MHz and 3 GHz (Eliasson and Kogelschatz 1991). The most common frequency is 2.45 GHz, which is used in microwave ovens. Large-volume nonequilibrium plasmas can be generated with the microwave-induced, surface-wave-sustained discharges (Chan et al. 1996), which provides a good option for applications including plasma chemical processing and lasing.

14.2.6 Plasma Jets

Plasma jets have become one of the most attractive discharges for biomedical applications because the plasma can be extended to regions not confined by electrodes or dielectrics. Based on aforementioned NTAP plasma excitation schemes, the "internally" generated NTAP plasma plumes or jets are pushed outside the device by gases flowing at rates that are sufficient to "extend" the plumes externally without disturbing or distinguishing the discharge. DC, pulsed, AC, RF, and microwave plasma jets have all been reported and applied for environmental and biomedical applications. Examples of atmospheric plasma jets can be found in a recent review by Laroussi and Akan (2007).

14.3 Biomedical Applications

14.3.1 Instrument and Biomaterial Sterilization

Proper sterilization of materials, equipment, and medical products is crucial to help prevent nosocomial infections and improve the quality of patient care (Mosley 2005). "Sterilization is the act or process, physical or chemical, that destroys or eliminates all forms of life, especially microorganisms" (Block 2001). The physical types of sterilization include sterilization by moist or dry heat, gamma, x-ray or electron-beam radiation, UV irradiation, filtration, and plasma sterilization. Chemical sterilization includes the use of disinfectants and antiseptics such as chlorine, iodine, oxygen peroxide, and their compounds, and oxidative gases such as ethylene oxide and ozone (Sykes 1965; Block 2001).

One of the earliest reports on plasma as a sterilization agent was in a 1968 US patent assigned to the Arthur D. Little Company. Menashi (1968) disclosed the use of a pulsed RF discharge for surface sterilization of contaminated glass or plastic containers in the patent. High-pressure argon plasmas were typically generated in the dielectric containers to be sterilized by delivering 6–7 kV peak root-mean-square (RMS) voltages to the surrounding coil, as in a typical inductively coupled discharge system. Surface sterilization was achieved by turning on the plasma for a very brief period of time, typically one-tenth of a second. The author reported a complete destruction of 4×10^6 bacterial spores per square inch after plasma treatment of less than 0.1 s. Due to the inherent characteristics of inductively coupled RF discharges, the gas temperature of these plasmas is usually high. Peeples and Anderson (1985) suggested that the decontamination effects of these early prototypes of pulsed RF discharges are due to microincineration in addition to UV radiation. Soon after the first report on sterilization using high-temperature plasmas, Ashman and Menashi (1972) at Arthur D. Little, Fraser et al. (1976) at the Boeing Company, Boucher (1980) at Biophysics Research & Consulting Corporation, and Bithell (1982) at Motorola demonstrated low-temperature plasma sterilization by using other gas discharge systems.

Although the study of plasmas for microorganism inactivation began in the 1950s, the commercialization of plasma-based technology for medical device sterilization did not occur until the early 1990s (Jacobs and Lin 2001). Even so, the first commercial sterilization process using nonthermal plasmas that received approval by the US Food and Drug Administration (FDA) is the STERRAD 100 Sterilizer, manufactured by Advanced Sterilization Products (ASP), Irvine, CA (Crow and Smith 1995; Jacobs and Lin 2001). This sterilizer is based on a two-step, low-temperature sterilization process developed by Jacobs and Lin (1987, 1988) at Surgikos, Inc., Johnson & Johnson. In the system, hydrogen peroxide vapors as the precursor chemical were injected into a reactor in which a vacuum (about 0.1–10 Torr) was created. RF discharges were generated in the second step, generating active species from the residual hydrogen peroxide for material and device sterilization and concurrently removing the residual hydrogen peroxide by converting it into nontoxic decomposition products. This two-step sterilization process was demonstrated to be effective against a broad spectrum of microorganisms, including resistant bacterial spores (Jacobs and Lin 1996). The hydroxyl radicals generated during the H_2O_2 microwave discharge in combination with H_2O_2 were considered the primary source of the sporicidal activities apart from UV radiation (Venugopalan and Shih 1981; Jacobs and Lin 1988).

As the development of low-temperature plasma as sterilization tools has entered the commercialization stage, the interest in this area continues to grow. RF and microwave plasmas, DBD plasmas, and plasma jets have all been demonstrated to sterilize biocompatible and heat-sensitive materials and medical devices. We list a few works for each type of discharge here. Additional work can be found in the review by Moreau et al. (2008).

14.3.1.1 RF and Microwave Plasmas

One of the early careful works was presented by Koulik et al. (1999), who used pulsed RF argon plasma jet or inductively coupled RF plasma in air at atmospheric pressure for rapid and uniform sterilization of dielectric surfaces including the inside surface of PET or glass bottles, polymer caps, and plastic tubes. The authors reported that the plasmas were effective in the destruction of microorganisms, including *Bacillus subtilis*, *Saccharomyces cerevisiae*, and *Staphylococcus aureus*, in vegetative or sporulent form. Park et al. (2003) studied the sterilization effects of a microwave-induced argon plasma at atmospheric pressure on filter papers contaminated with four bacterial strains (*B. subtilis*, *Escherichia coli*, *Pseudomonas aeruginosa*, and *Salmonella typhimurium*) and two fungal strains (*Aspergillus niger* and *Penicillium citrinum*). The system consisted of a 1 kW magnetron power supply operating at 2.45 GHz, a WR-284 copper waveguide, an applicator including a tuning section, and a nozzle section, as shown in Figure 14.1. The authors reported

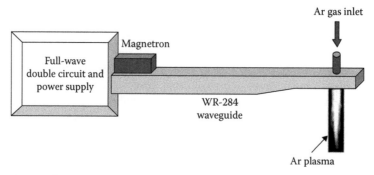

FIGURE 14.1   Schematic diagram of a 1 kW, 2.45 GHz microwave-induced argon plasma system. (From Park, B. J., et al., *Phys. Plasmas*, 10, 4539–44, 2003.)

that all the tested bacteria were fully sterilized within 20 s and that all the fungi were sterilized within 1 s. More recently, the same plasma was applied for bacterial biofilm removal from glass slides (Lee et al. 2009). *E. coli*, *S. epidermidis*, and methicillin-resistant *S. aureus* (MRSA) biofilms were cultured and tested. The authors reported that all bacterial biofilms were destroyed by plasma within 20 s, as shown in Figure 14.2. It was suggested that the sterilization effects of microwave-induced argon plasma on the microorganisms were caused by the generation of free radicals, UV light, and the etching process (Park et al. 2003).

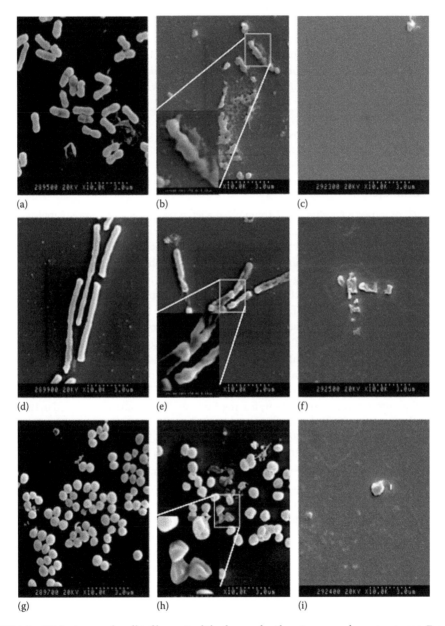

FIGURE 14.2 SEM micrographs of biofilms onto slide glasses after the microwave plasma treatment. *E. coli* (a–c), *S. epidermidis* (d–f), and MRSA (g–i). a, d, and g, untreated control groups; b, e, and h, plasma treated for 5 s; c, f, and i, plasma treated for 20 s. Damaged and ruptured morphologies were observed in the treated groups. Small holes and extensive leakage of cell constituents were observed for cells treated by plasma for 5 s. Biofilms were removed from the surface after plasma treatment for 20 s. (From Lee, M. H., et al., *New J. Phys.*, 11, 2009.)

14.3.1.2 DBD Plasmas

Birmingham and Hammerstrom (2000) used a packed barrier discharge device to deactivate bacterial spores and decompose mycotoxins from surfaces of materials such as steel, plastic, and cloth. The authors reported that the plasma treatment of contaminated metal or dielectric tubes with various times resulted in up to 99.9999% inactivation of *B. globigi* bacterial spores and 99.72% decomposition of the T-2 mycotoxin. In the same year, Montie et al. (2000) presented an atmospheric pressure glow discharge plasma in air generated with an RF-driven DBD configuration for sterilization of surfaces and materials. The water-cooled electrode systems and the specific range of audio RF-driving frequencies were required for the generation of stable, uniform plasmas. The authors reviewed the plasma treatment results for a collection of tested microorganisms including Gram-negative bacteria (*E. coli* K12, *E. coli* O157:H7, and *P. aeruginosa*), Gram-positive bacteria (*S. aureus*, *Deinococcus radiodurans*, *B. subtilus*, *B. stearothermophilus* endospores, *B. pumilus* spores, and *B. subtilis niger* spores), yeast (*S. cerevisiae*, *Candida albicans*), and viruses (*Bacteriophage phiX174*) on a variety of surfaces. It was reported that the plasma exposure resulted in log reductions of all the organisms tested, at least a 5-log (CFU) reduction in bacteria within 50–90 s of treatment. After 10–25 s of exposure, macromolecular leakage and bacterial fragmentation were observed. The authors also concluded that the nature of the surface influenced the degree of lethality, with microorganisms on polypropylene being most sensitive, followed by glass and cells embedded in agar.

14.3.1.3 Plasma Jets

Yu et al. (2007) proposed a brush-shaped argon plasma jet, powered by a DC power supply at low power (<20 W), for low-temperature sterilization of *E. coli* and *Micrococcus luteus* from filter paper, nutrient broth, or agar. The authors reported cell reduction of about 6 orders of magnitude for *E. coli* and *M. luteus* on nutrient broth or standard methods agar after less than 4 min of plasma treatment. Their results also indicated that bacteria grown on filter papers were less susceptible to plasma treatment. They demonstrated the bacterial inactivation effects of the argon plasma to be distinguishable from the possible synergetic effects of heat or fast gas blowing. Rather, the high-energy argon ions and the electronically excited argon neutrals presented in the plasmas were considered to account for the observed bactericidal effects. Laroussi et al. (2006) developed a DBD configuration-based, pulsed NTAP plasma jet and demonstrated the bactericidal effect of the cold plasma jet on *E. coli* grown on nutrient agar. Deng et al. (2007) used an AC-driven, atmospheric DBD jet to study plasma removal of proteinaceous matters from stainless-steel surfaces to demonstrate the ability of low-temperature plasma for sterilization of infectious protein-contaminated surgical instruments. The plasma system used in the study consisted of a dielectric tube wrapped with a metallic strip powered by 30 kHz sinusoidal voltages at a peak value of 8 kV and the ground electrode, essentially the sample holder at 10 mm from the dielectric tube. The schematic of the atmospheric DBD tube jet is shown in Figure 14.3a. He or He/(0.5%) O_2 mixture at a flow rate of 5 SLPM was used as the working gas. The authors employed a wide range of physical techniques including SEM images, EDX analysis, electrophoresis experiments (Figure 14.3b), fluorescence spectroscopy, and inactivation kinetics and showed that plasma-treated proteins were either removed from a stainless-steel surface as fragments or damaged significantly if retained on the surface (Deng et al. 2007).

14.3.2 Dental Treatment

According to the American Dental Association, dentistry is defined as the evaluation, diagnosis, prevention, and treatment of diseases, disorders, and conditions of the oral cavity, maxillofacial area, and the adjacent and associated structures and their impact on the human body.* There are several recognized dental specialties including periodontics, endodontics, prosthodontics, oral and maxillofacial

---

* American Dental Association, "General Dentistry and Interest Areas," available online: http://www.ada.org/6154.aspx

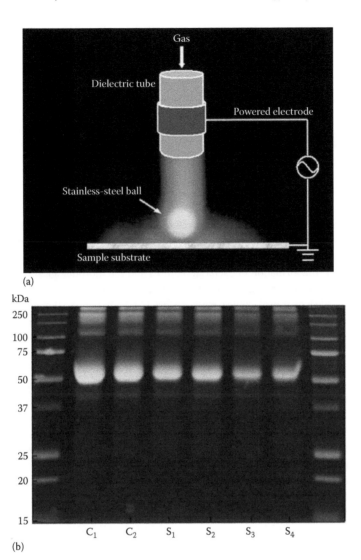

FIGURE 14.3 (a) Schematic of an atmospheric DBD tube jet with a stainless-steel ball immersed in the jet plume and (b) the electrophoresis images of six BSA samples. C_1 and C_2 are two control samples that were left in open air and treated with an unionized He–O_2 flow, both for 300 s. S_1, S_2, S_3, and S_4 are samples that were treated by the DBD tube jet for 4, 60, 180, and 300 s, respectively. When plasma treatment was applied for four different durations (e.g., 4, 60, 180, and 300 s), the 50 kDa band became progressively thinner and weaker, which suggested that the plasma-treated but unremoved proteins underwent considerable degradation. (From Deng, X. T. et al., *Appl. Phys. Lett.*, 90, 2007.)

pathology, radiology, and surgery, orthodontics and dentofacial orthopedics, pediatric dentistry, and dental public health.* Depending on the specialties, the common dental treatment may involve nonsurgical, surgical, or related procedures using mechanical methods, antibiotics, and, more recently, laser irradiation. Although different forms of plasmas have been frequently used for cleaning, preparation, and modification of biomaterial and implant surfaces (Aronsson et al. 1997) in dentistry, applications of plasmas for direct dental treatment as a therapeutic tool did not start till very recently. This is mostly due to the recent rapid development of atmospheric pressure cold plasmas and the urgent need in dentistry for less invasive, safe, effective, and cost-efficient disinfection tools.

---

* American Dental Association, "Specialty Definitions," available online: http://www.ada.org/495.aspx

The biggest challenge for conventional periodontal and endodontic treatments is to eliminate pathogenic microbial biofilms completely without causing damage or excessive loss of healthy tissues. Microbial biofilms, complex communities of microorganisms that are embedded in matrices of extracellular polymeric substance, are a common cause of numerous oral infections including dental caries, pulpitis, periodontitis, and periradicular lesions (Costerton et al. 1999; Estrela et al. 2009). Dental caries, for instance, is a localized destruction of tooth tissues by bacterial fermentation of dietary carbohydrates, and are typically induced by cariogenic dental plaque, an example of microbial biofilm with a diverse microbial composition (Moore and Moore 1994; Marsh 1999; Paster et al. 2001). The acids produced by the fermentation of these dietary carbohydrates demineralize the enamel, which results in the formation of cavities. Treatment of carious lesions and cavity preparation were often achieved by removal of the infected and demineralized soft and hard tissues with mechanical, chemo-mechanical methods, or laser ablation (Yip and Samaranayake 1998; Banerjee et al. 2000). To ensure the cavity free of bacteria, an excess of healthy tissue was often removed, which may have weakened the integrity of the remaining tooth (Banerjee et al. 2000). Root canal treatment is described as the reduction or elimination of intracanal microbes and their by-products from the root canal system, using endodontic instruments aided by antimicrobial agents, before filling (Schilder 1974; Buchanan 1994). The formation of bacterial biofilms and the morphological complexity of the root canal system are the primary factors accounted for the complication or failure of the root canal treatment (Chavez de Paz et al. 2003; Nair et al. 2005; Chavez de Paz 2007). Conventional methods include mechanical instrumentation, antimicrobial irrigation, and removal of the smear layer in order to eliminate intracanal bacteria and debris; however, they are not able to completely eliminate the postprocedure infection (Nair et al. 1990; Sjogren et al. 1997). Laser irradiation (Moritz et al. 1999) and laser-assisted photodynamic therapy (Soukos et al. 2006; Bergmans et al. 2008) were proposed in recent years as supplements to existing protocols for root canal disinfection, but they are not effective against all endodontic strains grown as biofilm phenotypes (Bergmans et al. 2006; Noiri et al. 2008); in addition, their capital cost is high, and they might cause tissue trauma in patients (Dederich and Bushick 2004).

Streptococcus mutans, *Lactobacillus acidophilus*, and *L. casei* are caries-associated bacteria and have often been used in periodontal studies (Ahmady et al. 1993; Babaahmady et al. 1998; Badet and Thebaud 2008). *Enterococcus faecalis*, a Gram-positive facultative bacterium, is a frequent and persistent isolate in teeth with failed root canal therapy and has been used extensively in endodontic studies (Sedgley et al. 2004). Yeast *C. albicans* is one of the predominant species isolated in different oral diseases including apical periodontitis and dental caries (Waltimo et al. 2003). In this section, we will focus on the application of cold plasmas for dental disinfection treatment of periodontal and endodontic diseases. The other plasma applications in dentistry such as biomaterial processing and tooth whitening (Lee et al. 2010) are not included here.

14.3.2.1 Periodontal Treatment

Stoffels et al. pioneered the development of an RF plasma source for periodontal applications. The authors initially proposed an RF "plasma needle" for low-temperature surface treatment of biomaterials (Stoffels et al. 2002). Sladek et al. of the same research group studied the antimicrobial activity of RF plasma against *Streptococcus mutans* biofilms and assessed the potential application of the nonthermal plasma for dental caries treatment. The so-called plasma needle in its later version for biomedical applications (Sladek et al. 2004, 2007) was a 0.3-mm diameter tungsten wire with a sharp tip, confined in a 4-mm inner diameter Perspex tube. The center needle served as the active electrode for the capacitively coupled RF discharges. The Perspex tube was filled with helium flow at 2 l min^{-1}. A 200 V peak-to-peak RF voltage at 13.56 MHz was typically applied to the center needle to initiate the plasma (about 0.1–1 mm long) at the tip of the needle. It was advised that the optimal working distance of the plasma range from 1 to 3 mm above the surface to be treated and the plasma power and treatment time not exceed 150 mW and 60 s, respectively, to avoid hypothermia heating and to maintain the sample temperature under 43°C. For the cariogenic bacterial biofilm *in vitro* study, *Strep. mutans* biofilms with and without 0.15% sucrose were grown on sterile cover glasses for 24 h, followed by plasma treatments and 0.2% chlorhexidine digluconate solution rinse. The latter served as a positive reference for the antimicrobial activity. It was found that a single plasma treatment for 1 min on biofilms cultured without sucrose caused no regrowth within the observation period. However, the growth of the biofilms cultured with 0.15% sucrose was only reduced after two consecutive days' plasma exposure (1 min per day). Goree et al. (2006) investigated a range of plasma parameters for the growth inhibition of *Strep. mutans* in nutrient agar. The presence of the radicals OH and O was verified using optical emission spectroscopy, implying these radicals may play an important role during the bactericidal process. Gonzalvo et al. (2006) used a molecular beam mass spectrometer (MBMS) system to conduct fractional number density measurements for the RF plasma "needle" operating at a range of discharge power up to 10 W. The density of NO generated from the plasma increased substantially after the discharge power value exceeded 2 W. However, the power condition used for the diagnostics does not align very well with the plasma conditions applied in the previous biomedical studies, and the role of NO in the plasma-induced antimicrobial activity could not be directly derived.

More recently, many other research groups have developed the periodontal application with plasma jet-based sources driven by various power sources. Yang et al. used a DC-powered, atmospheric pressure argon plasma brush (Duan et al. 2007) for oral bacterial deactivation (Yang et al. 2011). Oral bacteria of *Strep. mutans* and *L. acidophilus* with an initial bacterial population density of $1.0–5.0 \times 10^8$ CFU ml^{-1} were seeded on various supporting media including P5 filter papers, glass slides, and PTFE films. The plasma exposure time for a 99.9999% cell reduction was less than 15 s for *Strep. mutans* and within 5 min for *L. acidophilus*. It was found that the plasma deactivation efficiency was also dependent on the bacterial supporting media. Koban et al. (2010) tested the antifungal potential of different RF-driven plasma devices—an RF plasma jet (kINPen®09), a hollow electrode DBD, and volume DBD—against *C. albicans* biofilms on titanium disks *in vitro*. The plasma jet and the DBD plasmas were driven by high-frequency RF voltages (1.82 MHz for the jet, about 40 kHz for the DBDs) at power ranging typically from 3 to 16 W. Electrode cooling was needed to prevent overheating of the treated subject. Two-day-old *C. albicans* biofilms cultivated on titanium disks were subject to plasma, CHX, and NaOCl treatment for 10 min. The two chemical cleansers served as the positive controls. While the plasma jet resulted in only 1-log reduction for initial cell concentrations between 10^6 and 10^7 CFU ml^{-1}, the DBD plasma achieved a log reduction factor between 3.1 (hollow electrode DBD) and 5.2 (volume DBD). The authors noted that the DBD plasmas exceeded the antifungal effects of CHX and NaOCl. Nevertheless, the results of the above two experiments may not apply to the real oral environment where dentin or denture resin disks are better options as the substrate, which have cavities and porosities into which the biofilm could adhere.

Rupf et al. (2010) studied the antimicrobial effect of a microwave-powered nonthermal atmospheric plasma jet against adherent oral microorganisms to assess the potential of the cold plasma for dental surface disinfection. The plasma jet was powered by a modulated 2.45 GHz MW source with the peak and average power of 300 and 2.5 W, respectively. The working gas flow was He/O$_2$/N$_2$ at 2.0/1.2/1.5 l min^{-1}. Agar plates and dentin slices were inoculated with 6 log CFU cm^{-2} of *L. casei*, *Strep. mutans*, and *C. albicans*, with *E. coli* as a control. By placing the plasma nozzle 1.5 mm above the substrate with a motion of 11–30 mm s^{-1}, areas of 1 cm^2 on the agar plates or the complete dentin slices (8–10 mm^2) were irradiated with the plasma jet for 0.3–0.9 s mm^{-2}. A maximum temperature of 50.8°C was measured on the dentin surface without movement of the plasma jet over the sample. Approximately 4-log reduction was observed on the dentin slices for the bacteria *E. coli*, *L. casei*, and *Strep. mutans*, and even slightly higher killing rates for *C. albicans*, after the plasma treatment. SEM micrographs showed that the plasma treatment induced distinct morphological alterations but no complete removal of the microorganisms from the dentin surfaces. Killing of the microbes by plasma jet on the dentin slices was not as effective as treatment of agar plates. This might be because of penetration of the microorganisms into the dentin tubules. In addition, distinct differences in the susceptibilities of the four investigated microorganisms to cold plasma irradiation were detected in the present study. *Strep. mutans* revealed the strongest resistance to plasma jet irradiation on agar plates as well as on dentin slices.

Yamazaki et al. (2011) carried out experiments to evaluate the sterilization effects of a low-frequency atmospheric pressure plasma jet on oral pathogenic microorganisms (*Strep. mutans*, *C. albicans*, and

Ent. faecalis). The authors demonstrated that LF (16 kHz) plasma jet had sterilization effects, mainly through ROS, on oral pathogenic microorganisms present in both the solid and liquid phases. In particular, the role of superoxide anion radicals in the sterilization process was evaluated.

14.3.2.2 Endodontic Treatment

Jiang et al. (2008, 2009a) were among the earliest groups to develop a low-temperature pulsed "plasma dental probe" for root canal disinfection. A 2.5 cm long, 2 mm diameter pencil-like plasma jet was generated with a concentric tubular device that was typically powered with 100 ns, 4–8 kV voltage pulses at rates of 1–2 kHz. He/(1%) O_2 flow at 1–5 SLPM was passed through the tubular device to assist initiation and sustaining of the plasma plume. The schematic of a plasma device is shown in Figure 14.4. The plasma-mediated antimicrobial effect was assessed against *Ent. faecalis* biofilms grown on hydroxyapatite or bovine dentin disks *in vitro* and saliva-derived multispecies biofilms inoculated in human root canals *ex vivo* (Jiang et al. 2009b, 2012a). Treatment of dentin disks cultivated with *Ent. faecalis* monolayer biofilms with the plasma (average power ≤1 W) for 5 min resulted in 92.4% kill but no complete sterilization. Conspicuous biofilm disruption and cleared dentinal surfaces as well as partially biofilm-removed surfaces were observed in the human root canal after plasma treatment for 5 min, as shown in Figure 14.5. The authors suggested a thinner plasma jet, for example, ≤1 mm diameter, but with similar or longer length for better sterilization penetration in narrow and curving canals. They demonstrated a flexible plasma plume capable of filling up curving tubes, simulating root canals (Jiang and Schaudinn 2011). The feasibility of the plasma jet for root canal disinfection was tested using an

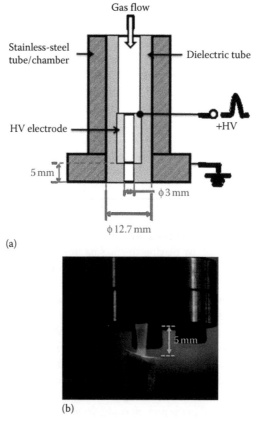

(a)

(b)

FIGURE 14.4 (a) Schematic of a plasma dental device and (b) the photograph of a He/(1%) O_2 plasma generated by the device impinging on a human tooth root 5 mm below the plasma device nozzle. (From Jiang, C. Q. et al., *Plasma Process. Polym.*, 6, 479–83, 2009a.)

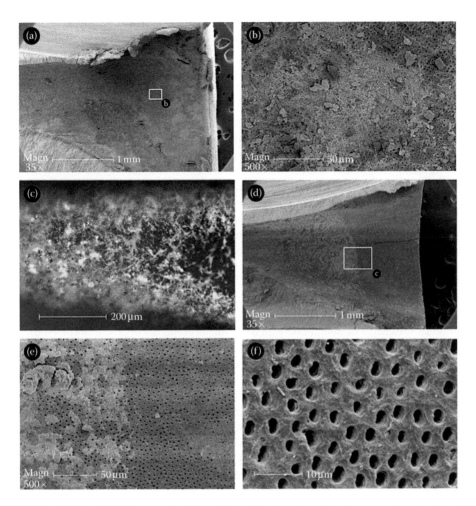

FIGURE 14.5 SEM and CLSM images of salivary biofilm-formed root canals after gas or plasma treatment: (a, b) the negative control (after 5 min gas treatment); (c) CLSM image of the same control: viable bacterial biofilms were observed for the control specimen; (d) a root canal surface after plasma treatment for 5 min; (e) magnified image of (d); and (f) a successful plasma-treated root canal surface revealing completely cleaned surface and open dentinal tubules. (From Jiang, C., et al., *Plasma for Bio-Decontamination, Medicine and Food Security*, eds., Machala, Z., K. Hensel and Y. Akishev, 2012a.)

Ent. faecalis biofilm model. The biofilms were grown inside the simulated root canals or hydroxyapatite disks and applied to the plasma jet treatment (Jiang and Schaudinn 2011; Jiang et al. 2012b). Although the bactericidal effects are prominent, complete biofilm removal was not achieved, which suggested that further improvement of the plasma jet-based techniques was needed. The authors concluded that reactive plasma species such as ROS were playing an important role during the bactericidal process.

Lu et al. (2009) used a syringe needle that was powered by repetitive submicrosecond voltage pulses for deactivation of *Ent. faecalis* bacteria. The authors demonstrated that the 3 cm long, 0.7 mm OD and about 200 μm ID syringe needle-based configuration allowed generation of the plasma inside the root canal. A preliminary work testing *Ent. faecalis* bacteria grown on nutrient agar was reported; about 2-log reduction but no complete sterilization was achieved after 10 min plasma exposure. Zhou et al. used a single electrode-based plasma jet system, also developed by Lu et al. (2008a), to study the antimicrobial activity of plasma on simulated root canals infected with *Ent. faecalis* bacteria. The experimental protocol was designed to test the growth inhibition effects by the cold plasma and by the combination of plasma and 5.25% NaOCl vapor. Sufficient number of single-rooted plastic resin blocks were instrumented and sterilized for aerobic *Ent. faecalis* inoculation (37°C for 72 h). The authors reported more than 4-log reduction of the bacteria by the plasma jet alone and almost complete inactivation by the plasma–bleach combined technique from an initial cell number of 10^6 in the canal models.

14.3.3 Wound Care

A wound is a disruption of normal anatomic structure and function, resulting from pathologic processes beginning internally or externally to the involved organ(s) (Lazarus et al. 1994). Wounds heal by various processes including coagulation, control of infection, resolution of inflammation, angiogenesis, fibroplasia, epithelialization, contraction, and remodeling (Lazarus et al. 1994; Robson 1997). Wounds may be classified as acute wounds that repair themselves or can be repaired in an orderly and timely process and chronic wounds that do not (Lazarus et al. 1994). Chronic wounds represent a worldwide problem, especially as the geriatric population increases. Institutional care of chronic wounds costs approximately $1000 per day, and more rapid healing of chronic wounds would result in potential savings of $11 billion in healthcare costs (Lazarus et al. 1994). Bacterial infection is known to be one of the leading causes of the pathobiology leading to wound chronicity and delay of healing (Tarnuzzer and Schultz 1996; Robson 1997). The NTAP with the demonstrated *in vitro* antimicrobial effects against a broad spectrum of microorganisms would therefore be a potential disinfection tool assisting wound healing. NTAPs may have additional advantages in wound care with their unique properties of promoting blood coagulation and enhancing endothelial cell proliferation. Here we focus on three aspects of the recent applications of NTAPs in wound healing: wound disinfection, blood coagulation, and cell stimulation.

14.3.3.1 Wound Disinfection

Different research groups have observed the antimicrobial effects of NTAP against skin-relevant microorganisms and microbial biofilms *in vitro* (Montie et al. 2000; Park et al. 2003; Lee et al. 2009). Fridman et al. (2006) applied a DBD air plasma for cadaver tissue sterilization to relate the results closer to that in treatment of human skins. The DBD plasma was generated between the insulated high-voltage electrode, the so-called floating-electrode, and the sample for treatment. The high-voltage electrode was insulated with a 1 mm thick quartz sheet, kept for 1.5 mm from the surface of the tissue or agar samples. Modulated AC voltages (20 kVp–p) at typically 12.5 kHz were delivered to the high-voltage electrode during the treatment. The authors conducted preliminary tests by growing human skin flora on blood agar and subject to DBD plasma irradiation. A visible microorganism-free surface was observed on the agar surface after less than 5 s of plasma treatment. The authors also reported a histology study of dead skin undergoing plasma treatment for 5 min without visible damage.

One of the earliest clinical trials was conducted by Isbary et al. (2010) on 38 chronic infected wounds in 36 patients with 5 min daily cold atmospheric argon plasma treatment in addition to standard wound care. The plasma was generated with a microwave-driven plasma device, called MicroPlaSter (originally developed by the Max Planck Institute, Germany), operated at 2.46 GHz, 86 W, with an Ar flow of 2.2 SLPM. The distance of the sample surface from the plasma nozzle was kept at 2 cm during treatment. The plasma treatment system in use for irradiation of a chronic infected wound and the schematic of the plasma device along with an end view of the plasma torch are shown in Figure 14.6. Prior to the plasma treatment, high-pressure water jet or scalpel was used to clean all wounds of debris. Bacterial load of all control and plasma-treated wounds were taken with swabs (for species identification) or nitrocellulose filters (for CFU counting) on different consecutive days. The plasma treatment was stopped if three consecutive daily negative bacterial swabs or nitrocellulose filters were obtained (wound was about to heal) or if the patient decided to stop the additional treatment. The authors detected diverse bacteria on wounds, among which *Enterobacteriaceae* and *Coagulase-negative streptococci* (normal skin flora) were the most prevalent with 18% each of the total sampled bacterial load, followed by *P. aeruginosa* (17%) and *S. aureus* (13%). MRSA was also observed

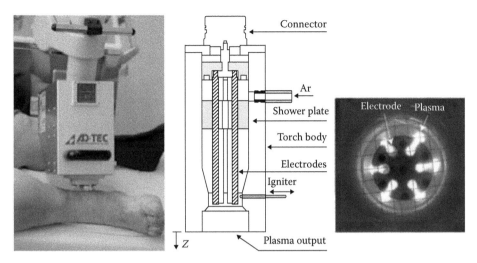

FIGURE 14.6 *Left*: A microwave (2.46 GHz, 86 W, Ar 2.2 SLPM) plasma torch, aka the MicroPlaSter, used in the treatment of a chronic infected wound. *Middle*: the schematic of the plasma device. *Right*: the end view of the plasma torch. (From Isbary, G. et al., *Brit. J. Dermatol.*, 163, 78–82, 2010.)

with a percentage of 3%. There was a highly significant reduction of about 34% in bacterial count in plasma-treated compared with untreated areas. The authors noted that although most patients were also treated with systemic antibiotics, the reduction in bacterial load was, in most cases, less in the control wound than in the plasma-treated area. These results demonstrated the substantial potential of cold atmospheric argon plasma treatment as a safe and painless new technique for wound disinfection and healing. Nevertheless, more clinical trials and long-term, follow-up studies are needed to assess the clinical efficacy and treatment safety of the plasma method.

14.3.3.2 Blood Coagulation

The use of argon plasma coagulation (APC) to assist open surgery was first introduced in the late 1970s (Morrison 1977) and has become the most commonly used endoscopic coagulation technique since its adapted use in endoscopy in the 1990s (Farin and Grund 1994; Grund et al. 1994; Vargo 2004; Raiser and Zenker 2006). Although the use of argon plasma was commonly considered to induce a thermal effect (cauterization) for devitalization of tissue and stanching of bleeding (hemostasis), the short duration (<1 μs) and low amplitude (<4 A) of the current pulses indicate that the APC plasma is a nonthermal plasma (Raiser and Zenker 2006). A typical APC device is based on an RF-driven DBD configuration, where the discharges are generated between a stainless-steel or tungsten electrode and the tissue to be treated (Raiser and Zenker 2006). The distance between the electrode and tissue may be 2–20 mm. The active electrode, usually a thin (typically submillimeter in diameter) wire or spatula-shaped tip, is powered by 4 kV AC voltages at 350 kHz, running at a burst mode of typically 20 kHz. Argon flows at rates between 0.5 and 7 l min^{-1} through a dielectric tube that surrounds the electrode. Commercially available endoscopic APC systems include the System 7550™ with ABC® from Conmed Corporation, Utica, NY, and the APC 2 unit from ERBE Electromedizin, Tübingen, Germany.

Recent studies by Fridman et al. (2006) suggested that a low-temperature air plasma achieved coagulation without inducing the thermal effect, rather by altering the protein or ion contents in blood, which promoted the coagulation process. The experiments were conducted by exposing human blood drops to a DBD air plasma, generated by the RF-driven floating electrode system as described above. The authors proposed a model to indirectly derive the time dependence for thrombin formation on the plasma treatment dosage. Thrombin is one of the proteins responsible for blood clot formation. They concluded that the thrombin formation time was reduced from ~30 s for normal blood to ~15 s for blood undergoing DBD plasma treatment.

14.3.3.3 Cell Stimulation

When plasma comes into contact with mammalian tissues/cells, reactive plasma species may cause any perturbations in the cellular structure, particularly the cell membrane, and induce degrees of structural and functional changes. Some changes may be encouraged for wound healing, while others may be potentially damaging to the host. The recent studies on NTAP interactions with different mammalian cells including fibroblasts and vascular cells have all indicated the dose dependence of the plasma methods and are important for establishing reliable and safe applications with cold plasma.

14.3.3.3.1 Fibroblasts

Kieft et al. (2004) and Stoffels et al. (2003) conducted studies on the interactions of an RF "plasma needle" with living cells in culture. Chinese hamster ovarian cells (CHO-K1)* and the human cells of lung carcinoma MR65[†] in suspension were subject to plasma treatment in different dosages by varying the RF power level and the concentrations of the molecular species. The authors showed that when applying the plasma under moderate conditions (e.g., 0.1 W and 3% air in 2 SLPM He flow), the plasma induced cell detachment between the adjacent cells or the substrates without inducing cell death. The authors suggested that the reactive nitrogen and ROS played a role in the plasma-induced damage of cell adhesion molecules, cadherins, and integrins and caused temporary loss of cell contact and cell detachment from the substrate. A study of the long-term effect of the mammalian cell reaction after exposure to plasma was conducted (Kieft et al. 2006), and 3T3 mouse fibroblast cells were used for the study. The authors reported that 24 h after plasma treatment more than 20% cells underwent apoptosis, that is, programmed cell death.

Lee et al. (2011) reported the suppression of hypertrophic scar generation in an animal model by plasma treatment. A DBD plasma powered by a 5 kV sinusoidal voltage at 4 kHz was ejected through two nozzles (20 mm × 1 mm) onto scars produced in a murine model. A mixture of He (3 SLPM) and air (50 SCCM) was used as the working gas for plasma generation. Eight-week-old C57/BL6 mice were subjected to a retractor to induce scars after the closure of the burn wound. The plasma treatment was performed 3 min each time and for three times a week over six weeks. The authors observed a positive effect on the suppression of vascularization by the plasma stimulation and higher apoptotic cell death levels for mouse hypertrophic scar fibroblasts (HTSF) than for normal fibroblasts. The authors further suggested that the plasma-induced scar suppression effects might be due to cellular apoptosis during the proliferative phase of wound healing and the early stage of scar generation, and the cellular apoptosis was caused by oxidative stress of the ROS in plasma.

14.3.3.3.2 Endothelial (Vascular) Cells

Using the same RF plasma needle source, Kieft et al. (2005) studied the interactions of the plasma and mammalian endothelial and smooth muscle cells that are prevalent in arterial walls. The specific cell types used in the study were bovine aortic endothelial cells (BAECs) and rat aortic smooth muscle cells. A cell viability assay using live/dead staining and fluorescence microscopy was used to examine the influence of the plasma treatment parameters on cells. The authors reported that the thickness of the liquid suspension layer covering the cells was the most important factor to achieve cell detachment and necrosis, compared to the plasma treatment time and the applied voltage. The results also suggested that decreasing the applied voltage could reduce the percentage of necrosis cells to below 10% while still disrupting cell adhesion. However, this detachment was not observed in a later report in which the cells were treated indirectly by exposure to plasma through a permeable membrane (Stoffels et al. 2008). Instead, a long-term behavior of vascular cells (endothelial and smooth muscle) after exposure to the RF plasma was observed: six to ten hours after indirect plasma treatment, apoptosis in low-dose plasma-treated smooth muscle cells was observed, but no immediate detachment occurred.

Kalghatgi et al. (2007) investigated the *in vitro* effect of a nonthermal DBD plasma on porcine aortic endothelial cells (PAECs) to assess the potential application of the plasma for dose-dependent blood vessel

---

* The CHO-K1 cells are basal type (fibroblasts), used as the first model to identify general cell responses.
[†] The MR65 cells are human epithelial cells or one of the common skin cell lines.

growth/regression control. The plasma source was based on a DBD configuration where the HV electrode was insulated with a thin layer of quartz dielectric that was separated from the sample by 2 mm. The sample consisted of PAEC in the culture media on a glass cover slip that was placed on a ground base. Modulated AC voltages (20 kVp–p) at frequencies between 0.5 and 1.5 kHz were delivered to the HV electrode for the plasma generation. The authors showed that low dose of plasma treatment (e.g., up to 30 s or 4 J cm^{-2}) enhanced endothelial cell proliferation by a factor of 2 when compared to the untreated control cells, and higher dose of plasma treatment (60 s and higher or 8 J cm^{-2}) led to cell necrosis. The authors suggested that the enhancement in cell proliferation was through fibroblast growth factor-2 (FGF2) release, which occurs only at cell injury or death, and the plasma induced the FGF2 release by the production of ROS including OH radicals, H_2O_2, and HO_2.

14.3.4 Additional Biomedical Applications

In addition to the above applications in medicine, there are many other biomedical or food industrial applications; some are ready for commercialization, and some are just at the stage of proof-of-principle study. They include the following:

- *Food processing.* It may include surface treatment of food packaging materials to prevent bacterial growth or to decontaminate the packaging surface (Kelly-Wintenberg et al. 1999; Kim et al. 2006; Yun et al. 2010) and direct treatment of food surfaces by deactivating pathogenic and/or spoilage microorganisms including *Listeria monocytogenes*, *E. coli*, *Salmonella*, *S. cerevisiae* (Vleugels et al. 2005; Critzer et al. 2007; Perni et al. 2008; Grzegorzewski et al. 2009; Misra et al. 2011; Fernandez et al. 2012; Niemira 2012; Rod et al. 2012). However, issues such as discoloring food through oxidation (Rod et al. 2012) and scale-up for commercial use require further study and development of the cold atmospheric pressure plasma-based technology prior to its applications in the food industry.
- *Tissue engineering.* This mainly involves plasma-assisted surface modification of polymer scaffolds for enhanced cell attachment and growth (Chim et al. 2003; Riekerink et al. 2003; Girard-Lauriault et al. 2005; Nelea et al. 2005; Aydin et al. 2006; Njatawidjaja et al. 2006). Fibroblast and human mesenchymal stem cells were among the studied biological cells for tissue engineering.
- *Cosmetic procedures.* They may include skin rejuvenation and resurfacing with plasma-induced thermal effects or ablation (Foest et al. 2007; Potter et al. 2007; Foster et al. 2008; Holcomb et al. 2009; Kono et al. 2009) and tooth whitening with cold plasma jet-assisted hydrogen peroxide treatment (Lee et al. 2010).
- *Cancer treatment.* This may be achieved via cold plasma-induced apoptosis of melanoma cells (Li et al. 2008; Sensenig et al. 2011). ROS generated by the plasma were considered to cause DNA double-strand breaks (DSBs) and apoptotic cell death (Li et al. 2008; Kim et al. 2011; O'Connell et al. 2011; Sensenig et al. 2011). However, the potential to damage the DNA of normal mammalian cells (Leduc et al. 2010) may be a problem, and the selectivity of plasma effects needs to be thoroughly studied before application of the technology for clinical use.

14.4 Technical Challenges

For any of the NTAP-based technologies to become a commercial product, the basic questions must be answered—does the technology work (efficacy), is it safe (toxicology), and has it obtained FDA approval?

14.4.1 Efficacy

The aforementioned research findings demonstrated that NTAP produces antimicrobial activity on microorganisms in planktonic forms or in biofilms and their spores, induces cell detachment or irreversible death, and enhances proliferation and coagulation. As different NTAP-based schemes are for different applications, the reality of the complex system must be considered to fully access the efficacy of the technology. For medical device sterilization, the sterilization techniques must be able to overcome the complexities of the instrument and evolve with the advancement of devices (Mosley 2005). For dental and wound treatments, quasi-clinical or clinical studies are required to evaluate the plasma-induced biological effects with influences of uncertainties in clinical settings, such as the physical condition of the patient and the presence of debris or excessive biofilms.

In addition, the efficacy has to be carefully assessed using sufficient and accurate biological assays. A study by Joaquin et al. (2009) showed that inactivated bacterial biofilms after 5 min plasma treatment were nonculturable but still alive. Many early studies assessed plasma antimicrobial activity with culture and CFU counting alone; they concluded that "complete sterilization," that is, "killing" of the organisms, was achieved, which is not accurate; newcomers should be aware of this. Nevertheless, an understanding of the efficacy of plasma technology in medicine is far from complete, but there has been some improvement thanks to the use of advanced microscopic imaging techniques and the involvement of not only plasma physicists but also microbiologists, dentists, and dermatologists.

14.4.2 Toxicology

Although drugs in medicine can be used effectively to cure diseases, they may have side effects or harmful effects when they are improperly used; it is the same with NTAP. Research found that the low-temperature plasma induced apoptosis (Kieft et al. 2006; Stoffels et al. 2008) or enhanced proliferation (Kalghatgi et al. 2009, 2010) in mammalian cells that could assist wound healing or treat cancer. On the other hand, they may also cause mutations to the healthy cell genome leading to the formation of cancer. Leduc et al. (2010) assessed the possible negative effects of direct and indirect plasma treatment on mammalian (HeLa) cells and naked DNA. Using a DBD plasma and an RF-driven plasma jet operating at power levels of <1 or 4 W, respectively, applied for 30 and 60 s, the authors reported that both types of plasmas induced oxidative stress to the cells, fragmented naked DNA, and caused DNA DSBs inside living cells without inducing mutation. The authors suggested that excited NO generated by the He/N_2 plasma might be responsible for the observed oxidative stress, and the reactive nature of nonthermal plasmas might have unintended negative biological effects on cells. After all, extensive and detailed studies on selectivity of the plasma-induced biological effects, plasma parameter (including the dosage, power level, and working gas) dependence, and long-term effects of plasma on living tissues are required to better define the role of NTAPs in specific applications. It is important to maintain controlled and localized plasma conditions during individual applications.

The other potential toxicology concern may be the occupational exposure of healthcare workers administering the treatments to reactive plasma species. The risks of long-term, low-level dosing with ozone, nitric oxides, and other reactive plasma species have to be reduced or eliminated for clinical acceptance of NTAP-based technology.

14.4.3 Regulatory Issues

Although the toxicology studies on NTAP-based medical devices may provide sufficient information on safety concerns, the release of any of these new medical devices to the public can only be possible/legal after they pass the scrutiny of regulatory authorities. Depending on the geography, the regulatory authorities and the detailed requirements regarding safety vary. We list here a few regulatory agencies in Asia, Europe, and America:

- China: the State Food and Drug Administration (SFDA)
- Germany: the Federal Institute for Drugs and Medical Devices (BfArM)
- The Netherlands: the Medicines Evaluation Board (MEB)
- The United Kingdom: the Medicines and Healthcare Products Regulatory Agency (MHRA)
- The United States: the FDA

Although the detailed responsibilities may vary, the common mission of these agencies is to protect and promote public health through the regulation and supervision of food and medicinal/medical products.

In the United States, the Center for Devices and Radiological Health (CDRH), a branch of the FDA, is responsible for the premarket approval (PMA) of all medical devices and oversees their manufacturing, performance, and safety.*

The consequence of failure to follow the legal procedure and obtain regulatory approval prior to the release of the product can be devastating. In 2006, the former AbTox, Inc., executives were convicted of selling unapproved hospital sterilizers,† the AbTox Plazlyte Sterilization System. The Plazlyte Sterilization System used a low-temperature argon, plasma-based sterilization technique in combination with a chemical pretreatment step for medical instrument sterilization (Moulton et al. 1992). In 1997, AbTox, Inc., introduced unapproved Plazlyte sterilization systems to a hospital for sterilization of ophthalmic surgery instruments, which caused a rare but serious disease, the toxic endothelial cell destruction syndrome, in six patients after intraocular ophthalmic surgery (Duffy et al. 2000). Until the company filed bankruptcy in 1998, they sold approximately 160 of the unapproved sterilizers to hospitals in the United States, and some caused catastrophic eye injuries in up to 18 patients who underwent eye surgeries at four different hospitals. The top executives at the company including one of the inventors of the plasma sterilization process, Captuto, were sentenced up to six years' imprisonment and more than a quarter million dollars' fine each.

Procedures toward medical device commercialization usually involve

- Fulfilling all the requirements specified by the safety regulations.
- Meeting the CE marking requirements for medical devices to allow the product to enter and move freely throughout the European markets. "CE marking on a product" indicates to governments that the product complies with the European Medical Device Directive and can be legally sold within the European Union and the European Free Trade Area.
- For the US market, acquiring clearance using the 510(k) pathway if the medical device is "substantially equivalent" to a device that is already legally marketed for the same use or acquiring approval through a PMA application by providing reasonable assurance of the device's safety and effectiveness.
- Conducting clinical trials.
- Pilot producing and manufacturing with validated and approved manufacturers.

14.5 Conclusion

The emerging biomedical applications of low-temperature plasmas have triggered research in both biomedical technology and plasma science. However, when it comes to the efficacy of these plasma-based technologies for clinical or hospital use, there are only few applications or areas of applications that successfully accepted and adopted plasma-based "novel" schemes or methods. Keeping this in mind, this chapter intends to provide the reader an extensive review of the NTAP plasmas and their emerging biomedical applications with a focus on areas of biomaterial and instrument sterilization, dental treatment, and wound care. The discussions on the technical challenges serve as a caution or guide for researchers and students who are interested in developing this fascinating technology for public use.

References

Ahmady, K., P. D. Marsh, et al. 1993. Distribution of *Streptococcus mutans* and *Streptococcus sobrinus* at sub-sites in human approximal dental plaque. *Caries Research* 27: 135–9.

Akishev, Y. S., A. A. Deryugin, et al. 1993. DC glow-discharge in air-flow at atmospheric-pressure in connection with waste gases treatment. *Journal of Physics D—Applied Physics* 26: 1630–7.

---

* http://www.fda.gov/AboutFDA/Transparency/Basics/ucm193731.htm
† https://www.va.gov/oig/pubs/press-releases/VAOIG-press-release-20060420.htm

Aronsson, B. O., J. Lausmaa, et al. 1997. Glow discharge plasma treatment for surface cleaning and modification of metallic biomaterials. *Journal of Biomedical Materials Research* 35: 49–73.

Ashman, L. E. and W. P. Menashi. 1972. Treatment of surface with low-pressure plasmas. U.S. Patent Office. Arthur D. Little, Inc. 3,701,628.

Aydin, H. M., M. Turk, et al. 2006. Attachment and growth of fibroblasts on poly(L-lactide/epsilon-caprolactone) scaffolds prepared in supercritical CO_2 and modified by polyethylene imine grafting with ethylene diamine-plasma in a glow-discharge apparatus. *International Journal of Artificial Organs* 29: 873–80.

Babaahmady, K. G., S. J. Challacombe, et al. 1998. Ecological study of *Streptococcus mutans, Streptococcus sobrinus* and *Lactobacillus* spp. at sub-sites from approximal dental plaque from children. *Caries Research* 32: 51–8.

Babaeva, N. Y., N. Ning, et al. 2012. Ion activation energy delivered to wounds by atmospheric pressure dielectric-barrier discharges: Sputtering of lipid-like surfaces. *Journal of Physics D—Applied Physics* 45.

Badet, C. and N. B. Thebaud. 2008. Ecology of lactobacilli in the oral cavity: A review of literature. *The Open Microbiology Journal* 2: 38–48.

Banerjee, A., T. F. Watson, et al. 2000. Dentine caries excavation: A review of current clinical techniques. *British Dental Journal* 188: 476–82.

Becker, K. H., K. H. Schoenbach, et al. 2006. Microplasmas and applications. *Journal of Physics D—Applied Physics* 39: R55–R70.

Bergmans, L., P. Moisiadis, et al. 2006. Bactericidal effect of Nd:YAG laser irradiation on some endodontic pathogens ex vivo. *International Endodontic Journal* 39: 547–57.

Bergmans, L., P. Moisiadis, et al. 2008. Effect of photo-activated disinfection on endodontic pathogens ex vivo. *International Endodontic Journal* 41: 227–39.

Birmingham, J. G. and D. J. Hammerstrom. 2000. Bacterial decontamination using ambient pressure nonthermal discharges. *IEEE Transactions on Plasma Science* 28: 51–5.

Bithell, R. M. 1982. Plasma pressure pulse sterilization. U.S. Patent Office. Motorola, Inc. 4,348,357.

Block, S. S., editor. 2001. Definition of Terms. In *Disinfection, Sterilization, and Preservation*, Ch 2, Philadelphia, PA: Lippincott Williams & Wilkins. pp. 19–28.

Boucher, R. M. G. 1980. Seeded gas plasma sterilization method. U.S. Patent Office. Biophysics Research & Consulting Corporation. 4,207,286.

Buchanan, L. S. 1994. Cleaning and shaping the root canal system: Negotiating canals to the termini. *Dentistry Today* 13: 76, 78–81.

Chan, C. M., T. M. Ko, et al. 1996. Polymer surface modification by plasmas and photons. *Surface Science Reports* 24: 3–54.

Chavez de Paz, L. E. 2007. Redefining the persistent infection in root canals: Possible role of biofilm communities. *Journal of Endodonotics* 33: 652–62.

Chavez de Paz, L. E., G. Dahlen, et al. 2003. Bacteria recovered from teeth with apical periodontitis after antimicrobial endodontic treatment. *International Endodonotics Journal* 36: 500–8.

Chim, H., J. L. Ong, et al. 2003. Efficacy of glow discharge gas plasma treatment as a surface modification process for three-dimensional poly (D,L-lactide) scaffolds. *Journal of Biomedical Materials Research Part A* 65A: 327–35.

Costerton, J. W., P. S. Stewart, et al. 1999. Bacterial biofilms: A common cause of persistent infections. *Science* 284: 1318–22.

Critzer, F. J., K. Kelly-Wintenberg, et al. 2007. Atmospheric plasma inactivation of foodborne pathogens on fresh produce surfaces. *Journal of Food Protection* 70: 2290–6.

Crow, S. and J. H. Smith. 1995. Gas plasma sterilization—Application of space-age technology. *Infection Control and Hospital Epidemiology* 16: 483–7.

Dederich, D. N. and R. D. Bushick. 2004. Lasers in dentistry: Separating science from hype. *Journal of the American Dental Association* 135: 204–12; quiz 29.

Deng, X. T., J. J. Shi, et al. 2007. Protein destruction by atmospheric pressure glow discharges. *Applied Physics Letters* 90: 013903.

Donohoe, K. G. and T. Wydeven. 1979. Plasma polymerization of ethylene in an atmospheric pressure-pulsed discharge. *Journal of Applied Polymer Science* 23: 2591–601.

Duan, Y. X., C. Huang, et al. 2007. Cold plasma brush generated at atmospheric pressure. *Review of Scientific Instruments* 78: 015104.

Duffy, R. E., S. E. Brown, et al. 2000. An epidemic of corneal destruction caused by plasma gas sterilization. The Toxic Cell Destruction Syndrome Investigative Team. *Archives of Ophthalmology* 118: 1167–76.

Eliasson, B. and U. Kogelschatz. 1991. Nonequilibrium volume plasma chemical-processing. *IEEE Transactions on Plasma Science* 19: 1063–77.

Estrela, C., G. B. Sydney, et al. 2009. Antibacterial efficacy of intracanal medicaments on bacterial biofilm: A critical review. *Journal of Applied Oral Science* 17: 1–7.

Farin, G. and K. E. Grund. 1994. Technology of argon plasma coagulation with particular regard to endoscopic applications. *Endoscopic Surgery and Allied Technologies* 2: 71–7.

Fernandez, A., N. Shearer, et al. 2012. Effect of microbial loading on the efficiency of cold atmospheric gas plasma inactivation of *Salmonella enterica* serovar Typhimurium. *International Journal of Food Microbiology* 152: 175–80.

Foest, R., E. Kindel, et al. 2007. RF capillary jet—A tool for localized surface treatment. *Contributions to Plasma Physics* 47: 119–28.

Foster, K. W., R. L. Moy, et al. 2008. Advances in plasma skin regeneration. *Journal of Cosmetic Dermatology* 7: 169–79.

Fraser, S. J., R. B. Gillette, et al. 1976. Sterilizing process and apparatus utilizing gas plasma. U.S. Patent Office. The Boeing Company. 3,948,601.

Fridman, G., M. Peddinghaus, et al. 2006. Blood coagulation and living tissue sterilization by floating-electrode dielectric barrier discharge in air. *Plasma Chemistry and Plasma Processing* 26: 425–42.

Girard-Lauriault, P. L., F. Mwale, et al. 2005. Atmospheric pressure deposition of micropatterned nitrogen-rich plasma-polymer films for tissue engineering. *Plasma Processes and Polymers* 2: 263–70.

Gonzalvo, Y. A., T. D. Whitmore, et al. 2006. Atmospheric pressure plasma analysis by modulated molecular beam mass spectrometry. *Journal of Vacuum Science and Technology A* 24: 550–3.

Goree, J., B. Liu, et al. 2006. Killing of S-mutans bacteria using a plasma needle at atmospheric pressure. *IEEE Transactions on Plasma Science* 34: 1317–24.

Grund, K. E., D. Storek, et al. 1994. Endoscopic argon plasma coagulation (APC) first clinical experiences in flexible endoscopy. *Endoscopic Surgery and Allied Technologies* 2: 42–6.

Grzegorzewski, F., O. Schluter, et al. 2009. Plasma-oxidative degradation of polyphenolics—Influence of non-thermal gas discharges with respect to fresh produce processing. *Czech Journal of Food Sciences* 27: S35–S9.

Holcomb, J. D., K. J. Kent, et al. 2009. Nitrogen plasma skin regeneration and aesthetic facial surgery: Multicenter evaluation of concurrent treatment. *Archives of Facial Plastic Surgery: Official Publication for the American Academy of Facial Plastic and Reconstructive Surgery, Inc. and the International Federation of Facial Plastic Surgery Societies* 11: 184–93.

Isbary, G., G. Morfill, et al. 2010. A first prospective randomized controlled trial to decrease bacterial load using cold atmospheric argon plasma on chronic wounds in patients. *The British Journal of Dermatology* 163: 78–82.

Jacobs, P. T. and S. M. Lin. 1987. Hydrogen peroxide plasma sterilization system. U.S. Patent Office. Surgikos, Inc. 4,643,876.

Jacobs, P. T. and S. M. Lin. 1988. Hydrogen peroxide plasma sterilization system. U.S. Patent Office. Surgikos, Inc. 4,756,882.

Jacobs, P. T. and S. M. Lin. 1996. Gas-plasma sterilization. *Irradiation of Polymers* 620: 216–39.

Jacobs, P. T. and S.-M. Lin. 2001. Sterilization processes utilizing low-temperature plasma. In *Disinfection, Sterilization, and Preservation*, edited by Block, S. S. Philadelphia, PA: Lippincott Williams & Wilkins. p. 747.

Jiang, C., C. Schaudinn, et al. 2012a. A sub-microsecond pulsed plasma jet for endodontic biofilm disinfection. In *Plasma for Bio-Decontamination, Medicine and Food Security*, edited by Machala, Z., K. Hensel, and Y. Akishev. Heidelberg, Germany: Springer.

Jiang, C., C. Schaudinn, et al. 2012b. In vitro antimicrobial effect of a cold plasma jet against *Enterococcus faecalis* biofilms. *ISRN Dentistry* 2012: 295736.

Jiang, C., P. T. Vernier, et al. 2008. Low energy nanosecond pulsed plasma sterilization for endodontic applications. *Proceedings of the 2008 IEEE International Power Modulators and High Voltage Conference.* pp. 77–9. IEEE Dielectrics and Electrical Insulation Society.

Jiang, C. Q., M. T. Chen, et al. 2009a. Nanosecond pulsed plasma dental probe. *Plasma Processes and Polymers* 6: 479–83.

Jiang, C. Q., M. T. Chen, et al. 2009b. Pulsed atmospheric-pressure cold plasma for endodontic disinfection. *IEEE Transactions on Plasma Science* 37: 1190–5.

Jiang, C. Q., A. A. H. Mohamed, et al. 2005. Removal of volatile organic compounds in atmospheric pressure air by means of direct current glow discharges. *IEEE Transactions on Plasma Science* 33: 1416–25.

Jiang, C. Q. and C. Schaudinn. 2011. A curving bactericidal plasma needle. *IEEE Transactions on Plasma Science* 39: 2966–7.

Joaquin, J. C., C. Kwan, et al. 2009. Is gas-discharge plasma a new solution to the old problem of biofilm inactivation? *Microbiology—Sgm* 155: 724–32.

Kalghatgi, S. U., G. Fridman, et al. 2007. Mechanism of blood coagulation by nonthermal atmospheric pressure dielectric barrier discharge plasma. *IEEE Transactions on Plasma Science* 35: 1559–66.

Kalghatgi, S. U., A. Fridman, et al. 2009. Cell proliferation following non-thermal plasma is related to reactive oxygen species induced fibroblast growth factor-2 release. *EMBC: 2009 Annual International Conference of the IEEE Engineering in Medicine and Biology Society.* Vols 1–20, pp. 6030–3.

Kalghatgi, S. U., G. Friedman, et al. 2010. Endothelial cell proliferation is enhanced by low dose non-thermal plasma through fibroblast growth factor-2 release. *Annals of Biomedical Engineering* 38: 748–57.

Kanazawa, S., M. Kogoma, et al. 1988. Stable glow plasma at atmospheric-pressure. *Journal of Physics D—Applied Physics* 21: 838–40.

Kedzierski, J., J. Engemann, et al. 2005. Atmospheric pressure plasma jets for 2D and 3D materials processing. *Solid State Phenomena* 107: 119–23.

Kelly-Wintenberg, K., T. C. Montie, et al. 1998. Room temperature sterilization of surfaces and fabrics with a one atmosphere uniform glow discharge plasma. *Journal of Industrial Microbiology and Biotechnology* 20: 69–74.

Kelly-Wintenberg, K., A. Hodge, et al. 1999. Use of a one atmosphere uniform glow discharge plasma to kill a broad spectrum of microorganisms. *Journal of Vacuum Science and Technology A: Vacuum Surfaces and Films* 17: 1539–44.

Kieft, I. E., J. L. V. Broers, et al. 2004. Electric discharge plasmas influence attachment of cultured CHO K1 cells. *Bioelectromagnetics* 25: 362–8.

Kieft, I. E., D. Darios, et al. 2005. Plasma treatment of mammalian vascular cells: A quantitative description. *IEEE Transactions on Plasma Science* 33: 771–5.

Kieft, I. E., M. Kurdi, et al. 2006. Reattachment and apoptosis after plasma-needle treatment of cultured cells. *IEEE Transactions on Plasma Science* 34: 1331–6.

Kim, J. H., G. M. Liu, et al. 2006. Deposition of stable hydrophobic coatings with in-line CH_4 atmospheric rf plasma. *Journal of Materials Chemistry* 16: 977–81.

Kim, K., J. D. Choi, et al. 2011. Atmospheric-pressure plasma-jet from micronozzle array and its biological effects on living cells for cancer therapy. *Applied Physics Letters* 98: 073701.

Koban, I., R. Matthes, et al. 2010. Treatment of *Candida albicans* biofilms with low-temperature plasma induced by dielectric barrier discharge and atmospheric pressure plasma jet. *New Journal of Physics* 12: 073039.

Kolb, J. F., A. A. H. Mohamed, et al. 2008. Cold atmospheric pressure air plasma jet for medical applications. *Applied Physics Letters* 92: 241501.

Kong, M. G., G. Kroesen, et al. 2009. Plasma medicine: an introductory review. *New Journal of Physics* 11: 115012.

Kono, T., W. F. Groff, et al. 2009. Treatment of traumatic scars using plasma skin regeneration (PSR) system. *Lasers in Surgery and Medicine* 41: 128–30.

Koulik, P., S. Begounov, et al. 1999. Atmospheric plasma sterilization and deodorization of dielectric surfaces. *Plasma Chemistry and Plasma Processing* 19: 311–26.

Laroussi, M. 1996. Sterilization of contaminated matter with an atmospheric pressure plasma. *IEEE Transactions on Plasma Science* 24: 1188–91.

Laroussi, M. 2002. Nonthermal decontamination of biological media by atmospheric-pressure plasmas: Review, analysis, and prospects. *IEEE Transactions on Plasma Science* 30: 1409–15.

Laroussi, M. 2005. Low temperature plasma-based sterilization: Overview and state-of-the-art. *Plasma Processes and Polymers* 2: 391–400.

Laroussi, M. and T. Akan. 2007. Arc-free atmospheric pressure cold plasma jets: A review. *Plasma Processes and Polymers* 4: 777–88.

Laroussi, M. and F. Leipold. 2004. Evaluation of the roles of reactive species, heat, and UV radiation in the inactivation of bacterial cells by air plasmas at atmospheric pressure. *International Journal of Mass Spectrometry* 233: 81–6.

Lazarus, G. S., D. M. Cooper, et al. 1994. Definitions and guidelines for assessment of wounds and evaluation of healing. *Archives of Dermatology* 130: 489–93.

Leduc, M., D. Guay, et al. 2010. Effects of non-thermal plasmas on DNA and mammalian cells. *Plasma Processes and Polymers* 7: 899–909.

Lee, D. H., J. O. Lee, et al. 2011. Suppression of scar formation in a murine burn wound model by the application of non-thermal plasma. *Applied Physics Letters* 99: 203701.

Lee, H. W., S. H. Nam, et al. 2010. Atmospheric pressure plasma jet composed of three electrodes: Application to tooth bleaching. *Plasma Processes and Polymers* 7: 274–80.

Lee, M. H., B. J. Park, et al. 2009. Removal and sterilization of biofilms and planktonic bacteria by microwave-induced argon plasma at atmospheric pressure. *New Journal of Physics* 11: 115022.

Li, G., H. P. Li, et al. 2008. Genetic effects of radio-frequency, atmospheric-pressure glow discharges with helium. *Applied Physics Letters* 92: 221504.

Lloyd, G., G. Friedman, et al. 2010. Gas plasma: Medical uses and developments in wound care. *Plasma Processes and Polymers* 7: 194–211.

Lu, X. P., Y. G. Cao, et al. 2009. An RC plasma device for sterilization of root canal of teeth. *IEEE Transactions on Plasma Science* 37: 668–73.

Lu, X. P., Z. H. Jiang, et al. 2008a. A single electrode room-temperature plasma jet device for biomedical applications. *Applied Physics Letters* 92: 151504.

Lu, X. P., T. Ye, et al. 2008b. The roles of the various plasma agents in the inactivation of bacteria. *Journal of Applied Physics* 104: 053309.

Marsh, P. D. 1999. Microbiologic aspects of dental plaque and dental caries. *Dental Clinics of North America* 43: 599–614, v–vi.

Menashi, W. P. 1968. Treatment of surfaces. U.S. Patent Office. Arthur D. Little, Inc. 3,383,163.

Misra, N. N., B. K. Tiwari, et al. 2011. Nonthermal plasma inactivation of food-borne pathogens. *Food Engineering Reviews* 3: 159–70.

Montie, T. C., K. Kelly-Wintenberg, et al. 2000. An overview of research using the one atmosphere uniform glow discharge plasma (OAUGDP) for sterilization of surfaces and materials. *IEEE Transactions on Plasma Science* 28: 41–50.

Moore, W. E. and L. V. Moore. 1994. The bacteria of periodontal diseases. *Periodontology 2000* 5: 66–77.

Moreau, M., N. Orange, et al. 2008. Non-thermal plasma technologies: New tools for bio-decontamination. *Biotechnology Advances* 26: 610–17.

Moritz, A., U. Schoop, et al. 1999. The bactericidal effect of Nd:YAG, Ho:YAG, and Er:YAG laser irradiation in the root canal: An in vitro comparison. *Journal of Clinical Laser Medicine and Surgery* 17: 161–4.

Morrison, J. C. F. 1977. Electrosurgical method and apparatus for initiating an electrical discharge in an inert gas flow. U.S. Patent Office. Valleylab, Inc. 4,040,426.

Mosley, G. 2005. Overcoming the complexities of instrument sterilization. *Materials Management in Health Care* 14: 26–8.

Moulton, K. A., B. A. Campbell, et al. 1992. Plasma sterilization process with pulsed antimicrobial agent Treatment. U.S. Patent Ofiice. AbTox, Inc. 5,084,239.

Nair, P. N., S. Henry, et al. 2005. Microbial status of apical root canal system of human mandibular first molars with primary apical periodontitis after "one-visit" endodontic treatment. *Oral Surgery, Oral Medicine, Oral Pathology, Oral Radiology, and Endodontics* 99: 231–52.

Nair, P. N., U. Sjogren, et al. 1990. Intraradicular bacteria and fungi in root-filled, asymptomatic human teeth with therapy-resistant periapical lesions: a long-term light and electron microscopic follow-up study. *Journal of Endodontics* 16: 580–8.

Nelea, V., L. Luo, et al. 2005. Selective inhibition of type X collagen expression in human mesenchymal stem cell differentiation on polymer substrates surface-modified by glow discharge plasma. *Journal of Biomedical Materials Research Part A* 75A: 216–23.

Niemira, B. A. 2012. Cold plasma reduction of *Salmonella* and *Escherichia coli* O157:H7 on almonds using ambient pressure gases. *Journal of Food Science* 77: M171–M5.

Njatawidjaja, E., M. Kodama, et al. 2006. Hydrophilic modification of expanded polytetrafluoroethylene (ePTFE) by atmospheric pressure glow discharge (APG) treatment. *Surface and Coatings Technology* 201: 699–706.

Noiri, Y., T. Katsumoto, et al. 2008. Effects of Er:YAG laser irradiation on biofilm-forming bacteria associated with endodontic pathogens in vitro. *Journal of Endodontics* 34: 826–9.

O'Connell, D., L. J. Cox, et al. 2011. Cold atmospheric pressure plasma jet interactions with plasmid DNA. *Applied Physics Letters* 98: 043701.

Park, B. J., D. H. Lee, et al. 2003. Sterilization using a microwave-induced argon plasma system at atmospheric pressure. *Physics of Plasmas* 10: 4539–44.

Paster, B. J., S. K. Boches, et al. 2001. Bacterial diversity in human subgingival plaque. *Journal of Bacteriology* 183: 3770–83.

Peeples, R. E. and N. R. Anderson. 1985. Microwave coupled plasma sterilization and depyrogenation II. Mechanisms of action. *Journal of Parenteral Science and Technology: A Publication of the Parenteral Drug Association* 39: 9–15.

Perni, S., D. W. Liu, et al. 2008. Cold atmospheric plasma decontamination of the pericarps of fruit. *Journal of Food Protection* 71: 302–8.

Potter, M. J., R. Harrison, et al. 2007. Facial acne and fine lines: transforming patient outcomes with plasma skin regeneration. *Annals of Plastic Surgery* 58: 608–13.

Raiser, J. and M. Zenker. 2006. Argon plasma coagulation for open surgical and endoscopic applications: State of the art. *Journal of Physics D—Applied Physics* 39: 3520–3.

Razer, Y. P. 1997. *Gas Discharge Physics*. New York: Springer-Verlag, LLC.

Riekerink, M. B. O., M. B. Claase, et al. 2003. Gas plasma etching of PEO/PBT segmented block copolymer films. *Journal of Biomedical Materials Research Part A* 65A: 417–28.

Robson, M. C. 1997. Wound infection. A failure of wound healing caused by an imbalance of bacteria. *The Surgical Clinics of North America* 77: 637–50.

Rod, S. K., F. Hansen, et al. 2012. Cold atmospheric pressure plasma treatment of ready-to-eat meat: Inactivation of *Listeria innocua* and changes in product quality. *Food Microbiology* 30: 233–8.

Rupf, S., A. Lehmann, et al. 2010. Killing of adherent oral microbes by a non-thermal atmospheric plasma jet. *Journal of Medical Microbiology* 59: 206–12.

Schilder, H. 1974. Cleaning and shaping the root canal. *Dental Clinics of North America* 18: 269–96.

Sedgley, C. M., S. L. Lennan, et al. 2004. Prevalence, phenotype and genotype of oral enterococci. *Oral Microbiology and Immunology* 19: 95–101.

Sensenig, R., S. Kalghatgi, et al. 2011. Non-thermal plasma induces apoptosis in melanoma cells via production of intracellular reactive oxygen species. *Annals of Biomedical Engineering* 39: 674–87.

Sjogren, U., D. Figdor, et al. 1997. Influence of infection at the time of root filling on the outcome of endodontic treatment of teeth with apical periodontitis. *International Endodontic Journal* 30: 297–306.

Sladek, R. E. J., S. K. Filoche, et al. 2007. Treatment of *Streptococcus mutans* biofilms with a nonthermal atmospheric plasma. *Letters in Applied Microbiology* 45: 318–23.

Sladek, R. E. J., E. Stoffels, et al. 2004. Plasma treatment of dental cavities: A feasibility study. *IEEE Transactions on Plasma Science* 32: 1540–3.

Soukos, N. S., P. S. Chen, et al. 2006. Photodynamic therapy for endodontic disinfection. *Journal of Endodontics* 32: 979–84.

Stark, R. H. and K. H. Schoenbach. 1999. Direct current glow discharges in atmospheric air. *Applied Physics Letters* 74: 3770–2.

Stoffels, E., A. J. Flikweert, et al. 2002. Plasma needle: A non-destructive atmospheric plasma source for fine surface treatment of (bio)materials. *Plasma Sources Science and Technology* 11: 383–8.

Stoffels, E., I. E. Kieft, et al. 2003. Superficial treatment of mammalian cells using plasma needle. *Journal of Physics D—Applied Physics* 36: 2908–13.

Stoffels, E., A. J. M. Roks, et al. 2008. Delayed effects of cold atmospheric plasma on vascular cells. *Plasma Processes and Polymers* 5: 599–605.

Sykes, G. 1965. *Disinfection and Sterilization: Theory and Practice*. Bath: The Pitman Press.

Tarnuzzer, R. W. and G. S. Schultz. 1996. Biochemical analysis of acute and chronic wound environments. *Wound Repair and Regeneration: Official Publication of the Wound Healing Society [and] The European Tissue Repair Society* 4: 321–5.

Vargo, J. J. 2004. Clinical applications of the argon plasma coagulator. *Gastrointestinal Endoscopy* 59: 81–8.

Venugopalan, M. and A. Shih. 1981. Reactions of hydrogen peroxide vapor dissociated in a microwave plasma. *Plasma Chemistry and Plasma Processing* 1: 191–9.

Vercammen, K. L. L., A. A. Berezin, F. Lox, and J.-S. Chang. 1997. Non-thermal plasma techniques for the reduction of volatile organic compounds in air streamers: A critical review. *Journal of Advanced Oxidation Technologies* 2: 17.

Vleugels, M., G. Shama, et al. 2005. Atmospheric plasma inactivation of biofilm-forming bacteria for food safety control. *IEEE Transactions on Plasma Science* 33: 824–8.

Waltimo, T. M., B. H. Sen, et al. 2003. Yeasts in apical periodontitis. *Critical Reviews in Oral Biology and Medicine* 14: 128–37.

Weltmann, K. D., E. Kindel, et al. 2010. Atmospheric-pressure plasma sources: Prospective tools for plasma medicine. *Pure and Applied Chemistry* 82: 1223–37.

Yamazaki, H., T. Ohshima, et al. 2011. Microbicidal activities of low frequency atmospheric pressure plasma jets on oral pathogens. *Dental Materials Journal* 30: 384–91.

Yang, B., J. R. Chen, et al. 2011. Oral bacterial deactivation using a low-temperature atmospheric argon plasma brush. *Journal of Dentistry* 39: 48–56.

Yip, H. K. and L. P. Samaranayake. 1998. Caries removal techniques and instrumentation: A review. *Clinical Oral Investigations* 2: 148–54.

Yu, Q. S., C. Huang, et al. 2007. Bacterial inactivation using a low-temperature atmospheric plasma brush sustained with argon gas. *Journal of Biomedical Materials Research Part B—Applied Biomaterials* 80B: 211–19.

Yun, H., B. Kim, et al. 2010. Inactivation of *Listeria monocytogenes* inoculated on disposable plastic tray, aluminum foil, and paper cup by atmospheric pressure plasma. *Food Control* 21: 1182–6.

15

Plasma Surface Engineering of Titanium-Based Materials for Osseointegration

Huiliang Cao
and Xuanyong Liu

15.1 Introduction

The orthopedic implant sector counts for much more in the worldwide biomedical industry. In the United States alone, orthopedic implant demand is forecast to rise 9.8% annually to $30.4 billion by the year 2015 (Freedonia Group 2008). Due to their medical benefits, including excellent mechanical properties, corrosion resistance, and biocompatibility, titanium and its alloys are widely used in the field. Titanium-based artificial hip, knee, shoulder, fracture fixations, spinal, and dental implants play an important role in improving the quality of life of aged or injured human beings. Nevertheless, a titanium implant may fail too. Lack of integration with bone tissue and infections are major complications of titanium implants (Neoh et al. 2012). In order to meet the long-term service demands, an implant should possess both good bulk mechanical properties and excellent surface properties, including biocompatibility, high corrosion and wear resistance, good osseointegration, and antibacterial activity (Geetha et al. 2009). However, titanium-based materials in use today can hardly satisfy all these requirements. Materials that possess favorable bulk properties such as strength frequently have inadequate surface biological properties such as osseointegration. Thus, surface modification gained an important and decisive place in the field of manufacturing orthopedic or dental implants. Recently, plasma-based technologies attracted much more attention in modifying the surface properties of various biomaterials (Chu et al. 2002; Liu et al. 2004a, 2010). This chapter focuses on utilizing such technologies for surface engineering of titanium-based materials for applications in osseointegration.

15.2 Biological Requirements for Titanium Surfaces

15.2.1 Biocompatibility and Bioactivity Issues

Once an artificial implant is placed *in vivo*, it induces a cascade of events in the biological microenvironment by interacting with body fluids, proteins, and various cells at different time intervals. This sequence of local events dominates the path and speed of bone healing and the long-term integration at the implant/tissue interface. It determines the actual significance of biocompatibility. The materials used as implantable devices are expected to be highly nontoxic and should not cause any inflammatory or allergic reactions in the human body. The historical goal of biomaterials was to "achieve a suitable combination of physical properties to match those of the replaced tissue with a minimal toxic response in the host" (Hench 1980). This led to the least chemically reactive devices being used in the human body during the years before 1980. A common feature of these implants is that they initiate the growth of a thin fibrous capsule that separates normal tissue from the implant. Thus, movement of the implant within the capsule occurs under stress, which undermines the long-term performance of implants. It is clinically evidenced that the service life of most "nearly inert" implants is much shorter than the patients require, and biologically active materials are becoming more widely used particularly in orthopedic and dental applications since the mid-1980s (Hench and Wilson 1984). The last few years have marked a substantial paradigm shift in design criteria for modern biomaterials as the concepts for bioactive and resorbable materials converged and the goal of biomaterials became to "design materials to stimulate specific cellular responses at the molecular level" (Hench and Polak 2002). Hence, research on the bioactivity of materials should focus on much more than crystallization of a biologically reactive hydroxyl-carbonate apatite layer from a simulated body solution (Bohner and Lemaitre 2009). Accordingly, the general definition of biocompatibility was modified to "the ability of a biomaterial to perform its desired function with respect to a medical therapy, without eliciting any undesirable local or systemic effects in the recipient or beneficiary of that therapy, but generating the most appropriate beneficial cellular or tissue response in that specific situation, and optimising the clinically relevant performance of that therapy" (Williams 2008). Naturally, tissue formation, homeostasis, and regeneration after damage are the result of a mazy spatial and temporal synergy of numerous individual cell-fate processes induced by a network of extracellular signals (as shown in Figure 15.1), including physical signals from insoluble hydrated macromolecules (such as fibronectin, laminin, and collagens), soluble signals from growth factors, chemokines, cytokines, and signal flow from neighboring cells in the form of cell–cell interactions (Bottaro et al. 2002; Lutolf and Hubbell 2005). In brief, the selection of a cell-fate process (replication, differentiation, migration, apoptosis, or other specific functions) is a coordinative response to the bidirectional interactions between cells and the signal network in their extracellular matrices (ECMs). And we may learn from the natural signal network and design biomimetic surface modification protocols for titanium-based materials.

Although titanium and its alloys earned a reputation of excellent corrosion resistance and biocompatibility, their long-term performance has raised concerns due to both formation of wear debris and leach of toxic metal ions from the substrate. Titanium-based materials tend to produce wear debris because of their high friction coefficient, resulting in inflammatory reaction and loosening of implants (Geetha et al. 2009). Further, leaching of metal ions from metallic prostheses is of much concern. All metals experience corrosion while in contact with the biological environment (Hallab et al. 2001), leading to release of metal ions that may induce adverse effects (Granchi et al. 2008;

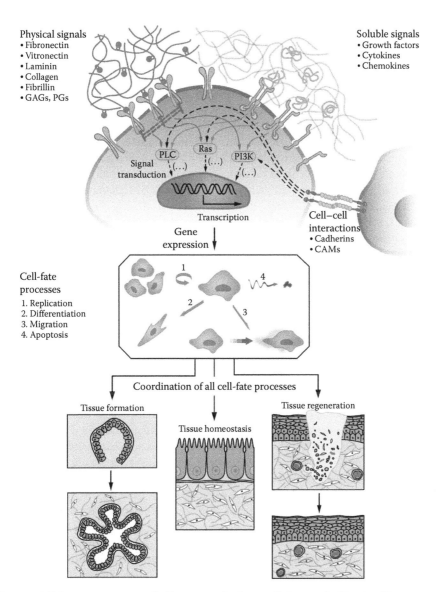

FIGURE 15.1 Cell-fate processes controlled by a network of extracellular signals. (Reprinted by permission from Macmillan Publishers Ltd. Lutolf, M.P. and Hubbell, J.A., *Nat. Biotechnol.*, 23, 47–55, copyright 2005.)

Manivasagam et al. 2010; Steinemann 1996). For example, it was reported that the release of aluminum or vanadium from a Ti6Al4V alloy induces problems in long-term health, such as neuropathy, Alzheimer's disease, and osteomalacia (Nag et al. 2005; Walker et al. 1989). The released metal ions may stimulate both immune system and bone metabolism through a series of direct and indirect pathways, contributing to aseptic loosening of metallic implants (Cadosch et al. 2009). However, on the other hand, metal ions are playing a fundamental role in life processes, functioning as osmotic regulators, as cofactors in enzymes, and as integrators of proteins (Elinder and Arhem 2004). The biological effects of metal ions in biomedical processes should be taken advantage of for developing novel therapeutic and diagnostic agents (Guo and Sadler 1999; Hambley 2007; Lippard 2006). At least 10 metal ions (Na, Mg, K, Ca, Mn, Fe, Co, Cu, Zn, and Mo) have been classified as essential because they occur in most biological tissues and cause reproducible physiological abnormalities on exclusion (Elinder and Arhem 2004). Metal ions, such as calcium, magnesium, strontium, and zinc, have an effect on skeletal cells. Many researchers have demonstrated that micronutrients are critical to or are able to alter bone growth, often during skeletal development (Habibovic and Barralet 2011). Numerous transition metals are accumulated by cells, and these metals have two uses: as required cofactors or as stimuli in cytotoxic reactions (Finney and Ohalloran 2003). Therefore, categorizing elements as essential and toxic should be done away with; instead, the dose–response concept and the overall health effects should be studied (Mertz 1993). Also, the concentration of a metal ion should not surpass normal levels and become toxic (Thompson and Orvig 2003). Besides, mammalian cells, as mentioned before, naturally reside in ECMs containing nanoscale collagen fibrils that interact with nanoscale receptors structured on their own membrane (Anselme et al. 2010; Bozec et al. 2007; Geiger et al. 2009). Mammalian cells respond to artificial topography through an indirect mechano-transduction process by which cells convert mechanical cues into chemical messages, influencing their local migration, polarization, and other functions (Bettinger et al. 2009; Wozniak and Chen 2009). Consequently, engineered nanoscale topography can be considered as another important signaling modality in controlling cell function. The biological benefits of both nanoscale topography and metal ions include eliciting the "appropriate beneficial cellular response." Nanoscale topography may be substitutes of natural physical signals to guide the cell adhesion; also, metal ions may be supplements for natural soluble signals for remotely controlling cell fate.

15.2.2 Bacterial Infection Issues

Biomedical device-associated infections (BDIs) are usually attributable to microbial contaminations. Bacterial species have the ability to adhere to all kinds of BDIs made of metals, ceramics, glasses, and polymers (Nazhat et al. 2009). The species, after adhering to the surface of a device, usually aggregate in a hydrated polymeric matrix resulting from their synthesis to form biofilms that answer for many persistent and chronic infections (Costerton et al. 1999). The colonization may occur at the time of surgery or through hematogenous paths (Long 2008). Most BDIs occurring within the first 3 months after implantation are due to perioperative exposure of bacterial species (Brady et al. 2009). This can happen due to commensal skin flora or physician's contaminations around the surgical site where clotted blood and compromised tissues for the colonization of bacteria can be present. Infections can also be triggered by an associated bacteremia through which bacteria are seeded into the vicinities of the implants and induce acute hematogenous infections (Rodríguez et al. 2009). Currently, both systemically administered and locally applied antibiotics are used to prevent BDIs. Most current bactericidal antimicrobials inhibit DNA, RNA, cell wall, or protein synthesis (Kohanski et al. 2010). At the same time, bacterial species may be resistant to antibiotics through various mechanisms, including impermeable membrane, efflux pumps, resistance mutations, and antibiotic inactivations (Allen et al. 2010). The outbreaks of antibiotic resistance in bacteria are universal life threats. The risk of the emergence and global spread of resistant genes looms large (Bush et al. 2011). For instance, the New Delhi metallo-beta-lactamase (NDM-1) resistance gene, which confers strong resistance against all antibiotics except colistin and tigecycline, has diffused rapidly after being reported in 2010 and has contributed to increasing mortality in hospitals (Kumarasamy et al. 2010). In order to tackle the antibiotic resistance storm, the use of antibiotics must be controlled and the development of alternatives to antibiotics should be encouraged (Bush et al. 2011). Therefore, inorganic material-based antibacterial coatings are attractive for orthopedic applications in combating BDIs (Simchi et al. 2011). Nonetheless, there are other issues we have to contend with. For example, silver nanoparticles (Ag NPs) have a strong action against a broad spectrum of bacterial and fungal species including antibiotic-resistant strains (Panáček et al. 2009; Prucek et al. 2011; Zhao et al. 2011), yet their ultrasmall size and high mobility raise concerns about their potential cytotoxicity (Brayner 2008; Park et al. 2011). Anyhow, surface modifications of implantable devices with antibacterial capability are likely feasible solutions to BDIs.

In summary, both osseointegration and antibacterial adhesion properties are important requirements for orthopedic implants. These requirements accompanied by stable mechanical properties are criteria for designing successful surface modification procedures.

15.3 Plasma Electrolytic Deposition on Titanium-Based Materials

15.3.1 Major Processes in Plasma Electrolytic Deposition

Plasma electrolytic deposition (PED), as a relatively new discipline of plasma electrolysis in surface engineering, is characterized by application of a different electrical potential between the workpieces and a counterelectrode in liquid electrolysis and the production of electrical discharges at (or in the vicinity of) the surface of these workpieces (Yerokhin et al. 1999). Due to the discharging-enhanced chemical reactions and diffusion processes on the electrode surfaces (primarily anode oxidation, solution electrolysis, and cation reduction), thermal and diffusion processes, new by-product chemical reactions and macroparticle transportation become possible in PED. Accordingly, PED can be used to fabricate oxide layers with enhanced mechanical properties on materials [termed plasma electrolytic oxidation (PEO)] and/or to saturate materials with various alloying elements of specific functions [termed plasma electrolytic saturation (PES)]. There are typically two kinds of current–voltage (I–U) diagrams for the plasma electrolytic process to develop discharge in the dielectric film or in the near-electrode area. As shown in Figure 15.2, a "type a" I–U plot represents a metal–electrolyte system with underlying gas liberation on either the anode or the cathode surface; a "type b" plot represents a system where oxide film formation occurs. At relatively low voltages the kinetics of the electrode processes for both systems conform to Faraday's laws and the I–U characteristics of the cell vary according to Ohm's law. However, beyond a certain critical voltage, the behavior of a particular system may change significantly. PES, such as plasma electrolytic nitriding (PEN) and plasma electrolytic carburizing

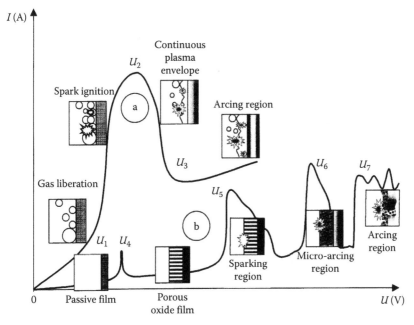

FIGURE 15.2 PED current (I)–voltage (U) diagrams classified according to the location where discharge phenomena developed: (a) in the near area of the workpiece and (b) in the dielectric coating on the workpiece surface. (Reprinted from *Surf. Coat. Technol.*, 122, Yerokhin, A.L. et al., 73–93, copyright 1999, with permission from Elsevier.)

(PEC) can be carried out at potentials above the breakdown voltage (U_2) of the gas bubbles formed on the surface of the workpiece, which may be either cathodic or anodic with respect to the electrolyte. At some critical voltage the vaporized electrolyte and/or gas released by oxidation/reduction processes on the electrode surface forms a continuous vapor envelope around the electrode, in which an arc plasma is established, providing reactive elements for substrate treatment. Because of the much larger current density achieved in such atmospheric discharges, compared with conventional vacuum plasma treatments, intense ion bombardment on the surface of the workpiece can occur, thus causing rapid heating and enhanced activation of the substrate surface. Therefore, only short treatment times (3–5 min) are required for fabrication of 200–500 mm thick diffusion/saturation layers, which have good mechanical and corrosion properties. PEO, also termed as "microplasma oxidation," "anode spark electrolysis," "plasma electrolytic anode treatment," or "anode oxidation under spark discharge," can be operated at potentials above the breakdown voltage (U_5) of an oxide film grown on the surface of a passivated metal anode and is characterized by multiple arcs moving rapidly over the treated surface. Complex compounds can be synthesized inside the high-voltage breakthrough channels formed across the growing oxide layer. These compounds are composed of oxides of both the substrate material (e.g., titanium) and electrolyte-borne modifying elements (e.g., calcium). Plasma thermochemical interactions in the multiple surface discharges result in a coating growing in both directions from the substrate surface.

15.3.2 Physical Properties of the PED Coatings

PED is an effective and rapid technique to improve the mechanical properties of titanium and its alloy by forming a ceramic oxide coating on their surfaces. The crystal structure is strongly influenced by the applied voltage. It was reported that the anatase phase usually forms at low applied voltage (350 V), whereas the rutile phase presents in the coating at high applied voltage (450 V) (Chen H. et al. 2010). Titania coatings consisting of almost pure crystal anatase and rutile phases can be synthesized by adding PED in a solution containing $Ti_2(SO_4)_3$ and H_2SO_4, and an increase in PED time leads to an increase in rutile content (Ragalevičius et al. 2008). It was found that the addition of $Ti_2(SO_4)_3$ extends the current density region used for oxidation of Ti and changes the discharge behavior from large, extensive, and long sparks to small, numerous, and short sparks, resulting in more uniform and relatively thick coatings. The microstructure of the oxide layer is greatly affected by the applied voltage and duration in PED. The oxide layer grows thicker and rougher with pores as the voltage and PED duration increase (Chen H. et al. 2009). The tribological modes of PED-synthesized titanium oxide (TiO_2) coatings (TOCs) vary with the load state. Zhou et al. investigated the tribological properties of PED-synthesized TOCs on a TC4 titanium alloy (Zhou et al. 2010). They found that the coatings are subject to abrasive wear at lower load, whereas fretting fatigue coexisting with abrasive wear mode is predominant at higher normal load. The corrosion resistance to a 3.5 wt.% NaCl solution for PED-treated Cp titanium can be improved notably. It was reported that the corrosion potential rose about 0.13 V and the corrosion current density decreased about an order of magnitude compared with that of bare titanium (Sun et al. 2010). Nanocrystalline TiO_2 films with primary particles size in the range of 10–20 nm can be synthesized by PED in an electrolyte containing calcium acetate monohydrate (Han et al. 2002). These kinds of coatings are expected to be significant for use in orthopedic/dental implants because the adhesion of osteoblast on nanocrystalline grain-sized surfaces can increase by 30% compared to the conventional coarse-grained ones (Webster et al. 1999).

The porous surface nature of a PED-fabricated ceramic coating is beneficial to bone tissue growth and enhances anchorage at the implant/bone interface. The pore size can be regulated by PED with selected applied voltage and electrolyte concentration. As mentioned above, the applied voltage has a direct role in controlling the location where electrical sparks commence, and as such it has the most critical role in the formation of structural pores. Therefore, as shown in Figure 15.3, no pore was observed in the absence of electric sparks (the applied voltage was less than 250 V), but the pore size increases when the voltage exceeds 250 V (Bayati et al. 2010). In addition to the voltage applied, electrolyte concentration also influences the pore formation process. Sun et al. (2008) studied the PED behavior of porous titanium in

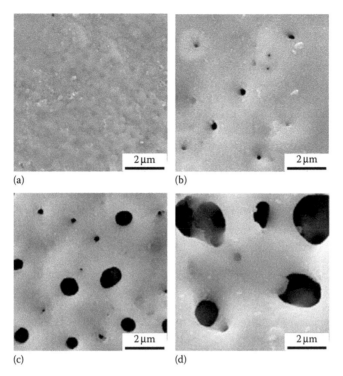

FIGURE 15.3 SEM morphology of the pores formed on titanium PED treated with various voltages in the same electrolyte: (a) 250 V; (b) 300 V; (c) 450 V; (d) 550 V. (Reprinted from *Mater. Chem. Phys.*, 120, Bayati, M.R. et al., 582–89, copyright 2010, with permission from Elsevier.)

the aqueous electrolytes containing 0.1 and 0.2 M sodium hydroxide (NaOH). They found that thin films with pore sizes of 20–60 nm were formed in both electrolytes. However, the film formed in the electrolyte containing 0.2 M NaOH presents higher roughness, more nanopores, and faster apatite-forming ability compared with the film formed in the electrolyte containing 0.1 M NaOH. They expected that the PED-formed bioactive porous titanium will be beneficial for both physical anchorage and chemical bonding at the implant/bone interface. Pure TOCs are usually inert to the host bone. However, bioactivity can be improved by PED in selected electrolytes. For example, Zhao et al. (2009) reported that bioactive oxide films on titanium containing the $Na_2Ti_6O_{13}$ phase, which can induce bone-like apatite, can be prepared by PED in a solution with a high concentration of NaOH. Byon et al. (2007) reported that PED in sulfuric acid can induce apatite-forming ability on a pure titanium surface. Besides, the bioactivity of the ceramic coating can be enhanced by combining PED with other surface modification techniques. It was reported that apatite nucleations can be rapidly induced on the titanium surface combination treated by PED and UV illumination (Han and Xu 2004). Also, the combination of PED with alkali-heat treatment (AH) can improve the cytocompatibility of the Ti–24Nb–4Zr–8Sn alloy (Han X. et al. 2011). Han et al. demonstrated that the growth rate of MC3T3-E1 cells on a PED- and AH-treated layer was significantly higher than that on just the PED-treated surface.

15.3.3 Incorporation of Inorganic Elements for Bioactivity and Biocompatibility

Different surface chemistry may stimulate distinctive bone tissue reactions at the implant/bone interface. The bioactivity and biocompatibility can be enhanced by doping the PED coatings with inorganic elements, including calcium, phosphate, silicon, and strontium. Calcium can be incorporated into the titanium surface (Ti–24Nb–4Zr–7.9Sn alloy) by PED in a calcium acetate [$(CH_2COO)_2Ca·H_2O$]

electrolyte (Tao et al. 2009). The x-ray photoelectron spectroscopy (XPS) results indicated that the calcium elements are incorporated into the oxide layer in the form of CaO. The CaO in the oxide layer favors the formation and nucleation of calcium hydroxide, hydroxyl radicals, and apatite in turn. Phosphorous (P)-containing anatase films can be prepared by PED treating the titanium (Ti) in an electrolyte containing β-glycerol phosphate disodium salt pentahydrate (β-GP, $C_3H_7Na_2O_6P\cdot5H_2O$) (Ryu et al. 2008). The P content in the films can be increased up to 8 at.% with an increasing applied voltage. The bioactivity of the synthesized films was evaluated in simulated body fluid (SBF), and they found that no apatite was induced on the as-prepared specimen for up to 28 days. However, apatite can be induced on the TiO_2 surfaces within 9 h of immersion after a hydrothermal treatment at 250°C, which is even faster than on Ca-containing PED surfaces. It is believed that the diffusion on the surface and hydrolyzation to form the hydrogen phosphate group (HPO_4^{2-}) in the films are responsible for the higher apatite-inducing ability of P-containing PED coatings after hydrothermal treatment. Sul (2003) pointed out that the incorporation of P and Ca into PED-treated implants increased the probability of biochemical bonding between the bone and the implant.

The biological properties of PED coatings incorporated with both Ca and P were also extensively explored. The apatite-inducing ability was closely related to phases containing either Ca and P, such as $CaTiO_3$, $CaTi_4(PO_4)_6$, $Ca_2P_2O_7$, $Ca_3(PO_4)_2$, and HA, or surface hydroxyl groups such as Ca–OH and Ti–OH. Huang et al. (2007) reported that TOCs incorporated with calcium and phosphorous can be prepared by PED in an electrolyte containing calcium acetate [$(CH_3COO)_2Ca\cdot H_2O$, 0.2 mol l^{-1}] and natrium-β-glycero-phosphate ($C_3H_7Na_2O_6P\cdot5H_2O$, 0.2 mol l^{-1}). The fabricated layer can rapidly induce the formation of apatite in SBF, and hydrolysis of the $CaTiO_3$ is a key factor controlling the process. Lim et al. (2009) demonstrated that the PED coating fabricated in an electrolyte consisting of $Ca(CH_3COO)_2\cdot H_2O$, $Ca(H_2PO_4)_2\cdot H_2O$, ethylene diamine tetraacetic acid (EDTA)–2Na, and NaOH can significantly increase the proliferation rate, alkaline phosphatase activity, and adhesion of osteoblast cell lines (SaOS-2). Zhao and Wen (2007) reported that $CaTi_4(PO_4)_6$ coatings can be directly fabricated on pure titanium substrate by PED in an electrolyte in which the calcium ions are complexed by sodium hexametaphosphate. The resulting coating possesses potential bioactivity and better stability. The apatite-inducing ability of Ca- and P-incorporated PED coating was investigated by Song W. et al. (2004). They oxidized the pure titanium substrate in an electrolytic solution containing β-glycerophosphate disodium salt pentahydrate ($C_3H_7Na_2O_6P\cdot5H_2O$) and calcium acetate monohydrate [$(CH_3COO)_2Ca\cdot H_2O$] and found Ca- and P-containing compounds such as $CaTiO_3$, $Ca_3(PO_4)_2$, and $Ca_2P_2O_7$ in the coatings only at higher applied voltages (>450 V). Carbonated hydroxyapatite (HA), when immersed in SBF, can be induced on the surfaces of the films oxidized at voltages above 450 V, which is closely related to these Ca- and P-containing phases. What is more, $Ca_3(PO_4)_2$-containing PED coatings, fabricated in a homogeneous alkaline electrolyte containing phosphate ions, calcium(II) complexes, and EDTA, show good *in vivo* osteogenic properties (Terleeva et al. 2010). The HA phase can be introduced into the coatings by dispersing HA particles in the PED electrolyte. Titanium oxide and HA coatings can be formed on a TC4 titanium alloy by PED in an electrolyte containing citric acid ($C_6H_8O_7$), ethylene diamine [$C_2H_4(NH_2)_2$], ammonium phosphate [$(NH_4)_3PO_4$], and dispersive HA NPs (Hong et al. 2011). The obtained PED coating demonstrates excellent bioactivity, cell attachment, and viability as well as good bonding strength to the substrate. However, this strategy for introduction of HA has the potential risk of poor distribution of Ca and P in the coating, which may induce macrodiversity in biological properties. In this respect, coupling PED with an electrophoretic deposition (EPD) process may aid in the uniform distribution of HA particles. This kind of procedure was initially reported by Kim et al. (2009). In recent years, Bai et al. (2010, 2011a, 2011b) paid a lot of attention to this process. They found that the direct incorporation of well-crystallized HA particles into the PED-formed TiO_2 layer can significantly improve the osteoblastic activity on the coating. However, for successful incorporation, it is essential that a suitable electrolyte solution, in which nanosized HA particles are homogeneously dispersed, be prepared (Kim et al. 2009). Thus, the process is usually not so easy. Hence, synthesis of HA during the PED treating process may be a better choice. It was reported that HA-containing TOCs can be formed by PED in an electrolyte containing calcium acetate monohydrate [$(CH_3COO)_2Ca \cdot H_2O$] and sodium phosphate monobasic dihydrate ($NaH_2PO_4 \cdot 2H_2O$) (Ni et al. 2008; Chen et al. 2006) or in electrolytes containing sodium glycerophosphate ($C_3H_7Na_2O_6P$) and calcium acetate [$Ca(CH_3COO)_2$] (Abbasi et al. 2011). The apatite-inducing advantages of Ca- and P-containing phases, such as $CaTiO_3$, can be made use of in designing PED processes. Kim et al. (2007) reported that nanocrystalline HA [$Ca_{10}(PO_4)_6(OH)_2$, HA] films can be directly fabricated onto the Ti surface by a single-step PED process in electrolytes containing calcium chloride ($CaCl_2$) and potassium phosphate monobasic (KH_2PO_4). The formation of an amorphous $CaTiO_3$ interlayer was identified between the HA and Ti substrates. It was suggested that high-density hydroxyl groups of $TiO(OH)_2$ formed by the reactions between the amorphous $CaTiO_3$ interlayer and the H^+ ions from the dissolution of KH_2PO_4 play a key role in nucleation and growth of the HA crystal by attracting calcium and phosphorous ions in the electrolytes during PED. Recently, based on a similar mechanism, a double-layer structure of porous TiO_2 covered with petal-like apatite (mainly composed of HA and carbonate apatite) was fabricated in an electrolyte in which $(CH_3COO)_2Ca \cdot H_2O$ and $NaH_2PO_4 \cdot 2H_2O$ were dissolved (Liu et al. 2011).

Beside calcium and phosphorous, other elements (such as silicon and strontium) were also incorporated into the PED coating to improve its biological properties. It was evidenced that silicon plays an important role in bone metabolism. Si deficiency may impede the normal growth of rats and cause disturbances in skull size and bone architecture (Schwarz and Milne 1972). Si, uniquely localized in active calcification sites in young bone, was found to be associated with calcium for calcification at an early stage (Carlisle 1970). Therefore, incorporating Si into the PED coating is an attractive and innovative option for enhancing the growth rate of bone tissue. This idea was first explored by Hu et al. (2010). They fabricated a kind of Si-incorporated TOC by PED in an electrolyte containing sodium silicate ($Na_2SiO_3 \cdot 9H_2O$) as the silicon source. The PED coatings containing Si showed higher proliferation rate and vitality of MG-63 cells than the Si-free coatings. Further studies proved that the incorporated Si and porous surface were able to promote adhesion behavior of MC3T3-E1 cells via the integrin b1-FAK signal transduction pathway (Zhang et al. 2011). Strontium (Sr) in human biology and pathology has attracted less attention than calcium, but the therapeutic potential of Sr in bone metabolism is significant. Research undertaken by Marie et al. showed that Sr has beneficial effects on bone cells and bone formation *in vivo* (Marie 2006; Marie et al 2001). Sr-based drugs, such as strontium renalate, have shown promise in treating osteoporosis and in enhancing bone healing (Marie et al. 2011; Hoppe et al. 2011). Sr can be incorporated into the titanium surface by carrying PED in an electrolyte containing $Sr(CH_3COO)_2$ or $Sr(OH)_2 \cdot 8H_2O$ as strontium sources. Nan et al. (2009) prepared a porous strontium-doped HA (Sr-HA) film on titanium substrates by PED in an electrolyte containing calcium acetate, strontium acetate, and β-glycerol phosphate disodium salt pentahydrate. Yan et al. (2010) fabricated bioactive Sr-riched TOCs on porous titanium by PED in a solution containing $Sr(CH_3COO)_2$. Kung et al. (2010a, 2010b, 2012) investigated the bioactivity, physicochemical characteristics, and mammalian cell response behaviors of a Sr-incorporated coating that was PED-formed in an electrolyte using strontium hydroxide 8-hydrate as the strontium source. The results indicated that the obtained coatings have good bioactivity, well-developed corrosion resistance, and excellent mechanical properties. But the response of human fetal osteoblastic cells (hFOB) on the coatings is strontium dependent. Cell proliferation at 14 days on coatings containing 1% or 5% strontium was higher than on strontium-free coatings, whereas the coating containing strontium higher than 10% was not beneficial to cell growth. These PED-fabricated Sr-loaded films are expected to be promising in medical applications.

15.3.4 Incorporation of Inorganic Elements for Antibacterial Applications

Fabrication of PED coating with selective bioactivity, that is, a coating that enhances mammalian cell activity but inhibits adhesion of pathogenic microbes, is highly desirable. As mentioned above, PED processes can easily fabricate TOC on titanium-based substrates. Accordingly, the physical property of TiO_2 can be taken advantage of in medical applications. Titanium oxide as an intrinsic n-type semiconductor has also been extensively investigated as antibacterial coatings for their photocatalytic effects (Chen et al. 2011; Fujishima and Honda 1972). The biocide efficiency of TiO_2 can be enhanced by introducing "electron traps" to reduce the recombination of photon-generated electron–hole pairs (Asahi et al. 2001). Thus, it is promising to enhance the photocatalytic activity (or antibacterial activity) of the TiO_2 layer by a bandgap engineering protocol. Chi et al. reported that photocatalytic activity of PED-fabricated TiO_2 layer can be enhanced by codoping of N and Eu to adapt its bandgap (Chi et al. 2011). However, these photon-stimulated processes cannot be easily implemented on implanted medical devices *in vivo* because they are not typically exposed to light. The porous structure of PED coatings is another physical property that may be taken advantage of. Research by Han et al. indicated that we may take use of the porous structure to locally administer antibiotic drugs. They created a nanoporous TiO_2 film by PED that can control the release of tetracycline hydrochloride (TCH) for up to 7 days (Han C. et al. 2011). This may prove effective in reducing the need for administration of antibiotic drugs.

An antibacterial PED coating can also be fabricated by incorporation of elements, such as fluorine, silver, and zinc, that are capable of biocidal activity. Li and Zhao (2012) reported that F-incorporated titanium substrate shows no cytotoxicity and has antibacterial efficacy against both *Staphylococcus aureus* and *Escherichia coli*. Silver as a nonspecific biocidal agent is able to act strongly against a broad spectrum of bacterial and fungal species including antibiotic-resistant strains (Agarwal et al. 2010). Silver (ions or NPs) incorporation has attracted a lot of attention in the synthesis of PED coatings. Song et al. (2009) compared the antibacterial activity and *in vitro* cytotoxicity of silver- or platinum-containing PED-fabricated calcium phosphate coatings on titanium substrates. They found that the calcium phosphate coatings obtained in an electrolyte with low Ag concentration [dissolved in the form of $AgNO_3$ (or CH_3COOAg)] exhibited excellent antibacterial activity and less cytotoxicity, whereas Pt-containing electrolyte demonstrated no apparent antibacterial activity. Silver can also be incorporated into the PED coating by dispersing Ag NPs in the electrolyte solution. The electron microscopy results obtained by Necula et al. (2011) confirmed that Ag NPs, delivered to the sites of coating growth through various pathways (open pores, cracks, or short-circuit channels), were embedded in the pore walls and on the surface of the coatings without any change in their morphology or chemistry. Zinc plays diverse roles in various biological systems, including in DNA synthesis, enzyme activity, and biomineralization, while it also possesses excellent antibacterial activity (Storrie and Stupp 2005; Tang Y. et al. 2009). Zn-incorporated TOCs can also be prepared by the PEO technique. Hu et al. (2012) fabricated Zn-incorporated TOCs on titanium by PED in an electrolyte containing calcium acetate monohydrate, glycerophosphate disodium salt pentahydrate, and zinc acetate [$Zn(CH_3COO)_2 \cdot 2H_2O$] with various concentrations. These Zn-incorporated TOCs were highly effective in inhibiting adhesion of both *S. aureus* and *E. coli* strains. As Figure 15.4 shows, although rod-shaped and

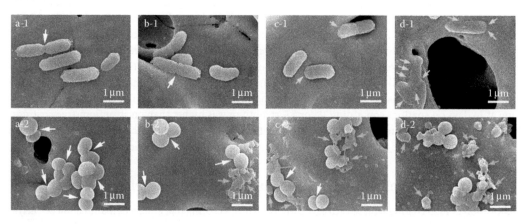

FIGURE 15.4 SEM morphology of the *E. coli* (i-1) and *S. aureus* (i-2) species seeded on the various surfaces (i = a represents the zinc-free coating, i = b, c, and d represent incorporated coatings of about 4.6, 7.1, and 9.3 wt.% zinc, respectively). (Reprinted from *Acta Biomater.*, 8, Hu, H. et al., 904–15, copyright 2012, with permission from Elsevier.)

undamaged binary fission *E. coli* cells can be detected on the coatings incorporated with less than 4.6 wt.% zinc (Figure 15.4b-1), they may become corrugated and may even merge into the surfaces incorporated with 7.1 and 9.3 wt.% zinc (Figure 15.4c-1 and d-1). The *S. aureus* cells were smooth and intact on zinc-free coatings (Figure 15.4a-2), while obvious cell debris and completely lysed cells were observed on coatings incorporated with zinc (Figure 15.4b-2, c-2, and d-2). What is more important is that the Zn-incorporated coatings exhibited obvious biological activity in promoting adhesion, proliferation, and differentiation of rat bMSC. These are encouraging results for fabrication of coatings with balanced antibacterial activity and osteoblast functions on surface-functionalized titanium.

15.4 Plasma Immersion Ion Implantation for Titanium-Based Materials

15.4.1 Plasma Immersion Ion Implantation Processes

Ion implantation is a very attractive method to modify the surface characters of various materials by forming new phases or alloys beyond the normal thermodynamic constraints (Meldrum et al. 2001). However, the application of ion implantation in industry is limited due to its high treatment costs and problems associated with treating three-dimensional (3D) workpieces, given that "line of sight" is the typical character of conventional ion implantation. Plasma immersion ion implantation (PIII), also known as plasma-based ion implantation (PBII) or plasma source ion implantation (PSII), has been developed to circumvent these limitations (Chu et al. 1996). The PIII technique was first proposed by Adler et al. (1985) and Conrad et al. (1987), and it offers an attractive process for simultaneous surface modification of complex-shaped 3D workpieces in large numbers. In PIII, a substrate is pulse-biased to a high negative potential relative to the chamber wall. When the substrate is immersed into a plasma, ions are extracted from the plasma sheath, accelerated by the bias potential, and implanted into the surface of the substrate. This process results in a mix of implanted ion species with the substrate material (Figure 15.5). In order to gain a uniform modification layer, it is necessary to immerse the workpiece in a plasma of high density (Gong et al. 2008). Therefore, PIII is usually carried out within isotropic gaseous plasmas resulting from radio frequency (RF) glow discharges. The power supply system in this hybrid process couples both pulsed RF and high-voltage pulses (Nishimura et al. 2002), When the RF pulses are turned off, negative high-voltage pulses are applied via the same feedthrough to the workpieces conducting PIII. In this respect, high-density plasma can be generated around the workpieces, which themselves act as antennas for RF power. This system is particularly suitable for surface modification of titanium-based substrates with gaseous materials, such as nitrogen, oxygen, and hydrogen. Vacuum arc processes can also be coupled with the PIII system. And thus different metal

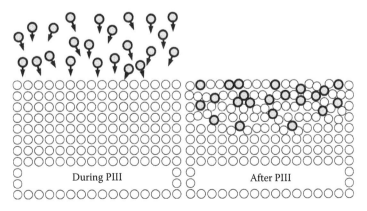

FIGURE 15.5 Illustration for a mix of the implanted ion species and the substrate material PIII treated.

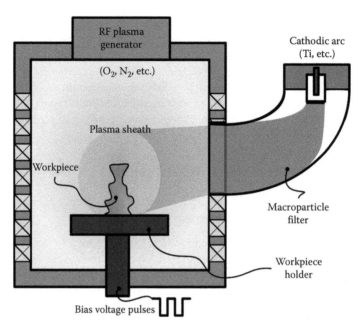

FIGURE 15.6 Coupling of RF plasma source with cathodic arc during PIII.

(include carbon) PIII and deposition (MePIIID) technologies can be developed (Anders 1997). Due to the condensation nature of the metal plasma, PIII with metal plasmas results in qualitatively distinct kind of surface engineering processes than with gaseous plasmas (Anders et al. 1994). Customized films of any solid metal, metal alloy, or carbon (amorphous or diamond) can be fabricated on the substrate surface by MePIIID using filtered vacuum arc plasma sources, and compounds such as oxides or nitrides can be formed by adding a gaseous flow of oxygen or nitrogen (Figure 15.6). PIII is a very versatile approach for modifying the surface characters of titanium-based materials given the wide range of ion species available.

15.4.2 PIII with Inorganic Elements for Osseointegration

15.4.2.1 Ion Implantation with Gaseous Materials

The biocompatibility of titanium is closely related to the thin (between 2 and 20 nm) native oxide layer spontaneously formed on its surface (Larsson et al. 1994). However, this thin oxide layer can be rapidly destroyed during relative movements between implant and tissue under loads. Disruption or destruction leads to increased titanium surface activity, where local electric fields can influence the tertiary and quaternary structure of proteins. Subsequent deformations of the steric atom positions lead to a nonrecognition of native proteins with an activation of the immune system (Mändl et al. 2003). Therefore, coating the titanium surface with a thick oxide layer may be a good solution to this limitation. A thick oxide layer can be obtained by oxygen PIII (O_2 PIII) at elevated temperatures or by annealing the O_2 PIII-treated titanium. Mändl et al. (2001; 2003) reported that it is possible to form hard, dense, and thick TiO_2 layers with excellent biocompatibility by O_2 PIII treating the titanium substrate at temperatures above 473 K. The thickness and crystallite size of the layers can be independently regulated by selecting proper temperature and duration for O_2 PIII. The TiO_2 can be present as rutile, anatase, or Magnéli phases, depending strongly on the temperature, ion energy, and dose of oxygen (Hammerl et al. 1999; Mändl et al. 2000; Rinner et al. 2000). However, the alloying elements presented in the substrate material can dramatically influence the oxidation or diffusion behavior of oxygen. A lamellar structure of TiO_2(rutile)/Ni_3Ti/NiTi can be formed by O_2 PIII treating the superelastic nitinol (NiTi) substrate at the temperature range 523–823 K, leading to the antisegregation of Ni at the outer surface (Mändl et al. 2005). The mechanical properties of pure titanium can be improved by adding vanadium and aluminum; titanium alloys, such as the TC4 alloy (Ti6Al4V), can be fabricated thus. When the alloy is subjected to O_2 PIII treatment, additional Al segregation toward the surface can be detected at elevated temperatures (Lutz et al. 2007). The phase structure of the TiO_2 layer can be regulated by postannealing procedures. Li et al. (2010) reported that although rutile is not formed in a TC4 surface O_2 PIII-treated at a low temperature (below 453 K), it precipitates after postannealing of the substrate at 773 K in vacuum. These postannealing procedures are simple and reproducible in controlling the phase structure in the O_2 PIII-treated titanium surfaces and improving the biocompatibility of titanium materials.

Furthermore, nitrogen, CO, water, and hydrogen were also used in ion implantation of titanium-based materials to enhance their mechanical, tribological, and biological properties. Plasma ion source nitriding of Ti and its alloys increases the wear and corrosion resistance (Byeli et al. 2012; Lei et al. 2011). N PIII at high temperatures was carried out to improve the surface properties of the TC4 titanium alloy (Silva et al. 2010). After N PIII, the alloy became better with respect to corrosion, mechanical, and wear resistances. The best results against corrosion attack were obtained with treatment for 2 h at 1033 K. The best result for mechanical resistance was obtained for samples implanted for 4 h at 933 K. It was reported that the osseointegration properties of TC4 dental implants were significantly improved by ion implantation of CO (Braceras et al. 2002). The bone implant contact percentage (%BIC) values for implants treated with CO are 61% and 62% at 3 and 6 months, respectively, which are statistically and significantly higher than that of both the commercially treated and the control implant groups (Maeztu et al. 2008). Water and hydrogen PIII have been extensively studied as a method to fabricate silicon-on-insulator (SOI) substrates in the semiconductor industry (Liu et al. 1995; Lu et al. 1997; Min et al. 1996), and so it is relatively straightforward to carry out the procedure on biomedical substrates. In water plasma, the dominant species are O^+, HO^+, and H_2O^+, which is in contrast to oxygen plasma consisting of O^+ and O_2^+. Water and hydrogen were plasma-implanted sequentially into titanium by Xie et al. (2005c). They found that water PIII can introduce damages in the near surface of the titanium substrate that acts as hydrogen traps during subsequent hydrogen PIII. Ti–OH functional groups were detected on the water and hydrogen ($H_2O + H_2$) PIII titanium substrate by both XPS and Fourier transform infrared (FTIR) spectroscopy. After incubating in SBFs, HA was precipitated on the ($H_2O + H_2$) PIII-treated titanium, whereas no apatite was detected on titanium treated by water or hydrogen PIII alone. Human osteoblast cells (OPC-1) exhibit good adhesion and growth activities on the ($H_2O + H_2$) PIII-treated titanium surface. These results suggest the practical significance of improving the bioactivity and cytocompatibility of titanium-based implants via the ($H_2O + H_2$) PIII process.

15.4.2.2 Ion Implantation with Nongaseous Materials

It is well known that surface chemistry plays an important role in mediating the interactions between an implant and the surrounding host tissue. Ion implantation with inorganic elements, such as calcium, sodium, magnesium, and phosphorus, leads to remarkable changes in chemical composition at the near-surface region of the biomedical materials, which are expected to influence the attachment and spreading characteristics of bone cells and subsequently their behaviors in signaling cascades and maturation. A number of studies have reported the physical, chemical, and biological properties of Ca-implanted titanium surfaces. Transmission electron microscopy (TEM) results indicated that the Ca-implanted titanium surface could be amorphous, and it increases the corrosion resistance under stationary conditions but undergoes pitting corrosion during anodic polarization (Krupa et al. 2001). The Ca-implanted Ti samples showed more active electrochemical behavior and easier apatite formation capability than the untreated samples, revealing their good bioactivity (Byon et al. 2005; Liu et al. 2005a). The performance of Ca-implanted titanium substrates was histologically investigated by inserting them into rat tibias (Hanawa et al. 1997).

The formation of new bone and mature bone with bone marrow can be detected on the Ca-implanted surface at 2 and 8 days after the insertion. Furthermore, part of the bone was found in direct contact with the Ca-implanted surface, whereas bone formation on the titanium surface without Ca implantation was delayed and the bone did not make contact with the surface. Nayab et al. (2005, 2007) studied the effects of Ca-implanted titanium surfaces on attachment, morphology, spreading, proliferation, and cell cycle of MG-63 cells at the molecular level. They found that Ca implantation, depending on the dose level of the implanted Ca, can qualitatively and quantitatively affect the adhesion of MG-63 cells. The Ca-implanted titanium surface, compared with the untreated titanium surface, markedly enhances MG-63 cells in expression of Ki-67, which is a nuclear antigen associated with cell proliferation. The MG-63 cells cultured on Ca-implanted titanium surfaces reenter and progress through the S and G2/M phases of the cell cycle more rapidly than on the untreated titanium surfaces. Other than calcium, sodium, magnesium, and phosphorus alone or accompanied by calcium were also introduced to enhance the osseointegration-related functions on titanium surfaces (Krischok et al. 2007; Krupa et al. 2002). The corrosion resistance and bioactivity of sodium-implanted titanium were examined by Krupa et al. They found that sodium implantation improves the corrosion resistance of titanium to a certain extent (Krupa et al. 2005). After exposure in SBF, calcium phosphates can be induced on the treated surfaces. This is inconsistent with the results reported by Maitz et al. (2002, 2005), who carried out the implantation both in the conventional ion beam and in the PIII system. The positive effect of magnesium implantation in promoting apatite nucleation and growth on titanium from SBF was also confirmed from multiple sources (Liang et al. 2007; Wan et al. 2006). Although not all the results mentioned above are obtained on a PIII-treated titanium surface, the advantages of a specific element affecting the biological responses of bone cells are valid with both conventional ion beam and PIII techniques. The superiority of the PIII process is mainly reflected in the uniformity obtained on treating 3D biomedical implants with complex shapes (Liu et al. 2005a; Maitz et al. 2005). Nevertheless, it is important to note that the biological benefits of these elements are highly dependent on their reactive nature, which, on the one hand, makes these chemicals available for bone cells or other biological systems, but, on the other hand, causes problems in preserving "the same surface" throughout the investigation because the chemical states of these elements change notably as they are exposed to air or water, and so it is hard to understand the exact mechanism affecting the functions of bone cells. For example, calcium hydroxide and calcium carbonate can be detected on the calcium PIII-treated titanium surface when it is exposed to water and carbon dioxide in the ambient atmosphere (Liu et al. 2005a). The present authors tracked this kind of changes in surface characters by measuring the water contact angle. We found that the contact angle on a Ca-implanted titanium surface was about 65°C after a 4 h exposure in air, and it could reach about 105°C after a 24 h exposure, indicating dramatic variation of the surface chemistry and the subsequent influences on biological properties. Because of this, special attention should be paid when handling PIII processes.

15.4.2.3 Ion Implantation with Both Gaseous and Nongaseous Materials

Titanium oxides or nitrides, as previously referred, can be deposited on titanium surfaces by adding a gaseous flow of oxygen or nitrogen into the MePIIID system. As shown in Figure 15.6, an RF plasma source was employed to create background plasmas of oxygen or nitrogen. It provides more active species for surface interaction (collisions) with titanium ions resulting from the impact of a cathodic arc during high-voltage pulses and thus allows the coating of large, nonflat substrates. Titanium oxide films can be produced in a plasma immersion ion implanter equipped with a metal cathodic arc and RF discharge plasma sources allowing generation of titanium and oxygen plasma at the same time (Chu et al. 2001; Tian et al. 2003). The synthesized TiO_2 films can be micropatterned using the argon plasma etching procedure (Jing et al. 2007). The resulting micropatterned surfaces with organized arrays of 25 μm^2 wells separated by 25 μm gaps exhibited good cytocompatibility with human umbilical vein endothelial (HUVE) cells, and demonstrated capability in modulating cell behavior in HUVE. It was demonstrated that TiO_2 thin films can be deposited by inserting an auxiliary RF plasma source in a MePIIID setup with a cathodic vacuum arc and high-voltage pulses (Gjevori et al. 2011). This auxiliary plasma source promotes growth rate at low gas flow ratios but does not increase the oxygen/titanium ratio. Due to impact effects of energetic ions, titanium nitride films deposited by high-voltage pulsing are also quite different from the gold-colored coatings produced by a native-energy titanium cathodic arc operating in a nitrogen atmosphere. When high-voltage pulsing is used, the color of the films changes to purple. There is also a distinct change in the preferred orientation of the crystallites in the film. For the usual arc-fabricated material, a <111> direction perpendicular to the plane of the film dominates, whereas with high-voltage pulsing the preferred direction becomes <200> (Bilek et al. 2002). Also, by additional supply of oxygen and nitrogen into the vacuum chamber of a PIII system, titanium–oxide–nitride (TiN_xO_y) can be synthesized (Tsyganov et al. 2007). The thrombocyte adhesion and fibrinogen adsorption of a TiN_xO_y film are lower than of TiO_2, which is correlated with the fact that its surface is less hydrophobic and has a higher polar component.

15.4.3 PIII with Inorganic Elements for Antibacterial Applications

Clinical practices have shown that systemic antibiotics are unable to provide effective treatment for all biomedical implant-related infections. Consequently, there is a need to find effective substitute therapy with antibacterial capabilities. Silver and copper stand out for their strong action against bacterial cells. It was reported that silver- or copper-incorporated materials are capable of reducing infections (Dan et al. 2005; Gosheger et al. 2004; Heidenau et al. 2005; Yoshida et al. 1999). Ion implantation was explored as a promising method to coat the biomedical materials with silver or copper for antibacterial applications. Wan et al. (2007a, 2007b) studied the antibacterial effects of titanium, Ti–Al–Nb, and 317L stainless steel, which were implanted with silver or copper using metal vapor vacuum arc (MEVVA) ion sources. The results obtained demonstrated that silver or copper implantation can enhance the antibacterial and wear performances of all these three substrates. Although the antibacterial efficiency of silver- or copper-implanted materials is notably high, research on how to control the antibacterial action of the implanted silver or copper is rare. Recent studies (Cao et al. 2011; Li J. et al. 2011) by our group on the antibacterial effects of Ag PIII-treated titanium indicate that the physicochemical property of substrate materials can be taken advantage of to influence precipitation and control the antibacterial actions of Ag PIII-fabricated Ag NPs. Ag NPs can result from the nonequilibrium conditions of the low (room)-temperature Ag PIII process, which produces local thermal spikes promoting precipitation of metallic silver particles and other interphases from substrate materials (such as pure titanium) with low silver solubility. This facilitates the fabrication of bound Ag NPs in the titanium surface. The results obtained through scanning electron microscopy (SEM), TEM, and high-resolution TEM (HRTEM), as shown in Figure 15.7 (Cao et al. 2011), proved that Ag NPs precipitate on (Figure 15.7b) and underneath (Figure 15.7c) the titanium surface after Ag PIII for 0.5 and 1.5 h, respectively, and the particles can be separated into small segments (Figure 15.7d) until the retained silver dose limitation in a titanium substrate is reached. The embedded Ag NPs may provide the basic components for microgalvanic couples. That is, the embedded Ag NPs act as the cathode and the α-Ti matrix serves as the anode due to the fact that the standard electrode potential of titanium ($E_{Ti}^{\circ} = -1.630$ V) is markedly more negative than that of silver ($E_{Ag}^{\circ} = 0.7996$ V). Accordingly, microgalvanic effects, as shown in Figure 15.8, can be easily triggered when these titanium surfaces (embedded with Ag NPs) are immersed in a physiological liquid that is abundant in protons (H^+) and chloride (Cl^-) ions. And these effects may cut off the proton electrochemical gradient (proton-motive force) in the intermembrane space of bacteria and interfere with the energy-dependent reactions in bacterial cells. The antibacterial efficiency of Ag PIII-treated titanium with various concentrations of Ag NPs is significantly corelated with the microgalvanic effects mentioned above (Cao et al. 2011). What is important is that the galvanic efficiency of a single

FIGURE 15.7 SEM and TEM morphology of Ag PIII-fabricated Ag NPs on and underneath titanium surface: (a) untreated Cp Ti (SEM); (b) Ag PIII treated for 0.5 h (SEM); (c) Ag PIII treated for 1.5 h (TEM); (d) Ag PIII treated for 1.5 h (HRTEM). (Reprinted from *Biomater.*, 32, Cao, H. et al., 693–705, copyright 2011, with permission from Elsevier.)

microgalvanic couple is determined basically by its galvanic current (Liu and Schlesinger 2009), I_g, which can be expressed theoretically as (Song G. et al. 2004):

$$I_g = (E_c - E_a)/(R_a + R_c + R_s + R_m),$$ (15.1)

where
    I_g is the galvanic current between the anode and the cathode
    E_c and E_a are the open circuit potentials of the cathode and anode, respectively
    R_c, R_a, R_s, and R_m are the resistances corresponding to the cathode, anode, solution path between the anode and the cathode, and metallic path from the anode surface to the cathode surface, respectively

Equation 15.1 provides basic approaches for "controlling" the antibacterial efficiency of titanium embedded with Ag NPs.

Additionally, TiO_2 is also considered as a promising antibacterial material because of its photodegradation effects, which can be used to lyse bacterial cells. A porous TiO_2 layer with increased total photodegradation area and light absorption efficiency can be fabricated by a series of PIII processes (Tang G. et al. 2009), which involves sequential steps: generating helium (He) bubbles in titanium substrate by helium (He PIII), oxidizing the layer by O_2 PIII, creating a crystal TiO_2 layer by annealing the O_2 PIII-treated layer in air, producing pores in the crystal TiO_2 layer by exposing the helium bubbles to the surface via an Ar ion sputtering process, and improving light absorption efficiency of the layer by nitrogen PIII (N_2 PIII). This procedure demonstrates fully the versatility of PIII in surface modification of titanium materials.

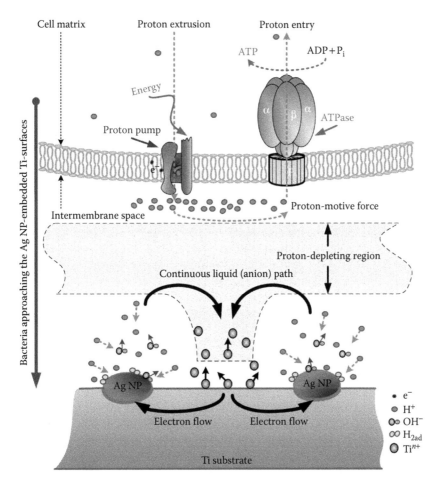

FIGURE 15.8 Figure illustration for the possible toxicity mechanism on the Ag NP-embedded titanium surfaces. (Reprinted from *Biomater.*, 32, Cao, H. et al., 693–705, copyright 2011, with permission from Elsevier.)

15.5 Plasma Spraying on Titanium-Based Materials

15.5.1 Plasma Spraying Processes

Plasma spraying as a part of thermal spraying involves a group of processes in which finely divided materials (metallic and nonmetallic) are deposited onto prepared substrates in a molten or semimolten state (Fauchais et al. 2001). Plasma spraying systems can be categorized by the spraying environment: atmospheric plasma spraying (APS) and controlled APS (CAPS) are basic categorizations. Variations of CAPS include high-pressure plasma spraying (HPPS) and low-pressure plasma spraying (LPPS). Vacuum plasma spraying (VPS) is the extreme case of LPPS. The high energy and density available in the plasma jet makes plasma spraying one of the most versatile surface modification techniques capable of fabricating coatings with almost all materials (ceramics, metals, and composites). The thermal plasma heat source with temperatures over 7723 K at atmospheric pressure allows the melting of any material (Fauchais 2004). In the plasma spraying process, powered materials are typically injected within a plasma jet where the particles are accelerated and melted (or partially melted) before they flatten and pile onto the substrates. The powders should be of spheroidal or similar shape and their size should be as uniform as possible (Liu et al. 2004a). The density, temperature, and velocity of the plasma beam are important for coat formation. The ideal situation would be when all the injected particles are uniformly heated (no vaporization), accelerated to velocities as high as possible but compatible with a completely melted state, and reach the substrates with a temperature over their melting point. High velocities reduce the residence times and thus the heating of the injected particles at a plasma jet. Due to the high deposition rate, low capital cost, and operating cost, plasma spraying has now been introduced into almost all industries including surface modification of biomedical implants.

15.5.2 Plasma Spraying Degradable Coatings for Osseointegration

As mentioned previously, inorganic ions (such as calcium, silicon, magnesium, and strontium) are involved in bone metabolism and play an important role in mineralization, angiogenesis, and growth of bone tissue (Hoppe et al. 2011). Thus, by coupling these elements in the fabrication of degradable ceramic coatings, their biological actions can be taken advantage of. Artificial HA is considered to be an ideal material in the surface modification of implants for rapid osseointegration. Plasma spraying of HA coatings was first proposed by Sumitomo Chemical Company Limited (Osaka, Japan) (Sumitomo Chemical Company Limited 1975). The HA-coated implants can induce earlier bone formation showing higher pushout strengths in the first months than uncoated implants (Wolke et al. 1992). It has been demonstrated that the positive effect of HA on osseointegration is highly related to its degradability in biological environments. It was reported that the deposition of biological apatite and subsequent bone formation on implants depends on the partial dissolution character of HA coating (Porter et al. 2002). The dissolution degree decreases as the crystallinity of HA increases. The crystallinity of As-sprayed HA coating can be altered by postheat treatments. It was found that thermal treatment is effective in converting amorphous HA coating to the crystalline phase (Chang et al. 1999; Khor et al. 2000). Vapor-flame treatment is a simple and efficient way to improve the crystallinity of the HA coating. The crystallinity of an As-sprayed HA coating can be raised from 53.5% to 98.7% by a 3–7 min treatment (Tao et al. 2000). Xue et al. (2004b) studied the bone bonding ability of these HA coatings with different crystallinity *in vivo* and found that the HA coating with high crystallinity (98%) possesses lower degradability but higher shear strengths than the HA coating with low crystallinity (56%). However, the short-term osseointegration property of a vapor-flame-treated HA coating may be diminished due to its high crystallinity (Xue et al. 2005a). Another concern with regard to a HA coating is its inferior mechanical property. Pure HA coating has been reported to have long-term side effects including flaking, foreign body inflammation, and implant loosening (Morscher et al. 1998; Reigstad et al. 2011). Hence, binary Ti/HA (Zheng et al. 2000), YSZ/HA (Fu et al. 2000), TiO_2/HA (Dimitrievska et al. 2011), and ternary YSZ/Ti–6Al–4V/HA (Gu et al. 2004), ZrO_2/Ti/HA (Ning et al. 2005), Al_2O_3/CNT/HA (Tercero et al. 2009) composite HA coatings were developed to enhance their mechanical properties. The reinforcement materials for HA coating should be carefully assessed because their release is very likely during degradation of the HA component, affecting the long-term biocompatibility of composite coatings.

CaO–SiO_2-based coatings, such as wollastonite ($CaSiO_3$, CS) and dicalcium silicate (Ca_2SiO_4, C_2S), have also been regarded as promising candidates for artificial implants due to their excellent bioactivity and biocompatibility (Liu et al. 2008). Liu et al. found that the formation of carbonate-containing HA on the plasma-sprayed wollastonite ($CaSiO_3$, CS) coating is related to the ionic exchange between H^+ in SBFs and Ca^{2+} released from the coating surface (Liu et al. 2001; Liu and Ding 2001, 2002a, 2002c). The apatite can only form on negatively charged surfaces with a functional group ($\equiv$Si–O^-). Increasing the calcium concentration in the SBF solution cannot affect the precipitation behavior of apatite (Liu et al. 2004b). The bioactivity of plasma-sprayed CS coatings was investigated *in vivo* by implanting in dog's muscle, cortical bone, and marrow. Results indicated that short-term osseointegration properties of the CS-coated implants were enhanced (Xue et al. 2005b). But the high dissolution rate of Ca–Si-based materials makes their long-term performance unstable. There are two ways that the stability of Ca–Si-based coatings can be improved. One way is the incorporation of Ti, Zn, or Zr by sol–gel methods and the formation of sphene ($CaTiSiO_5$) (Ramaswamy et al. 2009; Wu et al. 2008b), hardystonite ($Ca_2ZnSi_2O_7$)

(Wu et al. 2008a; Zreiqat et al. 2010), or baghdadite ($Ca_3ZrSi_2O_9$) (Ramaswamy et al. 2008) powders with improved chemical stability, which can be used as material sources for plasma spraying (Liang et al. 2010). Another way is by adding oxide powders, such as TiO_2 (Liu and Ding 2002b) or ZrO_2 (Liu and Ding 2003), to the CS coating, thus forming composite coatings with improved mechanical properties. Dicalcium silicate (Ca_2SiO_4, C_2S), another kind of CaO–SiO_2-based coating, was also under intensive study. The bioactivity of plasma-sprayed C_2S coating is highly related to the negatively charged Si three-ring sites activated during SBF incubation (Liu et al. 2002, 2005b). The ionic products of C_2S coating are beneficial for the proliferation and differentiation of MG-63 osteoblast-like cells (Sun et al. 2009a, 2009b). As with all the other degradable coatings, the application of C_2S coating is also limited by dissolution that cannot be controlled. To overcome this limitation, C_2S/Ti (Xie et al. 2005a, 2005d), C_2S/YSZ (Xie et al. 2005b), and C_2S/ZrO_2 (Xie et al. 2006) composite coatings were developed to improve the bonding strength and durability in biological environments. Although other degradable coatings, such as diopside (Xue et al. 2004a) and Mg_2SiO_4 (Xie et al. 2009), were developed and demonstrated good activity for rapid osseointegration, achieving a balance between osseointegration benefits and mechanical property deterioration is a challenge for this kind of coatings.

15.5.3 Plasma Spraying-Activatable TOC

Plasma-sprayed TiO_2 exhibits excellent long-term mechanical properties, but application in osseointegration is limited because of it bioinertness. Thus, various techniques, including UV illumination, alkali or acid treatment, and ion implantation processes, were developed to activate it for osseointegration. Due to the semiconductive nature of TiO_2, UV irradiation is usually applied to investigate the photocatalysis-related behavior of TiO_2 particles or films (Fujishima et al. 2008). It was evidenced that bone-like apatite, after incubating in SBF, can form on the plasma-sprayed TOC surface, which was previously UV illuminated in air for 24 h. The mechanism for this phenomenon can be narrated as follows (Liu et al. 2006): during UV illumination, oxygen vacancies are created at the two-coordinated bridging sites that transfer the Ti^{4+} sites into Ti^{3+} sites, which are favorable for dissociating water absorbed in atmosphere, resulting in abundance of Ti–OH groups at bridging oxygen sites. These Ti–OH groups play a key role in inducing nucleation and growth of the apatite layer. Furthermore, the surface nanostructures of plasma-sprayed TOC are also important for guaranteeing relative large surface areas and the required number of Ti–OH groups. This is why bone-like apatite was observed to precipitate on UV-irradiated TOC with a nanostructural surface, but not on the As-sprayed and UV-irradiated TOC without surface nanostructures. Apatite formation on a nano-TOC surface can be further improved by prolonging UV irradiation. What is more, the effect of UV irradiation strategy can be applied *in vivo* (Liu et al. 2008b). As shown in Figure 15.9, a yellow layer

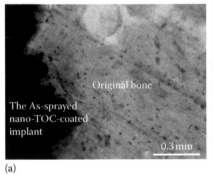
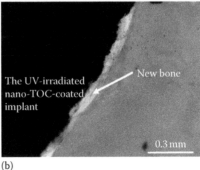

(a)   (b)

FIGURE 15.9 Histological morphology of the As-sprayed (a) and UV-irradiated (b) nano-TOC implanted in rabbit femoral condylus for 2 months, the undecalcified tissue marked by tetracycline. (Reprinted from *Acta Biomater.*, 4, Liu, X. et al., 544–52, copyright 2008, with permission from Elsevier.)

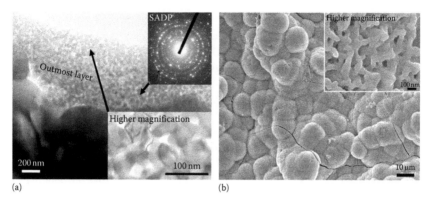

FIGURE 15.10 Cross-sectional TEM views of the As-sprayed nano-TOC surface (a), and surface SEM views of the hydrogen PIII nano-TOC after incubating in SBF for 2 weeks (b). (Reprinted from *Biomater.*, 26, Liu, X. et al., 6143–50, copyright 2005, with permission from Elsevier.)

(the undecalcified section stained by tetracycline), after implantation into the rabbit medial femoral condylus for 2 months, can be observed on the UV-irradiated nano-TOC, revealing formation of new bones. In contrast, the fluorescence view of the As-sprayed nano-TOC evidences no sign of new bone formation. These results indicate that UV irradiation is effective in activating the osseointegration ability of plasma-sprayed TOC.

Other activation methods for plasma-sprayed TOC include alkali treatment, acid immersion process, and PIII procedures. Alkali and acid treatment are regarded as the simplest methods to activate TOC coatings. Zhao et al. (2006) reported that bone-like apatite, after incubating in SBF, can be formed on alkali-treated TOC, whereas no apatite can be observed on As-sprayed TOC. This kind of bone-like apatite was also detected on acid-treated (hydrochloric or sulfuric acid at room temperature for 24 h) TOCs, which demonstrate good compatibility with human mesenchymal stem cells (MSCs) (Zhao et al. 2005, 2008). The inertness of plasma-sprayed TOC can also be changed by ion implantation processes. Liu et al. (2005c) and Zhao et al. (2007) studied the hydrogen PIII effects on the bioactivity of plasma-sprayed TOC. They fabricated a plasma-sprayed TOC with nanostructural surface revealed by cross-sectional TEM observations. As shown in Figure 15.10a, the outermost surface of the As-sprayed TOC consists of grains less than 50 nm in diameter. Then the TOC was transferred into the chamber of a PIII system where it was hydrogen PIII-treated. After incubating in SBF for 2 weeks, the hydrogen PIII-treated TOC surface was completely covered by a newly formed bone-like apatite layer (Figure 15.10b), suggesting good bioactivity for osseointegration. On the contrary, apatite can hardly form on either the As-sprayed TOC (both <50 nm grain size and >50 nm grain size) or the hydrogen PIII-treated TOC with grain size larger than 50 nm, or with other PIII procedures, including oxygen and ammonia PIII.

15.5.4 Plasma Spraying Antibacterial Coatings

In order to increase bone cell functions while decreasing bacterial adhesion (Shi et al. 2009), plasma-sprayed coatings incorporated with both bioactive and antibacterial materials were developed. It was reported that the plasma-sprayed TOC and CS coatings demonstrate good antibacterial and osteogenic properties when these coatings are grafted with collagen and antibiotics (Li et al. 2007, 2008a).

Recently, amidst serious concerns about antibiotic resistance, inorganic materials-based antibacterial strategies received significant attention (Simchi et al. 2011), and the safe and controlled antibacterial strategy that uses silver was widely studied (Prucek et al. 2011; Zhao et al. 2011). For example, silver-loaded CS coating was fabricated by soaking the As-sprayed coating in 5 wt.%

$AgNO_3$ solution at room temperature; the silver ions released from the coating strongly inhibit the growth of *S. aureus*, while they exhibit little side effects on the proliferation of osteoblastic cells (Li B. et al. 2008b). Silver-loaded TOC with excellent antibacterial property can be fabricated by first treating the plasma-sprayed TOC in alkali solution, and then immersing it in a calcification solution containing silver ions (Ag^+) (Chen Y. et al. 2009). Besides, silver can be loaded with various bioactive materials by coplasma-spraying processes. It was reported that coplasma-sprayed CS/silver coatings efficiently inhibit the growth of *E. coli* and possess good bioactivity and cytocompatibility (Li B. et al. 2009a). The nano-TOC/silver coatings fabricated by coplasma spraying also strongly inhibit the growth of *E. coli* and good UV activatable bioactivity (Li B. et al. 2009b). Chen Y. et al. (2010) compared the silver release kinetics based on the two loading strategies mentioned above. They found that both strategies could entrust the HA coating of good antibacterial property, although the release kinetics are highly related to the degradation nature of HA. As a result, both strategies can hardly lead to safe and controllable antibacterial actions due to the violent degradation nature of these matrix materials (HA, CS, and C_2S). This situation can be altered by introducing antibacterial elements into the crystal structure of bioactive materials. It was demonstrated that zinc-incorporated calcium silicate-based ceramic $Ca_2ZnSi_2O_7$ coating can be prepared by plasma spraying of the $Ca_2ZnSi_2O_7$ powder synthesized by a sol–gel method (Li K. et al. 2011). The obtained $Ca_2ZnSi_2O_7$ coating, compared with the pure CS coating, not only demonstrates improved chemical stability in Tris–HCl buffer solution, but also possesses good *in vitro* antibacterial activity, bioactivity, and cytocompatibility.

15.6 Summary

The ultimate goal of osseointegration is to achieve permanent stability of an implant by initiating controllable (or guidable) fine and fast osteogenesis leading to direct connections between living bone tissue and the implant surface that satisfy the structural and functional requirements of the patient. Osseointegration is manipulated by a cascade of extracellular and cellular biological events that commence at the bone/implant interface. It means that osseointegration should be regarded as "a system" and it runs well when all things are optimized. Therefore, "balancing and controlling all the aspects involved" may be a paradigm for engineering the surface of an implant with excellent osseointegration properties. Concretely, although all these three plasma-based techniques can be undertaken to modify an implant surface for osseointegration and antibacterial applications; they each have advantages and disadvantages, and technique selection depends on the characteristics needed for a particular implant. In addition, in the past few decades, researchers have tried many procedures and obtained encouraging results, but there is little known as to the specific mechanism involved. In this respect, PIII may go further than the other two techniques because of its accuracy in controlling the doping dose, which is critical in uncovering the exact biological actions of specific inorganic components. As bioactive and antibacterial properties impact a biomedical implant, a better understanding may inspire us to design "balancing strategies" for titanium-based materials that promote osteoblast functions but inhibit bacterial colonization. And exploring the synergetic effects of two (or more) different inorganic elements on adjusting the "balance point" may become the major field for future studies.

Acknowledgments

The work was jointly supported financially by National Basic Research Program of China (973 Program, 2012CB933601), National Natural Science Foundation of China (31100675 and 51071168), Shanghai Science and Technology R&D Fund under grant 11JC1413700, Science Foundation for Youth Scholar of State Key Laboratory of High Performance Ceramics and Superfine Microstructures (SKL201103), and Innovation Fund of SICCAS (Y26ZC3130G).

References

Abbasi, S., Bayati, M.R., Golestani-Fard, F., Rezaei, H.R., Zargar, H.R., Samanipour, F., and Shoaei-Rad, V. 2011. Micro arc oxidized Hap–TiO$_2$ nanostructured hybrid layers—part I: Effect of voltage and growth time. *Applied Surface Science* 257:5944–9.

Adler, R.J. and Picraux, S.T. 1985. Repetitively pulsed metal ion beams for ion implantation. *Nuclear Instruments and Methods in Physics Research Section B: Beam Interactions with Materials and Atoms* 6:123–8.

Agarwal, A., Weis, T.L., Schurr, M.J., Faith, N.G., Czuprynski, C.J., McAnulty, J.F., Murphy, C.J., and Abbott, N.L. 2010. Surfaces modified with nanometer-thick silver-impregnated polymeric films that kill bacteria but support growth of mammalian cells. *Biomaterials* 31:680–90.

Allen, H.K., Donato, J., Wang, H.H., Cloud-Hansen, K.A., Davies, J., and Handelsman, J. 2010. Call of the wild: Antibiotic resistance genes in natural environments. *Nature Reviews Microbiology* 8:251–9.

Anders, A. 1997. Metal plasma immersion ion implantation and deposition: A review. *Surface and Coatings Technology* 93:158–67.

Anders, A., Anders, S., Brown, I.G., Dickinson, M.R., and MacGill, R.A. 1994. Metal plasma immersion ion implantation and deposition using vacuum arc plasma sources. *Journal of Vacuum Science and Technology B* 12:815–20.

Anselme, K., Davidson, P., Popa, A.M., Giazzon, M., Liley, M., and Ploux, L. 2010. The interaction of cells and bacteria with surfaces structured at the nanometre scale. *Acta Biomaterialia* 6:3824–46.

Asahi, R., Morikawa, T., Ohwaki, T., Aoki, K., and Taga, Y. 2001. Visible-light photocatalysis in nitrogen-doped titanium oxides. *Science* 293:269–71.

Bai, Y., Kim, K., Park, I.S., Lee, S.J., Bae, T.S., and Lee, M.H. 2011a. In situ composite coating of titania–hydroxyapatite on titanium substrate by micro-arc oxidation coupled with electrophoretic deposition processing. *Materials Science and Engineering B* 176:1213–21.

Bai, Y., Park, I.S., Lee, S.J., Bae, T.S., Duncan, W., Swain, M., and Lee, M.H. 2011b. One-step approach for hydroxyapatite-incorporated TiO$_2$ coating on titanium via a combined technique of micro-arc oxidation and electrophoretic deposition. *Applied Surface Science* 257:7010–18.

Bai, Y., Park, I.S., Park, H.H., Bae, T.S., and Lee, M.H. 2010. Formation of bioceramic coatings containing hydroxyapatite on the titanium substrate by micro-arc oxidation coupled with electrophoretic deposition. *Journal of Biomedical Materials Research Part B: Applied Biomaterials* 95B:365–73.

Bayati, M.R., Golestani-Fard, F., and Moshfegh, A.Z. 2010. The effect of growth parameters on photocatalytic performance of the MAO-synthesized TiO$_2$ nano-porous layers. *Materials Chemistry and Physics* 120:582–9.

Bettinger, C.J., Langer, R., and Borenstein, J.T. 2009. Engineering substrate topography at the micro- and nanoscale to control cell function. *Angewandte Chemie International Edition* 48:5406–15.

Bilek, M.M.M., McKenzie, D.R., Tarrant, R.N., Lim, S.H.M., and McCulloch, D.G. 2002. Plasma-based ion implantation utilising a cathodic arc plasma. *Surface and Coatings Technology* 156:136–42.

Braceras, I., Alava, J.I., Onate, J.I., Brizuela, M., Garcia-Luis, A., Garagorri, N., Viviente, J.L., and de Maeztu, M.A. 2002. Improved osseointegration in ion implantation-treated dental implants. *Surface and Coatings Technology* 158–159:28–32.

Brady, R.A., Calhoun, J.H., Leid, J.G., and Shirtliff, M.E. 2009. Infections of orthopaedic implants and devices. In *The Role of Biofilms in Device-Related Infections*, eds. M.E. Shirtliff and J.G. Leid. Berlin: Springer-Verlag, pp. 15–56.

Bohner, M. and Lemaitre, J. 2009. Can bioactivity be tested in vitro with SBF solution. *Biomaterials* 30:2175–9.

Bottaro, D.P., Liebmann-Vinson, A., and Heidaran, M.A. 2002. Molecular signaling in bioengineered tissue microenvironments. *Annals of the New York Academy of Sciences* 961:143–53.

Bozec, L., Heijden, G., and Horton, M. 2007. Collagen fibrils: Nanoscale ropes. *Biophysical Journal* 92:70–5.

Brayner, R. 2008. The toxicological impact of nanoparticles. *Nano Today* 3:48–55.

Bush, K., Courvalin, P., Dantas, G., Davies, J., Eisenstein, B., Huovinen, P., Jacoby, G.A., et al. 2011. Tackling antibiotic resistance. *Nature Reviews Microbiology* 9:894–6.

Byeli, A.V., Kukareko, V.A., and Kononov, A.G. 2012. Titanium and zirconium based alloys modified by intensive plastic deformation and nitrogen ion implantation for biocompatible implants. *Journal of the Mechanical Behavior of Biomedical Materials* 6:89–94.

Byon, E., Jeong, Y., Takeuchi, A., Kamitakahara, M., and Ohtsuki, C. 2007. Apatite-forming ability of micro-arc plasma oxidized layer of titanium in simulated body fluids. *Surface and Coatings Technology* 201:5651–4.

Byon, E., Moon, T.S., Cho, S., Jeong, C., Jeong, Y., and Sul, Y. 2005. Electrochemical property and apatite formation of metal ion implanted titanium for medical implants. *Surface and Coatings Technology* 200:1018–21.

Cadosch, D., Chan, E., Gautschi, O.P., and Filgueira, L. 2009. Metal is not inert: Role of metal ions released by biocorrosion in aseptic loosening—Current concepts. *Journal of Biomedical Materials Research Part A* 91:1252–62.

Cao, H., Liu, X., Meng, F., and Chu, P.K. 2011. Biological actions of silver nanoparticles embedded in titanium controlled by micro-galvanic effects. *Biomaterials* 32:693–705.

Carlisle, E.M. 1970. Silicon: A possible factor in bone calcification. *Science* 167:279–80.

Chang, C., Huang, J., Xia, J., and Ding, C. 1999. Study on crystallization kinetics of plasma sprayed hydroxyapatite coating. *Ceramics International* 25:479–83.

Chen, H., Chung, C., Yang, T., Chiang, I., Tang, C., Chen, K., and He, J. 2010. Osteoblast growth behavior on micro-arc oxidized β-titanium alloy. *Surface and Coatings Technology* 205:1624–9.

Chen, H., Hsiao, C., Long, H., Chung, C., Tang, C., Chen, K., and He, J. 2009. Micro-arc oxidation of β-titanium alloy: Structural characterization and osteoblast compatibility. *Surface and Coatings Technology* 204:1126–31.

Chen, J., Shi, Y., Wang, L., Yan, F., and Zhang, F. 2006. Preparation and properties of hydroxyapatite-containing titania coating by micro-arc oxidation. *Materials Letters* 60:2538–43.

Chen, X., Liu, L., Yu, P.Y., and Mao, S.S. 2011. Increasing solar absorption for photocatalysis with black hydrogenated titanium dioxide nanocrystals. *Science* 331:746–50.

Chen, Y., Zheng, X., Xie, Y., Ji, H., and Ding, C. 2009. Antibacterial properties of vacuum plasma sprayed titanium coatings after chemical treatment. *Surface and Coatings Technology* 204:685–90.

Chen, Y., Zheng, X., Xie, Y., Ji, H., Ding, C., Li, H., and Dai, K. 2010. Silver release from silver-containing hydroxyapatite coatings. *Surface and Coatings Technology* 205:1892–6.

Chi, C., Choi, J., Jeong, Y., Lee, O. Y., and Oh, H.-J. 2011. Nitrogen and europium doped TiO_2 anodized films with applications in photocatalysis. *Thin Solid Films* 519:4676–80.

Chu, P.K., Chen, J.Y., Wang, L.P., and Huang, N. 2002. Plasma-surface modification of biomaterials. *Materials Science and Engineering R* 36:143–206.

Chu, P.K., Qin, S., Chan, C., Cheung, N.W., and Larson, L.A. 1996. Plasma immersion ion implantation semiconductor processing a fledgling technique for semiconductor processing. *Materials Science and Engineering R* 17:207–80.

Chu, P.K., Tang, B.Y., Wang, L.P., Wang, X.F., Wang, S.Y., and Huang, N. 2001. Third-generation plasma immersion ion implanter for biomedical materials and research. *Review of Scientific Instruments* 72:1660–5.

Conrad, J.R., Radtke, J.L., Dodd, R.A., Worzala, F.J., and Tran, N.C. 1987. Plasma source ion-implantation technique for surface modification of materials. *Journal of Applied Physics* 62:4591–6.

Costerton, J.W., Stewart, P.S., and Greenberg, E.P. 1999. Bacterial biofilms: A common cause of persistent infections. *Science* 284:1318–22.

Dan, Z.G., Ni, H.W., Xu, B.F., Xiong, J., and Xiong, P.Y. 2005. Microstructure and antibacterial properties of AISI 420 stainless steel implanted by copper ions. *Thin Solid Films* 492:93–100.

De Maeztu, M.A., Braceras, I., Alava, J.I., and Gay-Escoda, C. 2008. Improvement of osseointegration of titanium dental implant surfaces modified with CO ions: A comparative histomorphometric study in beagle dogs. *International Journal of Oral and Maxillofacial Surgery* 37:441–7.

Dimitrievska, S., Bureau, M.N., Antoniou, J., Mwale, F., Petit, A., Lima, R.S., and Marple, B.R. 2011. Titania–hydroxyapatite nanocomposite coatings support human mesenchymal stem cells osteogenic differentiation. *Journal of Biomedical Materials Research Part A* 98:576–88.

Elinder, F. and Arhem, P. 2004. Metal ion effects on ion channel gating. *Quarterly Reviews of Biophysics* 36:373–427.

Fauchais, P. 2004. Understanding plasma spraying. *Journal of Physics D: Applied Physics* 37:R86–108.

Fauchais, P., Vardelle, A., and Dussoubs, B. 2001. Quo vadis thermal spraying. *Journal of Thermal Spray Technology* 10:44–66.

Finney, L.A. and Ohalloran, T.V. 2003. Transition metal speciation in the cell: Insights from the chemistry of metal ion receptors. *Science* 300:931–6.

The Freedonia Group. 2008. *Orthopedic Implants to 2012: US Industry Study with Forecasts for 2012 and 2017*. Cleveland, OH: The Freedonia Group, Inc. www.freedoniagroup.com.

Fu, L., Khor, K.A., and Lim, J.P. 2000. Yttria stabilized zirconia reinforced hydroxyapatite coatings. *Surface and Coatings Technology* 127:66–75.

Fujishima, A. and Honda, K. 1972. Electrochemical photolysis of water at a semiconductor electrode. *Nature* 238:37–8.

Fujishima, A., Zhang, X., and Tryk, D.A. 2008. TiO_2 photocatalysis and related surface phenomena. *Surface Science Reports* 63:515–82.

Geetha, M., Singh, A.K., Asokamani, R., and Gogia, A.K. 2009. Ti based biomaterials, the ultimate choice for orthopaedic implants—A review. *Progress in Materials Science* 54:397–425.

Geiger, B., Spatz, J.P., and Bershadsky, A.D. 2009. Environmental sensing through focal adhesions. *Nature Reviews Molecular Cell Biology* 10:21–33.

Gjevori, A., Gerlach, J.W., Manova, D., Assmann, W., Valcheva, E., and Mändl, S. 2011. Influence of auxiliary plasma source and ion bombardment on growth of TiO_2 thin films. *Surface and Coatings Technology* 205:S232–4.

Gong, C., Tian, X., Yang, S., Fu, R.K.Y., and Chu, P.K. 2008. Direct coupling of pulsed radio frequency and pulsed high power in novel pulsed power system for plasma immersion ion implantation. *Review of Scientific Instruments* 79:043501.

Gosheger, G., Hardes, J., Ahrens, H., Streitburger, A., Buerger, H., Errenc, M., Gunsel, A., Kemper, F.H., Winkelmann, W., and von Eiff, C. 2004. Silver-coated megaendoprostheses in a rabbit model—An analysis of the infection rate and toxicological side effects. *Biomaterials* 25:5547–56.

Granchi, D., Cenni, E., Tigani, D., Trisolino, G., Baldini, N., and Giunti, A. 2008. Sensitivity to implant materials in patients with total knee arthroplasties. *Biomaterials* 29:1494–500.

Gu, Y.W., Khor, K.A., Pan, D., and Cheang, P. 2004. Activity of plasma sprayed yttria stabilized zirconia reinforced hydroxyapatite/Ti-6Al-4V composite coatings in simulated body fluid. *Biomaterials* 25:3177–85.

Guo, Z. and Sadler, P.J. 1999. Metals in medicine. *Angewandte Chemie International Edition* 38:1512–31.

Habibovic, P. and Barralet, J.E. 2011. Bioinorganics and biomaterials: Bone repair. *Acta Biomaterialia* 7:3013–26.

Hallab, N., Merritt, K., and Jacobs, J.J. 2001. Metal sensitivity in patients with orthopaedic implants. *Journal of Bone and Joint Surgery (American Volume)* 83-A:428–36.

Hambley, T.W. 2007. Metal-based therapeutics. *Science* 318:1392–3.

Hammerl, C., Renner, B., Rauschenbach, B., and Assmann, W. 1999. Phase formation in titanium after high-fluence oxygen ion implantation. *Nuclear Instruments and Methods in Physics Research Section B—Beam Interactions with Materials and Atoms* 148:851–7.

Han, C., Lee, E., Kim, H., Koh, Y., and Jang, J. 2011. Porous TiO_2 films on Ti implants for controlled release of tetracycline hydrochloride (TCH). *Thin Solid Films* 519:8074–6.

Han, X., Liu, H., Wang, D., Li, S., Yang, R., Tao, X., and Jiang, X. 2011. In vitro biological effects of Ti2448 alloy modified by micro-arc oxidation and alkali heatment. *Journal of Materials Science and Technology* 27:317–24.

Han, Y., Hong, S., and Xu, K. 2002. Synthesis of nanocrystalline titania films by micro-arc oxidation. *Materials Letters* 56:744–7.

Han, Y. and Xu, K. 2004. Photoexcited formation of bone apatite-like coatings on micro-arc oxidized titanium. *Journal of Biomedical Materials Research* 71A:608–14.

Hanawa, T., Kamiura, Y., Yamamoto, S., Kohgo, T., Amemiya, A., Ukai, H., Murakami, K., and Asaoka, K. 1997. Early bone formation around calcium-ion-implanted titanium inserted into rat tibia. *Journal of Biomedical Materials Research* 36:131–6.

Heidenau, F., Mittelmeier, W., Detsch, R., Haenle, M., Stenzel, F., Ziegler, G., and Gollwitzer, H. 2005. A novel antibacterial titania coating: Metal ion toxicity and in vitro surface colonization. *Journal of Materials Science: Materials in Medicine* 16:883–8.

Hench, L.L. 1980. Biomaterials. *Science* 208:826–31.

Hench, L.L and Polak, J.M. 2002. Third-generation biomedical materials. *Science* 295:1014–7.

Hench, L.L. and Wilson, J. 1984. Surface-active biomaterials. *Science* 226:630–36.

Hong, M., Lee, D., Kim, K., and Lee, Y. 2011. Study on bioactivity and bonding strength between Ti alloy substrate and TiO_2 film by micro-arc oxidation. *Thin Solid Films* 519:7065–70.

Hoppe, A., Güldal, N.S., and Boccaccini, A.R. 2011. A review of the biological response to ionic dissolution products from bioactive glasses and glass-ceramics. *Biomaterials* 32:2757–74.

Hu, H., Liu, X., and Ding, C. 2010. Preparation and cytocompatibility of Si-incorporated nanostructured TiO_2 coating. *Surface and Coatings Technology* 204:3265–71.

Hu, H., Zhang, W., Qiao, Y., Jiang, X., Liu, X., and Ding, C. 2012. Antibacterial activity and increased bone marrow stem cell functions of Zn-incorporated TiO_2 coatings on titanium. *Acta Biomaterialia* 8:904–15.

Huang, P., Xu, K., and Han, Y. 2007. Formation mechanism of biomedical apatite coatings on porous titania layer. *Journal of Materials Science: Materials in Medicine* 18:457–63.

Jing, F.J., Wang, L., Fu, R.K.Y., Leng, Y.X., Chen, J.Y., Huang, N., and Chu, P.K. 2007. Behavior of endothelial cells on micro-patterned titanium oxide fabricated by plasma immersion ion implantation and deposition and plasma etching. *Surface and Coatings Technology* 201:6874–7.

Khor, K.A., Dong, Z.L., Quek, C.H., and Cheang, P. 2000. Microstructure investigation of plasma sprayed HA: Ti6Al4V composites by TEM. *Materials Science and Engineering A* 28:221–8.

Kim, D., Kim, M., Kim, H., Koh, Y., Kim, H., and Jang, J. 2009. Formation of hydroxyapatite within porous TiO_2 layer by micro-arc oxidation coupled with electrophoretic deposition. *Acta Biomaterialia* 5:2196–205.

Kim, M., Ryu, J., and Sung, Y. 2007. One-step approach for nano-crystalline hydroxyapatite coating on titanium via micro-arc oxidation. *Electrochemistry Communications* 9:1886–91.

Kohanski, M.A., Dwyer, D.J., and Collins, J.J. 2010. How antibiotics kill bacteria: From targets to networks. *Nature Reviews Microbiology* 8:423–35.

Krischok, S., Blank, C., Engel, M., Gutt, R., Ecke, G., Schawohl, J., Spie, L., Schrempel, F., Hildebrand, G., and Liefeith, K. 2007. Influence of ion implantation on titanium surfaces for medical applications. *Surface Science* 601:3856–60.

Krupa, D., Baszkiewicz, J., Kozubowski, J.A., Barcz, A., Sobczak, J.W., Bilinski, A., Lewandowska-Szumie, M., and Rajchel, B. 2001. Effect of calcium-ion implantation on the corrosion resistance and biocompatibility of titanium. *Biomaterials* 22:2139–51.

Krupa, D., Baszkiewicz, J., Kozubowski, J.A., Barcz, A., Sobczak, J.W., Bilinski, A., Lewandowska-Szumie, M., and Rajchel, B. 2002. Effect of phosphorous-ion implantation on the corrosion resistance and biocompatibility of titanium. *Biomaterials* 23:3329–40.

Krupa, D., Baszkiewicz, J., Rajchel, B., Barcz, A., Sobczak, J.W., and Bilinski, A. 2005. Effect of sodium-ion implantation on the corrosion resistance and bioactivity of titanium. *Vacuum* 78:161–6.

Kumarasamy, K.K., Toleman, M.A., Walsh, T.R., Bagaria, J., Butt, F., Balakrishnan, R., Chaudhary, U., et al. 2010. Emergence of a new antibiotic resistance mechanism in India, Pakistan, and the UK: A molecular, biological, and epidemiological study. *The Lancet Infectious Diseases* 10:597–602.

Kung, K., Lee, T., Chen, J., and Lui, T. 2010a. Characteristics and biological responses of novel coatings containing strontium by micro-arc oxidation. *Surface and Coatings Technology* 205:1714–22.

Kung, K., Lee, T., and Lui, T. 2010b. Bioactivity and corrosion properties of novel coatings containing strontium by micro-arc oxidation. *Journal of Alloys and Compounds* 508:384–90.

Kung, K., Yuan, K., Lee, T., and Lui, T. 2012. Effect of heat treatment on microstructures and mechanical behavior of porous Sr–Ca–P coatings on titanium. *Journal of Alloys and Compounds* 515:68–73.

Larsson, C., Thomson, P., Lausmaa, J., Rodahl, M., Kasemo, B., and Ericson, L.E. 1994. Synthesis of hemocompatible materials. Part 1: Surface modification of polyurethanes based on poly(chloro alkyl vinyl ether)s by RGD fragments. *Biomaterials* 15:253–8.

Lei, M.K., Ou, Y.X., Wang, K.S., and Chen, L. 2011. Wear and corrosion properties of plasma-based low-energy nitrogen ion implanted titanium. *Surface and Coatings Technology* 205:4602–7.

Li, B., Liu, X., Cao, C., and Ding, C. 2007. Biocompatibility and antibacterial activity of plasma sprayed titania coating grafting collagen and gentamicin. *Journal of Biomedical Materials Research Part A* 83:923–30.

Li, B., Liu, X., Cao, C., Dong, Y., and Ding, C. 2009a. Biological and antibacterial properties of plasma sprayed wollastonite/silver coatings. *Journal of Biomedical Materials Research Part B Applied Biomaterials* 91:596–603.

Li, B., Liu, X., Cao, C., Dong, Y., Wang, Z., and Ding, C. 2008a. Biological and antibacterial properties of plasma sprayed wollastonite coatings grafting gentamicin loaded collagen. *Journal of Biomedical Materials Research Part A* 87:84–90.

Li, B., Liu, X., Cao, C., Meng, F., Dong, Y., Cui, T., and Ding, C. 2008b. Preparation and antibacterial effect of plasma sprayed wollastonite coatings loading silver. *Applied Surface Science* 255:452–4.

Li, B., Liu, X., Meng, F., Chang, J., and Ding, C. 2009b. Preparation and antibacterial properties of plasma sprayed nano-titania/silver coatings. *Materials Chemistry and Physics* 118:99–104.

Li, J., Qiao, Y., Ding, Z., and Liu, X. 2011. Microstructure and properties of Ag/N dual ions implanted titanium. *Surface and Coatings Technology* 205:5430–6.

Li, J., Sun, M., Ma, X., Li, X., and Song, Z. 2010. Effect of annealing on structure and hardness of oxygen-implanted layer on Ti6Al4V by plasma-based ion implantation. *Nuclear Instruments and Methods in Physics Research B* 268:135–9.

Li, J. and Zhao, Y. 2012. Biocompatibility and antibacterial performance of titanium by surface treatment. *Journal of Coatings Technology and Research* 9:223–8.

Li, K., Yu, J., Xie, Y., Huang, L., Ye, X., and Zheng, X. 2011. Chemical stability and antimicrobial activity of plasma sprayed bioactive $Ca_2ZnSi_2O_7$ coating. *Journal of Materials Science: Materials in Medicine* 22:2781–9.

Liang, H., Wan, Y.Z., He, F., Huang, Y., Xu, J.D., Li, J.M., Wang, Y.L., and Zhao, Z.G. 2007. Bioactivity of Mg-ion-implanted zirconia and titanium. *Applied Surface Science* 253:3326–33.

Liang, Y., Xie, Y., Ji, H., Huang, L., and Zheng, X. 2010. Excellent stability of plasma-sprayed bioactive $Ca_3ZrSi_2O_9$ ceramic coating on Ti–6Al–4V. *Applied Surface Science* 256:4677–81.

Lim, Y.W., Kwon, S.Y., Sun, D.H., Kim, H.E., and Kim, Y.S. 2009. Enhanced cell integration to titanium alloy by surface treatment with microarc oxidation. *Clinical Orthopaedics and Related Research* 467:2251–8.

Lippard, S.J. 2006. The inorganic side of chemical biology. *Nature Chemical Biology* 2:504–7.

Liu, J.B., Iyer, S.S.K., Hu, C.M., Cheung, N.W., Gronsky, R., Min, J., and Chu, P.K. 1995. Formation of buried oxide in silicon using separation by plasma implantation of oxygen (SPIMOX). *Applied Physics Letters* 67:2361–3.

Liu, L.J. and Schlesinger, M. 2009. Corrosion of magnesium and its alloys. *Corrosion Science* 51:1733–7.

Liu, S., Yang, X., Cui, Z., Zhu, S., and Wei, Q. 2011. One-step synthesis of petal-like apatite/titania composite coating on a titanium by micro-arc oxidation. *Materials Letters* 65:1041–4.

Liu, X., Chu, P.K., and Ding, C. 2004a. Surface modification of titanium, titanium alloys, and related materials for biomedical applications. *Materials Science and Engineering R* 47:49–121.

Liu, X., Chu, P.K., and Ding, C. 2010. Surface nano-functionalization of biomaterials. *Materials Science and Engineering R* 70:275–302.

Liu, X. and Ding, C. 2001. Phase compositions and microstructure of plasma sprayed wollastonite coating. *Surface and Coatings Technology* 141:269–74.

Liu, X. and Ding, C. 2002a. Characterization of plasma sprayed wollastonite powder and coatings. *Surface and Coatings Technology* 153:73–177.

Liu, X. and Ding, C. 2002b. Plasma sprayed wollastonite/TiO$_2$ composite coatings on titanium alloys. *Biomaterials* 23:4065–77.

Liu, X. and Ding, C. 2002c. Reactivity of plasma-sprayed wollastonite coating in simulated body fluid. *Journal of Biomedical Materials Research* 59:259–64.

Liu, X. and Ding, C. 2003. Plasma-sprayed wollastonite 2M/ZrO$_2$ composite coating. *Surface and Coatings Technology* 172:270–8.

Liu, X., Ding, C., and Chu, P.K. 2004b. Mechanism of apatite formation on wollastonite coatings in simulated body fluids. *Biomaterials* 25:1755–61.

Liu, X., Ding, C., and Wang, Z. 2001. Apatite formed on the surface of plasma-sprayed wollastonite coating immersed in simulated body fluid. *Biomaterials* 22:2007–12.

Liu, X., Morra, M., Carpi, A., and Li, B. 2008a. Bioactive calcium silicate ceramics and coatings. *Biomedicine and Pharmacotherapy* 62:526–9.

Liu, X., Poon, R.W.Y., Kwok, S.C.H., Chu, P.K., and Ding, C. 2005a. Structure and properties of Ca-plasma-implanted titanium. *Surface and Coatings Technology* 191:43–8.

Liu, X., Tao, S., and Ding, C. 2002. Bioactivity of plasma sprayed dicalcium silicate coatings. *Biomaterials* 23:963–8.

Liu, X., Xie, Y., Ding, C., and Chu, P.K. 2005b. Early apatite deposition and osteoblast growth on plasma-sprayed dicalcium silicate coating. *Journal of Biomedical Materials Research Part A* 74:356–65.

Liu, X., Zhao, X., Ding, C., and Chu, P.K. 2006. Light-induced bioactive TiO$_2$ surface. *Applied Physics Letters* 88:013905.

Liu, X., Zhao, X., Fu, R.K.Y., Ho, J.P.Y., Ding, C., and Chu, P.K. 2005c. Plasma-treated nanostructured TiO$_2$ surface supporting biomimetic growth of apatite. *Biomaterials* 26:6143–50.

Liu, X., Zhao, X., Li, B., Cao, C., Dong, Y., Ding, C., and Chu, P.K. 2008b. UV-irradiation-induced bioactivity on TiO$_2$ coatings with nanostructural surface. *Acta Biomaterialia* 4:544–52.

Lu, X., Iyer, S.S.K., Liu, J.B., Hu, C.M., Cheung, N.W., Min, J., and Chu, P.K. 1997. Separation by plasma implantation of oxygen to form silicon on insulator. *Applied Physics Letters* 70:1748–50.

Long, P.H. 2008. Medical devices in orthopedic applications. *Toxicologic Pathology* 36:85–91.

Lutolf, M.P. and Hubbell, J.A. 2005. Synthetic biomaterials as instructive extracellular microenvironments for morphogenesis in tissue engineering. *Nature Biotechnology* 23:47–55.

Lutz, T., Gerlach, J.W., and Mändl, S. 2007. Diffusion, phase formation and segregation effects in Ti6Al4V after oxygen PIII. *Surface and Coatings Technology* 201:6690–4.

Maitz, M.F., Pham, M.T., Matz, W., Reuther, H., Steiner, G., and Richter, E. 2002. Ion beam treatment of titanium surfaces for enhancing deposition of hydroxyapatite from solution. *Biomolecular Engineering* 19:269–72.

Maitz, M.F., Poon, R.W.Y., Liu, X.Y., Pham, M.T., and Chu, P.K. 2005. Bioactivity of titanium following sodium plasma immersion ion implantation and deposition. *Biomaterials* 26:5465–73.

Mändl, S., Gerlach, J.W., and Rauschenbach, B. 2005. Surface modification of NiTi for orthopaedic braces by plasma immersion ion implantation. *Surface and Coatings Technology* 196:293–7.

Mändl, S., Krause, D., Thorwarth, G., Sader, R., Zeilhofer, F., Horch, H.H., and Rauschenbach, B. 2001. Plasma immersion ion implantation treatment of medical implants. *Surface and Coatings Technology* 142–144:1046–50.

Mändl, S., Sader, R., Thorwarth, G., Krause, D., Zeilhofer, H.F., Horch, H.H., and Rauschenbach, B. 2003. Biocompatibility of titanium based implants treated with plasma immersion ion implantation. *Nuclear Instruments and Methods in Physics Research B* 206:517–21.

Mändl, S., Thorwarth, G., Schreck, M., Stritzker, B., and Rauschenbach, B. 2000. Raman study of titanium oxide layers produced with plasma immersion ion implantation. *Surface and Coatings Technology* 125:84–8.

Manivasagam, G., Dhinasekaran, D., and Rajamanickam, A. 2010. Biomedical implants: Corrosion and its prevention-A review. *Recent Patents on Corrosion Science* 2:40–54.

Marie, P.J. 2006. Strontium ranelate: A physiological approach for optimizing bone formation and resorption. *Bone* 38:S10–14.

Marie, P.J., Ammann, P., Boivin, G., and Rey, C. 2001. Mechanisms of action and therapeutic potential of strontium in bone. *Calcified Tissue International* 69:121–9.

Marie, P.J., Felsenberg, D., and Brandi, M.L. 2011. How strontium ranelate, via opposite effects on bone resorption and formation, prevents osteoporosis. *Osteoporosis International* 22:1659–67.

Meldrum, A., Haglund, R.F., Boatner, L.A., and White, C.W. 2001. Nanocomposite materials formed by ion implantation. *Advanced Materials* 13:1431–44.

Mertz, W. 1993. Essential trace metals: New definitions based on new paradigms. *Nutrition Reviews* 51:287–95.

Min, J., Chu, P.K., Cheng, Y.C., Liu, J., Iyer, S.S., and Cheung, N.W. 1996. Nucleation mechanism of SPIMOX (separation by plasma implantation of oxygen). *Surface and Coatings Technology* 85:60–3.

Morscher, E.W., Hefti, A., and Aebi, U. 1998. Severe osteolysis after third-body wear due to hydroxyapatite particles from acetabular cup coating. *Journal of Bone and Joint Surgery (British Volume)* 80:267–72.

Nag, S., Banerjee, R., and Fraser, H.L. 2005. Microstructural evolution and strengthening mechanisms in Ti–Nb–Zr–Ta, Ti–Mo–Zr–Fe and Ti–15Mo biocompatible alloys. *Materials Science and Engineering C* 25:357–62.

Nan, K., Wu, T., Chen, J., Jiang, S., Huang, Y., and Pei, G. 2009. Strontium doped hydroxyapatite film formed by micro-arc oxidation. *Materials Science and Engineering C* 29:1554–8.

Nayab, S.N., Jones, F.H., and Olsen, I. 2005. Effects of calcium ion implantation on human bone cell interaction with titanium. *Biomaterials* 26:4717–27.

Nayab, S.N., Jones, F.H., and Olsen, I. 2007. Modulation of the human bone cell cycle by calcium ion-implantation of titanium. *Biomaterials* 28:38–44.

Nazhat, S.N., Young, A.M., and Pratten, J. 2009. Sterility and infection. In *Biomedical Materials*, ed. R. Narayan. New York: Springer, pp. 239–60.

Necula, B.S., Apachitei, I., Tichelaar, F.D., Fratila-Apachitei, L.E., and Duszczyk, J. 2011. An electron microscopical study on the growth of TiO_2–Ag antibacterial coatings on Ti6Al7Nb biomedical alloy. *Acta Biomaterialia* 7:2751–7.

Neoh, K.G., Hu, X., Zheng, D., and Kang, E.T. 2012. Balancing osteoblast functions and bacterial adhesion on functionalized titanium surfaces. *Biomaterials* 33:2813–22.

Ni, J., Shi, Y., Yan, F., Chen, J., and Wang, L. 2008. Preparation of hydroxyapatite-containing titania coating on titanium substrate by micro-arc oxidation. *Materials Research Bulletin* 43:45–53.

Ning, C.Y., Wang, Y.J., Chen, X.F., Zhao, N.R., Ye, J.D., and Wu, G. 2005. Mechanical performances and microstructural characteristics of plasma-sprayed bio-functionally gradient HA–ZrO_2–Ti coatings. *Surface and Coatings Technology* 200:2403–8.

Nishimura, Y., Chayahara, A., Horino, Y., and Yatsuzuka, M. 2002. A new PBIID processing system supplying RF and HV pulses through a single feed-through. *Surface and Coatings Technology* 156:50–3.

Panáček, A., Kolář, M., Večeřova, R., Prucek, R., Soukupová, J., Kryštof, V., Hamal, P., Zbořil, R., and Kvítek, L. 2009. Antifungal activity of silver nanoparticles against Candida spp. *Biomaterials* 30:6333–40.

Park, M.V.D.Z., Neigh, A.M., Vermeulen, J.P., Fonteyne, L.J.J., Verharen, H.W., Briedé, J.J., Loveren, H., and Jong, W.H. 2011. The effect of particle size on the cytotoxicity, inflammation, developmental toxicity and genotoxicity of silver nanoparticles. *Biomaterials* 32:9810–17.

Porter, A.E., Hobbs, L.W., Benezra Rosen, V., and Spector, M. 2002. The ultrastructure of the plasma-sprayed hydroxyapatite–bone interface predisposing to bone bonding. *Biomaterials* 23:725–33.

Prucek, R., Tuček, J., Kilianová, M., Panáček, A., Kvítek, L., Filip, J., Kolář, M., Tománková, K., and Zbořil, R. 2011. The targeted antibacterial and antifungal properties of magnetic nanocomposite of iron oxide and silver nanoparticles. *Biomaterials* 32:4704–13.

Ragalevičius, R., Stalnionis, G., Niaura, G., and Jagminas, A. 2008. Micro-arc oxidation of Ti in a solution of sulfuric acid and Ti^{+3} salt. *Applied Surface Science* 254:1608–13.

Ramaswamy, Y., Wu, C., Dunstan, C.R., Hewson, B., Eindorf, T., Anderson, G.I., and Zreiqat, H. 2009. Sphene ceramics for orthopedic coating applications: An in vitro and in vivo study. *Acta Biomaterialia* 5:3192–204.

Ramaswamy, Y., Wu, C., Hummel, A.V., Combes, V., Grau, G., and Zreiqat, H. 2008. The responses of osteoblasts, osteoclasts and endothelial cells to zirconium modified calcium-silicate-based ceramic. *Biomaterials* 29:4392–402.

Reigstad, O., Johansson, C., Stenport, V., Wennerberg, A., Reigstad, A., and Rokkum, M. 2011. Different patterns of bone fixation with hydroxyapatite and resorbable CaP coatings in the rabbit tibia at 6, 12, and 52 weeks. *Journal of Biomedical Materials Research Part B Applied Biomaterials* 99:14–20.

Rinner, M., Gerlach, J., and Ensinger, W. 2000. Formation of titanium oxide films on titanium and Ti6Al4V by O$_2$-plasma immersion ion implantation. *Surface and Coatings Technology* 132:111–16.

Rodríguez, D., Pigrau, C., Euba, G., Cobo, J., García-Lechuz, J, Palomino, J., Riera, M., Del Toro, M.D., Granados, A., and Ariza, X. 2009. Acute hematogenous prosthetic joint infection: Prospective evaluation of medical and surgical management. *Clinical Microbiology Infection* doi:10.1111/j.1469- 0691.2009.03157.x.

Ryu, H.S., Song, W., and Hong, S. 2008. Biomimetic apatite induction of P-containing titania formed by micro-arc oxidation before and after hydrothermal treatment. *Surface and Coatings Technology* 202:1853–8.

Schwarz, K. and Milne, D.B. 1972. Growth-promoting effects of silicon in rats. *Nature* 239:333–4.

Shi, Z., Neoh, K.G., Kang, E.T., Poh, C.K., and Wang, W. 2009. Titanium with surface-grafted dextran and immobilized bone morphogenetic protein-2 for inhibition of bacterial adhesion and enhancement of osteoblast functions. *Tissue Engineering A* 15:417–26.

Silva, G., Ueda, M., Otani, C., Mello, C.B., and Lepienski, C.M. 2010. Improvements of the surface properties of Ti6Al4V by plasma based ion implantation at high temperatures. *Surface and Coatings Technology* 204:3018–21.

Simchi, A., Tamjid, E., Pishbin, F., and Boccaccini, A.R. 2011. Recent progress in inorganic and composite coatings with bactericidal capability for orthopaedic applications. *Nanomedicine: Nanotechnology, Biology, and Medicine* 7:22–39.

Song, G., Johannesson, B., Hapugoda, S., and St John, D. 2004. Galvanic corrosion of magnesium alloy AZ91D in contact with an aluminium alloy, steel and zinc. *Corrosion Science* 46:955–77.

Song, W., Jun, Y., Han, Y., and Hong, S. 2004. Biomimetic apatite coatings on micro-arc oxidized titania. *Biomaterials* 25:3341–9.

Song, W., Ryu, H.S., and Hong, S. 2009. Antibacterial properties of Ag (or Pt)-containing calcium phosphate coatings formed by micro-arc oxidation. *Journal of Biomedical Materials Research* 88A:246–54.

Steinemann, S.G. 1996. Metal implants and surface reactions. *Injury* 27:S/C16–22.

Storrie, H. and Stupp, S.I. 2005. Cellular response to zinc-containing organoapatite: An in vitro study of proliferation, alkaline phosphatase activity and biomineralization. *Biomaterials* 26:5492–9.

Sul, Y. 2003. The significance of the surface properties of oxidized titanium to the bone response: Special emphasis on potential biochemical bonding of oxidized titanium implant. *Biomaterials* 24:3893–907.

Sumitomo Chemical Company Limited (Osaka Japan). 1975. Implants for bones, joints and tooth roots. Japan, Patent application no. 50-158745.

Sun, C., Hui, R., Qu, W., Yick, S., Sun, C., and Qian, W. 2010. Effects of processing parameters on microstructures of TiO$_2$ coatings formed on titanium by plasma electrolytic oxidation. *Journal of Materials Science* 45:6235–41.

Sun, J., Han, Y., and Cui, K. 2008. Microstructure and apatite-forming ability of the MAO-treated porous titanium. *Surface and Coatings Technology* 202:4248–56.

Sun, J., Li, J., Liu, X., Wei, L., Wang, G., and Meng, F. 2009a. Proliferation and gene expression of osteoblasts cultured in DMEM containing the ionic products of dicalcium silicate coating. *Biomedicine and Pharmacotherapy* 63:650–7.

Sun, J., Wei, L., Liu, X., Li, J., Li, B., Wang, G., and Meng, F. 2009b. Influences of ionic dissolution products of dicalcium silicate coating on osteoblastic proliferation, differentiation and gene expression. *Acta Biomaterialia* 5:1284–93.

Tang, G., Li, J., Sun, M., and Ma, X. 2009. Fabrication of nitrogen-doped TiO_2 layer on titanium substrate. *Applied Surface Science* 255:9224–9.

Tang, Y., Chappell, H.F., Dove, M.T., Reeder, R.J., and Lee, Y.J. 2009. Zinc incorporation into hydroxylapatite. *Biomaterials* 30:2864–72.

Tao, S., Ji, H., and Ding, C. 2000. Effect of vapor-flame treatment on plasma sprayed hydroxyapatite coatings. *Journal of Biomedical Materials Research* 52:572–5.

Tao, X.J., Li, S.J., Zheng, C.Y., Fu, J., Guo, Z., Hao, Y.L., Yang, R., and Guo, Z.X. 2009. Synthesis of a porous oxide layer on a multifunctional biomedical titanium by micro-arc oxidation. *Materials Science and Engineering C* 29:1923–34.

Tercero, J.E., Namin, S., Lahiri, D., Balani, K., Tsoukias, N., and Agarwal, A. 2009. Effect of carbon nanotube and aluminum oxide addition on plasma-sprayed hydroxyapatite coating's mechanical properties and biocompatibility. *Materials Science and Engineering C* 29:2195–202.

Terleeva, O.P., Sharkeev, Y.P., Slonova, A.I., Mironov, I.V., Legostaeva, E.V., Khlusov, I.A., Matykina, E., Skeldon, P., and Thompson, G.E. 2010. Effect of microplasma modes and electrolyte composition on micro-arc oxidation coatings on titanium for medical applications. *Surface and Coatings Technology* 205:1723–9.

Thompson, K.H. and Orvig, C. 2003. Boon and bane of metal ions in medicine. *Science* 300:936–9.

Tian, X.B., Fu, R.K.Y., Chu, P.K., Anders, A., Gong, C.Z., and Yang, S.Q. 2003. Flexible system for multiple plasma immersion ion implantation–deposition processes. *Review of Scientific Instruments* 74:5137.

Tsyganov, I.A., Maitz, M.F., Richter, E., Reuther, H., Mashina, A.I., and Rustichelli, F. 2007. Hemocompatibility of titanium-based coatings prepared by metal plasma immersion ion implantation and deposition. *Nuclear Instruments and Methods in Physics Research B* 257:122–7.

Walker, P.R., Leblanc, J., and Sikorska, M. 1989. Effects of aluminum and other cations on the structure of brain and liver chromatin. *Biochemistry* 28:3911–15.

Wan, Y.Z., Huang, Y., He, F., Wang, Y.L., Zhao, Z.G., and Ding, H.F. 2006. Effect of Mg ion implantation on calcium phosphate formation on titanium. *Surface and Coatings Technology* 201:2904–9.

Wan, Y.Z., Raman, S., He, F., and Huang, Y. 2007a. Surface modification of medical metals by ion implantation of silver and copper. *Vacuum* 81:1114–18.

Wan, Y.Z., Xiong, G.Y., Liang, H., Raman, S., He, F., and Huang, Y. 2007b. Modification of medical metals by ion implantation of copper. *Applied Surface Science* 253:9426–9.

Webster, T.J., Siegel, R.W., and Bizios, R. 1999. Osteoblast adhesion on nanophase ceramics. *Biomaterials* 20:1221–7.

Williams, D.F. 2008. On the mechanisms of biocompatibility. *Biomaterials* 29:2941–53.

Wolke, J.G.C., de Blieck-Hogervorst, J.M.A., Dhert, W.J.A., Klein, C.P.A.T., and de Groot, K. 1992. Studies on the thermal spraying of apatite bioceramics. *Journal of Thermal Spray Technology* 1:75–82.

Wozniak, M.A. and Chen, C.S. 2009. Mechanotransduction in development: A growing role for contractility. *Nature Reviews Molecular Cell Biology* 10:34–43.

Wu, C., Ramaswamy, Y., Chang, J., Woods, J., Chen, Y., and Zreiqat, H. 2008a. The effect of Zn contents on phase composition, chemical stability and cellular bioactivity in Zn–Ca–Si system ceramics. *Journal of Biomedical Materials Research Part B* 87:346–53.

Wu, C., Ramaswamy, Y., Soeparto, A., and Zreiqat, H. 2008b. Incorporation of titanium into calcium silicate improved their chemical stability and biological properties. *Journal of Biomedical Materials Research Part A* 86:402–10.

Xie, Y., Liu, X., Chu, P.K., Zheng, X., and Ding, C. 2005a. Bioactive titanium-particle-containing dicalcium silicate coating. *Surface and Coatings Technology* 200:1950–3.

Xie, Y., Liu, X., Ding, C., and Chu, P.K. 2005b. Bioconductivity and mechanical properties of plasma-sprayed dicalcium silicate/zirconia composite coating. *Materials Science and Engineering C* 25:509–15.

Xie, Y., Liu, X., Huang, A., Ding, C., and Chu, P.K. 2005c. Improvement of surface bioactivity on titanium by water and hydrogen plasma immersion ion implantation. *Biomaterials* 26:6129–35.

Xie, Y., Liu, X., Zheng, X., and Ding, C. 2005d. Bioconductivity of plasma sprayed dicalcium silicate/titanium composite coatings on Ti–6Al–4V alloy. *Surface and Coatings Technology* 199:105–11.

Xie, Y., Liu, X., Zheng, X., Ding, C., and Chu, P.K. 2006. Improved stability of plasma-sprayed dicalcium silicate/zirconia composite coating. *Thin Solid Films* 515:1214–18.

Xie, Y., Zhai, W., Chen, L., Chang, J., Zheng, X., and Ding, C. 2009. Preparation and in vitro evaluation of plasma-sprayed Mg_2SiO_4 coating on titanium alloy. *Acta Biomaterialia* 5:2331–7.

Xue, W., Liu, X., Zheng, X., and Ding, C. 2004a. Plasma-sprayed diopside coatings for biomedical applications. *Surface and Coatings Technology* 185:340–5.

Xue, W., Liu, X., Zheng, X., and Ding, C. 2005a. Effect of hydroxyapatite coating crystallinity on dissolution and osseointegration in vivo. *Journal of Biomedical Materials Research Part A* 74:553–61.

Xue, W., Liu, X., Zheng, X., and Ding, C. 2005b. In vivo evaluation of plasma-sprayed wollastonite coating. *Biomaterials* 26:3455–60.

Xue, W., Tao, S., Liu, X., Zheng, X., and Ding, C. 2004b. In vivo evaluation of plasma sprayed hydroxyapatite coatings having different crystallinity. *Biomaterials* 25:415–21.

Yan, Y., Sun, J., Han, Y., Li, D., and Cui, K. 2010. Microstructure and bioactivity of Ca, P and Sr doped TiO_2 coating formed on porous titanium by micro-arc oxidation. *Surface and Coatings Technology* 205:1702–13.

Yerokhin, A.L., Nie, X., Leyland, A., Matthews, A., and Dowey, S.J. 1999. Plasma electrolysis for surface engineering. *Surface and Coatings Technology* 122:73–93.

Yoshida, K., Tanagawa, M., and Atsuta, M. 1999. Characterization and inhibitory effect of antibacterial dental resin composites incorporating silver-supported materials. *Journal of Biomedical Materials Research* 47:516–22.

Zhang, Z., Sun, J., Hu, H., Wang, Q., and Liu, X. 2011. Osteoblast-like cell adhesion on porous silicon-incorporated TiO_2 coating prepared by micro-arc oxidation. *Journal of Biomedical Materials Research Part B: Applied Biomaterials* 97B:224–34.

Zhao, L., Wang, H., Huo, K., Cui, L., Zhang, W., Ni, H., Zhang, Y., Wu, Z., and Chu, P.K. 2011. Antibacterial nano-structured titania coating incorporated with silver nanoparticles. *Biomaterials* 32:5706–16.

Zhao, X., Liu, X., and Ding, C. 2005. Acid-induced bioactive titania surface. *Journal of Biomedical Materials Research Part A* 75:888–94.

Zhao, X., Liu, X., Ding, C., and Chu, P.K. 2006. In vitro bioactivity of plasma-sprayed TiO_2 coating after sodium hydroxide treatment. *Surface and Coatings Technology* 200:5487–92.

Zhao, X., Liu, X., Ding, C., and Chu, P.K. 2007. Effects of plasma treatment on bioactivity of TiO_2 coatings. *Surface and Coatings Technology* 201:6878–81.

Zhao, X., Liu, X., You, J., Chen, Z., and Ding, C. 2008. Bioactivity and cytocompatibility of plasma-sprayed titania coating treated by sulfuric acid treatment. *Surface and Coatings Technology* 202:3221–6.

Zhao, Z., Chen, X., Chen, A., Shen, M., and Wen, S. 2009. Synthesis of bioactive ceramic on the titanium substrate by micro-arc oxidation. *Journal of Biomedical Materials Research* 90A:438–45.

Zhao, Z. and Wen, S. 2007. Direct preparation of CaTi4 (PO4)6 coatings on the surface of titanium substrate by micro arc oxidation. *Journal of Materials Science: Materials in Medicine* 18:2275–81.

Zheng, X., Huang, M., and Ding, C. 2000. Bond strength of plasma-sprayed hydroxyapatite/Ti composite coatings. *Biomaterials* 21:841–9.

Zhou, G., Ding, H., Zhang, Y., Liu, A., Lin, Y., and Zhu, Y. 2010. Fretting wear study on micro-arc oxidation TiO_2 coating on TC4 titanium alloys in simulated body fluid. *Tribology Letters* 40:319–26.

Zreiqat, H., Ramaswamy, Y., Wu, C., Paschalidis, A., Lu, Z., James, B., Birke, O., McDonald, M., Little, D., and Dunstan, C.R. 2010. The incorporation of strontium and zinc into a calcium–silicon ceramic for bone tissue engineering. *Biomaterials* 31:3175–84.

Index

9 780367 576363